Discovering Evolution Equations with Applications

Volume 2-Stochastic Equations

Published Titles

Advanced Differential Quadrature Methods, Zhi Zong and Yingyan Zhang

Computing with hp-ADAPTIVE FINITE ELEMENTS, Volume 1, One and Two Dimensional Elliptic and Maxwell Problems, Leszek Demkowicz

Computing with hp-ADAPTIVE FINITE ELEMENTS, Volume 2, Frontiers: Three Dimensional Elliptic and Maxwell Problems with Applications, Leszek Demkowicz, Jason Kurtz, David Pardo, Maciej Paszyński, Waldemar Rachowicz, and Adam Zdunek

CRC Standard Curves and Surfaces with Mathematica®*: Second Edition*, David H. von Seggern

Discovering Evolution Equations with Applications: Volume 1-Deterministic Equations, Mark A. McKibben

Discovering Evolution Equations with Applications: Volume 2-Stochastic Equations, Mark A. McKibben

Exact Solutions and Invariant Subspaces of Nonlinear Partial Differential Equations in Mechanics and Physics, Victor A. Galaktionov and Sergey R. Svirshchevskii

Fourier Series in Several Variables with Applications to Partial Differential Equations, Victor L. Shapiro

Geometric Sturmian Theory of Nonlinear Parabolic Equations and Applications, Victor A. Galaktionov

Green's Functions and Linear Differential Equations: Theory, Applications, and Computation, Prem K. Kythe

Introduction to Fuzzy Systems, Guanrong Chen and Trung Tat Pham

Introduction to non-Kerr Law Optical Solitons, Anjan Biswas and Swapan Konar

Introduction to Partial Differential Equations with MATLAB®, Matthew P. Coleman

Introduction to Quantum Control and Dynamics, Domenico D'Alessandro

Mathematical Methods in Physics and Engineering with Mathematica, Ferdinand F. Cap

Mathematical Theory of Quantum Computation, Goong Chen and Zijian Diao

Mathematics of Quantum Computation and Quantum Technology, Goong Chen, Louis Kauffman, and Samuel J. Lomonaco

Mixed Boundary Value Problems, Dean G. Duffy

Multi-Resolution Methods for Modeling and Control of Dynamical Systems, Puneet Singla and John L. Junkins

Optimal Estimation of Dynamic Systems, John L. Crassidis and John L. Junkins

Quantum Computing Devices: Principles, Designs, and Analysis, Goong Chen, David A. Church, Berthold-Georg Englert, Carsten Henkel, Bernd Rohwedder, Marlan O. Scully, and M. Suhail Zubairy

A Shock-Fitting Primer, Manuel D. Salas

Stochastic Partial Differential Equations, Pao-Liu Chow

CHAPMAN & HALL/CRC APPLIED MATHEMATICS
AND NONLINEAR SCIENCE SERIES

Discovering Evolution Equations with Applications

Volume 2-Stochastic Equations

Mark A. McKibben

Goucher College

Baltimore, Maryland

CRC Press
Taylor & Francis Group
Boca Raton London New York

CRC Press is an imprint of the
Taylor & Francis Group, an **informa** business
A CHAPMAN & HALL BOOK

CRC Press
Taylor & Francis Group
6000 Broken Sound Parkway NW, Suite 300
Boca Raton, FL 33487-2742

First issued in paperback 2017

© 2011 by Taylor & Francis Group, LLC
CRC Press is an imprint of Taylor & Francis Group, an Informa business

No claim to original U.S. Government works

ISBN 13: 978-1-138-11358-9 (pbk)
ISBN 13: 978-1-4200-9211-0 (hbk)

Visit the Taylor & Francis Web site at
http://www.taylorandfrancis.com

and the CRC Press Web site at
http://www.crcpress.com

Dedicated to my mother, Pat.

Contents

Preface   **xiii**

1 A Basic Analysis Toolbox   **1**
  1.1 Some Basic Mathematical Shorthand   1
  1.2 Set Algebra   2
  1.3 Functions   3
  1.4 The Space $(\mathbb{R}, |\cdot|)$ 5
    1.4.1 Order Properties   5
    1.4.2 Absolute Value   6
    1.4.3 Completeness Property of $(\mathbb{R}, |\cdot|)$ 7
    1.4.4 Topology of $\mathbb{R}$ 9
  1.5 Sequences in $(\mathbb{R}, |\cdot|)$ 12
    1.5.1 Sequences and Subsequences   12
    1.5.2 Limit Theorems   12
    1.5.3 Cauchy Sequences   19
    1.5.4 A Brief Look at Infinite Series   21
  1.6 The Spaces $\left(\mathbb{R}^N, \|\cdot\|_{\mathbb{R}^N}\right)$ and $\left(\mathbb{M}^N(\mathbb{R}), \|\cdot\|_{\mathbb{M}^N(\mathbb{R})}\right)$ 24
    1.6.1 The Space $\left(\mathbb{R}^N, \|\cdot\|_{\mathbb{R}^N}\right)$ 25
    1.6.2 The Space $\left(\mathbb{M}^N(\mathbb{R}), \|\cdot\|_{\mathbb{M}^N(\mathbb{R})}\right)$ 30
  1.7 Abstract Spaces   32
    1.7.1 Banach Spaces   33
    1.7.2 Hilbert Spaces   38
  1.8 Elementary Calculus in Abstract Spaces   42
    1.8.1 Limits   42
    1.8.2 Continuity   44
    1.8.3 The Derivative   47
  1.9 Some Elementary ODEs   48
    1.9.1 Separation of Variables   48
    1.9.2 First-Order Linear ODEs   49
    1.9.3 Higher-Order Linear ODEs   50
  1.10 A Handful of Integral Inequalities   51
  1.11 Fixed-Point Theory   53
  1.12 Guidance for Selected Exercises   56
    1.12.1 Level 1: A Nudge in a Right Direction   56
    1.12.2 Level 2: An Additional Thrust in a Right Direction   62

2 The Bare-Bone Essentials of Probability Theory   **69**

2.1   Formalizing *Randomness* 69

2.2 $\mathbb{R}$-Valued Random Variables 75

    2.2.1   Some Useful Statistics for Random Variables 79

    2.2.2   Some Common Random Variables 82

2.3   Introducing the Space $\mathfrak{L}^2(\Omega;\mathbb{R})$ 83

2.4 $\mathbb{R}^N$-Valued Random Variables 87

2.5   Conditional Probability and Independence 91

2.6   Conditional Expectation — A Very Quick Description 95

2.7   Stochastic Processes 97

2.8   Martingales 104

2.9   The Wiener Process 105

    2.9.1   A Common Simulation of Brownian Motion 106

    2.9.2   The Wiener Process 108

    2.9.3   Modeling with the Wiener Process 110

2.10 Summary of Standing Assumptions 112

2.11 Looking Ahead 113

2.12 Guidance for Selected Exercises 114

    2.12.1 Level 1: A Nudge in a Right Direction 114

    2.12.2 Level 2: An Additional Thrust in a Right Direction 116

3 Linear Homogenous Stochastic Evolution Equations in $\mathbb{R}$ **119**

3.1   Random Homogenous Stochastic Differential Equations 119

3.2   Introducing the Lebesgue and Itó Integrals 122

    3.2.1   The Lebesgue Integral for $\mathbb{R}$-Valued Stochastic Processes . . 122

    3.2.2   The Itó Integral for $\mathbb{R}$-Valued Stochastic Processes 127

    3.2.3   The Itó Formula in $\mathbb{R}$ 132

    3.2.4   Some Crucial Estimates 136

3.3   The Cauchy Problem — Formulation 140

3.4   Existence and Uniqueness of a Strong Solution 142

3.5   Continuous Dependence on Initial Data 152

3.6   Statistical Properties of the Strong Solution 153

    3.6.1   Mean and Variance 154

    3.6.2   Moment Estimates 156

    3.6.3   Continuity in the pth Moment 157

    3.6.4   The Distribution of a Strong Solution 157

    3.6.5   Markov Property 158

3.7   Some Convergence Results 158

3.8   A Brief Look at Stability 162

3.9   A Classical Example 165

3.10 Looking Ahead 166

3.11 Guidance for Selected Exercises 166

    3.11.1 Level 1: A Nudge in a Right Direction 166

    3.11.2 Level 2: An Additional Thrust in a Right Direction 168

4 Homogenous Linear Stochastic Evolution Equations in $\mathbb{R}^N$ **171**
4.1 Motivation by Models 171
4.2 Deterministic Linear Evolution Equations in $\mathbb{R}^N$ 179
    4.2.1 The Matrix Exponential 179
    4.2.2 The Homogenous Cauchy Problem 181
4.3 Exploring Two Models 182
4.4 The Lebesgue and Itó Integrals in $\mathbb{R}^N$ 183
    4.4.1 The Lebesgue Integral for $\mathbb{R}^N$-Valued Stochastic Processes . 183
    4.4.2 The Itó Integral for $\mathbb{R}^N$-Valued Stochastic Processes 185
    4.4.3 Some Crucial Estimates 188
    4.4.4 The Multivariable Itó Formula — Revisited 189
4.5 The Cauchy Problem — Formulation 191
4.6 Existence and Uniqueness of a Strong Solution 193
4.7 Continuous Dependence on Initial Data 195
4.8 Statistical Properties of the Strong Solution 196
4.9 Some Convergence Results 196
4.10 Looking Ahead 197
4.11 Guidance for Selected Exercises 199
    4.11.1 Level 1: A Nudge in a Right Direction 199
    4.11.2 Level 2: An Additional Thrust in a Right Direction 200

5 Abstract Homogenous Linear Stochastic Evolution Equations   **201**
5.1 Linear Operators 201
    5.1.1 Bounded versus Unbounded Operators 201
    5.1.2 Invertible Operators 205
    5.1.3 Closed Operators 206
    5.1.4 Densely Defined Operators 208
5.2 Linear Semigroup Theory — Some Highlights 208
5.3 Probability Theory in the Hilbert Space Setting 216
5.4 Random Homogenous Linear SPDEs 218
5.5 Bochner and Itó Integrals 227
    5.5.1 The Bochner Integral for $\mathscr{H}$-Valued Stochastic Processes . 227
    5.5.2 The Itó Integral for $\mathscr{H}$-Valued Stochastic Processes 229
5.6 The Cauchy Problem — Formulation 233
5.7 The Basic Theory 235
5.8 Looking Ahead 241
5.9 Guidance for Selected Exercises 242
    5.9.1 Level 1: A Nudge in a Right Direction 242
    5.9.2 Level 2: An Additional Thrust in a Right Direction 243

6 Nonhomogenous Linear Stochastic Evolution Equations   **245**
6.1 Finite-Dimensional Setting 245
    6.1.1 Motivation by Models 245
6.2 Nonhomogenous Linear SDEs in $\mathbb{R}$ 249
    6.2.1 The Cauchy Problem — Existence/Uniqueness Theory . . . 249

	6.2.2	Continuous Dependence Estimates	254
	6.2.3	Statistical Properties of the Solution	256
	6.2.4	Convergence Results	263
	6.2.5	Approximation by a Deterministic IVP	266
6.3		Nonhomogenous Linear SDEs in $\mathbb{R}^N$	267
6.4		Abstract Nonhomogenous Linear SEEs	270
	6.4.1	Motivation by Models	270
	6.4.2	The Cauchy Problem	271
6.5		Introducing Some New Models	272
6.6		Looking Ahead	277
6.7		Guidance for Selected Exercises	278
	6.7.1	Level 1: A Nudge in a Right Direction	278
	6.7.2	Level 2: An Additional Thrust in a Right Direction	279

7 Semi-Linear Stochastic Evolution Equations — **281**

7.1		Motivation by Models	281
	7.1.1	Some Models Revisited	281
	7.1.2	Introducing Two New Models	283
7.2		Some Essential Preliminary Considerations	286
7.3		Growth Conditions	287
7.4		The Cauchy Problem	291
	7.4.1	Problem Formulation	291
	7.4.2	Existence and Uniqueness Results	292
	7.4.3	Continuous Dependence Estimates	294
	7.4.4	p^{th} Moment Continuity	298
	7.4.5	Convergence of Yosida Approximations	300
	7.4.6	Convergence of Induced Probability Measures	303
	7.4.7	Zeroth-Order Approximation	307
7.5		Models Revisited	311
7.6		Theory for Non-Lipschitz-Type Forcing Terms	321
7.7		Looking Ahead	329
7.8		Guidance for Selected Exercises	331
	7.8.1	Level 1: A Nudge in a Right Direction	331
	7.8.2	Level 2: An Additional Thrust in a Right Direction	332

8 Functional Stochastic Evolution Equations — **335**

8.1		Motivation by Models	335
8.2		Functionals	339
8.3		The Cauchy Problem	346
	8.3.1	Problem Formulation	346
	8.3.2	Existence Results	347
	8.3.3	Convergence Results	353
	8.3.4	Zeroth-Order Approximation	358
8.4		Models — New and Old	361
8.5		Looking Ahead	379

8.6 Guidance for Selected Exercises 380
    8.6.1 Level 1: A Nudge in a Right Direction 380
    8.6.2 Level 2: An Additional Thrust in a Right Direction 381

9 Sobolev-Type Stochastic Evolution Equations   **383**
9.1 Motivation by Models 383
9.2 The Abstract Framework 386
9.3 Semi-Linear Sobolev Stochastic Equations 388
9.4 Functional Sobolev SEEs 393
9.5 Guidance for Selected Exercises 396
    9.5.1 Level 1: A Nudge in a Right Direction 396
    9.5.2 Level 2: An Additional Thrust in a Right Direction 397

10 Beyond Volume 2   **401**
10.1 Fully Nonlinear SEEs 401
10.2 Time-Dependent SEEs 404
10.3 Quasi-Linear SEEs 406
10.4 McKean-Vlasov SEEs 407
10.5 Even More Classes of SEEs 407

Bibliography   **409**

Index   **437**

Preface

The mathematical modeling of complex phenomena that evolve over time relies heavily on the analysis of a variety of systems of ordinary and partial differential equations. Such models are developed in very disparate areas of study, ranging from the physical and natural sciences and population ecology to economics, neural networks, and infectious disease epidemiology. Despite the eclectic nature of the fields in which these models are formulated, various groups of them share enough common characteristics that make it possible to study them within a unified theoretical framework. Such study is an area of functional analysis commonly referred to as the theory of evolution equations.

In the absence of "noise," the evolution equations are said to be deterministic. If noise is taken into account in these models, by way of perturbations of the operators involved or via a Wiener process, then the evolution equations become stochastic in nature. The development of the general theory is similar to the deterministic case, but a considerable amount of additional machinery is needed in order to rigorously handle the addition of noise, and questions regarding the nature of the solutions (which are now viewed as stochastic processes rather than deterministic mappings) need to be addressed due to the probabilistic nature of the equations.

One thread of development in this vast field is the study of evolution equations that can be written in an abstract form analogous to a system of finite-dimensional linear ordinary differential equations. The ability to represent the solution of such a finite-dimensional system by a variation of parameters formula involving the matrix exponential prompts one, by analogy, to identify the entity that plays the role of the matrix exponential in a more abstract setting. Depending on the class of equations, this entity can be interpreted as a linear C_0-semigroup, a nonlinear semigroup, a (co)sine family, etc. A general theory is then developed in each situation and applied, to the extent possible, to all models within its parlance.

The literature for the theory of evolution equations is massive. Numerous monographs and journal articles have been written, the total sum of which covers a practically insurmountable amount of ground. While there exist five-volume magna opi that provide excellent accounts of the big picture of aspects of the field (for instance, [105, 107, 418]), most books written on evolution equations tend to either provide a thorough treatment of a particular class of equations in tremendous depth for a beginner or focus on presenting an assimilation of materials devoted to a very particular timely research direction (see [11, 37, 38, 46, 47, 65, 90, 108, 131, 132, 133, 149, 159, 174, 178, 192, 206, 250, 252, 253, 290, 300, 305, 328, 329, 341, 365, 375, 381, 396, 407, 419]). The natural practice in such mathematics texts, given that they are written for readers trained in advanced mathematics, is to pay little attention to preliminary material or behind-the-scenes detail. Needless to say, initiating study in this field can be daunting for beginners. This begs the question, "How do newcomers obtain an overview of the field, in a reasonable amount of time, that prepares them to enter and initially navigate the research realm?" This is what prompted me to embark on writing the current volume. The purpose of this volume is to provide an engaging, accessible account of a rudimentary core of theoretical results that should be understood by anyone studying stochastic evolution equations in a way that gradually builds the reader's intuition. To accomplish this task, I have opted to write the book using a so-called discovery approach, the ultimate goal of which is to engage you, the reader, in the actual mathematical enterprise of studying stochastic evolution equations. Some characteristics of this approach that you will encounter in the text are mentioned below.

What are the "discovery approach" features of the text?

I have tried to extract the essence of my approaches to teaching this material to newcomers to the field and conducting my own research, and incorporate these features into the actual prose of the text. For one, I pose questions of all types throughout the development of the material, from verifying details and illustrating theorems with examples to posing (and proving) conjectures of actual results and analyzing broad strokes that occur within the development of the theory itself. At times, the writing takes the form of a conversation with you, by way of providing motivation for a definition, or setting the stage for the next step of a theoretical development, or prefacing an important theorem with a plain-English explanation of it. I sometimes pose rhetorical questions to you as a lead-in to a subsequent section of the text. The inclusion of such discussion facilitates "seeing the big picture" of a theoretical development that I have found naturally connects its various stages. You are not left guessing why certain results are being developed or why a certain path is being followed. As a result, the exposition in the text, at times, may lack the "polished style" of a mathematical monograph, and the language used will be colloquial English rather than the standard mathematical language that you would encounter in a journal article. But, this style has the benefit of encouraging you to not simply passively read the text, but rather work through it, which is essential to obtaining a meaningful grasp of the material.

I deliberately begin each chapter with a discussion of models, many of which are studied in several chapters and modified along the way to motivate the particular theory to be developed in a given chapter. The intent is to illustrate how taking into account natural additional complexity gives rise to more complicated initial-boundary value problems that, in turn, are formulated using more general abstract evolution equations. This connectivity among different fields and the centrality of the theory of evolution equations to their study are illustrated on the cover of the text.

The driving force of the discussion is the substantive collection of more than 500 questions and exercises dispersed throughout the text. I have inserted questions of all types directly into the development of the chapters with the intention of having you pause and either process what has just been presented or react to a rhetorical question posed. You might be asked to supply details in an argument, verify a definition or theorem using a particular example, create a counterexample to show why an extension of a theorem from one setting to another fails, or conjecture and prove a result of your own based on previous material, etc. The questions, in essence, constitute much of the behind-the-scenes detail that goes into actually formulating the theory. In the spirit of the conversational nature of the text, I have included a section entitled *Guidance for Selected Exercises* at the end of the first nine chapters that provides two layers of hints for selected exercises. Layer one, labeled as "A Small Nudge in a Right Direction" is intended to help you get started if you are stumped. The idea is that you will re-attempt the exercise using the hint. If you find this hint insufficient, the second layer of hints, labeled as "An Additional Thrust in a Right Direction" provides a more substantive suggestion as to how to proceed. In addition to this batch of exercises, you will encounter more questions or directives enclosed in parentheses throughout all parts of the text. The purpose of these less formal, yet equally important questions is to alert you to when details are being omitted or to call your attention to a specific portion of a proof to which I want you to pay close attention. You will likely view the occurrence of these questions to be, at times, disruptive. And, this is exactly the point of including them! The tendency is to gloss over details when working through material as technical as this, but doing so too often will create gaps in understanding. It is my hope that the inclusion of the combination of the two layers of hints for the formal exercises and this frequent questioning will reduce any reluctance you might have in working through the text.

Finally, most chapters conclude with a section in which some of the models used to motivate the chapter are revisited, but are now modified in order to account for an additional complexity. The impetus is to direct your thinking toward what awaits you in the next chapter. This short, but natural, section is meant to serve as a connective link between chapters.

For whom is this book accessible?

It is my hope that anybody possessing a basic familiarity with the real numbers and at least an exposure to the most elementary of differential equations, be it a student, engineer, scientist, or mathematician specializing in a different area, can work through this text to gain an initial understanding of stochastic evolution equations, how they are used in practice, and more than twenty different areas of study to which the theory applies. Indeed, while the level of the mathematics discussed in the text is conventionally viewed as a topic that a graduate student would encounter after studying stochastic and functional analysis, all of the underlying tools of stochastic and functional analysis necessary to intelligently work through the text are included, chapter by chapter as they arise. This, coupled with the conversational style in which the text is written, should make the material naturally accessible to a broad audience.

What material does this text cover, in broad strokes?

The present volume consists of ten chapters. The text opens with two substantive chapters devoted to creating basic real and stochastic analysis "toolboxes," the purpose of which is to arm you with the bare essentials of real and stochastic analysis needed to work through the rest of the book. If you are familiar with the topics in the chapter, I suggest you peruse the chapter to get a feel for the notation and terminology prior to moving on.

Chapter 3 is devoted to the development of the theory for homogenous one-dimensional stochastic ODEs, while Chapter 4 immediately extends this theory to systems of homogenous linear stochastic ordinary differential equations. These chapters act as a springboard into the development of its abstract counterpart in a more general separable Hilbert space. The discussion proceeds to the case of linear homogenous abstract stochastic evolution equations in Chapter 5, and subsequently in the next two chapters to the nonhomogenous and semi-linear cases. The case in which the forcing term is a functional (acting from one function space to another) is addressed in Chapter 8, followed by a discussion of Sobolev-type stochastic evolution equations in Chapter 9. These latter two chapters have been recent active research areas. Finally, the last chapter is devoted to a brief discussion of several different directions involving accessible topics of active research.

For each class of equations, a core of theoretical results concerning the following main topics is developed: the existence and uniqueness of solutions (in a variety of senses) under various growth assumptions, continuous dependence upon initial data and parameters, convergence results of various kinds, and elementary stability results (in a variety of senses).

A substantive collection of mathematical models arising in areas such as heat conduction, advection, fluid flow through fissured rocks, transverse vibrations in extensible beams, thermodynamics, population ecology, pharmacokinetics, spatial pattern formation, pheromone transport, neural networks, and infectious disease epidemiology are developed in stages throughout the text. In fact, the reason for studying the class of abstract equations of a given chapter is motivated by first considering modified versions of the model(s) discussed in the previous chapter, and subsequently formulating the batch of newly created initial-boundary value problems in the form of the abstract equation to be studied in that chapter.

In order to get the most out of this text, I strongly encourage you to read it alongside of volume 1 **[295]** and to make deliberate step-by-step comparisons of the theory in the deterministic and stochastic settings.

About the book cover

You might very well be wondering about the significance of the text cover. Would you believe that it embodies the main driving force behind the text? Indeed, the initial-value problem in the middle from which all arrows emanate serves as a theoretical central hub that mathematically binds the models depicted by the illustrations on the cover, to name just a few. Each of the eight pictures illustrates a scenario described by a mathematical model (involving partial differential equations) that is studied in this text. As you work through the text, you will discover that all of these models can be written abstractly in the form of the initial-value problem positioned in the middle of the cover. So, despite the disparate nature of the fields in which these models arise, they can all be treated under the same theoretical umbrella. This is the power of the abstract theory developed in this text.

Reading from left to right, and top to bottom, the fields depicted in the pictures are as follows: air pollution, infectious disease epidemiology, neural networks, chemical kinetics, combustion, population dynamics, spatial pattern formation, and soil mechanics.

Acknowledgments

Writing this text has been one of the most positive and fulfilling experiences of my professional career thus far. I found various stages of writing this book to be truly energizing and uplifting, while others required me to plumb the very depths of my patience and perseverance. I truly realize that a project like this would have never come to fruition without the constant support, encouragement, and good humor of many colleagues, students, friends, and family. While it is virtually impossible to acknowledge each and every individual who has, in some way, influenced me in a manner that helped either to steer me toward writing this book or to navigate murky (and at times very choppy) waters during the writing phase, I would like to acknowledge several at this time.

First and foremost, there are two notable women who have been indelible sources of encouragement, energy, and support for as long as they have been in my life. My wife, to whom the first volume is dedicated, prompted me for years to write this book. And, once I actually took her advice and began the process, she never abandoned my urgent and recurrent needs for technological help, editorial expertise, or idea-sounding, and has never begrudged me for the momentary lack of patience, uttering of angst-provoked witty remarks, or necessary "idea jotting" at 3 am. And my mother, to whom this second volume is dedicated, and who has supported all of my scholastic and professional endeavors for as long as I can remember. She relentlessly encourages me to temper hard work with balance; she always seems to know when to provide a good-humored story to lighten a stressed mood and when to politely remind me to "take a break!" This book would never have materialized if it were not for both of you; your unwavering support of my endeavors is a significant driving force!

Next, I am extremely lucky to have had two outstanding mentors during my college years. My first exposure to college-level mathematics, by way of analysis, took place nearly two decades ago in a one-on-one tutorial with Dr. David Keck. His unbridled enthusiasm for teaching and learning mathematics and well-timed witty humor have been infectious. Achieving the depth of his passion and honing my skills to mirror his innate ability to teach mathematics are goals to which I will continue to aspire for the duration of my academic career. And, my dissertation advisor, Dr. Sergiu Aizicovici, who took me under his wing as a graduate student and introduced me to various facets of the study of evolution equations and the world of mathematical research. His abilities to make the area come to life, to help a newcomer navigate the practically insurmountable literature with ease, and to tolerate and provide meaningful answers to even the most rudimentary of questions (which I admittedly asked quite often!) in a way that honed my intuition surrounding the subject matter are among the many reasons why I chose to pursue this area of research. You both have left indelible imprints on my development as a mathematician and educator.

Many people have been kind enough to provide honest feedback during various stages of this project. My colleagues Dr. Robert Lewand, Dr. Scott Sibley, Dr. Bernadette Tutinas, Dr. Cynthia Young, and Dr. Jill Zimmerman all provided valuable comments on portions of the prospectus that undoubtedly led to a stronger proposal. My colleagues Dr. Micah Webster and Dr. David Keck provided valuable feedback on their impression of the probability theory chapter, which led to improved transition to subsequent chapters of the text. My students Shana Lieberman, Jennifer Jordan, and Jordan Yoder endured various portions of the manuscript by way of independent study. I am proud to report that all three of them survived the experience (and seemed to enjoy it) and identified their fair share of errors in the early versions of the text! My colleague Dr. Tom Kelliher provided invaluable TEX help during early stages of the project; thank you for helping such a TEX neophyte! A special thanks to my wife, Jodi, for reading the prospectus and manuscript, and who will never let me live down the fact that she found a mathematical error in the prospectus.

I would like to thank the entire Taylor & Francis team. To my editor, Bob Stern, thank you for approaching me about writing this text and for patiently guiding me through the process from beginning to end. To my project coordinator, Jessica Vakili, who answered my production, stylistic, and marketing questions in a very helpful and timely manner. To my project editor, Karen Simon, for keeping the publication of this book on track. To Kevin Craig for designing the awesome cover of this book. And to Shashi Kumar, who helped me to overcome various LATEX issues throughout the typesetting process.

And, last but not least, I would like to thank you, the reader, for embarking on this journey with me through an amazingly rich field of mathematics. I hope your study is as fulfilling as mine has been thus far.

Mark A. McKibben

Chapter 1

A Basic Analysis Toolbox

Overview

The purpose of this chapter is to provide you with a succinct, hands-on introduction to elementary analysis that focuses on notation, main definitions and results, and the techniques with which you should be comfortable prior to working through this text. Additional topics will be introduced throughout the text whenever needed. Little is assumed beyond a working knowledge of the properties of real numbers, the "freshmen calculus," and a tolerance for mathematical rigor. Keep in mind that the presentation is not intended to be a complete exposition of real analysis. You are encouraged to refer to texts devoted to more comprehensive treatments of analysis (see **[17, 67, 196, 197, 234, 236, 250, 301, 353, 357, 372]**).

1.1 Some Basic Mathematical Shorthand

Symbolism is used heftily in mathematical exposition. Careful usage of some basic notation can streamline the verbiage. Some of the common symbols used are as follows.

Let P and Q be statements. (If the statement P changes depending on the value of some parameter x, we denote this dependence by writing $P(x)$.)

1.) The statement "not P," called the *negation of P*, is denoted by "$\neg P$."

2.) The statement "P or Q" is denoted by "$P \vee Q$," while the statement "P and Q" is denoted by "$P \wedge Q$."

3.) The statement "If P, then Q" is called an *implication*, and is denoted by "$P \Longrightarrow Q$" (read "P implies Q"). Here, P is called the *hypothesis* and Q is the *conclusion*.

4.) The statement "P if, and only if, Q" is denoted by "P iff Q" or "$P \Longleftrightarrow Q$." Precisely, this means "$(P \Longrightarrow Q) \wedge (Q \Longrightarrow P)$."

5.) The statement "$Q \Longrightarrow P$" is the *converse* of "$P \Longrightarrow Q$."

6.) The statement "$\neg Q \Longrightarrow \neg P$" is the *contrapositive* of "$P \Longrightarrow Q$." These two statements are equivalent.

7.) The symbol "$\exists$" is an existential quantifier and is read as "there exists" or "there is at least one."

8.) The symbol "$\forall$" is a universal quantifier and is read as "for every" or "for any."

Exercise 1.1.1. Let P, Q, R, and S be statements.
i.) Form the negation of "$P \wedge (Q \wedge R)$."
ii.) Form the negation of "$\exists\, x$ such that $P(x)$ holds."
iii.) Form the negation of "$\forall x$, $P(x)$ holds."
iv.) Form the contrapositive of "$(P \wedge Q) \Longrightarrow (\neg R \vee S)$."

Remark. Implication is a transitive relation in the sense that

$$((P \Longrightarrow Q) \wedge (Q \Longrightarrow R)) \Longrightarrow (P \Longrightarrow R).$$

For instance, a sequence of algebraic manipulations used to solve an equation is technically such a string of implications from which we conclude that the values of the variable obtained in the last step are the solutions of the original equation. Mathematical proofs are comprised of strings of implications, albeit of a somewhat more sophisticated nature.

1.2   Set Algebra

Informally, a *set* can be thought of as a collection of objects (e.g., real numbers, vectors, matrices, functions, other sets, etc.); the contents of a set are referred to as its *elements*. We usually label sets using uppercase letters and their elements by lowercase letters. Three sets that arise often and for whom specific notation will be reserved are

$$\mathbb{N} = \{1, 2, 3, ...\}$$
$$\mathbb{Q} = \text{the set of all rational numbers}$$
$$\mathbb{R} = \text{the set of all real numbers}$$

If P is a certain property and A is the set of all objects having property P, we write $A = \{x : x \text{ has } P\}$ or $A = \{x | x \text{ has } P\}$. A set with no elements is *empty*, denoted by $\varnothing$.

   If A is not empty and a is an element of A, we denote this fact by "$a \in A$." If a is not an element of A, a fact denoted by "$a \notin A$," where is it located? This prompts us to prescribe a universal set $\mathscr{U}$ that contains all possible objects of interest in our discussion. The following definition provides an algebra of sets.

Definition 1.2.1. Let A and B be sets.
i.) A is a *subset* of B, written $A \subset B$, whenever $x \in A \Longrightarrow x \in B$.
ii.) A *equals* B, written $A = B$, whenever $(A \subset B) \wedge (B \subset A)$.
iii.) The *complement of A relative to B*, written $B \setminus A$, is the set $\{x | x \in B \wedge x \notin A\}$.

Specifically, the complement relative to $\mathcal{U}$ is denoted by $\widetilde{A}$.
iv.) The *union of A and B* is the set $A \cup B = \{x | x \in A \vee x \in B\}$.
v.) The *intersection of A and B* is the set $A \cap B = \{x | x \in A \wedge x \in B\}$.
vi.) $A \times B = \{(a,b) | a \in A \wedge b \in B\}$.

Proving set equality requires that we show two implications. Use this fact when appropriate to complete the following exercises.

Exercise 1.2.1. Let A, B, and C be sets. Prove the following:
i.) $A \subset B$ iff $\widetilde{B} \subset \widetilde{A}$
ii.) $A = (A \cap B) \cup (A \setminus B)$
iii.) $A \cap (B \cup C) = (A \cap B) \cup (A \cap C)$ and $A \cup (B \cap C) = (A \cup B) \cap (A \cup C)$
iv.) $\widetilde{(A \cap B)} = \widetilde{A} \cup \widetilde{B}$ and $\widetilde{(A \cup B)} = \widetilde{A} \cap \widetilde{B}$

Exercise 1.2.2. Explain how you would prove $A \neq B$.

Exercise 1.2.3. Formulate an extension of Def. 1.2.1(iv) through (vi) that works for any finite number of sets.

It is often necessary to consider the union or intersection of more than two sets, possibly infinitely many. So, we need a succinct notation for unions and intersections of an arbitrary number of sets. Let $\Gamma \neq \emptyset$. (We think of the members of Γ as labels.) Suppose to each $\gamma \in \Gamma$, we associate a set A_γ. The collection of all these sets, namely $\mathscr{A} = \{A_\gamma | \gamma \in \Gamma\}$, is a *family of sets indexed by* Γ. We define

$$\bigcup_{\gamma \in \Gamma} A_\gamma = \{x | \exists \gamma \in \Gamma \text{ such that } x \in A_\gamma\}, \tag{1.1}$$

$$\bigcap_{\gamma \in \Gamma} A_\gamma = \{x | \forall \gamma \in \Gamma, x \in A_\gamma\}. \tag{1.2}$$

If $\Gamma = \mathbb{N}$, we write $\bigcup_{n=1}^{\infty}$ and $\bigcap_{n=1}^{\infty}$ in place of $\bigcup_{\gamma \in \Gamma}$ and $\bigcap_{\gamma \in \Gamma}$, respectively.

Exercise 1.2.4. Let A be a set and $\{A_\gamma | \gamma \in \Gamma\}$ a family of sets indexed by Γ. Prove
i.) $A \cap \bigcup_{\gamma \in \Gamma} A_\gamma = \bigcup_{\gamma \in \Gamma} (A \cap A_\gamma)$ and $A \cup \bigcap_{\gamma \in \Gamma} A_\gamma = \bigcap_{\gamma \in \Gamma} (A \cup A_\gamma)$
ii.) $\left(\bigcup_{\gamma \in \Gamma} A_\gamma\right)^{\sim} = \bigcap_{\gamma \in \Gamma} \widetilde{A_\gamma}$ and $\left(\bigcap_{\gamma \in \Gamma} A_\gamma\right)^{\sim} = \bigcup_{\gamma \in \Gamma} \widetilde{A_\gamma}$
iii.) $A \times \bigcup_{\gamma \in \Gamma} A_\gamma = \bigcup_{\gamma \in \Gamma} (A \times A_\gamma)$ and $A \times \bigcap_{\gamma \in \Gamma} A_\gamma = \bigcap_{\gamma \in \Gamma} (A \times A_\gamma)$
iv.) $\bigcap_{\gamma \in \Gamma} A_\gamma \subset A_{\gamma_0} \subset \bigcup_{\gamma \in \Gamma} A_\gamma, \forall \gamma_0 \in \Gamma$.

1.3 Functions

The concept of a function is central to the study of mathematics.

Definition 1.3.1. Let A and B be sets.

i.) A subset $f \subset A \times B$ satisfying

    a.) $\forall x \in A, \exists y \in B$ such that $(x,y) \in f$,

    b.) $(x,y_1) \in f \land (x,y_2) \in f \implies y_1 = y_2$,

    is called a *function from A into B*. We say f is *B-valued*, denoted by $f : A \to B$.

ii.) The set A is called the *domain* of f, denoted dom(f).

iii.) The *range* of f, denoted by rng(f), is given by $\mathrm{rng}(f) = \{f(x) | x \in A\}$.

Remarks.

1. Notation: When defining a function using an explicit formula, say $y = f(x)$, the notation $x \mapsto f(x)$ is often used to denote the function. Also, we indicate the general dependence on a variable using a dot, say $f(\cdot)$. If the function depends on two independent variables, we distinguish between them by using a different number of dots for each, say $f(\cdot, \cdot\cdot)$.

2. The term *mapping* is used synonymously with the term *function*.

3. $\mathrm{rng}(f) \subset B$.

Exercise 1.3.1. Precisely define what it means for two functions f and g to be equal.

The following classification plays a role in determining if a function is invertible.

Definition 1.3.2. $f : A \to B$ is called

i.) *one-to-one* if $f(x_1) = f(x_2) \implies x_1 = x_2$, $\forall x_1, x_2 \in A$;

ii.) *onto* whenever $\mathrm{rng}(f) = B$.

We sometimes wish to apply functions in succession in the following sense.

Definition 1.3.3. Suppose that $f : \mathrm{dom}(f) \to A$ and $g : \mathrm{dom}(g) \to B$ with $\mathrm{rng}(g) \subset \mathrm{dom}(f)$. The *composition of f with g*, denoted $f \circ g$, is the function $f \circ g : \mathrm{dom}(g) \to A$ defined by $(f \circ g)(x) = f(g(x))$.

Exercise 1.3.2. Show that, in general, $f \circ g \neq g \circ f$.

Exercise 1.3.3. Let $f : \mathrm{dom}(f) \to A$ and $g : \mathrm{dom}(g) \to B$ be such that $f \circ g$ is defined. Prove

i.) If f and g are onto, then $f \circ g$ is onto.

ii.) If f and g are one-to-one, then $f \circ g$ is one-to-one.

At times, we need to compute the functional values for all members of a subset of the domain, or perhaps determine the subset of the domain whose collection of functional values is a prescribed subset of the range. These notions are made precise below.

Definition 1.3.4. Let $f : A \to B$.

i.) For $X \subset A$, the *image of X under f* is the set $f(X) = \{f(x) | x \in X\}$.

ii.) For $Y \subset B$, the *pre-image of Y under f* is the set

$$f^{-1}(Y) = \{x \in \mathscr{A} \mid \exists y \in Y \text{ such that } y = f(x)\}.$$

The following related properties are useful.

Proposition 1.3.5. *Suppose $f : A \to B$ is a function, X, X_1, X_2, and X_γ, $\gamma \in \Gamma$, are all subsets of A and Y, Y_1, Y_2, and Y_γ, $\gamma \in \Gamma$, are all subsets of B. Then,*
i.) **a.)** $X_1 \subset X_2 \Longrightarrow f(X_1) \subset f(X_2)$
    b.) $Y_1 \subset Y_2 \Longrightarrow f^{-1}(Y_1) \subset f^{-1}(Y_2)$
ii.) **a.)** $f\left(\bigcup_{\gamma \in \Gamma} X_\gamma\right) = \bigcup_{\gamma \in \Gamma} f(X_\gamma)$
    b.) $f^{-1}\left(\bigcup_{\gamma \in \Gamma} Y_\gamma\right) = \bigcup_{\gamma \in \Gamma} f^{-1}(Y_\gamma)$
iii.) **a.)** $f\left(\bigcap_{\gamma \in \Gamma} X_\gamma\right) \subset \bigcap_{\gamma \in \Gamma} f(X_\gamma)$
    b.) $f^{-1}\left(\bigcap_{\gamma \in \Gamma} Y_\gamma\right) = \bigcap_{\gamma \in \Gamma} f^{-1}(Y_\gamma)$
iv.) **a.)** $X \subset f^{-1}(f(X)))$
    b.) $f\left(f^{-1}(Y)\right) \subset Y$

Exercise 1.3.4.
i.) Prove Prop. 1.3.5.
ii.) Impose conditions on f that would yield equality in Prop. 1.3.5(iv)(a) and (b).

We often consider functions whose domains and ranges are subsets of $\mathbb{R}$. For such functions, the notion of monotonicity is often a useful characterization.

Definition 1.3.6. Let $f : \text{dom}(f) \subset \mathbb{R} \to \mathbb{R}$ and suppose that $\varnothing \neq S \subset \text{dom}(f)$. We say that f is
i.) *nondecreasing on S* whenever $x_1, x_2 \in S$ with $x_1 < x_2 \Longrightarrow f(x_1) \leq f(x_2)$;
ii.) *nonincreasing on S* whenever $x_1, x_2 \in S$ with $x_1 < x_2 \Longrightarrow f(x_1) \geq f(x_2)$.

Remark. The prefix "non" in both parts of Def. 1.3.6 is removed when the inequality is strict.

The arithmetic operations of real-valued functions are defined in the natural way. For such functions, consider the following exercise.

Exercise 1.3.5. Suppose that $f : \text{dom}(f) \subset \mathbb{R} \to \mathbb{R}$ and $g : \text{dom}(g) \subset \mathbb{R} \to \mathbb{R}$ are nondecreasing (resp. nonincreasing) functions on their domains.
i.) Which of the functions $f + g$, $f - g$, $f \cdot g$, and $\frac{f}{g}$, if any, are nondecreasing (resp. nonincreasing) on their domains?
ii.) Assuming that $f \circ g$ is defined, must it be nondecreasing (resp. nonincreasing) on its domain?

1.4 The Space $(\mathbb{R}, |\cdot|)$

1.4.1 Order Properties

The basic arithmetic and order features of the real number system are likely familiar, even if you have not worked through its formal construction. For our purposes, we shall begin with a set $\mathbb{R}$ equipped with two operations, addition and multiplication, satisfying these algebraic properties:

(i) addition and multiplication are both commutative and associative,
(ii) multiplication distributes over addition,
(iii) adding zero to any real number yields the same real number,
(iv) multiplying a real number by 1 yields the same real number,
(v) every real number has a unique additive inverse, and
(vi) every nonzero real number has a unique multiplicative inverse.

Moreover, $\mathbb{R}$ equipped with the natural "<" ordering is an ordered field and obeys the following properties.

Proposition 1.4.1. (Order Features of $\mathbb{R}$)
For all $x, y, z \in \mathbb{R}$, the following are true:
i.) *Exactly one of the relationships $x = y$, $x < y$, or $y < x$ holds;*
ii.) $x < y \implies x + z < y + z$;
iii.) $(x < y) \wedge (y < z) \implies x < z$;
iv.) $(x < y) \wedge (c > 0) \implies cx < cy$;
v.) $(x < y) \wedge (c < 0) \implies cx > cy$;
vi.) $(0 < x < y) \wedge (0 < w < z) \implies 0 < xw < yz$.

The following is an immediate consequence of these properties and is often the underlying principle used when verifying an inequality.

Proposition 1.4.2. *If $x, y \in \mathbb{R}$ are such that $x < y + \varepsilon$, $\forall \varepsilon > 0$, then $x \leq y$.*

Proof. Suppose not; that is, $y < x$. Observe that for $\varepsilon = \frac{x-y}{2} > 0$, $y + \varepsilon = \frac{x+y}{2} < x$. (Why?) This is a contradiction. Hence, it must be the case that $x \leq y$. $\square$

Remark. The above argument is a very simple example of a proof by contradiction. The strategy is to assume that the conclusion is false and then use this additional hypothesis to obtain an obviously false statement or a contradiction of another hypothesis in the claim. More information about elementary proof techniques can be found in **[372]**.

Exercise 1.4.1.
i.) Let $x, y > 0$. Prove that $xy \leq \frac{x^2 + y^2}{2}$.
ii.) Show that if $0 < x < y$, then $x^n < y^n$, $\forall n \in \mathbb{N}$.

1.4.2　Absolute Value

The above is a heuristic description of the familiar algebraic structure of $\mathbb{R}$. When equipped with a distance-measuring artifice, a deeper topological structure of $\mathbb{R}$ can be defined and studied. This is done with the help of the absolute value function.

Definition 1.4.3. For any $x \in \mathbb{R}$, the *absolute value* of x, denoted $|x|$, is defined by

$$|x| = \begin{cases} x, x \geq 0, \\ -x, x < 0. \end{cases}$$

This can be viewed as a measurement of distance between real numbers within the context of a number line. For instance, the solution set of the equation "$|x - 2| = 3$" is the set of real numbers x that are "3 units away from 2," namely $\{-1, 5\}$.

Exercise 1.4.2. Determine the solution set for the following equations:
i.) $|x - 3| = 0$
ii.) $|x + 6| = 2$.

Proposition 1.4.4. *These properties hold for all $x, y, z \in \mathbb{R}$ and $a \geq 0$:*
i.) $-|x| = \min\{-x, x\} \leq x \leq \max\{-x, x\} = |x|$
ii.) $|x| \geq 0, \forall x \in \mathbb{R}$
iii.) $|x| = 0$ *iff* $x = 0$
iv.) $\sqrt{x^2} = |x|$
v.) $|xy| = |x||y|$
vi.) $|x| \leq a$ *iff* $-a \leq x \leq a$
vii.) $|x + y| \leq |x| + |y|$
viii.) $|x - y| \leq |x - z| + |z - y|$
ix.) $||x| - |y|| \leq |x - y|$
x.) $|x - y| < \varepsilon, \forall \varepsilon > 0 \implies x = y$

Exercise 1.4.3. Prove Prop. 1.4.4.

Exercise 1.4.4. Let $n \in \mathbb{N}$ and $x_1, x_2, \ldots, x_n, y_1, y_2, \ldots, y_n \in \mathbb{R}$. Prove:
i.) (Cauchy-Schwarz) $\sum_{i=1}^{n} x_i y_i \leq \left(\sum_{i=1}^{n} x_i^2\right) \left(\sum_{i=1}^{n} y_i^2\right)$

ii.) (Minkowski) $\left(\sum_{i=1}^{n} (x_i + y_i)^2\right)^{1/2} \leq \left(\sum_{i=1}^{n} x_i^2\right)^{1/2} + \left(\sum_{i=1}^{n} y_i^2\right)^{1/2}$

iii.) $\left|\sum_{i=1}^{n} x_i\right|^M \leq \left(\sum_{i=1}^{n} |x_i|\right)^M \leq n^{M-1} \sum_{i=1}^{n} |x_i|^M, \forall M \in \mathbb{N}$

1.4.3 Completeness Property of $(\mathbb{R}, |\cdot|)$

It turns out that $\mathbb{R}$ has a fundamental and essential property referred to as *completeness*, without which the study of analysis could not proceed. We introduce some terminology needed to state certain fundamental properties of $\mathbb{R}$.

Definition 1.4.5. Let $\varnothing \neq S \subset \mathbb{R}$.
i.) S is *bounded above* if $\exists u \in \mathbb{R}$ such that $x \leq u, \forall x \in S$;
ii.) $u \in \mathbb{R}$ is an *upper bound* of S (ub(S)) if $x \leq u, \forall x \in S$;
iii.) $u_0 \in \mathbb{R}$ is the *maximum* of S (max(S)) if u_0 is an ub(S) and $u_0 \in S$;
iv.) $u_0 \in \mathbb{R}$ is the *supremum* of S (sup(S)) if u_0 is an ub(S) and $u_0 \leq u$, for any other $u =$ub(S).

The following analogous terms can be defined by reversing the inequality signs in Def. 1.4.5: *bounded below, lower bound of S* (lb(S)), *minimum of S* (min(S)), and *infimum of S* (inf(S)).

Exercise 1.4.5. Formulate precise definitions of the above terms.

Exercise 1.4.6. Let $\varnothing \neq S \subset \mathbb{R}$.
i.) How would you prove that $\sup(S) = \infty$?
ii.) Repeat (i) for $\inf(S) = -\infty$.

Definition 1.4.6. A set $\varnothing \neq S \subset \mathbb{R}$ is *bounded* if $\exists M > 0$ such that $|x| \leq M, \forall x \in S$.

It can be formally shown that $\mathbb{R}$ possesses the so-called *completeness property*. The importance of this concept in the present and more abstract settings cannot be overemphasized. We state it in the form of a theorem to highlight its importance. Consult **[17, 234]** for a proof.

Theorem 1.4.7. *If $\varnothing \neq S \subset \mathbb{R}$ is bounded above, then $\exists u \in \mathbb{R}$ such that $u = \sup(S)$. We say $\mathbb{R}$ is complete.*

Remark. The duality between the statements concerning sup and inf leads to the formulation of the following alternate statement of the completeness property:

$$If \, \varnothing \neq T \subset \mathbb{R} \, is \, bounded \, below, then \, \exists v \in \mathbb{R} \, such \, that \, v = \inf(S). \qquad (1.3)$$

Exercise 1.4.7. Prove that (1.3) is equivalent to Thrm 1.4.7.

Proposition 1.4.8. (Properties of inf and sup) *Let $\varnothing \neq S, T \subset \mathbb{R}$.*
i.) *Assume $\exists \sup(S)$. Then, $\forall \varepsilon > 0, \exists x \in S$ such that $\sup(S) - \varepsilon < x \leq \sup(S)$.*
ii.) *If $S \subset T$ and $\exists \sup(T)$, then $\exists \sup(S)$ and $\sup(S) \leq \sup(T)$.*
iii.) *Let $S + T = \{s + t | s \in S \wedge t \in T\}$. If S and T are bounded above, then $\exists \sup(S + T)$ and it equals $\sup(S) + \sup(T)$.*
iv.) *Let $c \in \mathbb{R}$ and define $cS = \{cs | s \in S\}$. If S is bounded, then $\exists \sup(cS)$ given by*

$$\sup(cS) = \begin{cases} c \cdot \sup(S), & if \, c \geq 0, \\ c \cdot \inf(S), & if \, c < 0. \end{cases} \qquad (1.4)$$

v.) *Let $\varnothing \neq S, T \subset (0, \infty)$ and define $S \cdot T = \{s \cdot t | s \in S \wedge t \in T\}$. If S and T are bounded above, then $\exists \sup(S \cdot T)$ and it equals $\sup(S) \cdot \sup(T)$.*

Proof. We prove (iii) and leave the others for you to verify as an exercise.
    Because S and T are nonempty, $S + T \neq \varnothing$. Further, because

$$s + t \leq \sup(S) + \sup(T), \forall s \in S, t \in T, \qquad (1.5)$$

it follows that $\sup(S) + \sup(T)$ is an upper bound of $(S + T)$. (Why?) Hence, $\exists \sup(S + T)$ and

$$\sup(S + T) \leq \sup(S) + \sup(T). \qquad (1.6)$$

To establish the reverse inequality, let $\varepsilon > 0$. By Prop. 1.4.8, $\exists s_0 \in S$ and $t_0 \in T$ such that

$$\sup(S) - \frac{\varepsilon}{2} < s_0 \text{ and } \sup(T) - \frac{\varepsilon}{2} < t_0. \tag{1.7}$$

Consequently,

$$\sup(S) + \sup(T) - \varepsilon < s_0 + t_0 \leq \sup(S + T). \tag{1.8}$$

Thus, we conclude from Prop. 1.4.2 that

$$\sup(S) + \sup(T) \leq \sup(S + T). \tag{1.9}$$

Claim (iii) now follows from (1.6) and (1.9). (Why?) ☐

Exercise 1.4.8.
i.) Prove the remaining parts of Prop. 1.4.8.
ii.) Formulate statements analogous to those in Prop. 1.4.8 for infs. Indicate the changes that must be implemented in the proofs.

Remark. Prop 1.4.8(i) indicates that we can get "arbitrarily close" to $\sup(S)$ with elements of S. This is especially useful in convergence arguments.

1.4.4 Topology of $\mathbb{R}$

You have worked with open and closed intervals in calculus, but what do the terms *open* and *closed* mean? Is there any significant difference between them? The notion of an open set is central to the construction of a so-called *topology* on $\mathbb{R}$. Interestingly, many of the theorems from calculus are formulated on closed, bounded intervals for very good reason. As we proceed with our analysis of $\mathbb{R}$, you will see that many of these results are consequences of some fairly deep topological properties of $\mathbb{R}$ which, in turn, follow from the completeness property.

Definition 1.4.9. Let $S \subset \mathbb{R}$.
i.) x is an *interior point* (int pt) of S if $\exists \varepsilon > 0$ such that $(x - \varepsilon, x + \varepsilon) \subset S$.
ii.) x is a *limit point* (lim pt) of S if $\forall \varepsilon > 0$, $(x - \varepsilon, x + \varepsilon) \cap S$ is infinite.
iii.) x is a *boundary point* (bdry pt) of S if

$$\forall \varepsilon > 0, (x - \varepsilon, x + \varepsilon) \cap S \neq \emptyset \text{ and } (x - \varepsilon, x + \varepsilon) \cap \tilde{S} \neq \emptyset.$$

iv.) The *boundary* of S is the set $\partial S = \{x \in \mathbb{R} | x \text{ is a bdry pt of } S\}$.
v.) The *interior* of S is the set $\text{int}(S) = \{x \in \mathbb{R} | x \text{ is an int pt of } S\}$.
vi.) The *derived set* of S is the set $S' = \{x \in \mathbb{R} | x \text{ is a lim pt of } S\}$.
vii.) The *closure* of S is the set $\text{cl}_{\mathbb{R}}(S) = S \cup S'$.
viii.) S is *open* if every point of S is an int pt of S.
ix.) S is *closed* if S contains all of its lim pts.

Illustrating these concepts using a number line can facilitate your understanding of them. Do so when completing the following exercise.

Exercise 1.4.9. For each of these sets S, compute $\text{int}(S)$, S', and $\text{cl}_{\mathbb{R}}(S)$. Also, determine if S is open, closed, both, or neither.

i.) $[1,5]$

ii.) $\mathbb{Q}$

iii.) $\left\{\frac{1}{n} \big| n \in \mathbb{N}\right\}$

iv.) $\mathbb{R}$

v.) $\varnothing$

It is not difficult to establish the following duality between a set and its complement. It is often a useful tool when proving statements about open and closed sets.

Proposition 1.4.10. *Let $S \subset \mathbb{R}$. S is open iff $\widetilde{S}$ is closed.*

Exercise 1.4.10. Verify the following properties of open and closed sets.

i.) Let $n \in \mathbb{N}$. If $G_1,\ldots,G_n$ is a finite collection of open sets, then $\bigcap_{k=1}^{n} G_k$ is open.

ii.) Let $n \in \mathbb{N}$. If $F_1,\ldots,F_n$ is a finite collection of closed sets, then $\bigcup_{k=1}^{n} F_k$ is closed.

iii.) Let $\Gamma \neq \varnothing$. If G_γ is open, $\forall \gamma \in \Gamma$, then $\bigcup_{\gamma \in \Gamma} G_\gamma$ is open.

iv.) Let $\Gamma \neq \varnothing$. If F_γ is closed, $\forall \gamma \in \Gamma$, then $\bigcap_{\gamma \in \Gamma} F_\gamma$ is closed.

v.) If $S \subset T$, then $\text{int}(S) \subset \text{int}(T)$.

vi.) If $S \subset T$, then $\text{cl}_{\mathbb{R}}(S) \subset \text{cl}_{\mathbb{R}}(T)$.

Exercise 1.4.11. Let $\varnothing \neq S \subset \mathbb{R}$. Prove the following:

i.) If S is bounded above, then $\sup(S) \in \text{cl}_{\mathbb{R}}(S)$.

ii.) If S is bounded above and closed, then $\max(S) \in S$.

iii.) Formulate results analogous to (i) and (ii) assuming that S is bounded below.

Intuitively, S' is the set of points to which elements of S become arbitarily close. It is natural to ask if there are proper subsets of $\mathbb{R}$ that sprawl widely enough through $\mathbb{R}$ as to be sufficiently near every real number. Precisely, consider sets of the following type.

Definition 1.4.11. A set $\varnothing \neq S \subset \mathbb{R}$ is *dense* in $\mathbb{R}$ if $\text{cl}_{\mathbb{R}}(S) = \mathbb{R}$.

Exercise 1.4.12. Identify two different subsets of $\mathbb{R}$ that are dense in $\mathbb{R}$.

By way of motivation for the first major consequence of completeness, consider the following exercise.

Exercise 1.4.13. Provide examples, if possible, of sets $S \subset \mathbb{R}$ illustrating the following scenarios.

i.) S is bounded, but $S' = \varnothing$.

ii.) S is infinite, but $S' = \varnothing$.

iii.) S is bounded and infinite, but $S' = \varnothing$.

As you discovered in Exercise 1.4.13, the combination of bounded and infinite for a set S of real numbers implies the existence of a limit point of S. This is a consequence of the following theorem due to Bolzano and Weierstrass.

Theorem 1.4.12. (Bolzano-Weierstrass) *If S is a bounded, infinite subset of $\mathbb{R}$, then $S' \neq \emptyset$.*

Outline of Proof: Let $T = \{x \in \mathbb{R} \mid S \cap (x, \infty) \text{ is infinite}\}$. Then,

$$T \neq \emptyset. \text{ (Why?)} \tag{1.10}$$
$$T \text{ is bounded above. (Why?)} \tag{1.11}$$
$$\exists \sup(T); \text{ call it } t. \text{ (Why?)} \tag{1.12}$$
$$\forall \varepsilon > 0, \ S \cap (t - \varepsilon, \infty) \text{ is infinite. (Why?)} \tag{1.13}$$
$$\forall \varepsilon > 0, \ S \cap [t + \varepsilon, \infty) \text{ is finite. (Why?)} \tag{1.14}$$
$$\forall \varepsilon > 0, \ S \cap (t - \varepsilon, t + \varepsilon) \text{ is infinite. (Why?)} \tag{1.15}$$
$$t \in S'. \text{ (Why?)} \tag{1.16}$$

This completes the proof. $\qquad\qquad\qquad\qquad\qquad\qquad\qquad\qquad\qquad\qquad\square$

Exercise 1.4.14. Provide the details in the proof of Thrm. 1.4.12. Where was completeness used?

Another important concept is that of compactness. Some authors define this notion more generally using open covers (see **[17]**).

Definition 1.4.13. A set $S \subset \mathbb{R}$ is *compact* if every infinite subset of S has a limit point *in S*.

Remark. The "in S" portion of Def. 1.4.13 is crucial, and it distinguishes between the sets $(0, 1)$ and $[0, 1]$, for instance. (Why?) This is evident in Thrm. 1.4.14.

Exercise 1.4.15. Try to determine if the following subsets of $\mathbb{R}$ are compact.
i.) Any finite set.
ii.) $\{\frac{1}{n} \mid n \in \mathbb{N}\}$ versus $\{\frac{1}{n} \mid n \in \mathbb{N}\} \cup \{0\}$
iii.) $\mathbb{Q}$
iv.) $\mathbb{Q} \cap [0, 1]$
v.) $\mathbb{N}$
vi.) $\mathbb{R}$
vii.) $(0, 1)$ versus $[0, 1]$

Both the completeness property and finite dimensionality of $\mathbb{R}$ enter into the proof of the following characterization theorem for compact subsets of $\mathbb{R}$. The proof can be found in **[17]**.

Theorem 1.4.14. (Heine-Borel) *A set $S \subset \mathbb{R}$ is compact iff S is closed and bounded.*

Exercise 1.4.16. Revisit Exer. 1.4.15 in light of Thrm. 1.4.14.

1.5   Sequences in $(\mathbb{R}, |\cdot|)$

Sequences play a prominent role in analysis, especially in the development of numerical schemes used for approximation purposes.

1.5.1   Sequences and Subsequences

Definition 1.5.1. A *sequence* in $\mathbb{R}$ is a function $x : \mathbb{N} \to \mathbb{R}$. We often write x_n for $x(n)$, $n \in \mathbb{N}$, called the nth-*term* of the sequence, and denote the sequence itself by $\{x_n\}$ or by enumerating the range as $x_1, x_2, x_3, \ldots$.

The notions of monotonicity and boundedness given in Defs. 1.3.6 and 1.4.6 apply in particular to sequences. We formulate them in this specific setting for later reference.

Definition 1.5.2. A sequence is called
i.) *nondecreasing* whenever $x_n \leq x_{n+1}$, $\forall n \in \mathbb{N}$;
ii.) *increasing* whenever $x_n < x_{n+1}$, $\forall n \in \mathbb{N}$;
iii.) *nonincreasing* whenever $x_n \geq x_{n+1}$, $\forall n \in \mathbb{N}$;
iv.) *decreasing* whenever $x_n > x_{n+1}$, $\forall n \in \mathbb{N}$;
v.) *monotone* if any of (i)–(iv) are satisfied;
vi.) *bounded above* (resp. *below*) if $\exists M \in \mathbb{R}$ such that $x_n \leq M$ (resp. $x_n \geq M$), $\forall n \in \mathbb{N}$;
vii.) *bounded* whenever $\exists M > 0$ such that $|x_n| \leq M$, $\forall n \in \mathbb{N}$.

Exercise 1.5.1. Explain why a nondecreasing (resp. nonincreasing) sequence must be bounded below (resp. above).

Definition 1.5.3. If $x : \mathbb{N} \to \mathbb{R}$ is a sequence in $\mathbb{R}$ and $n : \mathbb{N} \to \mathbb{N}$ is an increasing sequence in $\mathbb{N}$, then the composition $x \circ n : \mathbb{N} \to \mathbb{R}$ is called a *subsequence* of x in $\mathbb{R}$.

Though this is a formal definition of a subsequence, let us examine carefully what this means using more conventional notation. Suppose that the terms of Def. 1.5.3 are represented by $\{x_n\}$ and $\{n_k\}$, respectively. Because $\{n_k\}$ is increasing, we know that $n_1 < n_2 < n_3 < \ldots$. Then, the official subsequence $x \circ n$ has values $(x \circ n)(k) = x(n(k))$, which, using our notation, can be written as x_{n_k}, $\forall k \in \mathbb{N}$. Thus, the integers n_k are just the indices of those terms of the original sequence that are retained in the subsequence as k increases, and roughly speaking, the remainder of the terms are omitted.

1.5.2   Limit Theorems

We now consider the important notion of convergence.

Definition 1.5.4. A sequence $\{x_n\}$ has *limit* L whenever $\forall \varepsilon > 0$, $\exists N \in \mathbb{N}$ (N depending in general on ε) such that

$$n \geq N \implies |x_n - L| < \varepsilon.$$

In such case, we write $\lim_{n\to\infty} x_n = L$ or $x_n \longrightarrow L$ and say that $\{x_n\}$ *converges* (or *is convergent*) to L. Otherwise, we say $\{x_n\}$ *diverges*.

If we paraphrase Def. 1.5.4, it would read: $\lim_{n\to\infty} x_n = L$ whenever given <u>any</u> open interval $(L-\varepsilon, L+\varepsilon)$ around L (that is, no matter how small the positive number ε is), it is the case that $x_n \in (L-\varepsilon, L+\varepsilon)$ for all but possibly finitely many indices n. That is, the "tail" of the sequence ultimately gets into every open interval around L. Also note that, in general, the smaller the ε, the larger the index N must be used (to get deeper into the tail) because ε is an error gauge, namely how far the terms are from the target. We say N must be chosen "sufficiently large" as to ensure the tail behaves in this manner for the given ε.

Exercise 1.5.2.
i.) Precisely define $\lim_{n\to\infty} x_n \neq L$.
ii.) Prove that $x_n \longrightarrow L$ iff $|x_n - L| \longrightarrow 0$.

Example. As an illustration of Def. 1.5.4, we prove that $\lim_{n\to\infty} \frac{2n^2+n+5}{n^2+1} = 2$.
Let $\varepsilon > 0$. We must argue that $\exists N \in \mathbb{N}$ such that

$$n \geq N \implies \left| \frac{2n^2+n+5}{n^2+1} - 2 \right| < \varepsilon. \tag{1.17}$$

To this end, note that $\exists N \in \mathbb{N}$ such that $N > 3$ and $N\varepsilon > 2$. (Why?) We show this N "works." Indeed, observe that $\forall n \geq N$,

$$\left| \frac{2n^2+n+5}{n^2+1} - 2 \right| = \left| \frac{2n^2+n+5-2n^2-2}{n^2+1} \right| = \frac{n+3}{n^2+1}. \tag{1.18}$$

Subsequently, by choice of N, we see that $n \geq N > 3$ and for all such n,

$$\frac{n+3}{n^2+1} < \frac{2n}{n^2+1} < \frac{2n}{n^2} = \frac{2}{n} < \frac{2}{N} < \varepsilon. \tag{1.19}$$

(Why?) Thus, by definition, it follows that $\lim_{n\to\infty} \frac{2n^2+n+5}{n^2+1} = 2$.   □

Exercise 1.5.3. Use Def. 1.5.4 to prove that $\lim_{n\to\infty} \frac{a}{n} = 0, \forall a \in \mathbb{R}$.

We now discuss the main properties of convergence. We mainly provide outlines of proofs, the details of which you are encouraged to provide.

Proposition 1.5.5. *If $\{x_n\}$ is a convergent sequence, then its limit is unique.*

Outline of Proof: Let $\lim_{n\to\infty} x_n = L_1$ and $\lim_{n\to\infty} x_n = L_2$ and suppose that, by way of contradiction, $L_1 \neq L_2$.
Let $\varepsilon = \frac{|L_1-L_2|}{2}$. Then, $\varepsilon > 0$. (Why?)
$\exists N_1 \in \mathbb{N}$ such that $n \geq N_1 \implies |x_n - L_1| < \varepsilon$. (Why?)

$\exists N_2 \in \mathbb{N}$ such that $n \geq N_2 \implies |x_n - L_2| < \varepsilon$. (Why?)
Choose $N = \max\{N_1, N_2\}$. Then, $|x_N - L_1| < \varepsilon$ and $|x_N - L_2| < \varepsilon$. (Why?)
Consequently, $2\varepsilon = |L_1 - L_2| \leq |x_N - L_1| + |x_N - L_2| < 2\varepsilon$. (Why?)
Thus, $L_1 = L_2$. (How?)
This completes the proof. □

Proposition 1.5.6. *If $\{x_n\}$ is a convergent sequence, then it is bounded.*

Outline of Proof: Assume that $\lim_{n \to \infty} x_n = L$. We must produce an $M > 0$ such that $|x_n| \leq M, \forall n \in \mathbb{N}$. Using $\varepsilon = 1$ in Def. 1.5.4, we know that

$$\exists N \in \mathbb{N} \text{ such that } n \geq N \implies |x_n - L| < \varepsilon = 1. \tag{1.20}$$

Using Prop. 1.4.4(ix) in (1.20) then yields

$$|x_n| < |L| + 1, \forall n \geq N. \tag{1.21}$$

(Tell how.) For how many values of n does x_n possibly not satisfy (1.21)? How do you use this fact to construct a positive real number M satisfying Def. 1.5.2(vii)? □

Proposition 1.5.7. (Squeeze Theorem) *Let $\{x_n\}, \{y_n\}$, and $\{z_n\}$ be sequences such that*

$$x_n \leq y_n \leq z_n, \forall n \in \mathbb{N}, \tag{1.22}$$

and

$$\lim_{n \to \infty} x_n = L = \lim_{n \to \infty} z_n. \tag{1.23}$$

Then, $\lim_{n \to \infty} y_n = L$.

Outline of Proof: Let $\varepsilon > 0$. From (1.23) we know that $\exists N_1, N_2 \in \mathbb{N}$ such that

$$|x_n - L| < \varepsilon, \forall n \geq N_1 \text{ and } |z_n - L| < \varepsilon, \forall n \geq N_2. \tag{1.24}$$

Specifically,

$$-\varepsilon < x_n - L, \forall n \geq N_1 \text{ and } z_n - L < \varepsilon, \forall n \geq N_2. \tag{1.25}$$

Choose $N = \max\{N_1, N_2\}$. Using (1.25) we see that

$$-\varepsilon < x_n - L, \text{ and } z_n - L < \varepsilon, \forall n \geq N. \text{ (Why?)}$$

Using this with (1.22) we can conclude that

$$n \geq N \implies -\varepsilon < y_n - L < \varepsilon. \text{ (Why?)}$$

Hence, $\lim_{n \to \infty} y_n = L$, as desired. □

Remark. The conclusion of Prop. 1.5.7 holds true if we replace (1.22) by

$$\exists N_0 \in \mathbb{N} \text{ such that } x_n \leq y_n \leq z_n, \forall n \geq N_0. \tag{1.26}$$

Suitably modify the way N is chosen in the proof of Prop. 1.5.7 to account for this more general condition. (Tell how.)

Proposition 1.5.8. *If* $\lim\limits_{n\to\infty} x_n = L$, *where* $L \neq 0$, *then* $\exists m > 0$ *and* $N \in \mathbb{N}$ *such that*

$$|x_n| > m, \forall n \geq N.$$

(In words, if a sequence has a nonzero limit, then its terms must be bounded away from zero for sufficiently large indices n.)

Outline of Proof:

Let $\varepsilon = \frac{|L|}{2}$. Then, $\varepsilon > 0$. (Why?)

$\exists N \in \mathbb{N}$ such that $|x_n - L| < \varepsilon = \frac{|L|}{2}, \forall n \geq N$. (Why?)

Thus, $||x_n| - |L|| < \frac{|L|}{2}, \forall n \geq N$. (Why?)

That is, $-\frac{|L|}{2} < |x_n| - |L| < \frac{|L|}{2}, \forall n \geq N$. (Why?)

So, $\frac{|L|}{2} < |x_n|, \forall n \geq N$.

The conclusion follows by choosing $m = \frac{|L|}{2}$. (Why?) $\qquad\qquad\square$

Proposition 1.5.9. *Suppose that* $\lim\limits_{n\to\infty} x_n = L$ *and* $\lim\limits_{n\to\infty} y_n = M$. *Then,*

i.) $\lim\limits_{n\to\infty} (x_n + y_n) = L + M;$

ii.) $\lim\limits_{n\to\infty} x_n y_n = LM.$

Outline of Proof:

<u>Proof of (i):</u> The strategy is straightforward. Because there are two sequences, we split the given error tolerance ε into two parts of size $\frac{\varepsilon}{2}$ each, apply the limit definition to each sequence with the $\frac{\varepsilon}{2}$ tolerance, and finally put the two together using the triangle inequality.

Let $\varepsilon > 0$. Then, $\frac{\varepsilon}{2} > 0$. We know that

$$\exists N_1 \in \mathbb{N} \text{ such that } |x_n - L| < \frac{\varepsilon}{2}, \forall n \geq N_1. \text{ (Why?)} \qquad (1.27)$$

$$\exists N_2 \in \mathbb{N} \text{ such that } |y_n - M| < \frac{\varepsilon}{2}, \forall n \geq N_2. \text{ (Why?)} \qquad (1.28)$$

How do you then select $N \in \mathbb{N}$ such that (1.27) and (1.28) hold simultaneously for all $n \geq N$? For such an N, observe that

$$n \geq N \implies |(x_n + y_n) - (L + M)| \leq |x_n - L| + |y_n - M| < \varepsilon. \qquad (1.29)$$

(Why?) Hence, we conclude that $\lim\limits_{n\to\infty} (x_n + y_n) = L + M$.

<u>Proof of (ii):</u> This time the strategy is a bit more involved. We need to show that $|x_n y_n - LM|$ can be made arbitrarily small for sufficiently large n using the hypotheses that $|x_n - L|$ and $|y_n - M|$ can each be made arbitrarily small for sufficiently large n. This requires two approximations, viz., making x_n close to L while simultaneously making y_n close to M. This suggests that we bound $|x_n y_n - LM|$ above by an expression involving $|x_n - L|$ and $|y_n - M|$. To accomplish this, we add and subtract

the same middle term in $|x_n y_n - LM|$ and apply certain absolute value properties. Precisely, observe that

$$
\begin{aligned}
|x_n y_n - LM| &= |x_n y_n - Mx_n + Mx_n - LM| \\
&= |x_n(y_n - M) + M(x_n - L)| \\
&\leq |x_n| |y_n - M| + |M| |x_n - L|.
\end{aligned}
\tag{1.30}
$$

(This trick is a workhorse throughout the text!) The tack now is to show that both terms on the right-hand side of (1.30) can be made less than $\frac{\varepsilon}{2}$ for sufficiently large n.

Let $\varepsilon > 0$. Proposition 1.5.6 implies that $\exists K > 0$ for which

$$
|x_n| \leq K, \forall n \in \mathbb{N}.
\tag{1.31}
$$

Also, because $\{y_n\}$ is convergent to M, $\exists N_1 \in \mathbb{N}$ such that

$$
|y_n - M| < \frac{\varepsilon}{2K}, \forall n \geq N_1.
\tag{1.32}
$$

In view of (1.31) and (1.32), we obtain

$$
n \geq N_1 \implies |x_n| |y_n - M| \leq K |y_n - M| < K \cdot \frac{\varepsilon}{2K} = \frac{\varepsilon}{2}.
\tag{1.33}
$$

This takes care of the first term in (1.30). Next, because $\{x_n\}$ is convergent to L, $\exists N_2 \in \mathbb{N}$ such that

$$
|x_n - L| < \frac{\varepsilon}{2(|M| + 1)}, \forall n \geq N_2.
\tag{1.34}
$$

Now, argue in a manner similar to (1.33) to conclude that

$$
n \geq N_2 \implies |M| |x_n - L| < \frac{\varepsilon}{2}. \text{ (Tell how.)}
\tag{1.35}
$$

Choose $N \in \mathbb{N}$ so that (1.33) and (1.35) hold simultaneously, $\forall n \geq N$. Then, use (1.30) through (1.35) to conclude that

$$
n \geq N \implies |x_n y_n - LM| < \varepsilon. \text{ (How?)}
$$

Hence, we conclude that $\lim_{n\to\infty} x_n y_n = LM$. This completes the proof. $\qquad\square$

Because we were unfolding the argument in somewhat reverse order for motivation, it would be better now to start with $\varepsilon > 0$ and reorganize the train of the suggested argument into a polished proof. (Do so!)

Exercise 1.5.4. Let $c \in \mathbb{R}$ and assume that $\lim_{n\to\infty} x_n = L$ and $\lim_{n\to\infty} y_n = M$. Prove that

i.) $\lim_{n\to\infty} cx_n = cL$,

ii.) $\lim_{n\to\infty} (x_n - y_n) = L - M$.

The following lemma can be proven easily using induction. (Tell how.)

Lemma 1.5.10. *If $\{n_k\} \subset \mathbb{N}$ is an increasing sequence, then $n_k \geq k$, $\forall k \in \mathbb{N}$.*

Proposition 1.5.11. *If $\lim\limits_{n\to\infty} x_n = L$ and $\{x_{n_k}\}$ is any subsequence of $\{x_n\}$, then $\lim\limits_{k\to\infty} x_{n_k} = L$. (In words, all subsequences of a sequence convergent to L also converge to L.)*

Outline of Proof: Let $\varepsilon > 0$. There exists $N \in \mathbb{N}$ such that

$$|x_n - L| < \varepsilon, \forall n \geq N.$$

Now, fix any $K_0 \geq N$ and use Lemma 1.5.10 to infer that

$$k \geq K_0 \implies n_k > k \geq K_0 \geq N \implies |x_{n_k} - L| < \varepsilon. \text{ (Why?)}$$

The conclusion now follows. (Tell how.) □

Exercise 1.5.5. Prove that if $\lim\limits_{n\to\infty} x_n = 0$ and $\{y_n\}$ is bounded, then $\lim\limits_{n\to\infty} x_n y_n = 0$.

Exercise 1.5.6.
i.) Prove that if $\lim\limits_{n\to\infty} x_n = L$, then $\lim\limits_{n\to\infty} |x_n| = |L|$.
ii.) Provide an example of a sequence $\{x_n\}$ for which $\exists \lim\limits_{n\to\infty} |x_n|$, but $\nexists \lim\limits_{n\to\infty} x_n$.

Exercise 1.5.7. Prove the following:
i.) If $\lim\limits_{n\to\infty} x_n = L$, then $\lim\limits_{n\to\infty} x_n^p = L^p$, $\forall p \in \mathbb{N}$.
ii.) If $x_n > 0$, $\forall n \in \mathbb{N}$, and $\lim\limits_{n\to\infty} x_n = L$, then $\lim\limits_{n\to\infty} \sqrt{x_n} = \sqrt{L}$.

Important connections between sequences and the derived set and closure of a set are provided in the following exercise.

Exercise 1.5.8. Let $\varnothing \neq S \subset \mathbb{R}$. Prove the following:
i.) $x \in S'$ iff $\exists \{x_n\} \subset S \setminus \{x\}$ such that $\lim\limits_{n\to\infty} x_n = x$.
ii.) $x \in cl_{\mathbb{R}}(S)$ iff $\exists \{x_n\} \subset S$ such that $\lim\limits_{n\to\infty} x_n = x$.

Proposition 1.5.12. *If $\{x_n\}$ is a bounded sequence in $\mathbb{R}$, then there exists a convergent subsequence $\{x_{n_k}\}$ of $\{x_n\}$.*

Outline of Proof: Let $\mathscr{R}_x = \{x_n | n \in \mathbb{N}\}$. We split the proof into two cases.

Case 1: $\mathscr{R}_x$ is a finite set, say $\mathscr{R}_x = \{y_1, y_2, \ldots, y_m\}$.
It cannot be the case that the set $x^{-1}(\{y_i\}) = \{n \in \mathbb{N} | x_n = y_i\}$ is finite, for every $i \in \{1, 2, \ldots, m\}$ because $\mathbb{N} = \bigcup_{i=1}^m x^{-1}(\{y_i\})$. (Why?) As such, there is at least one $i_0 \in \{1, 2, \ldots, m\}$ such that $x^{-1}(\{y_{i_0}\})$ is infinite. Use this fact to inductively construct a sequence $n_1 < n_2 < \ldots$ in $\mathbb{N}$ such that $x_{n_k} = y_{i_0}$, $\forall k \in \mathbb{N}$. (Tell how.) Observe that $\{x_{n_k}\}$ is a convergent subsequence of $\{x_n\}$. (Why?)

Case 2: $\mathscr{R}_x$ is infinite.

Because $\{x_n\}$ is bounded, it follows from Thrm. 1.4.12 that $\mathscr{R}'_x \neq \varnothing$, say $L \in \mathscr{R}'_x$. Use the definition of limit point to inductively construct a subsequence $\{x_{n_k}\}$ of $\{x_n\}$ such that $x_{n_k} \longrightarrow L$. How does this complete the proof?   □

The combination of the hypotheses of monotonicity and boundedness implies convergence, as the next result suggests.

Proposition 1.5.13. *If $\{x_n\}$ is a nondecreasing sequence that is bounded above, then $\{x_n\}$ converges and* $\lim\limits_{n \to \infty} x_n = \sup \{x_n | n \in \mathbb{N}\}$.

Outline of Proof: Because $\{x_n | n \in \mathbb{N}\}$ is a nonempty subset of $\mathbb{R}$ that is bounded above, $\exists \sup \{x_n | n \in \mathbb{N}\}$, call it L. (Why?) Let $\varepsilon > 0$. Then,

$$\exists N \in \mathbb{N} \text{ such that } L - \varepsilon < x_N. \text{ (Why?)}$$

Consequently,

$$n \geq N \Longrightarrow L - \varepsilon < x_N \leq x_n \leq L < L + \varepsilon \Longrightarrow |x_n - L| < \varepsilon.$$

(Why?) This completes the proof.   □

Exercise 1.5.9. Formulate and prove a result analogous to Prop. 1.5.13 for nonincreasing sequences.

Exercise 1.5.10.
i.) Let $\{x_k\}$ be a sequence of nonnegative real numbers. For every $n \in \mathbb{N}$, define $s_n = \sum_{k=1}^{n} x_k$. Prove that the sequence $\{s_n\}$ converges iff it is bounded above.
ii.) Prove that $\left\{ \frac{a^n}{n!} \right\}$ converges, $\forall a \in \mathbb{R}$. In fact, $\lim\limits_{n \to \infty} \frac{a^n}{n!} = 0$.

Now that we know about subsequences, it is convenient to introduce a generalization of the notion of the limit of a real-valued sequence. We make the following definition.

Definition 1.5.14. Let $\{x_n\} \subset \mathbb{R}$ be a sequence.
i.) We say that $\lim\limits_{n \to \infty} x_n = \infty$ whenever $\forall r > 0, \exists N \in \mathbb{N}$ such that $x_n > r, \forall n \geq N$. (In such case, we write $x_n \to \infty$.)
ii.) For every $n \in \mathbb{N}$, let $u_n = \sup \{x_k : k \geq n\}$. We define the *limit superior* of x_n by

$$\varlimsup_{n \to \infty} x_n = \inf \{u_n | n \in \mathbb{N}\} = \inf_{n \in \mathbb{N}} \left(\sup_{k \geq n} x_k \right).$$

iii.) The dual notion of *limit inferior*, denoted $\varliminf\limits_{n \to \infty} x_n$, is defined analogously with sup and inf interchanged in (ii), viz.,

$$\varliminf_{n \to \infty} x_n = \sup_{n \in \mathbb{N}} \left(\inf_{k \geq n} x_k \right).$$

Some properties of limit superior (inferior) are gathered below. The proofs are standard and can be found in standard analysis texts (see [234]).

Proposition 1.5.15. (Properties of Limit Superior and Inferior)
i.) $\overline{\lim}_{n\to\infty} x_n = p \in \mathbb{R}$ *iff* $\forall \varepsilon > 0,$
   a.) *There exist only finitely many n such that* $x_n > p + \varepsilon$, *and*
   b.) *There exist infinitely many n such that* $x_n > p - \varepsilon$;
ii.) $\overline{\lim}_{n\to\infty} x_n = p \in \mathbb{R}$ *iff p is the largest limit of any subsequence of* $\{x_n\}$;
iii.) $\overline{\lim}_{n\to\infty} x_n = \infty$ *iff* $\forall r \in \mathbb{R}, \exists$ *infinitely many n such that* $x_n > r$;
iv.) *If* $x_n < y_n, \forall n \in \mathbb{N}$, *then*
   a.) $\overline{\lim}_{n\to\infty} x_n \leq \overline{\lim}_{n\to\infty} y_n$,
   b.) $\underline{\lim}_{n\to\infty} x_n \leq \underline{\lim}_{n\to\infty} y_n$;
v.) $\overline{\lim}_{n\to\infty} (-x_n) = -\underline{\lim}_{n\to\infty} x_n$;
vi.) $\underline{\lim}_{n\to\infty} x_n \leq \overline{\lim}_{n\to\infty} x_n$;
vii.) $\lim_{n\to\infty} x_n = p$ *iff* $\underline{\lim}_{n\to\infty} x_n \leq \overline{\lim}_{n\to\infty} x_n = p$;
viii.) $\underline{\lim}_{n\to\infty} x_n + \underline{\lim}_{n\to\infty} y_n \leq \underline{\lim}_{n\to\infty} (x_n + y_n) \leq \overline{\lim}_{n\to\infty} (x_n + y_n) \leq \overline{\lim}_{n\to\infty} x_n + \overline{\lim}_{n\to\infty} y_n$;
ix.) *If* $x_n \geq 0$ *and* $y_n \geq 0, \forall n \in \mathbb{N}$, *then* $\overline{\lim}_{n\to\infty} (x_n y_n) \leq \left(\overline{\lim}_{n\to\infty} x_n \right) \left(\overline{\lim}_{n\to\infty} y_n \right)$, *provided the product on the right is not of the form* $0 \cdot \infty$.

1.5.3   Cauchy Sequences

Definition 1.5.16. A sequence $\{x_n\}$ is a *Cauchy sequence* if $\forall \varepsilon > 0, \exists N \in \mathbb{N}$ such that

$$n, m \geq N \implies |x_n - x_m| < \varepsilon.$$

Intuitively, the terms of a Cauchy sequence squeeze together as the index increases. Given any positive error tolerance ε, there is an index past which any two terms of the sequence, no matter how greatly their indices differ, have values within the tolerance of ε of one another. For brevity, we often write "$\{x_n\}$ is Cauchy" instead of "$\{x_n\}$ is a Cauchy sequence."

Exercise 1.5.11. Prove that the following statements are equivalent:
  (1) $\{x_n\}$ is a Cauchy sequence.
  (2) $\forall \varepsilon > 0, \exists N \in \mathbb{N}$ such that $|x_{N+p} - x_{N+q}| < \varepsilon, \forall p, q \in \mathbb{N}$.
  (3) $\forall \varepsilon > 0, \exists N \in \mathbb{N}$ such that $|x_n - x_N| < \varepsilon, \forall n \geq N$.
  (4) $\forall \varepsilon > 0, \exists N \in \mathbb{N}$ such that $|x_{N+p} - x_N| < \varepsilon, \forall p \in \mathbb{N}$.
  (5) $\lim_{n\to\infty} (x_{n+p} - x_n) = 0, \forall p \in \mathbb{N}$.

We could have included the statement "$\{x_n\}$ is a convergent sequence" in the above list and asked which others imply it or are implied by it. Indeed, which of the two statements

$$\{x_n\} \text{ is a convergent sequence}$$

or

$$\{x_n\} \text{ is a Cauchy sequence}$$

seems stronger to you? Which implies which, if either? We will revisit this question after the following lemma.

Lemma 1.5.17. (Properties of Cauchy Sequences in $\mathbb{R}$)
i.) *A Cauchy sequence is bounded.*
ii.) *If a Cauchy sequence $\{x_n\}$ has a subsequence $\{x_{n_k}\}$ that converges to L, then $\{x_n\}$ itself converges to L.*

Outline of Proof:
Proof of (i): Let $\{x_n\}$ be a Cauchy sequence. Then, by Def. 1.5.16, $\exists N \in \mathbb{N}$ such that

$$n, m \geq N \implies |x_n - x_m| < 1.$$

In particular,

$$n \geq N \implies |x_n - x_N| < 1.$$

Starting with the last statement, argue as in Prop. 1.5.6 that $|x_n| \leq M, \forall n \in \mathbb{N}$, where

$$M = \max\{|x_1|, |x_2|, \ldots, |x_{N-1}|, |x_N| + 1\}.$$

So, $\{x_n\}$ is bounded.

Proof of (ii): Let $\varepsilon > 0$. $\exists N_1 \in \mathbb{N}$ such that

$$n, m \geq N_1 \implies |x_n - x_m| < \frac{\varepsilon}{2} \tag{1.36}$$

and $\exists N_2 \in \mathbb{N}$ such that

$$k \geq N_2 \implies |x_{n_k} - L| < \frac{\varepsilon}{2}. \tag{1.37}$$

(Why?) Now, how do you select N so that (1.36) and (1.37) hold simultaneously? Let $n > N$ and choose any $k \in \mathbb{N}$ such that $k \geq N$. Then, $n \geq N_1$ and $n_k \geq N$. (Why?) As such,

$$n \geq N \implies |x_n - L| = |x_n - x_{n_k} + x_{n_k} - L| \leq |x_n - x_{n_k}| + |x_{n_k} - L| < \varepsilon.$$

(Why?) This completes the proof. $\square$

We now shall prove that convergence and Cauchy are equivalent notions in $\mathbb{R}$.

Theorem 1.5.18. (Cauchy Criterion in $\mathbb{R}$)
$\{x_n\}$ *is convergent* $\Longleftrightarrow \{x_n\}$ *is a Cauchy sequence.*

Outline of Proof:

Proof of $\Longrightarrow$): Suppose that $\lim\limits_{n\to\infty} x_n = L$ and let $\varepsilon > 0$. Then, $\exists N \in \mathbb{N}$ such that

$$n \geq N \Longrightarrow |x_n - L| < \frac{\varepsilon}{2}.$$

Thus,

$$n, m \geq N \Longrightarrow |x_n - x_m| \leq |x_n - L| + |x_m - L| < \frac{\varepsilon}{2} + \frac{\varepsilon}{2} = \varepsilon. \text{ (Why?)}$$

Thus, $\{x_n\}$ is a Cauchy sequence.

Proof of $\Longleftarrow$): Let $\{x_n\}$ be a Cauchy sequence. Then, $\{x_n\}$ is bounded (Why?) and so contains a convergent subsequence $\{x_{n_k}\}$. (Why?) Denote its limit by L. Then, it follows that, in fact, $\{x_n\}$ converges to L, as needed. (Why?) This completes the proof. $\qquad\square$

Remark. The proofs of the results:

   i.) A bounded sequence has a convergent subsequence,

   ii.) A bounded monotone sequence converges,

   iii.) Cauchy Criterion in $\mathbb{R}$,

all require the completeness property of $\mathbb{R}$, the first indirectly via Bolzano-Weierstrass, which in turn uses it, the second directly, and the third via use of the first. Actually, all three statements are not only consequences of the completeness property of $\mathbb{R}$, but are equivalent to it. In fact, in more general settings in which order is no longer available (cf. Sections 1.6 and 1.7), completeness of the space is *defined* to be the property that all Cauchy sequences converge <u>in the space</u>.

Exercise 1.5.12. For every $n \in \mathbb{N}$, define $s_n = \sum_{k=1}^{n} \frac{1}{k}$. Prove that $\{s_n\}$ diverges.

1.5.4   A Brief Look at Infinite Series

Sequences defined by forming partial sums using terms of a second sequence (e.g., see Exer. 1.5.12) often arise in applied analysis. You might recognize them by the name *infinite series*. We shall provide the bare essentials of this topic below. A thorough treatment can be found in **[236]**.

Definition 1.5.19. Let $\{a_n\}$ be a sequence in $\mathbb{R}$.

i.) The sequence $\{s_n\}$ defined by $s_n = \sum_{k=1}^{n} a_k$, $n \in \mathbb{N}$ is the *sequence of partial sums of $\{a_n\}$*.

ii.) The pair $(\{a_n\}, \{s_n\})$ is called an *infinite series,* denoted by $\sum_{n=1}^{\infty} a_n$ or $\sum a_n$.

iii.) If $\lim\limits_{n\to\infty} s_n = s$, then we say $\sum a_n$ *converges* and has *sum s*; we write $\sum a_n = s$. Otherwise, we say $\sum a_n$ *diverges*.

Remarks.

1. The sequence of partial sums can begin with an index n strictly larger than 1.

2. Suppose $\sum_{k=1}^{\infty} a_k = s$ and $s_n = \sum_{k=1}^{n} a_k$. Observe that

$$s_n + \underbrace{\sum_{k=n+1}^{\infty} a_k}_{\text{Tail}} = s. \tag{1.38}$$

Because $\lim_{n \to \infty} s_n = s$, it follows that $\lim_{n \to \infty} \sum_{k=n+1}^{\infty} a_k = 0$. (Why?)

Example. (Geometric Series)
Consider the series $\sum_{k=0}^{\infty} cx^k$, where $c, x \in \mathbb{R}$. For every $n \geq 0$, subtracting the expressions for s_n and xs_n yields

$$s_n = c\left[1 + x + x^2 + \ldots + x^n\right] \tag{1.39}$$
$$-xs_n = c\left[x + x^2 + \ldots + x^n + x^{n+1}\right] \tag{1.40}$$
$$(1-x)s_n = c\left[1 - x^{n+1}\right].$$

Hence,

$$s_n = \begin{cases} \frac{c[1-x^{n+1}]}{1-x}, & x \neq 1, \\ c(n+1), & x = 1. \end{cases}$$

If $|x| < 1$, then $\lim_{n \to \infty} x^{n+1} = 0$, so that $\lim_{n \to \infty} s_n = \frac{c}{1-x}$. Otherwise, $\lim_{n \to \infty} s_n$ does not exist. (Tell why.)

Exercise 1.5.13. Let $p \in \mathbb{N}$. Determine the values of x for which $\sum_{n=p}^{\infty} c(5x+1)^{3n}$ converges, and for such x, determine its sum.

Proposition 1.5.20. $\sum a_n$ *converges iff* $\{s_n\}$ *is Cauchy iff* $\forall \varepsilon > 0, \exists N \in \mathbb{N}$ *such that* $n \geq N \implies \left|\sum_{k=1}^{p} a_{n+k}\right| < \varepsilon$.

Outline of Proof: The first equivalence is immediate (Why?) and the second follows from Exer. 1.5.11. (Tell how.) □

Corollary 1.5.21. (n^{th} *-term test*) *If* $\sum a_n$ *converges, then* $\lim_{n \to \infty} a_n = 0$.

Outline of Proof: Take $p = 1$ in Prop. 1.5.20. □

Exercise 1.5.14. Prove that $\sum_{n=1}^{\infty} \frac{n!}{a^n}$ diverges, $\forall a > 0$.

Proposition 1.5.22. (Comparison Test)
If $a_n, b_n \geq 0$, $\forall n \in \mathbb{N}$, *and* $\exists c > 0$ *and* $N \in \mathbb{N}$ *such that* $a_n \leq cb_n$, $\forall n \geq N$, *then*
i.) $\sum b_n$ *converges* $\implies \sum a_n$ *converges;*
ii.) $\sum a_n$ *diverges* $\implies \sum b_n$ *diverges.*

Outline of Proof: Use Prop. 1.5.13 (Tell how.) □

Example. Consider the series $\sum_{n=1}^{\infty} \frac{5n}{3^n}$. Because $\lim_{n \to \infty} \frac{5n}{3^{n/2}} = 0$, $\exists N \in \mathbb{N}$ such that

$$n \geq N \implies \frac{5n}{3^{n/2}} < 1 \implies \frac{5n}{3^n} < \frac{1}{3^{n/2}} = \left(\frac{1}{\sqrt{3}} \right)^n. \tag{1.41}$$

(Why?) But, $\sum_{n=1}^{\infty} \left(\frac{1}{\sqrt{3}} \right)^n$ is a convergent geometric series. Thus, Prop. 1.5.22 implies that $\sum_{n=1}^{\infty} \frac{5n}{3^n}$ converges.

Definition 1.5.23. A series $\sum a_n$ is *absolutely convergent* if $\sum |a_n|$ converges.

It can be shown that rearranging the terms of an absolutely convergent series does not affect convergence (see [**236**]). So, we can regroup terms at will, which is especially useful when groups of terms simplify nicely.

Proposition 1.5.24. (Ratio Test)
Suppose $\sum a_n$ is a series with $a_n \neq 0$, $\forall n \in \mathbb{N}$. Let

$$r = \varliminf_{n \to \infty} \left| \frac{a_{n+1}}{a_n} \right| \text{ and } R = \varlimsup_{n \to \infty} \left| \frac{a_{n+1}}{a_n} \right|,$$

(where R could be ∞). Then,
i.) $R < 1 \implies \sum a_n$ converges absolutely;
ii.) $r > 1 \implies \sum a_n$ diverges;
iii.) If $r \leq 1 \leq R$, then the test is inconclusive.

Outline of Proof: We argue as in [**234**].
Proof of (i): Assume $R < 1$ and choose x such that $R < x < 1$. Observe that

$$\varlimsup_{n \to \infty} \left| \frac{a_{n+1}}{a_n} \right| = R < x \implies \exists N \in \mathbb{N} \text{ such that } \left| \frac{a_{n+1}}{a_n} \right| \leq x, \forall n \geq N$$

$$\implies |a_{n+1}| \leq |a_n| x, \forall n \geq N. \tag{1.42}$$

Thus,

$$|a_{N+1}| \leq |a_N| x$$
$$|a_{N+2}| \leq |a_{N+1}| x \leq |a_N| x^2$$
$$|a_{N+3}| \leq |a_{N+2}| x \leq |a_{N+1}| x^2 \leq |a_N| x^3$$
$$\vdots$$

(Why?) What can be said about the series

$$|a_N| \left(x + x^2 + x^3 + \ldots \right)?$$

Use Prop. 1.5.22 to conclude that $\sum |a_n|$ converges.

Proof of (ii): Next, assume $1 < r$. Observe that

$$\lim_{n \to \infty} \left| \frac{a_{n+1}}{a_n} \right| = r > 1 \implies \exists N \in \mathbb{N} \text{ such that } \left| \frac{a_{n+1}}{a_n} \right| \geq 1, \forall n \geq N$$

$$\implies |a_{n+1}| \geq |a_n| \geq |a_N| > 0, \forall n \geq N. \qquad (1.43)$$

(Why?) Thus, $a_n \not\to 0$. (So what?)

Proof of (iii): For both $\sum \frac{1}{n}$ and $\sum \frac{1}{n^2}$, $r = R = 1$, but $\sum \frac{1}{n}$ diverges and $\sum \frac{1}{n^2}$ converges. $\qquad \square$

Exercise 1.5.15. Determine if $\sum_{n=1}^{\infty} \frac{n^n}{n!}$ converges.

Finally, we will need to occasionally multiply two series in the following sense.

Definition 1.5.25. Given two series $\sum_{n=0}^{\infty} a_n$ and $\sum_{n=0}^{\infty} b_n$, define

$$c_n = \sum_{k=0}^{n} a_k b_{n-k}, \forall n \geq 0.$$

The series $\sum_{n=0}^{\infty} c_n$ is called the *Cauchy product* of $\sum_{n=0}^{\infty} a_n$ and $\sum_{n=0}^{\infty} b_n$.

To see why this is a natural definition, consider the partial sum $\sum_{n=0}^{p} c_n$ and form a grid by writing the terms $a_0, \ldots, a_p$ as a column and $b_0, \ldots, b_p$ as a row. Multiply the terms from each row and column pairwise and observe that the sums along the diagonals (formed left to right) coincide with $c_0, \ldots, c_p$. (Check this!)

The following proposition describes a situation when such a product converges. The proof of this and other related results can be found in **[17]**.

Proposition 1.5.26. *If $\sum_{n=0}^{\infty} a_n$ and $\sum_{n=0}^{\infty} b_n$ both converge absolutely, then the Cauchy product $\sum_{n=0}^{\infty} c_n$ converges absolutely and $\sum_{n=0}^{\infty} c_n = \left(\sum_{n=0}^{\infty} a_n \right) \left(\sum_{n=0}^{\infty} b_n \right)$.*

1.6   The Spaces $\left(\mathbb{R}^N, \|\cdot\|_{\mathbb{R}^N} \right)$ and $\left(\mathbb{M}^N(\mathbb{R}), \|\cdot\|_{\mathbb{M}^N(\mathbb{R})} \right)$

We now introduce two spaces of objects with which you likely have some familiarity, namely vectors and square matrices, as a first step in formulating more abstract spaces. The key observation is that the characteristic properties of $\mathbb{R}$ carry over to these spaces, and their verification requires minimal effort. As you work through this section, use your intuition about how vectors in two and three dimensions behave to help you understand the more abstract setting.

1.6.1 The Space $\left(\mathbb{R}^N, \|\cdot\|_{\mathbb{R}^N}\right)$

Definition 1.6.1. For every $N \in \mathbb{N}$, $\mathbb{R}^N = \underbrace{\mathbb{R} \times \cdots \times \mathbb{R}}_{N\,\text{times}}$ is the set of all ordered N-tuples of real numbers. This set is often loosely referred to as *N-space*.

A typical element of $\mathbb{R}^N$ (called a *vector*) is denoted by a boldface letter, say $\mathbf{x}$, representing the ordered N-tuple $\langle x_1, x_2, \ldots, x_N \rangle$. (Here, x_k is the kth *component* of $\mathbf{x}$.) The *zero element* in $\mathbb{R}^N$ is the vector $\mathbf{0} = \underbrace{\langle 0, 0, \ldots, 0 \rangle}_{N\,\text{times}}$.

The algebraic operations defined in $\mathbb{R}$ can be applied componentwise to define the corresponding operations in $\mathbb{R}^N$. Indeed, we have

Definition 1.6.2. (Algebraic Operations in $\mathbb{R}^N$)
Let $\mathbf{x} = \langle x_1, x_2, \ldots, x_N \rangle$ and $\mathbf{y} = \langle y_1, y_2, \ldots, y_N \rangle$ be elements of $\mathbb{R}^N$ and $c \in \mathbb{R}$,
i.) $\mathbf{x} = \mathbf{y}$ if and only if $x_k = y_k$, $\forall k \in \{1, \ldots, N\}$,
ii.) $\mathbf{x} + \mathbf{y} = \langle x_1 + y_1, x_2 + y_2, \ldots, x_N + y_N \rangle$,
iii.) $c\mathbf{x} = \langle cx_1, cx_2, \ldots, cx_N \rangle$.

The usual properties of commutativity, associativity, and distributivity of scalar multiplication over addition carry over to this setting by applying the corresponding property in $\mathbb{R}$ componentwise. For instance, because $x_i + y_i = y_i + x_i$, $\forall i \in \{1, \ldots, n\}$, it follows that

$$\begin{aligned}
\mathbf{x} + \mathbf{y} &= \langle x_1 + y_1, x_2 + y_2, \ldots, x_N + y_N \rangle \\
&= \langle y_1 + x_1, y_2 + x_2, \ldots, y_N + x_N \rangle \\
&= \mathbf{y} + \mathbf{x}.
\end{aligned} \tag{1.44}$$

Exercise 1.6.1. Establish associativity of addition and distributivity of scalar multiplication over addition in $\mathbb{R}^N$.

Geometric and Topological Structure

From the viewpoint of its geometric structure, what is a natural candidate for a distance-measuring artifice for $\mathbb{R}^N$? There is more than one answer to this question, arguably the most natural of which is the Euclidean distance formula, defined below.

Definition 1.6.3. Let $\mathbf{x} \in \mathbb{R}^N$. The *(Euclidean) norm* of $\mathbf{x}$, denoted $\|\mathbf{x}\|_{\mathbb{R}^N}$, is defined by

$$\|\mathbf{x}\|_{\mathbb{R}^N} = \sqrt{\sum_{k=1}^{N} x_k^2}. \tag{1.45}$$

We say that the *distance between x and y in* $\mathbb{R}^N$ is given by $\|\mathbf{x} - \mathbf{y}\|_{\mathbb{R}^N}$.

Remarks.
1. When referring to the norm generically or as a function, we write $\|\cdot\|_{\mathbb{R}^N}$.

2. There are other "equivalent" ways to define a norm on $\mathbb{R}^N$ that are more convenient to use in some situations. Indeed, a useful alternative norm is given by

$$\|\mathbf{x}\|_{\mathbb{R}^N} = \max_{1 \leq i \leq N} |x_i|. \tag{1.46}$$

By *equivalent*, we do not mean that the numbers produced by (1.45) and (1.46) are the same for a given $\mathbf{x} \in \mathbb{R}^N$. In fact, this is false in a big way! Rather, two norms $\|\cdot\|_1$ and $\|\cdot\|_2$ are *equivalent* if there exist constants $0 < \alpha < \beta$ such that

$$\alpha \|\mathbf{x}\|_1 \leq \|\mathbf{x}\|_2 \leq \beta \|\mathbf{x}\|_1, \forall \mathbf{x} \in \mathbb{R}. \tag{1.47}$$

Suffice it to say that you can choose whichever norm is most convenient to work with within a given series of computations, as long as you don't decide to use a different one halfway through! **By default, we use (1.45) unless otherwise specified.**

Exercise 1.6.2. Let $\varepsilon > 0$. Provide a geometric description of these sets:
i.) $A = \left\{ \mathbf{x} \in \mathbb{R}^2 \big| \|\mathbf{x}\|_{\mathbb{R}^2} < \varepsilon \right\}$,
ii.) $B = \left\{ \mathbf{y} \in \mathbb{R}^3 \big| \|\mathbf{y} - \langle 1, 0, 0 \rangle \|_{\mathbb{R}^3} \geq \varepsilon \right\}$,
iii.) $C = \left\{ \mathbf{y} \in \mathbb{R}^3 \big| \|\mathbf{y} - \mathbf{x}_0\|_{\mathbb{R}^3} = 0 \right\}$, where $\mathbf{x}_0 \in \mathbb{R}^3$ is prescribed.

The $\mathbb{R}^N$-norm satisfies similar properties as $|\cdot|$ (cf. Prop. 1.4.4), summarized below.

Proposition 1.6.4. *Let* $\mathbf{x}, \mathbf{y} \in \mathbb{R}^N$ *and* $c \in \mathbb{R}$. *Then,*
i.) $\|\mathbf{x}\|_{\mathbb{R}^N} \geq 0$,
ii.) $\|c\mathbf{x}\|_{\mathbb{R}^N} = |c| \|\mathbf{x}\|_{\mathbb{R}^N}$,
iii.) $\|\mathbf{x} + \mathbf{y}\|_{\mathbb{R}^N} \leq \|\mathbf{x}\|_{\mathbb{R}^N} + \|\mathbf{y}\|_{\mathbb{R}^N}$,
iv.) $\mathbf{x} = \mathbf{0}$ *iff* $\|\mathbf{x}\|_{\mathbb{R}^N} = 0$.

Exercise 1.6.3. Prove Prop. 1.6.4 using Def. 1.6.3. Then, redo it using (1.46).

Exercise 1.6.4. Let $M, p \in \mathbb{N}$. Prove the following string of inequalities:

$$\left\| \sum_{i=1}^{M} \mathbf{x}_i \right\|_{\mathbb{R}^N}^{p} \leq \left(\sum_{i=1}^{M} \|\mathbf{x}_i\|_{\mathbb{R}^N} \right)^{p} \leq M^{p-1} \sum_{i=1}^{M} \|\mathbf{x}_i\|_{\mathbb{R}^N}^{p} \tag{1.48}$$

The space $\left(\mathbb{R}^N, \|\cdot\|_{\mathbb{R}^N} \right)$ has an even richer geometric structure since it can be equipped with a so-called *inner product* that enables us to define orthonormality (or perpendicularity) and, by extension, the notion of angle in the space. Precisely, we have

Definition 1.6.5. Let $\mathbf{x}, \mathbf{y} \in \mathbb{R}^N$. The *inner product of* $\boldsymbol{x}$ *and* $\boldsymbol{y}$, denoted $\langle \mathbf{x}, \mathbf{y} \rangle_{\mathbb{R}^N}$, is defined by

$$\langle \mathbf{x}, \mathbf{y} \rangle_{\mathbb{R}^N} = \sum_{i=1}^{N} x_i y_i. \tag{1.49}$$

Note that taking the inner product of any two elements of $\mathbb{R}^N$ produces a real number. Also, $\langle \mathbf{x}, \mathbf{y} \rangle_{\mathbb{R}^N}$ is often written more compactly as $\mathbf{x}\mathbf{y}^T$, where $\mathbf{y}^T$ is the transpose of $\mathbf{y}$ (that is, $\mathbf{y}$ written as a column vector rather than as a row vector). Some of the properties of this inner product are as follows.

Proposition 1.6.6. (Properties of the Inner Product on $\mathbb{R}^N$)
Let $\mathbf{x}, \mathbf{y}, \mathbf{z} \in \mathbb{R}^N$ *and* $c \in \mathbb{R}$. *Then,*
i.) $\langle c\mathbf{x}, \mathbf{y} \rangle_{\mathbb{R}^N} = \langle \mathbf{x}, c\mathbf{y} \rangle_{\mathbb{R}^N} = c \langle \mathbf{x}, \mathbf{y} \rangle_{\mathbb{R}^N}$;
ii.) $\langle \mathbf{x} + \mathbf{y}, \mathbf{z} \rangle_{\mathbb{R}^N} = \langle \mathbf{x}, \mathbf{z} \rangle_{\mathbb{R}^N} + \langle \mathbf{y}, \mathbf{z} \rangle_{\mathbb{R}^N}$;
iii.) $\langle \mathbf{x}, \mathbf{x} \rangle_{\mathbb{R}^N} \geq 0$;
iv.) $\langle \mathbf{x}, \mathbf{x} \rangle_{\mathbb{R}^N} = 0$ *iff* $\mathbf{x} = \mathbf{0}$;
v.) $\langle \mathbf{x}, \mathbf{x} \rangle_{\mathbb{R}^N} = \|\mathbf{x}\|_{\mathbb{R}^N}^2$;
vi.) $\langle \mathbf{x}, \mathbf{z} \rangle_{\mathbb{R}^N} = \langle \mathbf{y}, \mathbf{z} \rangle_{\mathbb{R}^N}$, $\forall \mathbf{z} \in \mathbb{R}^N \Longrightarrow \mathbf{x} = \mathbf{y}$.

Verifying these properties is straightforward and will be argued in a more general setting in Section 1.7. (Try proving them here!) Property (v) is of particular importance because it asserts that an inner product generates a norm.

Exercise 1.6.5. Prove Prop. 1.6.6.

The following Cauchy-Schwarz inequality is very important.

Proposition 1.6.7. (Cauchy-Schwarz Inequality)
Let $\mathbf{x}, \mathbf{y} \in \mathbb{R}^N$. *Then,*

$$|\langle \mathbf{x}, \mathbf{y} \rangle_{\mathbb{R}^N}| \leq \|\mathbf{x}\|_{\mathbb{R}^N} \|\mathbf{y}\|_{\mathbb{R}^N} \tag{1.50}$$

Outline of Proof: For any $\mathbf{y} \in \mathbb{R}^N \setminus \{\mathbf{0}\}$,

$$0 \leq \left\langle \mathbf{x} - \left(\frac{\langle \mathbf{x}, \mathbf{y} \rangle_{\mathbb{R}^N}}{\|\mathbf{y}\|_{\mathbb{R}^N}^2} \right) \mathbf{y}, \mathbf{x} - \left(\frac{\langle \mathbf{x}, \mathbf{y} \rangle_{\mathbb{R}^N}}{\|\mathbf{y}\|_{\mathbb{R}^N}^2} \right) \mathbf{y} \right\rangle_{\mathbb{R}^N}.$$

(So what?) Why does (1.50) hold for $\mathbf{y} = \mathbf{0}$? □

The inner product can be used to formulate a so-called *orthonormal basis* for $\mathbb{R}^N$. Precisely, let

$$\mathbf{e}_1 = \langle 1, 0, \ldots, 0 \rangle, \, \mathbf{e}_2 = \langle 0, 1, 0, \ldots, 0 \rangle, \, \ldots, \, \mathbf{e}_n = \langle 0, \ldots, 0, 1 \rangle,$$

and observe that

$$\|\mathbf{e}_i\|_{\mathbb{R}^N} = 1, \, \forall i \in \{1, \ldots, N\}, \tag{1.51}$$
$$\left\langle \mathbf{e}_i, \mathbf{e}_j \right\rangle_{\mathbb{R}^N} = 0, \text{ whenever } i \neq j. \tag{1.52}$$

This is useful because it yields the following unique representation for the members of $\mathbb{R}^N$ involving the inner product.

Proposition 1.6.8. *For every* $\mathbf{x} \in \mathbb{R}^N$,

$$\mathbf{x} = \sum_{i=1}^{N} \langle \mathbf{x}, \mathbf{e}_i \rangle_{\mathbb{R}^N} \, \mathbf{e}_i. \tag{1.53}$$

If $\mathbf{x} = \langle x_1, x_2, \ldots, x_N \rangle$, then (1.53) is a succinct way of writing

$$\mathbf{x} = \langle x_1, 0, \ldots, 0 \rangle + \langle 0, x_2, \ldots, 0 \rangle + \langle 0, 0, \ldots, x_N \rangle. \tag{1.54}$$

(Tell why.) Heuristically, this representation indicates how much to "move" in the direction of each basis vector to arrive at $\mathbf{x}$.

For any $x_0 \in \mathbb{R}$ and $\varepsilon > 0$, an open interval centered at x_0 with radius ε is defined by

$$(x_0 - \varepsilon, x_0 + \varepsilon) = \{x \in \mathbb{R} | \, |x - x_0| < \varepsilon\}. \tag{1.55}$$

Because $\|\cdot\|_{\mathbb{R}^N}$ plays the role of $|\cdot|$ and shares its salient characteristics, it is natural to define an open N-ball centered at $\mathbf{x}_0$ with radius ε by

$$\mathfrak{B}_{\mathbb{R}^N}(\mathbf{x}_0; \varepsilon) = \{\mathbf{x} \in \mathbb{R}^N | \, \|\mathbf{x} - \mathbf{x}_0\|_{\mathbb{R}^N} < \varepsilon\}. \tag{1.56}$$

Exercise 1.6.6. Interpret (1.56) geometrically in $\mathbb{R}^2$ and $\mathbb{R}^3$.

The terminology and results developed for $(\mathbb{R}, |\cdot|)$ in Section 1.4 can be extended to $(\mathbb{R}^N, \|\cdot\|_{\mathbb{R}^N})$ with the only formal change being to replace $|\cdot|$ by $\|\cdot\|_{\mathbb{R}^N}$. Theorem 1.4.12 also holds, but a different approach is used to prove it because of the lack of ordering in $\mathbb{R}^N$. (See **[17]** for details.)

Exercise 1.6.7. Convince yourself of the validity of the generalization of the topological results to $\mathbb{R}^N$.

Sequences in $\mathbb{R}^N$

Definition 1.6.9. A function $\mathbf{x} : \mathbb{N} \longrightarrow \mathbb{R}^N$ is a *sequence* in $\mathbb{R}^N$.

The definitions of convergent and Cauchy sequences are essentially the same as in $\mathbb{R}$ and all the results carry over without issue, requiring only that we replace $|\cdot|$ by $\|\cdot\|_{\mathbb{R}^N}$.

Exercise 1.6.8. Convince yourself that the results from Sections 1.5.1 through 1.5.3 extend to the $\mathbb{R}^N$ setting.

Use the fact that the algebraic operations in $\mathbb{R}^N$ are performed componentwise to help you complete the following exercise.

Exercise 1.6.9.
i.) Consider the two real sequences $\{x_m\}$ and $\{y_m\}$ whose m^{th} terms are given by

$$x_m = \frac{1}{m^2}, \; y_m = \frac{2m}{4m+2}, \; m \in \mathbb{N}.$$

Show that $\lim\limits_{m\to\infty} \langle x_m, y_m \rangle = \langle 0, \frac{1}{2} \rangle$.

ii.) Generally, if $\lim\limits_{m\to\infty} x_m = p$ and $\lim\limits_{m\to\infty} y_m = q$, what can you conclude about $\{\langle x_m, y_m \rangle\}$ in $\mathbb{R}^2$?

iii.) Consider a sequence $\{\langle (x_1)_m, (x_2)_m, \ldots, (x_N)_m \rangle\}$ in $\mathbb{R}^N$. Establish a necessary and sufficient condition for this sequence to converge in $\mathbb{R}^N$.

A strategy similar to the one used in Exer. 1.6.9, coupled with Thrm. 1.5.18, is used to prove the following theorem.

Theorem 1.6.10. (Cauchy Criterion in $\mathbb{R}^N$)
$\{\mathbf{x}_n\}$ *converges in* $\mathbb{R}^N$ *iff* $\{\mathbf{x}_n\}$ *is a Cauchy sequence in* $\mathbb{R}^N$. *We say* $\left(\mathbb{R}^N, \|\cdot\|_{\mathbb{R}^N}\right)$ *is complete.*

Use the properties of sequences in $\mathbb{R}^N$ to complete the following exercises.

Exercise 1.6.10. Assume that $\{\mathbf{x}_m\}$ is a convergent sequence in $\mathbb{R}^N$. Prove that $\lim\limits_{m\to\infty} \|\mathbf{x}_m\|_{\mathbb{R}^N} = \left\|\lim\limits_{m\to\infty} \mathbf{x}_m\right\|_{\mathbb{R}^N}$. (We say that the norm $\|\cdot\|_{\mathbb{R}^N}$ is continuous.)

Exercise 1.6.11. Let $a \in \mathbb{R}$. Compute $\sup\left\{\left\|\frac{a\mathbf{x}}{\|\mathbf{x}\|_{\mathbb{R}^N}}\right\|_{\mathbb{R}^N} : \mathbf{x} \in \mathbb{R}^N \setminus \{\mathbf{0}\}\right\}$.

Exercise 1.6.12. Let $\delta, \varepsilon > 0$, $a, b \in \mathbb{R}$, and $\mathbf{x}_0 \in \mathbb{R}^N$ be prescribed. Compute $\sup\{\|\mathbf{z}\|_{\mathbb{R}^N} : \mathbf{z} \in \mathscr{A}\}$, where

$$\mathscr{A} = \{a\mathbf{x} + b\mathbf{y} : \mathbf{x} \in \mathfrak{B}_{\mathbb{R}^N}(\mathbf{x}_0; \varepsilon) \wedge \mathbf{y} \in \mathfrak{B}_{\mathbb{R}^N}(\mathbf{x}_0; \delta)\}.$$

Exercise 1.6.13. Let $\mathbf{x} \in \mathbb{R}^N \setminus \{\mathbf{0}\}$, $a \neq 0$, and $p \in \mathbb{N}$. Must the series $\sum_{m=p}^{\infty} \left\|\frac{\mathbf{x}}{a\|\mathbf{x}\|_{\mathbb{R}^N}}\right\|_{\mathbb{R}^N}^{2m}$ converge? If so, can you determine its sum?

Exercise 1.6.14. Let $\mathbf{x} \in \mathfrak{B}_{\mathbb{R}^2}\left(\mathbf{0}; \frac{1}{3}\right)$ and $p \in \mathbb{N}$. Must the series $\sum_{m=p}^{\infty} \left|\langle 2\mathbf{x}, \frac{1}{4}\mathbf{x}\rangle_{\mathbb{R}^2}\right|^{m/2}$ converge? If so, can you determine its sum?

Exercise 1.6.15. Let $\{c_n\}$ be a real sequence and $\mathbf{z} \in \mathbb{R}^N$. Assuming convergence of all series involved, prove that

$$\left\|\sum_{n=1}^{\infty} c_n \mathbf{z}\right\|_{\mathbb{R}^N} \leq \sum_{n=1}^{\infty} |c_n| \|\mathbf{z}\|_{\mathbb{R}^N}.$$

Exercise 1.6.16. Let $R > 0$ and $\langle a, b, c \rangle \in \partial\mathfrak{B}_{\mathbb{R}^3}(\mathbf{0}; R)$ and define the function $\mathbf{f} : \mathbb{R} \longrightarrow \mathbb{R}^3$ by $\mathbf{f}(t) = \langle a\sin\left(\frac{t}{\pi}\right), b\cos(2t + \pi), c \rangle$. Show that $\{\|\mathbf{f}(t)\|_{\mathbb{R}^3} : t \in \mathbb{R}\} < \infty$.

1.6.2   The Space $\left(\mathbb{M}^N(\mathbb{R}), \|\cdot\|_{\mathbb{M}^N(\mathbb{R})}\right)$

A mathematical description of certain scenarios involves considering vectors whose components are themselves vectors. Indeed, consider

$$\mathbf{A} = \langle \mathbf{x}_1, \mathbf{x}_2, \ldots, \mathbf{x}_N \rangle, \tag{1.57}$$

where

$$\mathbf{x}_i = \langle x_{i1}, x_{i2}, \ldots, x_{iN} \rangle, \ 1 \leq i \leq N. \tag{1.58}$$

Viewing $\mathbf{x}_i$ in column form enables us to express $\mathbf{A}$ more elegantly as the $N \times N$ matrix

$$\mathbf{A} = \begin{bmatrix} x_{11} & x_{12} & \cdots & x_{1N} \\ x_{21} & x_{22} & \cdots & x_{2N} \\ \vdots & \vdots & \vdots & \vdots \\ x_{N1} & x_{N2} & \cdots & x_{NN} \end{bmatrix}. \tag{1.59}$$

It is typical to write

$$\mathbf{A} = [x_{ij}], \text{ where } 1 \leq i, j \leq N, \tag{1.60}$$

and refer to x_{ij} as the ij^{th} *entry* of $\mathbf{A}$.

Definition 1.6.11. Let $N \in \mathbb{N} \setminus \{1\}$. $\mathbb{M}^N(\mathbb{R})$ is the set of all $N \times N$ matrices with real entries.

The following terminology is standard in this setting.

Definition 1.6.12. Let $\mathbf{A} \in \mathbb{M}^N(\mathbb{R})$.
i.) $\mathbf{A}$ is *diagonal* if $x_{ij} = 0$, whenever $i \neq j$;
ii.) $\mathbf{A}$ is *symmetric* if $x_{ij} = x_{ji}$, $\forall i, j \in \{1, \ldots, N\}$;
iii.) The *trace* of $\mathbf{A}$ is the real number $\text{trace}(\mathbf{A}) = \sum_{i=1}^{N} x_{ii}$;
iv.) The *zero matrix*, denoted $\mathbf{0}$, is the unique member of $\mathbb{M}^N(\mathbb{R})$ for which $x_{ij} = 0$, $\forall 1 \leq i, j \leq N$.
v.) The *identity matrix*, denoted $\mathbf{I}$, is the unique diagonal matrix in $\mathbb{M}^N(\mathbb{R})$ for which $x_{ii} = 1$, $\forall 1 \leq i \leq N$.
vi.) The *transpose* of $\mathbf{A}$, denoted $\mathbf{A}^T$, is the matrix $\mathbf{A}^T = [x_{ji}]$. (That is, the ij^{th} entry of $\mathbf{A}^T$ is x_{ji}.)

We assume a modicum of familiarity with elementary matrix operations and gather some basic ones below. Note that some, but not all, of the operations are performed entry-wise.

Definition 1.6.13. (**Algebraic Operations in** $\mathbb{M}^N(\mathbb{R})$)
Let $\mathbf{A} = [a_{ij}]$, $\mathbf{B} = [b_{ij}]$, and $\mathbf{C} = [c_{ij}]$ be in $\mathbb{M}^N(\mathbb{R})$ and $\alpha \in \mathbb{R}$.
i.) $\mathbf{A}^0 = \mathbf{I}$,
ii.) $\alpha \mathbf{A} = [\alpha a_{ij}]$,
iii.) $\mathbf{A} + \mathbf{B} = [a_{ij} + b_{ij}]$,
iv.) $\mathbf{AB} = \left[\sum_{r=1}^{N} a_{ir} b_{rj} \right]$.

Exercise 1.6.17. Consider the operations defined in Def. 1.6.13.
i.) Does $\mathbf{A} + \mathbf{B} = \mathbf{B} + \mathbf{A}$, $\forall \mathbf{A}, \mathbf{B} \in \mathbb{M}^N(\mathbb{R})$?
ii.) Does $\mathbf{AB} = \mathbf{BA}$, $\forall \mathbf{A}, \mathbf{B} \in \mathbb{M}^N(\mathbb{R})$?
iii.) Must either $(\mathbf{A} + \mathbf{B})\mathbf{C} = \mathbf{AC} + \mathbf{BC}$ or $\mathbf{C}(\mathbf{A} + \mathbf{B}) = \mathbf{CA} + \mathbf{CB}$ hold, $\forall \mathbf{A}, \mathbf{B}, \mathbf{C} \in \mathbb{M}^N(\mathbb{R})$?
iv.) Does $(\mathbf{AB})\mathbf{C} = \mathbf{A}(\mathbf{BC})$, $\forall \mathbf{A}, \mathbf{B}, \mathbf{C} \in \mathbb{M}^N(\mathbb{R})$?

We assume familiarity with the basic properties of determinants of square matrices (see [196]). They are used to define invertibility.

Proposition 1.6.14. *For any* $\mathbf{A} \in \mathbb{M}^N(\mathbb{R})$ *for which* $\det(\mathbf{A}) \neq 0$, *there exists a unique* $\mathbf{B} \in \mathbb{M}^N(\mathbb{R})$ *such that* $\mathbf{AB} = \mathbf{BA} = \mathbf{I}$. *We say* $\mathbf{A}$ *is invertible and write* $\mathbf{B} = \mathbf{A}^{-1}$.

The notion of an *eigenvalue* arises in the study of stability theory of ordinary differential equations (ODEs). Precisely, we have

Definition 1.6.15. Let $\mathbf{A} \in \mathbb{M}^N(\mathbb{R})$.
i.) A complex number λ_0 is an *eigenvalue* of $\mathbf{A}$ if $\det(\mathbf{A} - \lambda_0 \mathbf{I}) = 0$.
ii.) An eigenvalue λ_0 has *multiplicity* M if $\det(\mathbf{A} - \lambda_0 \mathbf{I}) = p(\lambda)(\lambda - \lambda_0)^M$; that is, $(\lambda - \lambda_0)^M$ divides evenly into $\det(\mathbf{A} - \lambda_0 \mathbf{I})$.

Exercise 1.6.18. Let $\mathbf{A} = \begin{bmatrix} a & 0 \\ 0 & b \end{bmatrix}$, where $a, b \neq 0$.
i.) Compute the eigenvalues of $\mathbf{A}$.
ii.) Compute $\mathbf{A}^{-1}$ and its eigenvalues.
iii.) Generalize the computations in (i) and (ii) to the case of a diagonal $N \times N$ matrix $\mathbf{B}$ whose diagonal entries are all nonzero. Fill in the blank:

If λ is an eigenvalue of $\mathbf{B}$, then _____ is an eigenvalue of $\mathbf{B}^{-1}$.

We can equip $\mathbb{M}^N(\mathbb{R})$ with various norms in the spirit of those used in $\mathbb{R}^N$. Let $\mathbf{A} \in \mathbb{M}^N(\mathbb{R})$. Three standard choices for $\|\mathbf{A}\|_{\mathbb{M}^N}$ are

$$\|\mathbf{A}\|_{\mathbb{M}^N} = \left[\sum_{i=1}^{N} \sum_{j=1}^{N} |a_{ij}|^2 \right]^{1/2}, \tag{1.61}$$

$$\|\mathbf{A}\|_{\mathbb{M}^N} = \sum_{i=1}^{N} \sum_{j=1}^{N} |a_{ij}|, \tag{1.62}$$

$$\|\mathbf{A}\|_{\mathbb{M}^N} = \max_{1 \leq i, j \leq N} |a_{ij}|. \tag{1.63}$$

It can be shown that (1.61) through (1.63) are equivalent in a sense similar to (1.47).

Exercise 1.6.19. Prove that each of (1.61) through (1.63) satisfies the properties in Prop. 1.6.4, appropriately extended to $\mathbb{M}^N(\mathbb{R})$.

Exercise 1.6.20. Let $\mathbf{A}, \mathbf{B} \in \mathbb{M}^N(\mathbb{R})$. Prove that $\|\mathbf{AB}\|_{\mathbb{M}^N} \leq \|\mathbf{A}\|_{\mathbb{M}^N} \|\mathbf{B}\|_{\mathbb{M}^N}$.

As in $\mathbb{R}^N$, a *sequence* in $\mathbb{M}^N(\mathbb{R})$ is an $\mathbb{M}^N(\mathbb{R})$-valued function whose domain is $\mathbb{N}$. If $\lim_{m \to \infty} \|\mathbf{A}_m - \mathbf{A}\|_{\mathbb{M}^N} = 0$, we say $\{\mathbf{A}_m\}$ *converges* to $\mathbf{A}$ in $\mathbb{M}^N(\mathbb{R})$ and write "$\mathbf{A}_m \longrightarrow \mathbf{A}$ in $\mathbb{M}^N(\mathbb{R})$." The similarity in the definitions of the norms used in $\mathbb{R}^N$ and $\mathbb{M}^N(\mathbb{R})$ suggests that checking convergence is performed entry-wise. The same is true of Cauchy sequences in $\mathbb{M}^N(\mathbb{R})$. (Convince yourself!) By extension of Thrm. 1.6.10, we can argue that $\left(\mathbb{M}^N(\mathbb{R}), \|\cdot\|_{\mathbb{M}^N(\mathbb{R})}\right)$ is complete with respect to any of the norms (1.61) through (1.63). (Tell why carefully.)

Let $\mathbf{A} \in \mathbb{M}^N(\mathbb{R})$ and $\mathbf{x} \in \mathbb{R}^N$. From the definition of matrix multiplication, if we view $\mathbf{x}$ as a $N \times 1$ column matrix, then because $\mathbf{A}$ is an $N \times N$ matrix, we know that $\mathbf{Ax}$ is a well-defined $N \times 1$ column matrix that can be identified as a member of $\mathbb{R}^N$. As such, the function $\mathbf{f_A} : \mathbb{R}^N \longrightarrow \mathbb{R}^N$ given by $\mathbf{f_A}(\mathbf{x}) = \mathbf{Ax}$ is well-defined. Such mappings are used frequently in Chapter 2.

Exercise 1.6.21. Prove that $\left\{ \left\| \begin{bmatrix} e^{-t} & 0 \\ 0 & e^{-2t} \end{bmatrix} \mathbf{x} \right\|_{\mathbb{R}^2} : \|\mathbf{x}\|_{\mathbb{R}^2} = 1 \wedge t > 0 \right\}$ is bounded.

Exercise 1.6.22. If $\left\{(x_{ij})_m\right\}$ is a real Cauchy sequence for each $1 \leq i, j \leq 2$, must the sequence $\{\mathbf{A}_m\}$, where $\mathbf{A}_m = \left[(x_{ij})_m\right]$, be Cauchy in $\mathbb{M}^2(\mathbb{R})$? Prove your claim.

Exercise 1.6.23. Let $\{\mathbf{A}_m\}$ be a sequence in $\mathbb{M}^N(\mathbb{R})$.
i.) If $\lim_{m \to \infty} \mathbf{A}_m = \mathbf{0}$, what must be true about each of the N^2 sequences formed using the entries of $\mathbf{A}_m$?
ii.) If $\lim_{m \to \infty} \mathbf{A}_m = \mathbf{I}$, what must be true about each of the N^2 sequences formed using the entries of $\mathbf{A}_m$?
iii.) More generally, if $\lim_{m \to \infty} \mathbf{A}_m = \mathbf{B}$, what must be true about each of the N^2 sequences formed using the entries of $\mathbf{A}_m$?

Exercise 1.6.24.
i.) If $\mathbf{A}_m \longrightarrow \mathbf{A}$ in $\mathbb{M}^N(\mathbb{R})$, must $\mathbf{A}_m\mathbf{x} \longrightarrow \mathbf{Ax}$ in $\mathbb{R}^N$, $\forall \mathbf{x} \in \mathbb{R}^N$?
ii.) If $\mathbf{x}_m \longrightarrow \mathbf{x}$ in $\mathbb{R}^N$, must $\mathbf{Ax}_m \longrightarrow \mathbf{Ax}$ in $\mathbb{R}^N$, $\forall \mathbf{A} \in \mathbb{M}^N(\mathbb{R})$?
iii.) If $\mathbf{A}_m \longrightarrow \mathbf{A}$ in $\mathbb{M}^N(\mathbb{R})$ and $\mathbf{x}_m \longrightarrow \mathbf{x}$ in $\mathbb{R}^N$, must $\{\mathbf{A}_m\mathbf{x}_m\}$ converge in $\mathbb{R}^N$? If so, what is its limit?

1.7 Abstract Spaces

Many other spaces possess the same salient features regarding norms, inner products, and completeness exhibited by $\left(\mathbb{R}^N, \|\cdot\|_{\mathbb{R}^N}\right)$ and $\left(\mathbb{M}^N(\mathbb{R}), \|\cdot\|_{\mathbb{M}^N(\mathbb{R})}\right)$. At the moment we would need to verify them for each such space that we encountered individually, which is inefficient. Rather, it would be beneficial to examine a more abstract structure possessing these characteristics and establish results directly for *them*. In turn, we would need only to verify that a space arising in an investigation had this basic structure and then invoke all concomitant results immediately. This will save us considerable work in that we will not need to reformulate all properties each time we introduce a new space. This section is devoted to the development of such abstract structures. (See **[243]** for a thorough treatment.)

1.7.1 Banach Spaces

We begin with the notion of a linear space over $\mathbb{R}$.

Definition 1.7.1. A *real linear space* X is a set equipped with addition and scalar multiplication by real numbers satisfying the following properties:
i.) $x + y = y + x, \forall x, y \in X$;
ii.) $x + (y + z) = (x + y) + z, \forall x, y, z \in X$;
iii.) There exists a unique element $0 \in X$ such that $x + 0 = 0 + x, \forall x \in X$;
iv.) For every $x \in X$, there exists a unique element $-x \in X$ such that

$$x + (-x) = (-x) + x = 0, \forall x \in X;$$

v.) $a(bx) = (ab)x, \forall x \in X$ and $a, b \in \mathbb{R}$;
vi.) $a(x + y) = ax + ay, \forall x, y \in X$ and $a \in \mathbb{R}$;
vii.) $(a + b)x = ax + bx, \forall x \in X$ and $a, b \in \mathbb{R}$.

Restricting attention to a subset $\mathscr{Y}$ of elements of a linear space $\mathscr{X}$ that possesses the same structure as the larger space leads to the following notion.

Definition 1.7.2. Let X be a real linear space. A subset $Y \subset X$, equipped with the same operations as X, is a *linear subspace* of X if
i.) $x, y \in Y \implies x + y \in Y$,
ii.) $x \in Y \implies ax \in Y, \forall a \in \mathbb{R}$.

Exercise 1.7.1.
i.) Verify that $(\mathbb{R}, |\cdot|)$, $\left(\mathbb{R}^N, \|\cdot\|_{\mathbb{R}^N}\right)$, and $\left(\mathbb{M}^N(\mathbb{R}), \|\cdot\|_{\mathbb{M}^N(\mathbb{R})}\right)$ are linear spaces.
ii.) Is $Y = \left\{\mathbf{A} \in \mathbb{M}^N(\mathbb{R}) : \mathbf{A} \text{ is diagonal}\right\}$ a linear subspace of $\mathbb{M}^N(\mathbb{R})$?

We can enhance the structure of a real linear space by introducing a topology so that limit processes can be performed. One way to accomplish this is to equip the space with a norm in the following sense.

Definition 1.7.3. Let X be a real linear space. A real-valued function $\|\cdot\|_X : X \longrightarrow \mathbb{R}$ is a *norm* on X if $\forall x, y \in X$ and $a \in \mathbb{R}$,
i.) $\|x\|_X \geq 0$,
ii.) $\|ax\|_X = |a| \|x\|_X$,
iii.) $\|x + y\|_X \leq \|x\|_X + \|y\|_X$,
iv.) $x = 0$ iff $\|x\|_X = 0$.

We say that the *distance* between x and y is $\|x - y\|_X$. We use this to obtain the following richer abstract structure.

Definition 1.7.4. A real linear space X equipped with a norm $\|\cdot\|_X$ is called a *(real) normed linear space.*

We know from our work in Sections 1.4 through 1.6 that $(\mathbb{R}, |\cdot|)$, $(\mathbb{R}^N, \|\cdot\|_{\mathbb{R}^N})$, and $\left(\mathbb{M}^N(\mathbb{R}), \|\cdot\|_{\mathbb{M}^N(\mathbb{R})}\right)$ are all normed linear spaces. Many of the normed linear spaces that we will encounter are collections of functions satisfying certain properties. Some standard function spaces (aka *Sobolev spaces*) and typical norms with which they are equipped are listed below. Momentarily, we assume an intuitive understanding of continuity, differentiability, and integrability. These notions will be defined more rigorously in Section 1.8. A detailed technical treatment of Sobolev spaces can be found in **[1]**.

<u>Some Common Function Spaces:</u> Let $I \subset \mathbb{R}$ and X be a normed linear space.

1.) $\mathbb{C}(I;X) = \{f : I \longrightarrow X \mid f \text{ is continuous on } I\}$ equipped with the *sup norm*

$$\|f\|_{\mathbb{C}} = \sup_{t \in I} \|f(t)\|_X. \tag{1.64}$$

2.) $\mathbb{C}^n(I;X) = \{f : I \longrightarrow X \mid f \text{ is } n \text{ times continuously differentiable on } I\}$ equipped with

$$\|f\|_{\mathbb{C}^n} = \sup_{t \in I} \left[\|f(t)\|_X + \|f'(t)\|_X + \ldots + \left\|f^{(n)}(t)\right\|_X \right]. \tag{1.65}$$

3.) Let $1 \leq p < \infty$. $\mathbb{L}^p(I;\mathbb{R}) = \{f : I \longrightarrow \mathbb{R} \mid f^p \text{ is integrable on } I\}$ equipped with

$$\|f\|_{\mathbb{L}^p} = \left[\int_I |f(t)|^p \, dt \right]^{\frac{1}{p}}. \tag{1.66}$$

4.) Let $1 \leq p < \infty$. $\mathbb{L}^p_{\text{loc}}(I;\mathbb{R}) = \{f : I \longrightarrow \mathbb{R} \mid f^p \text{ is integrable on compact subsets of } I\}$ equipped with (1.66).

5.) $\mathbb{H}^2(I;\mathbb{R}) = \{f \in \mathbb{L}^2(I;\mathbb{R}) \mid f', f'' \text{ exist and } f'' \in \mathbb{L}^2(I;\mathbb{R})\}$ equipped with

$$\|f\|_{\mathbb{H}^2} = \left[\int_I |f(t)|^2 \, dt \right]^{\frac{1}{2}}. \tag{1.67}$$

6.) $\mathbb{H}^1_0(a,b;\mathbb{R}) = \{f : (a,b) \longrightarrow \mathbb{R} \mid f' \text{ exists and } f(a) = f(b) = 0\}$ equipped with

$$\|f\|_{\mathbb{H}^1_0} = \left[\int_I \left(|f(t)|^2 + |f'(t)|^2 \right) dt \right]^{\frac{1}{2}}. \tag{1.68}$$

7.) Let $m \in \mathbb{N}$. $\mathbb{W}^{2,m}(I;\mathbb{R}) = \left\{f \in \mathbb{L}^2(I;\mathbb{R}) \mid f^{(k)} \in \mathbb{L}^2(I;\mathbb{R}), \forall k = 1, \ldots, m\right\}$ equipped with

$$\|f\|_{\mathbb{W}^{2,m}} = \left[\int_I \left(|f(t)|^2 + |f'(t)|^2 + \ldots + \left|f^{(m)}(t)\right|^2 \right) dt \right]^{\frac{1}{2}}. \tag{1.69}$$

Here, $f^{(k)}$ represents the kth-order derivative of f. (Technically, this is a generalized derivative defined in a distributional sense.)

The notions of convergent and Cauchy sequences extend to any normed linear space in the same manner as in Section 1.6. For example, we say that an X-valued sequence $\{x_n\}$ *converges to x in X* if $\lim\limits_{n \to \infty} \|x_n - x\|_X = 0$.

Exercise 1.7.2.
i.) Interpret the statement "$\lim\limits_{n \to \infty} x_n = x$ in X" for these specific choices of X:

   a.) (1.64)
   b.) (1.66)
   c.) (1.69)

ii.) Interpret the statement "$\{x_n\}$ is Cauchy in X" for the same choices of X.

When working with specific function spaces, knowing when Cauchy sequences in X must converge <u>in X</u> is often crucial. In other words, we need to know if a space is complete in the following sense.

Definition 1.7.5. (Completeness)
i.) A normed linear space X is *complete* if every Cauchy sequence in X converges to an element of X.
ii.) A complete normed linear space is called a *Banach space.*

We shall routinely work with sequences in $\mathbb{C}(I; \mathscr{X})$, where $\mathscr{X}$ is a Banach space. We now focus on the terminology and some particular results for this space.

Definition 1.7.6. Suppose that $\varnothing \neq S \subset D \subset \mathbb{R}$ and $f_n, f : D \longrightarrow \mathscr{X}, n \in \mathbb{N}$.
i.) $\{f_n\}$ *converges uniformly to f on S* whenever $\forall \varepsilon > 0, \exists N \in \mathbb{N}$ such that

$$n \geq N \implies \sup_{x \in S} \|f_n(x) - f(x)\|_{\mathscr{X}} < \varepsilon.$$

(We write "$f_n \longrightarrow f$ uniformly on S.")
ii.) $\{f_n\}$ *converges pointwise to f on S* whenever $\lim\limits_{n \to \infty} f_n(x) = f(x), \forall x \in S$.
iii.) $\{f_n\}$ is *uniformly bounded on S* whenever $\exists M > 0$ such that

$$\sup_{n \in \mathbb{N}} \left(\sup_{x \in S} \|f_n(x)\|_{\mathscr{X}} \right) \leq M.$$

iv.) $\sum_{k=1}^{\infty} f_k(x)$ *converges uniformly to $f(x)$ on S* whenever $s_n \longrightarrow f$ uniformly on S, where $s_n(x) = \sum_{k=1}^{n} f_k(x)$.

It can be shown that each of the function spaces listed above is complete with respect to the norm provided. Verification of this requires the use of various tools involving the behavior of sequences of functions and integrability. We consider the most straightforward one in the following exercise.

Exercise 1.7.3.
i.) Prove that $\mathbb{C}([a,b];\mathbb{R})$ equipped with the sup norm (1.64) is complete.
ii.) How does the argument change if $\mathbb{R}$ is replaced by a Banach space $\mathscr{X}$? Can $\mathscr{X}$ be any normed linear space, or must it be complete? Explain.

Strong Cautionary Remark! We have seen that a normed linear space can be equipped with different norms. As such, we must bear in mind that completeness is norm dependent. Indeed, equipping the spaces above with norms other than those specified by (1.64) through (1.69) could foresake completeness! For instance, $\mathbb{C}([a,b];\mathbb{R})$ equipped with (1.67) is NOT complete. (See **[160]**.)

The next Cauchy-like condition for checking the uniform convergence of a series of functions follows directly from the completeness of $\mathbb{C}(I;\mathscr{X})$.

Proposition 1.7.7. $\sum_{k=1}^{\infty} f_k(x)$ *converges uniformly on S iff* $\forall \varepsilon > 0, \exists N \in \mathbb{N}$ *such that*

$$n \geq N \wedge p \in \mathbb{N} \Longrightarrow \sup_{x \in S} \left\| \sum_{k=n+1}^{n+p} f_k(x) \right\|_{\mathscr{X}} < \varepsilon. \tag{1.70}$$

The following convergence result is useful in certain fixed-point arguments arising in Chapter 5.

Proposition 1.7.8. (Weierstrass M-Test)
Let $\{M_k\} \subset [0,\infty)$ *such that* $\forall k \in \mathbb{N}$,

$$\sup_{x \in S} \|f_k(x)\|_{\mathscr{X}} \leq M_k.$$

If $\sum_{k=1}^{\infty} M_k$ *converges, then* $\sum_{k=1}^{\infty} f_k(x)$ *converges uniformly on S.*

Outline of Proof: Let $\varepsilon > 0$. There exists $N \in \mathbb{N}$ such that

$$n \geq N \Longrightarrow \sum_{k=n+1}^{n+p} M_k < \varepsilon.$$

(Why?) For every $n \geq N$ and $p \in \mathbb{N}$, observe that $\forall x \in S$,

$$\left\| \sum_{k=n+1}^{n+p} f_k(x) \right\|_{\mathscr{X}} \leq \sum_{k=n+1}^{n+p} \|f_k(x)\|_{\mathscr{X}} \leq \sum_{k=n+1}^{n+p} M_k < \varepsilon.$$

Now, use the completeness of $\mathbb{C}(I;\mathscr{X})$ and Prop. 1.7.7. (Tell how.) $\square$

Exercise 1.7.4. Prove Props. 1.7.7 and 1.7.8.

Exercise 1.7.5. Assume that $g_n \longrightarrow g$ uniformly on $[a,b]$ and that f is uniformly continuous on $[a,b]$. Prove that $f(g_n) \longrightarrow f(g)$ uniformly on $[a,b]$.

Remark. Taylor series representations of infinitely differentiable functions are presented in elementary calculus. Some common examples are

$$e^x = \lim_{N \to \infty} \sum_{n=0}^{N} \frac{x^n}{n!}, x \in \mathbb{R}, \tag{1.71}$$

$$\sin(x) = \lim_{N \to \infty} \sum_{n=0}^{N} \frac{(-1)^n x^{2n+1}}{(2n+1)!}, x \in \mathbb{R}, \tag{1.72}$$

$$\cos(x) = \lim_{N \to \infty} \sum_{n=0}^{N} \frac{(-1)^n x^{2n}}{(2n)!}, x \in \mathbb{R}. \tag{1.73}$$

It can be shown that the convergence in each case is uniform on all compact subsets of $\mathbb{R}$. The benefit of such a representation is the uniform approximation of the function on the left-hand side by the sequence of nicely behaved polynomials on the right-hand side. Generalizations of these formulae to more abstract settings will be a key tool throughout the text.

The basic topological notions of open, closed, bounded, etc. carry over to normed linear spaces in the form of the metric topology defined using the norm $\|\cdot\|_{\mathscr{X}}$. We use the following notation:

$$\mathfrak{B}_{\mathscr{X}}(x_0; \varepsilon) = \{x \in \mathscr{X} \mid \|x - x_0\|_{\mathscr{X}} < \varepsilon\}, \tag{1.74}$$
$$\mathrm{cl}_{\mathscr{X}}(\mathscr{Z}) = \text{closure of } \mathscr{Z} \text{ (in the sense of } \|\cdot\|_{\mathscr{X}}).$$

Exercise 1.7.6. Describe the elements of the ball $\mathfrak{B}_{\mathbb{C}([0,2];\mathbb{R})}(x^2; 1)$.

Some topological results, like the Bolzano-Weierstrass and Heine-Borel Theorems, do not extend to the general Banach space setting because they rely on intrinsic properties of $\mathbb{R}^N$. This will present a minor obstacle in Chapter 5, at which time we will revisit the issue.

The need to restrict our attention to a particular subspace of a function space whose elements satisfy some special characteristic arises often. But, can we be certain that we remain in the subspace upon performing limiting operations involving its elements? Put differently, must a subspace of a Banach space be complete? The answer is provided easily by the following exercise.

Exercise 1.7.7. Let Y be the subspace $((0,2]; |\cdot|)$ of $\mathbb{R}$. Prove that $\left\{\frac{2}{n}\right\}$ is Cauchy in Y, but that there does not exist $y \in Y$ to which $\left\{\frac{2}{n}\right\}$ converges.

If the subspace had been closed in the topological sense, would it have made a difference? It turns out that it would have indeed, as suggested by:

Proposition 1.7.9. *A closed subspace $\mathscr{Y}$ of a Banach space $\mathscr{X}$ is complete.*

Exercise 1.7.8. Prove Prop. 1.7.9.

Exercise 1.7.9. Let $(\mathscr{X}, \|\cdot\|_{\mathscr{X}})$ and $(\mathscr{Y}, \|\cdot\|_{\mathscr{Y}})$ be real Banach spaces. Prove that $(\mathscr{X} \times \mathscr{Y}; \|\cdot\|_1)$ and $(\mathscr{X} \times \mathscr{Y}; \|\cdot\|_2)$ are also Banach spaces, where

$$\|(x,y)\|_1 = \|x\|_{\mathscr{X}} + \|y\|_{\mathscr{Y}}, \tag{1.75}$$

$$\|(x,y)\|_2 = \left(\|x\|_{\mathscr{X}}^2 + \|y\|_{\mathscr{Y}}^2\right)^{1/2}. \tag{1.76}$$

1.7.2 Hilbert Spaces

Equipping $\mathbb{R}^N$ with a dot product enhanced its structure by introducing the notion of orthogonality. This prompts us to define the general notion of an inner product on a linear space.

Definition 1.7.10. Let X be a real linear space. A real-valued function $\langle \cdot, \cdot \rangle_X : X \times X \longrightarrow \mathbb{R}$ is an *inner product on X* if $\forall x, y, z \in X$ and $a \in \mathbb{R}$,
i.) $\langle x, y \rangle_X = \langle y, x \rangle_X$,
ii.) $\langle ax, y \rangle_X = a \langle x, y \rangle_X$,
iii.) $\langle x + y, z \rangle_X = \langle x, z \rangle_X + \langle y, z \rangle_X$,
iv.) $\langle x, x \rangle_X > 0$ iff $x \neq 0$.
The pair $(X, \langle \cdot, \cdot \rangle_X)$ is called a *(real) inner product space.*

Some Common Inner Product Spaces:
1.) $\mathbb{R}^N$ equipped with (1.49).
2.) $\mathbb{C}([a,b];\mathbb{R})$ equipped with

$$\langle f, g \rangle_{\mathbb{C}} = \int_a^b f(t)g(t)dt. \tag{1.77}$$

3.) $\mathbb{L}^2(a,b;\mathbb{R})$ equipped with (1.77).
4.) $\mathbb{W}^{2,m}(a,b;\mathbb{R})$ equipped with

$$\langle f, g \rangle_{\mathbb{W}^{2,k}} = \int_a^b \left[f(t)g(t) + f'(t)g'(t) + \ldots + f^{(m)}(t)g^{(m)}(t) \right] dt. \tag{1.78}$$

Exercise 1.7.10. Verify that (1.77) and (1.78) are inner products.

An inner product on X induces a norm on X via the relationship

$$\langle x, x \rangle_X^{1/2} = \|x\|_X. \tag{1.79}$$

Exercise 1.7.11. Prove that the usual norms in $\mathbb{R}^N$, $\mathbb{C}([a,b];\mathbb{R})$, and $\mathbb{W}^{2,m}(a,b;\mathbb{R})$ can be obtained from their respective inner products (1.49), (1.77), and (1.78).

Propositions 1.6.6 and 1.6.7 actually hold for general inner products. We have

Proposition 1.7.11. *Let* $(X, \langle \cdot, \cdot \rangle_X)$ *be an inner product space and suppose* $\|\cdot\|_X$ *is given by (1.79). Then,* $\forall x, y \in X$ *and* $a \in \mathbb{R}$,

i.) $\langle x, ay \rangle_X = a \langle x, y \rangle_X$;
ii.) $\langle x, z \rangle_X = \langle y, z \rangle_X$, $\forall z \in X \Longrightarrow x = y$;
iii.) $\|ax\|_X = |a| \|x\|_X$;
iv.) (Cauchy-Schwarz) $|\langle x, y \rangle_X| \leq \|x\|_X \|y\|_X$;
v.) (Minkowski) $\|x + y\|_X \leq \|x\|_X + \|y\|_X$;
vi.) *If* $x_n \longrightarrow x$ *and* $y_n \longrightarrow y$ *in X, then* $\langle x_n, y_n \rangle_X \longrightarrow \langle x, y \rangle_X$.

Exercise 1.7.12. Prove Prop. 1.7.11.

Exercise 1.7.13. Interpret Prop. 1.7.11(iv) specifically for the space $\mathbb{L}^2(a, b; \mathbb{R})$. (This is a special case of the so-called *Hölder's Inequality*.)

Because inner product spaces come equipped with a norm, it makes sense to further characterize them using completeness.

Definition 1.7.12. A *Hilbert space* is a complete inner product space.

Both $\mathbb{R}^N$ and $\mathbb{L}^2(a, b; \mathbb{R})$ equipped with their usual norms are Hilbert spaces, while $\mathbb{C}([a, b]; \mathbb{R})$ equipped with (1.77) is not. Again, the underlying norm plays a crucial role.

The notion of a *basis* encountered in linear algebra can be made precise in the Hilbert space setting and plays a central role in formulating representation formulae for elements of the space. We begin with the following definition.

Definition 1.7.13. Let $\mathscr{H}$ be a Hilbert space and $\mathfrak{B} = \{e_n | n \in K \subset \mathbb{N}\}$.
i.) The *span of* $\mathfrak{B}$ is given by $\text{span}(\mathfrak{B}) = \{\sum_{n \in K} \alpha_n e_n | \alpha_n \in \mathbb{R}, \forall n \in K\}$;
ii.) If $\langle e_n, e_m \rangle_{\mathscr{H}} = 0$, then e_n and e_m are *orthogonal*;
iii.) The members of $\mathfrak{B}$ are *linearly independent* if

$$\sum_{n \in K} \alpha_n e_n = 0 \Longrightarrow \alpha_n = 0, \forall n \in K;$$

iv.) $\mathfrak{B}$ is an *orthonormal set* if
    a.) $\|e_n\|_{\mathscr{H}} = 1, \forall n \in K$,
    b.) $\langle e_n, e_m \rangle_{\mathscr{H}} = \begin{cases} 0, & \text{if } n \neq m, \\ 1, & \text{if } n = m; \end{cases}$
v.) $\mathfrak{B}$ is a *complete set* if $(\langle x, e_n \rangle_{\mathscr{H}} = 0, \forall n \in K) \Longrightarrow (x = 0, \forall x \in \mathscr{H})$;
vi.) A complete orthonormal subset of $\mathscr{H}$ is a *basis* for $\mathscr{H}$.

The utility of a basis $\mathfrak{B}$ of a Hilbert space $\mathscr{H}$ is that every element of $\mathscr{H}$ can be decomposed into a linear combination of the members of $\mathfrak{B}$. For general Hilbert spaces, specifically those that are not finite dimensional like $\mathbb{R}^N$, the existence of a basis is not guaranteed. There are, however, sufficiency results that indicate when a basis must exist. For instance, consider

Definition 1.7.14. An inner product space is *separable* if it contains a countable dense subset $\mathfrak{D}$.

Remark. A set $\mathfrak{D}$ is *countable* if a one-to-one function $f : \mathfrak{D} \to \mathbb{N}$ exists. In such case, the elements of $\mathfrak{D}$ can be matched in a one-to-one manner with those of $\mathbb{N}$. Intuitively, $\mathfrak{D}$ has no more elements than $\mathbb{N}$. A thorough treatment of countability can be found in **[234]**.

The proof of the following result can be found in **[243]**.

Theorem 1.7.15. *Any separable inner product space has a basis.*

Example. The set

$$\mathfrak{B} = \left\{ \frac{1}{\sqrt{2\pi}} \right\} \cup \left\{ \frac{\cos(nt)}{\sqrt{\pi}} \,\Big|\, n \in \mathbb{N} \right\} \cup \left\{ \frac{\sin(nt)}{\sqrt{\pi}} \,\Big|\, n \in \mathbb{N} \right\} \tag{1.80}$$

is an orthonormal, dense subset of $\mathbb{L}^2 (-\pi, \pi; \mathbb{R})$ equipped with inner product (1.77). (Here, $\cos(n\cdot)$ means "$\cos(nt), -\pi \leq t \leq \pi$.")

Exercise 1.7.14.
i.) Prove that $\|f\|_{\mathbb{L}^2} = 1, \forall f \in \mathfrak{B}$.
ii.) Prove that $\langle f, g \rangle_{\mathbb{L}^2} = 0, \forall f \neq g \in \mathfrak{B}$.
iii.) How would you adapt the set defined in (1.80) for $\mathbb{L}^2 (a, b; \mathbb{R})$, where $a < b$, such that properties (i) and (ii) remain true?

Proposition 1.7.16. (Properties of Orthonormal Sets)
Let $\mathscr{H}$ be an inner product space and $\mathscr{Y} = \{y_1, \ldots, y_n\}$ an orthonormal set in $\mathscr{H}$. Then,
i.) $\left\| \sum_{i=1}^{n} y_i \right\|_{\mathscr{H}}^2 = \sum_{i=1}^{n} \|y_i\|_{\mathscr{H}}^2$;
ii.) *The elements of $\mathscr{Y}$ are linearly independent.*
iii.) *If $x \in \mathrm{span}(\mathscr{Y})$, then $x = \sum_{i=1}^{n} \langle x, y_i \rangle_{\mathscr{H}} y_i$;*
iv.) *If $x \in H$, then $\left\langle x - \sum_{i=1}^{n} \langle x, y_i \rangle_{\mathscr{H}} y_i, y_k \right\rangle_{\mathscr{H}} = 0, \forall k \in \{1, \ldots, n\}$.*

Exercise 1.7.15. Prove Prop. 1.7.16.

The following result is the "big deal!"

Theorem 1.7.17. (Representation Theorem for a Hilbert Space)
Let $\mathscr{H}$ be a Hilbert space and $\mathfrak{B} = \{e_n | n \in \mathbb{N}\}$ a basis for $\mathscr{H}$. Then,
i.) *For every $x \in \mathscr{H}$, $\sum_{k=1}^{\infty} |\langle x, e_k \rangle_{\mathscr{H}}|^2 \leq \|x\|_{\mathscr{H}}^2$;*
ii.) $\lim_{N \to \infty} \left\| \sum_{k=1}^{N} \langle x, e_k \rangle_{\mathscr{H}} e_k - x \right\|_{\mathscr{H}} = 0$, *and we write $x = \sum_{k=1}^{\infty} \langle x, e_k \rangle_{\mathscr{H}} e_k$.*

Outline of Proof:

Proof of (i): Observe that $\forall N \in \mathbb{N}$,

$$0 \leq \left\| \sum_{k=1}^{n} \langle x, e_k \rangle_{\mathscr{H}} \, e_k - x \right\|_{\mathscr{H}}^2$$

$$= \left\langle \sum_{k=1}^{n} \langle x, e_k \rangle_{\mathscr{H}} \, e_k - x, \sum_{k=1}^{n} \langle x, e_k \rangle_{\mathscr{H}} \, e_k - x \right\rangle_H$$

$$= \|x\|_H^2 - \sum_{k=1}^{n} |\langle x, e_k \rangle_{\mathscr{H}}|^2.$$

(Why?) Thus, $\sum_{k=1}^{n} |\langle x, e_k \rangle_{\mathscr{H}}|^2 \leq \|x\|_{\mathscr{H}}^2$, $\forall n \in \mathbb{N}$. The result then follows because $\left\{ \sum_{k=1}^{n} |\langle x, e_k \rangle_{\mathscr{H}}|^2 : n \in \mathbb{N} \right\}$ is an increasing sequence bounded above. (Why and so what?)

Proof of (ii): For each $N \in \mathbb{N}$, let $S_N = \sum_{k=1}^{N} \langle x, e_k \rangle_{\mathscr{H}} \, e_k$. The fact that $\{S_N\}$ is a Cauchy sequence in $\mathscr{H}$ follows from Prop. 1.7.16 and part (i) of this theorem. (How?) Moreover, $\{S_N\}$ must converge because $\mathscr{H}$ is complete. The fact that the limit is x follows from the completeness of $\mathfrak{B}$. (Tell how.) □

Exercise 1.7.16. Provide the details in the proof of Thrm. 1.7.17(ii).

Remark. (Fourier Series)
An important application of Thrm. 1.7.17 occurs in the study of Fourier series. A technique often used to solve elementary partial differential equations is the *method of separation of variables.* This involves identifying a (sufficiently smooth) function with a unique series representation defined using a family of sines and cosines (cf. Section 3.2). To this end, we infer from the example directly following Thrm. 1.7.15 that every $f \in \mathbb{L}^2(-\pi, \pi; \mathbb{R})$ can be expressed uniquely as

$$f(t) = \sum_{n=1}^{\infty} \langle f(\cdot), e_n \rangle_{\mathbb{L}^2} \, e_n$$

$$= \left\langle f(\cdot), \frac{1}{\sqrt{2\pi}} \right\rangle_{\mathbb{L}^2} \frac{1}{\sqrt{2\pi}} + \sum_{n=1}^{\infty} \left\langle f(\cdot), \frac{\cos(n\cdot)}{\sqrt{\pi}} \right\rangle_{\mathbb{L}^2} \frac{\cos(nt)}{\sqrt{\pi}}$$

$$+ \sum_{n=1}^{\infty} \left\langle f(\cdot), \frac{\sin(n\cdot)}{\sqrt{\pi}} \right\rangle_{\mathbb{L}^2} \frac{\sin(nt)}{\sqrt{\pi}}, \quad -\pi \leq t \leq \pi, \tag{1.81}$$

where the convergence of the series is in the $\mathbb{L}^2$-sense. For brevity, let

$$a_0 = \left\langle f(\cdot), \frac{1}{\sqrt{2\pi}} \right\rangle_{\mathbb{L}^2} = \frac{1}{\sqrt{2\pi}} \int_{-\pi}^{\pi} f(t) dt, \tag{1.82}$$

$$a_n = \left\langle f(\cdot), \frac{\cos(n\cdot)}{\sqrt{\pi}} \right\rangle_{\mathbb{L}^2} = \frac{1}{\sqrt{\pi}} \int_{-\pi}^{\pi} f(t) \cos(nt) dt, \, n \in \mathbb{N}, \tag{1.83}$$

$$b_n = \left\langle f(\cdot), \frac{\sin(n\cdot)}{\sqrt{\pi}} \right\rangle_{\mathbb{L}^2} = \frac{1}{\sqrt{\pi}} \int_{-\pi}^{\pi} f(t) \sin(nt) dt, \, n \in \mathbb{N}. \tag{1.84}$$

Then, (1.81) can be written as

$$f(t) = \frac{a_0}{\sqrt{2\pi}} + \sum_{n=1}^{\infty} \left[a_n \frac{\cos(nt)}{\sqrt{\pi}} + b_n \frac{\sin(nt)}{\sqrt{\pi}} \right], \quad -\pi \leq t \leq \pi.$$

The utility of this representation will become apparent in Chapter 3. For additional details on Fourier series, see **[121, 301]**.

1.8 Elementary Calculus in Abstract Spaces

Convergent sequences and their properties play a central role in the development of the notions of limits, continuity, the derivative, and the integral. A heuristic discussion is often what is provided in an elementary calculus course, depicting the process visually by appealing to graphs and using sentences of the form, "As x gets closer to a from left or right, quantity A gets closer to quantity B." The intuition gained from such an exposition is helpful, but it needs to be formalized for the purposes of our study.

The plan of this section is to provide the formal definitions of these notions, together with their properties and important main results. The discussion we provide is a terse outline at best, and you are strongly encouraged to review these topics carefully to fill in the gaps (see **[243, 250]**). We cut to the chase and consider <u>abstract</u> functions at the onset because the development is very similar to that of real-valued functions. Of course, the drawback is that the graphical illustrations of these concepts that permeate a presentation of the calculus of real-valued functions is not available for general Banach space-valued functions. Nevertheless, retaining the mental association to the visual interpretation of the concepts is advantageous, by way of analogy.

Throughout the remainder of this chapter, $\mathscr{X}$ and $\mathscr{Y}$ are assumed to be real Banach spaces unless otherwise specified.

1.8.1 Limits

We begin with the extension of the notion of convergence (as defined for sequences) to the function setting.

Definition 1.8.1. A function $f : \text{dom}(f) \subset \mathscr{X} \to \mathscr{Y}$ has *limit L (in $\mathscr{Y}$) at $x = a \in (\text{dom}(f))'$* if for every sequence $\{x_n\} \subset \text{dom}(f)$ for which $\lim_{n \to \infty} \|x_n - a\|_{\mathscr{X}} = 0$, it is the case that $\lim_{n \to \infty} \|f(x_n) - L\|_{\mathscr{Y}} = 0$.

We write $\lim_{x \to a} f(x) = L$ or equivalently, "$\|f(x_n) - L\|_{\mathscr{Y}} \to 0$ as $x \to a$."

Loosely speaking, the interpretation of Def. 1.8.1 for $\mathscr{X} = \mathscr{Y} = \mathbb{R}$ is that as the inputs approach a in any manner possible (i.e., via any sequence in $\text{dom}(f)$ convergent to a), the corresponding functional values approach L. The benefit of this particular definition is that the limit rules follow easily from the corresponding sequence properties.

Exercise 1.8.1. Formulate and prove extensions of Prop. 1.5.5 - Prop. 1.5.7 and Prop. 1.5.9 to the present function setting.

An alternate definition equivalent to Def. 1.8.1, which is often more convenient to work with when involving certain norm estimates in an argument, is as follows:

Definition 1.8.2. A function $f : \text{dom}(f) \subset \mathscr{X} \to \mathscr{Y}$ has *limit L (in $\mathscr{Y}$) at $x = a \in (\text{dom}(f))'$* if $\forall \varepsilon > 0, \exists \delta > 0$ for which

$$x \in \text{dom}(f) \text{ and } 0 < \|x - a\|_{\mathscr{X}} < \delta \implies \|f(x) - L\|_{\mathscr{Y}} < \varepsilon. \qquad (1.85)$$

Remark. Interpreting Def. 1.8.2 verbally, we have $\lim_{x \to a} f(x) = L$ provided that a is a limit point of $\text{dom}(f)$ (so that points of the domain crowd against a) and for any $\varepsilon > 0$, given any open ball $\mathscr{B}_{\mathscr{Y}}(L; \varepsilon)$ around L, there is some sufficiently small so-called "deleted" open ball $\mathscr{B}_{\mathscr{X}}(a; \delta) \setminus \{a\}$ around a such that all members of this ball have images $f(x) \in \mathscr{B}_{\mathscr{Y}}(L; \varepsilon)$. That is, $\forall \varepsilon > 0, \exists \delta > 0$ such that

$$f\left(\text{dom}(f) \cap [\mathscr{B}_{\mathscr{X}}(a; \delta) \setminus \{a\}]\right) \subset \mathscr{B}_{\mathscr{Y}}(L; \varepsilon). \qquad (1.86)$$

The special case when $\mathscr{X} = \mathscr{Y} = \mathbb{R}$ shall arise often in our discussion, as will many related situations involving infinity. In particular, we have

Definition 1.8.3. Let $f : \text{dom}(f) \subset \mathbb{R} \to \mathbb{R}$.
i.) $\lim_{x \to a} f(x) = \infty$ means
   a.) $a \in (\text{dom}(f))'$,
   b.) $\forall M > 0, \exists \delta > 0$ such that

$$x \in \text{dom}(f) \text{ and } 0 < |x - a| < \delta \implies f(x) > M. \qquad (1.87)$$

ii.) $\lim_{x \to \infty} f(x) = L$ means
   a.) $\text{dom}(f) \cap (M, \infty) \neq \emptyset, \forall M > 0$,
   b.) $\forall \varepsilon > 0, \exists N > 0$ such that

$$x \in \text{dom}(f) \text{ and } x > N \implies |f(x) - L| < \varepsilon. \qquad (1.88)$$

Exercise 1.8.2.
i.) Interpret the terms in Def. 1.8.3 geometrically.
ii.) Formulate analogous definitions when ∞ is replaced by $-\infty$.

The notion of *one-sided limits* for real-valued functions arises occasionally, especially when limits are taken as the inputs approach the endpoints of an interval. Definition 1.8.2 can be naturally modified in such case, with the only changes occurring regarding which inputs near a are considered. (Form such extensions.) We denote the *right-limit at a* by $\lim\limits_{x \to a^+} f(x)$, meaning that all inputs chosen when forming sequences that approach a are comprised of values that are greater than or equal to a. Likewise, we denote the *left-limit at a* by $\lim\limits_{x \to a^-} f(x)$.

1.8.2   Continuity

Understanding the nature of continuous functions is crucial, as much of the work in this text is performed in the space $\mathbb{C}(I;X)$. To begin, we need only to slightly modify Defs. 1.8.1 and 1.8.2 to arrive at the following stronger notion of *(norm) continuity*.

Definition 1.8.4. A function $f : \text{dom}(f) \subset \mathscr{X} \to \mathscr{Y}$ is *continuous* at $a \in \text{dom}(f)$ if either of these two equivalent statements hold:

i.) For every sequence $\{x_n\} \subset \text{dom}(f)$ for which $\lim\limits_{n \to \infty} \|x_n - a\|_{\mathscr{X}} = 0$, it is the case that $\lim\limits_{n \to \infty} \|f(x_n) - f(a)\|_{\mathscr{Y}} = 0$. We write $\lim\limits_{x \to a} f(x_n) = f\left(\lim\limits_{n \to \infty} x_n\right) = f(a)$.

ii.) $\forall \varepsilon > 0, \exists \delta > 0$ for which

$$x \in \text{dom}(f) \wedge \|x - a\|_{\mathscr{X}} < \delta \implies \|f(x) - f(a)\|_{\mathscr{Y}} < \varepsilon. \qquad (1.89)$$

We say f is *continuous on* $S \subset \text{dom}(f)$ if f is continuous at every element of S.

"Continuity at a" is a strengthening of merely "having a limit at a" because the limit candidate being $f(a)$ requires that a be in the domain of f. It follows from Exer. 1.8.1 that the arithmetic combinations of continuous functions preserve continuity. (Tell how.)

Exercise 1.8.3. Prove that $f : \text{dom}(f) \subset \mathscr{X} \to \mathbb{R}$ defined by $f(x) = \|x\|_{\mathscr{X}}$ is continuous.

More complicated continuous functions can be built by forming compositions of continuous functions, as the following result indicates.

Proposition 1.8.5. *Let* $(\mathscr{X}, \|\cdot\|_{\mathscr{X}})$, $(\mathscr{Y}, \|\cdot\|_{\mathscr{Y}})$, *and* $(\mathscr{Z}, \|\cdot\|_{Z})$ *be Banach spaces and suppose that* $g : \text{dom}(g) \subset \mathscr{X} \to \mathscr{Y}$ *and* $f : \text{dom}(f) \subset \mathscr{Y} \to \mathscr{Z}$ *with* $\text{rng}(g) \subset \text{dom}(f)$. *If g is continuous at* $a \in \text{dom}(g)$ *and f is continuous at* $g(a) \in \text{dom}(f)$, *then* $f \circ g$ *is continuous at a. In such case, we write*

$$\lim\limits_{n \to \infty} f\left(g(x_n)\right) = f\left(\lim\limits_{n \to \infty} g(x_n)\right) = f\left(g\left(\lim\limits_{n \to \infty} x_n\right)\right) = f(g(a)).$$

Exercise 1.8.4. Prove Prop. 1.8.5 using both formulations of continuity in Def. 1.8.4.

We will frequently consider functions defined on a product space, such as $f : \mathscr{X}_1 \times \mathscr{X}_2 \to \mathscr{Y}$. Interpreting Def. 1.8.4 for such a function requires that we use $\mathscr{X}_1 \times \mathscr{X}_2$ as the space $\mathscr{X}$. This raises the question as to what is meant by the phrases "$\|(x_1, x_2) - (a, b)\|_{\mathscr{X}_1 \times \mathscr{X}_2} < \delta$" or "$\{(x_1^n, x_2^n)\} \to (a, b)$ in $\mathscr{X}_1 \times \mathscr{X}_2$." One typical product space norm is

$$\|(x_1, x_2)\|_{\mathscr{X}_1 \times \mathscr{X}_2} = \|x_1\|_{\mathscr{X}_1} + \|x_2\|_{\mathscr{X}_2}. \qquad (1.90)$$

Both conditions can be loosely interpreted by focusing on controlling each of the components of the members of the product space. (Make this precise.)

Different forms of continuity are used in practice. A weaker form of continuity is to require that the function be continuous in only a selection of the input variables and that such "section continuity" hold uniformly for all values of the remaining variables. For instance, saying that a function $f : \mathscr{X}_1 \times \mathscr{X}_2 \to \mathscr{Y}$ is "continuous in $\mathscr{X}_1$ uniformly on $\mathscr{X}_2$" means that for every fixed $x_2 \in \mathscr{X}_2$, the function $g : \mathscr{X}_1 \to \mathscr{Y}$ defined by $g(x_1) = f(x_1, x_2)$ satisfies Def. 1.8.4 with $\mathscr{X} = \mathscr{X}_1$, and that the choice of $x_2 \in \mathscr{X}_2$ does not affect the estimates or convergence of sequences arising in the continuity calculations involving g.

Exercise 1.8.5.
i.) Explain what it means for a function $f : [a,b] \times \mathscr{X} \times \mathscr{X} \to \mathscr{X}$ to be continuous on $\mathscr{X} \times \mathscr{X}$ uniformly on $[a,b]$.
ii.) Explain what it means for a mapping $\Phi : \mathbb{C}([a,b]; \mathscr{X}) \to \mathbb{C}([a,b]; \mathscr{X})$ to be continuous.
iii.) Interpret Def. 1.8.4 for functions of the form $f : [a,b] \to \mathbb{M}^N(\mathbb{R})$.

The notion of continuity for real-valued functions can be modified to give meaning to *left-* and *right-sided continuity* in a manner similar to one-sided limits. All continuity results also hold for one-sided continuity. A function that possesses both left- and right-limits at $x = a$, but for which these limits are different, is said to have a *jump discontinuity* at $x = a$.

Continuous functions enjoy interesting topological properties that lead to some rather strong results concerning boundedness and the existence of fixed-points. We list the essential results below, without proof, for later reference. (See **[17, 243]** for proofs.)

Proposition 1.8.6. (Properties of Continuous Functions)
Assume that $f : \mathrm{dom}(f) \subset \mathscr{X} \to \mathscr{Y}$ is continuous.
i.) *For every open set G in $\mathscr{Y}$, $f^{-1}(G)$ is open in $\mathscr{X}$.*
ii.) *For every compact set K in $\mathscr{X}$, $f(K)$ is compact in $\mathscr{Y}$.*
iii.) *Let $\mathscr{Y} = \mathbb{R}$. If K is a compact set in $\mathscr{X}$, then f is bounded on K and $\exists x_0, y_0 \in K$ such that $f(x_0) = \inf\{f(x) : x \in K\}$ and $f(y_0) = \sup\{f(x) : x \in K\}$.*
iv.) (Intermediate-Value Theorem) *Assume $\mathrm{dom}(f) = [a,b]$ and that $\mathscr{X} = \mathscr{Y} = \mathbb{R}$. If $f(a) \neq f(b)$, then for any z between $f(a)$ and $f(b)$, $\exists c_z \in (a,b)$ such that $f(c_z) = z$.*
v.) *If $f : \mathscr{X} \to \mathscr{Y}$ and $g : \mathscr{X} \to \mathscr{Y}$ are continuous and $f = g$ on a set $\mathscr{D}$ dense in $\mathscr{X}$, then $f = g$ on $\mathscr{X}$.*

Exercise 1.8.6.
i.) If $f : [a,b] \to [a,b]$ is continuous, prove that $\exists c \in [a,b]$ such that $f(c) = c$.
ii.) Show that the conclusion of (i) fails if $[a,b]$ is replaced by a half-open, open, or unbounded interval.

We now define a concept that is stronger than continuity in the sense that for a given $\varepsilon > 0$, there exists a <u>single</u> $\delta > 0$ that "works" for every point in the set.

Precisely,

Definition 1.8.7. A function $f : S \subset \text{dom}(f) \subset \mathscr{X} \to \mathscr{Y}$ is *uniformly continuous (UC) on S* provided that $\forall \varepsilon > 0, \exists \delta > 0$ for which

$$x, y \in S \wedge \|x - y\|_{\mathscr{X}} < \delta \implies \|f(x) - f(y)\|_{\mathscr{Y}} < \varepsilon.$$

Remark. The critical feature of uniform continuity on S is that the δ depends on the ε only, and not on the actual points $x, y \in S$ at which we are located. That is, given any $\varepsilon > 0$, there exists $\delta > 0$ such that $\|f(x) - f(y)\|_{\mathscr{Y}} < \varepsilon$, for any pair of points $x, y \in S$ with $\|x - y\|_{\mathscr{X}} < \delta$, no matter where they are located in S. In this sense, uniform continuity of f on S is a "global" property, whereas mere continuity at $a \in S$ is a "local" property.

Example. We claim that $f : (0, 1] \to \mathbb{R}$ defined by $f(x) = \frac{1}{x}$ is not UC on $(0,1]$. To see this, let $0 < \varepsilon < 1$ and suppose that δ is any positive real number. There exists $n \in \mathbb{N}$ such that $\frac{1}{2n} < \delta$. Let $x_0 = \frac{1}{n}$ and $y_0 = \frac{1}{2n}$. Then, x_0 and y_0 are points of $(0,1]$ such that

$$|x_0 - y_0| = \left| \frac{1}{n} - \frac{1}{2n} \right| = \frac{1}{2n} < \delta,$$

but

$$|f(x_0) - f(y_0)| = |n - 2n| = n > \varepsilon.$$

Hence, no choice of δ satisfies the definition of uniform continuity and thus, f is not UC on $(0,1]$.

This example illustrates the fact that continuity on S does not imply uniform continuity. However, if S is compact, the implication does hold, as the next result suggests. (See [**17**] for a proof.)

Proposition 1.8.8. *Let $f : \text{dom}(f) \subset \mathscr{X} \to \mathscr{Y}$ and K a compact subset of $\text{dom}(f)$. If f is continuous on K, then f is UC on K.*

Exercise 1.8.7.
i.) If $f : \text{dom}(f) \subset \mathbb{R} \to \mathbb{R}$ be UC on $S \subset \text{dom}(f)$. Prove that the image under f of any Cauchy sequence in S is itself a Cauchy sequence.
ii.) Prove that if f is UC on a bounded set $S \subset \text{dom}(f)$, then f is bounded on S.

The notion of absolute continuity, which involves controlling the total displacement of functional values across small intervals, arises in the definition of certain function spaces.

Definition 1.8.9. A function $f : [a, b] \to \mathbb{R}$ is *absolutely continuous (AC) on $[a, b]$* if $\forall \varepsilon > 0, \exists \delta > 0$ such that for any finite collection $\{(a_i, b_i) : i = 1, \ldots, n\}$ of pairwise disjoint open subintervals of $[a, b]$ for which $\sum_{k=1}^{n} |b_k - a_k| < \delta$, it is the case that $\sum_{k=1}^{n} |f(b_k) - f(a_k)| < \varepsilon$.

It can be shown that the usual arithmetic combinations of AC functions are also AC, and that AC functions are necessarily continuous. (Try showing this!)

1.8.3 The Derivative

Measuring the rate of change of one quantity with respect to another is central to the formulation and analysis of many mathematical models. The concept is formalized in the real-valued setting via a limiting process of quantities that geometrically resembles slopes of secant lines. We can extend this definition to $\mathscr{X}$-valued functions by making use of the norm on $\mathscr{X}$. This leads to

Definition 1.8.10. A function $f : (a,b) \to \mathscr{X}$ is *differentiable* at $x_0 \in (a,b)$ if there exists a member of $\mathscr{X}$, denoted by $f'(x_0)$, such that

$$\lim_{h \to 0} \left\| \frac{f(x_0 + h) - f(x_0)}{h} - f'(x_0) \right\|_{\mathscr{X}} = 0. \qquad (1.91)$$

The number $f'(x_0)$ is called the *derivative of f at x_0*. We say f is *differentiable on S* if f is differentiable at every element of S.

Exercise 1.8.8. Interpret Def. 1.8.10 using the formulation of limit given in Def. 1.8.2.

One-sided derivatives for real-valued functions $f : [a,b] \to \mathbb{R}$ are naturally defined using one-sided limits. We write $\frac{d^+}{dx} f(x)|_{x=c}$ to stand for the *right-sided derivative of f at c*, $\frac{d^-}{dx} f(x)|_{x=c}$ for the *left-sided derivative of f at c*, and $\frac{d}{dx} f(x)|_{x=c}$ the *derivative of f at c*, when they exist.

Exercise 1.8.9. Explain how you would show that $\frac{d^+}{dx} f(x)|_{x=c} = \infty$.

The notion of differentiability is more restrictive than continuity, a fact typically illustrated for real-valued functions by examining the behavior of $f(x) = |x|$ at $x = 0$. Indeed, differentiable functions are necessarily continuous, but not vice versa, in the abstract setting of $\mathscr{X}$-valued functions. Further, if the derivative of a function f is itself differentiable, we say f has a second derivative. Such a function has a "higher degree of regularity" than one that is merely differentiable. The pattern continues with each order of derivative, from which the following string of inclusions is derived, $\forall n \in \mathbb{N}$:

$$\mathbb{C}^n (I; \mathscr{X}) \subset \mathbb{C}^{n-1} (I; \mathscr{X}) \subset \ldots \subset \mathbb{C}^1 (I; \mathscr{X}) \subset \mathbb{C}(I; \mathscr{X}). \qquad (1.92)$$

Here, inclusion means that a space further to the left in the string is a closed linear subspace of all those occurring to its right.

We shall often work with real-valued differentiable functions. The arithmetic combinations of differentiable functions are again differentiable, although some care must be taken when computing the derivative of a product and composition. The following result provides two especially nice features of real-valued differentiable functions, the first of which is used to establish l'Hopital's rule (see [17]).

Proposition 1.8.11. (Properties of Real-Valued Differentiable Functions)
i.) (Mean Value Theorem) *If $f : [a,b] \rightarrow \mathbb{R}$ is differentiable on (a,b) and continuous on $[a,b]$, then $\exists c \in (a,b)$ for which*

$$f(b) - f(a) = f'(c)(b-a). \tag{1.93}$$

ii.) (Intermediate Value Theorem) *If $f : I \rightarrow \mathbb{R}$ is differentiable on $[a,b] \subset I$ and $f'(a) < f'(b)$, then $\forall z \in (f'(a), f'(b))$, $\exists c \in (a,b)$ such that $f'(c) = z$.*

Exercise 1.8.10. Assume that $f : [a,b] \rightarrow \mathbb{R}$ is continuous with $f'(x) = 0$, $\forall x \in [a,b]$. Prove that f is constant on $[a,b]$.

Remarks.
1. There are other ways of defining differentiability that are guided by different applications in which such calculations arise. For instance, there are extensions of the notion of a directional derivative, as well a weaker notion of differentiability defined using *distributions*. These topics are treated in **[1]**.
2. We shall be interested in bounded domains $\Omega \subset \mathbb{R}^N$ with a so-called *smooth boundary* $\partial \Omega$. This boundary is necessarily a curve in $\mathbb{R}^N$ and so, by *smooth* we mean that each of the N component functions used to define the curve is differentiable in the sense of Def. 1.8.10.

1.9 Some Elementary ODEs

Courses on elementary ordinary differential equations (ODEs) are chock full of techniques used to solve particular types of elementary differential equations. Within this vast toolbox are three particular scenarios that play a role in this text. We recall them informally here, along with some elementary exercises, to refresh your memory.

1.9.1 Separation of Variables

An ODE of the form $\frac{dy}{dx} = f(x)g(y)$, where $y = y(x)$, is called *separable* because symbolically the terms involving y can be gathered on one side of the equality and the terms involving x can be written on the other, thereby resulting in the equivalent equation (expressed in differential form) $\frac{1}{g(y)}dy = f(x)dx$. Integrating both sides yields an implicitly defined function $H(y) = G(x) + C$ that satisfies the original ODE on some set and is called the *general solution* of the ODE. This process can be made formal through the use of appropriate changes of variable.

Exercise 1.9.1. Determine the general solution of these ODEs:
i.) $\frac{dy}{dx} = e^{2x} \csc(\pi y)$;
ii.) $\frac{dy}{dx} = ax^n$, where $n \neq -1$ and $a \in \mathbb{R} \setminus \{0\}$;
iii.) $(1 - y^3) \frac{dy}{dx} = \sum_{i=1}^{N} a_i \sin(b_i x)$, where $a_i, b_i \in \mathbb{R}$ and $n \in \mathbb{N}$.

1.9.2 First-Order Linear ODEs

A *first-order linear ODE* is of the form

$$\frac{dy}{dx} + a(x)y = b(x), \text{ where } y = y(x). \tag{1.94}$$

We shall develop the solution of the initial-value problem (IVP) obtained by coupling (1.94) with the initial condition (IC)

$$y(x_0) = y_0 \tag{1.95}$$

using a simplified version of the so-called *variation of parameters method*.

<u>Step 1</u>: Solve the related homogenous equation $\frac{dy_h}{dx} + a(x)y_h = 0$.
  This equation is separable, so integrating both sides over the interval (x_0, x) yields

$$\frac{dy_h}{y_h} = -a(x)dx \implies \underbrace{\ln|y_h(x)| - \ln|y_h(x_0)|}_{=\ln\left|\frac{y_h(x)}{y_h(x_0)}\right|} = -\int_{x_0}^{x} a(s)ds$$

$$\implies y_h(x) = y_h(x_0)e^{-\int_{x_0}^{x} a(s)ds}. \tag{1.96}$$

<u>Step 2</u>: Determine $C(x)$ for which $y(x) = C(x)y_h(x)$ satisfies (1.94).
  Substitute this function into (1.94) to obtain

$$\frac{d}{dx}[C(x)y_h(x)] + a(x)[C(x)y_h(x)] = b(x)$$

$$C(x)\frac{dy_h}{dx} + y_h(x)\frac{dC}{dx} + a(x)C(x)y_h(x) = b(x)$$

$$C(x)\underbrace{\left[\frac{dy_h}{dx} + a(x)y_h(x)\right]}_{=0 \text{ by Step 1}} + y_h(x)\frac{dC}{dx} = b(x) \tag{1.97}$$

$$\frac{dC}{dx} = \frac{b(x)}{y_h(x)} = b(x)\left[y_h(x_0)e^{-\int_{x_0}^{x} a(s)ds}\right]^{-1}$$

$$C(x) = \int_{x_0}^{x} b(s)y_h^{-1}(x_0)e^{\int_{x_0}^{s} a(t)dt}ds + K,$$

where K is an integration constant.
<u>Step 3</u>: Substitute (1.97) into $y(x) = C(x)y_h(x)$ and apply (1.95) to find the general solution of the IVP.

$$y(x) = y_h(x_0)e^{-\int_{x_0}^{x} a(s)ds}\left[\int_{x_0}^{x} b(s)y_h^{-1}(x_0)e^{\int_{x_0}^{s} a(t)dt}ds + K\right]$$

$$= \int_{x_0}^{x} b(s)e^{-\int_{s}^{x} a(t)dt}ds + \underbrace{Ky_h(x_0)}e^{-\int_{x_0}^{x} a(t)dt}. \tag{1.98}$$

$$\text{Call this } \overline{K}$$

<u>Step 4</u>: Apply the IC (1.95) to determine the solution of the IVP.

Now, apply (1.95) to see that $y(x_0) = 0 + \overline{K} = y_0$. Hence, the solution of the IVP is

$$y(x) = y_0 e^{-\int_{x_0}^x a(t)dt} + \int_{x_0}^x b(s)e^{-\int_x^s a(t)dt} ds. \tag{1.99}$$

This formula is called the *variation of parameters formula.*

Exercise 1.9.2. Justify all steps in the derivation of (1.99).

Exercise 1.9.3. Solve the IVP:

$$\begin{cases} \frac{dy}{dx} + \frac{1}{2}y(x) = e^{-3x}, \\ y(0) = \frac{1}{2}. \end{cases}$$

1.9.3   Higher-Order Linear ODEs

Higher-order linear ODEs with constant coefficients of the form

$$a_n x^{(n)} + a_{n-1}x^{(n-1)} + \ldots + a_1 x' + a_0 x = 0, \tag{1.100}$$

where $a_i \in \mathbb{R}$, $a_n \neq 0$, and $n \in \mathbb{N}$, arise in Chapters 2 through 5. A more general version of the procedure outlined in Section 1.9.2 can be used to derive the general solution of (1.100) (see **[92, 93]**). We consider the special case

$$ax''(t) + bx'(t) + cx(t) = 0, \tag{1.101}$$

where $a \neq 0$ and $b, c \in \mathbb{R}$. Assuming that the solution of (1.101) is of the form $x(t) = e^{mt}$ yields

$$e^{mt}\left(am^2 + bm + c\right) = 0 \implies am^2 + bm + c = 0. \tag{1.102}$$

So, the nature of the solution of (1.101) is completely determined by the values of m. There are three distinct cases regarding the nature of the solution using (1.102):

Nature of the Roots of (1.102)	General Solution of (1.101)
$m_1 \neq m_2$ (real)	$x(t) = C_1 e^{m_1 t} + C_2 e^{m_2 t}$
$m_1 = m_2$ (real)	$x(t) = C_1 e^{m_1 t} + C_2 t e^{m_1 t}$
$m_1, m_2 = \alpha \pm i\beta$	$x(t) = C_1 e^{\alpha t} \sin(\beta t) + C_2 e^{\alpha t} \cos(\beta t)$

Exercise 1.9.4. For what values of m_1 and m_2 is it guaranteed that
i.) $\lim\limits_{t \to \infty} x(t) = 0$, $\forall C_1, C_2 \in \mathbb{R}$?
ii.) $x(\cdot)$ is a bounded function of t for a given $C_1, C_2 \in \mathbb{R}$?

The variation of parameters method can be extended to solve higher-order nonhomogenous linear ODEs (that is, when the right-hand side of (1.100) is not identically zero) as well (see **[92, 93]**).

1.10 A Handful of Integral Inequalities

Establishing *a priori* estimates is a crucial step in the proofs of most existence results. Such estimates often take the form of an upper bound of the state process in some function space. We begin with the following classical inequality from which many others are derived, proved as in [94].

Theorem 1.10.1. Gronwall's Lemma
Let $t_0 \in (-\infty, T)$, $x \in C([t_0, T]; \mathbb{R})$, $K \in C([t_0, T]; [0, \infty))$, and M be a real constant. If

$$x(t) \leq M + \int_{t_0}^{t} K(s)x(s)ds, \ t_0 \leq t \leq T, \tag{1.103}$$

then

$$x(t) \leq M e^{\int_{t_0}^{t} K(s)ds}, \ t_0 \leq t \leq T. \tag{1.104}$$

Proof. Define $y : [t_0, T] \to \mathbb{R}$ by

$$y(t) = M + \int_{t_0}^{t} K(s)x(s)ds.$$

Observe that $y \in C^1([t_0, T]; \mathbb{R})$ and satisfies the IVP

$$\begin{cases} y'(t) & = K(t)x(t), t_0 \leq t \leq T, \\ y(t_0) & = M. \end{cases} \tag{1.105}$$

(Tell why.) By assumption, $x(t) \leq y(t), \forall t_0 \leq t \leq T$. Multiplying both sides of this inequality by $K(t)$ and then substituting into (1.105) yields

$$\begin{cases} y'(t) \leq K(t)y(t), t_0 \leq t \leq T, \\ y(t_0) = M. \end{cases} \tag{1.106}$$

As such,

$$y'(t) - K(t)y(t) \leq 0, t_0 \leq t \leq T,$$

so that multiplying both sides by $e^{-\int_{t_0}^{t} K(s)ds}$ yields

$$\frac{d}{dt}\left[e^{-\int_{t_0}^{t} K(s)ds} y(t)\right] \leq 0, t_0 \leq t \leq T. \tag{1.107}$$

(Why?) Consequently,

$$e^{-\int_{t_0}^{t} K(s)ds} y(t) \leq y(t_0) = M, t_0 \leq t \leq T,$$

and so

$$y(t) \leq M e^{\int_{t_0}^{t} K(s)ds}, \ t_0 \leq t \leq T. \tag{1.108}$$

Because $x(t) \leq y(t), \forall t_0 \leq t \leq T$, the conclusion follows from (1.108). $\qquad \square$

This form of Gronwall's Lemma applies only for $t \geq t_0$, but we might need an estimate that holds for $t < t_0$. It is not difficult to establish such an estimate, as suggested by the following corollary.

Corollary 1.10.2. *Let $t_0 \in \mathbb{R}$, $x \in \mathbb{C}((-\infty, t_0]; \mathbb{R})$, $K \in \mathbb{C}((-\infty, t_0]; [0, \infty))$, and M be a real constant. If*

$$x(t) \leq M + \int_t^{t_0} K(s)x(s)ds, \quad -\infty < t \leq t_0, \tag{1.109}$$

then

$$x(t) \leq Me^{\int_t^{t_0} K(s)ds}, \quad -\infty < t \leq t_0. \tag{1.110}$$

Proof. Let $t \in (-\infty, t_0]$ and substitute $t = t_0 - s$ into (1.109) to obtain

$$x(t_0 - s) \leq M + \int_{t_0 - s}^{t_0} K(\tau)x(\tau)d\tau, \, s > 0. \tag{1.111}$$

Implementing the change of variable $\tau = t_0 - \xi$ in (1.111) yields

$$x(t_0 - s) \leq M + \int_0^s K(t_0 - \xi)x(t_0 - \xi)d\xi, \, s > 0. \tag{1.112}$$

(Tell why.) Now, let $\bar{x}(s) = x(t_0 - s)$ and $\overline{K}(s) = K(t_0 - s)$ in (1.112) to obtain

$$\bar{x}(s) \leq M + \int_0^s \overline{K}(\xi)\bar{x}(\xi)d\xi, \, s > 0.$$

Applying Thrm. 1.10.1 then yields the estimate

$$\bar{x}(s) \leq Me^{\int_0^s \overline{K}(\xi)d\xi}, \, s > 0. \tag{1.113}$$

Going back to the original variable then results in (1.110). □

While these integral inequalities are veritable workhorses in practice, it might not be possible (or feasible) to verify the hypotheses of the lemma. One natural question is what to do if the constant M were allowed to vary with t. In such case, the resulting estimate would depend on the regularity of M. One such result is as follows.

Proposition 1.10.3. *Let $t_0 \in (-\infty, T)$, $x \in \mathbb{C}([t_0, T]; \mathbb{R})$, $K \in \mathbb{C}([t_0, T]; [0, \infty))$, and $M : [t_0, T] \to [0, \infty)$. Assume that*

$$x(t) \leq M(t) + \int_{t_0}^t K(s)x(s)ds, \, t_0 \leq t \leq T. \tag{1.114}$$

i.) *If $M \in \mathbb{C}([t_0, T]; [0, \infty))$, then*

$$x(t) \leq M(t) + \int_{t_0}^t M(s)K(s)e^{\int_s^t K(\tau)d\tau}ds, \, t_0 \leq t \leq T. \tag{1.115}$$

ii.) *If $M \in \mathbb{C}^1\left((t_0, T); [0, \infty)\right)$, then*

$$x(t) \le e^{\int_{t_0}^t K(s)ds}\left[M(t_0) + \int_{t_0}^t M'(s)e^{-\int_{t_0}^s K(\tau)d\tau}ds\right], \quad t_0 \le t \le T. \tag{1.116}$$

A rich source of such inequalities is the text **[320]**. The following inequality, stated without proof, is useful when imposing more general growth conditions on the forcing term.

Proposition 1.10.4. *Let w, ψ_1, ψ_2, and $\psi_3 \in \mathbb{C}\left([0, \infty); [0, \infty)\right)$ and $w_0 \ge 0$. If, $\forall t > 0$,*

$$w(t) \le w_0 + \int_0^t \psi_1(s)w(s)ds + \int_0^t \psi_1(s)\left(\int_0^s \psi_2(\tau)w(\tau)d\tau\right)ds$$

$$+ \int_0^t \psi_1(s)\left(\int_0^s \psi_2(\tau)\left(\int_0^\tau \psi_3(\theta)w(\theta)d\theta\right)d\tau\right)ds, \tag{1.117}$$

then $\forall t > 0$,

$$w(t) \le w_0\left[1 + \int_0^t \left\{\psi_1(s)e^{\int_0^s \psi_1(\tau)d\tau}\left(1 + \right.\right.\right.$$

$$\left.\left.\left.\int_0^s \psi_2(\tau)e^{\int_0^\tau[\psi_2(\theta)+\psi_3(\theta)]d\theta}d\tau\right)\right\}ds\right]. \tag{1.118}$$

1.11   Fixed-Point Theory

One of our central strategies involves the use of so-called *fixed-point theory*. This broad approach is based on a very straightforward strategy whose utility is especially evident in the study of existence theory. We present several useful results.

Definition 1.11.1. A mapping $\Phi : \mathscr{X} \to \mathscr{X}$ is a *contraction* if $\exists 0 < \alpha < 1$ such that

$$\|\Phi x - \Phi y\|_{\mathscr{X}} \le \alpha \|x - y\|_{\mathscr{X}}, \quad \forall x, y \in \mathscr{X}. \tag{1.119}$$

Note that a contraction is automatically uniformly continuous. (Why?)

Theorem 1.11.2. Contraction Mapping Principle
If $\Phi : \mathscr{X} \to \mathscr{X}$ is a contraction, then there exists a unique $z^\star \in \mathscr{X}$ such that $\Phi(z^\star) = z^\star$. (We call $z^\star$ a fixed-point of Φ.)

Proof. Let $z_0 \in \mathscr{X}$ and define the sequence $\{z_n : n \in \mathbb{N}\} \subset \mathscr{X}$ by

$$z_n = \Phi(z_{n-1}), \quad n \in \mathbb{N}. \tag{1.120}$$

If $z_1 = z_0$, then we are done. (Why?) If not, then prove inductively that the following two statements hold, $\forall m, k \ge 0$.

$$\|z_{m+1} - z_m\|_{\mathscr{X}} \le \alpha^m \|z_1 - z_0\|_{\mathscr{X}}, \tag{1.121}$$

$$\|z_{m+k} - z_m\|_{\mathscr{X}} \le \sum_{j=0}^{k-1} \alpha^{n+j} \|z_1 - z_0\|_{\mathscr{X}} \le \frac{\alpha^m}{1-\alpha} \|z_1 - z_0\|_{\mathscr{X}}. \tag{1.122}$$

Let $\varepsilon > 0$. There exists $M \in \mathbb{N}$ such that

$$m \ge M \implies \alpha^m < \frac{(1-\alpha)\varepsilon}{\|z_1 - z_0\|_{\mathscr{X}}}. \tag{1.123}$$

As such, (1.122) and (1.123) together imply that

$$m \ge M \implies \|z_{m+k} - z_m\|_{\mathscr{X}} < \varepsilon, \ \forall k \ge 0. \tag{1.124}$$

So, $\{z_n : n \in \mathbb{N}\}$ is a Cauchy sequence in $\mathscr{X}$ and hence convergent to some $z^\star \in \mathscr{X}$. The fact that $z^\star$ is a fixed point of Φ follows from the fact that

$$\lim_{n \to \infty} \Phi(z_{n-1}) = \lim_{n \to \infty} z_n = z^\star.$$

As for uniqueness, suppose that both $x^\star$ and $y^\star$ are fixed-points of Φ and argue that $\|\Phi x - \Phi y\|_{\mathscr{X}} < \|x - y\|_{\mathscr{X}}$. (So what?) □

Corollary 1.11.3. *If* $\Phi : \mathscr{X} \to \mathscr{X}$ *is a mapping for which* $\exists 0 < \alpha < 1$ *such that for some* $n_0 \in \mathbb{N}$,

$$\|\Phi^{n_0} x - \Phi^{n_0} y\|_{\mathscr{X}} \le \alpha \|x - y\|_{\mathscr{X}}, \ \forall x, y \in \mathscr{X}, \tag{1.125}$$

then Φ *has a unique fixed-point in* $\mathscr{X}$.

Exercise 1.11.1. Prove Cor. 1.11.3

Remark. The above two results are particularly useful when establishing the existence and uniqueness of mild solutions of many classes of stochastic evolution equations. Often, it is beneficial to apply them on a closed subspace of $\mathscr{X}$, specifically a closed ball $\mathfrak{B}_{\mathscr{X}}(x_0; \varepsilon)$, rather than on the entire space $\mathscr{X}$. The theorems are still applicable because a closed metric subspace of a Banach space is complete.

Exercise 1.11.2. Was completeness of $\mathscr{X}$ an essential ingredient of Thrm. 1.11.2 and Cor. 1.11.3? Explain.

Not every continuous operator $\Phi : \mathscr{X} \to \mathscr{X}$ is a contraction. As such, Thrm. 1.11.2 is not always applicable. But, is it necessary for Φ to be a contraction in order for it to have a fixed-point?

Even in the one-dimensional case there is no shortage of continuous functions $f : \mathbb{R} \to \mathbb{R}$ that do not have fixed-points. For instance, take $f(x) = x - 1$. We also know from Exer. 1.8.6(i) that a continuous function $f : [a,b] \to [a,b]$ *must* possess a fixed-point, while continuity is not strong enough to ensure a function $g : (a,b) \to \mathbb{R}$ has a fixed-point. (Why?) This collection of examples suggests that the structure of the domain and range plays an important role. The following theorem due to Brouwer [160, 243] provides a complete answer for the $\mathbb{R}^N$-setting.

Theorem 1.11.4. Brouwer's Fixed-Point Theorem

Let $\mathscr{D} \subset \mathbb{R}^N$ be a closed, convex, and bounded set. If $\Psi : \mathscr{D} \to \mathscr{D}$ is continuous, then Ψ has at least one fixed-point in $\mathscr{D}$.

This theorem does not hold true in the infinite-dimensional setting because while a closed ball $\mathfrak{B}_{\mathbb{R}^N}(x_0;R)$ in $\mathbb{R}^N$ must be compact, it need not be in a general Banach space. This fact significantly enters into the proof of Thrm.1.11.4, as well as in the construction of a counterexample that illustrates the falsity of the theorem in an infinite-dimensional space. (For the latter, see [160].) In light of this failure, we must impose more stringent conditions on the nature of the set $\mathscr{D}$ in order to obtain the desired extension of Thrm. 1.11.4. This is where the notion of a *compact set* in a Banach space comes into play. Some of the following notions were introduced in Section 1.4.4 and are recalled here for convenience.

Definition 1.11.5. Let $(\mathscr{X}, \|\cdot\|_{\mathscr{X}})$ be a normed linear space. A set $\mathscr{K} \subset \mathscr{X}$ is called
i.) *convex* if $\forall 0 \leq t \leq 1$ and $x, y \in \mathscr{K}$, $tx + (1-t)y \in \mathscr{K}$;
ii.) *bounded* if $\exists M > 0$ such that $\|x\|_{\mathscr{X}} \leq M$, $\forall x \in \mathscr{K}$;
iii.) *compact* if every sequence $\{x_n\} \subset \mathscr{K}$ contains a subsequence $\{x_{n_k}\}$ that converges to a member of $\mathscr{K}$;
iv.) *precompact* if $\mathrm{cl}_{\mathscr{X}}(\mathscr{K})$ is a compact subset of $\mathscr{X}$.

A compact set $\mathscr{K} \subset \mathscr{X}$ must be closed and bounded (Why?), but not conversely, as shown in the following exercise.

Exercise 1.11.3. Let $\mathscr{K} = \left\{ \cos\left(\frac{2n\pi x}{b-a}\right) : n \in \mathbb{N} \wedge a \leq x \leq b \right\}$. Show that while $\mathscr{K}$ is a closed and bounded subset of $\mathbb{L}^2(a,b;\mathbb{R})$, it is not compact in $\mathbb{L}^2(a,b;\mathbb{R})$.

Verifying precompactness can be a tedious task, depending on the underlying topology of $\mathscr{X}$ because the closure is taken in $\mathscr{X}$. The so-called Arzela–Ascoli Theorem offers a method of attack. To state it, we begin with

Definition 1.11.6. A set $\mathscr{Z} \subset \mathbb{C}([a,b];\mathscr{Y})$ is *equicontinuous* at $t_0 \in [a,b]$ if $\forall \varepsilon > 0$, $\exists \delta > 0$ (depending on ε and t_0) such that

$$s \in [a,b] \text{ with } |s - t_0| < \delta \implies \|z(t_0) - z(s)\|_{\mathscr{X}} < \varepsilon, \ \forall z \in \mathscr{Z}.$$

Theorem 1.11.7. Arzela-Ascoli in $\mathbb{C}([a,b];\mathscr{Y})$

A set $\mathscr{Z} \subset \mathbb{C}([a,b];\mathscr{Y})$ is precompact if and only if the following hold, $\forall t \in [a,b]$:
i.) $\{z(t) \,|\, z \in \mathscr{Z}\}$ *is precompact in $\mathscr{Y}$.*
ii.) $\{z(t) \,|\, z \in \mathscr{Z}\}$ *is equicontinuous.*

Other commonly used fixed-point theorems using the notion of compactness are discussed in [70, 355].

1.12   Guidance for Selected Exercises

1.12.1   Level 1: A Nudge in a Right Direction

1.1.1 (i) Interpret this verbally. What must happen in order for this NOT to occur?
  (ii) $\forall x$, what happens?
  (iii) $\exists x$ for which what happens?
  (iv) Apply the definition of contrapositive directly.
1.2.1. (i) What is the contrapositive of "$x \in A \Longrightarrow x \in B$"?
  (ii) Start with $x \in (A \cap B) \cup (A \setminus B)$ and use Def. 1.2.1 to show $x \in A$. (Now what?)
  (iii) Use the approach from (ii). If P, Q, and R are statements, then $P \wedge (Q \vee R) \equiv$ __?__.
  (iv) Use the hint for Exer. 1.1.1(i). (How?)
1.2.2. Negate Def. 1.2.1(ii).
1.2.3. (iv) $\bigcup_{i=1}^{n} A_i = \{x \mid \exists i \in \{1, \ldots n\}$ such that __?__$\}$
  (v) This is similar to (iv), but the quantifier changes. (How?)
  (vi) This set must be comprised of what kind of elements?
1.2.4. The proofs of all of these statements are similar. Start with an x in the set on the left-hand side and, using the defining characteristics of that set, manipulate the expressions involved to argue that x must belong to the set on the right-hand side. Sometimes, these implications will reverse (thereby resulting in set equality), while for others they will not.
1.3.1. You must have $\mathrm{dom}(f) = \mathrm{dom}(g)$, and ....
1.3.2. Many example exist. Try $f(x) = 2x$ and $g(x) = x^2$.
1.3.3. Argue these in two stages. For (ii),

$$(f \circ g)(x_1) = (f \circ g)(x_2) \Longrightarrow f(\underline{g(x_1)}) = f(\underline{g(x_2)}).$$

What must be true about the underlined quantities? (Why? Now what?)
1.3.4. (i) Use the same approach as in Exer. 1.2.4. As an example, we prove (ii)(a):

$$y \in f\left(\bigcup_{\gamma \in \Gamma} X_\gamma\right) \Longleftrightarrow \exists x \in \bigcup_{\gamma \in \Gamma} X_\gamma \text{ such that } y = f(x)$$
$$\Longleftrightarrow \exists \gamma \in \Gamma \text{ such that } x \in X_\gamma \text{ and } y = f(x)$$
$$\Longleftrightarrow \exists \gamma \in \Gamma \text{ such that } y \in f(X_\gamma)$$
$$\Longleftrightarrow y \in \bigcup_{\gamma \in \Gamma} f(X_\gamma)$$

(ii) Identify which implications do not reverse.
1.3.5. (i) The arithmetic operation and ordering must work together in order for the arithmetic combination to retain the monotonicity of the functions used to form it. The sign of the output also contributes to the result. (How?)

(ii) If f and g are nondecreasing, then so is $f \circ g$. (Why?) But, something peculiar happens when f and g are nonincreasing? (What?)

1.4.1. (i) Compute $(x+y)^2$. (So what?)

(ii) Use induction with Prop. 1.4.1(vi).

1.4.2. Apply the result: $|x-a| = b$ iff $x-a = \pm b$.

1.4.3. (i) through **(vi)** are immediate consequences of Def. 1.4.3.

(vii) It follows from (i) that $-|x| \leq x \leq |x|$, $\forall x \in \mathbb{R}$. (So what?)

(viii) Use (vii). (How?)

(ix) Use $|x| = |(x-y)+y| \leq |x-y|+|y|$. (Now what?)

(x) Use Prop. 1.4.2.

1.4.4. (i) Note that $\sum_{i=1}^{n} (\alpha x_i - \beta y_i)^2 \geq 0$, $\forall \alpha, \beta \in \mathbb{R}$. (Now what?)

(ii) Use (i).

(iii) Apply Prop. 1.4.4(vii) for the first inequality.

1.4.5. This simply involves reversing the inequalities.

1.4.6. (i) Negate Def. 1.4.5(i).

(ii) Repeat (i) for sets bounded below.

1.4.7. Let T be a nonempty set bounded below. Apply Thrm. 1.4.7 to $S = -T$ and then appropriately reverse the inequalities to get back to T. (How?)

1.4.8. *Proof of (i)*: Argue by contradiction.

Proof of (iv): Use Prop. 1.4.1(iv) and (v) with Def. 1.4.5.

Proof of (v): Why is $\sup S + \sup T + 1 > 0$?

1.4.9. A partially completed table is as follows:

S	$\mathrm{int}(S)$	S'	$\mathrm{cl}_{\mathbb{R}}(S)$	Open?	Closed?
$[1,5]$	$(1,5)$	$[1,5]$	$[1,5]$	No	Yes
$\mathbb{Q}$	$\emptyset$		$\mathbb{R}$		
$\{\frac{1}{n}|n \in \mathbb{N}\}$		$\{0\}$			No
$\mathbb{R}$	$\mathbb{R}$			Yes	Yes
$\emptyset$					Yes

1.4.10. (i) Let $x \in \bigcap_{i=1}^{n} G_i$. Show $\exists \varepsilon > 0$ such that $(x-\varepsilon, x+\varepsilon) \subset \bigcap_{i=1}^{n} G_i$.

(ii) Use Prop. 1.4.10.

(iii) Use the same approach as in (i).

(iv) Use the same approach as in (ii).

(v) $\exists \varepsilon > 0$ such that $(x-\varepsilon, x+\varepsilon) \subset S \subset T$. (So what?)

(vi) Use Def. 1.4.9(vii).

1.4.11. (i) Apply Thrm. 1.4.7 and Prop. 1.4.8(i). (How?)

(ii) This follows from (i) because $\mathrm{cl}_{\mathbb{R}}(S) = \underline{\quad?\quad}$.

(iii) Implement standard changes involving bounded below and infs.

1.4.12. Think of Exer. 1.4.9.

1.4.13. (i) Any finite set will work. (Why?)

(ii) Think of an unbounded set with "gaps."

(iii) Keep trying....

1.4.14. These all follow from the definitions of T, $\sup T$, and limit point. Completeness is used in (1.12).

1.4.15 & 1.4.16. The only compact sets are (i), $\{\frac{1}{n}|n \in \mathbb{N}\} \cup \{0\}$, and $[0,1]$. Why are they compact? Why are the others not?

1.5.1. If $\{x_n|n \in \mathbb{N}\}$ is nondecreasing, then x_1 is a lb $\{x_n|n \in \mathbb{N}\}$. (Why?) Adapt this for nonincreasing sequences.

1.5.2. (i) Negate Def. 1.5.4.

   (ii) Use Def. 1.5.4 directly. (How?)

1.5.3. Let $\varepsilon > 0$. $\exists N \in \mathbb{N}$ such that $|a| < \varepsilon N$. (So what?)

1.5.4. (i) Let $\varepsilon > 0$. $\exists N \in \mathbb{N}$ such that $n \geq N \Longrightarrow |x_n - L| < \frac{\varepsilon}{|c|+1}$. (So what?)

   (ii) Apply Prop. 1.5.9(i) and Exer. 1.5.4(i).

1.5.5. $\exists M > 0$ such that $|y_n| \leq M$, $\forall n \in \mathbb{N}$. How do you control $|x_n y_n|$?

1.5.6. (i) Use Prop. 1.4.4(ix) with Def. 1.5.4.

   (ii) Use $x_n = (-1)^n$, $n \in \mathbb{N}$.

1.5.7. (i) Argue inductively using Prop. 1.5.9(ii).

   (ii) Proceed in two cases. First, assume $L \neq 0$. Observe that

$$\sqrt{x_n} - \sqrt{L} = \left(\sqrt{x_n} - \sqrt{L}\right) \frac{\left(\sqrt{x_n} + \sqrt{L}\right)}{\left(\sqrt{x_n} + \sqrt{L}\right)} = \frac{x_n - L}{\sqrt{x_n} + \sqrt{L}}.$$

Use this with Prop. 1.5.8. (How?) If $L = 0$, then $|x_n| < \varepsilon^2 \Longrightarrow \sqrt{x_n} < \varepsilon$. (Why?)

1.5.8. (i) Proof of $(\Longrightarrow)$: $\forall N \in \mathbb{N}$, show that you can choose

$$x_N \in (S \setminus \{x_1, \ldots, x_{N-1}\}) \cap \left(x - \frac{1}{N}, x + \frac{1}{N}\right).$$

   (ii) This is similar to (i), but now you can have a constant sequence. (Why? How does this alter the proof?)

1.5.9. If $\{x_n\}$ is a nonincreasing sequence bounded below, then $\{x_n\}$ converges and $\lim\limits_{n \to \infty} x_n = \inf\{x_n|n \in \mathbb{N}\}$. Now, prove it!

1.5.10. For both, apply Prop. 1.5.13 or its analogous version developed in Exer. 1.5.9.

1.5.11. The main trick is identifying the indices correctly when verifying each implication. Argue $(1) \Longrightarrow (2) \Longrightarrow (3) \Longrightarrow (4) \Longrightarrow (5) \Longrightarrow (1)$ to show all statements are equivalent.

1.5.12. Consider $s_{2m} - s_{2m-1}$, $\forall m \in \mathbb{N}$.

1.5.13. Apply the example directly.

1.5.14. Use Exer. 1.5.10(ii).

1.5.15. Apply Prop. 1.5.24.

1.6.1. Argue in a manner similar to (1.44). You will need to use associativity of addition of real numbers and distributivity of multiplication over addition.

1.6.2. Since this norm measures Euclidean distance, you should expect each of these to be related to circles or spheres, somehow.

1.6.3. (i), (ii), and (iv) follow from elementary radical properties. For (iii), apply Exer. 1.4.4(ii).

1.6.4. Argue as in Exer. 1.4.4(iii).

1.6.5. Most of these follow directly using the commutativity, associativity, and distributivity properties of $\mathbb{R}$. For (iv), use Hint 2 for Exer. 1.6.3. And for (vi), use $z = x - y$.

1.6.6. Revisit Exer. 1.6.2.

1.6.8. The computations go through without incident when replacing $|\cdot|$ by $\|\cdot\|_{\mathbb{R}^N}$.

1.6.9. $\|\mathbf{x}_n - \mathbf{L}\|^2_{\mathbb{R}^N} = \sum_{i=1}^{N}((x_i)_n - L_i)^2 \to 0$ as $n \to \infty$ iff what happens?

1.6.10. Argue that $\left| \|\mathbf{x}_m\|_{\mathbb{R}^N} - \left\| \lim_{p \to \infty} \mathbf{x}_p \right\|_{\mathbb{R}^N} \right| \to 0$ as $m \to \infty$.

1.6.11. $\left\| \frac{a\mathbf{x}}{\|\mathbf{x}\|_{\mathbb{R}^N}} \right\|_{\mathbb{R}^N} = \frac{|a| \|\mathbf{x}\|_{\mathbb{R}^N}}{\|\mathbf{x}\|_{\mathbb{R}^N}} = |a|, \forall \mathbf{x} \in \mathbb{R}^N \setminus \{0\}$. (Why?)

1.6.12. $\|a\mathbf{x} + b\mathbf{y}\|_{\mathbb{R}^N} \leq |a| \|\mathbf{x}\|_{\mathbb{R}^N} + |b| \|\mathbf{y}\|_{\mathbb{R}^N} \leq \underline{\quad?\quad}$. (Now what?)

1.6.13. $\left\| \frac{\mathbf{x}}{a\|\mathbf{x}\|_{\mathbb{R}^N}} \right\|^2_{\mathbb{R}^N} \leq \frac{1}{a^2}, \forall \mathbf{x} \in \mathbb{R}^N \setminus \{0\}$. (So what?)

1.6.14. $\left| \langle 2\mathbf{x}, \frac{1}{4}\mathbf{x} \rangle_{\mathbb{R}^2} \right|^{1/2} = \frac{1}{\sqrt{2}} \|\mathbf{x}\|_{\mathbb{R}^2}$. (Why? So what?)

1.6.15. Argue that $\left\| \sum_{n=1}^{p} c_n \mathbf{z} \right\|_{\mathbb{R}^N} \leq \sum_{n=1}^{p} |c_n| \|\mathbf{z}\|_{\mathbb{R}^N}, \forall p \in \mathbb{N}$. (Now what?)

1.6.16. Show that $\|\mathbf{f}(t)\|_{\mathbb{R}^3} \leq R^2, \forall t \in \mathbb{R}$. (Now what?)

1.6.17. (i) and (iii) hold; this is easily shown because the corresponding properties in $\mathbb{R}$ can be applied entrywise. (ii) rarely holds (Why?). And, (iv) is true, but the bookkeeping is a bit more tedious. (Try showing it.)

1.6.18. (i) Solve for λ : $\det(\mathbf{A} - \lambda\mathbf{I}) = \det \begin{bmatrix} a - \lambda & 0 \\ 0 & b - \lambda \end{bmatrix} = 0$.

(ii) $\mathbf{A}^{-1}$ is another diagonal matrix. What are its components?

(iii) Let $\mathbf{B}$ be a diagonal $N \times N$ matrix with nonzero entries $b_{11}, b_{22}, \ldots, b_{NN}$. The eigenvalues of a diagonal matrix are precisely the diagonal entries. (Now what?)

1.6.19. The only additional hitch with which we must contend is the *double* sum, but a moment's thought reveals that this can be expressed as a single sum. (Now what?)

1.6.20. Prove that

$$\sum_{i=1}^{N} \sum_{j=1}^{N} \left| \sum_{r=1}^{N} a_{ir} b_{rj} \right| \leq \left(\sum_{i=1}^{N} \sum_{j=1}^{N} |a_{ij}| \right) \left(\sum_{i=1}^{N} \sum_{j=1}^{N} |b_{ij}| \right).$$

1.6.21. Observe that

$$\left\| \begin{bmatrix} e^{-t} & 0 \\ 0 & e^{-2t} \end{bmatrix} \begin{bmatrix} x_1 \\ x_2 \end{bmatrix} \right\|^2_{\mathbb{R}^2} = \left\| \begin{bmatrix} e^{-t} x_1 \\ e^{-2t} x_2 \end{bmatrix} \right\|^2_{\mathbb{R}^2} = \left(e^{-t} x_1 \right)^2 + \left(e^{-2t} x_2 \right)^2.$$

1.6.22. Yes, naively because this boils down to entrywise calculation. Prove this formally.

1.6.23. Convergence is entrywise for all of these. So what?

1.6.24. (i) $\|\mathbf{A}_m \mathbf{x} - \mathbf{A}\mathbf{x}\|_{\mathbb{R}^N} = \|(\mathbf{A}_m - \mathbf{A})\mathbf{x}\|_{\mathbb{R}^N} \leq \|\mathbf{A}_m - \mathbf{A}\|_{\mathbb{M}^N} \|\mathbf{x}\|_{\mathbb{R}^N}$. So what?

(ii) This is similar to (i). Tell how.

(iii) Use (i) and (ii) combined with the triangle inequality.

1.7.1. These properties are known for $(\mathbb{R}, |\cdot|)$ and applying them componentwise enables you to argue that $\left(\mathbb{R}^N, \|\cdot\|_{\mathbb{R}^N} \right)$ and $\left(\mathbb{M}^N(\mathbb{R}), \|\cdot\|_{\mathbb{M}^N(\mathbb{R})} \right)$ are linear spaces.

1.7.2. For instance, (a) reads: $\forall \varepsilon > 0, \exists N \in \mathbb{N}$ such that

$$n \geq N \implies \sup_{t \in I} \|f_n(t) - f(t)\|_{\mathscr{X}} < \varepsilon$$

1.7.3. Let $\{f_n\} \subset \mathbb{C}([a,b];\mathbb{R})$ be Cauchy. Then, $\{f_n(x)\}$ is Cauchy in $\mathbb{R}$, $\forall x \in [a,b]$. (Why?) Hence, the function $f : [a,b] \to \mathbb{R}$ given by $f(x) = \lim_{n\to\infty} f_n(x)$ is well-defined. (Why? Now what?)

1.7.4. Prop. 1.7.7: $\{s_n\}$ converges in $\mathbb{C}$ iff $\{s_n\}$ is Cauchy in $\mathbb{C}$. (Why?) Now, use a modified version of Exer. 1.5.11 to conclude.

Prop. 1.7.8: Define $s_N = \sum_{k=1}^{N} M_k$. $\{s_N\}$ is Cauchy in $\mathbb{R}$. (Why? So what?)

1.7.6. This is the set of all $z \in \mathbb{C}([0,2];\mathbb{R})$ such that $\sup\{|z(x) - x^2| : x \in [0,2]\} < 1$. Interpret this geometrically.

1.7.7. Showing the sequence is Cauchy is easy. Note that $\lim_{n\to\infty} \frac{2}{n} = 0$. So, what is the issue?

1.7.8. Let $\{x_n\}$ be Cauchy in $\mathscr{Y}$. Then, $\{x_n\}$ converges in $\mathscr{X}$. (Why?) How do you prove that $\{x_n\}$ actually converges in $\mathscr{Y}$?

1.7.9. Let $\{(x_n, y_n)\}$ be Cauchy in $\mathscr{X} \times \mathscr{Y}$. Can you conclude that $\{x_n\}$ is Cauchy in $\mathscr{X}$ and $\{y_n\}$ is Cauchy in $\mathscr{Y}$? (How?)

1.7.10. Linearity of the integral is a key tool here.

1.7.11. This follows immediately.

1.7.12. (i) $\langle x, ay \rangle = \langle ay, x \rangle$. (Now what?)

(ii) The hypothesis implies that $\langle x - y, z \rangle = 0$, $\forall z \in \mathscr{X}$. Now, choose z appropriately to conclude. (How?)

(iii) $\|ax\|_{\mathscr{X}}^2 = \langle ax, ax \rangle_{\mathscr{X}}$. (So what?)

(iv) Argue as in Prop. 1.6.7.

(v) Apply the hint for (iii) with $x + y$ in place of ax.

(vi) $|\langle x_n, y_n \rangle - \langle x_n, y \rangle + \langle x_n, y \rangle - \langle x, y \rangle| \leq \ldots$ (Now what?)

1.7.13. $\left|\int_a^b f(x)g(x)dx\right| \leq \ldots$ (Now what?)

1.7.14. (i) Compute $\int_{-\pi}^{\pi} f^2(x)dx$, $\forall f \in \mathfrak{B}$, using trigonometric identities as needed.

(ii) Compute $\int_{-\pi}^{\pi} f(x)g(x)dx$ using a change of variable and trigonometric identity.

(iii) First, determine $C \in \mathbb{R}$ such that $\int_a^b C^2 dx = 1$. The other basis elements need to be replaced by

$$\mathfrak{B}^\star = \left\{c^\star \cos\left(\frac{2n\pi t}{b-a}\right) \,\middle|\, n \in \mathbb{N}\right\} \cup \left\{c^\star \sin\frac{2n\pi t}{b-a} \,\middle|\, n \in \mathbb{N}\right\}$$

for an appropriate choice of $c^\star \in \mathbb{R}$ that ensures $\int_a^b c^\star f^2(x)dx = 1$, $\forall f \in \mathfrak{B}^\star$.

1.7.15. (i) $\left\|\sum_{i=1}^n y_i\right\|_{\mathscr{H}}^2 = \langle \sum_{i=1}^n y_i, \sum_{i=1}^n y_i \rangle_{\mathscr{H}}$ (Now what?)

(ii) Take the inner product with y_i on both sides of $\alpha_1 y_1 + \ldots + \alpha_n y_n = 0$.

(iii) Take the inner product with y_i on both sides of $\alpha_1 y_1 + \ldots + \alpha_n y_n = x$.

(iv) Simplify using the properties of inner product to get $\langle x, y_i \rangle - \langle x, y_i \rangle = 0$. (Tell how.)

1.7.16. Proof of (i): Use Prop. 1.5.13.

Proof of (ii): Let $\varepsilon > 0$. $\exists N \in \mathbb{N}$ such that

$$m, n \geq N \implies \sum_{k=m}^{n} |\langle x, e_k \rangle|^2 < \varepsilon.$$

(So what?)

1.8.1. Prop. 1.5.5: If $\exists \lim_{x \to a} f(x)$, then it is unique.

Prop. 1.5.6: If $\exists \lim_{x \to a} f(x)$, then $\exists M > 0$ and $\delta > 0$ such that

$$|f(x)| \leq M, \forall x \in \text{dom}(f) \cap (a - \delta, a + \delta).$$

Prop. 1.5.7: If $f(x) \leq g(x) \leq h(x)$, for all "appropriate x near a" and $\exists \lim_{x \to a} f(x) = \lim_{x \to a} h(x) = L$, then $\exists \lim_{x \to a} g(x) = L$. (Make precise the phrase in quotes!)

Prop. 1.5.9: If $\exists \lim_{x \to a} f(x) = L$ and $\lim_{x \to a} g(x) = M$, then

(i) $\exists \lim_{x \to a} (f(x) + g(x)) = L + M$;

(ii) $\exists \lim_{x \to a} (f(x) \cdot g(x)) = L \cdot M$.

1.8.2. (i) These are formal ways of defining asymptotes. Which is which?

(ii) Certain inequalities will change since the inputs of interest are different. (How?)

1.8.3. Argue as in Exer. 1.6.10.

1.8.4. The proof using Def. 1.8.4(i) is suggested by the string of equalities in the statement of Prop. 1.8.5. Alternatively, using Def. 1.8.4(ii), let $\varepsilon > 0$. Find $\delta > 0$ such that

$$\|x - a\|_{\mathscr{X}} < \delta \implies \|f(g(x)) - f(g(a))\|_{\mathscr{X}} < \varepsilon.$$

1.8.5. (i) Mimic the statement in the paragraph preceding Exer. 1.8.5 with $\mathscr{X}_2 = [a, b]$ and $\mathscr{X}_1 = \mathscr{X} \times \mathscr{X}$.

(ii) It is more intuitive to use Def. 1.8.4(i). (Do so.)

(iii) Interpret this entrywise as suggested by Exer. 1.6.22.

1.8.6. (i) Consider the function $g(x) = f(x) - x$.

(ii) Align the function so that the fixed-point would occur at one of the endpoints that you are now excluding, or use asymptotes to your advantage.

1.8.7. (i) Let $\{x_n\}$ be a Cauchy sequence in S and $\varepsilon > 0$. Prove that $\{f(x_n)\}$ is a Cauchy sequence in $\mathbb{R}$.

(ii) If f is not bounded on S, then $\forall n \in \mathbb{N}, \exists x_n \in S$ such that $|f(x_n)| > n$. (Now what?)

1.8.8. For any real sequence $\{x_n\}$ such that $x_n \to 0$, ...

1.8.9. How do you show that a subset of $\mathbb{R}$ is unbounded? Adapt this.

1.8.10. Use Prop. 1.8.11(i) to show that $f(x) = f(a), \forall x \in [a, b]$.

1.9.1. (i) $\sin(\pi y) dy = e^{2x} dx$ ... Now, integrate.

(ii) $dy = ax^n dx$ ... Now, integrate.

(iii) $(1 - y^3) dy = \sum_{i=1}^{n} a_i \sin(b_i x) dx$ ... Now, integrate.

1.9.2. Use linearity and additivity of the integral.

1.9.3. Apply (1.99) directly with $y_0 = \frac{1}{2}$, $x_0 = 0$, $a(x) = \frac{1}{2}$, and $b(x) = e^{-3x}$.

1.9.4. (i) All exponential terms must go to zero.

(ii) Consider the case in which the roots are complex.

1.11.1. Φ^{n_0} has a fixed point x^*. (Why? So what?)

1.11.2. What is the definition of completeness? How does this help?

1.11.3. "$\{y_n\}$ converges to y in $L^2(a,b;\mathbb{R})$"0 means $\lim_{n\to\infty} \int_a^b |y_n(x) - y(x)|^2 dx = 0$. What does it mean for $\{y_n\}$ to be Cauchy in $L^2(a,b;\mathbb{R})$? Compute $\left\|\cos\left(\frac{2n\pi\cdot}{b-a}\right)\right\|_{L^2(a,b;\mathbb{R})}$, $\forall n \in \mathbb{N}$.

1.12.2 Level 2: An Additional Thrust in a Right Direction

1.1.1. (i) Use the fact that $\neg(P \wedge Q)$ is equivalent to $(\neg P) \vee (\neg Q)$. Interpret this.

(iv) Use (i) and an equivalent form of $\neg(P \vee Q)$ similar to (i).

1.2.1. (i) Use the contrapositive for both implications.

(ii) For the reverse inclusion, begin with $x \in A$. To which of the two sets on the right-hand side of the equality must x belong? (So what?)

(iii) $\ldots (P \wedge Q) \vee (P \wedge R)$. Similar reasoning applied to $P \vee (Q \wedge R)$ can be used to verify the related distributive law.

(iv) Negate $P \vee Q$.

1.2.2. This boils down to arguing that either A is not a subset of B, or vice versa.

1.2.3. (iv) Fill in the blank with "$x \in A_i$."

(v) $\bigcap_{i=1}^n A_i = \{x | x \in A_i, \forall i \in \{1,\ldots,n\}\}$

(vi) $\{(x_1,\ldots,x_n) | x_i \in A_i, \forall i \in \{1,\ldots,n\}\}$

1.2.4. As an example, we prove (ii):

$$x \in \left(\bigcup_{\gamma \in \Gamma} A_\gamma\right)^{\widetilde{}} \Longleftrightarrow x \notin \bigcup_{\gamma \in \Gamma} A_\gamma$$

$$\Longleftrightarrow \neg \left(\exists \gamma \in \Gamma \text{ such that } x \in A_\gamma\right)$$

$$\Longleftrightarrow \forall \gamma \in \Gamma, x \notin A_\gamma$$

$$\Longleftrightarrow \forall \gamma \in \Gamma, x \in \widetilde{A_\gamma}$$

$$\Longleftrightarrow x \in \bigcap_{\gamma \in \Gamma} \widetilde{A_\gamma}.$$

1.3.1. $\ldots$ and $f(x) = g(x)$, $\forall x \in \text{dom}(f) = \text{dom}(g)$.

1.3.3. Continuing, we have $g(x_1) = g(x_2) \Longrightarrow x_1 = x_2$, where the facts that f and g are one-to-one were used (in that order). The proof of (i) is similar.

1.3.4. (ii) Try using one-to-one for one of them, and onto for the other.

1.3.5. (i) The sum is the only one for which this holds. The product would have worked if the range were restricted to $(0,\infty)$. (Why?) Why don't the others work?

(ii) If f and g are both nonincreasing, then

$$x_1 < x_2 \Longrightarrow g(x_1) > g(x_2) \Longrightarrow f(g(x_1)) < f(g(x_2)).$$

(So what?)

1.4.1. (i) Note that $(x+y)^2 \geq 0$. Expand the left-hand side.

(ii) At the inductive step, use $(x^n < y^n) \wedge (0 < x < y) \Longrightarrow x^n x < y^n y$. (Now what?)

1.4.3. (vii) Apply this to both x and y and add. (Now what?) Alternatively, expand $(x+y)^2$ and apply (i), (v), and (vi).

(ix) Apply this to y also and subtract. (Now what?)

1.4.4. (i) Expand the expression and choose α, β appropriately. (How?)

(iii) For the second inequality, use Exer. 1.4.1(i) with

$$(a+b)^N = \sum_{k=0}^{N} \binom{N}{k} a^k b^{N-k}.$$

1.4.6. (i) Show that $\forall M > 0, \exists x \in S$ such that $x > M$.

(ii) Adapt (i) appropriately.

1.4.7. Now, argue similarly to show $(1.3) \Longrightarrow$ Thrm. 1.4.7.

1.4.8. Proof of (i): If the conclusion does not hold, then $\sup(S) - \varepsilon$ is an ub(S). Why is this a contradiction?

Proof of (v): Now use Prop. 1.4.8(i) with the number $\zeta = \frac{\varepsilon}{\sup(S)+\sup(T)+1}$ and mimic the argument of Prop. 1.4.8(iii). (Tell how.)

Regarding the proofs of the corresponding INF statements, all changes are straightforward and primarily involve inequality reversals and the appropriate modification of Prop. 1.4.8(i). (Supply the details.)

1.4.9. Make certain to supply the details.

S	int(S)	S'	cl$_{\mathbb{R}}(S)$	Open?	Closed?
$[1,5]$	$(1,5)$	$[1,5]$	$[1,5]$	No	Yes
$\mathbb{Q}$	Ø	$\mathbb{R}$	$\mathbb{R}$	No	No
$\{\frac{1}{n}|n \in \mathbb{N}\}$	Ø	$\{0\}$	$\{\frac{1}{n}|n \in \mathbb{N}\}\cup\{0\}$	No	No
$\mathbb{R}$	$\mathbb{R}$	$\mathbb{R}$	$\mathbb{R}$	Yes	Yes
Ø	Ø	Ø	Ø	Yes	Yes

1.4.10. (i) For every $i \in \{1,\ldots,n\}$, $\exists \varepsilon_i > 0$ such that $(x - \varepsilon_i, x + \varepsilon_i) \subset G_i$. (So what?)

(ii) Because $\tilde{F}_1, \ldots, \tilde{F}_n$ are open, (i) $\Longrightarrow \bigcap_{i=1}^{n} \tilde{F}_i$ is open. Now apply Exer. 1.2.4 (ii).

(iii) $\exists \gamma_0 \in \Gamma$ such that $x \in G_{\gamma_0}$, and because G_{γ_0} is open, $\exists \varepsilon > 0$ such that $(x - \varepsilon, x + \varepsilon) \subset G_{\gamma_0}$. Now, use Exer. 1.2.4 (iv).

1.4.11. (i) Note that $\forall \varepsilon > 0$, $(\sup(S) - \varepsilon, \sup(S) + \varepsilon) \cap S \neq$ Ø. Also, $\sup(S)$ need not be in S, but it must be in S'? (Why?)

(ii) cl$_{\mathbb{R}}(S) = S$

1.4.12. Some possibilities are $\mathbb{Q}, \tilde{\mathbb{Q}}, \mathbb{Q}\setminus A$, and $\tilde{\mathbb{Q}}\setminus A$, where A is any finite subset of $\mathbb{R}$.

1.4.13. (ii) $\mathbb{N}$

(iii) This is not possible.

1.4.15 & 1.4.16. Compute the closures of these sets and appeal to Thrm. 1.4.14.

1.5.2. (i) $\exists \varepsilon > 0$ such that no matter what $N \in \mathbb{N}$ is chosen, $\exists n \geq N$ for which $|x_n - L| \geq \varepsilon$.

(ii) $|x_n - L| = ||x_n - L| - 0|$. (So what?)

1.5.4. (i) $|cx_n - cL| = |c| |x_n - L| < \frac{\varepsilon |c|}{|c|+1}$. (So what?)

1.5.5. Let $\varepsilon > 0$. $\exists N \in \mathbb{N}$ such that $n \geq N \Longrightarrow |x_n - 0| < \frac{\varepsilon}{M+1}$. Now, argue as in Exer. 1.5.4(i).

1.5.6. (i) Let $\varepsilon > 0$. $\exists N \in \mathbb{N}$ such that $n \geq N \Longrightarrow ||x_n| - |L|| \leq |x_n - L| < \varepsilon$. (So what?)

1.5.7. (i) $\left(x_n^p \rightarrow L^p\right) \wedge \left(x_n \rightarrow L\right) \Longrightarrow x_n^p x_n \rightarrow L^p L$. (Why?)

(ii) Let $\varepsilon > 0$. $\exists M > 0$ and $N_1 \in \mathbb{N}$ such that $n \geq N_1 \Longrightarrow x_n > M$. Also, $\exists N_2 \in \mathbb{N}$ such that $n \geq N_2 \Longrightarrow |x_n - L| < \left(M + \sqrt{L}\right)\varepsilon$. So,

$$n \geq \max\{N_1, N_2\} \Longrightarrow \left| \frac{x_n - L}{\sqrt{x_n} + \sqrt{L}} \right| \leq \left(\frac{1}{M + \sqrt{L}} \right) |x_n - L| < \varepsilon.$$

1.5.8. (i) Proof of ($\Longleftarrow$) : Let $\varepsilon > 0$. $\exists N \in \mathbb{N}$ such that $n \geq N \Longrightarrow x_n \in (x - \varepsilon, x + \varepsilon)$. (So what?)

1.5.9. The proof is very similar to the proof of Prop. 1.5.13, but you need to use the fact that $\exists N \in \mathbb{N}$ such that $x_N < L + \varepsilon$. (Now what?)

1.5.10. The fact that $\lim\limits_{n \to \infty} \frac{a^n}{n!} = 0$ readily follows by applying Prop. 1.5.24 in conjunction with Cor. 1.5.21. (Revisit this when you reach this point.)

1.5.12. Use Thrm. 1.5.18. Find a real number ζ_0 for which $|s_{2m} - s_{2m-1}| \geq \zeta_0$, $\forall m \in \mathbb{N}$. Then, how do you conclude?

1.5.13. This series converges for any x such that $|5x + 1|^3 < 1$. For such x, the sum is $\frac{c(5x+1)^{3p}}{1 - (5x+1)^3}$.

1.5.14. If $x_n > 0$ and $x_n \rightarrow 0$, what can you say about $\left\{ \frac{1}{x_n} \right\}$? Use this with Cor. 1.5.21.

1.5.15. $\lim\limits_{n \to \infty} \frac{(n+1)^{n+1}}{(n+1)!} \cdot \frac{n!}{n^n} = \lim\limits_{n \to \infty} \left(1 + \frac{1}{n}\right)^n = e$. (So what?)

1.6.1. For instance,

$$\begin{aligned}
c\left(\mathbf{x} + \mathbf{y}\right) &= c\left\langle x_1 + y_1, x_2 + y_2, \ldots, x_N + y_N \right\rangle \\
&= \left\langle c\left(x_1 + y_1\right), c\left(x_2 + y_2\right), \ldots, c\left(x_N + y_N\right) \right\rangle \\
&= \left\langle cx_1 + cy_1, cx_2 + cy_2, \ldots, cx_N + cy_N \right\rangle \\
&= \left\langle cx_1, cx_2, \ldots, cx_N \right\rangle + \left\langle cy_1, cy_2, \ldots, cy_N \right\rangle \\
&= c\mathbf{x} + c\mathbf{y}.
\end{aligned}$$

1.6.2. (i) Open circle with radius ε centered at $(0,0)$.

(ii) Complement of an open sphere with radius ε centered at $(1,0,0)$.

(iii) The singleton set $\{\mathbf{x}_0\}$.

1.6.3. Use $\sqrt{z^2} = |z|$ and $\sum_{i=1}^N a_i^2 = 0 \Longleftrightarrow a_i = 0$, $\forall i \in \{1, \ldots, N\}$.

1.6.4. Adapt the hints provided for Exer. 1.4.4(iii).

1.6.5. For instance, the proof of (ii) is

$$\begin{aligned}
\left\langle \mathbf{x} + \mathbf{y}, \mathbf{z} \right\rangle_{\mathbb{R}^N} &= \sum_{i=1}^N \left(x_i + y_i\right) z_i = \sum_{i=1}^N \left(x_i z_i + y_i z_i\right) \\
&= \sum_{i=1}^N x_i z_i + \sum_{i=1}^N y_i z_i = \left\langle \mathbf{x}, \mathbf{z} \right\rangle_{\mathbb{R}^N} + \left\langle \mathbf{y}, \mathbf{z} \right\rangle_{\mathbb{R}^N}.
\end{aligned}$$

For (vi), expand $\langle \mathbf{x}, \mathbf{x} - \mathbf{y} \rangle_{\mathbb{R}^N} = \langle \mathbf{y}, \mathbf{x} - \mathbf{y} \rangle_{\mathbb{R}^N}$ to arrive at $\sum_{i=1}^{N} (x_i - y_i)^2 = 0$. (Now what?)

1.6.6. Open circle (or sphere) with radius ε centered at $\mathbf{x}_0$.

1.6.9. $(x_i)_n \to L_i$ as $n \to \infty$, $\forall i \in \{1, \dots, N\}$.

1.6.10. Using Prop. 1.6.4(iii) yields

$$0 \le \left| \|\mathbf{x}_m - \mathbf{L} + \mathbf{L}\|_{\mathbb{R}^N} - \left\| \lim_{p \to \infty} \mathbf{x}_p \right\|_{\mathbb{R}^N} \right|$$

$$\le \left| \|\mathbf{x}_m - \mathbf{L}\|_{\mathbb{R}^N} + \underbrace{\|\mathbf{L}\|_{\mathbb{R}^N} - \left\| \lim_{p \to \infty} \mathbf{x}_p \right\|_{\mathbb{R}^N}}_{=0} \right| \to 0.$$

1.6.11. What is the supremum of a singleton set?

1.6.12. Continuing, we conclude that $\eta = |a| \left(\|\mathbf{x}_0\|_{\mathbb{R}^N} + \varepsilon \right) + |b| \left(\|\mathbf{x}_0\|_{\mathbb{R}^N} + \delta \right)$ is an ub($\mathscr{A}$). Completeness ensures $\exists \sup(\mathscr{A})$. In order to prove that $\eta = \sup(\mathscr{A})$, let $0 < \zeta < \eta$. Produce $\mathbf{x} \in \mathfrak{B}_{\mathbb{R}^N}(\mathbf{x}_0; \varepsilon)$ and $\mathbf{y} \in \mathfrak{B}_{\mathbb{R}^N}(\mathbf{x}_0; \delta)$ such that $\zeta = \|a\mathbf{x} + b\mathbf{y}\|_{\mathbb{R}^N}$.

1.6.13. This is dominated by a geometric series that converges iff $|a| > 1$. (Why?)

1.6.14. $\left| \langle 2\mathbf{x}, \frac{1}{4}\mathbf{x} \rangle_{\mathbb{R}^2} \right|^{1/2} < 1$, $\forall \mathbf{x} \in \mathfrak{B}_{\mathbb{R}^2}(\mathbf{0}; \frac{1}{3})$. Use Prop. 1.5.22 to conclude. (How?)

1.6.15. Now, because $|c_n| \|\mathbf{z}\|_{\mathbb{R}^N}$, $\forall n \in \mathbb{N}$, we see that

$$\left\| \sum_{n=1}^{p} c_n \mathbf{z} \right\|_{\mathbb{R}^N} \le \sum_{n=1}^{\infty} |c_n| \|\mathbf{z}\|_{\mathbb{R}^N}, \ \forall p \in \mathbb{N},$$

from which the conclusion follows. (How?)

1.6.16. Thus, R is an ub($\{ \|\mathbf{f}(t)\|_{\mathbb{R}^3} : t \in \mathbb{R} \}$). In fact, R is the sup of this set; this can be shown using the continuity of the components of $\mathbf{f}(t)$.

1.6.18. (i) $\lambda = a, b$

(ii) $\mathbf{A}^{-1} = \begin{bmatrix} \frac{1}{a} & 0 \\ 0 & \frac{1}{b} \end{bmatrix}$. Compute the eigenvalues in the same manner as in (i).

(iii) The reciprocals of $b_{11}, b_{22}, \dots, b_{NN}$ are the eigenvalues of $\mathbf{B}^{-1}$.

1.6.20. Expand both sides and compare the terms. Replace some terms on the left by larger terms on the right to arrive at the right-hand side provided.

1.6.21. Continuing, we see that $\forall t > 0$,

$$\left(e^{-t} x_1 \right)^2 + \left(e^{-2t} x_2 \right)^2 = e^{-2t} x_1^2 + e^{-4t} x_2^2$$
$$\le e^{-2t} \left(x_1^2 + x_2^2 \right) + e^{-4t} \left(x_1^2 + x_2^2 \right)$$
$$\le e^{-2t} + e^{-4t}$$
$$\le 2.$$

(Note that the upper bound you end with depends on which $\mathbb{R}^N$ norm you use.)

1.6.22. Let $\varepsilon > 0$. $\forall i, j \in \{1, 2\}$, $\exists M_{ij} \in \mathbb{N}$ such that

$$n, m \ge M_{ij} \implies \left| (x_{ij})_n - (x_{ij})_m \right| < \frac{\varepsilon}{4}.$$

How do you use this to argue that $\{\mathbf{A}_m\}$ is a Cauchy sequence in $\mathbb{M}^2(\mathbb{R})$?

1.6.23. If $\mathbf{A}_m \to \mathbf{B}$ in $\mathbb{M}^N(\mathbb{R})$, then $\forall i, j \in \{1, \ldots, N\}$, $(a_{ij})_m \to b_{ij}$ as $m \to \infty$. Apply this to all parts of the exercise.

1.6.24. (i) Use the Squeeze Theorem with the inequality to conclude.

    (ii) $\|\mathbf{A}\mathbf{x}_m - \mathbf{A}\mathbf{x}\|_{\mathbb{R}^N} \le \|\mathbf{A}\|_{\mathbb{M}^N} \|\mathbf{x}_m - \mathbf{x}\|_{\mathbb{R}^N}$. (So what?)

    (iii) $\|\mathbf{A}_m\mathbf{x}_m - \mathbf{A}\mathbf{x}\|_{\mathbb{R}^N} \le \|\mathbf{A}_m\|_{\mathbb{M}^N} \|\mathbf{x}_m - \mathbf{x}\|_{\mathbb{R}^N} + \|\mathbf{A}_m - \mathbf{A}\|_{\mathbb{M}^N} \|\mathbf{x}\|_{\mathbb{R}^N}$. (So what?)

1.7.2. (b) reads: $\forall \varepsilon > 0$, $\exists N \in \mathbb{N}$ such that

$$n \ge N \implies \int_I |f_n(t) - f(t)|^p \, dt < \varepsilon^p.$$

(c) is formulated similarly. The modifications for Cauchy are obvious.

1.7.3. Observe that $|f_n(x) - f_m(x)| \le \|f_n - f_m\|_C \to 0$ as $n, m \to \infty$. Argue that $\|f_n - f\|_C \to 0$ as $n \to \infty$.

    (ii) The space $\mathscr{X}$ must be complete; otherwise, f would not be well-defined. The argument remains unchanged.

1.7.4. Prop. 1.7.7: Compute $\sup_{x \in S} \|s_{N+p}(x) - s_N(x)\|_{\mathscr{X}}$. (Now what?)

    Prop. 1.7.8: $\sum_{k=n+1}^{n+p} M_k \to 0$ as $n \to \infty$. So, $\forall \varepsilon > 0$, $\exists N \in \mathbb{N}$ such that $n \ge N \implies \sum_{k=n+1}^{n+p} M_k < \varepsilon$. Now, apply Prop. 1.7.7.

1.7.6. Construct a tube centered at the graph of $f(x) = x^2$ on the interval $[0, 2]$ by translating copies of the graph of f vertically up 1 unit and down 1 unit to form its boundaries. Any continuous function that remains strictly inside this tube is a member of this ball.

1.7.7. $\{\frac{2}{n}\}$ does not converge in the space $\mathscr{Y}$. So, $\mathscr{Y}$ is not complete.

1.7.8. Use an appropriately modified version of Exer. 1.5.8. (How?)

1.7.9. Yes, and in fact, the completeness of the respective spaces implies that $\{x_n\}$ converges in $\mathscr{X}$ and $\{y_n\}$ converges in $\mathscr{Y}$. So, $\{(x_n, y_n)\}$ converges in $\mathscr{X} \times \mathscr{Y}$ using either norm.

1.7.10. (i) Now use Def. 1.7.10(ii), then (i).

    (ii) Choose $z = x - y$ and use Def. 1.7.10(iv).

    (iii) $\langle ax, ax \rangle_{\mathscr{X}} = a \langle x, ax \rangle_{\mathscr{X}} = a \langle ax, x \rangle_{\mathscr{X}} = a^2 \langle x, x \rangle_{\mathscr{X}}$. (Now what?)

    (v) $\ldots = \|\mathbf{x}\|_{\mathscr{X}}^2 + 2|\langle x, y \rangle| + \|\mathbf{y}\|_{\mathscr{X}}^2 \le (\|\mathbf{x}\|_{\mathscr{X}} + \|\mathbf{y}\|_{\mathscr{X}})^2$. (Now what?)

    (vi) $\ldots = |\langle y_n - y, x_n \rangle + \langle x_n - x, y \rangle| \le \|y_n - y\|_{\mathscr{X}} \|x_n\|_{\mathscr{X}} + \|x_n - x\|_{\mathscr{X}} \|y\|_{\mathscr{X}}$ (Now what?)

1.7.11. $\ldots \le \left(\int_a^b f^2(x)dx \right)^{1/2} \left(\int_a^b g^2(x)dx \right)^{1/2}$.

1.7.12. (i) Use a double-angle formula.

    (ii) Use a product-to-sum formula.

    (iii) Use $C = \frac{1}{\sqrt{b-a}}$ and use a change of variable to find $c^\star$.

1.7.13. (i) Use properties of inner product to arrive at $\sum_{i=1}^n \sum_{j=1}^n \langle y_i, y_j \rangle_{\mathscr{H}}$. (Now what?)

    (ii) Conclude that $\alpha_i = 0$, $\forall i \in \{1, \ldots, n\}$. (How? So what?)

    (iii) Now, use $\langle x, y_i \rangle = \alpha_i$ in the definition of span($\mathscr{Y}$) to conclude.

1.7.14. Proof of (ii): To see why the limit is x, use

$$\left\langle x - \sum_{k=1}^{\infty} \langle x, e_k \rangle_{\mathscr{H}} e_k, e_j \right\rangle_{\mathscr{H}} = 0, \forall j \in \mathbb{N}.$$

(So what?)

1.8.1. The proofs mirror those of the corresponding results in the sequence setting with δ playing the role of N and the "tail of the sequence" corresponding to the "deleted neighborhood of a." Keep in mind that if there are several conditions involving different δ neighborhoods of a, then in order to ensure they hold simultaneously, take the MIN of the $\delta's$. (Why?)

1.8.2. (i) Def. 1.8.3(i) means that the graph of f has a vertical asymptote at a, while (ii) implies the existence of a horizontal asymptote.

   (ii) Alternatively, use $|\,\|x\|_{\mathscr{X}} - \|a\|_{\mathscr{X}}\,| \leq \|x - a\|_{\mathscr{X}}$ with Def. 1.8.4(ii).

1.8.4. Tackle the implication in two stages. Let $\varepsilon > 0$. First, find $\delta_1 > 0$ such that

$$\|y - g(a)\|_{\mathscr{X}} < \delta_1 \implies \|f(y) - f(g(a))\|_{\mathscr{X}} < \varepsilon.$$

Then, find $\delta_2 > 0$ such that $\|x - a\|_{\mathscr{X}} < \delta_2 \implies \|g(x) - g(a)\|_{\mathscr{X}} < \delta_1$. (Now what?)

1.8.6. (i) Show that at least one of these holds: The sign of $g(x)$ changes at some point within the interval $[a, b]$, $g(a) = 0$, or $g(b) = 0$.

1.8.7. (i) Use Def. 1.8.7 carefully to link control between the two Cauchy sequences.

   (ii) Because $\{x_n\}$ is bounded, it contains a convergent subsequence $\{x_{n_k}\}$. Now use part (i). (How? So what?)

1.8.8. Substitute x_n in for h and interpret.

1.8.9. Show that $\forall N \in \mathbb{N}, \exists h_n \in \mathbb{R}$ such that $\frac{f(x_o + h_n) - f(x_0)}{h_n} > N$.

1.8.10. Apply the Mean Value Theorem on $[a, x], \forall x \in [a, b]$.

1.9.1. The integration is standard. Note that the solutions of (i) and (ii) can be solved explicitly for y.

1.9.3. $y(x) = \frac{1}{2} e^{-\frac{1}{2}x} + \int_0^x e^{-3x} e^{-\frac{1}{2}(s-x)} ds$. Now, simplify.

1.9.4. (i) m_1 and m_2 are negative.

   (ii) $\alpha \pm i\beta = \pm i\beta, \forall \beta > 0$.

1.11.1. Use $\Phi^{n_0}(\Phi(x^\star)) = \Phi(\Phi^{n_0}(x^\star)) = \Phi(x^\star)$. (Now what?)

1.11.2. Where was the fact that a Cauchy sequence in a complete space must converge in the space used?

1.11.3. Does any subsequence of $\left\{ \|\cos\left(\frac{2n\pi \cdot}{b-a}\right)\|_{\mathbb{L}^2(a,b;\mathbb{R})} : n \in \mathbb{N} \right\}$ converge in $\mathscr{H}$ in the $\mathbb{L}^2$-sense?

Chapter 2

The Bare-Bone Essentials of Probability Theory

Overview

Initial-boundary value problems (IBVPs) involve differential equations whose terms describe certain features (physical, chemical, biological, etc.) of the phenomenon under investigation, as well as initial and boundary conditions that are often determined experimentally. The experiments that yield these parameters or conditions are conducted repeatedly and produce slightly different outcomes due to underlying noise. In deterministic settings, an average of these values is often used as an approximation to the parameter. Doing so effectively removes randomness from the IBVP. Such IBVPs were studied in Volume 1 [295]. The goal of this text is to develop an analogous abstract theory that enables us to study IBVPs *without* removing randomness from the model.

You cannot conduct a meaningful study of stochastic differential equations without having a reasonable understanding of probability theory. What often turns off a newcomer is the hefty dose of measure theory that is typically presented along the way. This chapter constitutes a bare-bones presentation of the necessary notions and theorems of elementary probability theory needed to work through this text — nothing more, nothing less. The discussion is not meant to be rigorous. Rather, it is intended to illustrate the probabilistic ideas very heuristically. A more thorough treatment of the material contained in this chapter can be found in [**20, 22, 64, 73, 80, 97, 116, 125, 126, 139, 142, 143, 163, 195, 212, 225, 257, 285, 302, 318, 350 - 352, 368, 371, 383, 390, 391, 404**].

2.1 Formalizing *Randomness*

You probably have an intuitive understanding of the term *randomness*. In fact, the previous sentence carries with it an informal quantification of certainty. But, a rigorous mathematical theory cannot be built on intuition alone. We need to develop a more formal framework in which to work so that randomness can be studied precisely. Specifically, we need to define a measure that quantifies the likelihood of the occurrence of all conceivable outcomes of an experiment, as well as any event of interest defined using them.

We are interested in studying experiments conducted repeatedly under the same conditions, but whose result can change from one trial to another due to some underlying noise. The prototypical familiar examples include tossing a coin or casting a die, recording the outcome, and assigning probabilities to each possible outcome. This is what you encounter in an introductory probability course. To illustrate and motivate the concepts at this early stage, we will first appeal to these elementary examples. Once we have attained comfortable familiarity with the general notions, we will extend the notions to more elaborate settings, and then apply them to the study of more complicated models of interest related to understanding, for instance, the behavior of temperature, population density, or concentration of a chemical or pollutant as time goes on under the influence of some source of randomness.

To begin, we gather all possible outcomes of an experiment in a set.

Definition 2.1.1. The set of all possible outcomes of an experiment is called the *sample space*, denoted by Ω. The individual outcomes, called *sample points*, are denoted by ω. Any subset A of Ω is called an *event*.

Exercise 2.1.1. Consider the experiment of rolling three typical six-sided, different-colored dice and recording the outcomes as ordered triples of the form (Die 1, Die 2, Die 3).
i.) Identify the sample space.
ii.) List the outcomes described by each of the following scenarios:
   a.) An even number appears on all three dice.
   b.) An even number appears on at least one of the dice.
   c.) The same number appears on all three dice.
   d.) The number "7" appears on at least one of the dice.

Exercise 2.1.2. Consider the experiment of flipping a fair coin N times and recording the outcomes as ordered N-tuples.
i.) Identify the sample space.
ii.) Suppose $N = 4$. List the outcomes described by each of the following scenarios:
   a.) The four tosses of the coin result in three heads or four tails.
   b.) The four tosses of the coin results in either no head or no tails.
   c.) The four tosses of the coin results in an odd number of heads and an odd number of tails.

Exercise 2.1.3. Consider the experiment of rolling a four-sided die (the sides of which are numbered 1, 2, 3, 4) until a "4" appears.
i.) Systematically describe the sample space. Does it contain only finitely many outcomes?
ii.) Now, suppose that you instead record the number of rolls of the die that it takes before a "4" appears.
   a.) Identify the sample space.

b.) To what outcome of this sample space do the outcomes (1, 4) and (2, 4) obtained when conducting the experiment described in (i) correspond?

The scenarios for each of the experiments described in the above exercises can be characterized as subsets of the sample space. For an experiment under consideration, we need to collect all such subsets of Ω "of interest" into a set $\mathscr{F}$. The usual set-theoretic combinations of events (e.g., unions, intersections, complements, etc.) arise naturally and must also belong to $\mathscr{F}$. For practical reasons, however, it is not in our interest to simply include *all* possible subsets of Ω in $\mathscr{F}$. Rather, we would like to construct $\mathscr{F}$ so that it contains the fewest sets possible for which the collection of events with which we are interested can be studied without ambiguity. Consider the following example.

Example. Suppose that, in the context of Exer. 2.1.1, we are interested only in the event $A = \{(1,1,1)\}$. Automatically, A is included in $\mathscr{F}$. But, what other events must be included in order to ensure that $\mathscr{F}$ is closed under set-theoretic combinations? For one, $\tilde{A}$ must belong to $\mathscr{F}$. (What outcomes belong to $\tilde{A}$?) The minute we include this event in $\mathscr{F}$, its union and intersection with any other member of $\mathscr{F}$ must also be included. As such, the events $A \cup \tilde{A} = \Omega$ and $A \cap \tilde{A} = \varnothing$ are also members of $\mathscr{F}$. At this point, you can check directly that $\mathscr{F} = \left\{\varnothing, A, \tilde{A}, \otimes\right\}$ is closed under set-theoretic combinations. (Do so!)

The set $\mathscr{F}$ in the above example is sufficient if we are only interested in one event, namely A. If later we decide that we would like to include event B in our study, and B is not one of the four members of $\mathscr{F}$, then we must carefully determine all other events that must be included to ensure that the new set $\mathscr{F}$ is closed under set-theoretic combinations; the set $\mathscr{F} \cup \{B\}$ simply does not work. (Why?)
Consider the next example.

Example. Suppose now that, in the context of Exer. 2.1.1, we are interested in the following two events:

$$A = \{(1,1,1)\},$$
$$B = \{(a,b,c) \,|\, a,b, \text{ and } c \text{ are all even natural numbers}\}.$$

We assert that

$$\mathscr{F} = \left\{\varnothing, A, \tilde{A}, B, \tilde{B}, A \cup B, \tilde{A} \cup \tilde{B}, \tilde{A} \cap \tilde{B}, A \cap B, \Omega\right\}. \tag{2.1}$$

Exercise 2.1.4. Verify (2.1).

By construction, the sets $\mathscr{F}$ in the above two examples contain the fewest events possible. Of course, we could proceed in this manner to construct the set $\mathscr{F}$ appropriate for the analysis of any finite collection of events. But, what if we are interested in *infinitely* many events? While actually constructing the set $\mathscr{F}$ as we did above is not be feasible, we impose the same requirement that $\mathscr{F}$ be closed under set-theoretic combinations in the sense of the following definition.

Definition 2.1.2. Let Ω be a sample space. We say that $\mathscr{F}$ is a *σ-algebra on Ω* if
i.) $\varnothing \in \mathscr{F}$,
ii.) $A \in \mathscr{F} \Longrightarrow \tilde{A} \in \mathscr{F}$,
iii.) $\{A_i \,|\, i \in I\} \subset \mathscr{F}$, where I is a countable set $\Longrightarrow \bigcup_{i \in I} A_i \in \mathscr{F}$.

Remark. Naturally, you might ask why we did not demand that intersections also belong to $\mathscr{F}$ as part of Def. 2.1.2. The reason is that by requiring complements and unions belong to $\mathscr{F}$, intersections are automatically contained in $\mathscr{F}$ due to DeMorgan's laws. Indeed,

$$A_i \in \mathscr{F}, \forall i \in I \Longrightarrow \bigcup_{i \in I} A_i \in \mathscr{F} \Longrightarrow \widetilde{\left(\bigcup_{i \in I} A_i\right)} = \bigcap_{i \in I} \tilde{A}_i \in \mathscr{F}.$$

Exercise 2.1.5. Assume that A, B, and C belong to $\mathscr{F}$. Show that the events $A \cap \tilde{B} \cap C$ and $\widetilde{(A \cup B \cup C)}$ both belong to $\mathscr{F}$.

Note that efficiency, in the sense of containing only those sets that are absolutely necessary in order to study a set of events identified a priori, is not built into the definition of a σ-algebra on Ω. The σ-algebras constructed in the above two examples are illustrations of the following special type of σ-algebra on Ω.

Definition 2.1.3. Let Ω be a sample space and consider the collection of events $\mathscr{A} = \{A_i \subset \Omega \,|\, i \in I\}$. The σ-algebra generated by $\mathscr{A}$, denoted by $\sigma(\mathscr{A})$, is defined by $\sigma(\mathscr{A}) = \bigcap_{\mathscr{A} \subset \mathscr{G}} \mathscr{G}$, where $\mathscr{G}$ is a σ-algebra on Ω.

Remark. By definition, $\sigma(\mathscr{A})$ is the smallest σ-algebra on Ω containing $\mathscr{A}$ in the sense that if $\mathscr{Y}$ is any σ-algebra on Ω containing $\mathscr{A}$, $\sigma(\mathscr{A}) \subset \mathscr{Y}$.

Exercise 2.1.6. Prove the above remark.

The following σ-algebra on $\mathbb{R}$ will be used often in our discussion.

Definition 2.1.4. Let $\mathscr{G}_1 = \{(a,b] \,|\, -\infty < a < b < \infty\}$. The *Borel class on $\mathbb{R}$* is $\sigma(\mathscr{G}_1)$.

Remarks.
1. The Borel class on $\mathbb{R}$ can be formed in different ways, one of which is the σ-algebra on $\mathbb{R}$ generated by the collection $\mathscr{G}_2 = \{(-\infty, x] \,|\, -\infty < x < \infty\}$.
2. The Borel class on $\mathbb{R}^N$ is $\sigma(\mathscr{G}_N)$, where

$$\mathscr{G}_N = \{(a_1, b_1] \times \ldots \times (a_N, b_N] \,|\, -\infty < a_i < b_i < \infty, \, i = 1, \ldots, N\}.$$

Exercise 2.1.7. Let $a \in \mathbb{R}$.
i.) Does $\{a\}$ belong to $\sigma(\mathscr{G}_1)$?

ii.) Does (a, b) belong to $\sigma(\mathcal{G}_1)$?

Suppose that for a given experiment, the pair $(\Omega, \mathcal{F})$ has been chosen. The next step is to systematically assign to each event $A \in \mathcal{F}$ a likelihood (or chance of occurrence), denoted $\mathcal{P}(A)$. In an elementary sense, you can think of repeatedly conducting the trials of an experiment and recording the outcome of the ith trial as $\omega_i \in \Omega$. For large values of N,

$$\mathcal{P}(A) \approx \frac{\text{number of } i \in \{1, \dots, N\} \text{ such that } \omega_i \in A}{N}$$

and $\mathcal{P}(A)$ equals the limiting value of this expression as $N \to \infty$.

Exercise 2.1.8. Based on this interpretation, answer the following questions:
i.) Why must $\mathcal{P}(\varnothing) = 0$ and $\mathcal{P}(\Omega) = 1$?
ii.) Why must $0 \le \mathcal{P}(A) \le 1, \forall A \in \mathcal{F}$?
iii.) If $A \cap B = \varnothing$, explain why $\mathcal{P}(A \cup B) = \mathcal{P}(A) + \mathcal{P}(B)$.
iv.) If $A \cap B \ne \varnothing$, explain how you would modify the formula in (iii).
v.) Explain why it is important for $\bigcup_{n=1}^{\infty} A_n \in \mathcal{F}$ in the context of Exer. 2.1.3. Can you formulate equalities comparable to those mentioned in (iii) and (iv) for two sets?

We make the notion of the function $\mathcal{P}$ precise in the following definition.

Definition 2.1.5. A function $\mathcal{P} : \mathcal{F} \to [0, 1]$ for which
i.) $\mathcal{P}(\Omega) = 1$,
ii.) $\mathcal{P}\left(\bigcup_{i=1}^{\infty} A_i\right) = \sum_{i=1}^{\infty} \mathcal{P}(A_i)$ whenever $A_i \in \mathcal{F}$ and $A_i \cap A_j, \forall i \ne j$,
is a *probability measure on* $(\Omega, \mathcal{F})$. We call $(\Omega, \mathcal{F}, \mathcal{P})$ a *probability space* and refer to the members of $\mathcal{F}$ as $(\mathcal{P}-)measurable$ *sets*.

We gather some useful properties of probabilitiy measures below, many of which you encounter in an elementary probability course.

Proposition 2.1.6. (Properties of Probability Measures)
Let $(\Omega, \mathcal{F}, \mathcal{P})$ *be a probability space and assume that* $A, A_i (i = 1, 2, \dots)$ *belong to* $\mathcal{F}$. *Then,*
i.) (Complements) $\mathcal{P}\left(\tilde{A}\right) = 1 - \mathcal{P}(A)$;
ii.) $\mathcal{P}(\varnothing) = 0$;
iii.) (Addition Law) $\mathcal{P}(A \cup B) = \mathcal{P}(A) + \mathcal{P}(B) - \mathcal{P}(A \cap B)$;
iv.) (Monotonicity) $A \subset B \implies \mathcal{P}(A) \le \mathcal{P}(B)$;
v.) (Partition Law) *If* $\Omega = \bigcup_{i=1}^{N} A_i$ *and* $A_i \cap A_j = \varnothing$ *whenever* $i \ne j$ *(that is, the sets* $A_1, \dots, A_N$ *are pairwise disjoint (pwd)), then* $\mathcal{P}(B) = \sum_{i=1}^{N} \mathcal{P}(B \cap A_i), \forall B \in \mathcal{F}$;
vi.) (Countable Subadditivity) *For any countable collection* $\{A_i : i \in I \subset \mathbb{N}\} \subset \mathcal{F}$, $\mathcal{P}\left(\bigcup_{i \in I} A_i\right) \le \sum_{i \in I} \mathcal{P}(A_i)$;
vii.) (Continuity) *If* $A_i \uparrow A$ *(i.e.,* $A_i \subset A_{i+1}, \forall i \in \mathbb{N}$*) and* $\bigcup_{i=1}^{\infty} A_i = A$, *then* $\mathcal{P}(A_i) \uparrow$ $\mathcal{P}(A)$ *as* $i \to \infty$. *Similarly, if* $A_i \downarrow A$ *(i.e.,* $A_i \supset A_{i+1}, \forall i \in \mathbb{N}$*) and* $\bigcup_{i=1}^{\infty} A_i = A$, *then* $\mathcal{P}(A_i) \downarrow \mathcal{P}(A)$ *as* $i \to \infty$;
viii.) (Subtractivity) $\mathcal{P}(A \setminus B) = \mathcal{P}(A) - \mathcal{P}(B)$.

Proof. We outline the proofs of several parts.

i.) Use $\widetilde{A} \cup A = \Omega$. (How?)

ii.) Use (i) and $\widetilde{\Omega} = \varnothing$. (How?)

iii.) Use $A = \left(A \cap \widetilde{B}\right) \cup (A \cap B)$ and $B = \left(B \cap \widetilde{A}\right) \cup (A \cap B)$. (How?)

iv.) Because $A \subset B$, $B = A \cup (B \setminus A)$. (So what?)

v.) Express B as a finite union of pairwise disjoint events. (How?)

vi.) $\bigcup_{i \in I} A_i = A_1 \cup \left(\widetilde{A_1} \cap A_2\right) \cup \left(\widetilde{A_1} \cap \widetilde{A_2} \cap A_3\right) \cup \ldots$ (Draw a diagram illustrating this. Now what?)

vii.) Assume that $A_i \uparrow A$. Observe that $\bigcup_{i=1}^{\infty} A_i$ can be partitioned by

$$\begin{cases} D_1 & = A_1 \\ D_i & = A_i \setminus A_{i-1}, i = 2, 3, \ldots \end{cases}$$

Observe that $\bigcup_{i=1}^{\infty} D_i = \bigcup_{i=1}^{\infty} A_i$ and $\{D_i : i \in \mathbb{N}\}$ are pwd. (Draw a picture.) Then,

$$\mathscr{P}\left(\lim_{i \to \infty} A_i\right) = \mathscr{P}(A) = \mathscr{P}\left(\bigcup_{i=1}^{\infty} D_i\right) = \sum_{i=1}^{\infty} \mathscr{P}(A_i) = \lim_{N \to \infty} \sum_{i=1}^{N} \mathscr{P}(A_i) =$$

$$= \lim_{N \to \infty} \mathscr{P}\left(\bigcup_{i=1}^{N} D_i\right) = \lim_{N \to \infty} A_N.$$

viii.) This follows from (iv). (Why?)   □

Exercise 2.1.9. Provide the details in the proof of Prop. 2.1.6.

Definition 2.1.7. Let $\{A_i : i \in \mathbb{N}\} \subset \mathscr{F}$.

i.) $\limsup A_i = \bigcap_{j=1}^{\infty} \bigcup_{i \geq j} A_i = \{\omega \in \Omega : \omega \in A_i \text{ for infinitely many } i \in \mathbb{N}\}$,

ii.) $\liminf A_i = \bigcup_{j=1}^{\infty} \bigcap_{i \geq j} A_i = \{\omega \in \Omega : \exists N(\omega) \in \mathbb{N} \text{ such that } i \geq N(\omega) \Longrightarrow \omega \in A_i\}$.

The following proposition plays an important role in certain convergence arguments.

Proposition 2.1.8. (Borel-Cantelli)
If $\{A_i : i \in \mathbb{N}\} \subset \mathscr{F}$ is such that $\sum_{i=1}^{\infty} \mathscr{P}(A_i) < \infty$, then $\mathscr{P}(\limsup A_i) = 0$.

Remarks.
1. Borel-Cantelli says that the set of outcomes that occur infinitely often in $\{A_i : i \in \mathbb{N}\}$ has probability zero.
2. An event with probability zero is called a $\mathscr{P}$-*null event.* At the other extreme, if $\mathscr{P}(A) = 1$, then we say the event A occurs *almost surely,* abbreviated as "a.s. $[\mathscr{P}]$."

Exercise 2.1.10. Let $\{A_i : i \in \mathbb{N}\} \subset \mathscr{F}$. Prove that
i.) $\mathscr{P}(\limsup A_i) \geq \limsup \mathscr{P}(A_i)$.
ii.) $\mathscr{P}(\liminf A_i) \leq \liminf \mathscr{P}(A_i)$.

In order to avoid theoretical problems, we need to make certain that any subset of a $\mathscr{P}$-null event belongs to our σ-algebra $\mathscr{F}$ and is also $\mathscr{P}$-null. This is handled by assuming that the probability space $(\Omega, \mathscr{F}, \mathscr{P})$ is complete in the following sense.

Definition 2.1.9. A probability space $(\Omega, \mathscr{F}, \mathscr{P})$ is *complete* if for any event A such that $A \subset B$ with $B \in \mathscr{F}$ and $\mathscr{P}(B) = 0$, it is the case that $A \in \mathscr{F}$ and $\mathscr{P}(A) = 0$.

Convention: We henceforth assume without further mention that all probability spaces under consideration in this text are complete.

2.2 $\mathbb{R}$-Valued Random Variables

Consider the experiment of rolling a die (whose faces are labeled 1, 2, 3, and 4) until a 4 appears (cf. Exer. 2.1.3). When conducting this experiment, it can take any number N of rolls in order for a 4 to appear, and we do not know a priori how many such rolls will occur. As such, if it takes N rolls in order for a 4 to appear, then letting 0 act as a place holder, a typical outcome ω is of the form $(\omega_1, \ldots, \omega_{N-1}, 4, 0, 0, \ldots)$, where $\omega_i \in \{1, 2, 3\}$, $\forall i \in \{1, \ldots, N-1\}$. The collection of all such outcomes constitutes the sample space Ω. Now assume, for simplicity, that $\mathscr{F} = \mathbb{P}(\Omega)$ (the set of all subsets of Ω) and define the function $X : \Omega \to \mathbb{R}$ by

$$X(\omega) = N, \text{ where } \omega_N = 4. \tag{2.2}$$

This function assigns a numerical value to every outcome in the sample space Ω.

Exercise 2.2.1.
i.) Compute $X^{-1}((-\infty, N])$, where $N \in \mathbb{N}$.
ii.) Extend the computation from (i) to $X^{-1}((-\infty, x])$, where $x \in \mathbb{R}$.
iii.) For what values of $x \in \mathbb{R}$ is $X^{-1}((-\infty, x]) \in \mathscr{F}$?

Guided by our discussion, defining a distribution function (or probability accumulation function) involving X requires that we consider probabilities of the form

$$\mathscr{P}(\{\omega \in \Omega : X(\omega) \le x\}), \text{ where } x \in \mathbb{R}.$$

Doing so requires that

$$\left\{ \omega \in \Omega : \underbrace{X(\omega) \le x}_{\text{i.e., } X(\omega) \in (-\infty, x]} \right\} \in \mathscr{F}, \forall x \in \mathbb{R},$$

or equivalently,

$$\{\omega \in \Omega : \omega \in X^{-1}((-\infty, x])\} = X^{-1}((-\infty, x]) \in \mathscr{F}, \forall x \in \mathbb{R}. \tag{2.3}$$

Remark. If (2.3) were true, then because $X^{-1}(\bigcup_i A_i) = \bigcup X^{-1}(A_i)$ and $X^{-1}(\bigcap_i A_i) = \bigcap X^{-1}(A_i)$, it would follow that $X^{-1}(B) \in \mathscr{F}$, for any Borel set B. (Convince yourself.)

Exercise 2.2.2. Let $a, b \in \mathbb{R}$. Explain how to compute $\mathscr{P}\left(\{\omega \in \Omega : a \leq X(\omega) \leq b\}\right)$.

We focus our attention on such mappings from Ω into $\mathbb{R}$, defined formally below.

Definition 2.2.1. Let $(\Omega, \mathscr{F}, \mathscr{P})$ be a complete probability space. A function $X : \Omega \to \mathbb{R}$ for which $X^{-1}\left((-\infty, x]\right) \in \mathscr{F}, \forall x \in \mathbb{R}$, is an $\mathbb{R}$-valued *random variable*.

If the codomain of a random variable of interest is an interval or $\mathbb{R}$ itself, it is called *continuous*. If the codomain is a countable set, then we say the random variable is *discrete*. Most random variables in practice, and especially in more elaborate settings, are *mixed*, meaning that the range is comprised of both intervals and discrete points. For simplicity, we restrict our discussion to simple examples. Some natural questions that arise are

1. Which operations on and combinations of random variables defined on $(\Omega, \mathscr{F}, \mathscr{P})$ produce another random variable defined on $(\Omega, \mathscr{F}, \mathscr{P})$?
2. What are some important statistical features of a random variable?
3. Can a calculus be developed for an appropriate collection of random variables?

The answer to the first question is addressed by the following proposition, which is easily proved. (Try it!)

Proposition 2.2.2. (Combinations of Random Variables)
Let X, Y, and $\{X_n : n \in \mathbb{N}\}$ be $\mathbb{R}$-valued random variables defined on $(\Omega, \mathscr{F}, \mathscr{P})$ and let $\alpha, \beta \in \mathbb{R}$. Then,
i.) $\alpha X \pm \beta Y, XY, \inf\{X_n : n \in \mathbb{N}\}, \liminf\{X_n : n \in \mathbb{N}\},$ *and* $\limsup\{X_n : n \in \mathbb{N}\}$ *are also $\mathbb{R}$-valued random variables.*
ii.) *If $f : \mathbb{R} \to \mathbb{R}$ is continuous, then $f \circ X$ is an $\mathbb{R}$-valued random variable.*

Remark. We must precisely define $\liminf\{X_n : n \in \mathbb{N}\}$ and $\limsup\{X_n : n \in \mathbb{N}\}$. Indeed, $\{X_n : n \in \mathbb{N}\}$ is a sequence of real-valued functions, not sets as in Def. 2.1.7. Using Def. 1.5.14, the function $\liminf\{X_n : n \in \mathbb{N}\}$ is shorthand notation for the function $F : \Omega \to \mathbb{R}$ given by $F(\omega) = \liminf\{X_n(\omega) : n \in \mathbb{N}\}, \forall \omega \in \Omega$. The function $\limsup\{X_n : n \in \mathbb{N}\}$ is defined in a similar fashion.

Exercise 2.2.3. Suppose that X is an $\mathbb{R}$-valued random variable defined on $(\Omega, \mathscr{F}, \mathscr{P})$.
i.) Prove that $\forall n \in \mathbb{N}$, the function $\omega \mapsto [X(\omega)]^n$ is also an $\mathbb{R}$-valued random variable.
ii.) Must the function $\omega \mapsto |X(\omega)|^n$ be an $\mathbb{R}$-valued random variable, $\forall n \in \mathbb{N}$?
iii.) More generally than (ii), must the function $\omega \mapsto |X(\omega)|^\gamma$ be an $\mathbb{R}$-valued random variable, $\forall \gamma > 0$?
iv.) More generally than (iii), must the function $\omega \mapsto |\sum_{i=1}^n \alpha_i X_i(\omega)|^\gamma$ be an $\mathbb{R}$-valued random variable, $\forall \gamma > 0$ and $n \in \mathbb{N}$?

The notion of a simple function, defined below, and pointwise and uniform limits thereof are important and will arise often in the development of the material.

Definition 2.2.3. i.) Let $A \in \mathscr{F}$. The *random characteristic function* $\chi_A : \Omega \to \{0,1\}$ is defined by

$$\chi_A(\omega) = \begin{cases} 0, & \omega \in A, \\ 1, & \omega \in \Omega \setminus A. \end{cases}$$

ii.) More generally, let $\{c_k : k = 1, \ldots, m\} \subset \mathbb{R}$ and $\{A_k : k = 1, \ldots, m\} \subset \mathscr{F}$ be such that $A_k \cap A_j = \varnothing$ whenever $k \neq j$ and $\Omega = \bigcup_{k=1}^{m} A_k$. The random variable $s : \Omega \to \mathbb{R}$ defined by $s(\omega) = \sum_{k=1}^{m} c_k \chi_{A_k}(\omega)$ is called a *random simple function*.

Exercise 2.2.4. Explain carefully why a simple function is indeed a random variable.

Exercise 2.2.4, together with the results concerning limsup and liminf, begs the question as to what random variables can be constructed using limits of sequences of simple functions. Certainly, such limits must themselves be random variables. (Why?) Such an approximation will have tremendous utility, as you will see.

The notion of an "information conduit" is central to our discussion throughout this text; such a conduit is formally built using a sequence of σ-algebras. Plainly speaking, given a sample space Ω, based on the features of our idealized model of the experiment, we identify a collection $\mathscr{E}$ of events that we are interested in studying. Then, we form $\sigma(\mathscr{E})$ to be the smallest collection of events containing $\mathscr{E}$ that must be considered in our discussion (whether directly of interest or not) in order to avoid ambiguity. We now extend this notion using the information gained through the use of a random variable.

Suppose that our experiment is modeled abstractly as $(\Omega, \mathscr{F}, \mathscr{P})$ and that we define a random variable $X : \Omega \to \mathbb{R}$. We are interested in those events formed using the information provided by the function X. Among the sets of immediate interest are those of the form

$$\mathscr{A}_{(a,b]} = \{\omega \in \Omega : -\infty < a < X(\omega) \leq b < \infty\} = X^{-1}(a,b]. \qquad (2.4)$$

The entire Borel class can be constructed using these sets. (Why?) Using (2.4), we define the σ-algebra generated by X, denoted $\sigma(X)$, by

$$\sigma(X) \equiv \sigma\left(\bigcup\{\mathscr{A}_{(a,b]} : -\infty < a < b < \infty\}\right). \qquad (2.5)$$

Note that $\sigma(X) \subset \mathscr{F}$ because we cannot escape the set of events used to model the experiment. The practical importance of $\sigma(X)$ is that it consists precisely of the collection $\mathscr{E}$ of events for which for any possible outcome $\omega \in \Omega$, the information provided by X is all that you need in order to decide whether or not $\omega \in \mathscr{E}$.

Of course, the downside is that information about the experiment is lost when computing a random variable. For instance, within the context of Exer. 2.1.3, knowing that $X(\omega) = 2$ simply tell us that $\omega \in \{(1,4,0,\ldots),(2,4,0,\ldots),(3,4,0,\ldots)\}$. We cannot deduce which of these three elements actually occurred. As such, while (2.5) is a rich set, it does not contain all the information about our experiment. That said, we can build onto this set to produce an even richer set. Indeed, we can use information provided by another random variable defined on Ω. Precisely, suppose that we are interested in the collection of all events that are completely determined by the information provided by two random variables X_1 and X_2. Then, $\sigma(\{X_1, X_2\})$ should be defined by

$$\sigma(\{X_1, X_2\}) = \sigma\left(\bigcup_{i \in \{1,2\}} \{\omega \in \Omega : -\infty < a < X_i(\omega) \leq b < \infty\}\right). \qquad (2.6)$$

Clearly, $\sigma(X_i) \subset \sigma(\{X_1, X_2\})$, for $i = 1, 2$. (Why?) As such, we intuitively think of $\sigma(\{X_1, X_2\})$ as carrying more information about the experiment than does $\sigma(X_i)$.

Later in our discussion, we will encounter the need to consider more elaborate collections of such random variables, say $\{X_\gamma : \gamma \in \Gamma\}$, where Γ is an uncountable index set. The σ-algebra $\sigma(\{X_\gamma : \gamma \in \Gamma\})$ is defined as in (2.6) with $\{1, 2\}$ replaced by Γ. In such case, observe that for $\Gamma = [\gamma_1, \gamma_2]$,

$$\gamma_1 < s < t \leq \gamma_2 \implies \sigma(\{X_\gamma : \gamma \in [\gamma_1, s]\}) \subset \sigma(\{X_\gamma : \gamma \in [\gamma_1, t]\}) \subset \mathscr{F}. \qquad (2.7)$$

The collection $\{\{X_\gamma : \gamma \in [\gamma_1, s]\} : s \in [\gamma_1, \gamma_2]\}$ is an increasing family of sub σ-algebras of $\mathscr{F}$ called a *filtration*. This provides an information conduit of sorts. We will revisit this notion later in the chapter.

By definition, to every random variable there corresponds a unique distribution function defined as follows.

Definition 2.2.4. Let $X : \Omega \to \mathbb{R}$ be a random variable. The function $F_X : \mathbb{R} \to [0, 1]$ defined by

$$F_X(x) = \mathscr{P}(\{\omega \in \Omega : X(\omega) \leq x\}) = \mathscr{P}(X^{-1}(x)) \qquad (2.8)$$

is the *distribution function* of X.

Naively, F_X can be thought of as an accumulation function. See **[212]** for a proof of the uniqueness of such a function. Convince yourself of the following properties.

Proposition 2.2.5. (Properties of Distribution Functions)
Let $F_X : \mathbb{R} \to [0, 1]$ be the distribution function of a random variable $X : \Omega \to \mathbb{R}$.
i.) F_X *is increasing.*
ii.) $\lim_{x \to \infty} F_X(x) = 1.$
iii.) $\lim_{x \to -\infty} F_X(x) = 0.$
iv.) F_X *is right-continuous.*
v.) $\mathscr{P}(\{\omega \in \Omega : a < X(\omega) \leq b\}) = F_X(b) - F_X(a).$

Remark. We write $\mathscr{P}(a < X \leq b)$ in place of $\mathscr{P}(\{\omega \in \Omega : a < X(\omega) \leq b\})$.

Exercise 2.2.5. Prove Prop. 2.2.5.

A random variable X is defined "uniquely" (i.e., equality with probability one) by its distribution function F_X in the sense that

$$F_X = F_Y \implies X(\omega) = Y(\omega) \text{ a.s.} [\mathscr{P}]. \qquad (2.9)$$

For instance, consider the prototypical example of tossing a coin N times and recording each time whether the coin lands heads or tails side up. Assuming that tosses are mutually independent (naively, this means that the result of any toss has no impact on any of the other tosses) and $\mathscr{P}$ (head) $= p$, for some $0 < p < 1$, for all tosses, this experiment is historically referred to as a *binomial* experiment. (What is the sample space Ω?) Define the random variable $X_p : \Omega \rightarrow \{0, 1, \ldots, N\}$ by

$$X_p(\omega) = \text{number of heads obtained in } N \text{ tosses.}$$

In such case, X_p is a *binomial random variable* and we write "X_p is $b(N, p)$." It is intuitive that if $p \neq q$, then $X_p \neq X_q$. More precisely, there is a set $\mathscr{A}$ of outcomes for which $\mathscr{P}(\mathscr{A}) > 0$ and $X_p(\omega) \neq X_q(\omega)$, $\forall \omega \in \mathscr{A}$.

Exercise 2.2.6. Let $N = 2$, $p = \frac{1}{2}$, and $q = \frac{1}{4}$. Show that $X_p \neq X_q$.

Assume that X is a continuous $\mathbb{R}$-valued random variable. The "accumulation function" interpretation of F_X suggests that

$$F_X(x) = \mathscr{P}(X \leq x) = \text{accumulation of probability up to } x. \quad (2.10)$$

In some situations, a function $f_X : \mathbb{R} \longrightarrow [0, \infty)$ (called the *probability density function of X*) can be identified such that

$$F_X(x) = \int_{-\infty}^{x} f_X(t)dt.$$

Typically, it is not feasible, or even possible, to explicitly determine such a function.

2.2.1   Some Useful Statistics for Random Variables

In elementary settings, the *mean* (or *expectation*) of a random variable X, denoted by μ_X or $E[X]$, can be thought of as a sort of weighted average and is computed using the probability density of X. For instance, if X is $b(N, p)$, then the mean of X is Np. However, given that a probability density need not exist, we need a definition that does not rely on it. To this end, we have the following.

Definition 2.2.6. The *expectation* of X, denoted $E[X]$, is defined by

$$E[X] = \int_{\Omega} X(\omega)d\mathscr{P}.$$

The integral $\int_{\Omega} X(\omega)d\mathscr{P}$ might seem a bit strange to you, especially if you are not familiar with the Lebesgue integral. After all, what do we mean by an integral with respect to a *measure* $\mathscr{P}$? We use a "building-block" approach to define such an integral, as follows.

Definition 2.2.7. "Building Block" Definition of $\int_\Omega X(\omega)\,d\mathscr{P}$

i.) <u>Step 1</u>: <u>Characteristic Function</u>
Let $A \in \mathscr{F}$ and $c \in \mathbb{R}$, and consider $X(\omega) = c\chi_A(\omega)$. We define

$$\int_\Omega c\chi_A(\omega)\,d\mathscr{P} \equiv c\left[1 \cdot \mathscr{P}(A) + 0 \cdot \mathscr{P}(\Omega \setminus A)\right] = c\mathscr{P}(A). \qquad (2.11)$$

ii.) <u>Step 2</u>: <u>Simple Function</u>
Let $\{A_k : k = 1, \dots, m\} \subset \mathscr{F}$ be such that $A_k \cap A_j = \varnothing$ whenever $k \neq j$ and $\Omega = \bigcup_{k=1}^m A_k$ and let $\{c_k : k = 1, \dots, m\} \subset \mathbb{R}$. Consider $X(\omega) = \sum_{k=1}^m c_k\chi_{A_k}(\omega)$. We define

$$\int_\Omega \left(\sum_{k=1}^m c_k\chi_{A_k}(\omega)\right) d\mathscr{P} \equiv \sum_{k=1}^m c_k \int_\Omega \chi_{A_k}(\omega)\,d\mathscr{P} = \sum_{k=1}^m c_k\mathscr{P}(A_k). \qquad (2.12)$$

iii.) <u>Step 3</u>: <u>General Positive Random Variable</u>
Let $X : \Omega \to [0, \infty)$ be a nonnegative random variable and consider a monotone increasing sequence of random simple functions $\{s_m : m \in \mathbb{N}\}$ such that

$$\lim_{m \to \infty} s_m(\omega) = X(\omega), \forall \omega \in \Omega.$$

We define

$$\int_\Omega X(\omega)\,d\mathscr{P} \equiv \lim_{m \to \infty} \int_\Omega s_m(\omega)\,d\mathscr{P}. \qquad (2.13)$$

iv.) <u>Step 4</u>: <u>General Random Variable</u>
Let $X : \Omega \to \mathbb{R}$ be a random variable. Using two new random variables $X^+, X^- : \Omega \to [0, \infty)$ by

$$X^+(\omega) = \max(X(\omega), 0), \ X^-(\omega) = \max(-X(\omega), 0), \qquad (2.14)$$

we define

$$\int_\Omega X(\omega)\,d\mathscr{P} \equiv \int_\Omega X^+(\omega)\,d\mathscr{P} - \int_\Omega X^-(\omega)\,d\mathscr{P}. \qquad (2.15)$$

Remarks.
1. The familiar properties of the integral, such as linearity and monotonicity, also hold for the integral defined in Def. 2.2.7 and are proven easily using the properties of the measure $\mathscr{P}$ with the help of the building-block approach. (See **[212, 404]** for a thorough discussion.)
2. We recover the interpretation that $E[X]$ is the "average value" of X over Ω by simply noting that $E[X] = \frac{1}{\mathscr{P}(\Omega)} \int_\Omega X(\omega)\,d\mathscr{P}$ because $\mathscr{P}(\Omega) = 1$. This quantity resembles the average value of a real-valued function f over the interval $[a, b]$, which is given by $\frac{1}{b-a} \int_a^b f(x)\,dx$.

The following properties of expectation are useful in all facets of our discussion.

Proposition 2.2.8. (Basic Properties of Expectation)
Let $X, X_1, X_2, \dots$ be $\mathbb{R}$-valued random variables defined on $(\Omega, \mathscr{F}, \mathscr{P})$ and $\{\alpha_k : k \in \mathbb{N}\} \subset \mathbb{R}$.

i.) **(Linearity)** $E\left[\sum_{k=1}^{m} \alpha_k X_k\right] = \sum_{k=1}^{m} \alpha_k E[X_k]$. *Further, if* $\{X_k : k \in \mathbb{N}\}$ *are non-negative random variables for which* $\sum_{k=1}^{\infty} E[X_k] < \infty$, *then* $E\left[\sum_{k=1}^{\infty} \alpha_k X_k\right] = \sum_{k=1}^{\infty} \alpha_k E[X_k]$.

ii.) **(Monotonicity)** $Y(\omega) \leq X(\omega), \forall \omega \in \Omega \implies E[Y] \leq E[X]$.

iii.) $|E[X]| \leq E[|X|]$.

iv.) **(Jensen's Inequality)** *If* $\varphi : \mathbb{R} \to \mathbb{R}$ *is convex,* $E[|X|] < \infty$, *and* $E[|\varphi(X)|] < \infty$, *then*

$$\varphi(E[X]) \leq E[\varphi(X)].$$

v.) **(Hölder's Inequality)** *Let* $1 \leq p \leq q < \infty$ *be such that* $\frac{1}{p} + \frac{1}{q} = 1$. *If* $E[|X|^p] < \infty$ *and* $E[|X|^q] < \infty$, *then*

$$|E[XY]| \leq (E[|X|^p])^{\frac{1}{p}} (E[|Y|^q])^{\frac{1}{q}}. \tag{2.16}$$

vi.) **(Minkowski's Inequality)** $\forall p > 1$, $(E[|X+Y|^p])^{\frac{1}{p}} \leq (E[|X|^p])^{\frac{1}{p}} + (E[|Y|^p])^{\frac{1}{p}}$.

vii.) *If* $E\left[|X|^2\right] = 0$, *then* $X = 0$ *a.s.* $[\mathscr{P}]$.

viii.) *If* $X(\omega) = Y(\omega)$ *a.s.* $[\mathscr{P}]$, *then* $E[X] = E[Y]$.

Remarks.

1. We have used brackets in the expression $E[X]$ to emphasize that the expectation is an operator with input X. That said, when considering $E[|X|]$, we shall streamline the notation and henceforth routinely write $E|X|$.

2. When $p = q = 2$, (2.16) is often referred to as the *Cauchy-Schwarz inequality*.

3. Remember, the phrase "$X = 0$ a.s. $[\mathscr{P}]$" means $\exists \mathscr{D} \in \mathscr{F}$ such that $\mathscr{P}(\mathscr{D}) = 0$ and $X(\omega) = 0, \forall \omega \in \Omega \setminus \mathscr{D}$.

4. Proposition 2.2.8(viii) is similar to the following familiar property of the Riemann integral: *If* $f(x) \geq 0, \forall x \in [a,b]$, *then* $\int_a^b f(x)dx = 0 \implies f(x) = 0, \forall x \in [a,b]$. (Interpret this visually.)

Establishing estimates involving the expectation of various random quantities will become standard practice in this text. Use the properties in Prop. 2.2.8 to complete the following exercises.

Exercise 2.2.7. Let $\alpha, \beta \in \mathbb{R}$ and $X, Y : \Omega \to \mathbb{R}$ be random variables. Prove

i.) $|E[\alpha X + \beta Y]| \leq |\alpha| E|X| + |\beta| E|Y|$;

ii.) $(E[\alpha X + \beta Y])^2 \leq \alpha^2 E[X^2] + 2\alpha\beta E[XY] + \beta^2 E[Y^2]$;

iii.) $\left(E\left[\left(\frac{X-\alpha}{\beta}\right) \cdot \left(\frac{Y-\alpha}{\beta}\right)\right]\right)^4 \leq \frac{1}{\beta^8} \left(E|X-\alpha|^4\right) \left(E|Y-\alpha|^{\frac{4}{3}}\right)^3$, provided $\beta \neq 0$.

Exercise 2.2.8. Suppose that $Z : \Omega \to \mathbb{R}$ is defined by $Z(\omega) = 1, \forall \omega \in \Omega$. Prove

i.) $E[Z] = 1$;

ii.) $E[\alpha] = \alpha, \forall \alpha \in \mathbb{R}$. (That is, the expectation of a constant is the constant.)

Another useful statistic that measures the dispersion of the values $X(\omega)$ around $E[X]$ is called the *variance*, defined below.

Definition 2.2.9. Let $X : \Omega \to \mathbb{R}$ be a random variable. The variance of X, denoted σ_X^2 or $Var[X]$, is defined by

$$Var[X] \equiv E[X - E[X]]^2 = E[X^2] - (E[X])^2. \qquad (2.17)$$

The standard deviation of X, denoted σ_X, is given by $\sigma_X = \sqrt{Var[X]}$.

Exercise 2.2.9. Let $\alpha, \beta \in \mathbb{R}$. Prove that $Var[\alpha X + \beta] = \alpha^2 Var[X]$.

Other useful statistics (such as skewness and kurtosis) can be defined in terms of $E|X|^p$, for $p \in \mathbb{N}$. These quantities arise when establishing important estimates and when studying stability.

Definition 2.2.10. Let $p \in \mathbb{N}$ and $X : \Omega \to \mathbb{R}$ be a random variable. The pth *moment of X is given by $E|X|^p$.*

A well-known lower bound for the pth moment is provided by Chebyshev's inequality, a version of which is as follows.

Proposition 2.2.11. (Chebyshev's Inequality)
Suppose that $X : \Omega \to [0, \infty)$ is a random variable. Then, $\forall \varepsilon, p > 0$,

$$E[X^p] \geq \varepsilon^p \mathscr{P}(\{\omega \in \Omega : X(\omega) \geq \varepsilon\}). \qquad (2.18)$$

Exercise 2.2.10. Let $Y : \Omega \to \mathbb{R}$ be such that $E[Y] < \infty$. Prove that $\forall \alpha > 0$,

$$\mathscr{P}(\{\omega \in \Omega : |Y(\omega) - E[Y]| \geq \alpha \sigma_Y\}) \leq \frac{1}{\alpha^2}.$$

At times, we want to compare two random variables X and Y. Two useful statistical measures in this regard are as follows.

Definition 2.2.12. Let $X, Y : \Omega \to \mathbb{R}$ be random variables.
i.) The *correlation of X and Y* is defined by $r(X,Y) = E[XY]$. If $r(X,Y) = 0$, we say X and Y are *orthogonal*.
ii.) The *covariance of X and Y* is defined by $Cov(X,Y) = E[(X - E[X])(Y - E[Y])]$.
iii.) The *correlation coefficient of X and Y* is defined by $\rho_{XY} = \frac{Cov(X,Y)}{\sigma_X \cdot \sigma_Y}$, provided that $\sigma_X \cdot \sigma_Y \neq 0$.

2.2.2   Some Common Random Variables

We already encountered one prototypical model, namely the binomial random variable. Another commonly-occurring random variable is one that assigns the same probability to each of its outcomes; this is called a *uniform random variable*. If such a random variable X is discrete, then its sample space must be finite. (Why?) If X is continuous, then there exist $\alpha, \beta \in \mathbb{R}$ such that $\alpha < \beta$ for which the probability density function equals $\frac{1}{\beta - \alpha}$ at all values within the interval $[\alpha, \beta]$ and zero outside this interval. (Why?) For this case, we write "X is $u(\alpha, \beta)$."

The single most commonly occurring continuous random variable is the *Gaussian*, or *normal*, random variable, defined as follows.

Definition 2.2.13. A random variable $X : \Omega \to \mathbb{R}$ is *Gaussian* if its probability density function is given by

$$n_{\mu,\sigma}(x) = \frac{1}{\sigma\sqrt{2\pi}} \exp\left[-\frac{1}{2}\left(\frac{X-\mu}{\sigma}\right)^2\right], \; x \in \mathbb{R}, \tag{2.19}$$

where $\mu \in \mathbb{R}$ and $\sigma > 0$. We write "X is $n\left(\mu, \sigma^2\right)$."

It can be shown that $E[X] = \mu$ and $Var[X] = \sigma^2$. (See **[212, 404]**.)

Proposition 2.2.14. (Properties of Gaussian Random Variables)
i.) (Standard Normal) *If X is $n\left(\mu, \sigma^2\right)$, then $Z = \frac{X-\mu}{\sigma}$ is $n(0,1)$.*
ii.) (Linear Transformation of Normal) *If X is $n\left(\mu, \sigma^2\right)$ and $\alpha, \beta \in \mathbb{R}$, then $\alpha X + \beta$ is $n\left(\alpha\mu + \beta, \beta^2\sigma^2\right)$.*

Exercise 2.2.11. Let $N \in \mathbb{N}$ and suppose $X_N : \Omega \to \mathbb{R}$ is $b\left(N, p\right)$.
i.) Compute $E[X_N]$ and $Var[X_N]$.
ii.) Suppose that $Y : \Omega \to \mathbb{R}$ is $n(Np, Np(1-p))$. Does the density for Y serve as a good approximation of the density of X_N? What happens to this approximation as $N \to \infty$?

2.3 Introducing the Space $\mathfrak{L}^2(\Omega; \mathbb{R})$

The collection of all random variables with finite second moment will be especially important in our development of stochastic calculus used in the analysis of stochastic evolution equations. We begin with the following space.

Definition 2.3.1. The space $\mathfrak{L}^2(\Omega; \mathbb{R})$ is defined by

$$\mathfrak{L}^2(\Omega; \mathbb{R}) = \left\{X : \Omega \to \mathbb{R} : X \text{ is a random variable with } E[X^2] < \infty\right\} \tag{2.20}$$

Naturally, we need to establish the structure of $\mathfrak{L}^2(\Omega; \mathbb{R})$, with an eye pointed toward showing that it is a Hilbert space when equipped with the appropriate inner product. The easiest question to answer is whether or not $\mathfrak{L}^2(\Omega; \mathbb{R})$ is a linear space (in the sense of Def. 1.7.1). Indeed, using earlier exercises in Section 2.2 with the properties of expectation, it is not difficult to conclude that it is.

Exercise 2.3.1. Prove that $\mathfrak{L}^2(\Omega; \mathbb{R})$ is a linear space.

Next, define $\langle \cdot, \cdots \rangle_{\mathfrak{L}^2} : \mathfrak{L}^2(\Omega; \mathbb{R}) \to \mathbb{R}$ by

$$\langle X, Y \rangle_{\mathfrak{L}^2} \equiv E\left|XY\right|. \tag{2.21}$$

Exercise 2.3.2. Prove that (2.21) defines an inner product on $\mathcal{L}^2(\Omega; \mathbb{R})$.

This inner product induces the following norm on $\mathcal{L}^2(\Omega; \mathbb{R})$:

$$\|X\|_{\mathcal{L}^2(\Omega;\mathbb{R})} \equiv \langle X, X \rangle_{\mathcal{L}^2}^{\frac{1}{2}} = \left(E|X|^2 \right)^{\frac{1}{2}}. \tag{2.22}$$

Exercise 2.3.3. Argue directly that (2.22) defines a norm on $\mathcal{L}^2(\Omega; \mathbb{R})$.

At this point, we conclude that $\mathcal{L}^2(\Omega; \mathbb{R})$ is an inner product space. But, does it earn the rank of Hilbert space, that is, is it complete with respect to the norm (2.22)? Before answering this question, we must make precise the notions of a convergent sequence and a Cauchy sequence in $\mathcal{L}^2(\Omega; \mathbb{R})$.

Definition 2.3.2. A sequence of $\mathbb{R}$-valued random variables $\{X_n\}$ is said to be
i.) *convergent to X in $\mathcal{L}^2(\Omega; \mathbb{R})$* (or *convergent in mean square*) if $\forall \varepsilon > 0, \exists N \in \mathbb{N}$ such that

$$n \geq N \implies \|X_n - X\|_{\mathcal{L}^2}^2 = E|X_n - X|^2 < \varepsilon.$$

We write $X_n \longrightarrow X$ in $\mathcal{L}^2$ (or equivalently, $\|X_n - X\|_{\mathcal{L}^2} \longrightarrow 0$).
ii.) *Cauchy in $\mathcal{L}^2(\Omega; \mathbb{R})$* (or *Cauchy in mean square*) if $\forall \varepsilon > 0, \exists N \in \mathbb{N}$ such that

$$m, n \geq N \implies \|X_m - X_n\|_{\mathcal{L}^2}^2 = E|X_m - X_n|^2 < \varepsilon.$$

Many of the following limit properties are standard and can be proven using arguments similar to those used in Section 1.5.2, but with the absolute value replaced by the $\mathcal{L}^2$-norm. (See [**212**], for instance, for more details.)

Proposition 2.3.3. (Limit Theorems for $\mathcal{L}^2(\Omega; \mathbb{R})$)
Let $\{X_n\}$ and $\{Y_n\}$ be sequences in $\mathcal{L}^2(\Omega; \mathbb{R})$ and let $\{\alpha_n\} \subset \mathbb{R}$.
i.) *If $\{X_n\}$ is convergent in $\mathcal{L}^2(\Omega; \mathbb{R})$, then $\{X_n\}$ is bounded in $\mathcal{L}^2(\Omega; \mathbb{R})$.*
ii.) *For every $Z \in \mathcal{L}^2(\Omega; \mathbb{R})$, if $\alpha_n \longrightarrow \alpha$ in $\mathbb{R}$, then $\alpha_n Z \longrightarrow \alpha Z$ in $\mathcal{L}^2(\Omega; \mathbb{R})$.*
iii.) *For every $\alpha, \beta \in \mathbb{R}$, if $X_n \longrightarrow X$ and $Y_n \longrightarrow Y$ in $\mathcal{L}^2(\Omega; \mathbb{R})$, then $(\alpha X_n \pm \beta Y_n) \longrightarrow \alpha X \pm \beta Y$ in $\mathcal{L}^2(\Omega; \mathbb{R})$.*
iv.) *If $X_n \longrightarrow X$ in $\mathcal{L}^2(\Omega; \mathbb{R})$, then $E[X_n] \longrightarrow E[X]$ in $\mathbb{R}$.*
v.) *If $X_n \longrightarrow X$ and $Y_n \longrightarrow Y$ in $\mathcal{L}^2(\Omega; \mathbb{R})$, then $E[X_nY_n] \longrightarrow E[XY]$ in $\mathbb{R}$.*

Proof. Linearity and the Cauchy-Schwarz inequality are the primary tools at work here. We shall prove (v) and leave the others for you to complete as an exercise. Let $\varepsilon > 0$. By (i) of this proposition, $\exists M > 0$ such that $\|Y_n\|_{\mathcal{L}^2} \leq M, \forall n \in \mathbb{N}$. By assumption, $\exists N_1, N_2 \in \mathbb{N}$ such that

$$n \geq N_1 \implies \|X_n - X\|_{\mathcal{L}^2} < \frac{\varepsilon}{2M}, \tag{2.23}$$

$$n \geq N_2 \implies \|Y_n - Y\|_{\mathcal{L}^2} < \frac{\varepsilon}{2(\|X\|_{\mathcal{L}^2} + 1)}. \tag{2.24}$$

Let $N = \max\{N_1, N_2\}$. Then, $n \geq N \implies$

$$
\begin{aligned}
|E[X_n Y_n] - E[XY]| &= |E[X_n Y_n] - E[XY_n] + E[XY_n] - E[XY]| \\
&= |E[(X_n - X)Y_n] + E[X(Y_n - Y)]| \\
&\leq |E[(X_n - X)Y_n]| + |E[X(Y_n - Y)]| \\
&\leq \left(E|X_n - X|^2\right)^{\frac{1}{2}} \left(E|Y_n|^2\right)^{\frac{1}{2}} + \\
&\quad \left(E|X|^2\right)^{\frac{1}{2}} \left(E|Y_n - Y|^2\right)^{\frac{1}{2}} \\
&< \varepsilon.
\end{aligned}
$$

$\square$

Exercise 2.3.4. Prove the remaining parts of Prop. 2.3.3.

Exercise 2.3.5. If $\forall i \in \{1, \ldots, N\}$, $(\alpha_i)_m \longrightarrow \alpha_i^\star$ in $\mathbb{R}$ and $(X_i)_m \longrightarrow X_i^\star$ in $\mathcal{L}^2(\Omega; \mathbb{R})$ as $m \longrightarrow \infty$, must $E\left[\sum_{i=1}^N (\alpha_i)_m (X_i)_m\right]$ have a limit in $\mathcal{L}^2(\Omega; \mathbb{R})$ as $m \longrightarrow \infty$? If so, what is the limit? Prove your assertion.

Exercise 2.3.6. Assume that $f : \mathbb{R} \to \mathbb{R}$ is uniformly continuous. Prove that if $\{X_n\}$ is Cauchy in $\mathcal{L}^2(\Omega; \mathbb{R})$, then $\{f(\|X_n\|_{\mathcal{L}^2})\}$ converges in $\mathbb{R}$.

Exercise 2.3.7. Prove that if $\{X_n\}$ converges in $\mathcal{L}^2(\Omega; \mathbb{R})$, then $\{X_n\}$ is Cauchy in $\mathcal{L}^2(\Omega; \mathbb{R})$. (In fact, this is true in *any* normed space.)

In order to prove that $\mathcal{L}^2(\Omega; \mathbb{R})$ is a Hilbert space with inner product (2.21), we need to establish the converse of Exer. 2.3.7, namely that every Cauchy sequence in $\mathcal{L}^2(\Omega; \mathbb{R})$ converges in $\mathcal{L}^2(\Omega; \mathbb{R})$. The proof of this fact resembles the proof that $(\mathbb{R}, |\cdot|)$ is complete. Summarizing, we have

Theorem 2.3.4. *The space* $\mathcal{L}^2(\Omega; \mathbb{R})$ *given by Def. 2.3.1 is a Hilbert space when equipped with the inner product (2.21) and induced norm (2.22).*

More generally, the space $\mathcal{L}^p(\Omega; \mathbb{R})$, where $p \geq 1$, is defined as in Def. 2.3.1 with the condition $E[X^2] < \infty$ replaced by the $E[X^p] < \infty$. While this space is not a Hilbert space unless $p = 2$, it is a Banach space when equipped with the norm

$$
\|X\|_{\mathcal{L}^p(\Omega; \mathbb{R})} \equiv (E|X|^p)^{\frac{1}{p}}. \tag{2.25}
$$

There are many other useful notions of convergence in the probabilistic setting. In fact, the richness of the structure enables us to define a wide variety of notions, each useful for different reasons. We mention a few of them below, along with some interpretation, and state, without proof, their interrelationships. (See **[212]** for a detailed discussion.) Of all of them, convergence in mean will be predominantly used in this text, with an occasional appeal to the other ones.

Definition 2.3.5. A sequence of random variables $\{X_n\}$ is said to be

i.) *convergent to X in p^{th} moment* $(p \geq 1)$ if $E|X_n|^p < \infty, \forall n \in \mathbb{N}, E|X|^p < \infty$, and $\forall \varepsilon > 0, \exists N \in \mathbb{N}$ such that

$$n \geq N \implies E|X_n - X|^p < \varepsilon.$$

We write $X_n \longrightarrow X$ in $\mathscr{L}^p(\Omega; \mathbb{R})$.

ii.) *convergent to X almost surely* (a.s.) if $\exists \mathscr{D} \in \mathscr{F}$ such that $\mathscr{P}(\mathscr{D}) = 0$ and $\forall \varepsilon > 0, \exists N \in \mathbb{N}$ such that $\forall \omega \in \Omega \setminus \mathscr{D}$,

$$n \geq N \implies |X_n(\omega) - X(\omega)| < \varepsilon.$$

We write $X_n \longrightarrow X$ a.s. $[\mathscr{P}]$.

iii.) *convergent to X in probability* if $\forall \varepsilon, \eta > 0, \exists N \in \mathbb{N}$ such that

$$n \geq N \implies \mathscr{P}(\{\omega \in \Omega : |X_n(\omega) - X(\omega)| \geq \varepsilon\}) < \eta.$$

We write $X_n \longrightarrow X$ in probability.

iv.) *convergent to X in distribution* if the sequence of distribution functions $\{F_{X_n}(x)\}$ converges pointwise (in $\mathbb{R}$) to the distribution function F_X at all points of continuity of F_X. We write $X_n \longrightarrow X$ in distribution.

Remarks.

1. Def. 2.3.5(ii) essentially means that the random sequence $\{X_n\}$ converges pointwise to X at all $\omega \in \Omega$ except for possibly those in a $\mathscr{P}$-null set. The condition can be written alternatively as

$$\mathscr{P}\left(\left\{\omega \in \Omega : \lim_{n \to \infty} X_n(\omega) = X(\omega)\right\}\right) = 1.$$

2. Some of the convergence types are easier to verify in practice than others. As such, it is often helpful to take advantange of the several interrelationships among the different types of convergence notions. Summarizing these relationships, assume that $1 \leq p \leq q \leq \infty$. The most straightforward string of relationships is as follows:

$$\text{(i) with } p \implies \text{(i) with } q \implies \text{(iii)} \implies \text{(iv)}. \tag{2.26}$$

Type (ii) can be woven into this string as well, in the sense that

$$\text{(ii)} \implies \text{(iii)} \implies \text{(iv)}. \tag{2.27}$$

3. Recall that if $\lim_{n \to \infty} f_n(t) = f(t)$ and $\lim_{n \to \infty} f_n(t) = g(t), \forall t \in \Omega$, (where all of the functions are real-valued and defined on Ω), then $f(t) = g(t), \forall t \in \Omega$, in the deterministic setting. This "uniqueness of limit" must be interpretted a bit more loosely in the probabilistic setting. Precisely, if f_n, f, and g are now random variables, then we interpret the uniqueness result as occurring "with probability 1," meaning that $f(t) = g(t)$ a.s.$[\mathscr{P}]$. As such, the equality need not occur $\forall t \in \Omega$, but rather $\exists \mathscr{D} \in \mathscr{F}$ such that $\mathscr{P}(\mathscr{D}) = 0$ and $f(t) = g(t), \forall t \in \Omega \setminus \mathscr{D}$.

Exercise 2.3.8. Assume that $1 < p \leq q < \infty$. Prove that if $X_n \longrightarrow X$ in $\mathfrak{L}^q(\Omega; \mathbb{R})$, then $X_n \longrightarrow X$ in $\mathfrak{L}^p(\Omega; \mathbb{R})$.

The corresponding "Cauchy variants" of the convergence types in Def. 2.3.5, as well as the existence of a subsequence that satisfies each condition, can be considered and appropriately woven into the interrelationship implications (2.26) and (2.27) to produce a massive implication diagram. (Try it!) Of these, we will need the following.

Proposition 2.3.6. *If $\{X_n\}$ is Cauchy in $\mathfrak{L}^2(\Omega; \mathbb{R})$, then there exists a subsequence $\{X_{n_k}\}$ for which $\exists Y \in \mathfrak{L}^2(\Omega; \mathbb{R})$ such that $\lim_{k \to \infty} X_{n_k}(\omega) = Y(\omega)$ a.s. $[\mathscr{P}]$.*

Before extending our discussion to $\mathbb{R}^N$-valued random variables, we state the following fact, which will be useful later in the text. (See **[20, 212]** for a proof.)

Proposition 2.3.7. *The mean square limit of a sequence of Gaussian random variables is itself a Gaussian random variable.*

2.4 $\mathbb{R}^N$-Valued Random Variables

As you work through this section, try to deliberately make connections to the corresponding development in $\mathbb{R}$ in Section 2.2 and pay particular attention to the nature of the modifications. Doing so will be helpful when extending the theory of stochastic evolution equations (SEEs) in $\mathbb{R}$ to the theory of SEEs in $\mathbb{R}^N$. Definition 2.2.1 can be extended in a natural way to the case when the codomain is $\mathbb{R}^N$, meaning that we now consider functions of the form $\mathbf{X} : \Omega \longrightarrow \mathbb{R}^N$, where

$$\mathbf{X}(\omega) = \langle X_1(\omega), \ldots, X_N(\omega) \rangle, \ \omega \in \Omega, \tag{2.28}$$

where the component functions $X_i \ (i = 1, \ldots, N)$ are random variables in the sense of Def. 2.2.1.

In order to make the definition precise, we need the inverse images of the "right sets" to belong to $\mathscr{F}$. For an $\mathbb{R}$-valued random variable X, we required that

$$X^{-1}(B) \in \mathscr{F}, \ \forall B \in \sigma\left((a, b] : a, b \in \mathbb{R}\right). \tag{2.29}$$

Exercise 2.4.1. Based on (2.29), what would a natural condition be for $\mathbb{R}^N$-valued random variables?

As it turns out, if we consider Cartesian products of the form

$$\mathscr{R} = (a_1, b_1] \times \ldots \times (a_N, b_N], \tag{2.30}$$

where $a_i, b_i \in \mathbb{R}$ with $a_i < b_i$, $\forall i \in \{1,\ldots,N\}$, then (2.29) translates as

$$\mathbf{X}^{-1}(\mathbf{B}) \in \mathscr{F}, \forall \mathbf{B} \in \sigma(\mathscr{G}), \tag{2.31}$$

where $\mathscr{G} = \{\mathbf{B} : \mathbf{B} \text{ is of the form (2.30)}\}$. Formally, we have

Definition 2.4.1. A function $\mathbf{X} : \Omega \longrightarrow \mathbb{R}^N$ for which (2.31) holds is an $\mathbb{R}^N$-*valued random variable* (or *random vector*). We often write $\mathbf{X} = \langle X_1,\ldots,X_N \rangle$, where the input ω (as written in (2.28)) has been suppressed.

We can also generalize this notion one step further to the case when the codomain is a separable Hilbert space. This will be discussed in Chapter 5.

Random vectors arise in practical situations in the same manner that nonrandom vectors do when more than one measurement is needed in order to describe a phenomenon. For instance, the trajectory of an object moving through a region in 3-space during a prescribed time interval can be described as a vector-valued function $t \mapsto \langle x(t), y(t), z(t) \rangle$. If the component functions are viewed as random variables (which is reasonable given that seemingly unpredictable wind gusts and other environmental factors can affect the motion of a moving object), then this function would be a *random* vector function $t \mapsto \langle x(t,\omega), y(t,\omega), z(t,\omega) \rangle$; and for any fixed time t_0, the vector $\langle x(t_0,\omega), y(t_0,\omega), z(t_0,\omega) \rangle$ is an $\mathbb{R}^3$-valued random variable.

The arithmetic operations of vectors in $\mathbb{R}^N$ and the calculus of $\mathbb{R}^N$-valued functions are all performed componentwise, as shown in Section 1.6.1. This suggests a natural extension of the notions developed for $\mathbb{R}$-valued random variables to the present setting. Using this fact, it is not difficult to deduce that the combinations of $\mathbb{R}$-valued random variables stated in Prop. 2.2.2 also hold for $\mathbb{R}^N$-valued random variables. (Convince yourself!) Also, the notion of a distribution function of an $\mathbb{R}^N$-valued random variable $\mathbf{X}$ is characterized in terms of its components $\{X_i\}$. Precisely, we have

Definition 2.4.2. Let $\mathbf{X} = \langle X_1,\ldots,X_N \rangle$ be an $\mathbb{R}^N$-valued random variable. The function $F_\mathbf{X} : \mathbb{R}^N \longrightarrow [0,1]$ defined by

$$F_\mathbf{X}(x_1,\ldots,x_N) = \mathscr{P}(\{\omega \in \Omega : X_1(\omega) \leq x_1 \text{ and } \ldots \text{ and } X_N(\omega) \leq x_N\}) \tag{2.32}$$

is the *(joint) distribution function* of $\mathbf{X}$.

As in the real-valued case, such a random vector $\mathbf{X}$ is defined uniquely by its distribution function $F_\mathbf{X}$ in the sense that

$$F_\mathbf{X} = F_\mathbf{Y} \implies \mathbf{X}(\omega) = \mathbf{Y}(\omega) \text{ a.s. } [\mathscr{P}],$$

which further simplifies to

$$X_i(\omega) = Y_i(\omega) \text{ a.s. } [\mathscr{P}], \forall i \in \{1,\ldots,N\}.$$

Of course, *proving* that $F_\mathbf{X} = F_\mathbf{Y}$ is generally nontrivial, unless you have a very nice representation formula for the distribution function.

Naturally, the definition of the expectation of an $\mathbb{R}^N$-valued random variable is as follows:

Definition 2.4.3. Let $\mathbf{X} = \langle X_1, \ldots, X_N \rangle$ be an $\mathbb{R}^N$-valued random variable.
i.) The *expectation* of $\mathbf{X}$, denoted $E[\mathbf{X}]$, is the nonrandom constant vector in $\mathbb{R}^N$ given by

$$E[\mathbf{X}] = \langle E[X_1], \ldots, E[X_N] \rangle. \tag{2.33}$$

ii.) The *variance* of $\mathbf{X}$, denoted $Var[\mathbf{X}]$, is the nonrandom constant vector in $\mathbb{R}^N$ given by

$$Var[\mathbf{X}] = \langle Var[X_1], \ldots, Var[X_N] \rangle. \tag{2.34}$$

Exercise 2.4.2.
i.) Explain carefully why Def. 2.4.3 is natural.
ii.) Explain why the relationship $Var[\mathbf{X}] = E\left[(\mathbf{X} - E[\mathbf{X}])^2\right]$ is valid when $\mathbf{X}$ is an $\mathbb{R}^N$-valued random variable.

Exercise 2.4.3. Assume that X is $n\left(0, \sigma_X^2\right)$ and Y is $n\left(0, \sigma_Y^2\right)$.
i.) Compute $E[\langle X, Y \rangle]$.
ii.) Compute $E[\langle \alpha_1 X + \beta_1, \alpha_2 Y + \beta_2 \rangle]$, where $\alpha_i, \beta_i \in \mathbb{R}$ $(i = 1, 2)$.
iii.) Compute $E[\langle X + Y, X - 2Y \rangle]$.
iv.) Explain why $\omega \mapsto \|\langle X(\omega), Y(\omega) \rangle\|_{\mathbb{R}^2}$ is an $\mathbb{R}$-valued random variable.
v.) How does $E[\|\langle X, Y \rangle\|_{\mathbb{R}^2}]$ compare to $\|E[\langle X, Y \rangle]\|_{\mathbb{R}^2}$?
vi.) Try to extract some properties of expectation for $\mathbb{R}^2$-valued random variables from these examples. Do you suspect these properties also apply to $\mathbb{R}^N$-valued random variables? (Compare to Prop. 2.2.8.)

The following properties are analogous to those established for $\mathbb{R}$-valued random variables (cf. Prop. 2.2.8.)

Proposition 2.4.4. (Properties of Expectation for $\mathbb{R}^N$-Valued Random Variables)

Let $\mathbf{X}$, $\mathbf{Y}$, $\mathbf{X}_1, \ldots, \mathbf{X}_m$ be $\mathbb{R}^N$-valued random variables on $(\Omega, \mathscr{F}, \mathscr{P})$ and $\{\alpha_k : k \in \mathbb{N}\}$ a sequence in $\mathbb{R}$.
i.) (Linearity) $E\left[\sum_{k=1}^m \alpha_k \mathbf{X}_k\right] = \sum_{k=1}^m \alpha_k E[\mathbf{X}_k]$.
ii.) $\|E[\mathbf{X}]\|_{\mathbb{R}^N} \leq E\|\mathbf{X}\|_{\mathbb{R}^N}$.
iii.) (Hölder's Inequality) *Let* $1 \leq p \leq q < \infty$ *be such that* $\frac{1}{p} + \frac{1}{q} = 1$. *If* $E\|\mathbf{X}\|_{\mathbb{R}^N}^p < \infty$ *and* $E\|\mathbf{X}\|_{\mathbb{R}^N}^q < \infty$, *then*

$$\left|E[\langle \mathbf{X}, \mathbf{Y} \rangle_{\mathbb{R}^N}]\right| \leq \left(E\|\mathbf{X}\|_{\mathbb{R}^N}^p\right)^{\frac{1}{p}} \left(E\|\mathbf{X}\|_{\mathbb{R}^N}^q\right)^{\frac{1}{q}}. \tag{2.35}$$

iv.) (Minkowski's Inequality) $\forall p > 1$, $E\left[\|\mathbf{X} + \mathbf{Y}\|_{\mathbb{R}^N}^p\right] \leq \left(E\|\mathbf{X}\|_{\mathbb{R}^N}^p\right)^{\frac{1}{p}} + \left(E\|\mathbf{Y}\|_{\mathbb{R}^N}^p\right)^{\frac{1}{p}}$.

v.) (Chebyshev's Inequality) $\forall \varepsilon, p > 0$, $E\|\mathbf{X}\|_{\mathbb{R}^N}^p \geq \varepsilon^p \mathscr{P}\left(\left\{\omega \in \Omega : \|\mathbf{X}(\omega)\|_{\mathbb{R}^N}^p \geq \varepsilon\right\}\right)$.

vi.) *If* $E\left[\|\mathbf{X}\|_{\mathbb{R}^N}^2\right] = 0$, *then* $X_i = 0$ *a.s.* $[\mathscr{P}]$, $\forall i \in \{1, \ldots, N\}$.

Exercise 2.4.4. Prove Prop. 2.4.4(i) and (v).

Exercise 2.4.5. Let $\alpha, \beta \in \mathbb{R}$ and $\mathbf{X}, \mathbf{Y}$ be $\mathbb{R}^N$-valued random variables. Prove:
i.) $\|E\left[\alpha \mathbf{X} + \beta \mathbf{Y}\right]\|_{\mathbb{R}^N} \leq |\alpha| E \|\mathbf{X}\|_{\mathbb{R}^N} + |\beta| E \|\mathbf{Y}\|_{\mathbb{R}^N}$.
ii.) Derive an estimate for $\|E\left[\alpha \mathbf{X} + \beta \mathbf{Y}\right]\|_{\mathbb{R}^N}^2$.

The following two statistical measures constitute an extension of Def. 2.2.12 that applies to $\mathbb{R}^N$-valued random variables.

Definition 2.4.5. Let $\mathbf{X} = \langle X_1, \ldots, X_N \rangle$ and $\mathbf{Y} = \langle Y_1, \ldots, Y_N \rangle$ be $\mathbb{R}^N$-valued random variables.
i.) The *correlation of* $\mathbf{X}$ *and* $\mathbf{Y}$ is the $N \times N$ matrix $\mathbf{r}(\mathbf{X}, \mathbf{Y})$ defined by

$$\mathbf{r}(\mathbf{X}, \mathbf{Y}) = \begin{bmatrix} E[X_1 Y_1] & \cdots & E[X_1 Y_N] \\ \vdots & \ddots & \vdots \\ E[X_N Y_1] & \cdots & E[X_N Y_N] \end{bmatrix} = \begin{bmatrix} r(X_1, Y_1) & \cdots & r(X_1, Y_N) \\ \vdots & \ddots & \vdots \\ r(X_N, Y_1) & \cdots & r(X_N, Y_N) \end{bmatrix}. \quad (2.36)$$

ii.) The *covariance of* $\mathbf{X}$ *and* $\mathbf{Y}$ is the $N \times N$ matrix $\mathbf{Cov}(\mathbf{X}, \mathbf{Y})$ is defined by

$$\mathbf{Cov}(\mathbf{X}, \mathbf{Y}) = \begin{bmatrix} Cov(X_1, Y_1) & \cdots & Cov(X_1, Y_N) \\ \vdots & \ddots & \vdots \\ Cov(X_N, Y_1) & \cdots & Cov(X_N, Y_N) \end{bmatrix}. \quad (2.37)$$

A Gaussian $\mathbb{R}$-valued random variable is characterized completely by its mean and variance. This extends in a natural way to an $\mathbb{R}^N$-valued Gaussian random variable $\mathbf{X}$, but we must account for all pairwise variances of the components of $\mathbf{X}$ because each contributes to its total variance.

Definition 2.4.6. An $\mathbb{R}^N$-valued random variable $\mathbf{X}$ is *Gaussian with mean* $\vec{\mu}$ *and covariance matrix* Σ, denoted by $n(\vec{\mu}, \Sigma)$, if its probability density is given by

$$f_{\mathbf{X}}(\mathbf{x}) = \frac{1}{(2\pi)^{\frac{N}{2}} \sqrt{\det \Sigma}} \exp\left(-\frac{1}{2}(\mathbf{x} - \vec{\mu}) \Sigma^{-1} (\mathbf{x} - \vec{\mu})^{\mathrm{T}}\right). \quad (2.38)$$

The *standard Gaussian* has mean $\vec{\mu} = \mathbf{0}$ and Σ is the identity element of $\mathbb{M}^N(\mathbb{R})$.

Exercise 2.4.6. Prove that if $\mathbf{X}$ is the standard Gaussian $\mathbb{R}^N$-valued random variable, then its density is the product of N one-dimensional $n(0, 1)$ densities.

Proposition 2.4.7. (Properties of an N-Dimensional Gaussian)
i.) *If* $\mathbf{X} : \Omega \longrightarrow \mathbb{R}^N$ *is* $n(\vec{\mu_X}, \Sigma_X)$, $\vec{\alpha}$ *is a constant* $m \times N$ *matrix, and* $\vec{\beta}$ *is an* $m \times 1$ *constant vector, then* $\mathbf{Y} = \vec{\alpha} \mathbf{X} + \vec{\beta}$ *is an m-dimensional Gaussian random variable wiht mean* $\vec{\mu_Y} = \vec{\alpha} \vec{\mu_X} + \vec{\beta}$ *and covariance* $\Sigma_Y = \vec{\alpha} (\vec{\alpha})^{\mathrm{T}} \Sigma_X$.
ii.) *An* $\mathbb{R}^N$-valued random variable $\mathbf{X} = \langle X_1, \ldots, X_N \rangle$ *is an N-dimensional Gaussian*

random variable iff $\forall k \in \{1, \ldots, N\}$, *there exist real constants* $a_k, (b_k)_1, \ldots, (b_k)_m$ *and* $\mathbb{R}$-*valued* $n(0,1)$ *random variables* $Z_1, \ldots, Z_n$ *such that* $\forall k \in \{1, \ldots, N\}$,

$$X_k = a_k + \sum_{j=1}^{m} (b_k)_j Z_j.$$

(That is, each component of $\mathbf{X}$ *is itself an* $\mathbb{R}$-*valued normal random variable.)*

Finally, the analog of $\mathcal{L}^2(\Omega; \mathbb{R})$ in this setting is constructed by making the usual change from the $\mathbb{R}$-norm to the $\mathbb{R}^N$-norm and making the appropriate computational modifications typical of those introduced in this section. Indeed, we have

Theorem 2.4.8. *The space* $\mathcal{L}^2\left(\Omega; \mathbb{R}^N\right)$ *given by*

$$\mathcal{L}^2\left(\Omega; \mathbb{R}^N\right) = \left\{ \mathbf{X} : \Omega \to \mathbb{R}^N : \mathbf{X} \text{ is a random variable with } E \left\| \mathbf{X} \right\|_{\mathbb{R}^N}^2 < \infty \right\} \quad (2.39)$$

equipped with the inner product

$$\langle \mathbf{X}, \mathbf{Y} \rangle_{\mathcal{L}^2\left(\Omega; \mathbb{R}^N\right)} \equiv E \left| \mathbf{X} \mathbf{Y}^T \right| \quad (2.40)$$

and induced norm

$$\left\| \mathbf{X} \right\|_{\mathcal{L}^2\left(\Omega; \mathbb{R}^N\right)} \equiv \left(E \left\| \mathbf{X} \right\|_{\mathbb{R}^N}^2 \right)^{\frac{1}{2}} \quad (2.41)$$

is a Hilbert space.

Exercise 2.4.7. Formulate and prove an extension of Prop. 2.3.3 for $\mathcal{L}^2\left(\Omega; \mathbb{R}^N\right)$.

All notions of convergence defined in Def. 2.3.5 can be formulated in a natural way for $\mathbb{R}^N$-valued random variables by simply replacing the $\mathbb{R}$-norm by the $\mathbb{R}^N$-norm, and interpretting the result componentwise. (Convince yourself that this is reasonable.) Moreover, all results stated in Section 2.3 extend without issue. We shall revisit these notions again in Chapter 5 when the need for a more general Hilbert space-valued random variable arises.

2.5 Conditional Probability and Independence

The probabilities associated with the events formed using the outcomes of an experiment can change if additional information is provided a priori and taken into account. Indeed, suppose that an event $B \in \mathscr{F}$ such that $\mathscr{P}(B) > 0$ is known to occur in the sense that any outcome $\omega \in \Omega$ is observed either to belong to B or not to belong to B. This is not as powerful as actually observing the outcome ω itself, but it does provide some information. A natural question to ask is for any $A \in \mathscr{F}$, what is the probability of A occurring now, given that we can now use information provided by knowing that B has occurred? Consider the following exercise.

Exercise 2.5.1. Consider the experiment of rolling two six-sided balanced dice with faces labeled as 1, 2, 3, 4, 5, and 6. Record the outcome as (Roll 1, Roll 2).
i.) What is the sample space Ω? For any $\omega \in \Omega$, compute $\mathscr{P}(\{\omega\})$.
ii.) Now, suppose that it is known a priori that a pair (i.e., the same number occurs on both dice) is NOT rolled.
   a.) Express this information as a specific event B and compute $\mathscr{P}(B)$.
   b.) In effect, knowing that B has occurred changes the sample space of our experiment to $\Omega \setminus B$. Why?
   c.) Let $A = \{\omega \in \Omega : \exists i \in \{1,\ldots,6\}$ such that $\omega = (1,i)\}$. Compute $\mathscr{P}(A)$ using $\Omega \setminus B$ as the new sample space. How does this compare to $\frac{\mathscr{P}(A \cap B)}{\mathscr{P}(B)}$, where the probabilities are computed using the entire sample space Ω?

   This suggests the following definition.

Definition 2.5.1. Let $A, B \in \mathscr{F}$ be such that $\mathscr{P}(B) > 0$. The *conditional probability of A given B*, denoted $\mathscr{P}(A|B)$, is defined as $\mathscr{P}(A|B) = \frac{\mathscr{P}(A \cap B)}{\mathscr{P}(B)}$.

Proposition 2.5.2. *If $B \in \mathscr{F}$ is such that $\mathscr{P}(B) > 0$, then $\mathscr{P}(\cdot|B)$ is a probability measure on $(\Omega, \mathscr{F})$.*

Proof. Certainly, $A \mapsto \mathscr{P}(A|B)$ is defined on $\mathscr{F}$ and is nonnegative, $\forall A \in \mathscr{F}$. Also, because $\Omega \cap B = B$, it follows that $\mathscr{P}(\Omega|B) = 1$. (Why?)
   Assume that $\{A_i : i \in I\}$ is a countable collection of pairwise disjoint events in $\mathscr{F}$. Observe that

$$
\begin{aligned}
\mathscr{P}\left(\bigcup_{i \in I} A_i \,\middle|\, B\right) &= \frac{\mathscr{P}\left(\left(\bigcup_{i \in I} A_i\right) \cap B\right)}{\mathscr{P}(B)} \\
&= \frac{\mathscr{P}\left(\bigcup_{i \in I} (A_i \cap B)\right)}{\mathscr{P}(B)} \text{ (Why?)} \\
&= \frac{\sum_{i \in I} \mathscr{P}(A_i \cap B)}{\mathscr{P}(B)} \text{ (Why?)} \\
&= \sum_{i \in I} \frac{\mathscr{P}(A_i \cap B)}{\mathscr{P}(B)} \\
&= \sum_{i \in I} \mathscr{P}(A_i|B),
\end{aligned}
$$

as needed.   □

Exercise 2.5.2. Let $(\Omega, \mathscr{F}, \mathscr{P})$ be a complete probability space and assume that Ω is finite. Let $\omega_0 \in \Omega$ and define the event $B = \{\omega_0\}$. Compute $\mathscr{P}(A|B)$, $\forall A \in \mathscr{F}$.

   Intuitively, two events A and B are "independent" if the knowledge of the event A having occurred does not change the probability of the occurrence of B, and vice versa. As such, $\mathscr{P}(B|A) = \mathscr{P}(B)$ and $\mathscr{P}(A|B) = \mathscr{P}(A)$. In such case, observe that

$$\mathscr{P}(A \cap B) = \mathscr{P}(A|B)\mathscr{P}(B) = \mathscr{P}(B|A)\mathscr{P}(A) = \mathscr{P}(A)\mathscr{P}(B).$$

This motivates the following definition.

Definition 2.5.3. Two events A and B are *independent* if $\mathscr{P}(A \cap B) = \mathscr{P}(A)\mathscr{P}(B)$.

This notion is sufficient when defining the independence of two specific events $A, B \in \mathscr{F}$, but if we want to know if the information "provided by a random variable X_1" has any influence on the likelihood of another random variable attaining its values, we need to develop a better definition that correctly identifies the meaning of such influence. To this end, we introduce the following notion.

Definition 2.5.4. Let $A, B \in \mathscr{F}$. The σ-algebras $\sigma(\{A\})$ and $\sigma(\{B\})$ are *independent* if $\forall C_A \in \sigma(\{A\})$ and $\forall C_B \in \sigma(\{B\})$, $\mathscr{P}(C_A \cap C_B) = \mathscr{P}(C_A)\mathscr{P}(C_B)$.

Clearly, Def. 2.5.4 implies Def. 2.5.3. (Why?) But, a moment's thought suggests that, in fact, the reverse implication also holds. Indeed, recall that $\sigma(\{A\}) = \left\{\varnothing, A, \widetilde{A}, \Omega\right\}$ and $\sigma(\{B\}) = \left\{\varnothing, B, \widetilde{B}, \Omega\right\}$. Verifying that Def. 2.5.3 implies Def. 2.5.4 entails verifying that the condition in Def. 2.5.3 holds for all sixteen pairwise combinations of sets in $\sigma(\{A\})$ and $\sigma(\{B\})$. For instance, consider the following exercise.

Exercise 2.5.3. Assuming that $\mathscr{P}(A \cap B) = \mathscr{P}(A)\mathscr{P}(B)$, show that $\mathscr{P}\left(\widetilde{A} \cap B\right) = \mathscr{P}(\widetilde{A})\mathscr{P}(B)$.

The other fifteen combinations are proven in a similar fashion. As such, we conclude that Definitions 2.5.3 and 2.5.4 are equivalent. Good!

Exercise 2.5.4. How would you extend Def. 2.5.3 to say that *three* events A_1, A_2, and A_3 are mutually independent?

We can generalize Def. 2.5.4 in a natural way to <u>any</u> finite collection of sub σ-algebras of $\mathscr{F}$, and they need not be generated by a single event. Precisely, we have

Definition 2.5.5. Suppose that $\{\mathscr{F}_i : i = 1, \ldots, m\}$ is a collection of σ-algebras on Ω such that $\mathscr{F}_i \subset \mathscr{F}$, $\forall i \in \{1, \ldots, m\}$. We say that $\{\mathscr{F}_i : i = 1, \ldots, m\}$ are *independent* if

$$\mathscr{P}\left(\bigcap_{i=1}^{m} A_i\right) = \prod_{i=1}^{m} \mathscr{P}(A_i), \forall A_i \in \mathscr{F}_i. \tag{2.42}$$

How can we use this to define the independence of two random variables? Well, remember that a random variable is a Borel measurable function and it is completely defined by its action on the Borel sets. Let $X_1, X_2 : \Omega \longrightarrow \mathbb{R}$ be random variables and let

$$\mathscr{F}_1 = \left\{ X_1^{-1}(B_1) : B_1 \in \sigma(\mathscr{G}) \right\},$$
$$\mathscr{F}_2 = \left\{ X_2^{-1}(B_2) : B_2 \in \sigma(\mathscr{G}) \right\}.$$

Certainly, $\mathscr{F}_1 \subset \mathscr{F}$ and $\mathscr{F}_2 \subset \mathscr{F}$. So, by Def. 2.5.5, $\mathscr{F}_1$ and $\mathscr{F}_2$ are independent iff

$$\mathscr{P}\left(\left\{ \omega \in \Omega : X_1^{-1}(B_1) \cap X_2^{-1}(B_2) \right\} \right) =$$
$$\mathscr{P}\left(\left\{ \omega \in \Omega : \omega \in X_1^{-1}(B_1) \right\} \right) \cdot \mathscr{P}\left(\left\{ \omega \in \Omega : \omega \in X_2^{-1}(B_2) \right\} \right),$$

$\forall B_1, B_2 \in \sigma(\mathscr{G})$. This is equivalent to saying

$$\mathscr{P}\left(\left\{ \omega \in \Omega : X_1(\omega) \in B_1 \text{ and } X_2(\omega) \in B_2 \right\} \right) =$$
$$\mathscr{P}\left(\left\{ \omega \in \Omega : X_1(\omega) \in B_1 \right\} \right) \cdot \mathscr{P}\left(\left\{ \omega \in \Omega : X_2(\omega) \in B_2 \right\} \right), \qquad (2.43)$$

$\forall B_1, B_2 \in \sigma(\mathscr{G})$. As such, we have the following definition.

Definition 2.5.6. Let $X_1, X_2 : \Omega \longrightarrow \mathbb{R}$ be random variables. We say that X_1 and X_2 are *independent* if (2.43) holds, $\forall B_1, B_2 \in \sigma(\mathscr{G})$.

Independence of random variables is a very important and powerful property in the underlying theory leading up to our study of stochastic evolution equations (SEEs). Two very important properties are stated below. (See **[404]** for a proof.)

Proposition 2.5.7. *Assume that $X_1, X_2 : \Omega \longrightarrow \mathbb{R}$ are independent random variables.*
i.) *If $E|X_1| < \infty$ and $E|X_2| < \infty$, then $E[X_1 X_2] = E[X_1]E[X_2]$.*
ii.) *If $E|X_1|^2 < \infty$ and $E|X_2|^2 < \infty$, then $Cov(X_1, X_2) = 0$.*

Exercise 2.5.5. Interpret Prop. 2.5.7(ii).

Exercise 2.5.6. Prove that if $X_1, \ldots, X_m$ are independent $\mathbb{R}$-valued random variables, then $Var\left(\sum_{i=1}^{m} X_i \right) = \sum_{i=1}^{m} Var(X_i)$.

Exercise 2.5.7. Assume that for each $i \in \{1, \ldots, m\}$, X_i is $n\left(\mu_i, \sigma_i^2 \right)$. Prove that if $\{X_i : i = 1, \ldots, m\}$ are independent, then $X = X_1 + \ldots + X_m$ is $n\left(\sum_{i=1}^{m} \mu_i, \sum_{i=1}^{m} \sigma_i^2 \right)$.

The following result is very powerful and used often in approximation arguments.

Theorem 2.5.8. (Central Limit Theorem)
Suppose that $\{X_i : i \in \mathbb{N}\}$ are independent $\mathbb{R}$-valued random variables all possessing the same distribution with mean μ and variance $\sigma^2 < \infty$. For every $n \in \mathbb{N}$, define $Z_n = \frac{\sum_{i=1}^{n} X_i - n\mu}{\sigma\sqrt{n}}$. Then,

$$\lim_{n \longrightarrow \infty} \underbrace{\mathscr{P}\left(\left\{ \omega \in \Omega : Z_n(\omega) \leq x \right\} \right)}_{=F_{Z_n}(x)} = \Phi(x), \forall x \in \mathbb{R},$$

where Φ is the distribution function for the one-dimensional standard Gaussian random variable.

An elementary illustration of Thrm. 2.5.8 is that the sequence of densities of standardized discrete $b(N,p)$ random variables converges to the density for the standard Gaussian (cf. Exer. 2.2.11).

2.6 Conditional Expectation — A Very Quick Description

For a more detailed discussion of the material in this section, refer to **[126, 302]**. We know that the expectation of X is the average value of X over the set of all possible outcomes and is given by

$$E[X] = \frac{1}{\mathscr{P}(\Omega)} \int_{\Omega} X(\omega) d\mathscr{P}. \tag{2.44}$$

In the absence of other information, this is the best estimate we have for $X(\omega)$, for any $\omega \in \Omega$. We saw in our discussion of conditional probability that the probability assignments to the events $A \in \mathscr{F}$ can change if it is known a priori that an event B with positive probability has occurred. Indeed, such knowledge enables you to shrink the sample space from Ω to B. As such, it stands to reason that such knowledge would affect the expected value of a random variable $X : \Omega \longrightarrow \mathbb{R}$. Indeed, based on (2.44), it is reasonable to compute the *conditional* expectation of X given that B has occurred by

$$E[X|B] = \frac{1}{\mathscr{P}(B)} \int_{B} X(\omega) d\mathscr{P}. \tag{2.45}$$

This quantity is just a real number, but it does provide a more informed approximation of X based on the information provided. But, what if the structure of the information known to have occurred is more complicated? For instance, we might know the result of calculating a different random variable $Y : \Omega \longrightarrow \mathbb{R}$ and this knowledge might help to provide a better idea of the expected value of X.

For definiteness, suppose that $\mathrm{rng}(Y) = \{y_1, y_2, y_3\}$. Then,

$$\Omega = Y^{-1}(\{y_1\}) \cup Y^{-1}(\{y_2\}) \cup Y^{-1}(\{y_3\}), \tag{2.46}$$

and the sets on the right-hand side of (2.46) are pairwise disjoint. (Why?) Now, suppose that conducting the experiment produces the outcome ω_0. We wish to estimate the value of $X(\omega_0)$. Does the information provided by Y (that is, the partition of Ω that its values induce) help refine the space in such a way as to improve our approximation of $X(\omega_0)$? In short, yes! Here's why. Note that $\omega_0 \in \Omega$ and belongs to precisely <u>one</u> of three sets on the right-hand side of (2.46), and having the information provided by Y a priori enables us to restrict the sample space to one of the pre-images $Y^{-1}(\{y_i\})$ $(i = 1,2,3)$. Precisely, $\forall i \in \{1,2,3\}$, if $\omega_0 \in Y^{-1}(\{y_i\})$, then

$$X(\omega_0) \approx \frac{1}{\mathscr{P}(Y^{-1}(\{y_i\}))} \int_{Y^{-1}(\{y_i\})} X(\omega) d\mathscr{P} = E\left[X \mid Y^{-1}(\{y_i\})\right]. \tag{2.47}$$

We summarize this by writing $E[X|Y]$. Note that the approximation in (2.47) changes depending on which pre-image $Y^{-1}(\{y_i\})$ the outcome ω_0 belongs. As such, $E[X|Y]$ is itself a random variable. Furthermore, it can be shown that it is the "best approximation of X" in the sense that

$$\|X - E[X|Y]\|_{\mathcal{L}^2(\Omega;\mathbb{R})} \leq \|X - Z\|_{\mathcal{L}^2(\Omega;\mathbb{R})}, \forall Z \in \mathcal{L}^2(\Omega;\mathbb{R}).$$

This only provides us with a glimpse into how one goes about defining the notion of conditional expectation. We will not need to compute with $E[X|Y]$ directly, although it is a powerful computational tool. Rather, for the purposes of our discussion, we shall be content with the following intuitive interpretation of conditional expectation: $E[X|\mathscr{A}]$ is a random variable that provides the best approximation of X based on the information provided by $\mathscr{A}$. Here, $\mathscr{A}$ can be a single event, a random variable, or a σ-algebra of events. We consider two extreme scenarios below.

Two Extreme Scenarios:
1. *The information provided need not improve the approximation.*
i.) $E[X|\Omega](\omega) = E[X], \forall \omega \in \Omega$, by (2.45). This is the crudest approximation of X. Note that the right-hand side is a real number independent of ω.
ii.) If X and Y are independent, then based on Def. 2.5.6, the information provided by Y cannot improve the approximation of X. As such, our best approximation of X remains $E[X]$. That is, $E[X|Y] = E[X]$. Moreover, because $\sigma(Y)$ provides no more information than Y (Why?), it follows that $E[X|\sigma(Y)] = E[X]$.
iii.) As an extension of (ii), if $\mathscr{A}$ is any σ-algebra that provides no information about X, then $E[X|\mathscr{A}] = E[X]$.

2. *The information provided renders the situation deterministic.*
If the values $\{(\omega, X(\omega)) : \omega \in \Omega\}$ can be extracted from the information with which you are provided a priori, then the random variable X is rendered a constant with respect to expectation. Indeed, in such case, the random variable that best approximates X is naturally X itself. This can occur as follows:
i.) $E[X|\sigma(X)](\omega) = X(\omega), \forall \omega \in \Omega$.
ii.) More generally, if we are trying to approximate an entire collection of random variables, indexed by time, say $\{X(t;\omega) : t \in [0,T], \omega \in \Omega\}$, then $\forall t \in [0,T]$,

$$E\left[X(t) \left| \bigcup_{0 \leq s \leq t} \sigma(X(s)) \right. \right] = X(t).$$

A more difficult question is determining the form of the best approximation when the information provided a priori falls somewhere in between these two extremes. Looking ahead, we will need to compute

$$E\left[X(t) \left| \sigma\left(\bigcup_{0 \leq \tau \leq s} X(\tau) \right) \right. \right],$$

where $0 < s < t$. We will revisit this in the next section.

We end this section with a list of some commonly used properties of conditional expectation. (Refer to any of the references mentioned at the beginning of this chapter for proofs and a more detailed discussion of conditional expectation.)

Proposition 2.6.1. (Properties of Conditional Expectation)
Let $(\Omega, \mathcal{F}, \mathcal{P})$ be a complete probability space and $X, Y : \Omega \longrightarrow \mathbb{R}$ random variables.
i.) (Linearity) *If $\alpha, \beta \in \mathbb{R}$, then $E[\alpha X + \beta Y | \mathcal{A}] = \alpha E[X | \mathcal{A}] + \beta E[Y | \mathcal{A}]$*
ii.) $E[X] = E[E[X | \mathcal{A}]]$
iii.) *If $\sigma(X) \subset \mathcal{A}$, then $E[XY | \mathcal{A}] = XE[Y | \mathcal{A}]$*
iv.) (Jensen) $|E[X | \mathcal{A}]| \leq E[|X| | \mathcal{A}]$

2.7 Stochastic Processes

For simplicity, the discussion is presented in the real-valued setting. The notions extend to the $\mathbb{R}^N$-setting by appealing to the discussion in Section 2.4 and making the natural modifications.

The focus of this text is the study of the evolution of the state of phenomena (e.g., concentration of chemicals, temperature of a medium, population density, etc.) whose behavior is governed by an evolution equation into which randomness has somehow been incorporated. In the deterministic setting studied in Volume 1, the solution process of such an equation was a function $u : [0, T] \longrightarrow \mathcal{H}$, where $\mathcal{H}$ was an appropriately chosen Hilbert space. (The case when $\mathcal{H} = \mathbb{R}$ corresponds to the most familiar case of an ordinary differential equation (ODE).) Introducing randomness into the evolution equation, possibly by way of modeling a physical parameter as a random quantity or by incorporating small random fluctuations into an external force acting on the system, will have the immediate effect of rendering each value $u(t)$ as a random variable $\omega \mapsto (u(t))(\omega)$ defined on some underlying probability space $(\Omega, \mathcal{F}, \mathcal{P})$ serving as an abstract framework of our experiment. As such, the study of <u>stochastic</u> evolution equations is quite a bit richer than the study of deterministic evolution equations in that the solution to an SEE is a collection of random variables $\{u(t; \cdot) : \Omega \longrightarrow \mathcal{H} \mid t \in [0, T]\}$ whose properties, both analytical and statistical, we seek to understand. The purpose of this section is to make precise the relevant notions of such collections of random variables and to formulate a calculus for them that naturally extends the operations of $\mathfrak{L}^2(\Omega; \mathbb{R})$. We begin with:

Definition 2.7.1. Let $(\Omega, \mathcal{F}, \mathcal{P})$ be a complete probability space. A *(real) stochastic process* is a collection of random variables $\mathcal{S} = \{X(t) : \Omega \longrightarrow \mathbb{R} \mid t \in I\}$, where $I \subset \mathbb{R}$. If I is a countable set, the stochastic process is called *discrete*; otherwise, it is said to be *continuous*.

It is customary, in our evolution equations context, to think of t as time. We will be interested in continuous stochastic processes the vast majority of the time. We will only consider the discrete case when formulating illustrative motivational examples.

Notation. For every $t \in I$, $X(t)$ is a random variable on Ω. As such, $X(t)$ assigns a real number to each $\omega \in \Omega$. We denote the corresponding output by $X(t)(\omega)$. In practice, the dependence on ω is typically suppressed with the understanding that the abbreviated notation $X(t)$ really means $X(t)(\omega)$. However, we shall often study the interplay between stochastic and deterministic evolution equations. As such, using this abbreviated notation can cause no end of confusion unless you are already familiar with the setting. To avoid such confusion, we shall view $\mathscr{S}$ as a function from $I \times \Omega$ into $\mathbb{R}$, meaning

$$\mathscr{S}(t, \omega) \equiv X(t)(\omega) \equiv X(t; \omega), \tag{2.48}$$

where t tells us which random variable in $\mathscr{S}$ to use and ω tells us the outcome at which to calculate $X(t)$. While more cumbersome, we will use the last term in (2.48) in place of just $X(t)$ when referring to that particular element of $\mathscr{S}$. When we refer to the stochastic process $\mathscr{S}$ as a single entity, we <u>will</u> suppress ω and typically just write X, for brevity.

Remarks.
1. For each fixed $t_0 \in I$, $\omega \mapsto X(t_0; \omega)$ is a random variable on Ω. As such, it has a distribution, expectation, variance, etc.
2. For each fixed $\omega_0 \in \Omega$, the real-valued function $t \mapsto X(t; \omega_0)$ is a *sample path* of the stochastic process $\mathscr{S}$. It represents the function obtained if the outcome $\omega_0 \in \Omega$ is chosen initially. It is one of (possibly infinitely) many realizations of the stochastic process $\mathscr{S}$.

Exercise 2.7.1. What does it mean for a sample path $t \mapsto X(t; \omega_0)$ to be continuous on I?

How do we go about studying the nature of a continuous stochastic process? As time moves from t to $t + \triangle t$, an entire continuum of random variables, each characterized by a massive collection of sample paths, is produced. While it might seem that having the knowledge of a density of $X(s)$, for every $s \in [t, t + \triangle t]$ would certainly suffice to completely characterize the stochastic process $\mathscr{S}$ on this interval, gaining such knowledge is generally intractable. Moreover, a moment's thought suggests that we really seek to understand how the properties change (or evolve) over time. We seek a distribution, of sorts, of the stochastic process $\mathscr{S}$. Certainly, this should involve the distribution of the individual random variables $X(s)$, but it should be constructed in a manner that covers all possibilities while remaining manageable. As it turns out, the collection of all finite-dimensional joint distribution functions, defined below, completely characterize the process when coupled with certain smoothness conditions, specified by Kolmogorov's theorem. (See [142] for a good discussion.)

Definition 2.7.2. Let $\mathscr{S} = \{X(t; \omega) \,|\, t \in I, \omega \in \Omega\}$, $m \in \mathbb{N}$, and $\{t_1, \dots, t_m\} \subset I$ be such that $t_1 < t_2 < \dots < t_m$. The collection of all joint probability distributions of

$\langle X(t_1; \cdot), X(t_2; \cdot), \ldots, X(t_m; \cdot) \rangle$, taken over all values of m and corresponding sets $\{t_1, \ldots, t_m\} \subset I$, is the collection of *finite-dimensional joint distributions of $\mathscr{S}$*.

Now that a stochastic process can be characterized by at least a somewhat verifiable condition, the next question to ask is how we can determine if two stochastic processes are "equivalent." Based on the notion of equality of two random variables (which coincides with usual function equality with the exception that inequality can occur on a $\mathscr{P}$-null set), we have the following definition:

Definition 2.7.3. Two stochastic processes $\mathscr{S} = \{X(t; \omega) \,|\, t \in I, \omega \in \Omega\}$ and $\mathscr{S}^* = \{X^*(t; \omega) \,|\, t \in I, \omega \in \Omega\}$ are *equivalent* if $\forall t \in I$,

$$\mathscr{P}(\{\omega \in \Omega : X(t; \omega) \neq X^*(t; \omega)\}) = 0. \qquad (2.49)$$

In such case, we say that $\mathscr{S}$ is a *version of $\mathscr{S}^*$*.

Exercise 2.7.2. Formulate and interpret Def. 2.7.3 for $\mathbb{R}^N$-valued stochastic processes.

A small miracle is that if a stochastic process $\mathscr{S} = \{X(t; \omega) \,|\, t \in I, \omega \in \Omega\}$ is sufficiently smooth (made precise in the theorem to follow), then you can find another stochastic process $\mathscr{S}^* = \{X^*(t; \omega) \,|\, t \in I, \omega \in \Omega\}$ such that almost all sample paths of $\mathscr{S}^*$ are uniformly continuous on I <u>and</u> $\mathscr{S}^*$ is a version of $\mathscr{S}$. Of course, the key is verifying the "smoothness condition," as stated below.

Theorem 2.7.4. (Kolmogorov's Criterion)
Let $\mathscr{S} = \{X(t; \omega) \,|\, t \in [a,b], \omega \in \Omega\}$ be a real stochastic process, where $[a,b] \subset [0, \infty)$. If $\exists \alpha, \beta, \eta > 0$ for which

$$E |X(t+h; \cdot) - X(t; \cdot)|^\alpha \leq \eta h^{1+\beta}, \qquad (2.50)$$

for all $h > 0$ for which $t + h \in [a,b]$, then there exists a stochastic process $\mathscr{S}^* = \{X^*(t; \omega) \,|\, t \in [a,b], \omega \in \Omega\}$ such that $t \mapsto X^*(t; \omega)$ is a continuous function on $[a,b]$, for almost all $\omega \in \Omega$. Moreover, $\mathscr{S}^*$ is equivalent to $\mathscr{S}$. (More succinctly, we say that $\mathscr{S}$ has a continuous version.)

We impose sufficiently regular conditions on all quantities upon which a stochastic process under investigation depends so that we are always assured of the existence of a continuous version. Showing this, of course, is nontrivial, and we will not be concerned with the details. Rather, we will provide references when appropriate.

There are various ways in which to categorize stochastic processes. Some common ones that will arise in our discussion are collected in the following definition.

Definition 2.7.5. A stochastic process $\mathscr{S} = \{X(t; \omega) \,|\, t \in [a,b], \omega \in \Omega\}$
i.) is *stationary* if $\langle X(t_1; \cdot), \ldots, X(t_n; \cdot) \rangle$ and $\langle X(t_1 + h; \cdot), \ldots, X(t_n + h; \cdot) \rangle$ have the same joint distribution, for all choices of $a \leq t_1 < \ldots < t_n \leq b$ and $h > 0$ for which $a \leq t_1 + h < \ldots < t_n + h \leq b$.
ii.) has *stationary increments* if for all t, s, h for which $t + h$ and $s + h$ are in $[a,b]$,

$X(t;\cdot) - X(s;\cdot)$ has the same distribution as $X(t+h;\cdot) - X(s+h;\cdot)$.

iii.) has *independent increments* if for all choices of $a \le t_1 < \ldots < t_n \le b$, $X(t_2;\cdot) - X(t_1;\cdot)$, $X(t_3;\cdot) - X(t_2;\cdot), \ldots, X(t_n;\cdot) - X(t_{n-1};\cdot)$ are mutually independent random variables.

iv.) is *Gaussian* if for all choices of $a \le t_1 < \ldots < t_n \le b$ and $\forall \{b_1, \ldots, b_n\} \subset \mathbb{R}$, $\sum_{i=1}^{n} b_i X(t_i;\cdot)$ is a Gaussian random variable. (That is, every finite linear combination of random variables in $\mathscr{S}$ is Gaussian.)

v.) is *Markov* if given the value $X(t_0;\cdot)$, $X(u;\cdot)$ is independent of $X(v;\cdot)$, $\forall u < t_0 < v$. (That is, our ability to predict the future behavior of the process given the present state is not enhanced by the knowledge of the past history.)

Exercise 2.7.3.
i.) Prove that if $\mathscr{S}$ has independent increments, then these increments must be orthogonal in $\mathcal{L}^2(\Omega;\mathbb{R})$ in the sense

$$E\left[(X(t_i;\cdot) - X(t_{i-1};\cdot))(X(t_j;\cdot) - X(t_{j-1};\cdot))\right] = \begin{cases} 0, & i \ne j, \\ E|X(t_i;\cdot) - X(t_{i-1};\cdot)|^2, & i = j. \end{cases}$$
$$(2.51)$$

ii.) Formulate (i) for $\mathbb{R}^N$-valued random variables.

Each element $X(t;\cdot)$ belonging to $\mathscr{S} = \{X(t;\omega) | t \in [a,b], \omega \in \Omega\}$ is itself a random variable. As such, it has a mean and variance, and the covariance of two members of $\mathscr{S}$ can be considered. In general, these values will change with t. In order to denote this fact, we introduce the following notions of the mean, variance, and covariance functions as they apply to stochastic processes.

Definition 2.7.6. Let $\mathscr{S} = \{X(t) | t \in I\}$ be a real-valued stochastic process.
i.) $\mu_{\mathscr{S}} : I \longrightarrow \mathbb{R}$ is defined by $\mu_{\mathscr{S}}(t) = E[X(t;\cdot)]$;
ii.) $Var_{\mathscr{S}} : I \longrightarrow \mathbb{R}$ is defined by $Var_{\mathscr{S}}(t) = Var[X(t;\cdot)]$;
iii.) $Cov_{\mathscr{S}} : I \times I \longrightarrow \mathbb{R}$ is defined by $Cov(t,s) = Cov(X(t;\cdot), X(s;\cdot))$.

Exercise 2.7.4. Interpret Def. 2.7.5(i) for a Gaussian process.

While the usual calculus applies to the sample paths because they are simply real-valued (or possibly $\mathbb{R}^N$-valued), we still need a calculus that takes into account the dependence on <u>both</u> t and ω. The structure of $\mathcal{L}^2(\Omega;\mathbb{R})$ has a natural influence here for reasons that will become apparent in upcoming sections. We begin with the most basic notion of a limit of a stochastic process at a particular value of $t \in I$. There are different ways to define this, but the most frequently used definition in this text is as follows.

Definition 2.7.7. A stochastic process $\{X(t;\omega) | t \in I, \omega \in \Omega\}$ has a *mean square limit* at $t_0 \in I$ if $\exists X^\star \in \mathcal{L}^2(\Omega;\mathbb{R})$ such that

$$\underbrace{\lim_{h \to 0} E|X(t_0+h;\cdot) - X^\star|^2}_{\text{A real-valued function of } h} = 0$$

(in the usual real-valued sense of Def. 1.8.2). We write $X(t;\cdot) \longrightarrow X^\star$ in $\mathcal{L}^2(\Omega;\mathbb{R})$ as $t \longrightarrow t_0$.

Remark. If t_0 belongs to the boundary of I (e.g., if $I = [a,b]$, then such a t_0 would be either a or b), we interpret Def. 2.7.7 as a one-sided limit.

Exercise 2.7.5.
i.) Formulate the $\varepsilon\delta$-form of Def. 2.7.7 (in the spirit of Def. 1.8.2).
ii.) Formulate Def. 2.7.7 for $\mathbb{R}^N$-valued random variables. Interpret verbally.

Exercise 2.7.6. Let $\{X(t;\omega)\,|\,t \in I, \omega \in \Omega\}$ and $\{Y(t;\omega)\,|\,t \in I, \omega \in \Omega\}$ be real-valued stochastic processes, $t_0 \in I$, and $\alpha, \beta \in \mathbb{R}$. Assume that $X(t;\cdot) \longrightarrow X^\star$ and $Y(t;\cdot) \longrightarrow Y^\star$ in $\mathcal{L}^2(\Omega;\mathbb{R})$ as $t \longrightarrow t_0$.
i.) Prove that $\alpha X(t;\cdot) + \beta Y(t;\cdot) \longrightarrow \alpha X^\star + \beta Y^\star$ in $\mathcal{L}^2(\Omega;\mathbb{R})$ as $t \longrightarrow t_0$.
ii.) If $g : I \longrightarrow \mathbb{R}$ is such that $\lim_{t \to t_0} g(t) = g^\star$ (in the sense of Def. 1.8.2), prove that
$g(t)X(t;\cdot) \longrightarrow g^\star X^\star$ in $\mathcal{L}^2(\Omega;\mathbb{R})$ as $t \longrightarrow t_0$.
iii.) Prove that the real-valued function $f : I \longrightarrow \mathbb{R}$ defined by $f(t) = E\,|X(t;\cdot)|$ is continuous on I.

We now consider several notions of continuity for stochastic processes.

Definition 2.7.8. Let $\mathscr{S} = \{X(t;\omega)\,|\,t \in I, \omega \in \Omega\}$ be a real-valued stochastic process. We say that $\mathscr{S}$ is
i.) *(right) left-continuous* at $t_0 \in I$ if the real-valued function $X(t;\omega)$ is (right) left-continuous at t_0 a.s. $[\mathscr{P}]$.
ii.) *continuous in probability* at $t_0 \in I$ if $\forall \varepsilon > 0$,

$$\lim_{t \to t_0} \mathscr{P}(\{\omega \in \Omega : |X(t;\omega) - X(t_0;\omega)| \geq \varepsilon\}) = 0.$$

iii.) *almost surely continuous* at $t_0 \in I$ if

$$\mathscr{P}\left(\left\{\omega \in \Omega : \lim_{t \to t_0} |X(t;\omega) - X(t_0;\omega)| = 0\right\}\right) = 1.$$

iv.) $\mathcal{L}^p$-*continuous* at $t_0 \in I$ $(p \geq 2)$ if

$$\lim_{h \to 0} E\,|X(t_0 + h) - X(t_0)|^p = 0.$$

Exercise 2.7.7.
i.) Interpret Def. 2.7.8 verbally.
ii.) Formulate Def. 2.7.8 for $\mathbb{R}^N$-valued random variables. Interpret verbally.

The notion of $\mathcal{L}^p$-continuity (called *mean square continuity* when $p = 2$) is the one with which we will work most often.

Exercise 2.7.8. Suppose that $1 < p < q < \infty$. If $\{X(t;\omega) \,|\, t \in I, \omega \in \Omega\}$ is $\mathfrak{L}^p$-continuous on I, must it also be $\mathfrak{L}^q$-continuous on I? How about conversely? Explain.

We also consider the notion of an $\mathfrak{L}^2$-derivative as follows.

Definition 2.7.9. A real-valued stochastic process $\{X(t;\omega) \,|\, t \in I, \omega \in \Omega\}$ is *mean square differentiable* at $t_0 \in I$ if $\exists X^\star \in \mathfrak{L}^2\,(\Omega;\mathbb{R})$ such that

$$\lim_{h \longrightarrow 0} E \left| \frac{X\,(t_0 + h;\cdot) - X\,(t_0;\cdot)}{h} - X^\star \right|^2 = 0.$$

Exercise 2.7.9. Formulate Def. 2.7.9 for $\mathbb{R}^N$-valued random variables.

The notion of accumulating information as time goes on when studying the evolution of a phenomenon described by a stochastic process is central to the study of SEEs. We first encountered this notion in Section 2.2. We now make it more formal. We shall state the definition for $I = [0,T]$ because this is the most common situation that we will encounter. The definition can be easily modified to account for different intervals $I \subset \mathbb{R}$.

Definition 2.7.10. Let $(\Omega, \mathscr{F}, \mathscr{P})$ be a complete probability space.
i.) A collection of sub σ-algebras $\{\mathscr{F}_t : t \in [0,T]\} \subset \mathscr{F}$ such that for all $0 \le s < t \le T$, $\mathscr{F}_s \subset \mathscr{F}_t \subset \mathscr{F}$, is called a *filtration* of $\mathscr{F}$.
ii.) $\{\mathscr{F}_t : t \in [0,T]\}$ is *right continuous* on $[0,T]$ if $\mathscr{F}_t = \bigcap_{s>t} \mathscr{F}_s, \forall t \in [0,T]$.
iii.) A stochastic process $\{X(t;\omega) \,|\, t \in [0,T], \omega \in \Omega\}$ is $\mathscr{F}_t$-*adapted* if $X(t;\cdot)$ is $\mathscr{F}_t$-measurable, $\forall t \in [0,T]$. That is,

$$\{\omega \in \Omega : X(t;\omega) \le x\} \in \mathscr{F}_t, \forall x \in \mathbb{R}, \forall t \in [0,T].$$

We need an intuitive understanding of what it means to be $\mathscr{F}_t$-adapted. Begin with $t = 0$. In a typical model, this corresponds to the time at which we start the experiment, so that $X(0;\cdot)$ is provided by initial data, or information about the system that is known. There is no randomness inherent to this information. So, saying "$X(0;\cdot)$ is $\mathscr{F}_0$-measurable" simply means that $X(0;\omega) = X_0(\omega)$ is known and events of the form $\{\omega \in \Omega : X_0(\omega) \le x\}$ are meaningful. As such, occurrence can be determined, without any guessing, at time 0. Said differently, $\mathscr{F}_0$ contains all the information necessary to describe $X(0;\cdot)$. (Why?)

Next, suppose that $t_0 > 0$. Saying "$X(t_0;\cdot)$ is $\mathscr{F}_{t_0}$-adapted" means that

$$\bigcup_{0 \le s \le t_0} \{\omega \in \Omega : X(s;\omega) \le x\} \subset \mathscr{F}_{t_0}, \forall x \in \mathbb{R}. \qquad (2.52)$$

(Why?) The left-hand side of (2.52) is comprised of all "basic events" formed using only those random variables $X(s;\cdot)$ in the stochastic process that correspond to times <u>prior or at</u> time t_0. As such, all the information about the stochastic process needed to describe $X\,(t_0;\cdot)$ and any events formed using the random variables in the set $\{X(t;\omega) \,|\, t \in [0,t_0], \omega \in \Omega\}$ is contained within $\mathscr{F}_{t_0}$.

Exercise 2.7.10. What is true about a stochastic process $\{X(t;\omega)\,|\,t \in [0,T], \omega \in \Omega\}$ if $X(t;\cdot)$ is $\mathscr{F}_0$-measurable, $\forall t \in [0,T]$?

Exercise 2.7.11. When is an event A $\mathscr{F}_t$-measurable?

The above discussion naturally leads to the following notion.

Definition 2.7.11. We say that $\{\mathscr{F}_t : t \in [0,T]\}$ is the *natural filtration* for $\mathscr{S} = \{X(t;\omega)\,|\,t \in [0,T], \omega \in \Omega\}$ if $\forall t \in [0,T]$,

$$
\mathscr{F}_t = \sigma \left(\underbrace{\bigcup_{0 \le s \le t} \sigma(X(s;\cdot))}_{\text{All the information about } \mathscr{S} \text{ up to time } t} \right).
$$
$$
\underbrace{\phantom{\mathscr{F}_t = \sigma \left(\bigcup_{0 \le s \le t} \sigma(X(s;\cdot)) \right)}}_{\text{All events that can be formed using this information}}
$$

This is the smallest filtration to which $\mathscr{S}$ is adapted.

When defining a stochastic integral in the next chapter, the following technical measurability notion will be needed.

Definition 2.7.12. A stochastic process $\{X(t;\omega)\,|\,t \in [0,T], \omega \in \Omega\}$ equipped with a filtration $\{\mathscr{F}_t : t \in [0,T]\}$ is *progressively measurable* with respect to this filtration if $\forall t \in [0,T]$, the random variables $X(s;\cdot)$, $0 \le s \le t$, are measurable functions on the product space $\mathscr{B}([0,t]) \times \mathscr{F}_t$, where $\mathscr{B}([0,t])$ is the collection of Borel sets contained within $[0,t]$.

Remark. The condition in Def. 2.7.12 is imposed to ensure double integrals in the probabilistic setting are well-defined and that the stochastic Fubini Theorem (used for switching the order of integration) will hold. As pointed out in **[161]**, it turns out that any $\mathscr{F}_t$-adapted stochastic process with continuous sample paths is automatically progressively measurable. And, moreover, this is precisely the type of stochastic process that we will encounter in this text. As such, we will not make explicit mention of this assumption of progressive measurable going forward because it is always satisfied.

Our study will naturally be restricted to stochastic processes possessing a certain degree of regularity, just as in the deterministic setting. We mention two such spaces in the following theorem.

Theorem 2.7.13. *Assume that $(\Omega, \mathscr{F}, \mathscr{P})$ is a complete probability space equipped with a right-continuous filtration $\{\mathscr{F}_t : t \in [0,T]\} \subset \mathscr{F}$. Let $p, r \ge 2$.*
***i.)** The space $\mathbb{C}([0,T]; \mathscr{L}^p(\Omega; \mathbb{R}))$ of stochastic processes $\{X(t;\omega)\,|\,t \in [0,T], \omega \in \Omega\}$ that are $\mathscr{F}_t$-adapted, progressively measurable with continuous sample paths, and*

are such that $\sup\{E\,|X(t;\cdot)|^p : 0 \le t \le T\} < \infty$ *is a Banach space when equipped with the norm*

$$\|X\|_{\mathbb{C}} \equiv \sup_{0 \le t \le T} (E\,|X(t;\cdot)|^p)^{\frac{1}{p}}. \tag{2.53}$$

ii.) *The space* $\mathbb{L}^r([0,T];\mathcal{L}^p(\Omega;\mathbb{R}))$ *of stochastic processes* $\{X(t;\omega)\,|t \in [0,T], \omega \in \Omega\}$ *that are $\mathcal{F}_t$-adapted, progressively measurable with continuous sample paths for which* $\sup\{E\,|X(t;\cdot)|^p : 0 \le t \le T\} < \infty$, *and* $\int_0^T \|X(t;\cdot)\|_{\mathbb{L}^p}^r\, dt < \infty$ *is a Banach space when equipped with the norm*

$$\|X\|_{\mathbb{L}^r} \equiv \left(\int_0^T \|X(t;\cdot)\|_{\mathbb{L}^p}^r\, dt \right)^{\frac{1}{r}}. \tag{2.54}$$

Exercise 2.7.12. Formulate analogous spaces to those in Thrm. 2.7.13 for $\mathbb{R}^N$-valued stochastic processes.

Exercise 2.7.13. Define precisely the notions of a convergent sequence and a Cauchy sequence for each of the two spaces in Thrm. 2.7.13. Pay particular attention to the form of the expressions involving the norms (2.53) and (2.54).

2.8   Martingales

At the end of Section 2.6, we posed the question

$$E\left[X(t;\cdot)\,\middle|\,\sigma\left(\bigcup_{0 \le \tau \le s} X(\tau;\cdot) \right) \right] = \underline{\quad?\quad}, \tag{2.55}$$

where $0 < s < t$. Based upon our very heuristic explanation of conditional expectation, we would *expect* this to be just $X(s;\cdot)$, end of story! But, this is not true in general. There are situations in which the best guess is not the value of the random variable at the "largest time" used to provide the a priori information. While this poses an interesting problem in stochastic processes, we will only be interested in those stochastic processes for which you <u>do</u> fill in the blank in (2.55) with $X(s)$. Such processes are defined as follows.

Definition 2.8.1. A real-valued stochastic process $\{X(t;\omega)\,|t \in I, \omega \in \Omega\}$ is a *martingale* with respect to the natural filtration $\{\mathcal{F}_t : t \in I\}$ if
i.) $X(t;\cdot)$ is $\mathcal{F}_t$-adapted, $\forall t \in I$,
ii.) $E\,|X(t;\cdot)| < \infty$, $\forall t \in I$, and
iii.) (Markov Property) $E[X(t;\cdot)|\mathcal{F}_s] = X(s;\cdot), \forall 0 \le s \le t$.

Proposition 2.8.2. *Let* $\{X(t;\omega)\,|t \in I, \omega \in \Omega\}$ *be a martingale with natural filtration* $\{\mathcal{F}_t : t \in I\}$.

i.) *There exists $M \in \mathbb{R}$ such that $E[X(t;\cdot)] = M, \forall t \in I$. (That is, the expectation of a martingale is constant.)*
ii.) *For every random variable Y for which $E|Y| < \infty$, $\{E[Y|\mathscr{F}_t] : t \in I\}$ is a martingale.*

Outline of Proof. (i) It suffices to show that $E[X(s;\cdot)] = E[X(t;\cdot)], \forall s,t \in I$. Use the Markov property with the fact that $E[E[X|\mathscr{A}]] = E[X]$. (Tell how.)
(ii) Verify the three conditions of Def. 2.8.1 directly. Condition (i) holds trivially because we are conditioning with respect to a natural filtration. To verify condition (ii), use Jensen's inequality together with the fact that $E[E[X|\mathscr{A}]] = E[X]$. And finally, condition (iii) holds because $E[E[Z|\mathscr{F}_t]|\mathscr{F}_s] = E[Z|\mathscr{F}_s]$. (Why?) $\square$

Exercise 2.8.1. Fill in the details in the proof of Prop. 2.8.2.

The following special case of Doob's Martingale Inequality (see **[404]**) comes in very handy when establishing estimates.

Theorem 2.8.3. (Doob's Martingale Property)
If $\{M(t;\omega)|t \in [0,T], \omega \in \Omega\}$ is a real-valued martingale for which the mapping $(t;\omega) \mapsto M(t;\omega)$ is almost surely continuous, then $\forall \delta > 0$,

$$\mathscr{P}\left(\left\{\omega \in \Omega : \sup_{0 \leq t \leq T} |M(t;\omega)| \geq \delta\right\}\right) \leq \frac{1}{\delta^2} E|M(T;\cdot)|^2. \qquad (2.56)$$

An analogous result holds for $\mathbb{R}^N$-valued random variables by simply replacing $|\cdot|$ by the $\mathbb{R}^N$-norm.

2.9 The Wiener Process

Physical parameters (e.g., reaction rates, diffusivity constants, densities) that are measured, external forces that act on a system, etc. are all subject to small random fluctuations, suitably referred to as "noise," that prevent us from obtaining a 100%-accurate deterministic description of them. But, how do we incorporate this vague notion into a rigorous mathematical framework that enables us to study the evolution of phenomena subject to such noise? The model presented in this section, the so-called *Wiener process*, is a mathematical construct used to simulate the motion first observed by Robert Brown in 1828 while looking through a microscope at a slide containing pollen grains encapsulated in water. Later, in 1905, Einstein studied this phenomenon and posited that the irregular movement was not due to the pollen, but rather to the molecular motion of the surrounding fluid. This motion has become known as *Brownian motion*.

2.9.1   A Common Simulation of Brownian Motion

There are many different ways to simulate Brownian motion (see **[161, 235, 302]**). One very intuitive approach is to define Brownian motion as the limit (of some sort) of a sequence of random variables arising from a description of random walks taken over successively more refined time steps. To get a feel for this, work through the following discussion.

Let $S : \Omega \longrightarrow \{-1, 1\}$ be a random variable for which its two possible outcomes are attained with equal probability; that is,

$$\mathscr{P}\left(\{\omega \in \Omega : S(\omega) = -1\}\right) = \mathscr{P}\left(\{\omega \in \Omega : S(\omega) = 1\}\right) = \frac{1}{2}.$$

Let us interpret this random variable as a model of "taking a single step of size 1 unit by moving either left or right from our current location on the real line." So, if we begin at location 0 on the real line, one application of S will require that we move either to position -1 or to position 1. Next, <u>from this position</u>, we take another step of size 1 unit, as described by S. Our current location on the real line has no bearing on the likelihood of moving left or right (Why?), so that successive steps are independent and identically distributed. The possible ending locations at this second round are $-2, 0, 2$. Similarly, if we take a third step, the possible ending locations are $-3, -1, 1, 3$. (Why?) Continuing in this manner, label the ith step as $S_i(\omega)$ with the understanding that $S_i(\omega)$ is nothing more than a copy of $S(\omega)$. Observe that the ending location on the real line after n such steps is given by $S_1(\omega) + \ldots + S_n(\omega)$. (Why?)

Exercise 2.9.1.
i.) Compute
    a.) $E[S]$
    b.) $Var[S]$
ii.) Let $n \in \mathbb{N}$. Compute
    a.) $E[S_1 + \ldots + S_n]$
    b.) $Var[S_1 + \ldots + S_n]$

Now, more generally, suppose that the step size is $0 < \varepsilon < 1$. Define a new random variable $S^* : \Omega \longrightarrow \{-\varepsilon, \varepsilon\}$ by $S^*(\omega) = \varepsilon S(\omega)$.

Exercise 2.9.2. Redo Exercise 2.9.1 for S^*.

We now apply this process on a fixed interval in a sequence of successively refined steps. Let $t > 0$ be fixed. (Think of t as having been measured in seconds.) Let $k \in \mathbb{N}$ and suppose that you take a step of size $\varepsilon = \frac{1}{\sqrt{k}}$ every $\frac{1}{k}$ seconds.

Exercise 2.9.3. Approximately how many such steps are taken during t seconds duration of time?

The location along the real line at time t is approximately

$$\beta_k(t;\omega) = \frac{1}{\sqrt{k}} \sum_{i=1}^{\lceil tk \rceil} S_i(\omega). \text{ (Why?)} \tag{2.57}$$

For simplicity, assume that $\beta_k(0;\omega) = 0$, $\forall k \in \mathbb{N}$. Observe that as k gets larger, more steps of a smaller size are taken. (How would you illustrate this pictorally?) Ultimately, for every $t \geq 0$, we are interested in determining $\lim_{k \to \infty} \beta_k(t;\omega)$, meaning that we would like to identify a stochastic process $\{\beta(t;\omega) : t \geq 0\}$ to which the sequence of stochastic processes $\{\beta_k(t;\omega) : t \geq 0\}$ converges (in an appropriate sense) as $k \longrightarrow \infty$ (i.e., as the step size shrinks to 0).

We consider the properties of $\{\beta_k(t;\omega) : t \geq 0\}$ below.

1. Stationary Increments. Let $0 \leq s < t$. Consider the increment $\beta_k(t) - \beta_k(s)$. Note that

$$\beta_k(t;\omega) - \beta_k(s;\omega) \approx \frac{1}{\sqrt{k}} \sum_{i=\lceil sk \rceil}^{\lceil tk \rceil} S_i(\omega). \text{ (Why?)} \tag{2.58}$$

Let $h \in \mathbb{R}$ be such that $t + h \geq 0$, $s + h \geq 0$ and consider

$$\beta_k(t+h;\omega) - \beta_k(s+h;\omega). \tag{2.59}$$

Exercise 2.9.4.
i.) How many copies of $S(\omega)$ are used to form the expression (2.58)?
ii.) How many copies of $S(\omega)$ are used to form the expression (2.59)? Is this true for all $h > 0$?
iii.) What can you conclude from this?

2. Independent Increments. Suppose that $(t_1, t_2] \cap (t_3, t_4] = \emptyset$ and that both intervals are contained within $[0, \infty)$.

Exercise 2.9.5. Explain why the random variables $\omega \mapsto (\beta_k(t_2;\omega) - \beta_k(t_1;\omega))$ and $\omega \mapsto (\beta_k(t_4;\omega) - \beta_k(t_3;\omega))$ are independent.

3. Limiting Distribution.

Exercise 2.9.6.
i.) Calculate $E[\beta_k(t;\cdot)]$ and $Var[\beta_k(t;\cdot)]$.
ii.) Explain why the distribution of the random variable

$$\omega \mapsto \frac{\sum_{i=1}^{\lceil tk \rceil} S_i(\omega) - \lceil tk \rceil}{\sqrt{\lceil tk \rceil}}$$

approaches the distribution of the standard Gaussian random variable as $k \longrightarrow \infty$.
iii.) From (ii), conclude that the distribution of $\omega \mapsto \beta_k(t;\omega)$ approaches the distribution of an $n(0,t)$ random variable.

iv.) Let $0 \leq s < t$. What is the distribution of $\omega \mapsto (\beta_k(t;\omega) - \beta_k(s;\omega))$?

4. Covariance. Let $0 \leq s < t$. Observe that

$$
\begin{aligned}
Cov_{B_k}(t,s) &= Cov\left(\beta_k(t;\cdot), \beta_k(s;\cdot)\right) \\
&= E\left[\beta_k(t;\cdot) \cdot \beta_k(s;\cdot)\right] \quad\quad\quad\quad\quad\quad\quad (2.60) \\
&= E\left[\beta_k(t;\cdot)\left(\beta_k(t;\cdot) - \beta_k(s;\cdot)\right) + \beta_k^2(s;\cdot)\right] \ \ (\text{Why?}) \\
&= E\left[\beta_k(t;\cdot)\right] \cdot E\left[\beta_k(t;\cdot) - \beta_k(s;\cdot)\right] + Var\left[\beta_k(t;\cdot)\right] \\
&= s \ (\text{Why?})
\end{aligned}
$$

Exercise 2.9.7. Fill in the missing details in (2.60).

5. Adaptedness. Let $0 \leq s < t$ and define

$$
\mathscr{F}_s = \sigma\left(\beta_k(u;\cdot) : 0 \leq u < s\right). \quad\quad\quad\quad (2.61)
$$

Exercise 2.9.8.
i.) Compute $E\left[\beta_k(t;\cdot) - \beta_k(s;\cdot)|\mathscr{F}_s\right]$.
ii.) Explain why $\beta_k(t;\cdot)$ is $\mathscr{F}_t$-adapted.

Exercise 2.9.9. Define a random variable $Y : I \times \Omega \longrightarrow \mathbb{R}$ by

$$
Y(t;\omega) = \sigma\beta_k(t;\omega) + \mu t.
$$

(For this random variable, μ is called the *drift*.)
i.) Compute $E\left[Y(t;\cdot)\right]$ and $Var\left[Y(t;\cdot)\right]$.
ii.) What is the approximate distribution of $\omega \mapsto Y(t;\omega)$?

2.9.2 The Wiener Process

We would like for the limiting stochastic process (obtained as $k \to \infty$) of the sequence $\{\beta_k(t;\omega) : t \geq 0, \omega \in \Omega\}$ to possess the same properties as those established in the previous subsection. The critical ones to include in our definition are (1), (2), and (3) because the others follow naturally from them. (Convince yourself.) We have the following definition.

Definition 2.9.1. An $\mathbb{R}$-*valued Wiener process* (or *one-dimensional Brownian motion*) is a stochastic process $\{\beta(t;\omega) : t \geq 0, \omega \in \Omega\}$ that satisfies the following properties:
i.) $\beta(0;\omega) = 0, \forall \omega \in \Omega$,
ii.) the increments are stationary and independent, and
iii.) $\beta(t;\cdot)$ is $n(0, \sigma^2 t)$.

Remarks.
1. Going forward, we will suppress the dependence of all Wiener processes on $\omega \in \Omega$ for notational simplicity. We use $\beta(t)$ and $W(t)$, the latter of which is in recoginition

of Wiener, interchangeably when referring to a one-dimensional Brownian motion.
2. Unless otherwise stated, we assume $\sigma = 1$, for simplicity. Then, $\{\beta(t) : t \geq 0\}$ is a standard Brownian motion.

Exercise 2.9.10. Assume that $\sigma > 0$ and $\{\beta(t) : t \geq 0\}$ is a Brownian motion. Must $\{\frac{1}{\sigma}\beta(t) : t \geq 0\}$ be a Brownian motion? Explain.

We summarize the main properties of the Wiener process below. They are extensions from the discussion in the previous subsection.

Theorem 2.9.2. (Properties of the Wiener Process)
Let $\{W(t) : t \geq 0\}$ be an $\mathbb{R}$-valued Wiener process. Then, $\forall 0 \leq s < t$,
i.) *$\{W(t) : t \geq 0\}$ is $\mathscr{F}_t$-adapted.*
ii.) *$\{W(t) : t \geq 0\}$ is a Gaussian process.*
iii.) *$Cov(W(t), W(s)) = s$.*
iv.) *$\lim_{t \to \infty} \frac{W(t)}{t} = 0$.*
v.) *$\{W(t) : t \geq 0\}$ is a martingale and $E[W(t) - W(s)|\mathscr{F}_s] = 0$.*
vi.) *The sample paths of $\{W(t) : t \geq 0\}$*
   a.) *are continuous a.s. $[\mathscr{P}]$,*
   b.) *are nowhere differentiable, and*
   c.) *do not have bounded variation on $[0, T]$.*

Outline of Proof:
(i) & (iii): These follow as before. (Tell why.)

(ii): Let $0 < t_1 < t_2 < \ldots < t_n$ be given. Observe that $\forall k \in \{1, \ldots n\}$,

$$W(t_k) = [W(t_k) - W(t_{k-1})] + [W(t_{k-1}) - W(t_{k-2})] \tag{2.62}$$

$$+ \ldots + \left[W(t_1) - \underbrace{W(0)}_{=0} \right].$$

The right-hand side of (2.62) is comprised of independent random variables (Why?), all of which are normal. (Why?) As such, by Exer. 2.5.7, the right-hand side (taken as a single random-variable) is normal. Thus, $W(t_k)$ is a normal random variable. Hence, $\langle W(t_1), \ldots, W(t_n) \rangle$ has a normal joint distribution (Why?). As such, we conclude that $\{W(t) : t \geq 0\}$ is Gaussian.

(iv): This property can be established using the *Strong Law of Large Numbers.* (See [285].)

(v): We verify the conditions of Def. 2.8.1. Note that conditions (i) and (ii) follow directly from the earlier properties. (Tell how.) To establish (iii), let $0 \leq s < t$ and

observe that

$$E\left[W(t)|\sigma\left(W(\tau)|\tau\le s\right)\right] \quad =$$
$$E\left[(W(t)-W(s))+W(s)|\sigma\left(W(\tau)|\tau\le s\right)\right] \quad = \quad (2.63)$$
$$\underbrace{E\left[(W(t)-W(s))|\sigma\left(W(\tau)|\tau\le s\right)\right]}_{=0}+\underbrace{E\left[W(s)|\sigma\left(W(\tau)|\tau\le s\right)\right]}_{=W(s)}. \ \text{(Why?)}$$

(vi) (a) Technically, we would need to verify that Kolmogorov's criterion is satisfied. (See **[161]**.)

(b) The idea is to show that for every sequence $\{t_n\}\subset[0,\infty)$ that converges to 0,

$$\mathscr{P}\left\{\omega\in\Omega:\limsup_{n\longrightarrow\infty}\left|\frac{W(t_n;\omega)-W(0;\omega)}{t_n-0}\right|=\infty\right\}=1.$$

(Why does this suffice? See **[302]** for a proof.)

(c) Said differently, this means that the total amount of "vertical movement" exhibited in the path of a Brownian motion on <u>any</u> interval, no matter how small, is infinite. You can think of this as the amount of vertical change experienced by the path of $\beta(t)$ as t moves through a given interval $[0,T]$. In order to be infinite, it must be the case that there is never any moment of smoothness in the graph, because this would temporarily (on a very small interval) allow the path to stop wiggling so wildly. This observation goes hand in hand with (b) and is the very reason why a different approach must be used when defining the so-called stochastic integral. This discussion awaits us in Chapter 3. $\square$

Exercise 2.9.11. Prove that $\{W(t):t\ge 0\}$ is continuous in probability at all $t_0\ge 0$.

Exercise 2.9.12. Let $T>0$ and $0<t_1<t_2<\ldots<t_n$. Show that the finite-dimensional joint distributions of $\{W(t):t\ge 0\}$ are such that $\left(T^{\frac{1}{2}}W(t_1),\ldots,T^{\frac{1}{2}}W(t_n)\right)$ has the same distribution as $(W(Tt_1),\ldots,W(Tt_n))$. We say that the Wiener process is *self-similar with Hurst parameter* $\frac{1}{2}$. If the parameter is replaced by some $H\in\left(\frac{1}{2},1\right]$, we get a so-called *fractional Brownian motion* (fBm).

Exercise 2.9.13. Let $0<t_1<t_2<\ldots<t_n$ be given. Show that for all $1\le i,j\le n$ for which $i\ne j$,

i.) $E\left[(W(t_i)-W(t_{i-1}))(W(t_j)-W(t_{j-1}))\right]=0$

ii.) (Quadratic Variation) $E\left[(W(t_i)-W(t_{i-1}))(W(t_i)-W(t_{i-1}))\right]=t_i-t_{i-1}$

2.9.3 Modeling with the Wiener Process

Disclaimer! The following discussion is purely heuristic and is meant simply to introduce some notions that will be more fully developed in the next chapter.

How does a Wiener process mysteriously end up as a term in a mathematical model? Suppose that the phenomenon in which we are interested can be modeled as the initial-value problem (IVP)

$$\begin{cases} \frac{dx}{dt} = \alpha x(t) + f(t), & t > 0, \\ x(0) = x_0, \end{cases}$$ (2.64)

where α is a real parameter of interest. Given that measurement is subject to error, regardless of how well-calibrated the equipment is, we would like to introduce randomness into the parameter itself, and in so doing, into IVP (2.64). Similarly, it is of interest to account for small random fluctuations in the modeling of the external force $f(t)$ impacting the system. Symbolically, we do this as follows:

$$\alpha \text{ is replaced by } \underbrace{\alpha_1 + \alpha_2 \frac{dW_1(t; \omega)}{dt}}_{\text{A stochastic process}}$$ (2.65)

$$f(t) \text{ is replaced by } \underbrace{f_1(t) + f_2(t) \frac{dW_2(t; \omega)}{dt}}_{\text{A stochastic process}}$$ (2.66)

Here, $\frac{dW_i(t; \omega)}{dt}$ stands for *white noise*, or formally the derivative (of some sort) of a Wiener process. But, this needs to be defined because, after all, we mentioned in Thrm. 2.9.2 that the sample paths of a Wiener process is NOWHERE differentiable! Technically, white noise is a "generalized" derivative in the sense of distributions. This is a delicate topic to treat correctly. (See **[1]**.) For our purposes, this is merely a stepping stone to get to a stochastic IVP, so we will not focus on the theoretical subtleties involved. Rather, from a stricly symbolic viewpoint, we substitute (2.65) and (2.66) into the differential equation portion of IVP (2.64) to obtain the stochastic differential equation:

$$\begin{aligned} \frac{dx}{dt} &= \left[\alpha_1 + \alpha_2 \frac{dW_1(t; \omega)}{dt} \right] x(t) + \left[f_1(t) + f_2(t) \frac{dW_2(t; \omega)}{dt} \right] \\ &= \alpha_1 x(t) + f_1(t) + \alpha_2 x(t) \frac{dW_1(t; \omega)}{dt} + f_2(t) \frac{dW_2(t; \omega)}{dt}. \end{aligned}$$ (2.67)

Now, multiply both sides of (2.67) by dt to obtain the differential

$$dx = (\alpha_1 x(t) + f_1(t)) dt + \alpha_2 x(t) dW_1(t; \omega) + f_2(t) dW_2(t; \omega).$$ (2.68)

Doing so yields a *stochastic differential equation* involving two sources of randomness. Equation (2.68) is said to be in *differential form*.

Going one step further, we integrate both sides of (2.68) over $(0, t)$ to obtain

$$x(t) - x_0 = \int_0^t (\alpha_1 x(s) + f_1(s)) \, ds$$ (2.69)

$$+ \int_0^t \alpha_2 x(s) dW_1(s; \omega) + \int_0^t f_2(s) dW_2(s; \omega).$$

Of course, performing these operations begs several questions, the most obvious of which is, "What is the precise meaning of the integrals on the right-hand side of (2.69)?" This question will be answered in the next chapter. And, in so doing, both (2.68) and (2.69) will become meaningful.

Another observation is that there are *two* Wiener processes in (2.68). Do they wipe out each other or amplify each other? How must they relate? A moment's thought suggests that they are introduced into the model via the assessment of very different quantities (e.g., a parameter and an unrelated external forcing term). As such, it is reasonable to assume that they are independent. Indeed, this provides us with a springboard into the following definition.

Definition 2.9.3. $\{\mathbf{W}(t) = \langle W_1(t), \ldots, W_M(t) \rangle : t \geq 0\}$ is an *M-dimensional Wiener process* if
i.) $\{W_i(t) : t \geq 0\}$ is a one-dimensional Wiener process, $\forall i \in \{1, \ldots, M\}$, and
ii.) $W_i(t)$ and $W_j(t)$ are independent whenever $i \neq j$.

We represent such a Wiener process as

$$\mathbf{W}(t) = \sum_{k=1}^{M} W_k(t)\mathbf{e}_k, \tag{2.70}$$

where $\{\mathbf{e}_k : k = 1, \ldots, M\}$ is an orthonormal basis for $\mathbb{R}^M$.

Exercise 2.9.14. Convince yourself that the same properties as for one-dimensional Wiener processes hold for M-dimensional Wiener processes, assuming appropriate modifications are made to account for the transition from $\mathbb{R}$ to $\mathbb{R}^M$.

Exercise 2.9.15. Let $0 \leq s < t$. Prove that

$$Cov_{\mathbf{W}}(t,s) = \underbrace{\begin{bmatrix} t & 0 & \cdots & 0 \\ 0 & \ddots & & \vdots \\ \vdots & & \ddots & 0 \\ 0 & \cdots & 0 & t \end{bmatrix}}_{\text{An } M \times M \text{ matrix}}. \tag{2.71}$$

2.10   Summary of Standing Assumptions

We gather a short list of standing assumptions used throughout the text in order to simplify the presentation going forward and to minimize, as much as is reasonable, the technical nature of the statement of the theorems. Some are listed loosely and suggest the nature of an assumption, the specific form of which will change slightly to accomodate a new setting.

(S.A.1) We always work with a complete probability space $(\Omega, \mathscr{F}, \mathscr{P})$ equipped with a right-continuous filtration $\{\mathscr{F}_t : t \geq 0\}$ to which all Wiener processes arising in the discussion are adapted.

(S.A.2) We assume, without further comment or justification, that we are able to find a continuous version of any stochastic process under consideration. (Showing this often requires one to verify the Kolmogorov criterion, which can be very technical.)

(S.A.3) All functions appearing as part of an integrand (of a Riemann, Lebesgue, Bochner, or stochastic integral) are assumed to be $\mathscr{F}_t$-adapted and progressively measurable in the appropriate space.

2.11 Looking Ahead

Armed with some rudimentary tools of analysis and the essentials of probability theory, we now ask what it means to account for noise in the mathematical modeling of a phenomenon, and in turn how it affects the differential equation(s) duly formed. Indeed, we caught a glimpse of a stochastic differential equation when describing how a Wiener process might end up as a term in a differential equation via "white noise." In preparation, we consider a more simplistic example of a random ordinary differential equation that can be solved using elementary methods.

Assume **(S.A. 1)** and let $a \in \mathbb{R}$ and $X_0 : \Omega \longrightarrow \mathbb{R}$ be a $n\left(\mu, \sigma^2\right)$ random variable. Consider the random IVP

$$\begin{cases} \frac{dX(t;\omega)}{dt} = aX(t;\omega), \ t > 0, \omega \in \Omega, \\ X(0;\omega) = X_0(\omega), \omega \in \Omega. \end{cases} \tag{2.72}$$

For each $\omega_0 \in \Omega$, note that $X_0(\omega_0)$ is simply a real number, so that (2.72) is nothing more than a standard IVP encountered in an elementary differential equations course.

Exercise 2.11.1.
i.) Solve (2.72) using separation of variables to show that

$$X(t;\omega_0) = e^{at}X_0(\omega_0). \tag{2.73}$$

ii.) Describe the possible behavior of the solution given in (2.73) as $t \longrightarrow \infty$ based on the nature of a.

This is fine, but if a different outcome $\omega_1 \in \Omega$ occurs instead of ω_0, then the solution given by (2.73) changes to

$$X(t; \omega_1) = e^{at} X_0(\omega_1).$$

We do not know a priori which outcome $\omega \in \Omega$ occurs, due to the fact that it is chosen randomly. As such, what *exactly* do we mean by a solution of (2.72)? Moreover, how does the normality assumption imposed on X_0 enter into the picture? We will investigate these issues, and much more, in Chapter 3.

2.12   Guidance for Selected Exercises

2.12.1   Level 1: A Nudge in a Right Direction

2.1.4. Begin with events A, B, and their complements. Then, use Exer. 1.2.1. (How?)

2.1.7. The answer is "yes" to both (i) and (ii). Express each as a set-theoretic combination of sets known to belong to $\sigma(\mathscr{G}_1)$.

2.1.10. (i) Let $B_j = \bigcup_{i \geq j} A_i$. Observe that $B_j \downarrow \limsup A_i$ as $j \to \infty$. (Why?) So, by Prop. 2.1.6(vii), $\mathscr{P}(B_j) \downarrow \mathscr{P}(\limsup A_i)$ as $j \to \infty$. Now what? The proof of (ii) is similar.

2.2.5. (i) Assume $x \leq y$. $F_X(y) = \mathscr{P}(X \leq x) + \mathscr{P}(x < X \leq y)$. (Now what?)

(ii) Use Prop. 2.1.6(vii). To what set does $\{\omega \in \Omega : X(\omega) \leq x\}$ converge as $x \to \infty$?

(iii) This is similar to (ii). To what set does $\{\omega \in \Omega : X(\omega) \leq x\}$ converge as $x \to -\infty$?

(iv) Observe that $\mathscr{P}\left(X \leq x + \frac{1}{n}\right) \downarrow \mathscr{P}(X \leq x)$ as $n \to \infty$. (Why?) What can you say about the set $\left\{F_X\left(x + \frac{1}{n}\right) : n \in \mathbb{N}\right\}$? (Now what?)

2.2.8. (iii) $\left(E\left[\left(\frac{X-\alpha}{\beta}\right) \cdot \left(\frac{Y-\alpha}{\beta}\right)\right]\right)^4 \leq \left(\left(E\left|\frac{X-\alpha}{\beta}\right|^4\right)^{\frac{1}{4}} \cdot \left(E\left|\frac{Y-\alpha}{\beta}\right|^{\frac{4}{3}}\right)^{\frac{3}{4}}\right)^4$. (Why? Now what?)

2.2.10. Use Prop. 2.2.11 with $X = |Y - E[Y]|$, $p = 2$, and $\varepsilon = \alpha \sigma_Y$.

2.3.3. (i) Use $\|Y_n\|_{\mathcal{L}^2}^2 \leq \|Y_n - Y\|_{\mathcal{L}^2}^2 + \|Y\|_{\mathcal{L}^2}^2$. (Now what?)

(ii) Note that $\|\alpha_n Z - \alpha Z\|_{\mathcal{L}^2}^2 = \|(\alpha_n - \alpha) Z\|_{\mathcal{L}^2}^2$. (Now what?)

(iii) Note that $\|(\alpha X_n \pm \beta Y_n) - (\alpha X \pm \beta Y)\|_{\mathcal{L}^2}^2 = \|\alpha(X_n - X) \pm \beta(Y_n - Y)\|_{\mathcal{L}^2}^2$. (Now what?)

(iv) Note that $|E[X_n] - E[X]| = |E[X_n - X]| \leq \left|E[X_n - X]^2\right|^{\frac{1}{2}} (E[1]^2)^{\frac{1}{2}}$. (Now what?)

2.3.6. Use Exer. 1.8.7 and the completeness of $\mathbb{R}$.

2.4.2. (i) Use the fact that the calculus is performed componentwise. (So what?)

(ii) Why is $\mathbf{X} - E[\mathbf{X}]$ an $\mathbb{R}^N$- random variable? How about its square?

2.4.4. (i) $E\left[\sum_{k=1}^m \alpha_k \mathbf{X}_k\right] = E[\alpha_1 \langle X_{1,1}, \ldots, X_{1,N} \rangle + \ldots + \alpha_m \langle X_{m,1}, \ldots, X_{m,N} \rangle]$. Now, combine the vectors. Then what?

(ii) Use Prop. 2.2.11 with an appropriate choice of random variable.

2.4.5. (i) Use Minkowski's inequality inside the expectation. Then, use monotonicity. (How?)

2.4.6. Observe that $f_{\mathbf{X}}(\mathbf{x}) = \dfrac{1}{\underbrace{(2\pi)^{\frac{1}{2}}\cdots(2\pi)^{\frac{1}{2}}}_{n \text{ times}}} \exp\left(-\frac{1}{2}\left(x_1^2+\ldots+x_N^2\right)\right)$. (Now what?)

2.4.7. (ii) Convergence of sequences in $\mathbb{R}^N$ is determined componentwise.

(iii) $E\left(\mathbf{X}_n\mathbf{Y}_n^{\mathsf{T}}\right) \longrightarrow E\left(\mathbf{X}\mathbf{Y}^{\mathsf{T}}\right)$

2.5.1. (ii) We can discard the outcomes contained in B from consideration because we know for certain that they did not occur. In effect, knowing that B has occurred removes a bit of the randomness of the experiment.

2.5.2. There are only two possibilities here. What are they?

2.5.3. $\mathscr{P}\left(\tilde{A}\cap B\right) = \mathscr{P}(B\setminus(A\cap B))$. (Now what?)

2.5.4. You must be able to compute the probabilities of all possible intersections of two <u>and three</u> of the events A_1, A_2, and A_3 as the product of the probabilities of the events used to form the intersection.

2.7.1. The definition of continuity of a real-valued function applies to this setting without issue. (Why?)

2.7.2. When are two $\mathbb{R}^N$-valued random variables $\mathbf{X}$ and $\mathbf{Y}$ not equal? (How do you use this?)

2.7.3. (i) If two random variables X and Y are independent, then $E[XY] = 0$. (So what?)

(ii) You must multiply the two differences in the sense of the $\mathbb{R}^N$ inner product. Otherwise, the definition is the same. (Tell why.)

2.7.4. Because a Gaussian random variable is characterized by its mean and covariance, the condition in Def. 2.7.5 (i) simplifies to $\mu_{\mathscr{S}}(t) = \mu_{\mathscr{S}}(t+h)$, $\forall t,h$. (So what?) Also, $Cov_{\mathscr{S}}(t,s) = Cov_{\mathscr{S}}(t+h,s+h)$. (So what?)

2.7.10. The stochastic process would actually be deterministic. (Why?)

2.7.11. A should belong to $\sigma(\{X(s)\,|s\in[0,t]\})$. (What does this mean?)

2.8.1. (i) $E[X(t)|\mathscr{F}_s] = X(s) \implies E[E[X(t)|\mathscr{F}_s]] = E[X(s)]$. (So what?)

2.9.1. (i)(a)0 **(b)**1 **(ii)(a)** 0 **(b)**n

2.9.2. (i)(a)0 **(b)**ε^2 **(ii)(a)** 0 **(b)**$n\varepsilon^2$

2.9.3. $\lfloor tk \rfloor$

2.9.4. (i) & (ii) $k(t-s)$

2.9.5. Both random variables are constructed using copies of $S(\omega)$ on nonoverlapping intervals.

2.9.6. (i) $E[B_k(t);\cdot] = 0$ and $Var[B_k(t);\cdot] = \frac{\lfloor tk\rfloor}{k} \approx t$.

(ii) Use the Central Limit Theorem. (How? Why does it apply?)

(iii) $n(0,t-s)$ (Why?)

2.9.7. (i) $\mu_{B_k} = 0$, $\forall t,s,k$.

(ii) Add and subtract $B_k(s)$.

(iii) $B_k(s)$ and $B_k(t) - B_k(s)$ are independent. (So what?) Also, what is the definition of $Var[B_k(t;\cdot)]$?

(iv) $E[B_k(s)] = 0$ and $Var[B_k(s;\cdot)] = s$. (Why?)

2.9.8. (i) $B_k(t) - B_k(s)$ is independent of $\mathscr{F}_s$. (So what?)

2.9.9. (i) $E\left[Y(t;\cdot)\right] = \mu t$ and $Var\left[Y(t;\cdot)\right] = \sigma^2 t$.

(ii) $n(\mu t, \sigma^2 t)$

2.9.10. Yes. Note that $\frac{1}{\sigma}\beta(t)$ is approximately $n(0,t)$.

2.9.11. $\mathscr{P}\left(|W(t) - W(s)| > \varepsilon\right) = \mathscr{P}\left(|W(t-s)| > \varepsilon\right) = 2\mathscr{P}\left(W(t-s) < -\varepsilon\right)$. (Now what?)

2.9.12. Argue each component separately. Observe that $W(Tt_i)$ is normal and thus, it is characterized by its mean and variance. (Now what?)

2.9.13 (i) Use the independence of the increments to conclude immediately.

(ii) Consult any of the standard references mentioned at the beginning of the chapter.

2.12.2   Level 2: An Additional Thrust in a Right Direction

2.1.4. Exer. 1.2.1. is useful when computing complements of certain unions of sets that must belong to $\mathscr{F}$.

2.1.7. (i) $\{a\} = [\overparen{(-\infty, a) \cup (a, \infty)}]$ and $(a,b) = \bigcup_{n=1}^{\infty} \left(a, b - \frac{1}{2^n}\right]$. Verify these.

2.1.10. (i) Because $\mathscr{P}(B_j) \geq \sup_{i \geq j} \mathscr{P}(A_i)$ (Why?), we see that

$$\mathscr{P}(\limsup A_i) \geq \lim_{j \to \infty} \left(\sup_{i \geq j} \mathscr{P}(A_i) \right) = \limsup \mathscr{P}(A_i).$$

(Tell how.)

2.2.5. (i) Note that $\mathscr{P}(x < X \leq y) \geq 0$. (So what?)

(ii) What is $\mathscr{P}(\Omega)$? (So what?)

(iii) What is $\mathscr{P}(\varnothing)$? (So what?)

(iv) The set $\left\{ F_X\left(x + \frac{1}{n}\right) : n \in \mathbb{N} \right\}$ is bounded below. (Why?) Using (i), we conclude that $\exists \lim_{n \to \infty} F_X\left(x + \frac{1}{n}\right) = \inf\left\{ F_X\left(x + \frac{1}{n}\right) : n \in \mathbb{N} \right\}$, which equals what?

2.2.8. (iii) Now, apply Prop. 2.2.8(i).

2.3.3. (i) Now, mimic the argument used in the real-valued case. See Prop. 1.5.6.

(ii) & (iii) Now, use the Cauchy-Schwarz inequality.

(iv) The right-hand side equals $\|X_n - X\|_{\mathscr{L}^2}$, which we know approaches zero as $n \to \infty$.

2.4.4. (i) $\ldots = E\left[\langle \alpha_1 X_{1,1} + \ldots + \alpha_m X_{m,1}, \ldots, \alpha_1 X_{1,N} + \ldots + \alpha_m X_{m,N} \rangle\right]$. Now, apply Def. 2.4.3(i), along with the properties of expectation in the real-valued case on each component. (Tell how.)

(ii) Use $\omega \mapsto \|X(\omega)\|_{\mathbb{R}^N}^p$ as X in Prop. 2.2.11. The result follows immediately.

2.4.6. Now, apply the exponent rule and regroup the terms in order to identify N copies of the $n(0,1)$ density.

2.5.2. If $\omega_0 \in A$, then $\mathscr{P}(A) = 1$. Otherwise, $\mathscr{P}(A) = 0$. (Explain this.)

2.5.3. $\mathscr{P}(B \setminus (A \cap B)) = \mathscr{P}(B) - \mathscr{P}(A)\mathscr{P}(B)$. (Now what?)

2.5.4. A reasonable condition is that $\mathscr{P}(C_1 \cap C_2 \cap C_3) = \mathscr{P}(C_1)\mathscr{P}(C_2)\mathscr{P}(C_3)$, $\forall C_i \in \sigma(A_i)$. (Why?)

2.7.2. If at least one pair of corresponding components X_i and Y_i are not equal random variables, then $\mathbf{X}$ is not equal to $\mathbf{Y}$. If, on the other hand, all pairs of corresponding components are equal except possibly on a $\mathscr{P}$-null set, then $\mathbf{X} = \mathbf{Y}$.

2.7.3. Note that $E[XY]$ is the inner product on $\mathcal{L}^2(\Omega; \mathbb{R})$.

2.7.4. This further implies that $\mu_{\mathscr{S}}(t) = \mu_{\mathscr{S}}(s)$, $\forall t, s$, so that $\mu_{\mathscr{S}}(\cdot)$ is a constant function. What can you say about $Cov_{\mathscr{S}}(\cdot, \cdot\cdot)$?

2.8.1. (i) Apply the fact that $E[E[X|\mathscr{A}]] = E[X]$.

2.9.11. $\ldots = 2\Phi\left(-\frac{\varepsilon}{\sqrt{t-s}}\right) \longrightarrow 0$ as $t \longrightarrow s$ because the distribution function is continuous and $-\frac{\varepsilon}{\sqrt{t-s}} \longrightarrow -\infty$.

2.9.12. $E\left[T^{\frac{1}{2}}W(t_i)\right] = T^{\frac{1}{2}}W(t_i) = 0$

$Var\left[T^{\frac{1}{2}}W(t_i)\right] = \left(T^{\frac{1}{2}}\right)^2 Var[W(t_i)] = Tt_i = Var[W(t_i)]$

Chapter 3

Linear Homogenous Stochastic Evolution Equations in $\mathbb{R}$

Overview

We begin our study of stochastic evolution equations with the simplest case of a stochastic ordinary differential equation in $\mathbb{R}$. It is here that you will get the lay of the land, learn about some typical questions that are of interest, and develop the machinery to answer them. Striking similarities and sharp differences from the deterministic setting will become apparent as you work through the chapter. Some standard references used throughout this chapter are **[11, 62, 178, 193, 244, 251, 287, 302, 318, 333, 375, 397, 399]**.

3.1 Random Homogenous Stochastic Differential Equations

Assume **(S.A.1).** Let $a \in \mathbb{R}$ and $X_0 : \Omega \longrightarrow \mathbb{R}$ be a given random variable, and consider the IVP

$$\begin{cases} \frac{dX}{dt}(t;\omega) = aX(t;\omega), 0 < t < T, \omega \in \Omega, \\ X(0;\omega) = X_0(\omega), \omega \in \Omega. \end{cases} \tag{3.1}$$

For a fixed $\omega_0 \in \Omega$, (3.1) is a typical IVP encountered in an elementary differential equations course. In fact, the separation of variables technique can be used to show that the solution of (3.1) is

$$X(t;\omega_0) = e^{at}X_0(\omega_0). \tag{3.2}$$

This solution trajectory changes depending on the random choice of ω_0. Because we do not know a priori which value of $\omega_0 \in \Omega$ occurs, every member of the collection of trajectories

$$\left\{ t \mapsto e^{at}X_0(\omega) : \omega \in \Omega \right\} \tag{3.3}$$

contributes to the solution of (3.1). As such, the solution of (3.1) is a stochastic process whose sample paths are $t \mapsto e^{at}X_0(\omega)$.

For every $\omega \in \Omega$, the sample paths are differentiable with respect to t. (Why?) In fact, the sample paths possess the same analytic properties as the solutions of the deterministic version of (3.1) (i.e., when there is no dependence on ω). But, the introduction of noise into the IVP by assuming that the initial condition (IC) $X_0(\omega)$ is a random variable enhances the mathematical description of any concrete model that can be written in the form (3.1). For instance, suppose that X_0 is $b(2, 0.8)$. Then, there are two possible outputs when X_0 is evaluated at a given $\omega \in \Omega$ with probabilities

$$\mathscr{P}(\{\omega \in \Omega : X_0(\omega) = 0\}) = 0.2,$$
$$\mathscr{P}(\{\omega \in \Omega : X_0(\omega) = 1\}) = 0.8. \tag{3.4}$$

Exercise 3.1.1.
i.) (3.3) is comprised of how many solution trajectories?
ii.) Describe the sample paths from (i).
iii.) Explain probabilistically how (3.4) affects the manner in which we interpret the solution process of (3.1).
iv.) Compute $E[X(t; \cdot)]$ and $Var[X(t; \cdot)]$. Interpret these quantities.
v.) Is $\{X(t; \cdot) : t \geq 0\}$ a Markov process? Explain.
vi.) Is $\{X(t; \cdot) : t \geq 0\}$ Gaussian? Explain.

Exercise 3.1.2. Suppose now that X_0 is $b(n, 0.5)$, for some $n \in \mathbb{N}$.
i.) (3.3) is comprised of how many solution trajectories?
ii.) Describe the sample paths from (i).
iii.) What percentage of the time would you expect each trajectory to arise, in the long run?
iv.) Compute $E[X(t; \cdot)]$ and $Var[X(t; \cdot)]$. Interpret these quantities.
v.) Is $\{X(t; \cdot) : t \geq 0\}$ a Markov process? Explain.
vi.) Is $\{X(t; \cdot) : t \geq 0\}$ Gaussian? Explain.
vii.) Describe the limiting behavior of the distribution of the solution process as $n \longrightarrow \infty$.

Exercise 3.1.3. Suppose now that X_0 is $n(0, 1)$. How does the interpretation of the solution trajectories change from the descriptions in Exercises 3.1.1 and 3.1.2?

Continuous dependence on the initial data is also of interest, but we need to accurately account for randomness. Consider (3.1) together with the following IVP:

$$\begin{cases} \frac{dY}{dt}(t; \omega) = aY(t; \omega), \ 0 < t < T, \omega \in \Omega, \\ Y(0; \omega) = Y_0(\omega), \ \omega \in \Omega, \end{cases} \tag{3.5}$$

where $\|X_0 - Y_0\|_{\mathcal{L}^2(\Omega; \mathbb{R})} < \delta$. A crude estimate of the difference between the solution processes of (3.4) and (3.5) in the $\mathcal{L}^2$-sense is provided in the following exercise.

Exercise 3.1.4. Prove that

$$\sup_{0 \leq t \leq T} \|X(t; \cdot) - Y(t; \cdot)\|_{\mathcal{L}^2(\Omega; \mathbb{R})} < \delta e^{aT}.$$

Continuous dependence can also be formulated as a convergence problem. Specifically, consider sequences $\{a_n : n \in \mathbb{N}\} \subset \mathbb{R}$ and $\{(X_0)_n : n \in \mathbb{N}\} \subset \mathcal{L}^2(\Omega; \mathbb{R})$ such that $a_n \longrightarrow a$ in $\mathbb{R}$ and $(X_0)_n \longrightarrow X_0$ in $\mathcal{L}^2(\Omega; \mathbb{R})$ as $n \longrightarrow \infty$. For each $n \in \mathbb{N}$, consider the IVP

$$\begin{cases} \frac{dX_n}{dt}(t; \omega) = a_n X_n(t; \omega), \ 0 < t < T, \omega \in \Omega, \\ X_n(0; \omega) = (X_0)_n(\omega), \ \omega \in \Omega. \end{cases} \tag{3.6}$$

Exercise 3.1.5. Does there exist an $\mathcal{L}^2(\Omega; \mathbb{R})$-limit of the sequence of solution processes $\{X_n\}$ of (3.6) as $n \longrightarrow \infty$? Prove your assertion.

The interval on which IVP (3.1) is considered can be extended to $[0, \infty)$, and in so doing, asking about the long-term behavior (as $t \longrightarrow \infty$) of the solution process becomes natural. This is a trivial question for the deterministic version of (3.1). (Why?)

Exercise 3.1.6. For a fixed $\omega_0 \in \Omega$, what are the possible long-term behaviors of (3.2), depending on the value of $a \in \mathbb{R}$?

Fixing an $\omega_0 \in \Omega$ simplified the situation. When this is not done, the question of long-term behavior of the stochastic version of (3.1) becomes more delicate.

Exercise 3.1.7. As a preliminary attack, try to formulate a natural notion of long-term stability in this setting.

We now make the transition to incorporating an additional source of noise into the model. Specifically, suppose that we replace the parameter a in (3.1) by

$$a_0 + a_1 \frac{dW}{dt}, \tag{3.7}$$

where $\frac{dW}{dt}$ is a white noise process. Proceeding as in our heuristic discussion from Section 2.9.3, we substitute (3.7) into (3.1) and subsequently recover the differential form (formally by multiplying both sides by the differential dt), as follows:

$$\begin{cases} \frac{dX}{dt}(t; \omega) = \left(a_0 + a_1 \frac{dW}{dt}\right) X(t; \omega), \ 0 < t < T, \omega \in \Omega, \\ X(0; \omega) = X_0(\omega), \omega \in \Omega, \end{cases} \tag{3.8}$$

and subsequently

$$\begin{cases} dX(t; \omega) = a_0 X(t; \omega) dt + a_1 X(t; \omega) dW(t), \ 0 < t < T, \omega \in \Omega, \\ X(0; \omega) = X_0(\omega), \ \omega \in \Omega. \end{cases} \tag{3.9}$$

More generally, we could introduce m independent sources of randomness into (3.1) by replacing the parameter a by

$$a_0 + \sum_{k=1}^{m} a_k \frac{dW_k(t)}{dt}. \tag{3.10}$$

Substituting (3.10) into (3.1) results in the more general stochastic IVP

$$\begin{cases} dX(t;\omega) = a_0 X(t;\omega)dt + \sum_{k=1}^{m} a_k X(t;\omega)dW_k(t), \ 0 < t < T, \omega \in \Omega, \\ X(0;\omega) = X_0(\omega), \ \omega \in \Omega. \end{cases} \quad (3.11)$$

We shall consider the integrated forms of (3.9) and (3.11), respectively given by

$$X(t;\omega) = X_0(\omega) + \int_0^t a_0 X(s;\omega)ds + \int_0^t a_1 X(s;\omega)dW(s), \quad (3.12)$$

$$X(t;\omega) = X_0(\omega) + \int_0^t a_0 X(s;\omega)ds + \sum_{k=1}^{m} \int_0^t a_k X(s;\omega)dW_k(s), \quad (3.13)$$

where $0 < t < T$ and $\omega \in \Omega$. Doing so merely replaces the solvability of one problem (an IVP) by another (a stochastic integral equation). What precisely do we mean by the integrals appearing on the right-hand sides of (3.12) and (3.13)? This question must be answered before going any further.

3.2   Introducing the Lebesgue and Itó Integrals

Assume **(S.A.1)**. We must first decide on the set of functions for which the integrals arising in (3.12) and (3.13) are to be defined. To this end, note that the integrands can depend on the solution process $X(t;\omega)$, which belongs to $\mathfrak{L}^2(\Omega;\mathbb{R})$. Also, for any $t \in [0,T]$, both integrals are themselves random variables and so it is natural to require that they belong to $\mathfrak{L}^2(\Omega;\mathbb{R})$. Furthermore, we must make certain that the integrands are $\mathscr{F}_t$-adapted and progressively measurable to ensure that the resulting solution process is $\mathscr{F}_t$-adapted and that Fubini's Theorem will be applicable, thereby enabling us to reverse the order of integration when necessary.

Suppose that the stochastic process $u : [0,t] \times \Omega \longrightarrow \mathbb{R}$ satisfies the following conditions:

$$\omega \mapsto u(s;\omega) \text{ is } \mathscr{F}_s - \text{adapted}, \forall s \in [0,t], \quad (3.14)$$

$$(s;\omega) \mapsto u(s;\omega) \text{ is progressively measurable on } [0,t] \times \Omega, \quad (3.15)$$

$$\int_0^t E|u(s;\cdot)|^2 ds < \infty. \quad (3.16)$$

The collection of all real-valued stochastic processes satisfying (3.14) through (3.16) shall be denoted by $\mathscr{U}$.

3.2.1   The Lebesgue Integral for $\mathbb{R}$-Valued Stochastic Processes

We have already encountered an integral different from the familiar Riemann integral when defining expectation. There, we integrated with respect to a probability measure rather than a real variable. The integral $\int_0^t u(s;\omega)ds$ arising in (3.12) and (3.13) need not be a Riemann integral, though it might not be apparent as to why. It can be a more general *Lebesgue integral*. This integral is defined using the same building block approach used in Def. 2.2.7.

Step 1: Random Characteristic Function
Let $c \in \mathcal{L}^2(\Omega; [0,\infty))$ and $[a,b] \subset [0,t]$. Define $u : [0,t] \times \Omega \longrightarrow [0,\infty)$ by

$$u(s;\omega) = c(\omega)\chi_{[a,b]}(s) = \begin{cases} c(\omega), & s \in [a,b], \\ 0, & s \in [0,t] \setminus [a,b]. \end{cases} \qquad (3.17)$$

Exercise 3.2.1. Why does u (given by (3.17)) belong to $\mathcal{U}$?

For a given $\omega_0 \in \Omega$, (3.17) is just an impulse function that is "on" with height $c(\omega_0)$ whenever the input is in the interval $[a,b]$, and "off" with height 0 outside of this interval. For such functions, we have the following definition:

$$\int_0^t u(s;\omega)ds \equiv c(\omega)(b-a). \qquad (3.18)$$

The next natural step is to consider those elements of $\mathcal{U}$ formed using finite combinations of random characteristic functions called *random step functions*.

Step 2: Random Step Functions
Let $0 = t_0 < t_1 < \ldots < t_{m-1} < t_m = t$ and $\{c_i : i = 1,\ldots,m\} \subset \mathcal{L}^2(\Omega; [0,\infty))$ be $\mathscr{F}_{t_i}$-adapted. Define $u : [0,t] \times \Omega \longrightarrow [0,\infty)$ by

$$u(s;\omega) = \sum_{i=1}^m c_i(\omega)\chi_{[t_{i-1},t_i)}(s). \qquad (3.19)$$

We denote the collection of all random step functions by $\mathbb{S}$.

Exercise 3.2.2. Why does u (given by (3.19)) belong to $\mathcal{U}$?

For any $u \in \mathbb{S}$, we have the following definition:

$$\int_0^t u(s;\omega)ds \equiv \sum_{i=1}^m c_i(\omega)(t_i - t_{i-1}). \qquad (3.20)$$

Remark. For a given $\omega_0 \in \Omega$, the picture of u is a typical step function. Pay particular attention to the nature of the intervals $[t_{i-1},t_i)$, specifically that the left endpoint is included and the right is excluded. How does this show up in the graph? What if the reverse were true? This choice has a significant impact when defining the other integral appearing in (3.12) and (3.13), especially regarding the $\mathscr{F}_t$-adaptedness. In analogy with the Riemann integral, you can loosely think of t_i as the sample point in the interval $[t_{i-1},t_i)$ at which we compute the integrand to obtain the height of that particular rectangle.

It turns out that the collection of all such step functions is dense in the space consisting of the positive-valued members of $\mathscr{U}$. (See **[161]**.) This can be used to show that $\forall u \in \mathscr{U}, \exists \{u_m\} \subset \mathbb{S}$ such that $u_m \longrightarrow u$ in $\mathcal{L}^2(\Omega; [0, \infty))$ as $m \to \infty$. Using this fact, the next step in the construction is natural.

Step 3: Positive Member of $\mathscr{U}$

Assume that $u : [0, t] \times \Omega \longrightarrow [0, \infty)$ belongs to $\mathscr{U}$. There exists a sequence of non-negative random step functions $\{u_m\} \subset \mathbb{S}$ such that $u_m \longrightarrow u$ in $\mathcal{L}^2(\Omega; [0, \infty))$ as $m \to \infty$. We have the following definition:

$$\int_0^t u(s; \omega)ds \equiv \lim_{m \longrightarrow \infty} \int_0^t u_m(s; \omega)ds, \qquad (3.21)$$

where the limit is taken in the mean square sense.

An arbitrary member of $\mathscr{U}$ can be written as the difference of two positive members of $\mathscr{U}$. Precisely, let $u \in \mathscr{U}$ and define $u^+ : [0, t] \times \Omega \longrightarrow [0, \infty)$ and $u^- : [0, t] \times \Omega \longrightarrow [0, \infty)$ by

$$u^+(s; \omega) = \begin{cases} u(s; \omega), & \text{if } u(s; \omega) \geq 0, \\ 0, & \text{if } u(s; \omega) < 0, \end{cases} \qquad (3.22)$$

$$u^-(s; \omega) = \begin{cases} 0, & \text{if } u(s; \omega) \geq 0, \\ -u(s; \omega), & \text{if } u(s; \omega) < 0. \end{cases} \qquad (3.23)$$

Observe that

$$u(s; \omega) = u^+(s; \omega) - u^-(s; \omega). \qquad (3.24)$$

Exercise 3.2.3. Why do u^+ and u^- belong to $\mathscr{U}$? Also, why does (3.24) hold?

Step 4: Arbitrary Member of $\mathscr{U}$

Let $u : [0, t] \times \Omega \longrightarrow \mathbb{R}$ belong to $\mathscr{U}$. Then, u can be written in the equivalent form (3.24) and so we have the following definition:

$$\int_0^t u(s; \omega)ds \equiv \int_0^t u^+(s; \omega)ds - \int_0^t u^-(s; \omega)ds. \qquad (3.25)$$

The integral defined via this construction is a *Lebesgue integral*. We say u is *Lebesgue integrable* if the right-hand side of (3.25) is defined.

This integral and the one arising when defining expectation are special cases of an "integral with respect to a measure"; refer to **[353]** for a more thorough discussion of the general case. For the most part in this text, you can think about Lebesgue integrals as if they were Riemann integrals (and quite often, they <u>are</u> Riemann integrals because the integrands are typically continuous). In fact, the properties of the

Lebesgue integral are similar to those of the Riemann integral. The key to verifying these properties is to appeal to the building block approach to first verify that the property holds for a characteristic function, and then proceed step-by-step through the construction.

Proposition 3.2.1. (Properties of the Lebesgue Integral of a Stochastic Process)
Assume that $u : [0,t] \times \Omega \longrightarrow \mathbb{R}$ *and* $v : [0,t] \times \Omega \longrightarrow \mathbb{R}$ *belong to* $\mathscr{U}$ *. Then, for any* $[a,b] \subset [0,t]$ *and* $\forall \omega \in \Omega$, *the following hold:*
i.) $\int_a^b u(s;\omega)ds = -\int_b^a u(s;\omega)ds.$
ii.) (Linearity)

$$\int_a^b [\alpha u(s;\omega) + \beta v(s;\omega)] \, ds = \alpha \int_a^b u(s;\omega)ds + \beta \int_a^b v(s;\omega)ds, \, \forall \alpha, \beta \in \mathbb{R}.$$

iii.) (Additivity) $\int_a^b u(s;\omega)ds = \int_a^c u(s;\omega)ds + \int_c^b u(s;\omega)ds, \, \forall c \in [a,b]$.
iv.) (Monotonicity)
   a.) *If* $|u(s;\omega)| \le |v(s;\omega)|$, *for almost all* $(s;\omega) \in [0,t] \times \Omega$, *then*

$$\int_a^b |u(s;\omega)| \, ds \le \int_a^b |v(s;\omega)| \, ds;$$

   b.) *If* $[c,d] \subset [a,b]$ *and* $u(s;\omega) \ge 0$, *for almost all* $(s;\omega) \in [0,t] \times \Omega$, *then*

$$\int_c^d u(s;\omega)ds \le \int_a^b u(s;\omega)ds.$$

v.) (Equality) *If* $u(s;\omega) = v(s;\omega)$, *for almost all* $(s;\omega) \in [0,t] \times \Omega$, *then*

$$\int_a^b u(s;\omega)ds = \int_a^b v(s;\omega)ds.$$

vi.) *If* $u \in \mathbb{C}([0,t] \times \Omega; \mathbb{R})$, *then* $\int_a^b u(s;\omega)ds$ *exists and*

$$\left| \int_a^b u(s;\omega)ds \right| \le \int_a^b |u(s;\omega)| \, ds. \tag{3.26}$$

vii.) (Absolute Continuity) *If* $u \in \mathbb{C}([0,t] \times \Omega; [0,\infty))$, *then* $\forall \varepsilon > 0$ *and* $\omega \in \Omega$, $\exists \delta > 0$ *such that* $\int_a^{a+\delta} u(s;\omega)ds < \varepsilon$.
viii.) (Hölder's Inequality)

$$\left| \int_a^b u(s;\omega)v(s;\omega)ds \right| \le \left[\int_a^b u^2(s;\omega)ds \right]^{1/2} \left[\int_a^b v^2(s;\omega)ds \right]^{1/2}. \tag{3.27}$$

More generally,

$$\left| \int_a^b u(s;\omega)v(s;\omega)ds \right| \le \left[\int_a^b u^p(s;\omega)ds \right]^{1/p} \left[\int_a^b v^q(s;\omega)ds \right]^{1/q}, \tag{3.28}$$

where $1 \le p, q < \infty$ *are such that* $\frac{1}{p} + \frac{1}{q} = 1$.
ix.) (Minkowski's Inequality)

$$
\left[\int_a^b (u(s; \omega) + v(s; \omega))^2 ds \right]^{1/2} \le \left[\int_a^b u^2(s; \omega) ds \right]^{1/2}
$$
$$
+ \left[\int_a^b v^2(s; \omega) ds \right]^{1/2}. \qquad (3.29)
$$

We defined the Lebesgue integral and stated its properties assuming the functions of interest were stochastic processes because that is typically the context in which the integral will arise. Equally as important, though, is the case when the integrand is a deterministic function, say $s \mapsto \hat{u}(s)$. Such a function can be viewed as the stochastic process $u : [0, t] \times \Omega \longrightarrow \mathbb{R}$ defined by $u(s; \omega) = \hat{u}(s), \forall \omega \in \Omega$. This can arise, for instance, when computing the integral of the expectation of a stochastic process. In such a case, it is customary to suppress the dependence on ω in the definition and properties.

Exercise 3.2.4. Assume that $f, u \in C([a, b]; \mathbb{R})$ and that $g : [a, b] \times \mathbb{R} \to \mathbb{R}$ is a continuous mapping for which there exist positive real numbers M_1 and M_2 such that

$$
|g(s, x)| \le M_1 |z| + M_2, \forall s \in [a, b], z \in \mathbb{R}. \qquad (3.30)
$$

i.) Prove that the set $\left\{ f(x) + \int_a^x g(z, u(z)) dz : x \in [a, b] \right\}$ is uniformly bounded above and provide an upper bound.
ii.) Let $N \in \mathbb{N}$. Determine upper bounds for these:
 a.) $\left| f(x) + \int_a^x g(z, u(z)) dz \right|^N$
 b.) $\int_a^b \left| f(x) + \int_a^x g(z, u(z)) dz \right|^N dx$

Exercise 3.2.5.
i.) Prove Prop. 3.2.1(viii) by appealing to the appropriate Hilbert space property as it specifically applies to $\mathbb{L}^2(a, b; \mathbb{R})$.
ii.) Prove that if f is Lebesgue integrable on $[0, T]$, then

$$
\left[\int_0^T f(x) dx \right]^2 \le \sqrt{T} \left[\int_0^T f^2(x) dx \right]^{1/2}.
$$

The following is an important property of Lebesgue integrals that is used frequently in convergence arguments.

Proposition 3.2.2. (Lebesgue Dominated Convergence (LDC))
Let $\{f_n\}$ *be a sequence of (Lebesgue) integrable real-valued functions defined on* $[a, b]$. *Assume that*
i.) $\lim_{n \to \infty} f_n(x) = f(x)$, *for almost all* $x \in [a, b]$, *and*

ii.) *there exists a Lebesgue integrable function* $g : [a,b] \to [0, \infty)$ *(called a dominator) for which*

$$|f_n(x)| \le g(x), \forall n \in \mathbb{N}, \text{ for almost all } x \in [a,b]. \tag{3.31}$$

Then, f is Lebesgue integrable and

$$\lim_{n \to \infty} \int_a^b f_n(x)dx = \int_a^b \lim_{n \to \infty} f_n(x)dx = \int_a^b f(x)dx. \tag{3.32}$$

We now turn our attention to the stochastic process $\mathscr{S} = \left\{ \int_0^t u(s;\omega)ds : 0 \le t \le T \right\}$. Observe that $\forall t \in [0,T]$, $\int_0^t u(s;\omega)ds \in \mathscr{L}^2(\Omega;\mathbb{R})$ (Why?), and (3.14) implies that $\mathscr{S}$ is $\mathscr{F}_t$-adapted. (Why?) The need to compute $E\left[\int_0^t u(s;\cdot)ds\right]$ arises often. If follows from the definition of expectation $E[\cdot]$ that

$$E\left[\int_0^t u(s;\cdot)ds\right] = \int_\Omega \int_0^t u(s;\omega)dsd\mathscr{P}. \tag{3.33}$$

Conditions (3.15) and (3.16) are sufficient to guarantee that Fubini's Theorem (see **[161]**) applies, so that the order of integration on the right-hand side of (3.33) can be interchanged, yielding

$$E\left[\int_0^t u(s;\cdot)ds\right] = \int_0^t \int_\Omega u(s;\omega)d\mathscr{P}ds = \int_0^t E[u(s;\cdot)]ds. \tag{3.34}$$

This result is particularly useful because it enables us to transfer the computation of the expectation of an integral to that of its integrand, of which we often have additional knowledge, typically in the form of a priori estimates.

Exercise 3.2.6. Verify (3.34) when u is a random characteristic function, and then when u is a random step function.

3.2.2   The Itó Integral for $\mathbb{R}$-Valued Stochastic Processes

The integral computed with respect to the Wiener process introduced in (3.9) and (3.11) is treated differently. We will apply essentially the same building block process to formulate a reasonable definition. In so doing, we must make certain that the stochastic process $\left\{ \int_0^t u(s;\omega)dW(s) : 0 \le t \le T \right\}$ is $\mathscr{F}_t$-adapted. Merely assuming that u satisfies (3.14) through (3.16) is insufficient because of the presence of the stochastic process $\{W(t) : 0 \le t \le T\}$. Indeed, we must further assume that

$$\sigma(\{W(u) : 0 \le u \le s\}) \subset \mathscr{F}_s, \forall 0 \le s \le t. \tag{3.35}$$

Which quantity is defined first, $\{W(u) : 0 \le u \le T\}$ or $\{\mathscr{F}_s : 0 \le s \le T\}$? By convention **(S.A.1)**, the very first thing we do is start with a complete probability space $(\Omega, \mathscr{F}, \mathscr{P})$ equipped with a filtration. But, this filtration is actually chosen based on the Wiener processes that are to be introduced into the stochastic differential equation being formulated. This is sensible because proceeding in the reverse order would run the risk of a Wiener process that we want to include in the model not being compatible with the filtration, thereby affecting the well-definedness of the Itó integral.

It is natural to expect that the integral $\int_0^t u(s;\omega)dW(s)$ would possess some of the same salient characteristics as the Wiener process $W(s)$. It will become apparent that this is indeed the case in the very first stage of the construction of the integral. Also, we shall employ the convention that for any partition arising in the construction of this integral, we will always choose the <u>left</u> endpoint of the subinterval as the sample point. Doing so will result in the so-called *Itó integral*. While different sample points (e.g., the midpoints or right endpoints of the subintervals) can be used to define meaningful stochastic integrals, the resulting constructions do not generate integrals equivalent to the Itó integral. (See **[161, 302]**.) This is an important departure from the Riemann integral, where the limiting process was independent of the choice of sample point. The various pathological problems and other constructions involved in defining a stochastic integral are very interesting, but we shall not pursue them here because our primary focus is the study of SEEs.

We now proceed with the construction of $\int_0^t u(s;\omega)dW(s)$. At each stage of the construction, we will consider the properties of the integral.

Step 1: <u>Random Characteristic Function</u>
Let $c \in \mathcal{L}^2(\Omega;\mathbb{R})$ and $[a,b] \subset [0,t]$. Define $u : [0,t] \times \Omega \longrightarrow [0,\infty)$ by $u(s;\omega) = c(\omega)\chi_{[a,b]}(s)$, as in (3.17). We define

$$\int_0^t u(s;\omega)dW(s) \equiv c(\omega)(W(b) - W(a)). \tag{3.36}$$

In particular, if $u(s;\omega) = c(\omega)\chi_{[0,t]}(s)$, then

$$\int_0^t u(s;\omega)dW(s) \equiv c(\omega)(W(t) - W(0)) = c(\omega)W(t). \tag{3.37}$$

So, for a fixed $\omega_0 \in \Omega$, this integral is a (random) constant multiple of a Wiener process. As such, the properties of the integral in this case follow almost exclusively from those of the Wiener process itself.

Proposition 3.2.3. (Properties of $\int_0^t c(\omega)dW(s)$)
Assume that $c, c_1, c_2 \in \mathcal{L}^2(\Omega;\mathbb{R})$.
i.) (Linearity) $\forall \alpha, \beta \in \mathbb{R}$,

$$\int_0^t [\alpha c_1(\omega) + \beta c_2(\omega)]dW(s) = \alpha \int_0^t c_1(\omega)dW(s) + \beta \int_0^t c_2(\omega)dW(s).$$

ii.) (Additivity) $\forall 0 < \tau < t$, $\int_0^t c(\omega)dW(s) = \int_0^\tau c(\omega)dW(s) + \int_\tau^t c(\omega)dW(s)$.
More generally, $\forall 0 = \tau_0 < \tau_1 < \ldots < \tau_m = t$, $\int_0^t c(\omega)dW(s) = \sum_{i=1}^m \int_{\tau_{i-1}}^{\tau_i} c(\omega)dW(s)$.

iii.) (Zero Expectation) *If* $E|c(\cdot)| < \infty$, *then* $E\left[\int_0^t c(\cdot)dW(s)\right] = 0$.

iv.) (Itó Isometry) $E \left| \int_0^t c(\cdot) dW(s) \right|^2 = \int_0^t E \left| c(\cdot) \right|^2 ds$.

v.) (Stochastic Properties) *Define* $F : [0,T] \times \Omega \longrightarrow \mathfrak{L}^2 (\Omega; \mathbb{R})$ *by*

$$F(t; \omega) = \int_0^t c(\omega) dW(s).$$

a.) $F(t; \cdot)$ *is* $\mathscr{F}_t$-*adapted,* $\forall 0 \leq t \leq T$;
b.) $\{F(t; \omega) : 0 \leq t \leq T, \, \omega \in \Omega\}$ *is a real martingale;*
c.) $F(t; \cdot)$ *has continuous sample paths a.s.* $[\mathscr{P}]$;
d.) *If* $c(\omega) = c^{\star}, \forall \omega \in \Omega$, *(i.e.,* $c(\cdot)$ *is deterministic), then* $F(t; \cdot)$ *is* $n(0,t)$.

Proof. (i)

$$\int_0^t \underbrace{[\alpha c_1(\omega) + \beta c_2(\omega)]}_{\text{Constant with respect to } s} dW(s) = [\alpha c_1(\omega) + \beta c_2(\omega)] \left(W(t) - \underbrace{W(0)}_{=0} \right)$$

$$= \alpha c_1(\omega) W(t) + \beta c_2(\omega) W(t)$$

$$= \alpha \int_0^t c_1(\omega) dW(s) + \beta \int_0^t c_2(\omega) dW(s).$$

(ii)

$$\int_0^t c(\omega) dW(s) = c_1(\omega) [W(t) - W(0)]$$

$$= c_1(\omega) [W(t) - W(\tau) + W(\tau) - W(0)]$$

$$= c_1(\omega) [W(\tau) - W(0)] + c_1(\omega) [W(t) - W(\tau)]$$

$$= \int_0^\tau c(\omega) dW(s) + \int_\tau^t c(\omega) dW(s).$$

(iii)

$$E \left[\int_0^t c(\cdot) dW(s) \right] = E\left[c(\cdot) W(t)\right] = E\left[c(\cdot)\right] E\left[W(t)\right] = E\left[c(\cdot)\right] (0) = 0.$$

(iv)

$$E \left| \int_0^t c(\cdot) dW(s) \right|^2 = E \left[\left| c(\cdot) W(t) \right|^2 \right]$$

$$= E \left[c^2(\cdot) W^2(t) \right]$$

$$= E \left[c^2(\cdot) \right] E \left[W^2(t) \right]$$

$$= E \left[c^2(\cdot) \right] t$$

$$= \int_0^t E \left[c^2(\cdot) \right] ds$$

$$= \int_0^t E \left| c(\cdot) \right|^2 ds.$$

(v) (a) For every $t \in [0,T]$, $W(t)$ is $\mathscr{F}_t$-measurable. Because constant multiples of $\mathscr{F}_t$-measurable functions are themselves $\mathscr{F}_t$-measurable (Why?), we conclude that $F(t;\omega) = c(\omega)W(t)$ is $\mathscr{F}_t$-measurable.

(b) The fact that $F(t;\cdot)$ has finite first and second moments follows from the properties of $c(\cdot)$. (Tell how.) The martingale property holds because $\forall 0 \le s \le t \le T$,

$$
\begin{aligned}
E[F(t)|\mathscr{F}_s] &= E[F(t) - F(s) + F(s)|\mathscr{F}_s] \\
&= E[F(t) - F(s)|\mathscr{F}_s] + E[F(s)|\mathscr{F}_s] \\
&= E[c(\cdot)(W(t) - W(s))|\mathscr{F}_s] + E[c(\cdot)W(s)|\mathscr{F}_s] \\
&= E[c(\cdot)(W(t) - W(s))] + c(\omega)W(s) \\
&= E[c(\cdot)]\underbrace{E[W(t) - W(s)]}_{=0} + \int_0^s c(\omega)dW(t) \\
&= 0 + F(s) = F(s).
\end{aligned}
$$

(c) For a given $\omega_0 \in \Omega$, $F(t;\omega_0) = c(\omega_0)W(t)$ is a constant multiple of a stochastic process possessing continuous sample paths a.s. $[\mathscr{P}]$. Hence, $F(t;\cdot)$ has continuous sample paths a.s. $[\mathscr{P}]$.

(d) If $c(\omega) = c^*, \forall \omega \in \Omega$, then $F(t;\omega) = c^*W(t)$, $\forall \omega \in \Omega$. Because $W(t)$ is $n(0,t)$, it follows that $F(t;\cdot)$ is also $n(0,t)$. $\qquad\square$

Exercise 3.2.7. Verify each step in the proof of Prop. 3.2.3. In particular, prove the more general version of the additivity property.

Exercise 3.2.8. Prove that $\{F(t;\cdot) : 0 \le t \le T\}$ has *orthogonal increments*, meaning that

$$
\forall 0 \le t_1 < t_2 < t_3 < t_4 \le T, \ E[(F(t_4;\cdot) - F(t_3;\cdot))(F(t_2;\cdot) - F(t_1;\cdot))] = 0.
$$

The next natural step in the construction of the Itó integral is to build upon Step 1 by considering random step functions.

Step 2: Random Step Functions
Assume that $u \in \mathscr{U}$ is given by (3.19) (with $[0,\infty)$ replaced by $\mathbb{R}$). The following definition is natural:

$$
\begin{aligned}
\int_0^t u(s;\omega)dW(s) &\equiv \sum_{i=1}^m \int_0^t c_i(\omega)\chi_{[t_{i-1},t_i)}(s)dW(s) \\
&= \sum_{i=1}^m \int_{t_{i-1}}^{t_i} c_i(\omega)dW(s) \hspace{3cm}(3.38) \\
&= \sum_{i=1}^m c_i(\omega)(W(t_i) - W(t_{i-1}))
\end{aligned}
$$

Using finite combinations of random characteristic functions should not severely impede our effort in verifying that (3.38) satisfies the properties listed in Prop. 3.2.3. Indeed, the nice properties of $E\,[\cdot]$ and $W(\cdot)$ should render the proofs straightforward.

Exercise 3.2.9. (**Mini-Project**) Verify that the integral defined by (3.38) satisfies the properties listed in Prop. 3.2.3.

The next step of the construction is where some more interesting analysis comes into play. For any $u \in \mathscr{U}$, we use the density of the step functions in $\mathscr{U}$ to construct a sequence of step functions that converges in the $\mathfrak{L}^2$-sense to u. Then, the stochastic integral is defined as the $\mathfrak{L}^2$-limit of the sequence of integrals of these step functions. (See **[161]** for details.)

Step 3: Arbitrary Member of $\mathscr{U}$
Let $u : [0,t] \times \Omega \longrightarrow \mathbb{R}$ belong to $\mathscr{U}$ and let $\{u_n\}$ be a sequence of step functions in $\mathscr{U}$ for which

$$\lim_{n \longrightarrow \infty} \int_0^t E\,|u_n(s;\cdot) - u(s;\cdot)|^2\,ds = 0. \tag{3.39}$$

Then, we define

$$\int_0^t u(s;\omega)dW(s) \equiv \lim_{n \longrightarrow \infty} \int_0^t u_n(s;\omega)dW(s), \tag{3.40}$$

where the limit is taken in the $\mathfrak{L}^2$-sense. This is the *Itó integral of u with respect to* $\{W(s) : 0 \le s \le t\}$ and we say u is *Itó integrable* whenever the limit on the right-hand side of (3.40) exists and is finite. Using some powerful convergence results of stochastic analysis, it can be shown that the Itó integral of such functions satisfies all of the properties in Prop. 3.2.3. This completes the construction.

Remark. If $u(s;\omega) = \widehat{u}(s)$, $\forall \omega \in \Omega$, (i.e., u is deterministic), then $\widehat{u}$ is automatically $\mathscr{F}_s$-adapted because it is constant with respect to ω and constant functions are measurable. In such case, the criterion used to define $\mathscr{U}$ (cf. (3.14) through (3.16)) reduces to $\int_0^t |\widehat{u}(s)|^2\,ds < \infty$. A common instance in which the integrand of a stochastic integral is, in fact, deterministic occurs with computations of the type $\int_0^t E\,|u(s;\cdot)|^2\,ds$.

As with the Lebesgue integral, we will need to compute $E\left[\int_0^t u(s;\cdot)dW(s)\right]$.

Exercise 3.2.10. Verify that

$$E\left[\int_0^t u(s;\cdot)dW(s)\right] = \int_0^t E\,[u(s;\cdot)]\,dW(s), \tag{3.41}$$

assuming first that u is a random characteristic function and then when u is a random step function. (In fact, it can be shown that (3.41) holds, $\forall u \in \mathscr{U}$.)

3.2.3   The Itó Formula in $\mathbb{R}$

Now, we are able to consider (3.8) and (3.11), or their equivalent integrated forms (3.12) and (3.13), respectively, more carefully.

Definition 3.2.4. Suppose that
i.) $u_1 : [0,T] \longrightarrow \mathbb{R}$ is $\mathscr{F}_t$-measurable and Lebesgue integrable;
ii.) $u_2 : [0,T] \longrightarrow \mathbb{R}$ is Itó integrable;
iii.) $Y_0 \in \mathcal{L}^2(\Omega;\mathbb{R})$ is an $\mathscr{F}_0$-measurable random variable independent of the Wiener process $\{W(t) : 0 \le t \le T\}$.
Then, the stochastic process $\{Y(t;\omega) : 0 \le t \le T, \omega \in \Omega\}$ for which

$$Y(t;\omega) = Y_0(\omega) + \int_0^t u_1(s;\omega)ds + \int_0^t u_2(s;\omega)dW(s), 0 \le t \le T, \qquad (3.42)$$

or equivalently,

$$\begin{cases} dY(t;\omega) = u_1(t;\omega)dt + u_2(t;\omega)dW(t), 0 < t < T, \omega \in \Omega, \\ Y(0;\omega) = Y_0(\omega), \omega \in \Omega, \end{cases} \qquad (3.43)$$

is called an *Itó process*.

Exercise 3.2.11. Prove that an Itó process $\{Y(t;\omega) : 0 \le t \le T, \omega \in \Omega\}$ is $\mathscr{F}_t$-measurable.

An important tool in obtaining explicit formulae for the solutions of some elementary SDEs, as well as some crucial a priori estimates, is the ability to compute stochastic differentials of the form $d[h(t,Y(t;\omega)]$, where $\{Y(t;\omega) : 0 \le t \le T, \omega \in \Omega\}$ is an Itó process. This suggests the need for a stochastic counterpart of the familiar chain rule for differentiation. We shall present a heuristic discussion that suggests the form of this rule. We begin with the deterministic setting, for comparison purposes.

Suppose that $x : [0,T] \longrightarrow \mathbb{R}$ is differentiable and that $h : [0,T] \times \mathbb{R} \longrightarrow \mathbb{R}$ has continuous first and second-order partials. We then know from elementary calculus that if $H(t) = h(t,x)$, where $x = x(t)$, then

$$\frac{dH}{dt} = \frac{\partial h}{\partial t} + \frac{\partial h}{\partial x}\frac{dx}{dt}, \qquad (3.44)$$

or in its equivalent differential form,

$$dH = \left(\frac{\partial h}{\partial t}\right)dt + \left(\frac{\partial h}{\partial x}\right)dx. \qquad (3.45)$$

By way of motivation for the stochastic chain rule, it is helpful to understand the formulation of (3.45). Recall that the two variable Taylor series for $h(t,x)$ centered at (t_0, x_0) is given by

$$
h(t,x) = h(t_0,x_0) + \frac{\partial h}{\partial t}(t_0,x_0)(t-t_0) + \frac{\partial h}{\partial x}(t_0,x_0)(x-x_0)
$$

$$
+ \frac{1}{2}\left[\frac{\partial^2 h}{\partial t^2}(t_0,x_0)(t-t_0)^2 + 2\frac{\partial^2 h}{\partial t \partial x}(t_0,x_0)(t-t_0)(x-x_0)\right.
$$

$$
\left. + \frac{\partial^2 h}{\partial x^2}(t_0,x_0)(x-x_0)^2\right] + [\text{Higher} - \text{order terms}].
\qquad (3.46)
$$

Let $\triangle x = x(t_0 + \triangle t) - x(t_0)$. We compute $\frac{dH}{dt}(t_0)$ as follows:

$$
\frac{dH}{dt}(t_0) = \lim_{\triangle t \rightarrow 0} \frac{H(t_0 + \triangle t) - H(t_0)}{\triangle t}
$$

$$
= \lim_{\triangle t \rightarrow 0} \frac{h(t_0 + \triangle t, x_0 + \triangle x) - h(t_0,x_0)}{\triangle t}
\qquad (3.47)
$$

$$
= \lim_{\triangle t \rightarrow 0} \frac{1}{\triangle t}\left\{\left[h(t_0,x_0) + \frac{\partial h}{\partial t}(t_0,x_0)(t_0 + \triangle t - t_0)\right.\right.
$$

$$
+ \frac{\partial h}{\partial x}(t_0,x_0)(x_0 + \triangle x - x_0) + \frac{1}{2}\left[\frac{\partial^2 h}{\partial t^2}(t_0,x_0)(t_0 + \triangle t - t_0)^2\right.
$$

$$
+ 2\frac{\partial^2 h}{\partial t \partial x}(t_0,x_0)(t_0 + \triangle t - t_0)(x_0 + \triangle x - x_0)
$$

$$
\left.\left.\left. + \frac{\partial^2 h}{\partial x^2}(t_0,x_0)(x_0 + \triangle x - x_0)^2\right] + [\text{Higher} - \text{order terms}]\right] - h(t_0,x_0)\right\}
$$

$$
= \lim_{\triangle t \rightarrow 0}\left\{\frac{\partial h}{\partial t}(t_0,x_0)\frac{\triangle t}{\triangle t} + \frac{\partial h}{\partial x}(t_0,x_0)\frac{\triangle x}{\triangle t}\right.
$$

$$
+ \frac{1}{2}\left[\frac{\partial^2 h}{\partial t^2}(t_0,x_0)\frac{(\triangle t)^2}{\triangle t} + 2\frac{\partial^2 h}{\partial t \partial x}(t_0,x_0)\frac{(\triangle x)(\triangle t)}{\triangle t} + \frac{\partial^2 h}{\partial x^2}(t_0,x_0)\frac{(\triangle x)^2}{\triangle t}\right]
$$

$$
\left. + \frac{(\triangle x)^m (\triangle t)^n}{\triangle t}[\text{Higher} - \text{order terms}]\right\},
$$

where $(m+n) \in \mathbb{N} \setminus \{1\}$.

All partial derivatives are computed at (t_0, x_0) (and hence are constant with respect to $\triangle t$). So, by continuity,

$$
\lim_{\triangle t \rightarrow 0} \triangle x = \lim_{\triangle t \rightarrow 0}[x(t_0 + \triangle t) - x(t_0)] = 0.
$$

As such, taking the limit as $\triangle t \rightarrow 0$ in (3.47) wipes out all terms except the first two, thereby resulting in (3.44).

Now, let us consider the stochastic case in which $\{x(t) : 0 \leq t \leq T\}$ is now an Itô process. The basic idea is to consider computations similar to those in (3.46) and (3.47), although the technical details are more complicated. We do not expect the first three terms of the Taylor formula to cause problems, but the fact that

$\{W(t) : 0 \le t \le T\}$ has quadratic variation, meaning $(W(t_0 + \triangle t) - W(t_0))^2$ (see Exer. 2.9.13 (ii)), creates a new wrinkle with which we must contend. But, in which terms would computing such a difference occur?

We must remember that now $x(t; \omega)$ is given by (3.42), which involves an Itó integral. Thus, when computing $x(t_0 + \triangle t; \omega) - x(t_0; \omega)$, the term $\int_{t_0}^{t_0 + \triangle t} u_2(s; \omega) dW(s)$ arises. The most simplistic case is when $u_2(s; \omega) = c(\omega) \chi_{[0,T)}$. Here,

$$\int_{t_0}^{t_0 + \triangle t} u_2(s; \omega) dW(s) = c(\omega) \left[W(t_0 + \triangle t) - W(t_0) \right]. \tag{3.48}$$

If $u_2(s; \omega)$ is a step function, then this integral is computed as a finite sum of the differences of the form occurring on the right-hand side of (3.48). Consequently, whenever we consider $(\triangle x)^n$, where $n \in \mathbb{N} \setminus \{1\}$, a term involving $(W(t_0 + \triangle t) - W(t_0))^2$ arises. However, when taking the limit as $\triangle t \longrightarrow 0$ in the stochastic version of (3.47), the only second-order that is not wiped out is $\frac{1}{2} \frac{\partial^2 h}{\partial x^2}(t_0, x_0)(\triangle x)^2$. (Why?) This, coupled with the fact that

$$\lim_{\triangle t \longrightarrow 0} \frac{(\triangle x)^m (\triangle t)^n}{\triangle t} = 0,$$

for all $m, n \in \mathbb{N}$ for which $(m + n) \in \mathbb{N} \setminus \{1\}$ (except for the combination $m = 0, n = 2$), is the basis for the following formula:

Proposition 3.2.5. (Itó's Formula in $\mathbb{R}$)
Suppose that $\{X(t; \omega) : 0 \le t \le T, \omega \in \Omega\}$ is an Itó process and let $h : [0, T] \times \mathbb{R} \longrightarrow \mathbb{R}$ have continuous first- and second-order partials. Then, the stochastic process $H : [0, T] \times \Omega \longrightarrow \mathbb{R}$ defined by $H(t; \omega) = h(t, X(t; \omega))$ is an Itó process and

$$dH(t; \omega) = \underbrace{\frac{\partial h}{\partial t}(t, X(t; \omega)) dt + \frac{\partial h}{\partial x}(t, X(t; \omega)) dX(t; \omega)}_{} \tag{3.49}$$

Familiar part

New part

$$+ \overbrace{\frac{1}{2} \frac{\partial^2 h}{\partial x^2}(t, X(t; \omega))(dX(t; \omega))^2}.$$

(See **[161]** for a formal proof.)

Remarks.
1. An infinitesimal version of the remarks directly preceding Prop. 3.2.5 is as follows:

$$dt \, dt = dt \, dW = dW \, dt = 0, \quad (dW)(dW) = dt. \tag{3.50}$$

2. Do not inadvertently confuse the term $(dX(t; \omega))^2$ for dt! Remember, we infer from (3.42) that

$$\begin{aligned}
(dX(t; \omega))^2 &= (u_1 dt + u_2 dW)^2 \\
&= u_1^2 (dt)^2 + 2u_1 u_2 dt \, dW + u_2^2 (dW)^2 \\
&= u_2^2 dt.
\end{aligned} \tag{3.51}$$

In a completely analogous manner, the following formula can be established using a three-variable Taylor series.

Proposition 3.2.6. (Multivariable Itó Formula)
Suppose that $\{X_i(t;\omega) : 0 \leq t \leq T, \omega \in \Omega\}$, $i = 1,\ldots,N$, *are Itó processes (involving the same Wiener process* $W(t)$*) and let* $h : [0,T] \times \mathbb{R}^N \longrightarrow \mathbb{R}$ *have continuous first- and second-order partials. Then, the stochastic process* $\{H(t;\omega) : 0 \leq t \leq T, \omega \in \Omega\}$ *defined by*

$$H(t;\omega) = h(t, X_1(t;\omega), \ldots, X_N(t;\omega))$$

is an Itó process and

$$
\begin{aligned}
dH(t;\omega) &= \frac{\partial h}{\partial t}(t, X_1(t;\omega), \ldots, X_N(t;\omega))\, dt + \\
&\quad \frac{\partial h}{\partial X_1}(t, X_1(t;\omega), \ldots, X_N(t;\omega))\, dX_1(t;\omega) + \ldots \\
&\quad + \frac{\partial h}{\partial X_N}(t, X_1(t;\omega), \ldots, X_N(t;\omega))\, dX_N(t;\omega) + \\
&\quad \frac{1}{2} \sum_{i=1}^{N} \sum_{j=1}^{N} \frac{1}{2} \frac{\partial^2 h}{\partial X_i \partial X_j}(t, X_1(t;\omega), \ldots, X_N(t;\omega))\, dX_i dX_j.
\end{aligned}
\tag{3.52}
$$

Exercise 3.2.12. Suppose that

$$dX_1 = u_1 dt + u_2 dW, \quad dX_2 = \widehat{u}_1 dt + \widehat{u}_2 dW. \tag{3.53}$$

Convince yourself that in the Taylor series computation leading to (3.52), all of the second-order terms not explicitly appearing in (3.52) go to zero as $\triangle t \longrightarrow 0$.

Exercise 3.2.13. **(Integration by Parts)**
Define $h : [0,T] \times \mathbb{R} \times \mathbb{R} \longrightarrow \mathbb{R}$ by $h(t, x_1, x_2) = x_1 x_2$.
i.) Compute all partial derivatives arising in (3.52) for this particular function h.
ii.) Apply (3.52) to derive a formula for $d(X_1(t;\omega)X_2(t;\omega))$, for Itó processes $\{X_1(t;\omega) : 0 \leq t \leq T, \omega \in \Omega\}$ and $\{X_2(t;\omega) : 0 \leq t \leq T, \omega \in \Omega\}$.

Exercise 3.2.14. **(Looking Ahead to Linear SDEs)**
Suppose that $\{X(t;\omega) : 0 \leq t \leq T, \omega \in \Omega\}$ is an Itó process satisfying

$$
\begin{cases}
dX(t;\omega) = AX(t;\omega)dt + BX(t;\omega)dW(t), 0 < t < T, \omega \in \Omega, \\
X(0;\omega) = X_0(\omega), \omega \in \Omega,
\end{cases}
\tag{3.54}
$$

where $A, B \in \mathbb{R}$. Compute $d[\ln X(t;\omega)]$.

Exercise 3.2.15. **(Looking Ahead to Nonhomogenous Linear SDEs)**
Suppose that $\{X(t;\omega) : 0 \leq t \leq T, \omega \in \Omega\}$ is an Itó process satisfying

$$
\begin{cases}
dX(t;\omega) = AX(t;\omega)dt + \varepsilon dW(t), 0 < t < T, \omega \in \Omega, \\
X(0;\omega) = X_0(\omega), \omega \in \Omega,
\end{cases}
\tag{3.55}
$$

where $A, \varepsilon \in \mathbb{R}$.

i.) Compute $d\left[e^{-At}X(t;\omega)\right]$.

ii.) Use the integrated form of the differential in (i) to derive a formula for $X(t;\omega)$.

3.2.4   Some Crucial Estimates

Certain estimates shall occur repeatedly in slightly different forms. We establish two general estimates that can be used as paradigms. Appealing to these results will help to streamline the computations in a given argument.

Lemma 3.2.7. *Let $p > 2$ and assume that $u \in \mathscr{U}$ with $E \int_0^T |u(s;\cdot)|^p\, ds < \infty$. Then, $\forall 0 \leq t' < t \leq T$,*

i.) $E \left| \int_{t'}^t u(s;\cdot)ds \right|^p \leq |t - t'|^{\frac{p}{q}} \int_{t'}^t E\, |u(s;\cdot)|^p\, ds$, *where* $\frac{1}{p} + \frac{1}{q} = 1$;

ii.) $E \left| \int_{t'}^t u(s;\cdot)dW(s) \right|^p \leq \zeta\,(t,t') \int_{t'}^t E\, |u(s;\cdot)|^p\, ds$, *where*

$$\zeta\,(t,t') = 2^p \left[\left(t^{\frac{p}{2}} + (t')^{\frac{p}{2}} \right) \left(\frac{p(p-1)}{2} \right)^{\frac{p}{2}} \right]. \tag{3.56}$$

Proof. (i) Apply Hölder's inequality and the properties of the integral, and then simplify as follows:

$$E \left| \int_{t'}^t u(s;\cdot)ds \right|^p \leq E \left| \left(\int_{t'}^t u^p(s;\cdot)ds \right)^{\frac{1}{p}} \underbrace{\left(\int_{t'}^t 1^q ds \right)^{\frac{1}{q}}}_{=(t-t')^{\frac{1}{q}}} \right|^p$$

$$\leq |t - t'|^{\frac{p}{q}} E \left| \int_{t'}^t u^p(s;\cdot)ds \right|$$

$$\leq |t - t'|^{\frac{p}{q}} E \int_{t'}^t |u^p(s;\cdot)|\, ds \tag{3.57}$$

$$= |t - t'|^{\frac{p}{q}} E \int_{t'}^t |u(s;\cdot)|^p\, ds.$$

(ii) We follow the presentation provided in **[203, 204, 285]**, with some modifications.
<u>Step 1:</u> First, assume that $t' = 0$.
Let $u \in \mathscr{U}$ and define $y : [0,T] \times \Omega \to \mathbb{R}$ by

$$y(t;\omega) = \int_0^t u(s;\omega)dW(s).$$

We shall apply Prop. 3.2.5 with $h(t,y) = |y(t)|^p$. To this end, we need the following:

$$\frac{\partial h}{\partial t} = 0$$

$$\frac{\partial h}{\partial y} = p|y(t;\omega)|^{p-1}\underbrace{\frac{d}{dy}|y(t;\omega)|}_{=\frac{y(t;\omega)}{|y(t;\omega)|}} = p|y(t;\omega)|^{p-2}y(t;\omega) \tag{3.58}$$

$$\frac{\partial^2 h}{\partial y^2} = p|y(t;\omega)|^{p-2}(1) + \left[p(p-2)|y(t;\omega)|^{p-3}\frac{y(t;\omega)}{|y(t;\omega)|}\right]y(t;\omega)$$

$$= p|y(t;\omega)|^{p-2} + p(p-2)|y(t;\omega)|^{p-2}$$

$$= p(p-1)|y(t;\omega)|^{p-2}.$$

Note that $d[y(t;\omega)] = u(t;\omega)dW(t)$. Hence, by Prop. 3.2.5, we have

$$d[|y(t;\omega)|^p] = 0 + p|y(t;\omega)|^{p-2}y(t;\omega)\underbrace{d[y(t;\omega)]}_{=u(t;\omega)dW(t)}$$

$$+\frac{1}{2}p(p-1)|y(t;\omega)|^{p-2}\cdot\underbrace{(d[y(t;\omega)])^2}_{=(u(t;\omega)dW(t))^2=u^2(t;\omega)dt} \tag{3.59}$$

$$= p|y(t;\omega)|^{p-2}y(t;\omega)u(t;\omega)dW(t)$$

$$+\frac{p(p-1)}{2}|y(t;\omega)|^{p-2}|u(t;\omega)|^2\,dt.$$

Note that

$$\int_0^t d[|y(s;\omega)|^p] = |y(t;\omega)|^p - \underbrace{|y(0;\omega)|^p}_{=0}. \tag{3.60}$$

As such, integrating both sides of (3.59) over $(0,t)$, followed by taking the expectation on both sides, yields

$$E|y(t;\cdot)|^p = \underbrace{E\left[\int_0^t p|y(s;\cdot)|^{p-2}y(s;\cdot)u(s;\cdot)dW(s)\right]}_{=0}$$

$$+E\left[\int_0^t \frac{p(p-1)}{2}|y(s;\cdot)|^{p-2}|u(s;\cdot)|^2\,ds\right] \tag{3.61}$$

$$+\frac{p(p-1)}{2}\int_0^t E\left[|y(s;\cdot)|^{p-2}|u(s;\cdot)|^2\right]\,ds.$$

An application of Hölder's inequality now yields the general estimate

$$E[XY] \le \left((E|X|^{p_*})^{\frac{1}{p_*}}(E|X|^{q_*})^{\frac{1}{q_*}}\right), \tag{3.62}$$

where $\frac{1}{p_*} + \frac{1}{q_*} = 1$ and X and Y are random variables for which the right-hand side of (3.62) is defined and finite. We apply this estimate with the following identifications:

$$X(s;\omega) = |y(s;\omega)|^{p-2}, \ Y(s;\omega) = |u(s;\omega)|^2, \ p_* = \frac{p}{p-2}, \ q_* = \frac{p}{2}.$$

(Note that we use this choice of q_* because we inevitably want the term $|u(s;\omega)|^p$.)
We now continue (3.61) as follows:

$$E|y(t;\cdot)|^p \leq \left(E\left[\left(|y(s;\cdot)|^{p-2} \right)^{\frac{p}{p-2}} \right] \right)^{\frac{p-2}{p}} \left(E\left[\left(|u(s;\cdot)|^2 \right)^{\frac{p}{2}} \right] \right)^{\frac{2}{p}}$$

$$= \left(E\left[|y(s;\cdot)|^p \right] \right)^{\frac{p-2}{p}} \left(E\left[|u(s;\cdot)|^p \right] \right)^{\frac{2}{p}}. \tag{3.63}$$

Thus, we infer from (3.61) that

$$E|y(t;\cdot)|^p \leq \frac{p(p-1)}{2} \int_0^t \left(E\left[|y(s;\cdot)|^p \right] \right)^{\frac{p-2}{p}} \left(E\left[|u(s;\cdot)|^p \right] \right)^{\frac{2}{p}} ds. \tag{3.64}$$

Observe that

$$0 \leq s \leq t \implies E|y(s;\cdot)|^p \leq E|y(t;\cdot)|^p. \tag{3.65}$$

(Why?) As such, we can bound the term $\left(E\left[|y(s;\cdot)|^p \right] \right)^{\frac{p-2}{p}}$ appearing inside the integrand on the right-hand side of (3.64) above by $\left(E\left[|y(t;\cdot)|^p \right] \right)^{\frac{p-2}{p}}$, which can be factored out of the integral using linearity. Thus, (3.64) implies that

$$E|y(t;\cdot)|^p \leq \frac{p(p-1)}{2} \left(E\left[|y(t;\cdot)|^p \right] \right)^{\frac{p-2}{p}} \int_0^t \left(E\left[|u(s;\cdot)|^p \right] \right)^{\frac{2}{p}} ds. \tag{3.66}$$

Next, note that $\forall a, b > 0$,

$$a \leq a^{\frac{p-2}{p}} b \implies \underbrace{a \cdot a^{-\left(\frac{p-2}{p} \right)}}_{=a^{\frac{2}{p}}} \leq b \implies a \leq b^{\frac{p}{2}}. \tag{3.67}$$

Applying (3.67) for each $t > 0$ in (3.66) yields

$$E|y(t;\cdot)|^p \leq \left(\frac{p(p-1)}{2} \right)^{\frac{p}{2}} \left(\int_0^t \left(E\left[|u(s;\cdot)|^p \right] \right)^{\frac{2}{p}} ds \right)^{\frac{p}{2}}. \tag{3.68}$$

Now, apply Hölder's inequality again, this time on the outside integral rather than on the expectation portion, to obtain

$$\int_0^t \left(\left(E\left[|u(s;\cdot)|^p \right] \right)^{\frac{2}{p}} \cdot 1 \right) ds \leq \left(\int_0^t \left(\left(E\left[|u(s;\cdot)|^p \right] \right)^{\frac{2}{p}} \right)^{\frac{p}{2}} ds \right)^{\frac{2}{p}} \left(\int_0^t 1^{\frac{p}{p-2}} ds \right)^{\frac{p-2}{p}}$$

$$= \left(\int_0^t E\left[|u(s;\cdot)|^p \right] ds \right)^{\frac{2}{p}} t. \tag{3.69}$$

Substituting (3.69) into (3.68) then yields

$$E\,|y(t;\cdot)|^p \leq \left(\frac{p(p-1)}{2}\right)^{\frac{p}{2}} \left(\left(\int_0^t E\,[|u(s;\cdot)|^p]\,ds\right)^{\frac{2}{p}} t\right)^{\frac{p}{2}}$$

$$= \left(\frac{tp(p-1)}{2}\right)^{\frac{p}{2}} \int_0^t E\,[|u(s;\cdot)|^p]\,ds, \qquad (3.70)$$

as needed.

<u>Step 2:</u> Now, assume that $0 < t' \leq t$.
Use additivity, followed by the inequality $(a+b)^p \leq 2^p\,(a^p+b^p)$ and the linearity of expectation, to obtain

$$E\left|\int_{t'}^t u(s;\cdot)dW(s)\right|^p = E\left|\int_0^t u(s;\cdot)dW(s) - \int_0^{t'} u(s;\cdot)dW(s)\right|^p$$

$$\leq 2^p E\left[\left|\int_0^t u(s;\cdot)dW(s)\right|^p + \left|\int_0^{t'} u(s;\cdot)dW(s)\right|^p\right] \quad (3.71)$$

$$= 2^p\left(E\left|\int_0^t u(s;\cdot)dW(s)\right|^p + E\left|\int_0^{t'} u(s;\cdot)dW(s)\right|^p\right).$$

Now, apply the result of Step 1 to each of the two integrals on the right-hand side of (3.71):

$$E\left|\int_{t'}^t u(s;\cdot)dW(s)\right|^p \leq 2^p\left[\left(\frac{tp(p-1)}{2}\right)^{\frac{p}{2}}\int_0^t E\,[|u(s;\cdot)|^p]\,ds+\right.$$

$$\left. + \left(\frac{t'p(p-1)}{2}\right)^{\frac{p}{2}}\int_0^{t'} E\,[|u(s;\cdot)|^p]\,ds\right]. \qquad (3.72)$$

Because $|u(s)|^p \geq 0$, it follows that $E\,|u(s;\cdot)|^p \geq 0$. (Why?) This fact, coupled with the monotonicity property of the integral, yields

$$\int_0^{t'} E\,[|u(s;\cdot)|^p]\,ds \leq \int_0^t E\,[|u(s;\cdot)|^p]\,ds.$$

Using this fact enables us to simplify (3.72) to obtain

$$E\left|\int_{t'}^t u(s;\cdot)dW(s)\right|^p \leq 2^p\left[t^{\frac{p}{2}}+(t')^{\frac{p}{2}}\right]\left(\frac{p(p-1)}{2}\right)^{\frac{p}{2}}\int_0^t E\,[|u(s;\cdot)|^p]\,ds$$

$$= \zeta\,(t,t')\int_{t'}^t E\,|u(s;\cdot)|^p\,ds,$$

as needed. This completes the proof. $\qquad\square$

Remark. We can apply the estimates in Lemma 3.2.7 to a stochastic process of the form $(s; \omega) \mapsto g(s, X(s; \omega))$, provided that $(s; \omega) \mapsto X(s; \omega)$ is nice enough (e.g., if it is an Itó process).

3.3 The Cauchy Problem — Formulation

Assume **(S.A. 1)**. The focus of our study in the remainder of this chapter will be SDEs of the form

$$\begin{cases} dX(t; \omega) = aX(t; \omega)dt + cX(t; \omega)dW(t), \, 0 < t < T, \omega \in \Omega, \\ X(0; \omega) = X_0(\omega), \omega \in \Omega, \end{cases} \quad (3.73)$$

for a single source of noise and, more generally,

$$\begin{cases} dX(t; \omega) = aX(t; \omega)dt + \sum_{k=1}^{m} c_k X(t; \omega)dW_k(t), \, 0 < t < T, \omega \in \Omega, \\ X(0; \omega) = X_0(\omega), \omega \in \Omega, \end{cases} \quad (3.74)$$

for multiple sources of noise. We assume that

(H3.1) $a, c, c_k (k = 1, \ldots, m)$ are real constants.
(H3.2) $\{W(t) : 0 \le t \le T\}$ is a Wiener process, or in the case of (3.74),

$$\{\{W_k(t) : 0 \le t \le T\} : k = 1, \ldots, m\}$$

is a collection of independent Wiener processes.
(H3.3) X_0 is an $\mathscr{F}_0$-measurable random variable in $\mathfrak{L}^2(\Omega; \mathbb{R})$ independent of the Wiener process(es) in **(H3.2)**.

Remark. We must assume that

$$\sigma(X_0, W(s) : 0 \le s \le t) \subset \mathscr{F}_t, \forall 0 \le t \le T, \quad (3.75)$$

when dealing with (3.73), and

$$\sigma(X_0, W_1(s), \ldots, W_m(s) : 0 \le s \le t) \subset \mathscr{F}_t, \forall 0 \le t \le T, \quad (3.76)$$

when dealing with (3.74).

A "solution" of (3.73) or (3.74), first and foremost, must be an $\mathbb{R}$-valued stochastic process $\{X(t; \omega) : 0 \le t \le T, \, \omega \in \Omega\}$, which we shall write succinctly as $X : [0, T] \times \Omega \longrightarrow \mathbb{R}$. But, what properties must it satisfy?

Exercise 3.3.1. Based on your experience with deterministic evolution equations and what you learned about stochastic processes in Chapter 2, what properties seem natural to expect a solution of (3.73) or (3.74) to possess?

Consider the following integrated forms of (3.73) and (3.74), respectively:

$$X(t;\omega) = X_0(\omega) + \int_0^t aX(s;\omega)ds + \int_0^t cX(s;\omega)dW(s), 0 < t < T \qquad (3.77)$$

$$X(t;\omega) = X_0(\omega) + \int_0^t aX(s;\omega)ds + \sum_{k=1}^m \int_0^t c_k X(s;\omega)dW_k(s), 0 < t < T \qquad (3.78)$$

First, the integrals on the right-hand sides of (3.77) and (3.78) must be defined. This requires X to belong to $\mathscr{U}$. Specifically, $X(t;\cdot)$ should be $\mathscr{F}_t$-adapted, $\forall 0 \le t \le T$, and at least $\int_0^T E|X(s;\cdot)|^2 ds < \infty$. Because $\mathscr{U}$ is a Banach space, constant multiples of X also satisfy these properties, so that the integrals appearing in (3.77) and (3.78) are defined.

Ultimately, we want to be able to apply the $\mathfrak{L}^2$-calculus to quantities involving the solution process $X : [0,T] \times \Omega \longrightarrow \mathbb{R}$ so that, for instance, we can estimate its moments. As such, we shall require that $\sup\left\{E|X(t;\cdot)|^2 : 0 \le t \le T\right\} < \infty$. This implies that $\int_0^T E|X(s;\cdot)|^2 ds < \infty$. (Why?)

Finally, it is reasonable to impose a certain degree of regularity on the solution process. The solution curves with which we dealt in the deterministic setting in Volume 1 were at least continuous, if not differentiable. Given that the worst part of the equation, namely the Wiener process, itself has continuous sample paths a.s. $[\mathscr{P}]$, it is not unreasonable to demand the same of the solution process.

Taking a step back, the above discussion can be summarized by requiring X to belong to the space $\mathbb{C}\left([0,T];\mathfrak{L}^2(\Omega;\mathbb{R})\right)$. A version of this condition will <u>always</u> be required for solution processes with which we are interested.

It remains to impose a connection between X and the equation (3.77) or (3.78). It is at this point where we will make different choices, depending on the behavior of the data. In this most basic setting, it will become apparent that it is reasonable to require that $X(t;\omega)$ satisfy (3.77) and (3.78) a.s. $[\mathscr{P}]$, $\forall 0 \le t \le T$. (To what requirement does this translate in the deterministic setting?) Later, we will encounter situations in which this is too restrictive and we will, as a result, weaken our definition of a solution to a *milder* version. This will occur in Chapter 7.

Summarizing, we seek solutions of (3.73) and (3.74) in the following sense:

Definition 3.3.1. A stochastic process $X : [0,T] \times \Omega \longrightarrow \mathbb{R}$ is a *strong solution* of (3.73) (resp. (3.74)) on $[0,T]$ if $X \in \mathbb{C}\left([0,T];\mathfrak{L}^2(\Omega;\mathbb{R})\right)$ and $X(t;\omega)$ satisfies (3.77) (resp. (3.78)) a.s. $[\mathscr{P}]$, $\forall 0 \le t \le T$.

3.4   Existence and Uniqueness of a Strong Solution

We shall state and prove the main result directly for (3.74) and recover (3.73) as a special case.

Theorem 3.4.1. *If* **(H3.1)** *through* **(H3.3)** *are satisfied, then (3.74) has a unique strong solution on* $[0, T]$.

Before launching into the proof, it is helpful to recall the strategy used in the deterministic case, and then determine to what extent the approach is applicable in the present setting. To this end, consider the following deterministic homogenous IVP in $\mathbb{R}$:

$$\begin{cases} U'(t) = aU(t), 0 \leq t \leq T, \\ U(0) = U_0, \end{cases} \tag{3.79}$$

where $U_0 \in \mathbb{R}$ and $a \in \mathbb{R}$. The desired solution $U(t) = e^{at} U_0$ is constructed using the iteration scheme described below.

We begin with the integrated form of (3.79) given by

$$U(t) = U_0 + \int_0^t aU(s)ds, \, 0 \leq t \leq T. \tag{3.80}$$

In order to overcome the self-referential nature of (3.80), we replace $U(s)$ on the right-hand side of (3.80) by an approximation of $U(s)$. Presently, the only knowledge about U that we have is its value at $t = 0$, namely U_0. So, naturally we use $U(s) = U_0$ as the initial approximation. Making this substitution yields the following crude approximation of $U(t)$:

$$U(t) \approx U_0 + \int_0^t aU_0 ds = U_0 + aU_0 t, \, 0 \leq t \leq T. \tag{3.81}$$

Now, let $U_1(t) = U_0 + aU_0 t$. In order to improve the approximation, we can replace $U(s)$ on the right-hand side of (3.80) by $U_1(s)$ to obtain

$$\begin{aligned} U(t) &\approx U_0 + \int_0^t aU_1(s)ds \\ &= U_0 + \int_0^t a\left[U_0 + \int_0^s aU_0 d\tau\right] ds \\ &= U_0 + \int_0^t aU_0 ds + \int_0^t a\left(\int_0^s aU_0 d\tau\right) ds \\ &= U_0 + aU_0 t + a^2 U_0 \frac{t^2}{2} \\ &= \left(a^0 t^0 + at + a^2 \frac{t^2}{2}\right) U_0. \end{aligned} \tag{3.82}$$

The above sequence of "successively better" approximations can be formally described by the recursive sequence

$$\mathbf{U}_m(t) = \mathbf{U}_0 + \int_0^t a\mathbf{U}_{m-1}(s)ds, \ m \in \mathbb{N}. \tag{3.83}$$

Proceeding as in (3.82) leads to the following explicit formula for $\mathbf{U}_m(t)$:

$$\mathbf{U}_m(t) = \sum_{k=0}^m \left(\frac{a^k t^k}{k!} \right) \mathbf{U}_0. \tag{3.84}$$

Moreover,

$$\lim_{m \to \infty} \left(\sup_{0 \le t \le T} \left| \mathbf{U}_m(t) - e^{at}\mathbf{U}_0 \right| = 0 \right), \tag{3.85}$$

(Why?) which then leads to the desired solution. (Tell how.)

The similarity between (3.79) and (3.74) suggests that considering the convergence (in an appropriate sense) of the sequence of stochastic processes recursively defined by

$$\begin{cases} X_0(t;\omega) = & X_0(\omega), \\ X_n(t;\omega) = & X_0(\omega) + \int_0^t aX_{n-1}(s;\omega)ds \\ & + \sum_{k=1}^m \int_0^t c_k X_{n-1}(s;\omega)dW_k(s), \ n \in \mathbb{N}, \end{cases} \tag{3.86}$$

for $0 \le t \le T$, might be a viable approach. In fact, the proof of Thrm. 3.4.1 very much resembles the proof in the deterministic case outlined above with appropriate modifications made to account for the probabilistic nature of (3.74), specifically the Itó integral.

Proof of Theorem 3.4.1: The proof of this theorem presented below is a blend of the approaches used in **[126, 142, 285]**. We divide the proof into several subclaims.

Existence of a Strong Solution:
<u>Claim 1</u>: Each term of the sequence $\{X_n\}$ defined in (3.86) is a well-defined martingale.
Proof. We must verify the following, $\forall n \ge 0$:
    i.) Both integrals in (3.86) are well-defined,
    ii.) the mapping $t \mapsto X_n(t;\omega)$ is continuous a.s. $[\mathscr{P}]$,
    iii.) $X_n(t)$ is $\mathscr{F}_t$-adapted, $\forall t \ge 0$.

We proceed inductively, beginning with $n = 0$. Observe that $X_0(t;\omega) = X_0(\omega)$ belongs to $\mathscr{U}$ (Why?) and is a constant random variable with respect to t. Hence, X_0 automatically has continuous sample paths a.s. $[\mathscr{P}]$. (Why?) As such, aX_0 and $c_k X_0$ ($k = 1, \ldots, m$) belong to $\mathscr{U}$ and so, the mappings $t \mapsto \int_0^t aX_0(\omega)ds$ and $t \mapsto \int_0^t c_k X_0(\omega)dW_k(s)$ ($k = 1, \ldots, m$) have continuous sample paths a.s. $[\mathscr{P}]$. (Why?) The $\mathscr{F}_t$-adaptedness follows because X_n is an Itó process. (Verify this!)

Finally, the martingale property holds trivially because X_0 is a constant random variable and the Itó integral (as a function of its upper limit) is a martingale. (Tell why.) This establishes the base case.

The inductive step follows easily. (How?) This completes the proof of Claim 1. $\diamondsuit$

<u>Claim 2</u>: The sequence $\{X_n\}$ is a uniformly bounded subset of $\mathbb{C}\left([0,T];\mathfrak{L}^2\left(\Omega;\mathbb{R}\right)\right)$.
Proof. We must show that $\exists M > 0$ such that

$$\sup_{n\geq 0}\left\{\sup_{0\leq t\leq T} E\,|X_n(t;\cdot)|^2\right\}\leq M. \tag{3.87}$$

Exercise 3.4.1. Upon what quantities do you suspect the bound M should depend?

Let $0 \leq t \leq T$ and $n \in \mathbb{N}$. The idea is to show that $\exists \zeta_1, \zeta_2 > 0$ (independent of n) such that

$$E\,|X_n(t;\cdot)|^2 \leq \zeta_1 + \zeta_2 \int_0^t E\,|X_{n-1}(s;\cdot)|^2\,ds. \tag{3.88}$$

If we can establish (3.88), then it follows that $\forall K \in \mathbb{N}$,

$$\sup_{1\leq n\leq K} E\,|X_n(t;\cdot)|^2 \leq \zeta_1 + \zeta_2\int_0^t \sup_{1\leq n\leq K} E\,|X_{n-1}(s;\cdot)|^2\,ds$$

$$\leq \zeta_1 + \zeta_2\int_0^t \left[E\,|X_0|^2 + \sup_{1\leq n\leq K} E\,|X_n(s;\cdot)|^2\right]ds \tag{3.89}$$

$$\leq \left(\zeta_1 + T\zeta_2\,\|X_0\|_{\mathfrak{L}^2}^2\right) + \zeta_2\int_0^t \sup_{1\leq n\leq K} E\,|X_n(s;\cdot)|^2\,ds.$$

Applying Gronwall's Lemma in (3.89) subsequently yields $\forall K \in \mathbb{N}$,

$$\sup_{1\leq n\leq K} E\,|X_n(t;\cdot)|^2 \leq \left(\zeta_1 + T\zeta_2\,\|X_0\|_{\mathfrak{L}^2}^2\right)e^{\zeta_2 t},$$

so that

$$\sup_{0\leq t\leq T}\left\{\sup_{1\leq n\leq K} E\,|X_n(t;\cdot)|^2\right\}\leq \underbrace{\left(\zeta_1 + T\zeta_2\,\|X_0\|_{\mathfrak{L}^2}^2\right)e^{\zeta_2 T}}_{\text{independent of } K}. \tag{3.90}$$

Because the bound on the right-hand side of (3.90) is independent of K, it is a suitable candidate for M in (3.87).

Now, we must verify the existence of $\zeta_1, \zeta_2 > 0$ in (3.88) in order for the above strategy to work. To this end, observe that

$$E\,|X_n(t;\cdot)|^2 = E\left|X_0(\cdot) + \int_0^t aX_{n-1}(s;\cdot)ds + \sum_{k=1}^m \int_0^t c_k X_{n-1}(s;\cdot)dW_k(s)\right|^2$$

$$\leq 3\left[\|X_0\|_{\mathfrak{L}^2}^2 + E\left|\int_0^t aX_{n-1}(s;\cdot)ds\right|^2\right. \tag{3.91}$$

$$\left. + m\sum_{k=1}^m E\left|\int_0^t c_k X_{n-1}(s;\cdot)dW_k(s)\right|^2\right].$$

Applying Lemma 3.2.7 then yields

$$E\left|\int_0^t aX_{n-1}(s;\cdot)ds\right|^2 \leq a^2T\int_0^t E\,|X_{n-1}(s;\cdot)|^2\,ds \tag{3.92}$$

$$E\left|\int_0^t c_kX_{n-1}(s;\cdot)dW_k(s)\right|^2 \leq 8Tc_k^2\int_0^t E\,|X_{n-1}(s;\cdot)|^2\,ds,\ k=1,\ldots,m. \tag{3.93}$$

Applying (3.92) and (3.93) in (3.91) yields the inequality

$$E\,|X_n(t;\cdot)|^2 \leq 3\,\|X_0\|_{\underline{\mathfrak{L}}^2}^2 + 3T\left(a^2 + 8m\sum_{k=1}^m c_k^2\right)\int_0^t E\,|X_{n-1}(s;\cdot)|^2\,ds, \tag{3.94}$$

so that the following are suitable choices for ζ_1, ζ_2:

$$\zeta_1 = 3\,\|X_0\|_{\underline{\mathfrak{L}}^2}^2\,,\ \zeta_2 = 3T\left(a^2 + 8m\sum_{k=1}^m c_k^2\right).$$

This completes the proof of Claim 2. ◇

<u>Claim 3:</u> There exists $\zeta > 0$ (independent of n) such that $\forall n \geq 0$ and $\forall 0 \leq t \leq T$,

$$E\,|X_{n+1}(t;\cdot) - X_n(t;\cdot)|^2 \leq \frac{(\zeta t)^{n+1}}{(n+1)!}. \tag{3.95}$$

Proof. We proceed by induction on n.
First, let $n = 0$. Observe that

$$E\,|X_1(t;\cdot) - X_0(t;\cdot)|^2 = E\left|\int_0^t aX_0(s;\cdot)ds + \sum_{k=1}^m \int_0^t c_kX_0(s;\cdot)dW_k(s)\right|^2$$

$$\leq 2\left(ta^2 + 8mt\sum_{k=1}^m c_k^2\right)\underbrace{\int_0^t E\,|X_0(s;\cdot)|^2\,ds}_{=\|X_0\|_{\underline{\mathfrak{L}}^2}^2 t}$$

$$\leq 2T\left(a^2 + 8m\sum_{k=1}^m c_k^2\right)\|X_0\|_{\underline{\mathfrak{L}}^2}^2\,t.$$

As such, it suffices to choose $\zeta = 2T\left(a^2 + 8m\sum_{k=1}^m c_k^2\right)\|X_0\|_{\underline{\mathfrak{L}}^2}^2$ to verify that (3.95) holds for the base case.

Next, assume $\exists \zeta^\star > 0$ (independent of n) such that $\forall 0 \leq t \leq T$,

$$E\,|X_n(t;\cdot) - X_{n-1}(t;\cdot)|^2 \leq \frac{(\zeta^\star t)^n}{n!}. \tag{3.96}$$

Observe that $\forall 0 \leq t \leq T$,

$$E\,|X_n(t;\cdot) - X_{n-1}(t;\cdot)|^2 = E\left|\int_0^t a\,(X_n(s;\cdot) - X_{n-1}(s;\cdot))\,ds\right.$$

$$\left. + \sum_{k=1}^m \int_0^t c_k\,(X_n(s;\cdot) - X_{n-1}(s;\cdot))\,dW_k(s)\right|^2$$

$$\leq 2\left(ta^2 + 8mt\sum_{k=1}^m c_k^2\right)\int_0^t E\,|X_n(s;\cdot) - X_{n-1}(s;\cdot)|^2\,ds$$

$$\leq 2T\left(a^2 + 8m\sum_{k=1}^m c_k^2\right)\int_0^t \frac{(\zeta^\star s)^n}{n!}\,ds \qquad (3.97)$$

$$= 2T\left(a^2 + 8m\sum_{k=1}^m c_k^2\right)\frac{(\zeta^\star)^n t^{n+1}}{(n+1)!}.$$

Both $2T\left(a^2 + 8m\sum_{k=1}^m c_k^2\right)$ and $\zeta^\star$ are less than or equal to

$$\zeta^{\star\star} = \max\left\{2T\left(a^2 + 8m\sum_{k=1}^m c_k^2\right),\,\zeta^\star\right\}.$$

Hence, we can continue the string of inequalities in (3.97) to conclude that for every $0 \leq t \leq T$,

$$E\,|X_n(t;\cdot) - X_{n-1}(t;\cdot)|^2 \leq \frac{(\zeta^{\star\star})^{n+1}\,t^{n+1}}{(n+1)!},$$

as needed. This proves Claim 3.$\Diamond$

Claim 4: There exists a stochastic process $X : [0,T] \times \Omega \longrightarrow \mathbb{R}$ such that

$$\mathscr{P}\left(\left\{\omega \in \Omega : \sup_{0 \leq t \leq T} |X_n(t;\omega) - X(t;\omega)| \longrightarrow 0 \text{ as } n \longrightarrow \infty\right\}\right) = 1. \qquad (3.98)$$

In such a case, we write $\lim_{n \to \infty} X_n(t;\omega) = X(t;\omega)$ a.s. $[\mathscr{P}]$.

Proof. We first outline the basic strategy. Observe that $\forall n \in \mathbb{N}$,

$$X_n(t;\omega) = X_0(\omega) + \sum_{j=0}^{n-1}(X_{j+1}(t;\omega) - X_j(t;\omega)). \qquad (3.99)$$

As in the deterministic case, we wish to apply the Weierstrass M-test. Doing so requires that we prove

$$\sup_{0 \leq t \leq T}\sum_{n=0}^\infty |X_{n+1}(t;\omega) - X_n(t;\omega)| < \infty, \text{ a.s. } [\mathscr{P}]. \qquad (3.100)$$

One way to show (3.100) is to prove that the tail of the series is dominated by a convergent geometric series; that is, show $\exists N = N(\omega)$ and $0 < \varepsilon < 1$ such that

$$\sup_{0 \le t \le T} \sum_{n=N}^{\infty} |X_{n+1}(t;\omega) - X_n(t;\omega)| \le \sum_{n=N}^{\infty} \varepsilon^n < \infty \text{ a.s. } [\mathscr{P}]. \tag{3.101}$$

This reduces to showing that for a given $0 < \varepsilon < 1$, $\exists N = N(\omega)$ such that

$$n \ge N \implies \sup_{0 \le t \le T} |X_{n+1}(t;\omega) - X_n(t;\omega)| < \varepsilon^n \text{ a.s. } [\mathscr{P}]. \tag{3.102}$$

(Why?) This suggests that the Borel-Cantelli Lemma (cf. Prop. 2.1.8) might be useful. (Why?)

We now present the argument. Let $0 < \varepsilon < 1$. For every $n \in \mathbb{N} \cup \{0\}$, define the event

$$A_n = \left\{ \omega \in \Omega : \sup_{0 \le t \le T} |X_{n+1}(t;\omega) - X_n(t;\omega)| > \varepsilon^n \right\}. \tag{3.103}$$

We claim that $\exists \overline{\zeta} > 0$ (independent of n) such that

$$\mathscr{P}(A_n) \le \frac{\left(\overline{\zeta}\right)^{n+1}}{(n+1)!}. \tag{3.104}$$

To see this, recall from Claim 1 that X_n is a martingale, $\forall n \in \mathbb{N} \cup \{0\}$, so that we can apply Doob's Martingale Property (cf. Thrm. 2.8.3) to the stochastic process

$$M(t;\omega) = X_{n+1}(t;\omega) - X_n(t;\omega)$$

(Why?) to conclude that

$$\begin{aligned}
\mathscr{P}(A_n) &\le \frac{1}{\varepsilon^{2n}} E\, |X_{n+1}(T;\cdot) - X_n(T;\cdot)|^2 \\
&\le \frac{1}{\varepsilon^{2n}} \sup_{0 \le t \le T} E\, |X_{n+1}(t;\cdot) - X_n(t;\cdot)|^2 \\
&\le \frac{1}{\varepsilon^{2n}} \sup_{0 \le t \le T} \frac{(\zeta t)^{n+1}}{(n+1)!} \\
&\le \frac{\varepsilon^2 \left(\varepsilon^{-2} \zeta t\right)^{n+1}}{(n+1)!}.
\end{aligned} \tag{3.105}$$

Observe that

$$\sum_{n=0}^{\infty} \frac{\varepsilon^2 \left(\varepsilon^{-2} \zeta T\right)^{n+1}}{(n+1)!} = \varepsilon^2 e^{\varepsilon^{-2} \zeta T} < \infty. \tag{3.106}$$

Hence, the Borel-Cantelli Lemma implies that $\exists \Omega_0 \in \mathscr{F}$ and $N_0 = N_0(\omega)$ such that $\mathscr{P}(\Omega_0) = 1$ and

$$n \ge N_0(\omega) \implies \sup_{0 \le t \le T} |X_{n+1}(t;\omega) - X_n(t;\omega)| \le \varepsilon^n, \forall \omega \in \Omega_0. \tag{3.107}$$

As such, $\forall \omega \in \Omega_0$,

$$\sum_{n=0}^{\infty} |X_{n+1}(t;\omega) - X_n(t;\omega)| = \underbrace{\sum_{n=0}^{N_0-1} |X_{n+1}(t;\omega) - X_n(t;\omega)|}_{\text{Bounded because it is a finite sum}} + \underbrace{\sum_{n=N_0}^{\infty} \underbrace{|X_{n+1}(t;\omega) - X_n(t;\omega)|}_{\leq \varepsilon^n}}_{\text{A convergent geometric series}}.$$

Hence, taking the supremum over $[0,T]$ in the above expression enables us to conclude that (3.100) holds, $\forall \omega \in \Omega_0$. Thus, by the Weierstrass M-test, we conclude that (3.98) holds, where $X(t;\omega)$ is given by

$$X(t;\omega) = X_0(\omega) + \sum_{j=0}^{\infty} \left(X_{j+1}(t;\omega) - X_j(t;\omega) \right), \qquad (3.108)$$

as needed. This proves Claim 4. $\diamond$

Claim 5: The stochastic process X defined by (3.108) belongs to $\mathbb{C}\left([0,T]; \mathcal{L}^2(\Omega;\mathbb{R})\right)$.
Proof. The facts that $X_n \in \mathbb{C}\left([0,T]; \mathcal{L}^2(\Omega;\mathbb{R})\right)$, $\forall n \in \mathbb{N}$, and X is the uniform limit of $\{X_n\}$ together constitute the driving force behind this argument.

First, observe that in light of (3.86), every term X_n can be ultimately expressed entirely in terms of a, c_k, $X_0(\omega)$, and the integral thereof. All thusly formed portions of this expression for $X_n(t;\cdot)$ are $\mathscr{F}_t$-adapted martingales with continuous sample paths a.s. $[\mathscr{P}]$. As such, $X(t;\cdot)$ inherits these properties, being the uniform limit of $\{X_n\}$.

It remains to show that $\sup\left\{ E|X(t;\cdot)|^2 : 0 \leq t \leq T \right\} < \infty$. To begin, Claim 3 guarantees that $\forall t \in [0,T]$, $\{X_n(t;\cdot)\}$ is a Cauchy sequence in $\mathcal{L}^2(\Omega;\mathbb{R})$. (Tell how.) Hence, by the completeness of $\mathcal{L}^2(\Omega;\mathbb{R})$, we know that $\forall t \in [0,T]$, $\{X_n(t;\cdot)\}$ is actually a convergent sequence in $\mathcal{L}^2(\Omega;\mathbb{R})$. Moreover, we know that $\{X_n\}$ converges uniformly to X on $[0,T]$. Thus, we can apply the LDC (where $E(\cdot)$ is the Lebesgue integral arising in the situation to which the theorem is being applied) to obtain

$$\sup_{0 \leq t \leq T} E\left[(X(t;\cdot))^2 \right] = \sup_{0 \leq t \leq T} E\left[\left(\lim_{n \to \infty} X_n(t;\cdot) \right) \right]^2 \quad \text{(Why?)}$$

$$= \sup_{0 \leq t \leq T} E\left[\lim_{n \to \infty} (X_n(t;\cdot))^2 \right]$$

$$= \sup_{0 \leq t \leq T} \left(\lim_{n \to \infty} E\left[(X_n(t;\cdot))^2 \right] \right) \quad \text{(by LDC)}$$

$$\leq M < \infty, \quad \text{(by Claim 1)}$$

as needed. This proves Claim 5. $\diamond$

Claim 6: The stochastic process $X(t;\cdot)$ defined by (3.108) satisfies (3.86) a.s. $[\mathscr{P}]$.
Proof. Define $Z : [0,T] \times \Omega \to \mathbb{R}$ by

$$Z(t;\omega) = X_0(\omega) + \int_0^t aX(s;\omega)ds + \sum_{k=1}^{m} \int_0^t c_k X(s;\omega)dW_k(s). \qquad (3.109)$$

Claim 5 ensures that (3.109) is a well-defined stochastic process in $\mathbb{C}\left([0,T];\mathcal{L}^2\left(\Omega;\mathbb{R}\right)\right)$. (Why?) We will show that

$$\lim_{n\to\infty}\left(\sup_{0\leq t\leq T}E\,|X_n(t;\cdot)-Z(t;\cdot)|^2\right)=0. \tag{3.110}$$

Indeed, $\forall 0\leq t\leq T$, using Lemma 3.2.7 yields

$$\sup_{0\leq t\leq T}E\,|X_n(t;\cdot)-Z(t;\cdot)|^2 \leq \sup_{0\leq t\leq T}E\left|\int_0^t a\left(X_n(s;\cdot)-X(s;\cdot)\right)ds\right.$$

$$\left.+\sum_{k=1}^m\int_0^t c_k\left(X_n(s;\cdot)-X(s;\cdot)\right)dW_k(s)\right|^2 \tag{3.111}$$

$$\leq\left(a^2+4m\sum_{k=1}^m c_k^2\right)\int_0^T E\,|X_n(s;\cdot)-X(s;\cdot)|^2\,ds.$$

Let $\varepsilon>0$. There exists $N\in\mathbb{N}$ such that

$$n\geq N\implies \sup_{0\leq s\leq T}E\,|X_n(s;\cdot)-X(s;\cdot)|^2<\frac{\varepsilon}{T\left(a^2+4m\sum_{k=1}^m c_k^2\right)}. \tag{3.112}$$

Because $\lim_{n\to\infty}E\,|X_n(s;\cdot)-X(s;\cdot)|^2=0,\ \forall s\in[0,T]$, we can apply LDC on the right-hand side of (3.111). As such, we can use (3.112) in (3.111) to conclude that

$$n\geq N\implies \sup_{0\leq t\leq T}E\,|X_n(t;\cdot)-Z(t;\cdot)|^2. \tag{3.113}$$

This shows (3.110).

We have shown that the sequence $\{X_n\}$ has two $\mathcal{L}^2\left(\Omega;\mathbb{R}\right)$-limits, namely X and Z. The uniqueness of $\mathcal{L}^2\left(\Omega;\mathbb{R}\right)$-limits guarantees that

$$Z(t;\omega)=X(t;\omega),\ \forall t\in[0,T],\ \text{a.s.}\ [\mathscr{P}]. \tag{3.114}$$

Making this substitution on the left-hand side of (3.109) proves that $X(t;\omega)$ satisfies (3.86) a.s. $[\mathscr{P}]$, as needed. This proves Claim 6. $\Diamond$

This completes the existence portion of the proof. $\Diamond$

Uniqueness of a Strong Solution:
Suppose that two stochastic processes $X:[0,T]\times\Omega\longrightarrow\mathbb{R}$ and $X^\star:[0,T]\times\Omega\longrightarrow\mathbb{R}$ satisfy Def. 3.3.1. We must show that

$$\mathscr{P}\left(\{\omega\in\Omega:X(t;\omega)=X^\star(t;\omega),\ \forall t\in[0,T]\}\right)=1. \tag{3.115}$$

To this end, we begin by showing that

$$E\,|X(t;\cdot)-X^\star(t;\cdot)|=0,\ \forall t\in[0,T]. \tag{3.116}$$

Let $0 \leq t \leq T$. Arguing exactly as in Claim 3 (specifically when establishing (3.97)) yields

$$E|X(t;\cdot) - X^{\star}(t;\cdot)|^2 = 2T\left(a^2 + 8m\sum_{k=1}^{m} c_k^2\right)\int_0^t E|X(s;\cdot) - X^{\star}(s;\cdot)|^2\, ds. \quad (3.117)$$

Thus, applying Gronwall's Lemma yields

$$E|X(t;\cdot) - X^{\star}(t;\cdot)|^2 = 0, \forall t \in [0,T]. \quad (3.118)$$

This is close to what we need, but is not yet quite there. (Why?)

We can use Prop. 2.2.8 (vii) to infer from (3.118) that

$$|X(t;\omega) - X^{\star}(t;\omega)| = 0, \forall t \in [0,T], \text{ a.s. } [\mathscr{P}]. \quad (3.119)$$

This means that $\forall t \in [0,T]$, there exists an event $\mathscr{D}_t \in \mathscr{F}$ such that

$$\mathscr{P}(\mathscr{D}_t) = 0 \text{ and } |X(t;\omega) - X^{\star}(t;\omega)| = 0, \forall \omega \in \Omega \setminus \mathscr{D}_t. \quad (3.120)$$

Using this information, we must come to the conclusion that

$$\mathscr{P}\left(\left\{\omega \in \Omega: \sup_{0 \leq t \leq T}|X(t;\omega) - X^{\star}(t;\omega)| = 0\right\}\right) = 1. \quad (3.121)$$

We cannot, however, simply use (3.120) <u>for all</u> $0 \leq t \leq T$ because while $\mathscr{P}(\mathscr{D}_t) = 0$, $\forall t \in [0,T]$, it need not be the case that $\mathscr{P}(\bigcup\{\mathscr{D}_t : 0 \leq t \leq T\}) = 0$. (This can be easily seen, for instance, by noting that while the Lebesgue measure of a singleton set $\{x\}$ is zero, it does not follow that the Lebesgue measure of $\bigcup\{\{x\} : 0 \leq x \leq 1\}$ is zero. In fact, this union is the interval $[0,1]$, which has Lebesgue measure 1.) The problem is that the collection $\{\mathscr{D}_t : 0 \leq t \leq T\}$ consists of too many sets. If we could argue that it was sufficient to only use countably many sets from this collection, say $\{\mathscr{D}_t : t \in [0,T] \cap \mathbb{Q}\}$, then the countable subadditivity of the probability measure (cf. Prop. 2.1.6 (vi)) would imply that

$$\mathscr{P}\left(\bigcup\{\mathscr{D}_t : t \in [0,T] \cap \mathbb{Q}\}\right) = 0. \text{ (Why?)} \quad (3.122)$$

Because both X and $X^{\star}$ both have continuous sample paths a.s. $[\mathscr{P}]$, Prop. 1.8.6 (v) implies that it is sufficient to note that (3.119) holds, $\forall t \in [0,T] \cap \mathbb{Q}$. Then, using (3.122), we conclude that

$$|X(t;\omega) - X^{\star}(t;\omega)| = 0, \forall t \in [0,T] \cap \mathbb{Q}, \omega \in \Omega \setminus \bigcup\{\mathscr{D}_t : t \in [0,T] \cap \mathbb{Q}\}. \quad (3.123)$$

Consequently, because both $X(t;\cdot)$ and $X^{\star}(t;\cdot)$ are continuous and equal on a dense subset of $[0,T]$, we conclude from Prop. 1.8.6(v) that (3.123) actually holds for all $0 \leq t \leq T$.

This completes the uniqueness portion of the theorem. $\Diamond$

This completes the proof of Thrm. 3.4.1. □

Summarizing, we have managed to prove the existence and uniqueness of a strong solution of (3.74). We now ask if it is possible to formulate a nice representation formula for this solution. We actually caught a glimpse of this in Exer. 2.11.1. In the absence of the Itô integral, as in (3.1) or its deterministic counterpart, we would apply the separation of variables method to "solve" the IVP. As an illustration, suppose that $a \neq 0$. We proceed as follows to solve the deterministic version of (3.1):

$$\frac{dx(t)}{dt} = ax(t)$$

$$\frac{dx(t)}{x(t)} = adt$$

$$\int_0^t \frac{d}{ds} \ln(x(s)) \, ds = \int_0^t \frac{1}{x(s)} \frac{dx(s)}{ds} \, ds = \int_0^t \frac{1}{x(s)} dx(s) = \int_0^t ads$$

$$\underbrace{}_{=d(\ln(x(s)))}$$

$$\ln(x(s))|_{s=0}^{s=t} = at \tag{3.124}$$

$$\ln(x(t)) - \ln(x(0)) = at$$

$$\ln(x(t)) = \ln(x(0)) + at$$

$$x(t) = e^{\ln(x(0)) + at}$$

$$x(t) = e^{at} x_0.$$

We would like to mimic this process for (3.73). Paying particular attention to the third line in (3.124) suggests that we need to compute the integral $\int_0^t d\left(\ln X(s; \omega)\right)$. It is here where Itô's formula comes into play! Indeed, because a strong solution $X(t; \cdot)$ is an Itô process, Prop. 3.2.5 is applicable. To this end, let $h(t, x) = \ln x$ and observe that

$$\frac{\partial h}{\partial t} = 0, \ \frac{\partial h}{\partial x} = \frac{1}{x}, \ \frac{\partial^2 h}{\partial x^2} = -\frac{1}{x^2}.$$

Hence, applying Itô's formula (with $dtdt = dtdW = dWdt = 0$, $(dW)^2 = dt$) yields

$$
\begin{aligned}
d\left(\ln X(t; \cdot)\right) &= 0dt + \frac{1}{X(t; \cdot)} dX(t; \cdot) + \frac{1}{2}\left(-\frac{1}{X^2(t; \cdot)}\right)(dX(t; \cdot))^2 \\
&= \frac{1}{X(t; \cdot)}\left[aX(t; \cdot)dt + cX(t; \cdot)dW(t)\right] \\
&\quad - \frac{1}{2X^2(t; \cdot)}\left[aX(t; \cdot)dt + cX(t; \cdot)dW(t)\right]^2 \\
&= adt + cdW - \frac{1}{2X^2(t; \cdot)}\left[c^2 X^2(t; \cdot)dt\right] \\
&= \left(a - \frac{1}{2}c^2\right)dt + cdW.
\end{aligned}
\tag{3.125}
$$

Integrating both sides of (3.125) over $(0,T)$ then yields

$$\int_0^t d\left(\ln X(s;\cdot)\right) ds = \int_0^t \left(a - \frac{1}{2}c^2\right) ds + \int_0^t cdW(s)$$

$$\ln(X(t;\cdot)) - \ln(X(0;\cdot)) = \left(a - \frac{1}{2}c^2\right) t + cW(t)$$

$$\ln(X(t;\cdot)) = \ln(X_0(\cdot)) + \left(a - \frac{1}{2}c^2\right) t + cW(t).$$

As such, the representation formula for a strong solution of (3.73) is

$$
\begin{aligned}
X(t;\cdot) &= e^{\ln(X_0(\cdot)) + \left(a - \frac{1}{2}c^2\right)t + cW(t)} \\
&= e^{at} X_0(\cdot) \underbrace{e^{-\frac{1}{2}c^2 t + cW(t)}}_{\text{New term!}}.
\end{aligned}
\tag{3.126}
$$

Remark. When $\{W(t) : 0 \le t \le T\}$ is a Brownian motion, the stochastic process $\left\{ e^{-\frac{1}{2}c^2 t + cW(t)} : 0 \le t \le T \right\}$ is called a *geometric Brownian motion*.

Exercise 3.4.2. Develop the representation formula for a strong solution of (3.74).

3.5 Continuous Dependence on Initial Data

Consider (3.74) and the related IVP

$$
\begin{cases}
dY(t;\omega) = aY(t;\omega)dt + \sum_{k=1}^m c_k Y(t;\omega)dW_k(t), \ 0 < t < T, \omega \in \Omega, \\
Y(0;\omega) = Y_0(\omega), \omega \in \Omega,
\end{cases}
\tag{3.127}
$$

both under hypotheses **(H3.1)** through **(H3.3)**. If the initial data $X_0(\cdot)$ and $Y_0(\cdot)$ are "close" in the $\mathcal{L}^2(\Omega;\mathbb{R})$-sense, we would like to establish an upper bound for the distance (in the $\mathcal{L}^2(\Omega;\mathbb{R})$-norm) between the corresponding strong solutions of (3.74) and (3.127) in terms of $\|X_0 - Y_0\|_{\mathcal{L}^2(\Omega;\mathbb{R})}$. Then, we could conclude that the closer the initial data $X_0(\cdot)$ and $Y_0(\cdot)$ are, the closer the corresponding solutions $X(\cdot)$ and $Y(\cdot)$ are. This is addressed by the following proposition.

Proposition 3.5.1. *For all $0 \le t \le T$,*

$$E|X(t;\cdot) - Y(t;\cdot)|^2 \le 3e^{3T\left(a^2 + m\sum_{k=1}^m c_k^2\right)t} \|X_0 - Y_0\|^2_{\mathcal{L}^2(\Omega;\mathbb{R})}. \tag{3.128}$$

Proof. Let $0 \leq t \leq T$. Subtracting the integrated forms of (3.74) and (3.127) yields

$$
E\,|X(t;\cdot) - Y(t;\cdot)|^2 = E\,\bigg|(X_0(\cdot) - Y_0(\cdot)) + \int_0^t a\,(X(s;\cdot) - Y(s;\cdot))\,ds
$$

$$
+ \sum_{k=1}^m \int_0^t c_k\,(X(s;\cdot) - Y(s;\cdot))\,dW_k(s)\bigg|^2 \tag{3.129}
$$

$$
\leq 3\bigg[\|X_0 - Y_0\|_{\mathfrak{L}^2(\Omega;\mathbb{R})}^2 + a^2 t \int_0^t E\,|X(s;\cdot) - Y(s;\cdot)|^2\,ds
$$

$$
+ m \sum_{k=1}^m c_k^2 t \int_0^t E\,|X(s;\cdot) - Y(s;\cdot)|^2\,ds\bigg]
$$

$$
\leq 3\,\|X_0 - Y_0\|_{\mathfrak{L}^2(\Omega;\mathbb{R})}^2
$$

$$
+ 3\bigg(a^2 + m \sum_{k=1}^m c_k^2\bigg) T \int_0^t E\,|X(s;\cdot) - Y(s;\cdot)|^2\,ds.
$$

Applying Gronwall's Lemma in (3.129) results in (3.128), as desired. $\qquad\square$

Exercise 3.5.1. Let $0 < \varepsilon < 1$. How small would $\|X_0 - Y_0\|_{\mathfrak{L}^2(\Omega;\mathbb{R})}^2$ need to be to ensure that $\|X - Y\|_{\mathbb{C}([0,T];\mathfrak{L}^2(\Omega;\mathbb{R}))} < \varepsilon$?

Exercise 3.5.2. Replace a in (3.127) by $a^\star$ and assume that $\exists \delta_1, \delta_2 > 0$ such that

$$
|a - a^\star| < \delta_1 \text{ and } \|X_0 - Y_0\|_{\mathfrak{L}^2(\Omega;\mathbb{R})} < \delta_2. \tag{3.130}
$$

Establish an estimate for $E\,|X(t;\cdot) - Y(t;\cdot)|^2$ as in Prop. 3.5.1. Comment on where the differences occur and additional assumptions, if needed, to overcome them.

3.6   Statistical Properties of the Strong Solution

Recovering the probability density for a solution process $X : [0,T] \times \Omega \to \mathbb{R}$ can generally be difficult. The random variables $X(t;\cdot)$ comprising the stochastic process change with t, and so do their densities. While understanding the probability density for the stochastic process, as a single entity, is wrought with difficulty, it is more reasonable to obtain information about the statistical properties of the individual random variables $\omega \mapsto X(t;\omega)$, for each fixed t. The specific properties of interest are defined in Section 2.7; each of these properties tells us something about the density of $\omega \mapsto X(t;\omega)$. We study these properties for the strong solution of (3.73) and suggest that you do the same for (3.74) as an exercise.

3.6.1   Mean and Variance

Let $0 \le t \le T$ and consider the integrated form (3.77) of (3.73). We shall derive two deterministic ODEs to which $t \mapsto \mu_X(t)$ and $t \mapsto Var_X(t)$ are solutions. To this end, taking the expectation on both sides of (3.77) yields

$$
\begin{aligned}
\mu_X(t) &= E\left[X_0(\cdot)\right] + E\left[\int_0^t aX(s;\cdot)ds\right] + E\left[\int_0^t cX(s;\cdot)dW(s)\right] \\
&= E\left[X_0(\cdot)\right] + E\left[\int_0^t aX(s;\cdot)ds\right] \\
&= E\left[X_0(\cdot)\right] + \int_0^t aE\left[X(s;\cdot)\right]ds \qquad\qquad (3.131) \\
&= \mu_X(0) + a\int_0^t \mu_X(s)ds.
\end{aligned}
$$

It follows from (3.131) that $t \mapsto \mu_X(t)$ satisfies the deterministic IVP

$$
\begin{cases}
\frac{d\mu_X(t)}{dt} = a\mu_X(t),\ 0 < t < T, \\
\mu_X(0) = E\left[X_0(\cdot)\right],
\end{cases}
\qquad\qquad (3.132)
$$

the unique solution of which is

$$
\mu_X(t) = e^{at}E\left[X_0(\cdot)\right]. \qquad\qquad (3.133)
$$

(Why?)

Exercise 3.6.1. Interpret the meaning of (3.133) in the context of the entire solution process $X : [0,T] \times \Omega \to \mathbb{R}$. Specifically, what if $a < 0$? How about if $a = 0$?

Alternatively, we can calculate $E\left[X(t;\cdot)\right]$ using the representation formula (3.126). Doing so yields

$$
\begin{aligned}
\mu_X(t) &= E\left[X(t;\cdot)\right] = E\left[e^{at}X_0(\cdot)e^{\left(-\frac{1}{2}c^2t+cW(t)\right)}\right] \\
&= e^{at}E\left[X_0(\cdot)e^{\left(-\frac{1}{2}c^2t+cW(t)\right)}\right]. \qquad\qquad (3.134)
\end{aligned}
$$

Formulae (3.133) and (3.134) must agree. As such, we must conclude that

$$
E\left[X_0(\cdot)\right] = E\left[X_0(\cdot)e^{\left(-\frac{1}{2}c^2t+cW(t)\right)}\right]. \qquad\qquad (3.135)
$$

Being skeptical about (3.135) is certainly natural. We must verify that (3.135) is indeed valid. To do so, define $\zeta : [0,T] \times \Omega \to \mathbb{R}$ by

$$
\zeta(t;\omega) = X_0(\omega)e^{\left(-\frac{1}{2}c^2t+cW(t)\right)}. \qquad\qquad (3.136)
$$

Exercise 3.6.2. Prove that $\zeta : [0,T] \times \Omega \to \mathbb{R}$ is the unique strong solution of the stochastic IVP

$$\begin{cases} d\zeta(t;\omega) = c\zeta(t;\omega)dW(t), 0 < t < T, \omega \in \Omega, \\ \zeta(0;\omega) = X_0(\omega), \omega \in \Omega. \end{cases} \tag{3.137}$$

The integrated form of (3.137) is

$$\zeta(t;\omega) = X_0(\omega) + c \int_0^t \zeta(s;\omega)dW(s), 0 \le t \le T. \tag{3.138}$$

Taking the expectation on both sides of (3.138) yields $E[\zeta(t;\cdot)] = E[X_0(\cdot)]$, as desired. (Why?)

Next, observe that

$$\begin{aligned} Var_X(t) &= E[X^2(t;\cdot)] - (E[X(t;\cdot)])^2 \\ &= E[X^2(t;\cdot)] - (e^{at}E[X_0(\cdot)])^2. \end{aligned} \tag{3.139}$$

In order to simplify (3.139), we need a concise expression for $X^2(t;\cdot)$.

Exercise 3.6.3.
i.) Show that

$$d(X^2(t;\cdot)) = (2a + c^2)X^2(t;\cdot)dt + 2cX^2(t;\cdot)dW(t). \tag{3.140}$$

ii.) Deduce from (i) that

$$X^2(t;\cdot) = X_0^2(\cdot) + (2a + c^2)\int_0^t X^2(s;\cdot)ds + 2c\int_0^t X^2(s;\cdot)dW(s). \tag{3.141}$$

Taking the expectation on both sides of (3.141) yields

$$E[X^2(t;\cdot)] = E[X_0^2(\cdot)] + (2a + c^2)\int_0^t E[X^2(s;\cdot)]ds. \tag{3.142}$$

(Why?) Similar reasoning that led to (3.132) shows that the deterministic function $t \mapsto E[X^2(t;\cdot)]$ satisfies the IVP

$$\begin{cases} \frac{d(E[X^2(t;\cdot)])}{dt} = (2a + c^2)E[X^2(t;\cdot)], 0 < t < T, \\ E[X^2(0;\cdot)] = E[X_0^2(\cdot)]. \end{cases} \tag{3.143}$$

Hence, we conclude that

$$E[X^2(t;\cdot)] = E[X_0^2(\cdot)]e^{(2a+c^2)t}. \tag{3.144}$$

Substituting (3.144) into (3.139) yields the following concise formula:

$$Var_X(t) = e^{2at}\left(E[X_0^2(\cdot)]e^{c^2t} - (E[X_0(\cdot)])^2\right). \tag{3.145}$$

Remark. Note that if $c = 0$, then (3.73) is no longer driven by a Wiener process and (3.145) simplifies to

$$Var_X(t) = Var(X_0(\cdot)) e^{2at}.$$

Exercise 3.6.4. Compute $\mu_X(t)$ and $Var_X(t)$ for the strong solution of (3.74).

Exercise 3.6.5. Compute the covariance function $Cov_X(t,s)$ for the strong solutions of (3.73) and (3.74). Simplify the resulting expressions using Itó's formula.

3.6.2 Moment Estimates

Higher-order moments are useful when studying other statistical properties of a random variable, such as kurtosis and skewness, and are used heavily in stability arguments. There are various ways to proceed. We begin with a result that follows from a straightforward application of Itó's formula.

Proposition 3.6.1. *Let $p \in \mathbb{N}$ and X be the strong solution of (3.73). If $E\left[X_0^{2p}(\cdot)\right] < \infty$, then*

$$E\left[X^{2p}(t;\cdot)\right] \leq E\left[X_0^{2p}(\cdot)\right] e^{p(2a+(2p-1)c^2)t}, 0 \leq t \leq T. \tag{3.146}$$

Proof. An application of Itó's formula (with $h(t,x) = x^{2p}$) yields

$$d(X^{2p}(t;\omega)) = 2pX^{2p-1}(t;\omega)d(X(t;\omega)) + \frac{1}{2}(2p(2p-1)X(t;\omega))^{2p-2}d(X^2(t;\omega))$$

$$= p\left(2a + (2p-1)c^2\right)X^{2p}(t;\omega)dt + 2pcX^{2p}(t;\omega)dW(t). \tag{3.147}$$

(Tell why carefully.) Integrating (3.147) over $(0,t)$ yields

$$X^{2p}(t;\omega) = X_0^{2p}(\omega) + \int_0^t p\left(2a + (2p-1)c^2\right)X^{2p}(s;\omega)ds$$

$$+ \int_0^t 2pcX^{2p}(s;\omega)dW(s), \tag{3.148}$$

and subsequently taking the expectation in (3.148) yields

$$E\left[X^{2p}(t;\cdot)\right] = E\left[X_0^{2p}(\cdot)\right] + \int_0^t p\left(2a + (2p-1)c^2\right)E\left[X^{2p}(s;\cdot)\right]ds.$$

(Why?) An application of Gronwall's Lemma finally results in (3.146). $\qquad\square$

Exercise 3.6.6. Let $p \in \mathbb{N} \setminus \{1\}$ and X be the strong solution of (3.73). If $E|X_0(\cdot)|^p < \infty$, then

$$E|X(t;\cdot)|^p \leq E|X_0(\cdot)|^p e^{\left(p\left(a - \frac{c^2}{2}\right) + \frac{p^2c^2}{2}\right)t}, 0 \leq t \leq T. \tag{3.149}$$

Exercise 3.6.7. Establish estimates in the spirit of (3.146) and (3.149) for a strong solution of (3.74).

3.6.3 Continuity in the pth Moment

We already know that a strong solution X of (3.73) is $\mathcal{L}^2(\Omega;\mathbb{R})$-continuous. Now, we extend this to $\mathcal{L}^p(\Omega;\mathbb{R})$-continuity in the following sense.

Proposition 3.6.2. *Let* $p \in \mathbb{N} \setminus \{1\}$. *If* $E|X_0(\cdot)|^p < \infty$, *then there exists a continuous function* $\zeta_p : [0,T] \times [0,T] \longrightarrow \mathbb{R}$ *such that* $\zeta_p(\tau,\tau') \longrightarrow 0$ *as* $(\tau - \tau') \longrightarrow 0$ *and*

$$E|X(\tau;\cdot) - X(\tau';\cdot)|^p \leq \zeta_p(\tau,\tau'). \tag{3.150}$$

Proof. Let $p \in \mathbb{N} \setminus \{1\}$ and $0 \leq \tau' < \tau \leq T$. Using (3.77) yields

$$E|X(\tau;\cdot) - X(\tau';\cdot)|^p = E\left|\int_{\tau'}^{\tau} aX(s;\cdot)ds + \int_{\tau'}^{\tau} cX(s;\cdot)dW(s)\right|^p \tag{3.151}$$

$$\leq 2^{p-1}\left[E\left|\int_{\tau'}^{\tau} aX(s;\cdot)ds\right|^p + E\left|\int_{\tau'}^{\tau} cX(s;\cdot)dW(s)\right|^p\right].$$

Applying Lemma 3.2.7 in (3.151) now yields

$$E|X(\tau;\cdot) - X(\tau';\cdot)|^p \leq 2^{p-1}\left[|\tau - \tau'|^{\frac{p}{q}}\int_{\tau'}^{\tau} E|aX(s;\cdot)|^p ds\right.$$

$$\left. + 2^{p+1}T^{\frac{p}{2}}\left(\frac{p(p-1)}{2}\right)^{\frac{p}{2}}\int_{\tau'}^{\tau} E|bX(s;\cdot)|^p ds\right] \tag{3.152}$$

$$\leq 2^{p-1}\left[|a|^p\underbrace{|\tau - \tau'|^{\frac{p}{q}}}_{\leq 2T^{\frac{p}{q}}} + 4|b|^p T^{\frac{p}{2}}\left(\frac{p(p-1)}{2}\right)^{\frac{p}{2}}\right]$$

$$\times \int_{\tau'}^{\tau} E|X(s;\cdot)|^p ds.$$

By Exer. 3.6.6, we know that

$$\int_{\tau'}^{\tau} E|X(s;\cdot)|^p ds \leq |\tau - \tau'|E|X_0(\cdot)|^p e^{\left(p\left(a - \frac{c^2}{2}\right) + \frac{p^2 c^2}{2}\right)T}. \tag{3.153}$$

Using (3.153) in (3.152) shows that $E|X(\tau;\cdot) - X(\tau';\cdot)|^p$ is dominated by a function of the form $\zeta_p(\tau,\tau') = \eta|\tau - \tau'|$, where η is a positive constant. This completes the proof. $\qquad\square$

Exercise 3.6.8. Formulate a result analogous to Prop. 3.6.2 for (3.74).

3.6.4 The Distribution of a Strong Solution

Generally speaking, the strong solutions of (3.73) and (3.74) are not Gaussian even if the initial data $\omega \mapsto X_0(\omega)$ is Gaussian. However, in the absence of the Itô integral term, if the initial data $\omega \mapsto X_0(\omega)$ is Gaussian, then we can immediately conclude that the solution process $\{X(t;\omega) = X_0(\omega)e^{at} : 0 \leq t \leq T, \omega \in \Omega\}$ is Gaussian.

(Why?) In such case, the formulae (3.133) and (3.145) completely characterize the solution process of (3.73). (A similar result holds for (3.74); fill in the details.)

3.6.5   Markov Property

Heuristically speaking, the Markov property can be described verbally as, "When the present is known, the future of a process is independent of the past." This basically means that for a stochastic process possessing this property, only the information at the present time is of any practical use when trying to predict the future behavior of the process. Certainly, not all stochastic processes behave in this manner, but for those that do the literature concerning their properties is massive. (See [125].)

A formal investigation involves rather technical computations involving the conditional expectation of the process based on the information provided by certain σ-algebras. We shall only discuss this on an intuitive level. We proceed as in [83].

Let $0 \le t \le T$ and $\varepsilon > 0$. We must verify that $\forall x \in \mathbb{R}$,

$$\mathscr{P}\left(\{\omega \in \Omega : X(t+\varepsilon;\omega) \le x\} \,|\, \mathscr{F}_t\right) = \mathscr{P}\left(\{\omega \in \Omega : X(t+\varepsilon;\omega) \le x\} \,|\, X(t;\omega)\right). \tag{3.154}$$

We need an expression for $X(t+\varepsilon;\omega)$ in terms of $X(t;\omega)$. This follows by integrating (3.73) over the interval $(t, t+\varepsilon)$, as follows:

$$\begin{aligned} X(t+\varepsilon;\omega) &= X(t;\omega) + \int_t^{t+\varepsilon} aX(s;\omega)ds + \int_t^{t+\varepsilon} cX(s;\omega)dW(s), \; 0 \le t \le T \\ &= I_1 + I_2 + I_3. \end{aligned} \tag{3.155}$$

The question we must address is whether or not the right-hand side of (3.155) depends on any information provided by $\mathscr{F}_t$ other than $X(t;\omega)$ itself. To this end, note that certainly the term I_1 is fine. (Why?) Also, because the integrand of the term I_2 does not extend into the past beyond time t and the integrand is a constant multiple of $X(s;\omega)$, this term depends only on $X(t;\omega)$ itself. Finally, note that term I_3 is independent of $X_0(\omega)$ (Why?) and the increments $W(t+\varepsilon) - W(t)$ and $W(t)$ are independent. (Why?) Because the integrand does not depend on $X(s;\omega)$ for $0 < s < t$, we can conclude that the right-hand side of (3.155) depends only on information provided by $X(t;\omega)$ itself. As such, we conclude that (3.154) holds.

Exercise 3.6.9. Convince yourself that the same holds true for the strong solution of (3.74).

3.7   Some Convergence Results

The notion of continuous dependence on initial data is naturally related to the convergence scheme introduced in this section.

For every $n \in \mathbb{N}$, consider the IVP

$$\begin{cases} dX_n(t;\omega) = a_n X_n(t;\omega)dt + \sum_{k=1}^{m}(c_k)_n X_n(t;\omega)dW_k(t), \\ X_n(0;\omega) = (X_0)_n(\omega), \omega \in \Omega \end{cases} \tag{3.156}$$

where $0 < t < T$, $\omega \in \Omega$. Assume the following:

(H3.4) There exists $a \in \mathbb{R}$ such that $\lim_{n \to \infty} |a_n - a| = 0$.

(H3.5) For every $k \in \{1, \ldots, m\}$, $\lim_{n \to \infty} |(c_k)_n - c_k| = 0$.

(H3.6) There exists $X_0 \in \mathscr{L}_0^2(\Omega;\mathbb{R})$ such that $\lim_{n \to \infty} \|(X_0)_n - X_0\|_{\mathscr{L}_0^2(\Omega;\mathbb{R})} = 0$.

Proposition 3.7.1. *If* **(H3.1)** *-* **(H3.6)** *hold, then* $\lim_{n \to \infty} \|X_n - X\|_{\mathbb{C}([0,T];\mathscr{L}^2(\Omega;\mathbb{R}))} = 0$, *where X is the strong solution of (3.74).*

Proof. The strong solutions of (3.74) and (3.156) (for a given $n \in \mathbb{N}$) are given by

$$X(t;\omega) = X_0(\omega) + \int_0^t aX(s;\omega)ds + \sum_{k=1}^{m}\int_0^t c_k X(s;\omega)dW_k(s), \tag{3.157}$$

$$X_n(t;\omega) = (X_0)_n(\omega) + \int_0^t a_n X_n(s;\omega)ds \tag{3.158}$$
$$+ \sum_{k=1}^{m}\int_0^t (c_k)_n X_n(s;\omega)dW_k(s).$$

Let $\varepsilon > 0$. There exists $M_0 > 0$ such that

$$E|X(t;\cdot)|^2 \le M_0, \forall t \in [0,T]. \tag{3.159}$$

(Why?) Because convergent sequences are bounded, **(H3.4)** and **(H3.5)** guarantee the existence of $M_1, \ldots, M_{m+1} > 0$ such that

$$|a_n|^2 \le M_1, \forall n \in \mathbb{N}, \tag{3.160}$$
$$|(c_k)_n|^2 \le M_{k+1}, \forall n \in \mathbb{N}, k \in \{1, \ldots, m\}. \tag{3.161}$$

Let $\eta = 2T\sum_{k=1}^{m+1} M_k$. There exist $N_1, \ldots, N_{m+2} \in \mathbb{N}$ such that

$$n \ge N_1 \implies |a_n - a|^2 < \frac{\varepsilon^2}{(2T^2 M_0)(m+2)\eta}, \tag{3.162}$$

$$n \ge N_{k+1} \implies |(c_k)_n - c_k|^2 < \frac{\varepsilon^2}{(6mT^2 M_0)(m+2)\eta}, k = 1, \ldots, m, \tag{3.163}$$

$$n \ge N_{m+2} \implies \|(X_0)_n - X_0\|^2_{\mathscr{L}_0^2(\Omega;\mathbb{R})} < \frac{\varepsilon^2}{3(m+2)\eta}. \tag{3.164}$$

Let $N = \max\{N_1, \ldots, N_{m+2}\}$. Then, (3.162) through (3.164) hold, $\forall n \geq N$. Observe that

$$E\,|X_n(t;\cdot) - X(t;\cdot)|^2 = E\,\left|((X_0)_n - X_0) + \int_0^t (a_n X_n(s;\cdot) - aX(s;\cdot))\,ds\right.$$

$$+ \sum_{k=1}^m \int_0^t \left.((c_k)_n X_n(s;\omega) - c_k X(s;\cdot))\,dW_k(s)\right|$$

$$\leq 3\left[\|(X_0)_n - X_0\|^2_{\mathscr{L}_0^2(\Omega;\mathbb{R})} + T\int_0^t E\,|a_n X_n(s;\cdot) - aX(s;\cdot)|^2\,ds\right.$$

$$+ 3mT\sum_{k=1}^m \int_0^t E\,|(c_k)_n X_n(s;\cdot) - c_k X(s;\cdot)|^2\,ds\Bigg]. \qquad (3.165)$$

Using the triangle inequality, together with (3.159), (3.160), and (3.162), yields

$$\int_0^t E\,|a_n X_n(s;\cdot) - aX(s;\cdot)|^2\,ds \leq$$

$$\int_0^t E\,|a_n X_n(s;\cdot) - a_n X(s;\cdot) + a_n X(s;\cdot) - aX(s;\cdot)|^2\,ds \leq \qquad (3.166)$$

$$\int_0^t 2\left[|a_n|^2 E\,|X_n(s;\cdot) - X(s;\cdot)|^2 + |a_n - a|^2 E\,|X(s;\cdot)|^2\right]ds \leq$$

$$2M_1 \int_0^t E\,|X_n(s;\cdot) - X(s;\cdot)|^2\,ds + \frac{\varepsilon^2}{T(m+2)\eta}.$$

Similarly, using (3.159), (3.161), and (3.163) yields, $\forall k \in \{1, \ldots, m\}$,

$$\int_0^t E\,|(c_k)_n X_n(s;\cdot) - c_k X(s;\cdot)|^2\,ds \leq$$

$$\int_0^t 2\left[|(c_k)_n|^2 E\,|X_n(s;\cdot) - X(s;\cdot)|^2 + |(c_k)_n - c_k|^2 E\,|X(s;\cdot)|^2\right]ds \leq \quad (3.167)$$

$$2M_{k+1}\int_0^t E\,|X_n(s;\cdot) - X(s;\cdot)|^2\,ds + \frac{\varepsilon^2}{3mT(m+2)\eta}.$$

Substituting (3.166) and (3.167) into (3.165) yields

$$E\,|X_n(t;\cdot) - X(t;\cdot)|^2 < \frac{\varepsilon^2}{\eta} + \left(2T\sum_{k=1}^{m+1} M_k\right)\int_0^t E\,|X_n(s;\cdot) - X(s;\cdot)|^2\,ds. \quad (3.168)$$

As such, applying Gronwall's Lemma shows that

$$E\,|X_n(t;\cdot) - X(t;\cdot)|^2 < \frac{\varepsilon^2}{\eta}e^{\left(2T\sum_{k=1}^{m+1} M_k\right)t},\ 0 \leq t \leq T. \qquad (3.169)$$

Hence, upon taking the supremum over $[0,T]$ on both sides of (3.169), we conclude that

$$n \geq N \implies \|X_n - X\|_{\mathbb{C}([0,T];\mathscr{L}^2(\Omega;\mathbb{R}))} < \varepsilon,$$

as needed. This completes the proof. $\qquad\qquad\qquad\qquad\qquad\qquad\qquad\square$

Exercise 3.7.1. Argue that $\forall 0 \le t \le T$,

$$\lim_{n \to \infty} \mu_{X_n}(t) = \mu_X(t),$$
$$\lim_{n \to \infty} Var_{X_n}(t) = Var_X(t).$$

We have introduced noise into linear ODEs and have established the existence and uniqueness of a strong solution of the stochastic IVP, as well as its analytical and statistical properties. We now seek to approximate the stochastic IVP by the original deterministic IVP. Precisely, let $0 < \varepsilon < 1$ and consider the following two IVPs:

$$\begin{cases} dX_\varepsilon(t; \omega) = a_\varepsilon X_\varepsilon(t; \omega)dt + \sum_{k=1}^m (c_k)_\varepsilon X_\varepsilon(t; \omega)dW_k(t), \\ X_\varepsilon(0; \omega) = X_0(\omega), \omega \in \Omega \end{cases} \quad (3.170)$$

where $0 < t < T$, $\omega \in \Omega$, and

$$\begin{cases} dX(t) = aX(t)dt, 0 < t < T, \\ X(0) = X_0. \end{cases} \quad (3.171)$$

We intend to compare the solution of the stochastic IVP (3.170) to the solution of the deterministic IVP (3.171), and this requires us to view a deterministic function $X(t)$ as a random variable. This is easily done because a deterministic function is simply a constant random variable in the sense that

$$X(t) \equiv X(t; \omega), \forall \omega \in \Omega.$$

We assume the following:
(H3.7) $\lim_{\varepsilon \to 0^+} |a_\varepsilon - a| = 0$,
(H3.8) For every $k \in \{1, \dots, m\}$, $\lim_{\varepsilon \to 0^+} (c_k)_\varepsilon = 0$.

We proceed in basically the same manner as in the previous convergence argument to prove the following result.

Proposition 3.7.2. *Let $p > 2$ and X_ε and X be the strong solutions of (3.170) and (3.171), respectively. Then, $\exists \zeta > 0$ and a function $\psi : I \subset [0,1] \longrightarrow (0, \infty)$ for which* $\lim_{\varepsilon \to 0^+} \psi(\varepsilon) = 0$ *and*

$$E|X_\varepsilon(t; \cdot) - X(t)|^p \le \zeta \psi(\varepsilon), \quad (3.172)$$

$\forall 0 \le t \le T$ *and $\varepsilon > 0$ sufficiently small. (That is, $\lim_{\varepsilon \to 0^+} \|X_\varepsilon - X\|_{\mathbb{C}([0,T]; \mathscr{L}^p(\Omega; \mathbb{R}))} = 0$.)*

Exercise 3.7.2. Prove Prop. 3.7.2.

As expected, as the noise term diminishes, the noise has less impact on the solution process and so, in the limit, the solution process tends to resemble the deterministic case.

3.8 A Brief Look at Stability

There are several different notions of stability, which is reasonable because there are various notions of convergence (cf. Def. 2.3.5). Some natural questions are:

1. If the solution of a deterministic IVP behaves in a certain manner, what natural conditions can be imposed on the stochastic terms to ensure that the same result holds in the presence of noise?

2. What are the relationships among the different notions of stability?

Entire volumes are devoted to this subject. (See **[105, 139, 161, 267, 268, 285, 414]**.) We only explore a few ideas to introduce the topic. Consider (3.74) on $[0, \infty)$, together with the IVP

$$\begin{cases} dY(t;\omega) = aY(t;\omega)dt + \sum_{k=1}^{m} c_k Y(t;\omega)dW_k(t), 0 < t < T, \omega \in \Omega, \\ Y(0;\omega) = Y_0(\omega), \omega \in \Omega. \end{cases} \tag{3.173}$$

Definition 3.8.1. A strong solution of (3.74) is said to be
i.) *asymptotically exponentially p^{th} moment stable* $(p \geq 2)$ if $\exists \zeta, \lambda > 0$ such that for any strong solution of (3.173),

$$E|X(t;\cdot) - Y(t;\cdot)|^p \leq \zeta e^{-\lambda t} E|X_0(\cdot) - Y_0(\cdot)|^p, \forall t \geq 0. \tag{3.174}$$

ii.) *almost surely exponentially stable* if $\exists \lambda > 0$ such that for any solution Y of (3.173),

$$\lim_{t \to \infty} \frac{1}{t} \log |X(t;\omega) - Y(t;\omega)| \leq -\lambda \text{ a.s. } [\mathscr{P}], \tag{3.175}$$

or equivalently,

$$|X(t;\omega) - Y(t;\omega)| \leq e^{-\lambda t}, \forall t \geq 0, \text{ a.s. } [\mathscr{P}]. \tag{3.176}$$

Here, $-\lambda$ is called the *Lyapunov exponent*.

In particular, for $Y_0(\cdot) = 0$, $Y \equiv 0$ is the so-called *equilibrium solution* of (3.173) and Def. 3.8.1 can be interpretted directly for this solution. An extensive discussion of stability can be found in **[20, 285, 414]**.

Remarks.
1. Definition 3.8.1(i) is given in terms of a deterministic function (namely, the pth-moments), while Def. 3.8.1 (ii) governs the behavior of the sample paths.

2. The Lyapunov exponent λ is significant because it is the rate of exponential decay (when $\lambda < 0$) of sample paths of the solution process to the equilibrium position.

For simplicity, consider (3.73). The pth-moments are given by

$$E\,|X(t;\cdot)|^p = E\,|X_0(\cdot)|^p\,e^{\left(p\left(a-\frac{c^2}{2}\right)+\frac{p^2c^2}{2}\right)t}. \tag{3.177}$$

Computing the pth-moment of the difference between (3.73) and the strong solution of the version of (3.173) where $cY(t;\omega)dW(t)$ is in place of $\sum_{k=1}^{m} c_k Y(t;\omega)dW_k(t)$ yields

$$E\,|X(t;\cdot) - Y(t;\cdot)|^p = E\,|X_0(\cdot) - Y_0(\cdot)|^p\,e^{\left(p\left(a-\frac{c^2}{2}\right)+\frac{p^2c^2}{2}\right)t}. \tag{3.178}$$

(Why?) As such, a necessary and sufficient condition to ensure that the strong solution of (3.73) satisfies Def. 3.8.1 (i) is that

$$p\left(a-\frac{c^2}{2}\right) + \frac{p^2c^2}{2} < 0. \tag{3.179}$$

Note that when $c = 0$ (that is, the deterministic case), (3.179) simplifies to $a < 0$, which is exactly what you would expect. (Why?)

Exercise 3.8.1. Determine a condition like (3.179) that guarantees that the strong solution of (3.74) satisfies Def. 3.8.1(i).

Exercise 3.8.2. For given values of a and c, does $\exists p_0 \in \mathbb{N}$ such that (3.179) does not hold for $p \geq p_0$. This would imply that some moments are stable (in the sense of Def. 3.8.1(i)), while others are not.

The notion of *robustness* with respect to perturbation in important in applications. Precisely, consider (3.73) and the following perturbed variant:

$$\begin{cases} dZ(t;\omega) &= [aZ(t;\omega)dt + cZ(t;\omega)dW(t)] \\ &\quad + [a^\star Z(t;\omega)dt + c^\star Z(t;\omega)dW(t)]\,,\, 0 < t < T,\, \omega \in \Omega, \\ Z(0;\omega) &= Z_0(\omega),\, \omega \in \Omega. \end{cases} \tag{3.180}$$

Assume that condition (3.179) holds. For a given $p \in \mathbb{N}$, we would like to determine which perturbations of (3.73) remain asymptotically exponentially p^{th}-moment stable. This requires that we determine when the strong solution of (3.180) satisfies Def. 3.8.1 (i). This is easily done because (3.180) is equivalent to the IVP

$$\begin{cases} dZ(t;\omega) = (a+a^\star)Z(t;\omega)dt + (c+c^\star)Z(t;\omega)dW(t), \\ Z(0;\omega) = Z_0(\omega),\, \omega \in \Omega \end{cases} \tag{3.181}$$

where $0 < t < T, \omega \in \Omega$. Arguing as above reveals that a necessary and sufficient condition for this to occur is

$$p\left((a+a^\star) - \frac{(c+c^\star)^2}{2}\right) + \frac{p^2(c+c^\star)^2}{2} < 0. \tag{3.182}$$

How do we interpret this condition? For the moment, assume that $c^\star = 0$. Then, condition (3.182) simplifies to

$$\underbrace{p\left(a - \frac{c^2}{2}\right) + \frac{p^2 c^2}{2}}_{<0 \text{ by assumption}} + pa^\star < 0. \tag{3.183}$$

As such, $a^\star$ can be any value for which the sum of $pa^\star$ and the expression in (3.179) remains negative. That is, $a^\star$ satisfies

$$a^\star < -\left(\frac{p\left(a - \frac{c^2}{2}\right) + \frac{p^2 c^2}{2}}{p}\right). \tag{3.184}$$

The interpretation when $c^\star \neq 0$ is more involved to treat analytically. But, for given values of p, a, and c, it suffices to consider the surface defined by $z =$ the left-hand side of (3.182) and determine the subset of the $a^\star c^\star$-plane in which the surface lies below the $a^\star c^\star$-plane.

Exercise 3.8.3. Establish such a robustness result for (3.74).

Finally, we consider Def. 3.8.1 (ii). Recall that the formula for the strong solution of (3.73) is given by

$$X(t; \omega) = X_0(\omega) e^{\left((a - \frac{1}{2}c^2)t + cW(t)\right)}. \tag{3.185}$$

Taking the absolute value on both sides of (3.185), followed by taking the log, yields

$$\log |X(t; \omega)| = \log |X_0(\omega)| + \left(a - \frac{1}{2}c^2\right)t + cW(t). \tag{3.186}$$

Dividing both sides of (3.186) by $t > 0$ and using the Law of Large Numbers results in

$$\lim_{t \to \infty} \frac{\log |X(t; \omega)|}{t} = \underbrace{\lim_{t \to \infty} \frac{\log |X_0 \omega)|}{t}}_{=0} + \underbrace{\lim_{t \to \infty} \frac{\left(a - \frac{1}{2}c^2\right)t}{t}}_{= \left(a - \frac{1}{2}c^2\right)} + \underbrace{\lim_{t \to \infty} \frac{cW(t)}{t}}_{=0}$$

$$= a - \frac{1}{2}c^2 \text{ a.s. } [\mathscr{P}]. \tag{3.187}$$

As such, the strong solution of (3.73) is almost surely exponentially stable if $a - \frac{1}{2}c^2 < 0$.

Exercise 3.8.4. Derive such a condition for the almost surely exponential stability of the strong solution of (3.74).

Exercise 3.8.5. Investigate the robustness of this type of stability.

Exercise 3.8.6. Prove that if the strong solution of (3.73) is asymptotically exponentially p^{th}-moment stable, then it is almost surely exponentially stable, but not conversely in general.

3.9   A Classical Example

Presently, we consider a classical example of an SDE arising in physics. The theory presented in this chapter can be applied directly.

Example 3.9.1. (The Ornstein-Uhlenbeck Process)
Consider a particle that moves through space merely under the influence of friction. Its position and velocity functions are naturally vector-valued. For simplicity, we consider a single component of each of these vectors, denoted by $s(t;\omega)$ and $v(t;\omega)$. Assuming that the acceleration is subject to white noise $\frac{dW}{dt}$, Ornstein and Uhlenbeck **[285]** formulated the following system of SDEs governing the motion of such a particle:

$$\begin{cases} dv(t;\omega) = -\alpha v(t;\omega)dt + \beta dW(t), t > 0, \omega \in \Omega, \\ v(0;\omega) = v_0(\omega), \omega \in \Omega, \end{cases} \tag{3.188}$$

$$\begin{cases} ds(t;\omega) = v(t;\omega)dt, t > 0, \omega \in \Omega, \\ s(0;\omega) = s_0(\omega), \omega \in \Omega, \end{cases} \tag{3.189}$$

where $\alpha > 0$, $\beta \in \mathbb{R}$, and $s_0(\cdot)$ and $v_0(\cdot)$ are $\mathscr{F}_0$-measurable members of $\mathfrak{L}^2(\Omega;\mathbb{R})$ that are normally distributed.

Exercise 3.9.1.
i.) Derive a representation formula for the strong solution of (3.188).
ii.) Solve (3.189). Write down its representation formula using (i).
iii.) Argue that both $\{v(t;\omega) : 0 \le t \le T, \omega \in \Omega\}$ and $\{s(t;\omega) : 0 \le t \le T, \omega \in \Omega\}$ are Gaussian processes.
iv.) Calculate the following: $\mu_v(t)$, $\mu_s(t)$, $Var_v(t)$, and $Var_s(t)$.
v.) Derive the pth-moments for (3.188) and (3.189) directly using the representation formula from (i).

3.10   Looking Ahead

Armed with some rudimentary tools of stochastic analysis, we are ready to embark on our journey through the world of stochastic evolution equations. We will begin by investigating some models whose mathematical description involves an extension of (3.1) to vector form. Specifically, it is natural to ask what is meant by a strong solution of the stochastic IVP:

$$\begin{cases} \frac{d\mathbf{X}}{dt}(t;\omega) = \mathbf{A}\mathbf{X}(t;\omega),\ 0 < t < T, \omega \in \Omega, \\ \mathbf{X}(0;\omega) = \mathbf{X}_0(\omega), \omega \in \Omega, \end{cases} \tag{3.190}$$

where $\mathbf{X}(t;\omega) = \begin{bmatrix} X_1(t;\omega) \\ \vdots \\ X_N(t;\omega) \end{bmatrix}$ and $\mathbf{A} \in \mathbb{M}^N(\mathbb{R})$. It is tempting to write $\mathbf{X}(t;\omega) = e^{\mathbf{A}t}\mathbf{X}_0(\omega)$, but what does this really mean? What are its properties? More generally, if we introduce randomness by means of white noise, the stochastic IVP

$$\begin{cases} d\mathbf{X}(t;\omega) = \mathbf{A}\mathbf{X}(t;\omega)dt + \mathbf{C}\mathbf{X}(t;\omega)dW(t),\ 0 < t < T, \omega \in \Omega, \\ \mathbf{X}(0;\omega) = \mathbf{X}_0(\omega), \omega \in \Omega, \end{cases} \tag{3.191}$$

arises. Interpreting this IVP requires that we broaden our definition of the Itô integral, to say the least. We address these questions and much more in Chapter 4.

3.11   Guidance for Selected Exercises

3.11.1   Level 1: A Nudge in a Right Direction

3.1.4. Integrate both sides of (3.5) corresponding to each initial condition. (Now what?)

3.1.5. Consider integrated versions of (3.6) and proceed as in Exer. 3.1.4.

3.2.1. Use the fact that $c \in \mathcal{L}^2(\Omega; [0, \infty))$.

3.2.2. Use the fact that $\mathcal{U}$ is a linear space. (So what?)

3.2.3. Use Step 3 of the building block process.

3.2.4. (i) f is continuous on a compact set. (So what?) Also, use Prop. 3.2.1(iv) and (vi)(a). (How?)

   (ii) Use the upper bound from (i).

   (iii) Use the upper bound from (ii).

3.2.5. (i) Use the Cauchy-Schwarz inequality for (i) and triangle inequality for (ii).

   (ii) Apply Prop. 3.2.1(viii) with $g(x) = 1$.

3.2.9. (i) You must be careful because the step functions u and v need not be defined in terms of characteristic functions using the same partition of $[0,t]$. Let $0 = t_0 < t_1 < \ldots < t_{m-1} < t_m = t$ and $0 = \overline{t_0} < \overline{t_1} < \ldots < \overline{t_{n-1}} < \overline{t_n} = t$ be partitions of $[0,t]$, and $\{b_i : i = 1, \ldots, n\} \cup \{c_i : i = 1, \ldots, m\} \subset \mathcal{L}^2(\Omega; [0, \infty))$ be $\mathscr{F}_{t_i}$-adapted. Define $u, v : [0,t] \times \Omega \longrightarrow [0, \infty)$ by

$$u(s; \omega) = \sum_{i=1}^{m} c_i(\omega) \chi_{[t_{i-1}, t_i)}(s),$$

$$v(s; \omega) = \sum_{j=1}^{n} b_j(\omega) \chi_{[\overline{t_{j-1}}, \overline{t_j})}(s).$$

Then,

$$(u+v)(s; \omega) = \sum_{i=1}^{m} \sum_{j=1}^{n} (c_i(\omega) + b_j(\omega)) \chi_{[t_{i-1}, t_i) \cap [\overline{t_{j-1}}, \overline{t_j})}(s).$$

Also, note that constant multiples do not affect the partitions used to define a step function. (Now what?)

(ii) Apply additivity to each member of the sum. Then, pull apart. (Now what?)

(iii) Use the linearity of $E[\cdot]$ with the analogous property established in Step 1 of the discussion.

(iv) Note that $W(t_j) - W(t_{j-1})$ is independent of $c_i(\cdot) c_j(\cdot)(W(t_i) - W(t_{i-1}))$ for all $i < j$. (Why?) So,

$$E\left[\int_0^T u(s; \cdot) dW(s)\right]^2 = \sum_{i=1}^{m} \sum_{j=1}^{n} E\left[c_i(\cdot) c_j(\cdot)(W(t_i) - W(t_{i-1}))(W(t_j) - W(t_{j-1}))\right]$$

$$= \sum_{i=1}^{m} E\left[c_i^2(\cdot)(W(t_i) - W(t_{i-1}))^2\right].$$

(Now what?)

(v) (a) The sum of $\mathscr{F}_t$-measurable functions is $\mathscr{F}_t$-measurable. Moreover, each term in the sum is $\mathscr{F}_t$-measurable.

(b) Apply on the subintervals of the partition used to define the step function. (Then what?)

(c) The sum of functions that are continuous a.s. $[\mathscr{P}]$ is again continuous a.s. $[\mathscr{P}]$.

(d) Use Exer. 2.5.7 to conclude that the sum in normal. (Why can this be applied?)

3.2.11. Because the sum of $\mathscr{F}_t$-measurable random variables is again $\mathscr{F}_t$-measurable, it suffices to argue that each of the three terms on the right-hand side of (3.42) is $\mathscr{F}_t$-measurable. First, Y_0 is $\mathscr{F}_0$-measurable and $\mathscr{F}_0 \subset \mathscr{F}_t, \forall t$. (So what? How about the other two terms?)

3.2.13. (i) $\frac{\partial h}{\partial t} = 0$, $\frac{\partial h}{\partial x_1} = x_2$, $\frac{\partial h}{\partial x_2} = x_1$. (How about the second partials?)

(ii) $d(X_1(t) X_2(t)) = X_2 dX_1 + X_1 dX_2 + \underbrace{\frac{1}{2}[dX_1 dX_2 + dX_2 dX_1]}_{= dX_1 dX_2}$

3.2.14. Use $h(t,x) = \ln x$ in (3.49). Observe that $\frac{\partial h}{\partial t} = 0$, $\frac{\partial h}{\partial x} = \frac{1}{x}$, $\frac{\partial^2 h}{\partial x^2} = -\frac{1}{x^2}$. Now, compute $d\left[\ln X(t)\right]$.

3.2.15. (i) Use $h(t,x) = e^{-At}x$. Use Itó's formula to show that

$$d\left[e^{-At}X(t;\omega)\right] = \varepsilon e^{-At}dW(t).$$

(ii) Observe that

$$e^{-At}X(t;\omega) - X_0(\omega) = \int_0^t \varepsilon e^{-As}dW(s),$$

and so

$$X(t;\omega) = X_0(\omega)e^{At} + \varepsilon \int_0^t e^{-A(t-s)}dW(s).$$

3.4.1. $\|X_0\|_{\mathscr{L}^2}$, T, a, c_k $(k = 1,\dots.m)$, and m

3.4.2. The process is essentially identical to the one used in (3.125) and (3.126). You need only determine how the specific terms on the right-hand side of (3.126) change.

3.5.1. Use (3.128) directly. (How?)

3.5.2. Add and subtract $a^\star X(t;\cdot)$ in (3.129). Separate the terms appropriately. What do you need in order to be able to apply Gronwall's Lemma?

3.6.3. (i) Use the integration by parts formula to see that

$$
\begin{aligned}
d\left(X(t;\cdot)X(t;\cdot)\right) &= 2X(t;\cdot)d(X(t;\cdot)) + (d(X(t;\cdot)))^2 \\
&= 2X(t;\cdot)\left(aX(t;\cdot)dt + cX(t;\cdot)dW(t)\right) \\
&\quad + (aX(t;\cdot)dt + cX(t;\cdot)dW(t))^2.
\end{aligned}
$$

Now, continue simplifying.

3.6.6. Use Itó's formula with $h(t,x) = x^p$. Argue as in Prop. 3.6.1.

3.7.2. Use the representation formulae for the solutions rather than the integrated forms of the IVPs.

3.8.6. We know that $p\left(a - \frac{c^2}{2}\right) + \frac{p^2c^2}{2} < 0$. Since $\frac{p^2c^2}{2} > 0$, we see that

$$p\left(a - \frac{c^2}{2}\right) \le p\left(a - \frac{c^2}{2}\right) + \frac{p^2c^2}{2} < 0.$$

You can construct a counterexample to show that the reverse implication does not hold simply by choosing appropriate values for a, p, c. (Tell how.)

3.9.1. (i) $v(t;\omega) = e^{-\alpha t}v_0(\omega) + \beta \int_0^t e^{-\alpha(t-s)}dW(s)$.

(ii) $s(t;\omega) = \frac{1}{\alpha}\left(1 - e^{-\alpha t}\right)v_0(\omega) + s_0(\omega) + \frac{\beta}{\alpha}\int_0^t\left(1 - e^{-\alpha(t-s)}dW(s)\right)$

(iv) $\mu_v(t) = e^{-\alpha t}E\left[v_0(\cdot)\right]$, $\mu_s(t) = \frac{1}{\alpha}\left(1 - e^{-\alpha t}\right)E\left[v_0(\cdot)\right] + E\left[s_0(\cdot)\right]$, and
$Var_s(t) = e^{-2\alpha t}Var\left[v_0(\cdot)\right] + \frac{\beta^2}{2\alpha}\left(1 - e^{-2\alpha t}\right).$

3.11.2   Level 2: An Additional Thrust in a Right Direction

3.1.4. Use Gronwall's Lemma. (How?)

3.2.4. (i) $\exists M^\star > 0$ such that $|f(x)| \le M^\star$, $\forall x \in [a,b]$. One natural upper bound for the given set is

$$\zeta = M^\star + (b - a)(M_1 M^{\star\star} + M_2),$$

where $M^{\star\star} = \sup\{|u(z)| : z \in [a,b]\}$. (Verify this.)

  (ii) ζ^N (Why?)

  (iii) $\zeta^N(b-a)$ (Why?)

3.2.9. (i) Now, apply the linearity of a finite sum and regroup the terms of u and v.

(iv) Continue the previous string of equalities as follows:

$$\ldots = \sum_{i=1}^{m} E\left[c_i^2(\cdot)\right] E\left[(W(t_i) - W(t_{i-1}))^2\right]$$

$$= E\left[\sum_{i=1}^{m} c_i^2(\cdot)(t_i - t_{i-1})\right]$$

$$= \int_0^T E\left[u^2(s;\cdot)\right] ds$$

(v) (b) Now, use linearity of $E[\cdot|\cdot\cdot]$ with the property from Step 1.

(d) The increments on the right-hand side are independent and each term is normal by the definition of $W(\cdot)$.

3.2.11. Because $u_1(t)$ is $\mathscr{F}_t$-measurable, it follows immediately that $\int_0^t u_1(s;\omega)ds$ is $\mathscr{F}_t$-measurable. Also, $\int_0^t u_2(s;\omega)dW(s)$ is $\mathscr{F}_t$-measurable by Prop. 3.2.3(v)(a). This completes the proof.

3.2.13. (i) $\frac{\partial^2 h}{\partial x_1^2} = \frac{\partial^2 h}{\partial x_2^2} = 0$, $\frac{\partial^2 h}{\partial x_1 \partial x_2} = 1 = \frac{\partial^2 h}{\partial x_2 \partial x_1}$

3.2.14. Observe that

$$d\left[\ln X(t;\omega)\right] = 0dt + \frac{1}{X(t;\omega)}dX(t;\omega) + \frac{1}{2}\left(-\frac{1}{X^2(t;\omega)}\right)(dX(t;\omega))^2$$

$$= \vdots$$

$$= \left(A - \frac{1}{2}B^2\right)dt + BdW.$$

(We will revisit this problem later in the chapter.)

3.4.2. The formula is $X(t;\omega) = e^{at}X_0(\omega)e^{-\frac{1}{2}\sum_{k=1}^{m}c_k^2 t + \sum_{k=1}^{m}c_k W_k(t)}$.

3.5.1. We need $\|X_0 - Y_0\|_{\mathscr{L}^2(\Omega;\mathbb{R})}^2 < \frac{\varepsilon}{3e^{3T^2\left(a^2 + m\sum_{k=1}^{m}c_k^2\right)}}$. (Why?)

3.5.2. You will need $E|X(t;\cdot)|^2 \leq C, \forall 0 \leq t \leq T$. (Why? Now what?)

3.6.6. Note that $\frac{\partial h}{\partial x} = p|x|^{p-1}\frac{x}{|x|}$.

Chapter 4

Homogenous Linear Stochastic Evolution Equations in $\mathbb{R}^N$

Overview

The question of how phenomena evolve over time is central to a broad range of fields within the social, natural, and physical sciences. The behavior of such phenomena is governed by established laws of the underlying field that typically describe the rates at which it and related quantities evolve over time. The measurement of all parameters is subject to noise. As such, a precise mathematical description involves the formulation of so-called *stochastic evolution equations* whose complexity depends to a large extent on the realism of the model. We focus in this chapter on models in which the evolution equation is generated by a system of ordinary differential equations with finitely many independent sources of multiplicative noise. We are in search of an abstract paradigm into which all of these models are subsumed as special cases. Once established, we will study the rudimentary properties of the abstract paradigm and subsequently apply the results to each model. Some standard references used throughout this chapter are **[11, 62, 178, 193, 244, 251, 287, 302, 318, 333, 375, 397, 399]**.

4.1 Motivation by Models

Throughout this chapter, the formulation of all models begins with a complete probability space $(\Omega, \mathscr{F}, \mathscr{P})$ equipped with a natural filtration $\{\mathscr{F}_t : t \geq 0\}$ to which all Wiener processes are adapted. We motivate the theoretical development presented in this chapter with a discussion of some elementary models.

Model I.1 Chemical Kinetics

Chemical substances are transformed via a sequence of reactions into other products. The use of stochastic differential equations facilitates the understanding the chemical kinetics of such reactions. (See **[59, 72, 128, 270, 303, 390, 391]**.) For instance, consider a first-order reaction $Y \xrightarrow{\alpha_1} Z$ in which a substance Y, whose concentration at time t is denoted by $[Y](t)$, is transformed into a product Z at a rate α_1. The rate at which $[Y](t)$ reacts and is converted into Z is described by an initial-value problem.

We first assume that noise is introduced via the initial data only in the sense that we assume that the initial concentration $[Y]_0$ is an $\mathscr{F}_0$-measurable random variable with $E\left|[Y]_0\right|^2 < \infty$. Doing this automatically renders both $[Y](t)$ and $[Z](t)$ as stochastic processes. For each fixed $\omega \in \Omega$, the random IVP governing this scenario is given by

$$\begin{cases} \frac{d[Y](t;\omega)}{dt} = -\alpha_1 [Y](t;\omega),\ t > 0,\ \omega \in \Omega, \\ \frac{d[Z](t;\omega)}{dt} = \alpha_1 [Z](t;\omega),\ t > 0,\ \omega \in \Omega, \\ [Y](0;\omega) = [Y]_0(\omega),\ [Z](0;\omega) = 0,\ \omega \in \Omega. \end{cases} \quad (4.1)$$

The system (4.1) is easily converted into the equivalent matrix form:

$$\begin{cases} \begin{bmatrix} [Y](t;\omega) \\ [Z](t;\omega) \end{bmatrix}' = \begin{bmatrix} -\alpha_1 & 0 \\ 0 & \alpha_1 \end{bmatrix} \begin{bmatrix} [Y](t;\omega) \\ [Z](t;\omega) \end{bmatrix},\ t > 0,\ \omega \in \Omega, \\ \begin{bmatrix} [Y](0;\omega) \\ [Z](0;\omega) \end{bmatrix} = \begin{bmatrix} [Y]_0(\omega) \\ 0 \end{bmatrix},\ \omega \in \Omega. \end{cases} \quad (4.2)$$

For a given $\omega \in \Omega$, (4.2) can be treated as in the deterministic case using the matrix exponential. We will revisit this in the next section.

A more interesting manner in which noise can be introduced into the model is in the measurement of the rate constant α_1. This parameter is measured experimentally and, hence, is subject to randomness. Precisely, we assume that

$$\alpha_1 = \overline{\alpha_1} + \overline{\alpha_2} \frac{dW(t)}{dt}, \quad (4.3)$$

where $\frac{dW(t)}{dt}$ formally represents a white noise process (cf. Section 2.9.3). Formally substituting (4.3) into (4.2) and rewriting the equation in differential form yields the following stochastic IVP:

$$\begin{cases} d[Y](t;\omega) = -\overline{\alpha_1}[Y](t;\omega)dt - \overline{\alpha_2}[Y](t;\omega)dW(t),\ t > 0,\ \omega \in \Omega, \\ d[Z](t;\omega) = \overline{\alpha_1}[Z](t;\omega)dt + \overline{\alpha_2}[Z](t;\omega)dW(t),\ t > 0,\ \omega \in \Omega, \\ [Y](0;\omega) = [Y]_0(\omega),\ [Z](0;\omega) = 0,\ \omega \in \Omega, \end{cases} \quad (4.4)$$

where $[Y]_0(\cdot)$ is independent of $\{W(t) : 0 \leq t \leq T\}$. We would like to write (4.4) in an equivalent matrix form similar to (4.2). Here, the natural approach is to mimic how the right-hand side of (4.2) is formulated. Doing so yields

$$\begin{cases} d\begin{bmatrix} [Y](t;\omega) \\ [Z](t;\omega) \end{bmatrix} = \begin{bmatrix} -\overline{\alpha_1} & 0 \\ 0 & \overline{\alpha_1} \end{bmatrix} \begin{bmatrix} [Y](t;\omega) \\ [Z](t;\omega) \end{bmatrix} dt \\ \qquad\qquad + \begin{bmatrix} -\overline{\alpha_2} & 0 \\ 0 & \overline{\alpha_2} \end{bmatrix} \begin{bmatrix} [Y](t;\omega) \\ [Z](t;\omega) \end{bmatrix} dW(t),\ t > 0,\ \omega \in \Omega, \\ \begin{bmatrix} [Y](0;\omega) \\ [Z](0;\omega) \end{bmatrix} = \begin{bmatrix} [Y]_0(\omega) \\ 0 \end{bmatrix},\ \omega \in \Omega. \end{cases} \quad (4.5)$$

(Check this!)

A related reversible reaction $Y \xrightarrow{\alpha} Z$, $Y \xleftarrow{\beta} Z$ in which part of the product resulting from the forward reaction is converted back into the original substance at a certain rate can also be considered. Suppose that both parameters α and β are modeled as white noise processes, say

$$\alpha = \overline{\alpha_1} + \overline{\alpha_2} \frac{dW_1(t)}{dt}, \tag{4.6}$$

$$\beta = \overline{\beta_1} + \overline{\beta_2} \frac{dW_2(t)}{dt}, \tag{4.7}$$

where $\{W_1(t) : 0 \leq t \leq T\}$ and $\{W_2(t) : 0 \leq t \leq T\}$ are independent, real-valued Wiener processes. The resulting stochastic IVP is given by

$$\begin{cases} d\,[Y]\,(t;\omega) = \left(-\overline{\alpha_1}\,[Y]\,(t;\omega) + \overline{\beta_1}\,[Z]\,(t;\omega)\right) dt - \overline{\alpha_2}\,[Y]\,(t;\omega)dW_1(t) \\ \quad + \overline{\beta_2}\,[Z]\,(t;\omega)dW_2(t),\ t > 0,\ \omega \in \Omega, \\ d\,[Z]\,(t;\omega) = \left(\overline{\alpha_1}\,[Y]\,(t;\omega) - \overline{\beta_1}\,[Z]\,(t;\omega)\right) dt + \overline{\alpha_2}\,[Y]\,(t;\omega)dW_1(t) \\ \quad - \overline{\beta_2}\,[Z]\,(t;\omega)dW_2(t),\ t > 0,\ \omega \in \Omega, \\ [Y]\,(0;\omega) = [Y]_0\,(\omega),\ [Z]\,(0;\omega) = [Z]_0\,(\omega),\ \omega \in \Omega, \end{cases} \tag{4.8}$$

where both random variables $[Y]_0\,(\cdot)$ and $[Z]_0\,(\cdot)$ are $\mathscr{F}_0$-measurable with finite second moments and are independent of the Wiener processes $\{W_1(t) : 0 \leq t \leq T\}$ and $\{W_2(t) : 0 \leq t \leq T\}$.

We want to express (4.8) in an equivalent matrix form, but the right-hand sides of the SDEs in IVP (4.8) are more complicated than those appearing in (4.4). How do we proceed? When converting (4.4) into (4.5), we inserted "zero place holders" without deliberately announcing as much. Doing so enabled us to formally define a second coefficient matrix that, when multiplied by the solution process $\begin{bmatrix} [Y]\,(t;\omega) \\ [Z]\,(t;\omega) \end{bmatrix}$ and then by $dW(t)$, accurately captured all of the noise terms. We shall proceed in the same manner for both equations in (4.8) and then appropriately combine them. To this end, observe that for every $t > 0$, $\omega \in \Omega$,

$$d\,[Y]\,(t;\omega) = \left(-\overline{\alpha_1}\,[Y]\,(t;\omega) + \overline{\beta_1}\,[Z]\,(t;\omega)\right)dt + \left(-\overline{\alpha_2}\,[Y]\,(t;\omega) + 0\,[Z]\,(t;\omega)\right)dW_1(t)$$

$$+ \left(0\,[Y]\,(t;\omega) + \overline{\beta_2}\,[Z]\,(t;\omega)\right)dW_2(t), \tag{4.9}$$

$$d\,[Z]\,(t;\omega) = \left(\overline{\alpha_1}\,[Y]\,(t;\omega) - \overline{\beta_1}\,[Z]\,(t;\omega)\right)dt + \left(\overline{\alpha_2}\,[Y]\,(t;\omega) + 0\,[Z]\,(t;\omega)\right)dW_1(t)$$

$$+ \left(0\,[Y]\,(t;\omega) - \overline{\beta_2}\,[Z]\,(t;\omega)\right)dW_2(t). \tag{4.10}$$

Now, gather all the $dW_1(t)$ terms together. We apply the same approach used to formulate the matrix version of (4.4) to obtain the matrix expression

$$\begin{bmatrix} -\overline{\alpha_2} & 0 \\ \overline{\alpha_2} & 0 \end{bmatrix} \begin{bmatrix} [Y]\,(t;\omega) \\ [Z]\,(t;\omega) \end{bmatrix} dW_1(t) = \begin{bmatrix} -\overline{\alpha_2}\,[Y]\,(t;\omega)dW_1(t) \\ \overline{\alpha_2}\,[Y]\,(t;\omega)dW_1(t) \end{bmatrix}. \tag{4.11}$$

Similarly, we have

$$\begin{bmatrix} 0 & \overline{\beta_2} \\ 0 & -\overline{\beta_2} \end{bmatrix} \begin{bmatrix} [Y](t;\omega) \\ [Z](t;\omega) \end{bmatrix} dW_2(t) = \begin{bmatrix} \overline{\beta_2}[Z](t;\omega)dW_2(t) \\ -\overline{\beta_2}[Z](t;\omega)dW_2(t) \end{bmatrix}. \tag{4.12}$$

Moreover, observe that

$$\begin{bmatrix} -\overline{\alpha_2} & 0 \\ \overline{\alpha_2} & 0 \end{bmatrix} \begin{bmatrix} [Y](t;\omega) \\ [Z](t;\omega) \end{bmatrix} dW_1(t) + \begin{bmatrix} 0 & \overline{\beta_2} \\ 0 & -\overline{\beta_2} \end{bmatrix} \begin{bmatrix} [Y](t;\omega) \\ [Z](t;\omega) \end{bmatrix} dW_2(t) \tag{4.13}$$

$$= \begin{bmatrix} -\overline{\alpha_2}[Y](t;\omega)dW_1(t) + \overline{\beta_2}[Z](t;\omega)dW_2(t) \\ \overline{\alpha_2}[Y](t;\omega)dW_1(t) - \overline{\beta_2}[Z](t;\omega)dW_2(t) \end{bmatrix}. \tag{4.14}$$

As such, we can write (4.8) in the equivalent matrix form:

$$\begin{cases} d\begin{bmatrix} [Y](t;\omega) \\ [Z](t;\omega) \end{bmatrix} = \begin{bmatrix} -\overline{\alpha_1} & \overline{\beta_1} \\ \overline{\alpha_1} & -\overline{\beta_1} \end{bmatrix} \begin{bmatrix} [Y](t;\omega) \\ [Z](t;\omega) \end{bmatrix} dt \\ + \begin{bmatrix} -\overline{\alpha_2} & 0 \\ \overline{\alpha_2} & 0 \end{bmatrix} \begin{bmatrix} [Y](t;\omega) \\ [Z](t;\omega) \end{bmatrix} dW_1(t) + \begin{bmatrix} 0 & \overline{\beta_2} \\ 0 & -\overline{\beta_2} \end{bmatrix} \begin{bmatrix} [Y](t;\omega) \\ [Z](t;\omega) \end{bmatrix} dW_2(t), \tag{4.15} \\ \begin{bmatrix} [Y](0;\omega) \\ [Z](0;\omega) \end{bmatrix} = \begin{bmatrix} [Y]_0(\omega) \\ 0 \end{bmatrix}, \ \omega \in \Omega, \end{cases}$$

where $t > 0$, $\omega \in \Omega$. What are the concentrations of both substances at time t, and do they approach an equilibrium state as $t \to \infty$?

Model II.1 Pharmacokinetics
The field of pharmacokinetics is concerned with studying the evolution of a substance (e.g., drugs, toxins, nutrients) administered to a living organism (by consumption, inhalation, absorption, etc.) and its effects on the organism. Models can be formed by partitioning portions of the organism into compartments, each of which is treated as a single unit; improvements of these models can be made by refining the compartments in various ways. We consider a rudimentary model motivated by the work discussed in **[120, 309]**. Let

$$a = \text{absorption rate into the bloodstream,}$$
$$b = \text{rate at which the drug is eliminated from the blood.}$$

As in Model I.1, we introduce noise into both of these parameters by assuming that

$$a = a_1 + a_2 \frac{dW_1(t)}{dt},$$
$$b = b_1 + b_2 \frac{dW_2(t)}{dt},$$

where $a_i, b_i > 0$ $(i = 1, 2)$, and $\{W_1(t) : 0 \le t \le T\}$ and $\{W_2(t) : 0 \le t \le T\}$ are independent real-valued Wiener processes. Further, let

$$y(t;\omega) = \text{concentration of the drug in the GI tract,}$$
$$z(t;\omega) = \text{concentration of the drug in the blood,}$$

and assume that $y_0 \in \mathcal{L}^2(\Omega; \mathbb{R})$ is independent of both Wiener processes. The following system is obtained based on an elementary "rate in *minus* rate out" model:

$$\begin{cases} dy(t; \omega) = -a_1 y(t; \omega)dt - a_2 y(t; \omega)dW_1(t), \\ dz(t; \omega) = (a_1 y(t; \omega) - b_1 z(t; \omega)) dt + a_2 y(t; \omega)dW_1(t) \\ \qquad\qquad -b_2 z(t; \omega)dW_2(t), \\ y(0; \omega) = y_0(\omega), \ z(0; \omega) = 0, \omega \in \Omega, \end{cases} \quad (4.16)$$

where $t > 0, \omega \in \Omega$.

Exercise 4.1.1. Reformulate (4.16) in an equivalent matrix form similar to (4.15).

We can further assume that the measurement of the parameter a is subject to <u>two</u> independent sources of noise, say

$$a = a_1 + a_2 \frac{dW_1(t)}{dt} + a_3 \frac{dW_3(t)}{dt}, \quad (4.17)$$

where $\{W_1(t) : 0 \le t \le T\}$, $\{W_2(t) : 0 \le t \le T\}$, and $\{W_3(t) : 0 \le t \le T\}$ are independent, and the initial condition $y_0(\cdot)$ is independent of all three Wiener processes. Doing so gives rise to the following modification of (4.16):

$$\begin{cases} dy(t; \omega) = -a_1 y(t; \omega)dt - a_2 y(t; \omega)dW_1(t) - a_3 y(t; \omega)dW_3(t), \\ dz(t; \omega) = (a_1 y(t; \omega) - b_1 z(t; \omega)) dt + a_2 y(t; \omega)dW_1(t) \\ \qquad +a_3 y(t; \omega)dW_3(t), -b_2 z(t; \omega)dW_2(t), \\ y(0; \omega) = y_0(\omega), \ z(0; \omega) = 0, \end{cases} \quad (4.18)$$

where $t > 0$ and $\omega \in \Omega$. As before, it is desirable to write (4.18) in an equivalent matrix form, but this time we must account for an $\mathbb{R}^3$-valued Wiener process $\mathbf{W}(t) = \langle W_1(t), W_2(t), W_3(t) \rangle$.

Exercise 4.1.2. Rewrite (4.18) in an equivalent matrix form similar to (4.15).

At what time is a certain level of the substance reached within each compartment in the body? When is the substance concentration among different compartments in equilibrium? What happens if our measurements of the parameters or the initial condition y_0 are a little off; are the resulting solutions drastically different?

Model III.1 Spring-Mass Systems

A second-order ordinary differential equation (ODE) describing the position $x(t)$ at time t of a mass m attached to an oscillating spring with spring constant k with respect to an equilibrium position can be derived using Newton's second law. Initially, let us assume that the movement of the mass is one-dimensional and that there is no damping factor or external driving force. Assuming that the initial data are random variables and that this is the only source of noise being introduced into the model, Hooke's law describes the force acting on the spring, resulting in the stochastic IVP:

$$\begin{cases} \frac{d^2x}{dt^2}(t;\omega) + \alpha^2 x(t;\omega) = 0, \ t > 0, \ \omega \in \Omega, \\ x(0;\omega) = x_0(\omega), \ \frac{dx}{dt}(0;\omega) = x_1(\omega), \ \omega \in \Omega. \end{cases} \tag{4.19}$$

Here, $\alpha^2 = \frac{k}{m}$, $x_0(\cdot)$ is the initial position of the mass with respect to the equilibrium, and $x_1(\cdot)$ is the initial speed. Both $x_0(\cdot)$ and $x_1(\cdot)$ are assumed to belong to $\mathcal{L}^2(\Omega;\mathbb{R})$. This is the most basic example of a *stochastic harmonic oscillator.* (See **[144, 170].**) This IVP can be converted into a system of first-order random ODEs by way of the change of variable

$$\begin{cases} y = x, \\ z = \frac{dx}{dt} \end{cases} \text{ so that } \begin{cases} \frac{dy}{dt} = \frac{dx}{dt} = z \\ \frac{dz}{dt} = \frac{d^2x}{dt^2} = -\alpha^2 x = -\alpha^2 y. \end{cases} \tag{4.20}$$

Then, (4.20) can be written in the equivalent matrix form:

$$\begin{cases} \frac{d}{dt}\begin{bmatrix} y(t;\omega) \\ z(t;\omega) \end{bmatrix} = \begin{bmatrix} 0 & 1 \\ -\alpha^2 & 0 \end{bmatrix} \begin{bmatrix} y(t;\omega) \\ z(t;\omega) \end{bmatrix}, \ t > 0, \ \omega \in \Omega, \\ \begin{bmatrix} y(0;\omega) \\ z(0;\omega) \end{bmatrix} = \begin{bmatrix} x_0(\omega) \\ x_1(\omega) \end{bmatrix}, \ \omega \in \Omega. \end{cases} \tag{4.21}$$

Certainly, $z(t;\omega)$ is redundant (Why?), but this transformation is useful because it enables us to consider a second-order random ODE in the same form as the previous two models. As before, **IVP** (4.21) can be handled using the deterministic theory.

We now go one step further and assume that

$$\alpha^2 = (\overline{\alpha_1})^2 + (\overline{\alpha_2})^2 \frac{dW(t)}{dt}. \tag{4.22}$$

(Actually, a more realistic stochastic spring mass system is obtained by introducing noise via an external driving force. This will be explored in Chapter 5.) Substituting (4.22) into (4.19) yields the equation

$$\frac{d^2x}{dt^2}(t;\omega) + (\overline{\alpha_1})^2 x(t;\omega) = -(\overline{\alpha_2})^2 x(t;\omega)\frac{dW(t)}{dt}, t > 0, \ \omega \in \Omega. \tag{4.23}$$

Performing a similar change of variable, we see that

$$\begin{cases} y = x, \\ z = \frac{dx}{dt} \end{cases} \text{ so that } \begin{cases} \frac{dy}{dt} = \frac{dx}{dt} = z \\ \frac{dz}{dt} = \frac{d^2x}{dt^2} = -(\overline{\alpha_1})^2 \underbrace{x}_{=y} - (\overline{\alpha_2})^2 \underbrace{x}_{=y} \frac{dW}{dt}. \end{cases} \tag{4.24}$$

Hence, converting (4.23) to differential form yields

$$\begin{cases} dy(t;\omega) = z(t;\omega)dt, t > 0, \ \omega \in \Omega, \\ dz(t;\omega) = -(\overline{\alpha_1})^2 y(t;\omega)dt - (\overline{\alpha_2})^2 y(t;\omega)dW, t > 0, \ \omega \in \Omega, \\ y(0;\omega) = x_0(\omega), \ z(0;\omega) = x_1(\omega), \ \omega \in \Omega. \end{cases} \tag{4.25}$$

Inserting "zero place holders" then yields

$$
\begin{cases}
dy = (0y + z)\,dt + (0y + 0z)\,dW, \\
dz = \left(-(\overline{\alpha_1})^2 y + 0z\right)dt + \left(-(\overline{\alpha_2})^2 y + 0z\right)dW, \\
y(0) = x_0,\ z(0) = x_1.
\end{cases}
\tag{4.26}
$$

(Note that we have suppressed the dependence on t and ω in (4.26).) Subsequently, this results in the following matrix form:

$$
\begin{cases}
\dfrac{d}{dt}\begin{bmatrix} y(t;\omega) \\ z(t;\omega) \end{bmatrix}
= \begin{bmatrix} 0 & 1 \\ -(\overline{\alpha_1})^2 & 0 \end{bmatrix}
\begin{bmatrix} y(t;\omega) \\ z(t;\omega) \end{bmatrix} dt \\
\qquad + \begin{bmatrix} 0 & 1 \\ -(\overline{\alpha_2})^2 & 0 \end{bmatrix}
\begin{bmatrix} y(t;\omega) \\ z(t;\omega) \end{bmatrix} dW(t), t > 0, \omega \in \Omega, \\
\begin{bmatrix} y(0;\omega) \\ z(0;\omega) \end{bmatrix}
= \begin{bmatrix} x_0(\omega) \\ x_1(\omega) \end{bmatrix}, \omega \in \Omega.
\end{cases}
\tag{4.27}
$$

Remark. The trick of converting a higher-order SDE into a system of first-order SDEs is very useful and arises in the study of several different models, such as wave equations and equations governing the dynamics of beams.

Next, consider the scenario in which there are two springs, one to which a mass m_A is affixed and the other to which a mass m_B is attached. One of these springs is fastened to the mass of the other to form a system of coupled springs. The deterministic IVP governing the positions $x_A(t)$ and $x_B(t)$ of the masses m_A and m_B, respectively, with respect to the equilibrium state is given by

$$
\begin{cases}
m_A x_A''(t) + (k_A + k_B)x_A(t) + k_A x_B(t) = 0,\ t > 0, \\
m_B x_B''(t) - k_B x_A(t) + k_B x_B(t) = 0,\ t > 0, \\
x_A(0) = x_{0,A},\ x_A'(0) = x_{1,A}, \\
x_B(0) = x_{0,B},\ x_B'(0) = x_{1,B}.
\end{cases}
\tag{4.28}
$$

Exercise 4.1.3. Formulate a random version of (4.28) analogous to (4.19) by introducing noise via the initial data. Then, convert the resulting system to an equivalent matrix form.

Exercise 4.1.4. Suppose that

$$
k_A = k_A^\star + \overline{k_A}\frac{dW_1}{dt},
\tag{4.29}
$$

$$
k_B = k_B^\star + \overline{k_B}\frac{dW_2}{dt},
\tag{4.30}
$$

where $\{W_1(t) : 0 \le t \le T\}$ and $\{W_2(t) : 0 \le t \le T\}$ are independent Wiener processes, and all the initial conditions are independent of both Wiener processes. Incorporate (4.29) and (4.30) into (4.28) to obtain a system of second-order SDEs. Then, convert this into a first-order system.

Common Theme: Although these applications arise in vastly different contexts, the nature of the IVPs used to describe the scenarios and the questions posed are strikingly similar. We encountered two different types of stochastic systems. The simpler of the two is when the noise is incorporated into the model only via the initial conditions. Such an IVP can be written compactly in vector form as

$$\begin{cases} \mathbf{U}'(t;\omega) = \mathbf{A}\mathbf{U}(t;\omega), t > 0, \omega \in \Omega, \\ \mathbf{U}(0;\omega) = \mathbf{U}_0(\omega), \omega \in \Omega, \end{cases} \tag{4.31}$$

where $\mathbf{U} : [0,\infty) \times \Omega \to \mathbb{R}^N$ is the solution stochastic process, $\mathbf{A}$ is an $N \times N$ constant matrix, and $\mathbf{U}_0 \in \mathfrak{L}^2\left(\Omega;\mathbb{R}^N\right)$ is the vector containing the initial conditions. It would be efficient to initially focus our attention on the abstract IVP (4.31) and answer as many rudimentary questions as possible regarding existence and uniqueness of a solution, continuous dependence of the solutions with respect to initial data, etc. Then, in turn, we could apply those results to any model that could be viewed as a special case of (4.31). In analogy with the first-order case, $\forall \omega \in \Omega$, we can treat (4.31) as a deterministic ODE and apply the theory of homogenous linear systems developed in **[295]** and outlined in Section 3.2. Indeed, we expect the solution process of (4.31) to be $\mathbf{U}(t;\omega) = e^{\mathbf{A}t}\mathbf{U}_0(\omega)$, a.s. $[\mathscr{P}]$, where $e^{\mathbf{A}t}$ is the matrix exponential. For such an IVP, a complete theory already has been developed and can be applied directly with the caveat that all statements hold a.s. $[\mathscr{P}]$.

The stochastic systems formed by introducing white noise processes into the description of the parameters is more involved, but still resembles (3.73) and (3.74). Specifically, all these IVPs can be written in the matrix form

$$\begin{cases} d\mathbf{U}(t;\omega) = \mathbf{A}\mathbf{U}(t;\omega)dt + \sum_{k=1}^{m} \mathbf{B}_k\mathbf{U}(t;\omega)dW_k(t), t > 0, \omega \in \Omega, \\ \mathbf{U}(0;\omega) = \mathbf{U}_0(\omega), \omega \in \Omega, \end{cases} \tag{4.32}$$

where $\mathbf{U} : [0,\infty) \times \Omega \to \mathbb{R}^N$ is the solution stochastic process; $\mathbf{A}, \mathbf{B}_1,\ldots,\mathbf{B}_m$ are $N \times N$ constant matrices; $\{W_i(t) : 0 \leq t \leq T\}, i = 1,\ldots,m$ are independent $\mathbb{R}$-valued Wiener processes; and $\mathbf{U}_0 \in \mathfrak{L}^2\left(\Omega;\mathbb{R}^N\right)$ is an $\mathscr{F}_0$-measurable random variable independent of $\{W_i(t) : 0 \leq t \leq T\}, i = 1,\ldots,m$. The striking similarity between (4.32) and (3.74) suggests that the matrix exponential might be of some utility. And it is, but not without certain restrictions. It is reasonable to expect that the theory should be able to be extended from the one-dimensional case to the N-dimensional setting without tremendous difficulty, just as in the deterministic theory, due to the structural properties of $\mathbb{R}^N$ and the matrix exponential.

Performing this generalization requires that we understand how to work with the matrix exponential and that we appropriately extend the notions of the Lebesgue and Itô integrals to $\mathbb{R}^N$-valued stochastic processes. To this end, we first recall the basics of the matrix exponential from **[295]** and investigate random IVPs of the form (4.31). Then, we define the two integrals and proceed with a more substantive development of the theory of Itô SDEs of the form (4.32) in $\mathbb{R}^N$.

4.2 Deterministic Linear Evolution Equations in $\mathbb{R}^N$

We provide the following terse outline of the basic definitions and results from the deterministic theory for finite-dimensional linear systems of ODEs that we need in order to develop an analogous theory in the stochastic setting. A thorough development of this material can be found in Volume 1.

4.2.1 The Matrix Exponential

A natural definition of e^{at} that is independent of a geometric context is the Taylor representation $e^{at} = \sum_{k=0}^{\infty} \frac{(at)^k}{k!}$, which converges $\forall t \in \mathbb{R}$. We can define $e^{\mathbf{A}t}$ in a similar manner.

Definition 4.2.1. For any $\mathbf{A} \in \mathbb{M}^N(\mathbb{R})$ and $t \in \mathbb{R}$, the *matrix exponential* $e^{\mathbf{A}t}$ is the unique member of $\mathbb{M}^N(\mathbb{R})$ defined by

$$e^{\mathbf{A}t} = \sum_{k=0}^{\infty} \frac{(\mathbf{A}t)^k}{k!}.$$

For instance, consider the following example.

Example. Let $\mathbf{A} = \begin{bmatrix} a & 0 \\ 0 & b \end{bmatrix}$, where $a, b \in \mathbb{R}$. Observe that

$$e^{\begin{bmatrix} a & 0 \\ 0 & b \end{bmatrix} t} = \sum_{k=0}^{\infty} \frac{\left(\begin{bmatrix} a & 0 \\ 0 & b \end{bmatrix} t \right)^k}{k!} = \sum_{k=0}^{\infty} \frac{\begin{bmatrix} at & 0 \\ 0 & bt \end{bmatrix}^k}{k!} = \sum_{k=0}^{\infty} \frac{\begin{bmatrix} (at)^k & 0 \\ 0 & (bt)^k \end{bmatrix}}{k!}$$

$$= \sum_{k=0}^{\infty} \begin{bmatrix} \frac{(at)^k}{k!} & 0 \\ 0 & \frac{(bt)^k}{k!} \end{bmatrix} = \begin{bmatrix} \sum_{k=0}^{\infty} \frac{(at)^k}{k!} & 0 \\ 0 & \sum_{k=0}^{\infty} \frac{(bt)^k}{k!} \end{bmatrix} = \begin{bmatrix} e^{at} & 0 \\ 0 & e^{bt} \end{bmatrix}.$$

Exercise 4.2.1. Compute $e^{\mathbf{A}t}$ where $\mathbf{A}$ is a diagonal $N \times N$ matrix with real diagonal entries $a_{11}, a_{22}, \ldots, a_{NN}$. Justify all steps.

Exercise 4.2.2. Let $\mathbf{A} \in \mathbb{M}^N(\mathbb{R})$ and assume that $\exists \mathbf{D}, \mathbf{P} \in \mathbb{M}^N(\mathbb{R})$ such that $\mathbf{D}$ is diagonal, $\mathbf{P}$ is invertible, and $\mathbf{A} = \mathbf{P}^{-1}\mathbf{D}\mathbf{P}$. (In such case, $\mathbf{A}$ is said to be *diagonalizable*.) Show that $\forall t \in \mathbb{R}$, $e^{\mathbf{A}t} = \mathbf{P}^{-1}e^{\mathbf{D}t}\mathbf{P}$. (Note: For nondiagonalizable matrices $\mathbf{A}$, even though $e^{\mathbf{A}t}$ exists, computing it can be tedious. In such case, the *Putzer algorithm* is a useful tool.)

We gather some important properties below about $e^{\mathbf{A}t}$ and its relationship to $\mathbf{A}$.

Proposition 4.2.2. *Let* $\mathbf{A} \in \mathbb{M}^N(\mathbb{R})$ *and define* $\mathscr{B} : \mathrm{dom}(\mathscr{B}) \subset \mathbb{R}^N \to \mathbb{R}^N$ *by* $\mathscr{B}\mathbf{x} = \mathbf{A}\mathbf{x}$. *Also, for every* $t \in \mathbb{R}$, *define* $S_t : \mathrm{dom}(S_t) \subset \mathbb{R}^N \to \mathbb{R}^N$ *by* $S_t\mathbf{x} = e^{\mathbf{A}t}\mathbf{x}$. *Then,*

i.) $\text{dom}(\mathscr{B}) = \mathbb{R}^N$;

ii.) *For every* $\alpha, \beta \in \mathbb{R}$ *and* $\mathbf{x}, \mathbf{y} \in \mathbb{R}^N$, $\mathscr{B}(\alpha\mathbf{x} + \beta\mathbf{y}) = \alpha\mathscr{B}\mathbf{x} + \beta\mathscr{B}\mathbf{y}$;

iii.) *For every* $\mathbf{x} \in \mathbb{R}^N$, $\|\mathscr{B}\mathbf{x}\|_{\mathbb{R}^N} \leq \|\mathbf{A}\|_{\mathbb{M}^N} \|\mathbf{x}\|_{\mathbb{R}^N}$;

iv.) $\text{dom}(S_t) = \mathbb{R}^N$;

v.) *For every* $\alpha, \beta \in \mathbb{R}$ *and* $\mathbf{x}, \mathbf{y} \in \mathbb{R}^N$, $S_t(\alpha\mathbf{x} + \beta\mathbf{y}) = \alpha S_t\mathbf{x} + \beta S_t\mathbf{y}$;

vi.) *For every* $\mathbf{x} \in \mathbb{R}^N$, $\|S_t\mathbf{x}\|_{\mathbb{R}^N} = \|e^{\mathbf{A}t}\mathbf{x}\|_{\mathbb{R}^N} \leq \|e^{\mathbf{A}t}\|_{\mathbb{M}^N} \|\mathbf{x}\|_{\mathbb{R}^N} \leq e^{t\|\mathbf{A}\|_{\mathbb{M}^N}} \|\mathbf{x}\|_{\mathbb{R}^N}$.

Definition 4.2.3. Let $\mathbf{A} \in \mathbb{M}^N(\mathbb{R})$ and $t \geq 0$. The family of operators $\{e^{\mathbf{A}t} : t \geq 0\}$ is *contractive* if $\|e^{\mathbf{A}t}\mathbf{x}\|_{\mathbb{R}^N} \leq \|\mathbf{x}\|_{\mathbb{R}^N}, \forall \mathbf{x} \in \mathbb{R}^N, t \geq 0$.

Exercise 4.2.3.
i.) Give an example of $\mathbf{A} \in \mathbb{M}^N(\mathbb{R})$ for which $\{e^{\mathbf{A}t} : t \geq 0\}$ is not contractive.

ii.) Determine a sufficient condition that could be imposed on a diagonal matrix $\mathbf{A} \in \mathbb{M}^N(\mathbb{R})$ to ensure that $\{e^{\mathbf{A}t} : t \geq 0\}$ is contractive.

Proposition 4.2.4. *Let* $\mathbf{A}, \mathbf{B} \in \mathbb{M}^N(\mathbb{R})$ *and* $t \geq 0$.

i.) $e^{\mathbf{0}} = \mathbf{I}$, *where* $\mathbf{0}$ *is the zero matrix and* $\mathbf{I}$ *is the identity matrix in* $\mathbb{M}^N(\mathbb{R})$;

ii.) *For every* $t, s \geq 0$, $e^{\mathbf{A}(t+s)} = e^{\mathbf{A}t}e^{\mathbf{A}s} = e^{\mathbf{A}s}e^{\mathbf{A}t}$;

iii.) *For every* $\mathbf{A} \in \mathbb{M}^N(\mathbb{R})$ *and* $t \geq 0$, $e^{\mathbf{A}t}$ *is invertible and* $\left(e^{\mathbf{A}t}\right)^{-1} = e^{-\mathbf{A}t}$;

iv.) *If* $\mathbf{AB} = \mathbf{BA}$, *then* $e^{\mathbf{B}}e^{\mathbf{A}} = e^{\mathbf{A}}e^{\mathbf{B}} = e^{\mathbf{A}+\mathbf{B}}$;

v.) $\lim\limits_{t \to 0^+} \|e^{\mathbf{A}t} - \mathbf{I}\|_{\mathbb{M}^N} = 0$;

vi.) *For every* $\mathbf{x}_0 \in \mathbb{R}^N$, $\mathbf{g} : [0, \infty) \to \mathbb{R}^N$ *defined by* $\mathbf{g}(t) = e^{\mathbf{A}t}\mathbf{x}_0$ *is continuous*;

vii.) *For every* $t_0 \geq 0$, $\mathbf{h} : \mathbb{R}^N \to \mathbb{R}^N$ *defined by* $\mathbf{h}(\mathbf{x}) = e^{\mathbf{A}t_0}\mathbf{x}$ *is continuous*;

viii.) *For every* $\mathbf{x}_0 \in \mathbb{R}^N$ *and* $t_0 \geq 0$, $\lim\limits_{h \to 0^+} \frac{1}{h}\int_{t_0}^{t_0+h} e^{\mathbf{A}s}\mathbf{x}_0 ds = e^{\mathbf{A}t_0}\mathbf{x}_0$;

ix.) *For every* $\mathbf{x}_0 \in \mathbb{R}^N$, $\lim\limits_{h \to 0^+} \frac{(e^{\mathbf{A}h} - \mathbf{I})\mathbf{x}_0}{h} = \mathbf{A}\mathbf{x}_0$;

x.) *For every* $t \geq 0$, $\mathbf{A}e^{\mathbf{A}t} = e^{\mathbf{A}t}\mathbf{A}$;

xi.) *For every* $\mathbf{x}_0 \in \mathbb{R}^N$, $\mathbf{g} : (0, \infty) \to \mathbb{R}^n$ *defined by* $\mathbf{g}(t) = e^{\mathbf{A}t}\mathbf{x}_0$ *is in* $\mathbb{C}^1\left((0, \infty); \mathbb{R}^N\right)$ and

$$\frac{d}{dt}e^{\mathbf{A}t}\mathbf{x}_0 = \mathbf{A}e^{\mathbf{A}t}\mathbf{x}_0 = e^{\mathbf{A}t}\mathbf{A}\mathbf{x}_0;$$

xii.) *For every* $t_0 \geq 0$ *and* $\mathbf{x}_0 \in \mathbb{R}^N$,

$$\mathbf{A}\int_0^{t_0} e^{\mathbf{A}s}\mathbf{x}_0 ds = e^{\mathbf{A}t_0}\mathbf{x}_0 - \mathbf{x}_0;$$

xiii.) *For every* $0 < s \leq t < \infty$ *and* $\mathbf{x}_0 \in \mathbb{R}^N$,

$$e^{\mathbf{A}t}\mathbf{x}_0 - e^{\mathbf{A}s}\mathbf{x}_0 = \int_s^t e^{\mathbf{A}u}\mathbf{A}\mathbf{x}_0 du = \int_s^t \mathbf{A}e^{\mathbf{A}u}\mathbf{x}_0 du.$$

Definition 4.2.5. Let $\mathbf{A} \in \mathbb{M}^N(\mathbb{R})$. The function $\mathscr{B} : \mathbb{R}^N \to \mathbb{R}^N$ defined by

$$\mathscr{B}\mathbf{z} = \lim_{h \to 0^+} \frac{\left(e^{\mathbf{A}h} - \mathbf{I}\right)}{h}\mathbf{z}$$

is the *generator* of the family $\{e^{\mathbf{A}t} : t \geq 0\}$.

By Prop. 4.2.4, the operator $\mathscr{B}$ generates $\left\{e^{\mathbf{A}t} : t \geq 0\right\}$. For brevity, we say "**A** generates $\left\{e^{\mathbf{A}t} : t \geq 0\right\}$." As such, to every $\mathbf{A} \in \mathbb{M}^N(\mathbb{R})$ there is associated at least one exponential family.

4.2.2 The Homogenous Cauchy Problem

The deterministic homogenous linear IVP in $\mathbb{R}^N$ is given by

$$\begin{cases} \mathbf{U}'(t) = \mathbf{A}\mathbf{U}(t), t > 0, \\ \mathbf{U}(0) = \mathbf{U}_0. \end{cases} \tag{4.33}$$

Definition 4.2.6. A *classical solution* of (4.33) is a function $\mathbf{U} \in \mathbb{C}\left([0,\infty);\mathbb{R}^N\right) \cap \mathbb{C}^1\left((0,\infty);\mathbb{R}^N\right)$ that satisfies the ODE and IC in (4.33).

The properties of $e^{\mathbf{A}t}$ established in Section 2.2 enable us to establish the following result.

Theorem 4.2.7. *For every* $\mathbf{U}_0 \in \mathbb{R}^N$, *the IVP (4.33) has a unique classical solution given by* $\mathbf{U}(t) = e^{\mathbf{A}t}\mathbf{U}_0$.

A natural change of variable can be used to show the following.

Corollary 4.2.8. *For every* $t_0 > 0$, *the IVP*

$$\begin{cases} \mathbf{U}'(t) = \mathbf{A}\mathbf{U}(t), t > t_0, \\ \mathbf{U}(t_0) = \mathbf{U}_0 \end{cases} \tag{4.34}$$

has a unique classical solution $\mathbf{U} \colon [t_0,\infty) \to \mathbb{R}^N$ *given by* $\mathbf{U}(t) = e^{\mathbf{A}(t-t_0)}\mathbf{U}_0$.

Because the classical solution is $\mathbf{U}(t) = e^{\mathbf{A}t}\mathbf{U}_0$, these questions ultimately concern the nature of $\left\{e^{\mathbf{A}t} : t \geq 0\right\}$, which in turn is directly linked to the matrix **A**. How, specifically, does the relationship between **A** and $\left\{e^{\mathbf{A}t} : t \geq 0\right\}$ translate into different long-term behavior? The following exercises shed some light on this situation.

Exercise 4.2.4. Let $\mathbf{A} = \begin{bmatrix} \alpha & 0 \\ 0 & \beta \end{bmatrix}$, where $\alpha, \beta \in \mathbb{R}$. Answer the following three questions for each description of α and β to follow.

(**I**) Let $t \geq 0$. Compute $e^{\mathbf{A}t}$ and determine an upper bound for $\left\|e^{\mathbf{A}t}\right\|_{\mathbb{M}^N}$.

(**II**) Does $\exists\lim\limits_{t\to\infty}\left\|e^{\mathbf{A}t}\mathbf{x}\right\|_{\mathbb{R}^N}$? Does it depend on the choice of $\mathbf{x} \in \mathbb{R}^N$?

(**III**) Address each of the four questions posed at the beginning of this section for the classical solution of the corresponding IVP.

i.) $\alpha, \beta < 0$;

ii.) $\alpha, \beta > 0$;

iii.) Exactly one of α, β is equal to zero and the other is strictly positive;

iv.) Exactly one of α, β is equal to zero and the other is strictly negative;

v.) $\alpha < 0 < \beta$.

The eigenvalues of the matrix $\mathbf{A}$ significantly impact the structure of $e^{\mathbf{A}t}$ (cf. Prop. 4.2.4). Every $N \times N$ matrix has N complex eigenvalues, including multiplicity. As such, it is reasonable to expect a connection between them and the long-term behavior of $\{e^{\mathbf{A}t} : t \geq 0\}$. This is apparent in the following simple scenario.

Exercise 4.2.5. Consider the matrix $\mathbf{A} = \begin{bmatrix} \alpha & 0 \\ 0 & \beta \end{bmatrix}$, whose eigenvalues are α and β. For each case listed in Exer. 4.2.4, associate the nature of the eigenvalues with the long-term behavior of the classical solution $\mathbf{U}(t)$ of (4.34). (For instance, if α, $\beta < 0$, then the eigenvalues of $\mathbf{A}$ are both negative and in such case, $\lim_{t \to \infty} \|\mathbf{U}(t)\|_{\mathbb{R}^2} = 0$.)

4.3   Exploring Two Models

As mentioned earlier, the deterministic theory discussed in Section 3.2 can be applied for fixed $\omega \in \Omega$ to study stochastic IVPs of the form (4.31). Consider the following exercises.

Exercise 4.3.1. Consider the chemical kinetics IVP (4.1).
i.) Derive a formula for the solution of this IVP and calculate $m_X(t)$ and $Var_X(t)$.
ii.) If $[Y]_0$ is $b(3, 0.5)$, how many different solution vectors are there? Compute $m_X(t)$ and $Var_X(t)$ for each of them.
iii.) If $[Y]_0$ is $n(0, 1)$, how do your responses in (ii) change?
iv.) Suppose that $\alpha_1^\star > 0$ is such that $|\alpha_1 - \alpha_1^\star| > 0$. Denote the solutions of (4.1) corresponding to α_1 and $\alpha_1^\star$ by $\begin{bmatrix} [Y](t; \cdot) \\ [Z](t; \cdot) \end{bmatrix}$ and $\begin{bmatrix} [Y^\star](t; \cdot) \\ [Z^\star](t; \cdot) \end{bmatrix}$, respectively. Determine an estimate for $\left\| \begin{bmatrix} [Y](t; \cdot) \\ [Z](t; \cdot) \end{bmatrix} - \begin{bmatrix} [Y^\star](t; \cdot) \\ [Z^\star](t; \cdot) \end{bmatrix} \right\|_{\mathcal{L}^2(\Omega; \mathbb{R})}^2$, $\forall t > 0$.

v.) For what values of α_1 does $\left\| \begin{bmatrix} [Y] \\ [Z] \end{bmatrix} \right\|_{\mathbb{C}([0,T]; \mathcal{L}^2(\Omega; \mathbb{R}))} \longrightarrow 0$ as $t \longrightarrow \infty$?

Exercise 4.3.2. Consider the spring-mass system described by the IVP (4.19).
i.) Use the definition of the matrix exponential to show that

$$e^{\begin{bmatrix} 0 & 1 \\ -\alpha^2 & 0 \end{bmatrix} t} = \begin{bmatrix} \cos(\alpha t) & \frac{1}{\alpha} \sin(\alpha t) \\ -\alpha \sin(\alpha t) & \cos(\alpha t) \end{bmatrix}. \tag{4.35}$$

ii.) Use (i) to derive an explicit formula for the solution of (4.19).
iii.) Assume that x_0 is $b(N, 0.5)$. Compute $m_X(t)$ and $Var_X(t)$.
iv.) Assume that x_0 is $n(\mu, \sigma^2)$. Compute $m_X(t)$ and $Var_X(t)$.

v.) Consider a sequence $\{a_n : n \in \mathbb{N}\} \subset \mathbb{R}$ for which $\lim\limits_{n \to \infty} a_n = a$. For every $n \in \mathbb{N}$, find a formula for the classical solution $\mathbf{U}_n$ of (2.5). Does there exist $\mathbf{U}^\star$ for which $\mathbf{U}_n \longrightarrow \mathbf{U}^\star$ uniformly as $n \to \infty$?

vi.) Error can occur in the measurement of any parameter, including the initial conditions. Consider two versions of IVP (2.5), one for which the initial data are $y(0) = y_0, z(0) = 0$ and the other for which $y(0) = \overline{y_0}, z(0) = \overline{z_0}$. Denote the classical solutions by $\mathbf{U}(t)$ and $\overline{\mathbf{U}}(t)$, respectively, and assume that $E\left|y_0 - \overline{y_0}\right|^2 < \delta_1$ and $E\left|\overline{z_0}\right|^2 < \delta_2$. For any $T > 0$, determine $\sup\left\{E\left\|\mathbf{U}(t) - \overline{\mathbf{U}}(t)\right\|_{\mathbb{R}^N}^2 : 0 \le t \le T\right\}$.

vii.) Determine an estimate for $E\left|x(t; \cdot)\right|^p$, for $p > 2$.

4.4   The Lebesgue and Itó Integrals in $\mathbb{R}^N$

Assume **(S.A.1)**. Developing a theory for SDEs in $\mathbb{R}^N$ of the form (4.32) analogous to the theory developed in Chapter 3 requires that we extend the definitions of the Lebesgue and Itó integrals to handle $\mathbb{R}^N$-valued random variables. As in Chapter 3, we restrict out attention to stochastic processes $\mathbf{u} : [0, t] \times \Omega \longrightarrow \mathbb{R}^N$ satisfying the following conditions:

$$\mathbf{u}(s; \cdot) \text{ is } \mathscr{F}_s - \text{adapted}, \forall s \in [0, t], \tag{4.36}$$

$$\mathbf{u}(s; \omega) \text{ is progressively measurable on } [0, t] \times \Omega, \tag{4.37}$$

$$\int_0^t E\left\|\mathbf{u}(s; \cdot)\right\|_{\mathbb{R}^N}^2 ds < \infty. \tag{4.38}$$

The collection of all $\mathbb{R}^N$-valued stochastic processes satisfying (4.36) through (4.38) is denoted by $\mathscr{U}_N$.

Exercise 4.4.1. Interpret (4.36) and (4.37) in terms of the components of $\mathbf{u}$.

4.4.1   The Lebesgue Integral for $\mathbb{R}^N$-Valued Stochastic Processes

The standard calculus operations for deterministic matrix-valued functions are performed componentwise. The benefit is that the operation (e.g., limit, derivative, or integral) is performed on real-valued functions for which we already have a working definition and properties. It is reasonable to define these calculus operations for matrix-valued stochastic processes in a similar manner. To this end, we have the following.

Definition 4.4.1. Let $\mathbf{u} : [0, t] \times \Omega \longrightarrow \mathbb{R}^N$ be a stochastic process, whose component form is $\mathbf{u}(t; \omega) = [u_1(t; \omega), \dots, u_N(t; \omega)]^T$, that belongs to $\mathscr{U}_N$. The *Lebesgue integral* *of* **u** *on* $(0,t)$ *is given by*

$$\int_0^t \mathbf{u}(s;\omega)ds = \begin{bmatrix} \int_0^t u_1(s;\omega)ds \\ \vdots \\ \int_0^t u_N(s;\omega)ds \end{bmatrix} = \left[\int_0^t u_1(s;\omega)ds, \cdots, \int_0^t u_N(s;\omega)ds \right]^T, \quad (4.39)$$

where each of the component integrals is as defined in Section 3.2.1.

Remarks.
1. Observe that for a given $t > 0$, the Lebesgue integral of an $\mathbb{R}^N$-valued stochastic process is an $\mathbb{R}^N$-valued random variable.
2. The properties listed in Prop. 3.2.1 carry over to the present setting by making the obvious modifications, namely replacing the absolute value by the $\mathbb{R}^N$-norm, interpreting the integrands as $\mathbb{R}^N$-valued random variables, etc. Each of these properties can be verified by applying the corresponding one-dimensional version of the property to the components and appealing to the arithmetic of vectors in $\mathbb{R}^N$ and using the properties of the $\mathbb{R}^N$-norm.

Exercise 4.4.2.
i.) State the analogous form of Prop. 3.2.1 for $\mathscr{U}_N$. Take careful note of the precise nature of all modifications.
ii.) Prove the linearity and additivity properties of the Lebesgue integral for stochastic processes in $\mathscr{U}_N$.

Recall that the expectation of an $\mathbb{R}^N$-valued random variable is itself a (deterministic) vector in $\mathbb{R}^N$. Fubini's theorem applies equally well for members of $\mathscr{U}_N$. Indeed, let $\mathbf{u} \in \mathscr{U}_N$ and observe that

$$\begin{aligned}
E\left[\int_0^t \mathbf{u}(s;\omega)ds \right] &= E\left[\int_0^t u_1(s;\omega)ds, \ldots, \int_0^t u_N(s;\omega)ds \right]^T \\
&= \left[E\int_0^t u_1(s;\omega)ds, \ldots, E\int_0^t u_N(s;\omega)ds \right]^T \\
&= \left[\int_0^t E\left[u_1(s;\omega)\right]ds, \ldots, \int_0^t E\left[u_N(s;\omega)\right]ds \right]^T \quad (4.40) \\
&= \int_0^t \left[E\left[u_1(s;\omega)\right], \ldots, E\left[u_{N1}(s;\omega)\right] \right]^T ds \\
&= \int_0^t E\left[u_1(s;\omega), \ldots, u_N(s;\omega)\right]^T ds \\
&= \int_0^t E\left[\mathbf{u}(s;\omega)\right]ds,
\end{aligned}$$

as needed.

4.4.2 The Itó Integral for $\mathbb{R}^N$-Valued Stochastic Processes

An examination of (4.32) suggests that we will need to consider the following integral at the very onset of our analysis:

$$\int_0^t \sum_{k=1}^m \mathbf{B}_k \mathbf{U}(s;\omega) dW_k(s),$$ (4.41)

where $0 \leq t \leq T$; $\{W_1(t) : 0 \leq t \leq T\}, \ldots, \{W_m(t) : 0 \leq t \leq T\}$ are independent real-valued Wiener processes; $\mathbf{B}_k \in \mathbb{M}^N(\mathbb{R})$ $(k = 1, \ldots, m)$; and $\mathbf{U} : [0,T] \times \Omega \longrightarrow \mathbb{R}^N$.

The independence of the Wiener processes arising in (4.41) enables us to gather the Wiener noise terms together as the m-dimensional Wiener process

$$\mathbf{W}(t) = \langle W_1(t), \ldots, W_m(t)\rangle^T.$$ (4.42)

We must assume that $\forall s \in [0,T]$,

$$\sigma(\mathbf{W}(u)|0 \leq u \leq s) \subset \mathscr{F}_s.$$ (4.43)

(Why?) We seek to define the Itó integral

$$\int_0^t \mathbf{g}(s;\omega) d\mathbf{W}(s)$$ (4.44)

in a manner consistent with the development in Section 3.2.2 and for which the integral (4.41) can be subsumed as a special case by appropriately rearranging terms. The question, of course, is for which stochastic processes $\mathbf{g}$ is (4.44) meaningful? In the present form, it is insufficient to simply require that $\mathbf{g} \in \mathscr{U}_N$. (Why?)

Answering this question requires that we think about the form of the noise terms that can occur in any of the N SDEs comprising the system (4.32). Momentarily ignoring the precise nature of the coefficients of dW_j $(j = 1, \ldots, m)$ in the SDEs, we note that $\forall 1 \leq i \leq N$, the form of the equation governing $X_i = X_i(t;\omega)$ is

$$dX_1 = (a_{11}X_1 + \ldots + a_{1N}X_N) dt + (g_{11}dW_1(t) + \ldots + g_{1m}dW_m(t)),$$

$$\vdots$$ (4.45)

$$dX_N = (a_{N1}X_1 + \ldots + a_{NN}X_N) dt + (g_{N1}dW_1(t) + \ldots + g_{Nm}dW_m(t)),$$

where each g_{ij} is an appropriately behaved linear expression of $X_1, \ldots, X_N$. (This will be made precise in Section 4.5.) Now, writing (4.45) in matrix form gives rise to the following modified form of (4.32):

$$d\mathbf{X}(t;\omega) = \mathbf{A}\mathbf{X}(t;\omega)dt + \begin{bmatrix} g_{11}(t) & \cdots & g_{1m}(t) \\ \vdots & \ddots & \vdots \\ g_{N1}(t) & \cdots & g_{Nm}(t) \end{bmatrix} \begin{bmatrix} dW_1(t) \\ \vdots \\ dW_m(t) \end{bmatrix}$$

$$= \mathbf{A}\mathbf{X}(t;\omega)dt + \underbrace{\mathbf{G}(t)}_{N \times m \text{ matrix}} \cdot \underbrace{d\mathbf{W}(t)}_{m \times 1 \text{ matrix}}.$$ (4.46)

Forming the integrated version of (4.46) suggests that we define the Itó integral with respect to an m-dimensional Wiener process for $\mathbb{M}^{N \times m}$-valued stochastic processes possessing similar characteristics to members of $\mathscr{U}_N$. It is helpful to view the matrices $\mathbf{G}(t)$ as operators, especially in the next chapter when we make the transition to the Hilbert space setting. To this end, note that for a given $\mathbf{G} \in \mathbb{M}^{N \times m}$, we can define the operator $\mathfrak{B} : \mathbb{R}^m \to \mathbb{R}^N$ by

$$\mathfrak{B}(\underbrace{\mathbf{y}}_{m \times 1}) = \underbrace{\underbrace{G}_{N \times m} \underbrace{y}_{m \times 1}}_{N \times 1} . \tag{4.47}$$

Let $\mathbf{G} = [g_{ij}]$, $1 \le i \le N$, $1 \le j \le m$. We equip $\mathbb{M}^{N \times m}$ with the so-called *Hilbert-Schmidt norm* defined by

$$\|\mathbf{G}\|^2_{\mathbb{M}^{N \times m}} \equiv \text{trace} \left(\mathbf{G}^T \mathbf{G}\right) = \sum_{i=1}^{N} \sum_{j=1}^{m} g_{ij}^2 < \infty. \tag{4.48}$$

Remark. Observe that $\forall \alpha, \beta \in \mathbb{R}$ and $\mathbf{y}, \mathbf{z} \in \mathbb{R}^m$,

$$\|\mathfrak{B}(\mathbf{y})\|_{\mathbb{R}^N} \le \|\mathbf{G}\|_{\mathbb{M}^{N \times m}} \|\mathbf{y}\|_{\mathbb{R}^m} = \left(\text{trace}\left(\mathbf{G}^T \mathbf{G}\right)\right)^{\frac{1}{2}} \|\mathbf{y}\|_{\mathbb{R}^m}, \tag{4.49}$$

$$\mathfrak{B}(\alpha \mathbf{y} + \beta \mathbf{z}) = \mathbf{G}(\alpha \mathbf{y} + \beta \mathbf{z}) = \alpha \mathbf{G} \mathbf{y} + \beta \mathbf{G} \mathbf{z} = \alpha \mathfrak{B}(\mathbf{y}) + \beta \mathfrak{B}(\mathbf{z}). \tag{4.50}$$

We say that $\mathfrak{B}$ is a bounded linear operator with finite trace or, equivalently, that $\mathfrak{B}$ is a *Hilbert-Schmidt operator from* $\mathbb{R}^m$ *into* $\mathbb{R}^N$. We denote the collection of all such operators by $\mathscr{B}_0\left(\mathbb{R}^m, \mathbb{R}^N\right)$. It can be shown that $\mathscr{B}_0\left(\mathbb{R}^m, \mathbb{R}^N\right)$ is a Banach space when equipped with the norm

$$\|\mathfrak{B}\|_{\mathscr{B}_0} \equiv \left(\text{trace}\left(\mathbf{G}^T \mathbf{G}\right)\right)^{\frac{1}{2}}, \tag{4.51}$$

where $\mathbf{G}$ is defined in (4.46). In order to avoid cumbersome notation, it is customary to treat the operator $\mathfrak{B}$ and the matrix $\mathbf{G}$ in (4.47) as being indistinguishable and refer to $\mathbf{G}$ itself as the Hilbert-Schmidt operator with norm $\|\mathbf{G}\|_{\mathscr{B}_0}$ given by the right-hand side of (4.51).

Exercise 4.4.3. Let $\mathbf{G} = [g_{ij}]$, $1 \le i \le N$, $1 \le j \le m$. Verify that

$$\text{trace}\left(\mathbf{G}^T \mathbf{G}\right) = \sum_{i=1}^{N} \sum_{j=1}^{m} g_{ij}^2.$$

Remark. Observe that $\text{trace}\left(\mathbf{G}^T \mathbf{G}\right) < \infty$, $\forall \mathbf{G} \in \mathbb{M}^{N \times m}$. (Why?) However, it is not true when either $\mathbb{R}^m$ or $\mathbb{R}^N$ are replaced by general separable Hilbert spaces. The requirement that "$\text{trace}\left(\mathbf{G}^T \mathbf{G}\right) < \infty$" becomes nontrivial in such case.

Next, we define the Itó integral for $\mathscr{B}_0\left(\mathbb{R}^m, \mathbb{R}^N\right)$-valued stochastic processes.

Definition 4.4.2. Let $\mathbf{u} : [0,T] \times \Omega \longrightarrow \mathscr{B}_0 \left(\mathbb{R}^m, \mathbb{R}^N \right)$ be a stochastic process whose component form is $\mathbf{u}(t;\omega) = [u_{ij}(t;\omega)]$, $1 \le i \le N$, $1 \le j \le m$. Assume that $\mathbf{u}$ is $\mathscr{F}_t$-adapted, progressively measurable, and is such that $\int_0^t E \|\mathbf{u}(s;\cdot)\|_{\mathscr{B}_0}^2 \, ds < \infty$. The *Itó integral of* $\mathbf{u}$ *over* $(0,t)$ *with respect to* $\{\mathbf{W}(s) : 0 \le s \le t\}$ is given by

$$\int_0^t \mathbf{u}(s;\omega) d\mathbf{W}(s) = \begin{bmatrix} \sum_{j=1}^m \int_0^t u_{1j}(s;\omega) dW_j(s) \\ \vdots \\ \sum_{j=1}^m \int_0^t u_{Nj}(s;\omega) dW_j(s) \end{bmatrix} \tag{4.52}$$

$$= \left[\sum_{j=1}^m \int_0^t u_{1j}(s;\omega) dW_j(s), \; \cdots, \; \sum_{j=1}^m \int_0^t u_{Nj}(s;\omega) dW_j(s) \right]^T,$$

where the integrals used to define the components are as defined in Section 3.2.2.

As with the Lebesgue integral, componentwise arguments can be used to show that the properties of the Itó integral of $\mathbb{R}$-valued stochastic processes carry over to the present setting. It would be instructive for you to verify these properties directly.

Exercise 4.4.4.
i.) State the analogous form of Prop. 3.2.3 for $\mathscr{B}_0 \left(\mathbb{R}^m, \mathbb{R}^N \right)$-valued stochastic processes. Make careful note of the precise nature of all modifications.
ii.) Prove the linearity and additivity properties of the Itó integral for $\mathscr{B}_0 \left(\mathbb{R}^m, \mathbb{R}^N \right)$-valued stochastic processes.

We verify that the Itó isometry holds. Observe that for all $\mathscr{B}_0 \left(\mathbb{R}^m, \mathbb{R}^N \right)$-valued stochastic processes $\mathbf{u}$ satisfying the conditions of Def. 4.4.2,

$$E \left\| \int_0^t \mathbf{u}(s;\cdot) d\mathbf{W}(s) \right\|_{\mathbb{R}^m}^2 = E \left[\sum_{i=1}^N \left(\sum_{j=1}^m \int_0^t u_{ij}(s;\cdot) dW_j(s) \right)^2 \right]$$

$$= \sum_{i=1}^N E \left[\sum_{j=1}^m \left(\int_0^t u_{ij}(s;\cdot) dW_j(s) \right)^2 \right.$$

$$\left. + \underbrace{\sum_j \sum_{k \neq j} \left(\int_0^t u_{ij}(s;\cdot) dW_j(s) \right) \left(\int_0^t u_{ik}(s;\cdot) dW_k(s) \right)}_{\text{All of the cross-terms}} \right]$$

$$= \sum_{i=1}^N E \left[\sum_{j=1}^m \left(\int_0^t u_{ij}(s;\cdot) dW_j(s) \right)^2 \right] \tag{4.53}$$

$$= \sum_{i=1}^{N} \sum_{j=1}^{m} E \left(\int_0^t u_{ij}(s;\cdot) dW_j(s) \right)^2$$

$$= \sum_{i=1}^{N} \sum_{j=1}^{m} \int_0^t E \left[u_{ij}^2(s;\cdot) \right] ds$$

$$= \int_0^t \underbrace{\sum_{i=1}^{N} \sum_{j=1}^{m} E \left[u_{ij}^2(s;\cdot) \right]}_{=\text{trace}\left(\mathbf{u}^T(s;\cdot) \mathbf{u}(s;\cdot) \right)} ds$$

$$= \int_0^t \| \mathbf{u}(s;\omega) \|^2_{\mathscr{B}_0(\mathbb{R}^m, \mathbb{R}^N)} ds.$$

4.4.3   Some Crucial Estimates

The following generalization of Lemma 3.2.7 will be useful.

Lemma 4.4.3. *Let $p > 2$.*
i.) *Assume that $\mathbf{u} \in \mathscr{U}_N$ with $E \int_0^T \| \mathbf{u}(s;\cdot) \|^p_{\mathbb{R}^N} ds < \infty$. Then, $\forall 0 \le t' < t \le T$,*

$$E \left\| \int_{t'}^t \mathbf{u}(s;\cdot) ds \right\|^p_{\mathbb{R}^N} \le |t - t'|^{\frac{p}{q}} \int_{t'}^t E \| \mathbf{u}(s;\cdot) \|^p_{\mathbb{R}^N} ds, \qquad (4.54)$$

where $\frac{1}{p} + \frac{1}{q} = 1$.
ii.) *Let $\mathbf{u} : [0, T] \times \Omega \longrightarrow \mathscr{B}_0 \left(\mathbb{R}^m, \mathbb{R}^N \right)$ be as in Def. 4.4.2 with $E \int_0^T \| \mathbf{u}(s;\cdot) \|^p_{\mathscr{B}_0} ds < \infty$. Then, $\forall 0 \le t' < t \le T$,*

$$E \left\| \int_{t'}^t \mathbf{u}(s;\cdot) d\mathbf{W}(s) \right\|^p_{\mathbb{R}^N} \le \zeta_{\mathscr{B}_0}(t,t') \int_{t'}^t E \| \mathbf{u}(s;\cdot) \|^p_{\mathscr{B}_0} ds, \qquad (4.55)$$

where

$$\zeta_{\mathscr{B}_0}(t,t') = 2^p N^{\frac{p+1}{2}} m^p \left[\left(t^{\frac{p}{2}} + (t')^{\frac{p}{2}} \right) \left(\frac{p(p-1)}{2} \right)^{\frac{p}{2}} \right]. \qquad (4.56)$$

Proof. It is more convenient to use the max-norm for $\mathbb{R}^N$ (cf. (1.46)). Doing so, observe that (i) holds because

$$E \left\| \int_{t'}^t \mathbf{u}(s;\cdot) ds \right\|^p_{\mathbb{R}^N} = E \left[\max_{1 \le i \le N} \left| \int_{t'}^t u_i(s;\cdot) ds \right|^p \right]$$

$$\le \max_{1 \le i \le N} \left(|t - t'|^{\frac{p}{q}} \int_{t'}^t E |u_i(s;\cdot)|^p ds \right)$$

$$\le \max_{1 \le i \le N} \left(|t - t'|^{\frac{p}{q}} \int_{t'}^t E \left[\max_{1 \le j \le N} |u_j(s;\cdot)|^p \right] ds \right)$$

$$\le |t - t'|^{\frac{p}{q}} \int_{t'}^t E \| \mathbf{u}(s;\cdot) \|^p_{\mathbb{R}^N} ds.$$

Next, we see that (ii) holds because

$$
E\left\|\int_{t'}^{t} \mathbf{u}(s;\cdot)d\mathbf{W}(s)\right\|_{\mathbb{R}^N}^{p} = E\left[\left\|\int_{t'}^{t} \mathbf{u}(s;\cdot)d\mathbf{W}(s)\right\|_{\mathbb{R}^N}^{2}\right]^{\frac{p}{2}}
$$

$$
= E\left[\sum_{i=1}^{N}\left(\sum_{j=1}^{m}\int_{t'}^{t} u_{ij}(s;\cdot)dW_j(s)\right)^{2}\right]^{\frac{p}{2}}
$$

$$
\leq N^{\frac{p-1}{2}}E\left[\sum_{i=1}^{N}\left(\sum_{j=1}^{m}\int_{t'}^{t} u_{ij}(s;\cdot)dW_j(s)\right)^{p}\right] \qquad (4.57)
$$

$$
\leq N^{\frac{p-1}{2}}m^{p-1}\sum_{i=1}^{N}\sum_{j=1}^{m}E\left|\int_{t'}^{t} u_{ij}(s;\cdot)dW_j(s)\right|^{p}
$$

$$
\leq 2^{p}N^{\frac{p-1}{2}}m^{p-1}\left[\left(t^{\frac{p}{2}}+(t')^{\frac{p}{2}}\right)\left(\frac{p(p-1)}{2}\right)^{\frac{p}{2}}\right]
$$

$$
\times \sum_{i=1}^{N}\sum_{j=1}^{m}\int_{t'}^{t} E\left|u_{ij}(s;\cdot)\right|^{p}ds.
$$

Note that

$$
E\left|u_{ij}(s;\cdot)\right|^{p} = E\left[\left(\left|u_{ij}(s;\cdot)\right|^{2}\right)^{\frac{p}{2}}\right] \leq \left[\left(\|\mathbf{u}\|_{\mathscr{B}_0}^{2}\right)^{\frac{p}{2}}\right]. \qquad (4.58)
$$

Using (4.58) in (4.57) enables us to continue the string of inequalities to conclude that (4.55) holds. This completes the proof. $\qquad\square$

Remark. The numerical estimate $\zeta_{\mathscr{B}_0}(t,t')$ can be improved by following the proof of Lemma 3.2.7(ii) and replacing the absolute value by the $\mathbb{R}^N$-norm. Given that this is a real-valued stochastic process, the same form of Itó's formula applies. The existence of such a bound $\zeta_{\mathscr{B}_0}(t,t')$ is all that we will need in our qualitative arguments.

4.4.4   The Multivariable It Formula — Revisited

We introduced Itó's formula for mappings $h:[0,T]\times\mathbb{R}^N\to\mathbb{R}$ in Prop. 3.2.6. This formula is sufficient for our purposes because we merely apply it componentwise to derive formulae for slightly more general mappings $\mathbf{h}:[0,T]\times\mathbb{R}^{N_1}\to\mathbb{R}^{N_2}$. But, by way of motivation for the extension of Prop. 3.2.6 to functions that map one separable Hilbert space into another in Chapter 5, we present the formula in an equivalent manner that is more amenable to generalization.

Suppose that $\mathbf{X}:[0,T]\times\Omega\to\mathbb{R}^N$ is an Itó process given by

$$
\begin{cases} d\mathbf{X}(t;\omega) = \mathbf{F}(t)dt + \mathbf{G}(t)d\mathbf{W}(t), & 0<t<T, \omega\in\Omega, \\ \mathbf{X}(0;\omega) = \mathbf{X}_0(\omega), \omega\in\Omega, \end{cases} \qquad (4.59)
$$

where $\mathbf{F} : [0,T] \to \mathbb{R}^N$, $\mathbf{G} : [0,T] \to \mathscr{B}_0\left(\mathbb{R}^m, \mathbb{R}^N\right)$, and $\mathbf{W}(t) = \langle W_1(t), \ldots, W_m(t) \rangle^T$ is an m-dimensional Wiener process. The ith component of $d\mathbf{X}(t;\omega)$ is given by

$$dX_i(t;\omega) = F_i(t)dt + G_{i1}(t)dW_1(t) + \ldots + G_{im}(t)dW_m(t). \quad (4.60)$$

We consider each term in the Itó formula (3.52) individually. Observe that

$$\frac{\partial h}{\partial X_i} dX_i(t;\omega) = \frac{\partial h}{\partial X_i}\left[F_i(t)dt + G_{i1}(t)dW_1(t) + \ldots + G_{im}(t)dW_m(t)\right],$$

so that

$$\sum_{i=1}^n \frac{\partial h}{\partial X_i} dX_i(t;\omega) = (\nabla h \cdot \mathbf{F})(t)dt + \nabla h \cdot (\mathbf{G}(t)d\mathbf{W}(t)), \quad (4.61)$$

where $\nabla h = \left\langle \frac{\partial h}{\partial X_1}, \ldots, \frac{\partial h}{\partial X_N} \right\rangle$ and "$\cdot$" represents the usual $\mathbb{R}^N$ dot product.

Next, $\forall 1 \le i, j \le N$, observe that

$$\begin{aligned}
dX_i(t;\omega)dX_j(t;\omega) &= (F_i(t)dt + G_{i1}(t)dW_1(t) + \ldots + G_{im}(t)dW_m(t)) \cdot \\
&\quad (F_j(t)dt + G_{j1}(t)dW_1(t) + \ldots + G_{jm}(t)dW_m(t)) \\
&= G_{i1}(t)G_{j1}(t)(dW_1(t))^2 + \ldots + G_{im}(t)G_{jm}(t)(dW_m(t))^2 \\
&= \sum_{k=1}^m G_{ik}(t)G_{jk}(t)dt \quad (4.62) \\
&= \left(\mathbf{G}(t)\mathbf{G}^T(t)\right)_{ij} dt,
\end{aligned}$$

where $\left(\mathbf{G}(t)\mathbf{G}^T(t)\right)_{ij}$ is the $(ij)^{\text{th}}$ component of the $n \times n$ matrix $\mathbf{G}(t)\mathbf{G}^T(t)$. As such, we have

$$\sum_{i=1}^N \sum_{j=1}^N \frac{\partial^2 h}{\partial X_i \partial X_j} dX_i(t;\omega)dX_j(t;\omega) = \sum_{i=1}^N \sum_{j=1}^N \frac{\partial^2 h}{\partial X_i \partial X_j}\left(\mathbf{G}(t)\mathbf{G}^T(t)\right)_{ij} dt. \quad (4.63)$$

Going one step further reveals that the right-hand side of (4.63) is

$$\text{trace}\left(\underbrace{\begin{bmatrix} \frac{\partial^2 h}{\partial X_1 \partial X_1} & \cdots & \frac{\partial^2 h}{\partial X_1 \partial X_N} \\ \vdots & \ddots & \vdots \\ \frac{\partial^2 h}{\partial X_N \partial X_1} & \cdots & \frac{\partial^2 h}{\partial X_N \partial X_N} \end{bmatrix}}_{= \mathscr{J}(h)} \mathbf{G}(t)\mathbf{G}^T(t) \right), \quad (4.64)$$

where $\mathscr{J}(h)$ is the Jacobian matrix of h.

Exercise 4.4.5. Show that (4.64) equals the right-hand side of (4.63).

Using (4.63) and (4.64) in (3.52) yields the equivalent formula

$$\begin{aligned}
d(h(t, \mathbf{X}(t;\omega))) &= \frac{\partial h}{\partial t}dt + (\nabla h \cdot \mathbf{F})(t)dt + \frac{1}{2}\text{trace}\left(\mathscr{J}(h)\mathbf{G}(t)\mathbf{G}^T(t)\right) \quad (4.65) \\
&\quad + \nabla h \cdot (\mathbf{G}(t)d\mathbf{W}(t)).
\end{aligned}$$

4.5   The Cauchy Problem — Formulation

Assume **(S.A.1)**. The focus of our study is linear systems of N stochastic ODEs of the general form

$$
\begin{cases}
dX_1 = (a_{11}X_1 + \ldots + a_{1N}X_N)\,dt + \left(b^1_{11}X_1 + \ldots + b^1_{1N}X_N\right) dW_1(t) \\
\quad + \ldots + \left(b^1_{m1}X_1 + \ldots + b^1_{mN}X_N\right) dW_m(t), \\
\vdots \\
dX_N = (a_{N1}X_1 + \ldots + a_{NN}X_N)\,dt + \left(b^N_{11}X_1 + \ldots + b^N_{1N}X_N\right) dW_1(t) \\
\quad + \ldots + \left(b^N_{m1}X_1 + \ldots + b^N_{mN}X_N\right) dW_m(t), \\
X_1(0;\omega) = (X_0)_1\,(\omega), \ldots, X_N(0;\omega) = (X_0)_N\,(\omega), \omega \in \Omega,
\end{cases}
\tag{4.66}
$$

where $0 < t < T, \omega \in \Omega$ and $X_i = X_i(t;\omega)\,(i = 1,\ldots,N)$. The system (4.66) can be written in the concise form (4.32). Indeed, let $\mathbf{X} : [0,T] \times \Omega \to \mathbb{R}^N$ be given by

$$
\mathbf{X}(t;\omega) = \begin{bmatrix} X_1(t;\omega) \\ \vdots \\ X_N(t;\omega) \end{bmatrix} \quad \text{and} \quad \mathbf{X}_0(\omega) = \begin{bmatrix} (X_0)_1\,(\omega) \\ \vdots \\ (X_0)_N\,(\omega) \end{bmatrix}.
\tag{4.67}
$$

We identify the constant coefficient matrix of the "dt" portion of the right-hand side of (4.66) as

$$
\mathbf{A} = \begin{bmatrix} a_{11} & \cdots & a_{1N} \\ \vdots & \ddots & \vdots \\ a_{N1} & \cdots & a_{NN} \end{bmatrix}.
\tag{4.68}
$$

Also, for each $1 \le j \le m$, group all terms involving dW_j and rewrite in matrix form in a manner that retains the individual contributions to each of the N equations in (4.66). Doing so results in

$$
\begin{bmatrix} b^1_{j1} & \cdots & b^1_{jN} \\ \vdots & \ddots & \vdots \\ b^N_{j1} & \cdots & b^N_{jN} \end{bmatrix} \begin{bmatrix} X_1(t;\omega) \\ \vdots \\ X_N(t;\omega) \end{bmatrix} dW_j(t) = B_j\mathbf{X}(t;\omega)dW_j(t).
\tag{4.69}
$$

Using (4.67) through (4.69) shows that (4.66) can be rewritten as (4.32).

We assume the following:

(H4.1) a_{ij}, b^k_{ij} are real constants, $\forall 1 \le i,j,k \le N$.

(H4.2) $\mathbf{W}(t) = \langle W_1(t), \ldots, W_m(t) \rangle^T, 0 \le t \le T$, is an m-dimensional Wiener process.

(H4.3) $\mathbf{X}_0$ is an $\mathscr{F}_0$-measurable random variable in $\mathcal{L}^2\left(\Omega; \mathbb{R}^N\right)$ that is independent of $\{\mathbf{W}(t) : 0 \le t \le T\}$. Moreover, $\forall t \in [0,T]$,

$$
\sigma\left(\mathbf{X}_0, \mathbf{W}(s) | 0 \le s \le t\right) \subset \mathscr{F}_t.
$$

The integral form of (4.32) is given by

$$\mathbf{X}(t;\omega) = \mathbf{X}_0(\omega) + \int_0^t \mathbf{A}\mathbf{X}(s;\omega)ds \tag{4.70}$$

$$+ \sum_{k=1}^m \int_0^t \mathbf{B}_k\mathbf{X}(s;\omega)dW_k(s), \, 0 \le t \le T, \omega \in \Omega.$$

Following the discussion in Section 3.3 regarding the well-posedness of (3.78), we must make certain that the right-hand side of (4.70) is well-defined. And, requiring that $\mathbf{X} \in \mathbb{C}\left([0,T];\mathcal{L}^2\left(\Omega;\mathbb{R}^N\right)\right)$ does the trick. Indeed, in such case, the Lebesgue integral in (4.70) is certainly defined. (Why?) We must argue that the Itô integral terms can be expressed in the form defined in Section 4.4.3. This follows immediately by defining

$$G_{ij}(s;\omega) = \sum_{k=1}^N b^i_{jk}X_k(s;\omega), \, \forall 1 \le i, j \le N, 0 \le s \le T, \omega \in \Omega, \tag{4.71}$$

so that

$$\int_0^t \mathbf{G}(s;\omega)d\mathbf{W}(s) = \begin{bmatrix} \sum_{j=1}^m \int_0^t \left(\sum_{k=1}^N b^1_{jk}X_k(s;\omega) \right) dW_j(s) \\ \vdots \\ \sum_{j=1}^m \int_0^t \left(\sum_{k=1}^N b^N_{jk}X_k(s;\omega) \right) dW_j(s) \end{bmatrix}. \tag{4.72}$$

Because the matrices $\mathbf{G}(s;\omega)$ are Hilbert-Schmidt operators, it follows that the integral defined in (4.72) is well-defined. (Why?) Moreover, (4.70) is equivalent to

$$\mathbf{X}(t;\omega) = \mathbf{X}_0(\omega) + \int_0^t \mathbf{A}\mathbf{X}(s;\omega)ds + \int_0^t \mathbf{G}(s;\omega)d\mathbf{W}(s), \, 0 \le t \le T, \omega \in \Omega. \tag{4.73}$$

Exercise 4.5.1. Carefully show that (4.70) and (4.73) are equivalent.

As such, seeking a solution of (4.71) in the following sense is reasonable.

Definition 4.5.1. A stochastic process $\mathbf{X} : [0,T] \times \Omega \longrightarrow \mathbb{R}^N$ is a *strong solution* of (4.70) on $[0,T]$ if $\mathbf{X} \in \mathbb{C}\left([0,T];\mathcal{L}^2\left(\Omega;\mathbb{R}^N\right)\right)$ and $\mathbf{X}(t;\omega)$ satisfies (4.70) a.s. $[\mathscr{P}]$, $\forall 0 \le t \le T$.

Now, take a step back and compare the setting of (4.70) to that of (3.74). Specifically, the main differences are that (i) the quantities that are constants in (3.74) are now constant matrices in (4.70), and (ii) the solution process is now an $\mathbb{R}^N$-valued stochastic process exhibiting characteristics that resemble those of the one-dimensional case almost verbatim.

Exercise 4.5.2. Formally show that (3.74) is a special case of (4.70).

This connection served as the basis for the development of the matrix exponential and the ensuing theory in the deterministic case. Looking back, the proofs of the theoretical results in the N-dimensional setting were virtually identical to those of the analogous one-dimensional results due to the fact that the properties of the matrix exponential so closely resembled the properties of the real-valued exponential function. In fact, other than a change of norm and being careful about the commutativity of products (because we are now dealing with matrices), the proofs go through unaltered, with the exception that the actual value of some of the numerical estimates are necessarily different. This renders the remainder of our work in this chapter rather simple; we simply need to carefully check line by line that the proofs work in the present setting. Because the proofs of the results in the more general case of multiple sources of noise were often left as exercises in Chapter 3, actually working through the proofs in this chapter for (4.70) will be especially instructive. In what follows, much of the detail-checking will be left as an exercise. A more complete discussion will be provided for those topics for which the details are new.

4.6   Existence and Uniqueness of a Strong Solution

The main existence/uniqueness result for (4.70) is as follows.

Theorem 4.6.1. *If (**H4.1**) through (**H4.3**) hold, then (4.70) has a unique strong solution on* $[0,T]$.

Exercise 4.6.1. Prove Thrm. 4.6.1 by carefully modifying the proof of Thrm. 3.4.1.

We can use the properties of the matrix exponential to derive a representation formula for a strong solution of (4.70), as long as all matrices $\mathbf{A}, \mathbf{B}_1, \ldots, \mathbf{B}_m$ are pairwise commutative. Precisely, we have

Proposition 4.6.2. *Assume that* $\mathbf{A}, \mathbf{B}_1, \ldots, \mathbf{B}_m$ *are pairwise commutative. Then, the strong solution of (4.70) is given by*

$$\mathbf{X}(t;\omega) = \left[e^{\left(\mathbf{A} - \frac{1}{2}\sum_{k=1}^m \mathbf{B}_k^2\right)t + \sum_{k=1}^m \mathbf{B}_k dW_k(t)} \right] \mathbf{X}_0(\omega), \ 0 \le t \le T, \omega \in \Omega. \tag{4.74}$$

Proof. Define $\mathbf{Y} : [0,T] \times \Omega \longrightarrow \mathbb{M}^N(\mathbb{R})$ by

$$\mathbf{Y}(t;\omega) = \int_0^t \left(\mathbf{A} - \frac{1}{2} \sum_{k=1}^m \mathbf{B}_k^2 \right) ds + \int_0^t \sum_{k=1}^m \mathbf{B}_k dW_k(s). \tag{4.75}$$

We must argue that the stochastic process $\mathbf{Z} : [0,T] \times \Omega \longrightarrow \mathbb{R}^N$ defined by

$$\mathbf{Z}(t;\omega) = e^{\mathbf{Y}(t;\omega)} \mathbf{X}_0(\omega)$$

satisfies (4.70). Applying Prop. 3.2.6 with $h(z) = e^z$ (so that $\frac{\partial h}{\partial t} = 0$, $\frac{\partial h}{\partial z} = \frac{\partial^2 h}{\partial z^2} = e^z$) yields

$$
\begin{aligned}
d\left(\mathbf{Z}(t;\omega)\right) &= d\left[e^{\mathbf{Y}(t;\omega)}\mathbf{X}_0(\omega)\right] \\
&= d\left(\mathbf{Y}(t;\omega)\right)\left(e^{\mathbf{Y}(t;\omega)}\mathbf{X}_0(\omega)\right) \qquad\qquad (4.76)\\
&\quad + \frac{1}{2}\left(d\left(\mathbf{Y}(t;\omega)\right)\right)^2\left(e^{\mathbf{Y}(t;\omega)}\mathbf{X}_0(\omega)\right) \\
&= d\left(\mathbf{Y}(t;\omega)\right)\mathbf{Z}(t;\omega) + \frac{1}{2}\left(d\left(\mathbf{Y}(t;\omega)\right)\right)^2\mathbf{Z}(t;\omega).
\end{aligned}
$$

Using

$$
\begin{cases}
d\left(\mathbf{Y}(t;\omega)\right) = \left(\mathbf{A} - \frac{1}{2}\sum_{k=1}^m \mathbf{B}_k^2\right)dt + \sum_{k=1}^m \mathbf{B}_k dW_k(t),\ 0 < t < T, \omega \in \Omega, \\
\mathbf{Y}(0;\omega) = 0, \omega \in \Omega,
\end{cases}
\qquad (4.77)
$$

in (4.76) and simplifying yields

$$
\begin{aligned}
d\left(\mathbf{Z}(t;\omega)\right) &= \left[\left(\mathbf{A} - \frac{1}{2}\sum_{k=1}^m \mathbf{B}_k^2\right)dt + \sum_{k=1}^m \mathbf{B}_k dW_k(t)\right]\mathbf{Z}(t;\omega) \\
&\quad + \frac{1}{2}\left(\sum_{k=1}^m \mathbf{B}_k^2 dt\right)\mathbf{Z}(t;\omega) \qquad\qquad (4.78)\\
&= \mathbf{A}\mathbf{Z}(t;\omega)dt + \left(\sum_{k=1}^m \mathbf{B}_k dW_k(t)\right)\mathbf{Z}(t;\omega) \\
&= \mathbf{A}\mathbf{Z}(t;\omega)dt + \sum_{k=1}^m \mathbf{B}_k\mathbf{Z}(t;\omega)dW_k(t).
\end{aligned}
$$

Also, observe that

$$
\mathbf{Z}(0;\omega) = e^{\mathbf{Y}(0;\omega)}\mathbf{X}_0(\omega) = \mathbf{I}\mathbf{X}_0(\omega) = \mathbf{X}_0(\omega).
$$

Consequently, $\mathbf{Z}(t;\omega)$ satisfies (4.70). By uniqueness, we conclude that $\mathbf{X}(t;\omega) = \mathbf{Z}(t;\omega)$, a.s. $[\mathscr{P}]$, $\forall 0 \le t \le T$, thereby resulting in (4.74). This completes the proof. $\qquad\square$

Exercise 4.6.2.
i.) Where was the commutativity of $\mathbf{A}, \mathbf{B}_1, \ldots, \mathbf{B}_m$ used in the proof of Prop. 4.6.2?
ii.) Explain how Itó's formula is being applied to a matrix-valued process $\mathbf{Y}(t;\omega)$.

The formula (4.74) simplifies considerably when $\mathbf{A}, \mathbf{B}_1, \ldots, \mathbf{B}_m$ are diagonal matrices. In such case, the N equations are all decoupled, all matrices necessarily commute (Why?), and the resulting formula is an N-dimensional stochastic process whose i^{th} component is precisely the representation formula (of the form (3.126)) of the solution of the ith equation in the system. Use this observation to complete the following exercises.

Exercise 4.6.3. Derive a simplified formula for the strong solution of (4.15).

Exercise 4.6.4. Use Putzer's algorithm to aid you in deriving a simplified formula for the strong solution of (4.27). Then, interpret the components of the solution process.

4.7 Continuous Dependence on Initial Data

Consider (4.70) and the related IVP

$$\begin{cases} d\mathbf{Y}(t;\omega) = \mathbf{A}\mathbf{Y}(t;\omega)dt + \sum_{k=1}^{m} \mathbf{B}_k \mathbf{Y}(t;\omega)dW_k(t), \, 0 < t < T, \omega \in \Omega, \\ \mathbf{Y}(0;\omega) = \mathbf{Y}_0(\omega), \omega \in \Omega, \end{cases} \tag{4.79}$$

both under hypotheses **(H4.1)** through **(H4.3)**.

Exercise 4.7.1. Before proceeding, try to derive an estimate for $E \, \|\mathbf{X}(t;\cdot) - \mathbf{Y}(t;\cdot)\|_{\mathbb{R}^N}^2$ in the spirit of (3.128). What differences do you encounter in this $\mathbb{R}^N$ setting?

Proposition 4.7.1. *For all* $0 \le t \le T,$

$$E \, \|\mathbf{X}(t;\cdot) - \mathbf{Y}(t;\cdot)\|_{\mathbb{R}^N}^2 \le 3 \, \|\mathbf{X}_0 - \mathbf{Y}_0\|_{\mathcal{L}^2}^2 \, e^{3t\left(T\|\mathbf{A}\|_{\mathbb{M}^N}^2 + m\zeta_{\mathscr{B}_0} \sum_{k=1}^{m} \|\mathbf{B}_k\|_{\mathbb{M}^N}^2\right)}. \tag{4.80}$$

Proof. Let $0 \le t \le T$. Subtracting (4.70) and the integrated form of (4.79) and using Lemma 4.4.3 yields

$$E \, \|\mathbf{X}(t;\cdot) - \mathbf{Y}(t;\cdot)\|_{\mathbb{R}^N}^2 \le 3 \left[\|\mathbf{X}_0 - \mathbf{Y}_0\|_{\mathcal{L}^2}^2 + E \left\| \int_0^t \mathbf{A}\left(\mathbf{X}(s;\cdot) - \mathbf{Y}(s;\cdot)\right) \right\|_{\mathbb{R}^N}^2 \right.$$

$$\left. + m \sum_{k=1}^{m} E \left\| \int_0^t \mathbf{B}_k\left(\mathbf{X}(s;\cdot) - \mathbf{Y}(s;\cdot)\right) dW_k(s) \right\|_{\mathbb{R}^N}^2 \right] \tag{4.81}$$

$$\le 3 \left[\|\mathbf{X}_0 - \mathbf{Y}_0\|_{\mathcal{L}^2}^2 + t \int_0^t E \, \|\mathbf{A}\left(\mathbf{X}(s;\cdot) - \mathbf{Y}(s;\cdot)\right)\|_{\mathbb{R}^N}^2 \, ds \right.$$

$$\left. + m\zeta_{\mathscr{B}_0} \sum_{k=1}^{m} \int_0^t E \, \|\mathbf{B}_k\left(\mathbf{X}(s;\cdot) - \mathbf{Y}(s;\cdot)\right)\|_{\mathscr{B}_0}^2 \, ds \right]$$

$$\le 3 \, \|\mathbf{X}_0 - \mathbf{Y}_0\|_{\mathcal{L}^2}^2 + 3 \left(T \|\mathbf{A}\|_{\mathbb{M}^N}^2 + m\zeta_{\mathscr{B}_0} \sum_{k=1}^{m} \|\mathbf{B}_k\|_{\mathbb{M}^N}^2 \right)$$

$$\cdot \int_0^t E \, \|\mathbf{X}(s;\cdot) - \mathbf{Y}(s;\cdot)\|_{\mathbb{R}^N}^2 \, ds.$$

Applying Gronwall's Lemma in (4.81) yields (4.80). This completes the proof. ☐

Exercise 4.7.2. Formulate (4.80) directly for IVPs (4.5) and (4.15).

Exercise 4.7.3. Replace $\mathbf{A}$ in (4.79) by $\mathbf{A}^\star$ and assume $\exists \delta_1, \delta_2 > 0$ such that

$$\|\mathbf{A} - \mathbf{A}^\star\|_{\mathbb{M}^N} < \delta_1 \text{ and } \|\mathbf{X}_0 - \mathbf{Y}_0\|_{\mathfrak{L}^2} < \delta_2. \tag{4.82}$$

Establish an estimate for $E \|\mathbf{X}(t;\cdot) - \mathbf{Y}(t;\cdot)\|_{\mathbb{R}^N}^2$ as in Prop. 4.7.1.

4.8 Statistical Properties of the Strong Solution

The same statistical properties for the $\mathbb{R}$-valued stochastic processes developed in Section 3.6 are of interest here. The computations closely resemble those in Section 3.6 and are therefore left as an exercise.

Exercise 4.8.1. Assume that **(H4.1)** through **(H4.3)** hold. Let $\mathbf{X} : [0, T] \times \Omega \longrightarrow \mathbb{R}^N$ be the strong solution of (4.70).
i.) Show that $\mu_{\mathbf{X}}(t) = e^{\mathbf{A}t} E[\mathbf{X}_0]$.
ii.) Compute $\mu_{\mathbf{X}}(t)$ directly for the IVPs in Section 4.1.
iii.) Assume that $m = 1$ in (4.70) and write the Itó term as $\int_0^t \mathbf{BX}(s;\cdot) d\mathbf{W}(s)$.
  a.) Develop a formula for $Var_{\mathbf{X}}(t)$.
  b.) Compute $Var_{\mathbf{X}}(t)$ directly for IVP (4.5).
  c.) Compute $Cov_{\mathbf{X}}(t, s)$ for the IVP (4.5).
iv.) Assume that $E \|\mathbf{X}_0(\cdot)\|_{\mathbb{R}^N}^p < \infty, \forall p \geq 2$.
  a.) Use Itó's formula to derive an estimate for $E \|\mathbf{X}(t;\cdot)\|_{\mathbb{R}^N}^p$.
  b.) Prove that there exists a continuous function $\eta : [0, T] \times [0, T] \longrightarrow \mathbb{R}$ such that

$$E \|\mathbf{X}(\tau;\cdot) - \mathbf{X}(\tau';\cdot)\|_{\mathbb{R}^N}^p \leq \eta_p(\tau, \tau').$$

  c.) Interpret the results in (a) and (b) specifically for the IVPs in Section 4.1.
v.) Convince yourself that $\{\mathbf{X}(t;\omega) : 0 \leq t \leq T, \omega \in \Omega\}$ is a Markov process.

4.9 Some Convergence Results

These results are natural extensions of those formulated in Section 3.7. You are encouraged to provide the details.
  For every $n \in \mathbb{N}$, consider the IVP

$$\begin{cases} d\mathbf{X}_n(t;\omega) = \mathbf{A}_n \mathbf{X}_n(t;\omega) dt + \sum_{k=1}^m (\mathbf{B}_k)_n \mathbf{X}_n(t;\omega) dW_k(t), \\ \mathbf{X}_n(0;\omega) = (\mathbf{X}_0)_n(\omega), \omega \in \Omega. \end{cases} \tag{4.83}$$

where $0 < t < T, \omega \in \Omega$. Complete the following exercise to establish a generalization of Prop. 3.7.1.

Exercise 4.9.1.
i.) Formulate hypotheses that correspond to **(H3.4)** through **(H3.6)** for (4.83).
ii.) Subtract (4.70) from the integrated version of (4.83) and use Lemma 4.4.3 to obtain an initial estimate for $E \|\mathbf{X}_n(t;\cdot) - \mathbf{X}(t;\cdot)\|^2_{\mathbb{R}^N}$.
iii.) Let $\varepsilon > 0$. Mimicking the proof of Prop. 3.7.1, scout ahead to determine estimates in the spirit of (3.162) through (3.164) so that the end result of the argument is the statement

$$n \geq N \Longrightarrow \|\mathbf{X}_n - \mathbf{X}\|_{\mathbb{C}\left([0,T];\mathscr{L}^2\left(\Omega;\mathbb{R}^N\right)\right)} < \varepsilon. \tag{4.84}$$

Exercise 4.9.2. Argue that $\forall 0 \leq t \leq T,$

$$\lim_{n\to\infty} \mu_{\mathbf{X}_n}(t) = \mu_{\mathbf{X}}(t), \ \lim_{n\to\infty} Var_{\mathbf{X}_n}(t) = Var_{\mathbf{X}}(t). \tag{4.85}$$

Exercise 4.9.3. Interpret the results from Exercises 4.9.1 and 4.9.2 specifically for the IVPs in Section 4.1.

In a similar fashion, we can approximate the solution of the deterministic IVP

$$\begin{cases} \mathbf{X}'(t) = \mathbf{A}\mathbf{X}(t), t > 0, \\ \mathbf{X}(0) = \mathbf{X}_0, \end{cases} \tag{4.86}$$

by a sequence of strong solutions of appropriately chosen stochastic IVPs of the form (4.32). Following the development of Prop. 3.7.2, complete the following exercise.

Exercise 4.9.4.
i.) Let $0 < \varepsilon < 1$. What stochastic IVP plays the role of (3.170) in this setting?
ii.) Formulate the hypotheses that should replace **(H3.7)** and **(H3.8)**.
iii.) Formulate the generalization of Prop. 3.7.2 for this setting.
iv.) Argue as in the proof of Prop. 3.7.2 to prove this proposition.

Exercise 4.9.5.
i.) Devise an approximation scheme of the type studied in Exer. 4.9.4 specifically for (4.16).
ii.) Repeat (i) for (4.19).

4.10   Looking Ahead

The content of this chapter establishes the well-posedness of the IVPs (4.31) and (4.32) (that is, the existence and uniqueness of a strong solution that depends continuously on the initial data) for any $N \times N$ matrices $\mathbf{A}, \mathbf{B}_1, \ldots, \mathbf{B}_m$ and any initial condition $\mathbf{U}_0$. This theory can be used in the description of numerous applications. However, not all phenomena can be described using a linear system of stochastic ODEs. In fact, as soon as the description depends on more than one variable (e.g., both time *and* position), the use of stochastic *partial* differential equations (SPDEs) is required. For instance, consider the following deterministic initial-boundary value problem (IBVP), which is a classical model of heat conduction in a one-dimensional rod of length L:

$$\begin{cases} \frac{\partial u}{\partial t}(z,t) = k\frac{\partial^2 u}{\partial z^2}(z,t), \ 0 \le z \le L, t > 0, \\ u(z,0) = u_0(z), \ 0 \le z \le L, \\ u(0,t) = u(L,t) = 0, \ t > 0. \end{cases} \tag{4.87}$$

Here, the constant k is a physical parameter involving the density of the rod and thermal conductivity, and $u(z,t)$ represents the temperature at position z along the rod at time t. A loose comparison of (4.87) to (4.31) suggests that we can identify $\mathbf{U}$ with the unknown $u : [0,L] \times [0,\infty) \to \mathbb{R}$, the left-hand side with $\frac{\partial u}{\partial t}$, and the right-hand side somehow with the differential operator $A = \frac{\partial^2}{\partial z^2}$. But, this certainly does not fall under the parlance of the theory developed in this chapter because $\frac{\partial^2}{\partial z^2}$ cannot be identified with any member of $\mathbb{M}^N(\mathbb{R})$. Still, expressing (4.87) in an abstract form similar to the deterministic version of (4.31) is not unreasonable, although doing so requires the use of linear semigroup theory.

This theory applies to a random version of (4.87) in which noise is incorporated strictly through the initial conditions. We can even derive an explicit formula for the solution that holds a.s. $[\mathscr{P}]$. But, the more interesting scenario is when noise is incorporated into the model from a source other than the IC, thereby resulting in an IBVP of the form

$$\begin{cases} \partial u(z,t;\omega) = \left(k\frac{\partial^2 u}{\partial z^2}(z,t;\omega)\right) dt + Bu(z,t;\omega)dW(t), \ 0 \le z \le L, t > 0, \omega \in \Omega, \\ u(z,0;\omega) = u_0(z;\omega), \ 0 \le z \le L, \omega \in \Omega, \\ u(0,t;\omega) = u(L,t;\omega) = 0, \ t > 0, \omega \in \Omega. \end{cases}$$

$$\tag{4.88}$$

Can (4.88) be transformed into an IVP of the *form* (4.32), albeit perhaps in some more abstract sense? If so, can the theory we have developed for stochastic ODEs be generalized to handle abstract stochastic ODEs with the help of linear semigroup theory? These questions will be considered in Chapter 5.

4.11 Guidance for Selected Exercises

4.11.1 Level 1: A Nudge in a Right Direction

4.2.1. Mimic the steps used to compute $e^{\begin{bmatrix} a & 0 \\ 0 & b \end{bmatrix} t}$. Compute $\mathbf{A}t$, $(\mathbf{A}t)^k$, and $\frac{(\mathbf{A}t)^k}{k!}$, where

$$\mathbf{A} = \begin{bmatrix} a_{11} & 0 & 0 & \cdots & 0 \\ 0 & a_{22} & 0 & \cdots & 0 \\ \vdots & \vdots & \ddots & & \vdots \\ 0 & 0 & & \ddots & 0 \\ 0 & 0 & 0 & \cdots & a_{NN} \end{bmatrix} \tag{4.89}$$

(Now what?)

4.2.2. Show $\mathbf{P}^{-1}(\mathbf{A}+\mathbf{B})\mathbf{P} = \mathbf{P}^{-1}\mathbf{A}\mathbf{P} + \mathbf{P}^{-1}\mathbf{B}\mathbf{P}$ and $(\mathbf{P}^{-1}\mathbf{D}\mathbf{P})^k = \mathbf{P}^{-1}\mathbf{D}^k\mathbf{P}$, $\forall k \in \mathbb{N}$.

4.2.3. (i) Use $\left\| e^{\mathbf{A}t}\mathbf{x} \right\|_{\mathbb{R}^N} \le \left\| e^{\mathbf{A}t} \right\|_{\mathbb{M}^N} \left\| \mathbf{x} \right\|_{\mathbb{R}^N}$. (How?)

(ii) Use Exer. 4.2.1.

(iii) Let $\mathbf{A}$ be given by (4.89), where $a_{ii} < 0$, $\forall i \in \{1,\ldots,N\}$. Compute $\|\mathbf{A}\|_{\mathbb{M}^N}$.

4.2.4. I) $e^{\mathbf{A}t}$ has the same form for (i) through (v), and $\left\| e^{\mathbf{A}t} \right\|_{\mathbb{M}^N} \le e^{t\|\mathbf{A}\|_{\mathbb{M}^N}}$, $\forall t$. (So what?)

II) Compute the limits entrywise. When does the term e^{at} have a limit as $t \to \infty$?

III) (1) Only for those for which the limit in (II) exists. (Why?)

(2) Does $\exists p > 0$ such that $e^{a(t+p)} = e^{at}$, $\forall t \ge 0$?

(3) For which values of $a \in \mathbb{R}$ is the set $\{e^{at}|t \ge 0\}$ bounded?

(4) What must be true about the graph of $y = e^{at}$ in order for this to occur?

4.2.5. The eigenvalues of $\mathbf{A}$ are α and β. Now, the conclusions follow immediately from Exer. 4.2.4. (How?)

4.5.2. Let $\mathbf{g} = \langle g_1,\ldots,g_m \rangle$ and $\mathbf{W}(t) = \langle W_1(t),\ldots,W_m(t) \rangle^T$. Then, what is the dimension of $\mathbf{g}d\mathbf{W}$?

4.6.1. You will need to use Lemma 4.4.3; so, the constants will change. Also, keep in mind that all stochastic processes are now $\mathbb{R}^N$-valued, so calculus operations and convergence arguments will be performed componentwise.

4.6.2. (i) $e^{\mathbf{A}+\mathbf{B}} = e^{\mathbf{A}}e^{\mathbf{B}} = e^{\mathbf{B}}e^{\mathbf{A}}$ only when $\mathbf{A}$ and $\mathbf{B}$ commute. (So what?)

(ii) An $N \times N$ matrix can be viewed as an operator from $\mathbb{R}^N$ into itself. Prop. 3.2.6 applies to such mappings.

4.7.1. Use the integrated form of (4.80) and then Gronwall's Lemma.

4.7.3. Modify the estimate (4.82) to account for the additional approximation of the operator A.

4.11.2   Level 2: An Additional Thrust in a Right Direction

4.2.1. If $\mathbf{A}$ is given by (4.89), then

$$
e^{\mathbf{A}t} = \begin{bmatrix}
e^{a_{11}t} & 0 & 0 & \cdots & 0 \\
0 & e^{a_{22}t} & 0 & \cdots & 0 \\
\vdots & \vdots & \ddots & & \vdots \\
0 & 0 & & \ddots & 0 \\
0 & 0 & 0 & \cdots & e^{a_{NN}t}
\end{bmatrix}
$$

4.2.2. Now apply the Taylor representation formula.

4.2.3. (i) Choose $\mathbf{A}$ such that $\left\| e^{\mathbf{A}t} \right\|_{\mathbb{M}^N} > 1$.

   (ii) Calculate $\|\mathbf{A}\|_{\mathbb{M}^N}$ using (1.63). What must be true about the eigenvalues?

   (iii) Using the Cauchy-Schwarz inequality, we see that

$$
|\langle \mathbf{A}\mathbf{x}, \mathbf{x} \rangle| \leq \|\mathbf{A}\mathbf{x}\|_{\mathbb{R}^N} \|\mathbf{x}\|_{\mathbb{R}^N} \leq \|\mathbf{A}\|_{\mathbb{M}^N} \|\mathbf{x}\|_{\mathbb{R}^N}^2 .
$$

4.2.4. I) See the example following Def. 2.2.3 for $e^{\mathbf{A}t}$. Also, $\left\| e^{\mathbf{A}t} \right\|_{\mathbb{M}^N} \leq e^{\max\{\alpha,\beta\}t}$, $\forall t$.

II) There is a uniform limit of $\mathbf{0}$ in (i) and (iv); the others do not have limit functions.

III) (2) None are periodic. (Can you use Exer. 1.9.4 to determine conditions under which such a solution *would* be periodic?)

   (3) (i) and (iv) only.

   (4) None exhibit this behavior because no component has a vertical asymptote.

4.5.2. $\mathbf{g}d\mathbf{W}$ is a 1×1 matrix.

Chapter 5

Abstract Homogenous Linear Stochastic Evolution Equations

Overview

Partial differential equations are often an important component of the mathematical description of phenomena. Guided by our study of (4.32) in Chapter 4, it is natural to ask whether or not certain classes of initial-boundary value problems could also be subsumed under some theoretical umbrella in the spirit of (4.32), albeit in a more elaborate sense. The quick answer is yes, provided we interpret the pieces correctly. Our work in this chapter focuses on extending the theoretical framework from Chapter 4 to a more general setting. Some standard references used throughout this chapter are **[78, 102, 107, 133, 155, 159, 206, 219, 221, 222, 252, 265, 300]**.

5.1   Linear Operators

Up to now, we have dealt only with mappings from $\mathbb{R}^N$ to $\mathbb{R}^M$, where $N, M \in \mathbb{N}$. Extending the theory from Chapter 2 to more elaborate settings will require the use of more general mappings between Banach spaces $\mathscr{X}$ and $\mathscr{Y}$. We begin with a preliminary discussion of linear operators. A thorough treatment of the topics discussed in this section can be found in **[118, 160, 215, 243, 256, 418]**.

5.1.1   Bounded versus Unbounded Operators

Definition 5.1.1. Let $(\mathscr{X}, \|\cdot\|_{\mathscr{X}})$ and $(\mathscr{Y}, \|\cdot\|_{\mathscr{Y}})$ be real Banach spaces.
i.) A *bounded linear operator* from $(\mathscr{X}, \|\cdot\|_{\mathscr{X}})$ into $(\mathscr{Y}, \|\cdot\|_{\mathscr{Y}})$ is a mapping $\mathscr{F} : \mathscr{X} \to \mathscr{Y}$ such that
    a.) (linear) $\mathscr{F}(\alpha x + \beta y) = \alpha \mathscr{F}(x) + \beta \mathscr{F}(y), \forall \alpha, \beta \in \mathbb{R}$ and $x, y \in \mathscr{X}$,
    b.) (bounded) There exists $m \geq 0$ such that $\|\mathscr{F}(x)\|_{\mathscr{Y}} \leq m \|x\|_{\mathscr{X}}, \forall x \in \mathscr{X}$.
We denote the set of all such operators by $\mathbb{B}(\mathscr{X}, \mathscr{Y})$. If $\mathscr{X} = \mathscr{Y}$, we write $\mathbb{B}(\mathscr{X})$ and refer to its members as "bounded linear operators on $\mathscr{X}$."
ii.) Let $\mathscr{F} \in \mathbb{B}(\mathscr{X}, \mathscr{Y})$. The *operator norm* of $\mathscr{F}$, denoted by $\|\mathscr{F}\|_{\mathbb{B}(\mathscr{X}, \mathscr{Y})}$ or more

succinctly as $\|\mathscr{F}\|_{\mathbb{B}}$, is defined by

$$\|\mathscr{F}\|_{\mathbb{B}} = \inf\{m : m > 0 \wedge \|\mathscr{F}(x)\|_{\mathscr{Y}} \le m\|x\|_{\mathscr{X}}, \forall x \in \mathscr{X}\}$$
$$= \sup\{\|\mathscr{F}(x)\|_{\mathscr{Y}} : \|x\|_{\mathscr{X}} = 1\}.$$

iii.) If there does not exist $m \ge 0$ such that $\|\mathscr{F}(x)\|_{\mathscr{Y}} \le m\|x\|_{\mathscr{X}}, \forall x \in \mathscr{X}$, we say that $\mathscr{F}$ is *unbounded*.

The terminology "bounded operator" may seem to be somewhat of a misnomer in comparison to the notion of a bounded real-valued function, but the name arose because such operators map norm-bounded subsets of $\mathscr{X}$ into norm-bounded subsets of $\mathscr{Y}$. We must simply contend with this nomenclature issue on a contextual basis. Also, regarding the notation, the quantity $\mathscr{F}(x)$ is often written more succinctly as $\mathscr{F}x$ (with parentheses suppressed) as in the context of matrix multiplication.

Exercise 5.1.1.
i.) Prove that $\|\cdot\|_{\mathbb{B}}$ is a norm on $\mathbb{B}(\mathscr{X}, \mathscr{Y})$.
ii.) Let $\mathscr{F} \in \mathbb{B}(\mathscr{X}, \mathscr{Y})$. Prove that $\|\mathscr{F}x\|_{\mathscr{Y}} \le \|\mathscr{F}\|_{\mathbb{B}}\|x\|_{\mathscr{X}}, \forall x \in \mathscr{X}$.

Exercise 5.1.2. Explain how to prove that an operator $\mathscr{F} : \mathscr{X} \to \mathscr{Y}$ is unbounded.

Some Examples.
1. The identity operator $I : \mathscr{X} \to \mathscr{X}$ is in $\mathbb{B}(\mathscr{X})$ with $\|I\|_{\mathbb{B}} = 1$.
2. Let $\mathbf{A} \in \mathbb{M}^N(\mathbb{R})$. The operator $\mathfrak{B} : \mathbb{R}^N \to \mathbb{R}^N$ defined by $\mathfrak{B}\mathbf{x} = \mathbf{A}\mathbf{x}$ is in $\mathbb{B}(\mathbb{R}^N)$ with $\|\mathfrak{B}\|_{\mathbb{B}} = \|\mathbf{A}\|_{\mathbb{M}^N}$ (cf. Prop. 4.2.2).
3. Assume that $g : [a,b] \times [a,b] \to \mathbb{R}$ is continuous. Let $x \in \mathbb{C}([a,b];\mathbb{R})$ and define $y : [a,b] \to \mathbb{R}$ by $y(t) = \int_a^b g(t,s)x(s)\,ds$. The operator $\mathscr{F} : \mathbb{C}([a,b];\mathbb{R}) \to \mathbb{C}([a,b];\mathbb{R})$ defined by $\mathscr{F}(x) = y$ is in $\mathbb{B}(\mathbb{C}([a,b];\mathbb{R}))$, where $\mathbb{C}([a,b];\mathbb{R})$ is equipped with the sup norm.

Exercise 5.1.3.
i.) Provide the details in Example 3 and identify an upper bound for $\|\mathscr{F}\|_{\mathbb{B}}$.
ii.) Assume that $x \in \mathbb{C}([a,b];\mathbb{R})$. If $\mathscr{F}$ is, instead, viewed as an operator on $\mathbb{L}^2(a,b;\mathbb{R})$, is $\mathscr{F} \in \mathbb{B}(\mathbb{L}^2(a,b;\mathbb{R}))$?

Exercise 5.1.4. Assume that $g : [a,b] \times [a,b] \times \mathbb{R} \to \mathbb{R}$ is a continuous mapping for which $\exists m_g > 0$ such that

$$|g(x,y,z)| \le m_g|z|, \forall x,y \in [a,b] \text{ and} z \in \mathbb{R}. \tag{5.1}$$

For every $x \in [a,b]$, define the operator $\mathscr{F} : \mathbb{C}([a,b];\mathbb{R}) \to \mathbb{C}([a,b];\mathbb{R})$ by

$$\mathscr{F}(z)[x] = \int_a^b g(x,y,z(y))\,dy. \tag{5.2}$$

Is $\mathscr{F}$ linear? bounded?

Exercise 5.1.5. Define $\mathscr{F} : C^1((0,a);\mathbb{R}) \to C([0,a];\mathbb{R})$ by $\mathscr{F}(g) = g'$.
i.) Certainly, $\mathscr{F}$ is linear. Show that $\mathscr{F}$ is unbounded if $C([0,a];\mathbb{R})$ is equipped with the sup norm.
ii.) If $C([0,a];\mathbb{R})$ is equipped with the $\mathbb{L}^2$-norm, show that $\mathscr{F}$ is in $\mathbb{B}(\mathbb{L}^2(0,a;\mathbb{R}))$ with $\|\mathscr{F}\|_{\mathbb{B}}^2 \leq a$.

Important Note. Exercise 5.1.5 illustrates the fact that changing the underlying norm (not to mention the function space) can drastically alter the nature of the operator. This has ramifications in the theoretical development, and to an even greater extent the application of the theory to actual IBVPs. Often, choosing the correct closed subspace equipped with the right norm is critical in establishing the existence of a solution to an IBVP. We will revisit this issue frequently in what is to come.

Exercise 5.1.6. Define $\mathscr{F} : C^2((0,a);\mathbb{R}) \to C([0,a];\mathbb{R})$ by $\mathscr{F}(g) = g''$. Certainly, $\mathscr{F}$ is linear. Show that $\mathscr{F}$ is unbounded if $C([0,a];\mathbb{R})$ is equipped with the sup norm.

Exercise 5.1.7. Let $\mathscr{F} \in \mathbb{B}(\mathscr{X},\mathscr{Y})$ and take $\{x_n\} \subset \mathscr{X}$ such that $\lim_{n\to\infty} \|x_n - x\|_{\mathscr{X}} = 0$.
i.) Show that $\lim_{n\to\infty} \|\mathscr{F}x_n - \mathscr{F}x\|_{\mathscr{Y}} = 0$. (Thus, bounded linear operators are continuous.)
ii.) Explain why $\mathscr{X}$ need not be complete in order for (i) to hold.

The notions of domain and range are the same for any mapping. One nice feature of an operator $\mathscr{F} \in \mathbb{B}(\mathscr{X},\mathscr{Y})$ is that both $\mathrm{dom}(\mathscr{F})$ and $\mathrm{rng}(\mathscr{F})$ are vector subspaces of $\mathscr{X}$ and $\mathscr{Y}$, respectively. The need to compare two operators and to consider the restriction of a given operator to a subset of its domain arise often. These notions are made precise below.

Definition 5.1.2. Let $\mathscr{F}, \mathscr{G} \in \mathbb{B}(\mathscr{X},\mathscr{Y})$.
i.) We say $\mathscr{F}$ *equals* $\mathscr{G}$, written $\mathscr{F} = \mathscr{G}$, if
    a.) $\mathrm{dom}(\mathscr{F}) = \mathrm{dom}(\mathscr{G})$,
    b.) $\mathscr{F}x = \mathscr{G}x$, for all x in the common domain.
ii.) Let $\mathscr{Z} \subset \mathrm{dom}(\mathscr{F})$. The operator $\mathscr{F}|_{\mathscr{Z}} : \mathscr{Z} \subset \mathscr{X} \to \mathscr{Y}$ defined by $\mathscr{F}|_{\mathscr{Z}}(z) = \mathscr{F}(z), \forall z \in \mathscr{Z}$, is called the *restriction* of $\mathscr{F}$ to $\mathscr{Z}$.

The following claim says that if two bounded linear operators "agree often enough," then they agree everywhere. This is not difficult to prove. (Compare this to Prop. 1.8.6(v).)

Proposition 5.1.3. *Let* $\mathscr{F}, \mathscr{G} \in \mathbb{B}(\mathscr{X})$ *and* $\mathscr{D}$ *a dense subset of* $\mathscr{X}$. *If* $\mathscr{F}x = \mathscr{G}x$, $\forall x \in \mathscr{D}$, *then* $\mathscr{F} = \mathscr{G}$.

All operators arising in Chapter 4 were members of the Banach space $\mathbb{M}^N(\mathbb{R})$. The structure inherent to a Banach space was essential to ensure that sums of $N \times N$ matrices and limits of convergent sequences of $N \times N$ matrices were well-defined. The very act of forming a more general theory suggests that spaces playing a comparable role in the present setting will need to possess a similar structure. While many of the operators arising in our models will be unbounded, the underlying theory relies heavily on the space $\mathbb{B}(\mathscr{X},\mathscr{Y})$. In accordance with intuition, it turns out that if $\mathscr{X}$ and $\mathscr{Y}$ are sufficiently nice, then $\mathbb{B}(\mathscr{X},\mathscr{Y})$ is also. Precisely, we have

Proposition 5.1.4. *If $\mathscr{X}$ and $\mathscr{Y}$ are Banach spaces, then $\mathbb{B}(\mathscr{X},\mathscr{Y})$ equipped with the norm $\|\cdot\|_{\mathbb{B}}$ is a Banach space.*

For a proof, see Volume 1. We frequently need to apply two operators in succession in the following sense.

Proposition 5.1.5. *Let $\mathscr{F},\mathscr{G} \in \mathbb{B}(\mathscr{X})$ where $\mathrm{rng}(\mathscr{G}) \subset \mathrm{dom}(\mathscr{F})$. The composition operator $\mathscr{F}\mathscr{G} : \mathscr{X} \to \mathscr{X}$ defined by $(\mathscr{F}\mathscr{G})(x) = \mathscr{F}(\mathscr{G}(x))$ is in $\mathbb{B}(\mathscr{X})$ and satisfies $\|\mathscr{F}\mathscr{G}\|_{\mathbb{B}} \leq \|\mathscr{F}\|_{\mathbb{B}}\|\mathscr{G}\|_{\mathbb{B}}$. (If $\mathscr{F} = \mathscr{G}$, then $\mathscr{F}^2$ is written in place of $\mathscr{F}\mathscr{G}$.)*

Exercise 5.1.8. Prove Prop. 5.1.5.

Convergence properties are important, especially when establishing numerical results for computational purposes. Different types of convergence can be defined by equipping the underlying spaces with different topologies. Presently, we focus only on the norm-topology.

Definition 5.1.6. Let $\{\mathscr{F}_n\} \subset \mathbb{B}(\mathscr{X},\mathscr{Y})$. We say that $\{\mathscr{F}_n\}$ is

i.) *uniformly convergent* to $\mathscr{F}$ in $\mathbb{B}(\mathscr{X},\mathscr{Y})$ if $\lim\limits_{n\to\infty} \|\mathscr{F}_n - \mathscr{F}\|_{\mathbb{B}} = 0$.

ii.) *strongly convergent* to $\mathscr{F}$ if $\lim\limits_{n\to\infty} \|\mathscr{F}_n x - \mathscr{F}x\|_{\mathscr{Y}} = 0, \forall x \in \mathscr{X}$.

We write $\mathscr{F}_n \xrightarrow{\text{uni}} \mathscr{F}$ and $\mathscr{F}_n \xrightarrow{\text{s}} \mathscr{F}$, respectively.

Exercise 5.1.9. Show (i) $\implies$ (ii) in Def. 5.1.6.

Convergence issues will arise when defining different types of solutions of an IVP. In general, the characteristic properties of the limit operator are enhanced when the type of convergence used is strengthened. We will revisit this notion as the need arises. Note that we do not assume $\mathscr{F} \in \mathbb{B}(\mathscr{X},\mathscr{Y})$ in Def. 5.1.6(ii). It actually turns out to be true, but not without some work. To show this, we make use of the following powerhouse theorem.

Theorem 5.1.7. (Principle of Uniform Boundedness)
Let $I \subset \mathbb{R}$ be a (possibly uncountable) index set and let $\{\mathscr{F}_i\} \subset \mathbb{B}(\mathscr{X},\mathscr{Y})$ be such that $\forall x \in \mathscr{X}$, $\exists m_x > 0$ for which $\sup\{\|\mathscr{F}_i x\|_{\mathscr{Y}} : i \in I\} \leq m_x$. Then, $\exists m^\star > 0$ such that $\sup\left\{\|\mathscr{F}_i\|_{\mathbb{B}(\mathscr{X},\mathscr{Y})} : i \in I\right\} \leq m^\star$.

Consequently, we have

Corollary 5.1.8. *If $\{\mathscr{F}_n : n \in \mathbb{N}\} \subset \mathbb{B}(\mathscr{X},\mathscr{Y})$ and $\mathscr{F}_n \xrightarrow{\text{s}} \mathscr{F}$ as $n \longrightarrow \infty$, then $\mathscr{F} \in \mathbb{B}(\mathscr{X},\mathscr{Y})$.*

Proof. Because $\mathscr{F}_n \xrightarrow{\text{s}} \mathscr{F}$, it follows that $\{\mathscr{F}_n x : n \in \mathbb{N}\}$ is a bounded subset of $\mathscr{Y}$, $\forall x \in \mathscr{X}$. So, by Thrm. 5.1.7, $\exists m^\star > 0$ such that $\sup\{\|\mathscr{F}_n\|_{\mathbb{B}} : n \in \mathbb{N}\} \leq m^\star$. Hence, $\forall x \in \mathscr{X}$,

$$\|\mathscr{F}_n x\|_{\mathscr{Y}} \leq \|\mathscr{F}_n\|_{\mathbb{B}} \|x\|_{\mathscr{X}} \leq m^\star \|x\|_{\mathscr{X}},$$

and so

$$\|\mathscr{F} x\|_{\mathscr{Y}} = \left\|\lim_{n\to\infty} \mathscr{F}_n x\right\|_{\mathscr{Y}} = \lim_{n\to\infty} \|\mathscr{F}_n x\|_{\mathscr{Y}} \leq m^\star \|x\|_{\mathscr{X}}.$$

$\square$

5.1.2 Invertible Operators

The notions of one-to-one and onto for operators, and their relationships to invertibility, coincide with the usual elementary notions for real-valued functions.

Definition 5.1.9. i.) A linear operator $\mathscr{F} : \text{dom}(\mathscr{F}) \subset \mathscr{X} \to \mathscr{Y}$ is
a.) *one-to-one* if $\mathscr{F} x = \mathscr{F} y \Longrightarrow x = y$, $\forall x, y \in \text{dom}(\mathscr{F})$.
b.) *onto* if $\text{rng}(\mathscr{F}) = \mathscr{Y}$.
ii.) An operator $\mathscr{G} : \text{rng}(\mathscr{F}) \to \text{dom}(\mathscr{F})$ that satisfies

$$\mathscr{G}(\mathscr{F}(x)) = x, \ \forall x \in \text{dom}(\mathscr{F}),$$
$$\mathscr{F}(\mathscr{G}(y)) = y, \ \forall y \in \text{rng}(\mathscr{F})$$

is an *inverse* of $\mathscr{F}$ and is denoted by $\mathscr{G} = \mathscr{F}^{-1}$. If $\mathscr{G}$ exists, we say that $\mathscr{F}$ is *invertible*.

As with real-valued functions, $\mathscr{F} \in \mathbb{B}(\mathscr{X}, \mathscr{Y})$ is invertible iff $\mathscr{F} : \text{dom}(\mathscr{F}) \to \text{rng}(\mathscr{F})$ is one-to-one. (Why?) The following characterization is useful.

Proposition 5.1.10. *Let* $\mathscr{F} \in \mathbb{B}(\mathscr{X}, \mathscr{Y})$.
i.) $\mathscr{F}$ *is invertible iff* $\mathscr{F} x = 0 \Longrightarrow x = 0$.
ii.) *If* $\mathscr{F}$ *is invertible, then* $\mathscr{F}^{-1}$ *is a linear operator.*

Exercise 5.1.10. Prove Prop. 5.1.10.

The following notion of a convergent series in a Banach space is needed in the next result.

Definition 5.1.11. Let $\mathscr{X}$ be a Banach space and $\{x_n\} \subset \mathscr{X}$. We say that $\sum_{n=1}^{\infty} x_n$ *converges absolutely* if $\sum_{n=1}^{\infty} \|x_n\|_{\mathscr{X}}$ converges (in the sense of Def. 1.5.19).

Exercise 5.1.11. Let $\mathscr{X}$ be a Banach space and $\{x_n\} \subset \mathscr{X}$. For each $n \in \mathbb{N}$, define $S_n = \sum_{k=1}^{n} x_k$. Prove that if $\sum_{n=1}^{\infty} \|x_n\|_{\mathscr{X}}$ converges, then $\{S_n\}$ is a strongly convergent sequence in $\mathscr{X}$.

The identity operator I on $\mathscr{X}$ is clearly invertible, as are bounded linear operators that are "sufficiently close to I" in the following sense.

Proposition 5.1.12. *Let $\mathscr{A} \in \mathbb{B}(\mathscr{X})$. If $\|\mathscr{A}\|_{\mathbb{B}} < 1$, then*

i.) $I - \mathscr{A}$ *is invertible with inverse* $(I - \mathscr{A})^{-1}$.

ii.) $(I - \mathscr{A})^{-1}$ *is bounded.*

iii.) $\left\|(I - \mathscr{A})^{-1}\right\|_{\mathbb{B}} \leq \frac{1}{1 - \|\mathscr{A}\|_{\mathbb{B}}}$.

Proof. Because $\|\mathscr{A}\|_{\mathbb{B}} < 1$, $\sum_{n=0}^{\infty} \|\mathscr{A}\|_{\mathbb{B}}^n$ is a convergent geometric series with sum $\frac{1}{1-\|\mathscr{A}\|_{\mathbb{B}}}$. By Exer. 5.1.11, $\{\sum_{k=0}^{n} \mathscr{A}^k : n \in \mathbb{N}\}$ is strongly convergent in $\mathbb{B}(\mathscr{X})$ with limit $\sum_{k=0}^{\infty} \mathscr{A}^k$. (Why?) Further, because

$$(I - \mathscr{A})\left(\sum_{k=0}^{\infty} \mathscr{A}^k\right) = I = \left(\sum_{k=0}^{\infty} \mathscr{A}^k\right)(I - \mathscr{A}),$$

it follows that $I - \mathscr{A}$ is invertible with inverse $(I - \mathscr{A})^{-1} = \sum_{k=0}^{\infty} \mathscr{A}^k$ in $\mathbb{B}(\mathscr{X})$. Finally,

$$\left\|(I - \mathscr{A})^{-1}\right\|_{\mathbb{B}} \leq \left\|\sum_{k=0}^{\infty} \mathscr{A}^k\right\|_{\mathbb{B}} \leq \sum_{k=0}^{\infty} \|\mathscr{A}\|_{\mathbb{B}}^k = \frac{1}{1 - \|\mathscr{A}\|_{\mathbb{B}}}.$$

This concludes the proof. $\qquad\square$

Exercise 5.1.12. Let $\mathscr{A} \in \mathbb{B}(\mathscr{X})$. Show that if $\mathscr{A}$ is invertible, then $\alpha\mathscr{A}$ is invertible, $\forall \alpha \neq 0$.

5.1.3 Closed Operators

Recall what it means for a set $\mathscr{D}$ to be closed in $\mathscr{X}$, assuming that $\mathscr{X}$ is equipped with the norm topology. This notion can be extended to operators in $\mathbb{B}(\mathscr{X}, \mathscr{Y})$, but we must first make precise the notions of a product space and the graph of an operator.

Proposition 5.1.13. *Let $(\mathscr{X}, \|\cdot\|_{\mathscr{X}})$ and $(\mathscr{Y}, \|\cdot\|_{\mathscr{Y}})$ be real Banach spaces. The product space*

$$\mathscr{X} \times \mathscr{Y} = \{(x,y) : x \in \mathscr{X} \wedge y \in \mathscr{Y}\}$$

equipped with the operations

$$(x_1, y_1) + (x_2, y_2) = (x_1 + x_2, y_1 + y_2), \ \forall (x_1, y_1), (x_2, y_2) \in \mathscr{X} \times \mathscr{Y},$$
$$\alpha(x, y) = (\alpha x, \alpha y), \ \forall \alpha \in \mathbb{R}, x \in \mathscr{X}, y \in \mathscr{Y},$$

and the so-called graph norm

$$\|(x,y)\|_{\mathscr{X} \times \mathscr{Y}} = \|x\|_{\mathscr{X}} + \|y\|_{\mathscr{Y}}, \tag{5.3}$$

is a Banach space.

Exercise 5.1.13. Prove Prop. 5.1.13.

Definition 5.1.14. $\mathscr{F} : \mathscr{X} \to \mathscr{Y}$ is a *closed operator* if its graph, defined by

$$\text{graph}(\mathscr{F}) = \{(x, \mathscr{F}x) : x \in \text{dom}(\mathscr{F})\},$$

is a closed set in $\mathscr{X} \times \mathscr{Y}$.

Proving that an operator $\mathscr{F} : \mathscr{X} \to \mathscr{Y}$ is closed requires that $\forall \{x_n\} \subset \text{dom}(\mathscr{F})$,

$$\left(\|x_n - x^\star\|_{\mathscr{X}} \to 0 \wedge \|\mathscr{F}x_n - y^\star\|_{\mathscr{Y}} \to 0\right) \Longrightarrow (x^\star \in \text{dom}(\mathscr{F}) \wedge y^\star = \mathscr{F}x^\star).$$

We have already encountered one example of a closed linear operator in Chapter 4. So, we know that the set of closed operators intersects $\mathbb{B}(\mathscr{X}, \mathscr{Y})$. However, neither set is contained within the other. Indeed, let $\mathscr{X}$ be a Banach space and $\mathscr{V}$ a linear subspace of $\mathscr{X}$ that is not closed. Then, the operator $I|_{\mathscr{V}}$ is bounded but not closed. (Why?) Next, consider the operator $\mathscr{F} : \mathbb{C}^1((a,b); \mathbb{R}) \to \mathbb{C}([a,b]; \mathbb{R})$ defined by $\mathscr{F}(g) = g'$ (cf. Exer. 5.1.5). We assert that $\mathscr{F}$ is a closed operator. To see this, let $\{f_n\} \subset \mathbb{C}^1((a,b); \mathbb{R})$ be such that $f_n \to g$ in $\mathbb{C}^1((a,b); \mathbb{R})$ and $\mathscr{F}(f_n) = f_n' \to h$ in $\mathbb{C}([a,b]; \mathbb{R})$. We must argue that $g \in \text{dom}(\mathscr{F})$ and $g' = h$. Indeed, $\forall t \in [a,b]$,

$$\int_a^t h(s)\,ds = \int_a^t \lim_{n\to\infty} f_n'(s)\,ds = \lim_{n\to\infty} \int_a^t f_n'(s)\,ds$$
$$= \lim_{n\to\infty} [f_n(t) - f_n(a)] = g(t) - g(a). \tag{5.4}$$

Because h is continuous, we know that g is differentiable (Why?) and $\frac{d}{dt}\int_a^t h(s)\,ds = h(t)$, which equals $g'(t)$, as desired.

Exercise 5.1.14. Define $\mathscr{B} : \text{dom}(\mathscr{B}) \subset \mathbb{C}^2((a,b); \mathbb{R}) \to \mathbb{C}([a,b]; \mathbb{R})$ by $\mathscr{B}(g) = -g''$, where $\text{dom}(\mathscr{B})$ is given by

$$\{g \in \mathbb{C}^2((a,b); \mathbb{R}) \,|\, g, g'' \in \mathbb{C}([a,b]; \mathbb{R}) \wedge g(a) = g(b) = g''(a) = g''(b) = 0\}. \tag{5.5}$$

Prove that $\mathscr{B}$ is a closed operator.

Exercise 5.1.15. Prove that if $\mathscr{B} : \text{dom}(\mathscr{B}) \subset \mathscr{X} \to \mathscr{Y}$ is a closed invertible operator, then $\mathscr{B}^{-1}$ is a closed operator.

The next theorem is another useful powerhouse result.

Theorem 5.1.15. (Closed Graph Theorem)
Let $\mathscr{F} : \text{dom}(\mathscr{F}) \subset \mathscr{X} \to \mathscr{Y}$ be a closed linear operator. If $\text{dom}(\mathscr{F})$ is closed in $\mathscr{X}$, then $\mathscr{F} \in \mathbb{B}(\mathscr{X}, \mathscr{Y})$.

Remark. In particular, if $\text{dom}(\mathscr{F}) = \mathscr{X}$, then Thrm. 5.1.15 reduces to saying "if $\mathscr{F}$ is a closed, linear operator, then $\mathscr{F}$ must be bounded."

5.1.4 Densely Defined Operators

Recall that a set $\mathscr{D} \subset \mathscr{X}$ is dense in $\mathscr{X}$ if $\mathrm{cl}_{\mathscr{X}}(\mathscr{D}) = \mathscr{X}$. Intuitively, we can get arbitrarily close (in the sense of the $\mathscr{X}$−norm) to any member of $\mathscr{X}$ with elements of $\mathscr{D}$. This notion can be used to define a so-called *densely defined operator*, as follows.

Definition 5.1.16. $\mathscr{F} : \mathrm{dom}(\mathscr{F}) \subset \mathscr{X} \to \mathscr{Y}$ is *densely defined* if $\mathrm{cl}_{\mathscr{X}}(\mathrm{dom}(\mathscr{F})) = \mathscr{X}$.

Exercise 5.1.16. Explain why, for normed spaces, it is sufficient to produce a set $\mathscr{D} \subset \mathrm{dom}(\mathscr{F})$ for which $\mathrm{cl}_{\mathscr{X}}(\mathscr{D}) = \mathscr{X}$ in order to prove that $\mathscr{F}$ is densely defined.

We will encounter this notion exclusively when $\mathscr{X}$ and $\mathscr{Y}$ are function spaces. The underlying details can sometimes be involved, but we will primarily focus on tame examples. For instance, consider the following.

Proposition 5.1.17. *The operator $\mathscr{B}$ defined in Exer. 5.1.14, now with domain $\mathbb{C}^1((a,b);\mathbb{R}) \cap \mathbb{C}([a,b];\mathbb{R})$, is linear, closed, and densely defined.*

Proof. The linearity of $\mathscr{B}$ is clear and the fact that $\mathscr{B}$ is closed can be shown as in Exer. 5.1.14. In order to argue that $\mathscr{B}$ is densely defined, we produce a set $\mathscr{D} \subset \mathrm{dom}(\mathscr{B})$ for which $\mathrm{cl}_{\mathbb{C}([a,b];\mathbb{R})}(\mathscr{D}) = \mathbb{C}([a,b];\mathbb{R})$. This is easily done because every $h \in \mathbb{C}^1((a,b);\mathbb{R}) \cap \mathbb{C}([a,b];\mathbb{R})$ has a unique Fourier representation given by

$$h(x) = \sum_{n=0}^{\infty} \lambda_n \cos\left(\frac{n\pi x}{2(b-a)}\right), \, x \in [a,b],$$

where $\{\lambda_n\} \subset \mathbb{R}$ are the Fourier coefficients of h (cf. Section 1.7.2). Because the set

$$\left\{ \sum_{n=0}^{N} \lambda_n \cos\left(\frac{n\pi x}{2(b-a)}\right) \middle| \{\lambda_n\} \subset \mathbb{R}, N \in \mathbb{N} \right\}$$

is dense in $\mathbb{C}^1((a,b);\mathbb{R}) \cap \mathbb{C}([a,b];\mathbb{R})$, we conclude that $\mathscr{B}$ is densely defined. This completes the proof. $\qquad\square$

Exercise 5.1.17. Consider the operator $\mathscr{B} : \mathrm{dom}(\mathscr{B}) \subset \mathbb{L}^2((a,b);\mathbb{R}) \to \mathbb{L}^2((a,b);\mathbb{R})$ defined by $\mathscr{B}(g) = -g''$, where

$$\mathrm{dom}(\mathscr{B}) = \left\{ g \in \mathbb{W}_2^2((a,b);\mathbb{R}) \middle| g(a) = g(b) = 0 \right\}. \qquad (5.6)$$

Is $\mathscr{B}$ densely defined on $\mathbb{L}^2((a,b);\mathbb{R})$? Compare this to Prop. 5.1.17.

5.2 Linear Semigroup Theory — Some Highlights

Simplistically speaking, the utility of the theory of strongly continuous linear semigroups in studying abstract Cauchy problems mirrors the utility of the matrix exponential in solving systems of linear ODEs. We provide a brief outline of definitions and theorems for use in the remainder of the text. Some standard references on semi-group theory are [**46, 90, 108, 123, 396**]. A thorough development of this material is provided in Volume 1.

Definition 5.2.1. A family of operators $\{S(t) : t \geq 0\}$, satisfying
i.) $S(t) \in \mathbb{B}(\mathscr{X})$, $\forall t \geq 0$,
ii.) $S(0) = I$, where I is the identity operator on $\mathscr{X}$, and
iii.) $S(t+s) = S(t)S(s)$, $\forall t, s \geq 0$ (called the *semigroup property*),
iv.) $\lim\limits_{t \to 0^+} \left\| e^{At}x - x \right\|_{\mathscr{X}} = 0$, $\forall x \in \mathscr{X}$,
is called a *strongly continuous semigroup of bounded linear operators on* $\mathscr{X}$.
(For brevity, we say that $\{S(t) : t \geq 0\}$ is a *linear C_0-semigroup on* $\mathscr{X}$.)

A linear semigroup $\{S(t) : t \geq 0\}$ as defined in Def. 5.2.1 bears a striking resemblance to the matrix exponential $\{e^{At} : t \geq 0\}$ used in Chapter 4. This is no accident, but what plays the role of the generator $\mathbf{A}$ in the present setting? Guided by Def. 4.2.5, we proceed as follows.

Let $\{S(t) : t \geq 0\}$ be a linear semigroup on $\mathscr{X}$ and define

$$\mathscr{D} = \left\{ x \in \mathscr{X} \, \middle| \, \exists \lim_{h \to 0^+} \frac{(S(h) - I)x}{h} \right\}. \tag{5.7}$$

It follows directly from the linearity of the semigroup $\{S(t) : t \geq 0\}$ that $\mathscr{D}$ is a linear subspace of $\mathscr{X}$. (Tell why.) Now, define the operator $A : \mathscr{D} \subset \mathscr{X} \to \mathscr{X}$ by

$$Ax = \lim_{h \to 0^+} \frac{(S(h) - I)x}{h}. \tag{5.8}$$

Observe that A is linear on $\mathscr{D}$. (Why?) Note that this is precisely how the generator of $\{e^{At} : t \geq 0\}$ was defined when $\mathbf{A} \in \mathbb{M}^N(\mathbb{R})$, with the notable difference being that in the finite-dimensional setting, $\mathscr{D} = \mathbb{R}^N = \mathscr{H}$, whereas in the infinite-dimensional setting we will find that $\mathscr{D}$ need only be a dense linear subspace of $\mathscr{X}$ and, in general, is not the entire space. Henceforth, we write $\mathrm{dom}(A)$ in place of $\mathscr{D}$. Formally, we have the following definition.

Definition 5.2.2. A linear operator $A : \mathrm{dom}(A) \subset \mathscr{X} \to \mathscr{X}$ defined by

$$Ax = \lim_{h \to 0^+} \frac{(S(h) - I)x}{h},$$

$$\mathrm{dom}(A) = \left\{ x \in \mathscr{X} \, \middle| \, \exists \lim_{h \to 0^+} \frac{(S(h) - I)x}{h} \right\}$$

is called an *(infinitesimal) generator* of $\{S(t) : t \geq 0\}$. We say that A *generates* $\{S(t) : t \geq 0\}$.

The similarity between the above discussion and the analogous one in Chapter 4 prompts us to introduce the notation $S(t) \equiv e^{At}$ for a linear semigroup on $\mathscr{X}$ whose generator is A.

Theorem 5.2.3. (Properties of C₀-Semigroups and their Generators)

Let $\{e^{At} : t \geq 0\}$ be a C_0-semigroup with generator $A : \mathrm{dom}(A) \subset \mathscr{X} \to \mathscr{X}$. Then,

i.) $\mathrm{dom}(A)$ *is a linear subspace of $\mathscr{X}$.*

ii.) *The mapping $t \mapsto \left\|e^{tA}\right\|_{\mathbb{B}(\mathscr{X})}$ is bounded on bounded subsets of $[0, \infty)$. Moreover, there exist constants $\omega \in \mathbb{R}$ and $M \geq 1$ such that*

$$\left\|e^{tA}\right\|_{\mathbb{B}(\mathscr{X})} \leq Me^{\omega t}, \ \forall t \geq 0. \tag{5.9}$$

iii.) *For every $x_0 \in \mathscr{X}$, the function $g: [0, \infty) \to \mathscr{X}$ defined by $g(t) = e^{At}x_0$ is continuous.*

iv.) *For every $t_0 \geq 0$, the function $h: \mathscr{X} \to \mathscr{X}$ defined by $h(x) = e^{At_0}x$ is continuous.*

v.) *For every $x_0 \in \mathscr{X}$ and $t_0 \geq 0$, $\lim\limits_{h \to 0^+} \frac{1}{h} \int_{t_0}^{t_0+h} e^{As}x_0 ds = e^{At_0}x_0$.*

vi.) *For every $x \in \mathrm{dom}(A)$ and $t \geq 0$, $e^{At}x \in \mathrm{dom}(A)$. Moreover, $\forall x \in \mathrm{dom}(A)$, the function $g: (0, \infty) \to \mathscr{X}$ defined by $g(t) = e^{At}x_0$ is in $\mathbb{C}^1\left((0, \infty); \mathrm{dom}(A)\right)$ and*

$$\frac{d}{dt}e^{At}x_0 = \underbrace{Ae^{At}x_0 = e^{At}Ax_0}_{\text{i.e., } A \text{ and } e^{At} \text{ commute}}. \tag{5.10}$$

vii.) *For every $x_0 \in \mathscr{X}$ and $t_0 \geq 0$,*

$$A\int_0^{t_0} e^{As}x_0 ds = e^{At_0}x_0 - x_0. \tag{5.11}$$

viii.) *For every $x_0 \in \mathrm{dom}(A)$ and $0 < s \leq t < \infty$,*

$$e^{At}x_0 - e^{As}x_0 = \int_s^t e^{Au}Ax_0 du = \int_s^t Ae^{Au}x_0 du. \tag{5.12}$$

ix.) *A is a densely defined operator.*

x.) *A is a closed operator.*

xi.) *If A generates two C_0-semigroups $\{S(t) : t \geq 0\}$ and $\{T(t) : t \geq 0\}$ on $\mathscr{X}$, then $S(t) = T(t)$, $\forall t \geq 0$.*

The abstract IVP

$$\begin{cases} u'(t) = Au(t), \ t > 0, \\ u(0) = u_0, \end{cases} \tag{5.13}$$

is referred to as an *abstract homogenous Cauchy problem in $\mathscr{X}$*. When does (5.13) have a solution? In fact, what do we even mean by a solution? We precisely define one notion of a solution below.

Definition 5.2.4. A function $u : [0, \infty) \to \mathscr{X}$ is a *classical solution* of (5.13) if

i.) $u \in \mathbb{C}\left([0, \infty); \mathscr{X}\right)$,

ii.) $u(t) \in \mathrm{dom}(A)$, $\forall t > 0$,

iii.) $u \in \mathbb{C}^1\left((0, \infty); \mathrm{dom}(A)\right)$,

iv.) (5.13) is satisfied.

The properties of Thrm. 5.2.3 enable us to establish the existence and uniqueness of a classical solution of (5.13) just as in the finite-dimensional case.

Theorem 5.2.5. *If $A : \mathrm{dom}(A) \subset \mathscr{X} \to \mathscr{X}$ generates a C_0-semigroup $\{e^{At} : t \geq 0\}$ on $\mathscr{X}$, then $\forall u_0 \in \mathrm{dom}(A)$, (5.13) has a unique classical solution given by $u(t) = e^{At} u_0$.*

It is necessary to assume that $u_0 \in \mathrm{dom}(A)$ in Thrm. 5.2.5. Indeed, the function $u(t) = e^{At} u_0$ might not be differentiable if $u_0 \in \mathscr{X} \setminus \mathrm{dom}(A)$. In such case, u could not formally satisfy (5.13) in the sense of Def. 5.2.4. Consider the following exercise.

Exercise 5.2.1. *Let $\mathscr{X} = \mathbb{C}_B((0,\infty);\mathbb{R})$ denote the space of continuous real-valued functions that remain bounded on $(0,\infty)$. This is a Banach space when equipped with the usual sup norm. Define $A : \mathrm{dom}(A) \subset \mathscr{X} \to \mathscr{X}$ by*

$$Af = f',$$
$$\mathrm{dom}(A) = \{f \in \mathbb{C}_B((0,\infty);\mathbb{R}) \,|\, f' \in \mathbb{C}_B((0,\infty);\mathbb{R})\}.$$

Choose $u_0 \in \mathbb{C}_B((0,\infty);\mathbb{R})$ for which u_0' does not exist for at least some $t_0 \in (0,\infty)$. *(Why is such a choice possible?) Show that for such a choice of u_0, (5.13) does not have a classical solution in the sense of Def. 5.2.4.*

The fact that we have been using $t = 0$ as the starting time of IVP (5.13) is not essential, but it does suffice, as illustrated by the following corollary. As a result, we assume henceforth that all abstract IVPs are equipped with ICs evaluated at $t = 0$.

Corollary 5.2.6. *Assume the conditions of Theorem 5.2.5. For any $t_0 > 0$, the IVP*

$$\begin{cases} u'(t) = Au(t), \ t > t_0, \\ u(t_0) = u_0 \end{cases} \tag{5.14}$$

has a unique classical solution $u : [t_0,\infty) \to \mathscr{X}$ given by $u(t) = e^{A(t-t_0)} u_0$.

Remark. The regularity of u_0 directly affects the regularity of u.

Once we reformulate the IBVP as the abstract evolution equation (5.13), we must determine whether or not the operator A generates a C_0-semigroup on our choice of space $\mathscr{X}$. If we can manage to conclude that A *does* generate a C_0-semigroup, then Thrm. 5.2.5 ensures the existence and uniqueness of a suitably regular classical solution of (5.13). As such, we need criteria that ensure an unbounded linear operator A generates a C_0-semigroup. This is given by some fairly deep theorems, the most notable of which is the *Hille-Yosida theorem*. In order to state this theorem, we need the following definition.

Definition 5.2.7. Let $A : \mathrm{dom}(A) \subset \mathscr{X} \to \mathscr{X}$ be a linear operator.
i.) The *resolvent set* of A, denoted $\rho(A)$, is the set of all complex numbers λ for which there exists an operator $R_\lambda(A) \in \mathbb{B}(\mathscr{X})$ such that

a.) For every $\forall y \in \mathscr{X}$, $R_\lambda(A)y \in \mathrm{dom}(A)$ and $(\lambda I - A)R_\lambda(A)y = y$; and
b.) for every $x \in \mathrm{dom}(A)$, $R_\lambda(A)(\lambda I - A)x = x$.
ii.) For any $\lambda \in \rho(A)$, $R_\lambda(A) = (\lambda I - A)^{-1}$ is the *resolvent operator of A*.

The underlying strategy of the proof is to approximate the operator A by a sequence of "nicer" bounded linear operators that converges to A in a sufficiently strong sense. To this end, we introduce the following sequence.

Definition 5.2.8. Let $A : \mathrm{dom}(A) \subset \mathscr{X} \to \mathscr{X}$ be a linear operator such that
i.) A is closed and densely defined, and
ii.) For every $\lambda > 0$, $\lambda \in \rho(A)$ and $\|R_\lambda(A)\|_{\mathbb{B}(\mathscr{X})} \le \frac{1}{\lambda}$.
The collection of operators $A_\lambda : \mathrm{dom}(A_\lambda) \subset \mathscr{X} \to \mathscr{X}$ defined by

$$A_\lambda x = \lambda A R_\lambda(A)x, \tag{5.15}$$

where $\lambda > 0$, is called the *Yosida approximation* of A.

The following list of properties involving the Yosida approximation of A are needed to establish the very important *Hille-Yosida theorem*, and are useful in some convergence arguments later in the text.

Proposition 5.2.9. (Properties of Resolvents and Yosida Approximations)
i.) *For every* $\lambda > 0$, $\mathrm{dom}(A_\lambda) = \mathscr{X}$.
ii.) *For every* $\lambda, \mu > 0$,
    a.) $R_\lambda(A)R_\mu(A) = R_\mu(A)R_\lambda(A)$.
    b.) *For every* $x \in \mathrm{dom}(A)$, $AR_\lambda(A)x = R_\lambda(A)Ax$.
    c.) $A_\lambda A_\mu = A_\mu A_\lambda$.
iii.) *For every* $x \in \mathscr{X}$, $\lim\limits_{\lambda \to \infty} \lambda R_\lambda(A)x = x$.
iv.) *For every* $x \in \mathscr{X}$, $A_\lambda x = \lambda^2 R_\lambda(A)x - \lambda x$.
v.) *For every* $x \in \mathrm{dom}(A)$, $\lim\limits_{\lambda \to \infty} \|A_\lambda x - Ax\|_{\mathscr{X}} = 0$.
vi.) *For every* $\lambda > 0$, A_λ *generates a U.C. contractive semigroup* $\{e^{A_\lambda t} : t \ge 0\}$ *on* $\mathscr{X}$. *As such,* $\|e^{A_\lambda t}\|_{\mathbb{B}(\mathscr{X})} \le 1$, $\forall \lambda > 0$ *and* $t \ge 0$.
vii.) *For every* $x \in \mathscr{X}$, $t \ge 0$, *and* $\lambda, \mu > 0$, $\|e^{A_\lambda t}x - e^{A_\mu t}x\|_{\mathscr{X}} \le t\|A_\lambda x - A_\mu x\|_{\mathscr{X}}$.
viii.) *For every* $t \ge 0$, $\exists T(t) \in \mathbb{B}(\mathscr{X})$ *such that* $\forall x \in \mathscr{X}$,
    $T(t)x = \lim\limits_{\lambda \to \infty} e^{A_\lambda t}x$ *uniformly on bounded intervals (in t) of* $[0, \infty)$.
ix.) $\{T(t) : t \ge 0\}$ *is a contractive* C_0-*semigroup on* $\mathscr{X}$.

We focus only the case of a contractive semigroup, for simplicity.

Theorem 5.2.10. (Hille-Yosida Theorem)
A linear operator $A : \mathrm{dom}(A) \subset \mathscr{X} \to \mathscr{X}$ *generates a contractive* C_0-*semigroup* $\{e^{At} : t \ge 0\}$ *on* $\mathscr{X}$ *if and only if*
i.) A *is closed and densely defined, and*
ii.) *For every* $\lambda > 0$, $\lambda \in \rho(A)$ *and* $\|R_\lambda(A)\|_{\mathbb{B}(\mathscr{X})} \le \frac{1}{\lambda}$.

Of course, not all C_0-semigroups on a given space $\mathscr{X}$ arising in practice are contractive. Unfortunately, Thrm. 5.2.10 does not apply to them. But, we still have the growth estimate $\left\|e^{At}\right\|_{\mathbb{B}(\mathscr{X})} \leq Me^{\omega t}$ (for some $M \geq 1$ and $\omega > 0$) that can be exploited to form an extension of the Hille-Yosida theorem that applies to *any* C_0-semigroup. Such a more general characterization result, formulated by Feller, Miyadera, and Phillips, is similar in spirit to the Hille-Yosida Theorem, but the lack of contractivity necessitates that hypothesis (ii) be replaced by a more technical counterpart. Consequently, the proof is somewhat more technical.

Theorem 5.2.11. (Feller-Miyadera-Phillips Theorem)
A linear operator $A : \text{dom}(A) \subset \mathscr{X} \to \mathscr{X}$ *generates a* C_0-*semigroup* $\left\{e^{At} : t \geq 0\right\}$
on $\mathscr{X}$ *(for which* $\left\|e^{At}\right\|_{\mathbb{B}(\mathscr{X})} \leq Me^{\omega t}$, *for some* $M \geq 1$ *and* $\omega > 0$*) if and only if*
i.) *A is closed and densely defined, and*
ii.) *For every* $\lambda > \omega$, $\lambda \in \rho(A)$ *and* $\left\|R_\lambda^n(A)\right\|_{\mathbb{B}(\mathscr{X})} \leq \frac{M}{(\lambda - \omega)^n}$, $\forall n \in \mathbb{N}$.

Monotonicity is a useful tool in analysis. This notion can be extended in a natural way to operators in a more abstract setting, like a Hilbert space or Banach space. As it turns out, this notion is closely related to the range condition (ii) of the Hille-Yosida Theorem and is often easier to verify in applications. Moreover, it is central to the development of the theory when A is a *nonlinear* unbounded operator. We begin with the following definitions connected to monotonicity formulated in a real Hilbert space.

Definition 5.2.12. An operator $B : \text{dom}(B) \subset \mathscr{H} \to \mathscr{H}$ is
i.) *accretive* if $\|Bx - By\|_{\mathscr{H}} \geq \|x - y\|_{\mathscr{H}}$, $\forall x, y \in \text{dom}(B)$,
ii.) *nonexpansive* if $\|Bx - By\|_{\mathscr{H}} \leq \|x - y\|_{\mathscr{H}}$, $\forall x, y \in \text{dom}(B)$,
iii.) *monotone* if $\langle Bx - By, x - y \rangle_{\mathscr{H}} \geq 0$, $\forall x, y \in \text{dom}(B)$,
iv.) *dissipative* if $\langle Bx, x \rangle_{\mathscr{H}} \leq 0$, $\forall x \in \text{dom}(B)$,
v.) *m-dissipative* if B is dissipative and $\forall \alpha > 0$, $\text{rng}(I - \alpha B) = \mathscr{H}$.

Examples.
1.) The operator $\mathscr{A} : \mathbb{R}^N \to \mathbb{R}^N$ defined by

$$\mathscr{A}\mathbf{x} = \begin{bmatrix} \alpha_1 & 0 & \cdots & 0 \\ 0 & \alpha_2 & & 0 \\ \vdots & & \ddots & \vdots \\ 0 & 0 & \cdots & \alpha_n \end{bmatrix} \mathbf{x}$$

where $\alpha_i \geq 0$, $1 \leq i \leq N$, is a monotone operator on $\mathbb{R}^N$. (Why?) In general, if $\mathbf{B} \in \mathbb{M}^N(\mathbb{R})$ has eigenvalues that are all nonnegative, then the operator $\mathscr{C} : \mathbb{R}^N \to \mathbb{R}^N$ defined by $\mathscr{C}\mathbf{x} = \mathbf{B}\mathbf{x}$ is monotone. (Why?)
2.) If $\mathscr{D} : \text{dom}(\mathscr{D}) \subset \mathscr{H} \to \mathscr{H}$ is nonexpansive, then $I - \mathscr{D}$ is monotone. (Why?)
3.) The operator $\mathscr{G} : \text{dom}(\mathscr{G}) \subset \mathbb{L}^2(0, a; \mathbb{R}) \to \mathbb{L}^2(0, a; \mathbb{R})$ defined by

$$\mathscr{G}f = -\frac{d^2 f}{dx^2}$$

$$\text{dom}(\mathscr{G}) = \left\{ f \in \mathbb{L}^2(0, a; \mathbb{R}) \left| \frac{df}{dx}, \frac{d^2 f}{dx^2} \in \mathbb{L}^2(0, a; \mathbb{R}) \wedge f(0) = f(a) = 0 \right. \right\}$$

is monotone because

$$\langle f, \mathcal{G} f\rangle_{L^2(0,a;\mathbb{R})} = -\int_0^a f(x)\cdot \frac{d^2 f}{dx^2}(x)dx = \int_0^a \left(\frac{df}{dx}(x)\right)^2 dx \geq 0.$$

More generally, the Laplacian $\triangle : \mathrm{dom}(\triangle) \subset L^2\left(\prod_{i=1}^n (0,a_i);\mathbb{R}\right) \to L^2\left(\prod_{i=1}^n (0,a_i);\mathbb{R}\right)$ defined by

$$\triangle f = -\left(\frac{\partial^2 f}{\partial x_1^2} + \ldots + \frac{\partial^2 f}{\partial x_n^2}\right)$$

$$\mathrm{dom}(\triangle) = \left\{ f \in L^2\left(\prod_{i=1}^n (0,a_i);\mathbb{R}\right) \,\middle|\, \frac{\partial f}{\partial x_i}, \frac{\partial^2 f}{\partial x_i^2} \in L^2\left(\prod_{i=1}^n (0,a_i);\mathbb{R}\right) \right.$$

$$\left. i = 1,\ldots,n \wedge f = 0 \,\mathrm{on}\, \partial\left(\prod_{i=1}^n (0,a_i)\right) \right\}$$

is monotone. (Prove this.)

4.) If $\mathbf{I}$ denotes the identity operator on $\mathbb{R}^N$, then $-\mathbf{I}$ is dissipative because $\forall \mathbf{x} \in \mathbb{R}^N$,

$$\langle \mathbf{x}, -\mathbf{I}\mathbf{x}\rangle_{\mathbb{R}^N} = \mathbf{x}\cdot(-\mathbf{x}) = -\|\mathbf{x}\|_{\mathbb{R}^N}^2 \leq 0.$$

5.) The operator $\mathcal{J} : \mathrm{dom}(\mathcal{J}) \subset L^2(0,a;\mathbb{R}) \to L^2(0,a;\mathbb{R})$ defined by

$$\mathcal{J} f = f',$$
$$\mathrm{dom}(\mathcal{J}) = \{f \in L^2(0,a;\mathbb{R}) \,|\, f(0) = 0\}$$

is dissipative because

$$\langle f, \mathcal{J} f\rangle_{L^2(0,a;\mathbb{R})} = \int_0^a f(s)f'(s)ds = -\frac{1}{2}(f(0))^2 = 0.$$

The following result is an equivalent formulation of the Hille-Yosida Theorem. Its benefit lies in the fact that the range condition (ii) in the Hille-Yosida Theorem is replaced by an easier-to-verify dissipativity one.

Theorem 5.2.13. (Lumer-Phillips Theorem)
A linear operator $A : \mathrm{dom}(A) \subset \mathscr{H} \to \mathscr{H}$ generates a contractive C_0-semigroup $\{e^{At} : t \geq 0\}$ on $\mathscr{H}$ if and only if
i.) *A is densely defined, and*
ii.) *A is m-dissipative.*

Suppose that a linear operator A generates a C_0-semigroup on $\mathscr{X}$ and a "sufficiently well-behaved" operator B is added to it. Intuitively, as long as $A + B$ is "relatively close" to A, it seems that $A + B$ ought to generate a C_0-semigroup on $\mathscr{X}$. Is this true? Some related questions that naturally arise are
1.) If both A and B generate C_0-semigroups on $\mathscr{X}$, say $\{e^{At} : t \geq 0\}$ and $\{e^{Bt} : t \geq 0\}$, respectively, must $A + B$ also generate a C_0-semigroup on $\mathscr{X}$?

2.) Must B generate a C_0-semigroup on $\mathscr{X}$ in order for $A + B$ to generate one?

The answer to (1) in the $\mathbb{R}^N$-setting is yes, because every bounded linear operator on $\mathbb{R}^N$ is identified with a member of $\mathbb{M}^N(\mathbb{R})$, and we know the matrix exponential is defined for all members of $\mathbb{M}^N(\mathbb{R})$. The answer to (2) is no. The combination of these two observations leads to the following proposition.

Proposition 5.2.14. (A Perturbation Result)
If $A : \operatorname{dom}(A) \subset \mathscr{X} \to \mathscr{X}$ generates a C_0-semigroup $\left\{e^{At} : t \geq 0\right\}$ on $\mathscr{X}$ and $B \in \mathbb{B}(\mathscr{X})$, then $A + B$ generates a C_0-semigroup $\left\{e^{(A+B)t} : t \geq 0\right\}$ on $\mathscr{X}$ given by $e^{(A+B)t} = \sum_{n=0}^{\infty} u_n(t)$, where $\{u_n\}$ is defined recursively by

$$\begin{cases} u_n(t) = \int_0^t e^{A(t-s)} B u_{n-1}(s) ds, \\ u_0(t) = e^{At}. \end{cases}$$

For operators A and B as in Prop. 5.2.14, it follows with the help of Thrm. 5.2.5 that the IVP

$$\begin{cases} u'(t) = (A+B)u(t), \ t > 0, \\ u(0) = u_0 \end{cases} \tag{5.16}$$

has a unique classical solution given by $u(t) = e^{(A+B)t} u_0, \ \forall t \geq 0$.

The following approximation results are useful when establishing certain convergence schemes.

Corollary 5.2.15. *Suppose that a linear operator $A : \operatorname{dom}(A) \subset \mathscr{X} \to \mathscr{X}$ generates a contractive C_0-semigroup $\left\{e^{At} : t \geq 0\right\}$ on $\mathscr{X}$. Then*

$$e^{At}x = \lim_{\lambda \to \infty} e^{A_\lambda t} x, \ \forall x \in \mathscr{X}, \tag{5.17}$$

where $\{A_\lambda : \lambda \geq 0\}$ is the Yosida approximation of A defined in (5.15).

One notable theorem connecting the convergence of a sequence of C_0-semigroups to the convergence of the sequence of their respective generators is the *Trotter-Kato Approximation* Theorem, stated below.

Theorem 5.2.16. (Trotter-Kato Approximation Theorem)
Assume that A and $\{A_n : n \in \mathbb{N}\}$ are linear operators on $\mathscr{X}$ for which $\exists M \geq 1$ and $\omega \in \mathbb{R}$ such that $\forall t \geq 0$,
***i.)** $A : \operatorname{dom}(A) \subset \mathscr{X} \to \mathscr{X}$ generates a C_0-semigroup $\left\{e^{At} : t \geq 0\right\}$ on $\mathscr{X}$ such that $\left\|e^{At}\right\|_{\mathbb{B}(\mathscr{X})} \leq M e^{\omega t}$, and*
***ii.)** For each $n \in \mathbb{N}$, $A_n : \operatorname{dom}(A_n) \subset \mathscr{X} \to \mathscr{X}$ generates a C_0-semigroup $\left\{e^{A_n t} : t \geq 0\right\}$ on $\mathscr{X}$ such that $\left\|e^{A_n t}\right\|_{\mathbb{B}(\mathscr{X})} \leq M e^{\omega t}$.*
Then, the following are equivalent:
***a.)** For all $x \in \mathscr{X}$ and $\lambda > \omega$, $\lim_{n \to \infty} R_\lambda (A_n) x = R_\lambda (A) x$.*
***b.)** For all $x \in \mathscr{X}$ and $t \geq 0$, $\lim_{n \to \infty} e^{A_n t} x = e^{At} x$ uniformly on bounded intervals in t.*

5.3 Probability Theory in the Hilbert Space Setting

Our approach to studying a PDE subject to randomness will be to reformulate it as an ODE defined on an appropriate function space. Specifically, a mapping of the form $z : [0,T] \times \mathbb{R} \times \Omega \to \mathbb{R}$ will be reformulated as another mapping $U : [0,T] \times \Omega \to \mathcal{H}$. Symbolically, the mapping

$$(t,x;\omega) \mapsto z(t,x;\omega)$$

is identified with

$$(t;\omega) \mapsto z(t,\cdot;\omega) = U(t;\omega).$$

So, $\forall t \in [0,T]$, we view $z(t,\cdot;\omega)$ as a function belonging to the space $\mathcal{H}$. As such, we must define what is meant by an $\mathcal{H}$-valued random variable and indicate how to modify the key concepts from Chapter 2 for such random variables. We provide a brief highlight of the necessary material in this section.

We begin with a complete probability space $(\Omega, \mathcal{F}, \mathcal{P})$ equipped with a natural filtration $\{\mathcal{F}_t | t \geq 0\}$ and assume that $\mathcal{H}$ is a separable Hilbert space equipped with an inner product $\langle \cdot, \cdot \rangle_{\mathcal{H}}$ and basis $\{\mathbf{e}_i | i \in \mathbb{N}\}$.

Definition 5.3.1. A function $X : \Omega \to \mathcal{H}$ is an $\mathcal{H}$-*valued random variable* if $\forall z \in \mathcal{H}$, $\omega \mapsto \langle X(\omega), z \rangle_{\mathcal{H}}$ is a real-valued random variable. The associated *probability distribution of X* is defined by

$$F_X(B) = \mathcal{P}(\{\omega \in \Omega | X(\omega) \in B\}), \forall B \in \mathcal{B}(\mathcal{H}). \tag{5.18}$$

The *expectation of X* is defined as before, namely as

$$E[X] = \int_\Omega X(\omega) d\mathcal{P}. \tag{5.19}$$

Because the integrand in (5.19) is Hilbert space-valued, we must carefully define what is meant by such an integral. Essentially the same building block used to define the integral for real-valued random variables applies here. To this end, let $A \in \mathcal{B}(\mathcal{H})$ and consider the random variable $X(\omega) = c\chi_A(\omega)$, where $c \in \mathcal{H}$. Certainly, X is a well-defined $\mathcal{H}$-valued random variable (Why?) and it is reasonable to define

$$E[X] = c\mathcal{P}(A) + 0\mathcal{P}(\Omega \setminus A) = c\mathcal{P}(A). \tag{5.20}$$

The second step of the building block process is also straightforward. Indeed, let $\{A_k : k = 1, \ldots, m\} \subset \Omega$ be a pairwise disjoint collection of events for which $\bigcup_{k=1}^m A_k = \Omega$, and let $\{c_k : k = 1, \ldots, m\} \subset \mathcal{H}$. Define the $\mathcal{H}$-valued random step function $X : \Omega \to \mathcal{H}$ by

$$X(\omega) = \sum_{k=1}^m c_k \chi_{A_k}(\omega). \tag{5.21}$$

Exercise 5.3.1. Explain why X given by (5.21) is a well-defined $\mathcal{H}$-valued random variable. How would you define $E[X]$ in this case?

Finally, because $\mathcal{H}$ is separable, it can be shown that there exists a sequence of $\mathcal{H}$-valued random step functions such that

$$\lim_{n \to \infty} \|X_n(\omega) - X(\omega)\|_{\mathcal{H}} = 0, \forall \omega \in \Omega. \tag{5.22}$$

In such case, we define $E[X]$ by

$$E[X] = \lim_{n \to \infty} \int_{\Omega} X_n(\omega) d\mathscr{P} = \int_{\Omega} X(\omega) d\mathscr{P}. \tag{5.23}$$

(Refer to **[105]** for details.)

Remarks.
1. Let X be an $\mathcal{H}$-valued random variable and $p > 2$. We often consider the expression $E\left[\|X(\cdot)\|_{\mathcal{H}}^p\right]$. Note that $\|X(\omega)\|_{\mathcal{H}}^p \in \mathbb{R}, \forall \omega \in \Omega$, and so, we are really applying the expectation in the sense of a real-valued random variable here. In such case, all properties established in Chapter 2 apply directly. In contrast, the quantity $E[X(\cdot)]$ is interpreted in the sense of (5.23).
2. The covariance can be defined for $\mathcal{H}$-valued random variables. (See **[203, 204]**.)
3. The heuristic explanation of the notion of conditional expectation provided in the one-dimensional setting applies in this more general setting with the caveat that the integral is understood in a more general sense. (Further discussion of the integral appears in the next section.)
4. An $\mathcal{H}$-valued random variable X is *Gaussian* if $\forall i \in \mathbb{N}$, the real-valued random variable $\omega \mapsto \langle X(\omega), \mathbf{e}_i \rangle_{\mathcal{H}}$ is Gaussian (in the sense of Def. 2.2.13).

It is not difficult to show that for $p \geq 2$, the space $\mathcal{L}^p(\Omega; \mathcal{H})$ given by

$$\mathcal{L}^p(\Omega; \mathcal{H}) = \left\{ X : \Omega \to \mathcal{H} \,\middle|\, E\|X(\cdot)\|_{\mathcal{H}}^p < \infty \right\} \tag{5.24}$$

equipped with the norm

$$\|X\|_{\mathcal{L}^p(\Omega; \mathcal{H})} = \left(E\|X(\cdot)\|_{\mathcal{H}}^p \right)^{\frac{1}{p}} \tag{5.25}$$

is a Banach space.

The notions of stochastic processes and martingales and their calculus also extend to this Hilbert space setting. All notions of convergence can be reformulated using $\|\cdot\|_{\mathcal{H}}$ in place of $|\cdot|$. (Convince yourself!) As such, the spaces $\mathbb{C}([0,T]; \mathcal{L}^p(\Omega; \mathcal{H}))$ and $\mathbb{L}^r((0,T); \mathcal{L}^p(\Omega; \mathcal{H}))$, for $r, p \geq 2$, can be defined using $\|\cdot\|_{\mathcal{H}}$ in place of $|\cdot|$.

Finally, we remark about more general Wiener processes. For simplicity, we shall only consider PDEs driven by $\mathbb{R}^m$-valued Wiener processes. The notion of a $\mathcal{K}$-valued Wiener process, where $\mathcal{K}$ is a separable Hilbert space, certainly exists, but the subsequent development of the Itó integral in such case becomes more delicate and requires one to appeal to more technical functional analysis that lies outside the pervue of this text. Suffice it to say that upon completion of this construction, all of the nice properties and estimates carry over as in the $\mathbb{R}^m$-setting.

5.4 Random Homogenous Linear SPDEs

We now consider some elementary models often discussed in an introductory PDEs course. An interesting recent account of progress made in the study of PDEs is provided in [66]. As we progress through the text, we will encounter increasingly more complex versions of these models and many others.

Model IV.1 Advection Equation - Pollution and Traffic Flow
Suppose we wish to study the concentration levels of a certain air pollutant over time at every point z in some enclosed region $\mathscr{D}$ of space. Let $c(z,t;\omega)$ denote the concentration of this pollutant at position z in $\mathscr{D}$ at time $t > 0$; and assume that for a given $\omega \in \Omega$, the initial distribution of pollutant concentration throughout $\mathscr{D}$ is described by the function $c(z,0;\omega) = c_0(z;\omega)$. Assume momentarily that we only account for the effect of the wind $v(z,t)$ on the concentration levels throughout $\mathscr{D}$ over time and ignore any effects due to diffusion or other atmospheric, chemical, or physical factors. Then, intuitively it would seem that as time goes on, the wind would simply "push" the initial profile $c_0(z;\omega)$ through $\mathscr{D}$ without changing its shape. How do we formally describe this phenomenon?

For simplicity, we reduce the above scenario to the one-dimensional case and take $\mathscr{D} = [0,\infty)$. Assume that the concentration is zero along the boundary of $\mathscr{D}$ (which is $\{0\}$) and, for ease of computation, assume that the wind is represented by the constant V, $\forall (z,t;\omega) \in \mathscr{D} \times [0,\infty) \times \Omega$. For any $\omega \in \Omega$, this scenario can be described by the IBVP

$$\begin{cases} \frac{\partial}{\partial t}c(z,t;\omega) &= V\frac{\partial}{\partial z}c(z,t;\omega), \ z > 0, t > 0, \omega \in \Omega, \\ c(z,0;\omega) &= c_0(z;\omega), \ z > 0, \omega \in \Omega, \\ c(0,t;\omega) &= 0, \ t > 0, \omega \in \Omega, \end{cases} \quad (5.26)$$

where $c_0(z;\cdot) \in \mathcal{L}^2(\Omega;\mathbb{R})$, $\forall z > 0$. As expected, the solution $c : \mathscr{D} \times [0,\infty) \times \Omega \to \mathbb{R}$ is given by

$$c(z,t;\omega) = c_0(z + Vt;\omega), \forall z \in \mathscr{D}, t \geq 0, \omega \in \Omega. \quad (5.27)$$

Exercise 5.4.1. Verify that (5.27) satisfies (5.26), $\forall \omega \in \Omega$, using the multivariable chain rule. Why is this solution sensible based on the underlying assumptions?

Exercise 5.4.2. Suppose that $c_0(z;\omega) = \sin(z)Y(\omega)$.
i.) If Y is $n(0,1)$, must the solution (5.27) be normally distributed, for every $t > 0$?
ii.) If Y is $b\left(3,\frac{1}{4}\right)$, what are the possible trajectories in (5.27)?

Exercise 5.4.3. Let $\{V_n : n \in \mathbb{N}\}$ be a real sequence for which $\lim_{n\to\infty} V_n = V$. For each $n \in \mathbb{N}$, consider (5.26) with V replaced by V_n. Assume that $c_0(z;\omega) = \sin(z)Y(\omega)$, where Y is $b\left(3,\frac{1}{4}\right)$. Does $\exists c \in \mathbb{C}\left([0,T];\mathcal{L}^2(\Omega;\mathbb{R})\right)$ such that $\lim_{n\to\infty} \|c_n - c\|_{\mathbb{C}} = 0$?

Explain.

Our present goal is to reformulate the IBVP (5.26) as an abstract stochastic IVP (called an *abstract stochastic evolution equation*) of the form

$$\begin{cases} \frac{d}{dt}\left(X(t;\omega)\right) = A\left(X(t;\omega)\right), \, t > 0, \omega \in \Omega, \\ X(0;\omega) = X_0(\omega), \omega \in \Omega \end{cases} \tag{5.28}$$

for some operator A in an appropriate separable Hilbert space $\mathscr{H}$. We begin with the following naive pairing of terms between (5.26) and (5.28). We will then analyze each identification in turn.

	IBVP (5.26)	Abstract IVP (5.28)	
Solution	$c : \mathscr{D} \times [0,\infty) \times \Omega \to \mathbb{R}$	$X : [0,\infty) \times \Omega \to \mathscr{H}$	(5.29)
	given by $c(z,t;\omega)$	given by $X(t;\omega)$	
Initial Condition	$c_0(z;\omega)$	$X_0(\omega)$	(5.30)
Left Side	$\frac{\partial}{\partial t}\underbrace{(\,\cdot\,)}_{\text{function of } z,t}$	$\frac{d}{dt}\underbrace{(\,\cdot\,)}_{\text{function of } t}$	(5.31)
Right Side	$V\frac{\partial}{\partial z}\underbrace{(\,\cdot\,)}_{\text{function of } z,t}$	$A\underbrace{(\,\cdot\,)}_{\text{function of } t}$	(5.32)
Boundary Condition	$c(0,t;\omega) = 0$	None	(5.33)

First, because we are attempting to reformulate a stochastic *partial* differential equation (whose solution by its very nature depends upon t, ω, and at least one other variable) as an abstract stochastic *ordinary* differential equation (whose solution depends only on t and ω), identification (5.29) suggests that for each $t_0 \geq 0$ and $\omega \in \Omega$, the term $X(t_0;\omega)$ must "contain" the information for the entire trajectory $\{c(z,t_0;\omega) : z \geq 0\}$. As such, $X(t_0;\omega)$ must itself be a function of z. We write

$$X(t_0;\omega)[z] = c(z,t_0;\omega), \, \forall z \geq 0. \tag{5.34}$$

Note that (5.30) follows from (5.34) because

$$X_0[z;\omega] = X(0;\omega)[z] = c(z,0;\omega) = c_0(z;\omega), \, \forall z \geq 0. \tag{5.35}$$

It follows from (5.34) and (5.35) that the space $\mathscr{H}$ mentioned in (5.29) must be a space of functions. But, *which* space exactly? This is a critical issue because our choice of the space $\mathscr{H}$ directly impacts the smoothness (also called *regularity*) of the solution process, as discussed in Volume 1. We will not presently linger on the subtleties involved in making this choice because using the underlying characteristics inherent to the model (physical, ecological, economical, etc.) inevitably enter into making the appropriate choice in a nontrivial manner. When information specific to the model is unavailable, we choose a convenient space for $\mathscr{H}$ that ensures all identifications involved in expressing the IBVP in the abstract form (5.28) are meaningful. The discussion of the deterministic version of (5.26) involved using the space

$$\{f \in C([0,\infty); \mathbb{R}) \, | f(0) = 0\}. \tag{5.36}$$

However, in order to study a stochastic version of (5.26) obtained by introducing noise through a Wiener process within the confines of our current theoretical framework, the space we use to reformulate the problem must be a separable Hilbert space. As such, we shall use the following larger space instead of (5.36):

$$\{f \in \mathbb{L}^2((0,\infty); \mathbb{R}) \, | f(0) = 0\}. \tag{5.37}$$

Based on the discussion leading to (5.34) and (5.35), it is quite natural that the *partial* derivative (in t) for the real-valued function c should be transformed into an *ordinary* derivative (also in t) for the $\mathscr{H}$-valued function X. So, (5.31) is reasonable (Why?), and the derivative is also an $\mathscr{H}$-valued function. (Why?)

Finally, we must handle (5.32) and (5.33). Judging from (5.34), the boundary condition

$$c(0,t;\omega) = 0, \ \forall t > 0, \ \omega \in \Omega \tag{5.38}$$

is easily transformed into

$$X(t;\omega)[0] = 0, \ \forall t > 0, \ \omega \in \Omega. \tag{5.39}$$

However, the expression (5.39) is nowhere to be found in (5.28), yet it must be accounted for in the transformation of (5.26) into (5.28). Consequently, we must define the operator A in a manner that satisfies

$$A(X(t;\omega))[\cdot] = V \frac{\partial}{\partial z} c(\cdot, t; \omega), \ \forall t > 0, \omega \in \Omega \tag{5.40}$$

<u>and</u> accounts for (5.38). It is apparent from (5.40) that the inputs of A and the corresponding outputs $A(X)$ are functions in the space $\mathscr{H}$. As such, A must be an operator from $\mathscr{H}$ into $\mathscr{H}$. But, not *all* functions in $\mathscr{H}$ should be included in dom(A) because there exist $\mathbb{L}^2$-functions f such that $f(0) \neq 0$, and such functions are not helpful in our search for a solution of (5.28). Therefore, it makes sense to restrict the domain of A to include only those functions for which (5.38) holds <u>and</u> (5.40) is defined, namely

$$A[f] = V \frac{d}{dz}[f],$$
$$\text{dom}(A) = \left\{ f \in \mathscr{H} \left| \frac{df}{dz} \in \mathscr{H} \text{ and} f(0) = 0 \right. \right\}. \tag{5.41}$$

Using (5.34) through (5.41) enables us to successfully reformulate (5.26) as the abstract evolution equation (5.28) in the space defined in (5.37), for each $\omega \in \Omega$. The benefit of doing this is that the theory of abstract Cauchy problems outlined in

Section 5.3 is applicable, as long as $(A, \text{dom}(A))$ generates a C_0-semigroup on $\mathscr{H}$. This turns out to be true (see [47, 311]) and, in fact,

$$X(t; \omega) = e^{At} X_0(\omega), \text{ a.s. } [\mathscr{P}], \tag{5.42}$$

where

$$e^{At} f[z; \omega] = f(z + Vt; \omega).$$

Model V.1 The Many Faces of Diffusion — Heat Conduction

Classical diffusion theory originated in 1855 with the work of physiologist Adolf Fick. The premise is simply that a diffusive substance (e.g., heat, gas, virus) will move from areas of high level of concentration toward areas of lower concentration. As such, in the absence of other factors (like advection or external forcing terms), we expect that if the substance diffuses only over a bounded region, its concentration would, over time, become uniformly distributed throughout the region.

This phenomenon arises in many disparate settings. Some common areas include intersymbol distortion of a pulse transmitted along a cable [273, 408], pheromone transport (emitted by certain species to identify mates) [48], migratory patterns of moving herds [137, 241, 319, 337], the spread of infectious disease through populated areas [205, 416], and the dispersion of salt through water [255]. We will consider different interpretations of diffusion (with added complexity) as the opportunity arises. We begin with a well-known classical model of heat conduction in one and two dimensions.

Consider a one-dimensional rod of length a with uniform properties and cross-sections. Assuming that no heat is generated and the surface is insulated, the homogenous heat equation describes the evolution of temperature throughout the rod over time. This equation, coupled with the initial profile, yields the IVP

$$\begin{cases} \frac{\partial}{\partial t} z(x, t; \omega) = k \frac{\partial^2}{\partial x^2} z(x, t; \omega), \ 0 < x < a, t > 0, \omega \in \Omega, \\ z(x, 0; \omega) = z_0(x; \omega), \ 0 < x < a, \omega \in \Omega, \end{cases} \tag{5.43}$$

where $z(x, t; \omega)$ represents the temperature at position x along the rod at time t and k is a proportionality constant depending on the thermal conductivity and material density, for a fixed $\omega \in \Omega$, and $z_0(x; \cdot) \in \mathcal{L}^2(\Omega; \mathbb{R})$, $\forall x \in [0, a]$. A very readable account of the derivation of this deterministic heat equation from basic physical principles can be found in [336].

A complete description of this phenomenon requires that we prescribe what happens to the temperature on the boundary of the rod. This can be done in many naturally occurring ways, some of which are described below.

1. Temperature is held constant along the boundary of the rod:

$$z(0, t; \omega) = C_1 \text{ and } z(a, t; \omega) = C_2, \ \forall t > 0, \omega \in \Omega. \tag{5.44}$$

2. Temperature is controlled along the boundary of the rod, but changes with time:

$$z(0, t; \omega) = C_1(t; \omega) \text{ and } z(a, t; \omega) = C_2(t; \omega), \ \forall t > 0, \omega \in \Omega. \tag{5.45}$$

3. Heat flow rate is controlled along the boundary of the rod:

$$\frac{\partial z}{\partial x}(0,t;\omega) = C_1(t;\omega) \text{ and } \frac{\partial z}{\partial x}(a,t;\omega) = C_2(t;\omega), \ \forall t > 0, \ \omega \in \Omega. \quad (5.46)$$

4. Convection (governed by Newton's law of heating and cooling):

$$C_1 z(0,t;\omega) + C_2 \frac{\partial z}{\partial x}(0,t;\omega) = C_3(t;\omega), \ \forall t > 0, \ \omega \in \Omega,$$

$$\overline{C_1} z(a,t;\omega) + \overline{C_2} \frac{\partial z}{\partial x}(a,t;\omega) = \overline{C_3}(t;\omega), \ \forall t > 0, \ \omega \in \Omega. \quad (5.47)$$

Boundary conditions (BCs) of the forms (5.44) and (5.45) are called *Dirichlet* BCs, while those of type (5.46) are called *Neumann* BCs. If the constants/functions C_i are zero, the BCs are called *homogenous*; otherwise, they are *nonhomogenous*. We can use a mixture of the types of BCs in the formulation of an IBVP. For instance, a homogenous Dirichlet BC can be imposed at one end of the rod and a nonhomogenous Neumann BC at the other end.

We first consider the IBVP formed by coupling (5.43) with the homogenous Dirichlet BCs

$$z(0,t;\omega) = z(a,t;\omega) = 0, \ \forall t > 0, \ \omega \in \Omega. \quad (5.48)$$

For any $\omega \in \Omega$, the solution can be constructed as in the deterministic case using the standard *separation of variables method* involving Fourier series (cf. Section 1.7.2 and **[111, 127, 138, 296, 336]**). As in Volume 1, doing so yields the solution of this IBVP as

$$z(x,t;\omega) = \sum_{m=1}^{\infty} \left(\frac{2}{a} \int_0^a z_0(y;\omega) \sin\left(\frac{m\pi}{a}y\right) dy \right) e^{-\left(\frac{m\pi}{a}\right)^2 kt} \sin\left(\frac{m\pi}{a}x\right) \quad (5.49)$$

$$= \sum_{m=1}^{\infty} \frac{2}{a} e^{-\left(\frac{m\pi}{a}\right)^2 kt} \left\langle z_0(\cdot;\omega), \sin\left(\frac{m\pi}{a}\cdot\right) \right\rangle_{\mathbb{L}^2(0,a;\mathbb{R})} \sin\left(\frac{m\pi}{a}x\right),$$

where $0 < x < a, t > 0$, and $\omega \in \Omega$.

We now transform this IBVP into an abstract stochastic evolution equation of the form (5.28). To this end, let $\mathscr{H} = \mathbb{L}^2(0,a;\mathbb{R})$, assume that $z_0(\cdot) \in \mathcal{L}^2(\Omega;\mathscr{H})$, and identify the solution and IC, respectively, by

$$X(t;\omega)[x] = z(x,t;\omega), \ 0 < x < a, t > 0, \ \omega \in \Omega, \quad (5.50)$$

$$X_0[x;\omega] = X(0)[x;\omega] = z(x,0;\omega) = z_0(x;\omega), \ 0 < x < a, \ \omega \in \Omega. \quad (5.51)$$

Define the operator $A : \text{dom}(A) \subset \mathbb{L}^2(0,a;\mathbb{R}) \rightarrow \mathbb{L}^2(0,a;\mathbb{R})$ by

$$A[f] = k \frac{d^2}{dx^2}[f], \quad (5.52)$$

$$\text{dom}(A) = \left\{ f \in \mathscr{H} \left| \exists \frac{df}{dx}, \frac{d^2 f}{dx^2}, \frac{d^2 f}{dx^2} \in \mathscr{H}, \text{ and } f(0) = f(a) = 0 \right. \right\}.$$

Identifying the time derivatives in the same manner as in Model IV.I, we see that using (5.50) through (5.52) yields a reformulation of the given IBVP into the form (5.28) in the separable Hilbert space $\mathbb{L}^2(0,a;\mathbb{R})$.

Remark. The domain specified in (5.52) is often written more succinctly using the Sobolev space $\mathbb{H}^2(0,a)$ (cf. (1.67)); indeed, it can be expressed equivalently as

$$\text{dom}(A) = \{f \in \mathbb{H}^2(0,a;\mathbb{R}) \,|\, f(0) = f(a) = 0\}.$$

Exercise 5.4.4.
i.) Use the separation of variables method to show that for any fixed $\omega \in \Omega$, the solution of the IBVP obtained by coupling (5.43) instead with the homogenous Neumann BCs

$$\frac{\partial z}{\partial x}(0,t;\omega) = \frac{\partial z}{\partial x}(a,t;\omega) = 0, \; \forall t > 0, \tag{5.53}$$

is given by

$$z(x,t;\omega) = \sum_{m=0}^{\infty} \left(\frac{2}{a}\int_0^a z_0(y;\omega)\cos\left(\frac{m\pi}{a}y\right)dy\right)e^{-\left(\frac{m\pi}{a}\right)^2 kt}\cos\left(\frac{m\pi}{a}x\right). \tag{5.54}$$

ii.) Simplify (5.54) when $z_0(x;\omega) = xY(\omega)$, where $Y(\cdot)$ is $b\left(2,\frac{1}{2}\right)$.
iii.) Transform the IBVP described in (i) into an abstract stochastic evolution equation of the form (5.28). Clearly define all identifications.

Next, we consider a similar model for heat conduction in a two-dimensional rect-angular plate composed of an isotropic, uniform material. Assuming that the temper-ature is zero along the boundary of the rectangle, the IBVP describing the transient temperature at every point on the plate over time is given by

$$\begin{cases} \frac{\partial z}{\partial t}(x,y,t;\omega) = k\left(\frac{\partial^2 z}{\partial x^2}(x,y,t;\omega) + \frac{\partial^2 z}{\partial y^2}(x,y,t;\omega)\right), \\ z(x,y,0;\omega) = z_0(x,y;\omega), \\ z(x,0,t;\omega) = 0 = z(x,b,t;\omega), \\ z(0,y,t;\omega) = 0 = z(a,y,t;\omega), \end{cases} \tag{5.55}$$

where $0 < x < a, 0 < y < b, t > 0, \omega \in \Omega$; $z(x,y,t;\omega)$ represents the temperature at the point (x,y) on the plate at time t corresponding to $\omega \in \Omega$; and $z_0(x,y,\cdot) \in \mathscr{L}^2(\Omega;\mathbb{R})$, $\forall 0 < x < a, 0 < y < b$. Again, using the separation of variables method (now as it applies to the two-dimensional setting) yields

$$z_0(x,y) = z(x,y,0) = \sum_{m=1}^{\infty}\sum_{n=1}^{\infty} b_{mn}\sin\left(\frac{m\pi}{a}x\right)\sin\left(\frac{n\pi}{b}y\right)e^{-\left(\left(\frac{m\pi}{a}\right)^2+\left(\frac{n\pi}{b}\right)^2\right)kt}, \tag{5.56}$$

$\forall 0 < x < a, 0 < y < b$, a.s. $[\mathscr{P}]$, where, assuming that $z_0(\cdot)$ is sufficiently smooth,

$$b_{mn} = \frac{4}{ab}\int_0^b\int_0^a z_0(v_1,v_2;\omega)\sin\left(\frac{m\pi}{a}v_1\right)\sin\left(\frac{n\pi}{b}v_2\right)dv_1 dv_2, \; m,n \in \mathbb{N}. \tag{5.57}$$

Exercise 5.4.5.

i.) Formulate (5.55) as an abstract stochastic evolution equation. Proceed by making suitable modifications to the approach used in the one-dimensional case.

ii.) Obtain an estimate for $E \|X(t;\cdot)\|_{\mathscr{H}}^p$, where $p > 2$, where X is a mild solution of the IVP in (i).

We again make the connection between the semigroup generated by A on $\mathscr{H}$ and the solution of the IBVP. We know that the form of the solution is (5.45). We would like a nice representation formula for e^{At}. To this end, for each $t \geq 0$, define the operator $e^{At} : \mathbb{L}^2(0,a;\mathbb{R}) \to \mathbb{L}^2(0,a;\mathbb{R})$ by

$$e^{At}[f][x] = \sum_{m=0}^{\infty} \left(\frac{2}{a} \int_0^a f(y) \cos\left(\frac{m\pi}{a}y\right) dy \right) e^{-\left(\frac{m\pi}{a}\right)^2 kt} \cos\left(\frac{m\pi}{a}x\right). \qquad (5.58)$$

We claim that $\left\{ e^{At} : t \geq 0 \right\}$ is a linear C_0-semigroup on $\mathbb{L}^2(0,a;\mathbb{R})$.

First, we show that $e^{At} \in \mathbb{B}(\mathbb{L}^2(0,a;\mathbb{R}))$. Let $t \geq 0$. For any $f,g \in \mathbb{L}^2(0,a;\mathbb{R})$, applying the linearity of the integral and convergent series immediately yields

$$e^{At}[f+g][x] = e^{At}[f][x] + e^{At}[g][x], \ 0 < x < a.$$

This proves linearity of e^{At}. As for boundedness, let $f \in \mathbb{L}^2(0,a;\mathbb{R})$. Using standard inequalities from Section 1.10 and properties of convergent series with the fact that $\sup \left\{ \left| \cos\left(\frac{m\pi}{a}y\right) \right| : y \in [0,a] \right\} \leq 1$ yields

$$\|e^{At}[f]\|_{\mathbb{L}^2(0,a;\mathbb{R})}^2 \leq 2 \int_0^a \left[\frac{2}{a} \|f\|_{\mathbb{L}^2(0,a;\mathbb{R})}^2 + \left(\frac{2}{a}\right)^2 \|f\|_{\mathbb{L}^2(0,a;\mathbb{R})}^2 M \sum_{m=1}^{\infty} e^{-\left(\frac{m\pi}{a}\right)^2 2kt} \right] dx$$

$$\leq \overline{M} \|f\|_{\mathbb{L}^2(0,a;\mathbb{R})}^2 < \infty, \qquad (5.59)$$

for some positive constants M and $\overline{M}$ depending on a,k,t, and the convergent series $\sum_{m=1}^{\infty} e^{-\left(\frac{m\pi}{a}\right)^2 2kt}$. This establishes boundedness, so that Def. 5.2.1(i) has been shown.

Next, let $f \in \mathbb{L}^2(0,a;\mathbb{R})$. We use the fact that $\left\{ \sqrt{\frac{2}{a}} \cos\left(\frac{m\pi}{a}\cdot\right) : m \in \mathbb{N} \cup \{0\} \right\}$ is an orthonormal basis of $\mathbb{L}^2(0,a;\mathbb{R})$ to see that

$$e^{A(0)}[f][x] = \sum_{m=0}^{\infty} \left(\frac{2}{a} \int_0^a f(y) \cos\left(\frac{m\pi}{a}y\right) dy \right) \cos\left(\frac{m\pi}{a}x\right)$$

$$= \sum_{m=0}^{\infty} \left\langle f(\cdot), \sqrt{\frac{2}{a}} \cos\left(\frac{m\pi}{a}\cdot\right) \right\rangle_{\mathbb{L}^2(0,a;\mathbb{R})} \sqrt{\frac{2}{a}} \cos\left(\frac{m\pi}{a}x\right)$$

$$= f(x).$$

Hence, $e^{A(0)} = I$, where I is the identity operator on $\mathbb{L}^2(0,a;\mathbb{R})$.

Finally, using the fact that $e^{-\left(\frac{m\pi}{a}\right)^2 k(t_1+t_2)} = e^{-\left(\frac{m\pi}{a}\right)^2 kt_1} \cdot e^{-\left(\frac{m\pi}{a}\right)^2 kt_2}$, it follows immediately that $e^{A(t_1+t_2)}[f][x] = e^{At_2}\left(e^{At_1}[f][x]\right), \forall f \in \mathbb{L}^2(0,a;\mathbb{R})$. This establishes the

semigroup property. The strong continuity is not difficult to verify. So, we have shown that $\{e^{At} : t \geq 0\}$ is a strongly continuous linear semigroup on $\mathbb{L}^2(0,a;\mathbb{R})$.
□

Exercise 5.4.6. Equip (5.55) with the homogenous Neumann BCs

$$\frac{\partial z}{\partial x}(0,y,t;\omega) = \frac{\partial z}{\partial x}(a,y,t;\omega) = 0, 0 < y < b, t > 0, \omega \in \Omega,$$

$$\frac{\partial z}{\partial y}(x,0,t;\omega) = \frac{\partial z}{\partial y}(x,b,t;\omega) = 0, 0 < x < a, t > 0, \omega \in \Omega. \tag{5.60}$$

i.) Solve the resulting IBVP using the separation of variables method.
ii.) Formulate the IBVP as an abstract stochastic evolution equation.

Exercise 5.4.7. Construct a stochastic IBVP for heat conduction on an n-dimensional rectangular plate $[0,a_1] \times \ldots \times [0,a_n]$ equipped with homogenous Neumann BCs. Without going through all of the computations, conjecture a form of the solution. How would you formulate this IBVP as an abstract stochastic evolution equation?

Remark. The operators A used to formulate all of the above heat conduction IB-VPs abstractly are forms of the *Laplacian operator* and are often denoted using the symbol $\triangle$.

Model VI.1 Fluid Flow Through Porous Media
The following model is a special case of a so-called *Sobolev-type* IBVP arising in the study of thermodynamics **[216]**, fluid flow through fissured rocks **[39]**, soil mechanics **[167, 237]**, and consolidation of clay **[307, 378, 403, 408]**. We shall investigate such models more thoroughly in Chapter 9. For now, we consider

$$\begin{cases} \frac{\partial}{\partial t}\left(z(x,t;\omega) - \frac{\partial^2}{\partial x^2}z(x,t;\omega)\right) = \frac{\partial^2}{\partial x^2}z(x,t;\omega), 0 < x < \pi, t > 0, \omega \in \Omega, \\ z(x,0;\omega) = z_0(x;\omega), 0 < x < \pi, \omega \in \Omega, \\ z(0,t;\omega) = z(\pi,t;\omega) = 0, t > 0, \omega \in \Omega, \end{cases}$$

$$\tag{5.61}$$

where $z_0(x,t;\cdot) \in \mathcal{L}^2(\Omega;\mathbb{R})$, $\forall x \in [0,\pi], t > 0$. The main difference between (5.61) and (5.43) is the presence of the term $-\frac{\partial^2 z}{\partial x^2}(x,t;\omega)$, which initially hinders our effort to transform (5.61) into the abstract form (5.28). As before, let $\mathcal{H} = \mathbb{L}^2(0,\pi;\mathbb{R})$ and define the operators $A : \mathrm{dom}(A) \subset \mathcal{H} \to \mathcal{H}$ and $B : \mathrm{dom}(B) \subset \mathcal{H} \to \mathcal{H}$ as follows:

$$A[f] = f'', \mathrm{dom}(A) = \left\{f \in \mathbb{H}^2(0,\pi;\mathbb{R}) \,|\, f(0) = f(\pi) = 0\right\},$$

$$B[f] = f - f'', \mathrm{dom}(B) = \mathrm{dom}(A). \tag{5.62}$$

Making the identification $X(t)[x;\omega] = z(x,t;\omega)$ enables us to reformulate (5.61) as the following abstract stochastic evolution equation in $\mathbb{L}^2(0,\pi;\mathbb{R})$:

$$\begin{cases} \frac{d}{dt}(BX(t;\omega)) = A(X(t;\omega)), t > 0, \omega \in \Omega, \\ X(0;\omega) = X_0(\omega), \omega \in \Omega. \end{cases} \tag{5.63}$$

Exercise 5.4.8. Intuitively, what would be the natural thing to *try* to do in order to further express (5.63) in the form (5.28)? What conditions are needed to justify such a transformation?

We will study the intricacies of such problems in Chapter 9. For the moment, let us just say A and B must be compatible in order to facilitate the further transition to the form (5.28). Moreover, the solution of (5.61) is given by

$$z(x,t;\omega) = \sum_{m=1}^{\infty} e^{-\left(\frac{m^2}{m^2+1}\right)^2 t} \frac{-m^4}{m^2+1} \left\langle z_0(\cdot;\omega), \sqrt{\frac{2}{\pi}} \sin(m\cdot) \right\rangle_{L^2(0,\pi;\mathbb{R})} \sqrt{\frac{2}{\pi}} \sin(mx),$$

(5.64)

where $0 < x < \pi, t > 0$, and $\omega \in \Omega$.

Exercise 5.4.9. Consider the solution of (5.61). For each $t \geq 0$, define the operator $e^{At} : L^2(0,\pi;\mathbb{R}) \to L^2(0,\pi;\mathbb{R})$ by

$$e^{At}[f][x;\omega] = \sum_{m=1}^{\infty} e^{-\left(\frac{m^2}{m^2+1}\right)^2 t} \left\langle f(\cdot), \sqrt{\frac{2}{\pi}} \sin(m\cdot) \right\rangle_{L^2(0,\pi;\mathbb{R})} \sqrt{\frac{2}{\pi}} \sin(mx).$$

Prove that $\{e^{At} : t \geq 0\}$ is a linear C_0-semigroup on $L^2(0,\pi;\mathbb{R})$.

Summarizing, a viable approach to studying linear homogenous PDEs where randomness is introduced through the initial conditions is to apply the usual deterministic theory for each $\omega \in \Omega$ and to interpret the resulting solution a.s. $[\mathscr{P}]$. That said, the more interesting scenario is when randomness is introduced via a white noise process. For instance, a simple advection model now perturbed by a one-dimensional white noise process $c(z,t;\omega)\frac{dW}{dt}(t;\omega)$ could be described by the following stochastic IBVP:

$$\begin{cases} \frac{\partial}{\partial t}c(z,t;\omega) = V\frac{\partial}{\partial z}c(z,t;\omega) + c(z,t;\omega)\frac{dW}{dt}(t;\omega), \ z > 0, t > 0, \omega \in \Omega, \\ c(z,0;\omega) = c_0(z;\omega), \ z > 0, \omega \in \Omega, \\ c(0,t;\omega) = 0, \ t > 0, \omega \in \Omega. \end{cases}$$

(5.65)

Likewise, incorporating two independent white noise processes into the equation portion of the two-dimensional diffusion IBVP (5.55) yields the stochastic PDE

$$\frac{\partial z}{\partial t}(x,y,t;\omega) = k\Delta z(x,y,t;\omega) + \alpha_1 z(x,y,t;\omega)\frac{dW_1}{dt} + \alpha_2 z(x,y,t;\omega)\frac{dW_2}{dt}.$$

(5.66)

We are interested in the integrated forms of (5.65) and (5.66). Formally, the differential form of (5.65) is

$$\partial c = \left(V\frac{\partial c}{\partial z}\right)\partial t + cdW,$$

(5.67)

and the subsequent integrated form of (5.67) on $(0,t)$ is given by

$$c(z,t;\omega) = c_0(z;\omega) + \int_0^t V\frac{\partial c}{\partial z}(z,s;\omega)ds + \int_0^t c(z,s;\omega)dW(s).$$

(5.68)

Similarly, the integrated form of (5.66) is given by

$$z(x,y,t;\omega) = z_0(x,y;\omega) + k \int_0^t \triangle z(x,y,s;\omega)ds \tag{5.69}$$

$$+\alpha_1 \int_0^t z(x,y,s;\omega)dW_1(s) + \alpha_2 \int_0^t z(x,y,s;\omega)dW_2(s).$$

Because we want to subsume the study of such stochastic PDEs as special cases of our theory of abstract SEEs similar to the one developed in the previous two chapters (but now in a Hilbert space), we must consider the abstract formulation of each of these PDEs in some Hilbert space $\mathscr{H}$. Loosely speaking, using the same identifications obtained when reformulating the IBVPs (5.26) and (5.43) as the abstract stochastic evolution equation (5.28) prompts us to express (5.68) abstractly as

$$X(t;\omega) = X_0(\omega) + \int_0^t AX(s;\omega)ds + \int_0^t X(s;\omega)dW(s) \tag{5.70}$$

in $\mathscr{H} = \mathbb{L}^2\left(0,\infty;\mathscr{L}^2\left(\Omega;\mathbb{R}\right)\right)$, and (5.69) as

$$X(t;\omega) = X_0(\omega) + \int_0^t AX(s;\omega)ds + \sum_{i=1}^2 \int_0^t \alpha_i X(s;\omega)dW_i(s) \tag{5.71}$$

in $\mathscr{H} = \mathbb{L}^2\left((0,a) \times (0,b);\mathscr{L}^2\left(\Omega;\mathbb{R}\right)\right)$. Note that in both cases X is an $\mathscr{H}$-valued random variable. So, we are now confronted with the same questions as in the $\mathbb{R}^N$-setting, namely how precisely are the integrals on the right-hand sides of (5.70) and (5.71) defined? We answer this question in the next section.

5.5 Bochner and It Integrals

In this section, we precisely define integrals $\int_0^t g(s;\omega)ds$ and $\int_0^t g(s;\omega)dW(s)$, where $g : [0,T] \times \Omega \to \mathscr{H}$ is an $\mathscr{H}$-valued random variable.

5.5.1 The Bochner Integral for $\mathscr{H}$-Valued Stochastic Processes

Defining $\int_0^t g(s;\omega)ds$ involves constructing a generalization of the Lebesgue integral (and so, the Riemann integral as well) to one that is applicable to $\mathscr{H}$-valued stochastic processes. A thorough treatment reveals that this integral satisfies the same basic properties as the Riemann integral. Indeed, for a stochastic process $g : [0,T] \times \Omega \to \mathscr{H}$, the process used to define $\int_0^t g(s;\omega)ds$ is as follows.

Step 1 (Partition): Let $n \in \mathbb{N}$. Divide $[0,T]$ into n subintervals using $0 = t_0 < t_1 <$

$\ldots < t_{n-1} < t_n = T$. The set $\mathscr{P} = \{t_0, t_1, \ldots, t_n\}$ is a *partition* of $[0,T]$. For convenience, let

$$\Delta t_i = t_i - t_{i-1}, \forall 1 \leq i \leq n,$$
$$\|\mathscr{P}\| = \max\{\Delta t_i : 1 \leq i \leq n\}.$$

<u>Step 2 (Approximation)</u>: For every $i \in \{1, \ldots, n\}$, choose $t_i^\star \in [t_{i-1}, t_i]$ and form the approximation

$$\underbrace{\underbrace{g(t_i^\star; \omega)}_{\text{in } \mathscr{H}} \underbrace{\Delta t_i}_{\text{in } \mathbb{R}}}_{\text{in } \mathscr{H}}, \forall \omega \in \Omega. \tag{5.72}$$

<u>Step 3 (Sum)</u>: Sum the approximations in (5.72) over $i \in \{1, \ldots, n\}$ to obtain

$$\underbrace{\sum_{i=1}^{n} \underbrace{g(t_i^\star; \omega) \Delta t_i}_{\text{in } \mathscr{H}}}_{\text{in } \mathscr{H}}, \forall \omega \in \Omega. \tag{5.73}$$

<u>Step 4 (Limit)</u>: Take the limit as $\|\mathscr{P}\| \to 0$ in (5.73) to obtain

$$\underbrace{\lim_{\|\mathscr{P}\| \to 0} \underbrace{\sum_{i=1}^{n} g(t_i^\star; \omega) \Delta t_i}_{\text{in } H}}_{\text{in } \mathscr{H}}, \forall \omega \in \Omega. \tag{5.74}$$

Definition 5.5.1. Let $g : [0,T] \times \Omega \to \mathscr{H}$ be an $\mathscr{H}$-valued random variable. If the limit in (5.74) exists, then it belongs to $\mathscr{H}$ (by completeness) and we say that g is *(Bochner) integrable* on $[0,T]$. We denote the limiting value by $\int_0^t g(s; \omega) ds$ and call it the *Bochner integral of g on* $[0,T]$.

Verifying the last step in the above construction is delicate. (Refer to [**37**] for details.) Suffice it to say that the process is well-defined and that the properties listed in Prop. 3.2.1 hold in this more general setting.

Remarks.
1. We will frequently consider integrals of the form $\int_0^t \|g(s; \omega)\|_{\mathscr{H}} ds$. Note that because the norm $\|\cdot\|_{\mathscr{H}}$ is a real-valued function, this integral is really just a one-dimensional Lebesgue integral, <u>not</u> a Bochner integral.
2. Fubini's Theorem applies in this setting as well, resulting in

$$E \int_0^t g(s; \cdot) ds = \int_0^t E[g(s; \cdot)] ds. \tag{5.75}$$

3. Let $\mathscr{L} \in \mathbb{B}(\mathscr{H})$. It can be shown that

$$\mathscr{L} \int_0^t g(s; \omega) ds = \int_0^t \mathscr{L}[g(s; \omega)] ds. \tag{5.76}$$

(Try arguing this using the building block approach, together with the linearity of $\mathscr{L}$.)

4. The following estimate is the analog of (4.54):

$$E\left\|\int_0^t g(s;\cdot)ds\right\|_{\mathscr{H}}^p \leq E\left[\left(\underbrace{\int_0^t \|g(s;\cdot)\|_{\mathscr{H}}\,ds}_{Real-valued\ Lebesgue\ integral}\right)^p\right]$$

$$\leq t^{\frac{p}{q}} E \int_0^t \|g(s;\cdot)\|_{\mathscr{H}}^p\,ds \qquad (5.77)$$

$$= t^{\frac{p}{q}} \int_0^t E\,\|g(s;\cdot)\|_{\mathscr{H}}^p\,ds.$$

5.5.2   The Itó Integral for $\mathscr{H}$-Valued Stochastic Processes

A thorough discussion of this construction can be found in **[105, 203, 204, 267, 338]**.

The formal construction of the Itó integral driven by a Hilbert space-valued Wiener process is rather technical. For our purposes, we simply need assurance that the main properties and estimates of the Itó integral carry over to the present setting. As such, we shall proceed by analogy with the construction of the Itó integral in the $\mathbb{R}^N$-setting and only highlight the main ideas involved.

Let $\mathbf{W}(t) = \langle W_1(t), \ldots, W_m(t)\rangle$ be an m-dimensional Wiener process and suppose for the moment that $\mathbf{g} : [0,t] \times \Omega \to \mathscr{B}_0\left(\mathbb{R}^m, \mathbb{R}^N\right)$ belongs to $\mathscr{U}_N$. Then, the integral $\int_0^t \mathbf{g}(s;\omega)d\mathbf{W}(s)$ is well-defined and satisfies the following two properties.

1. (Itó Isometry)

$$E\left\|\int_0^t \mathbf{g}(s;\cdot)d\mathbf{W}(s)\right\|_{\mathbb{R}^N}^2 = \int_0^t E\,\|\mathbf{g}(s;\cdot)\|_{\mathscr{B}_0\left(\mathbb{R}^m,\mathbb{R}^N\right)}^2\,ds. \qquad (5.78)$$

2. (p^{th} Moment Estimate) For every $p > 2$, there exists $\zeta_{\mathbf{g}}(t,p) > 0$ such that

$$E\left\|\int_0^t \mathbf{g}(s;\cdot)d\mathbf{W}(s)\right\|_{\mathbb{R}^N}^p \leq \zeta_{\mathbf{g}}(t,p)\int_0^t E\,\|\mathbf{g}(s;\cdot)\|_{\mathscr{B}_0\left(\mathbb{R}^m,\mathbb{R}^N\right)}^p\,ds. \qquad (5.79)$$

If we want these two properties to hold in a more general setting, how do we define the appropriate space $\mathscr{B}_0$ when $\mathbf{g}$ is $\mathscr{H}$-valued and what plays the role of the space $\mathscr{U}_N$?

Recall that the space $\mathscr{B}_0\left(\mathbb{R}^m, \mathbb{R}^N\right)$ consists of the bounded linear operators $\mathbf{g} : \mathbb{R}^m \to \mathbb{R}^N$ for which

$$\|\mathbf{g}\|_{\mathscr{B}_0} = \text{trace}\left(\mathbf{g}\mathbf{g}^T\right) = \sum_{i=1}^N \sum_{j=1}^m g_{ij}^2 = \sum_{j=1}^m E\,\|\mathbf{g}(\mathbf{e}_j)\|_{\mathbb{R}^N}^2 < \infty, \qquad (5.80)$$

where $\{\mathbf{e}_j : j = 1, \ldots, m\}$ is an orthonormal basis of $\mathbb{R}^m$. How do we modify this space to accommodate more general random variables $\mathbf{g} : \mathbb{R}^m \to \mathscr{H}$? Well, note that we still only consider an m-dimensional Wiener process $\mathbf{W}(t)$, for simplicity, but we need to replace $\mathbb{R}^N$ by the separable Hilbert space $\mathscr{H}$. Mimicking the above construction, we define the space

$$\mathscr{B}_0(\mathbb{R}^m, \mathscr{H}) = \left\{ \mathbf{g} : \mathbb{R}^m \to \mathscr{H} \,|\, \mathbf{g} \text{ is linear and } \sum_{j=1}^{m} \|\mathbf{g}(\mathbf{e}_j)\|_{\mathscr{H}}^2 < \infty \right\}, \qquad (5.81)$$

where $\{\mathbf{e}_j : j = 1, \dots, m\}$ is an orthonormal basis of $\mathbb{R}^m$. It turns out that $\mathscr{B}_0(\mathbb{R}^m, \mathscr{H})$ equipped with the inner product

$$\langle g, h \rangle_{\mathscr{B}_0} = \sum_{j=1}^{m} \langle g(\mathbf{e}_j), h(\mathbf{e}_j) \rangle_{\mathscr{H}} \qquad (5.82)$$

and the induced norm

$$\|\mathbf{g}\|_{\mathscr{B}_0}^2 = \sum_{j=1}^{m} \|\mathbf{g}(\mathbf{e}_j)\|_{\mathscr{H}}^2 \qquad (5.83)$$

is a separable Hilbert space.

Remark. More generally, if we were to consider a $\mathscr{K}$-valued Wiener process, where $\mathscr{K}$ is a separable Hilbert space, then the space $\mathscr{B}_0(\mathscr{K}, \mathscr{H})$ would be defined similarly, but now the norm condition in (5.81) would be replaced by

$$\sum_{j=1}^{\infty} \|\mathbf{g}(\mathbf{e}_j)\|_{\mathscr{H}}^2 < \infty, \qquad (5.84)$$

$\{\mathbf{e}_j : j \in \mathbb{N}\}$ is an orthonormal basis of $\mathscr{K}$. Unlike the condition in (5.81), (5.84) does not hold trivially. Also, in general, the space $\mathscr{B}_0(\mathscr{K}, \mathscr{H})$ is NOT a separable Hilbert space (see [105]) and so, we must refine our definition of an operator-valued random variable to take into account a weaker sense of measurability. Moreover, the construction of a $\mathscr{K}$-valued Wiener process is more delicate and, in fact, the covariance operator Q not only must satisfy certain technical conditions (among them being trace$Q < \infty$), it also enters in a nontrivial way into the very definition of the space $\mathscr{B}_0(\mathscr{K}, \mathscr{H})$. Beyond this, there are even more general constructions (for cylindrical Wiener processes). A very comprehensive discussion of such constructions can be found in [105, 338]. Because this text is primarily intended to introduce a beginner to the subject, we shall restrict our discussion to m-dimensional Wiener processes.

In view of how $\mathscr{B}_0(\mathbb{R}^m, \mathscr{H})$ has been defined, it is reasonable to restrict our attention, as in the case of $\mathbb{R}^N$-valued random variables, to mappings $g : [0,t] \times \Omega \to \mathscr{B}_0(\mathbb{R}^m, \mathscr{H})$ such that g is $\mathscr{F}_s$-adapted, progressively measurable, and

$$\int_0^t E \|g(s; \cdot)\|_{\mathscr{B}_0(\mathbb{R}^m, \mathscr{H})}^2 \, ds < \infty. \qquad (5.85)$$

The collection of all such operator-valued mappings shall be denoted $\mathscr{U}_{\mathscr{H}}$. Now, how do we proceed with the construction of the Itó integral $\int_0^t g(s; \omega) d\mathbf{W}(s)$ for such mappings g? The end result of the construction must belong to $\mathscr{H}$. As such, the construction is not as simple as it was when extending the definition from the case when g was a real-valued stochastic process to the case when it was an $\mathbb{R}^N$-valued stochastic process (via components). Indeed, we shall revert back to the building block approach used to construct the one-dimensional Itó integral. To this end, we start with a step function in the following operator sense.

Definition 5.5.2. A random variable $g : [0,t] \times \Omega \to \mathscr{B}_0(\mathbb{R}^m, \mathscr{H})$ is a *random operator-valued step function* if

$$g(t; \omega) = \sum_{k=0}^{n} g_k(\omega)\chi_{[t_k, t_{k+1}]}(t), \tag{5.86}$$

where $g_k : [0,T] \times \Omega \to \mathscr{B}_0(\mathbb{R}^m, \mathscr{H})$ is $\mathscr{F}_{t_k}$-measurable such that rng (g_k) is finite, $\forall k \in \{0,1,\dots,n\}$, and $0 = t_0 < t_1 < \dots < t_n = t$.

The Itó integral for such a step function g is defined by

$$\int_0^t g(s; \cdot) d\mathbf{W}(s) = \sum_{k=0}^{n} \underbrace{\underbrace{g_k(\omega)}_{\text{operator}} \underbrace{(\mathbf{W}(t_{k+1}; \omega) - \mathbf{W}(t_k; \omega))}_{\text{applied to input in } \mathbb{R}^m}}_{\text{in } \mathscr{H}, \forall k \in \{0,1,\dots,n\}}. \tag{5.87}$$

$$\underbrace{}_{\text{in } \mathscr{H} \text{ because it is a linear space}}$$

Then, we can use the fact that the collection of such step functions is dense in $\mathscr{B}_0(\mathbb{R}^m, \mathscr{H})$ (see **[105]**) to define the stochastic integral for more general members of $\mathscr{U}_\mathscr{H}$. It can be shown that the same properties from Prop. 3.2.3 hold here.

Exercise 5.5.1. Compile a list of properties analogous to Prop. 3.2.3, making the appropriate notational modifications.

The following properties will be of particular utility in the remainder of the text.

Proposition 5.5.3. *Assume that $g : [0,t] \times \Omega \to \mathscr{B}_0(\mathbb{R}^m, \mathscr{H})$ belongs to $\mathscr{U}_\mathscr{H}$ and $\mathbf{W}$ is an m-dimensional Wiener process. Then,*
i.) $E \int_0^t g(s; \cdot) d\mathbf{W}(s) = \int_0^t E[g(s; \cdot)] d\mathbf{W}(s);$
ii.) *If $\mathscr{L} \in \mathbb{B}(\mathscr{H})$, then* $\mathscr{L} \int_0^t g(s; \cdot) d\mathbf{W}(s) = \int_0^t \mathscr{L}[g(s; \cdot)] d\mathbf{W}(s);$
iii.) *For any $p > 2$, if $\int_0^t E \|g(s; \cdot)\|_{\mathscr{B}_0(\mathbb{R}^m, \mathscr{H})}^p ds < \infty$, then there exists $\zeta_g(t, p) > 0$ such that*

$$E \left\| \int_0^t g(s; \cdot) d\mathbf{W}(s) \right\|_\mathscr{H}^p \leq \zeta_g(t, p) \int_0^t E \|g(s; \cdot)\|_{\mathscr{B}_0(\mathbb{R}^m, \mathscr{H})}^p ds. \tag{5.88}$$

The General Itó Formula
The Itó formula in a Hilbert space is strikingly similar to the multivariable Itó formula. Indeed, suppose that $X : [0,T] \to \mathscr{H}$ is an $\mathscr{H}$-valued Itó process given by

$$\begin{cases} dX(t; \omega) = f(t)dt + g(t)d\mathbf{W}(t), 0 < t < T, \omega \in \Omega, \\ X(0; \omega) = X_0(\omega), \omega \in \Omega, \end{cases} \tag{5.89}$$

where $f : [0, T] \to \mathcal{H}$ is $\mathcal{F}_t$-adapted, $\int_0^T \|f(t)\|_{\mathcal{H}} dt < \infty$, $g : [0, T] \to \mathcal{B}_0(\mathbb{R}^m, \mathcal{H})$ belongs to $\mathcal{U}_{\mathcal{H}}$, $\mathbf{W}$ is an m-dimensional Wiener process, and $X_0 \in \mathcal{L}^2(\Omega; \mathcal{H})$ is an $\mathcal{F}_0$-measurable random variable independent of $\mathbf{W}$. Further, suppose that $H : [0, T] \times \mathcal{H} \to \mathcal{K}$, where $\mathcal{K}$ is a separable Hilbert space, is a sufficiently nice mapping and that we wish to compute $d(H(t, X(t)))$. If we were, as in the one-dimensional case, to write down the Taylor expansion, the same format of terms arises and because we are still using an m-dimensional Wiener process, it is reasonable to expect that the terms whose order is at least three will vanish. Of course, we need a more general notion of differentiability and the products must be defined properly because we are now "multiplying" members of a Hilbert space, not just real numbers.

Comparing this to (4.65) and its development, the first thing to note is that we are now dealing with $H(t, z)$, where z belongs to the <u>Hilbert space</u> $\mathcal{H}$. So, we need to define what is meant by $\frac{\partial H}{\partial z}$. The so-called *Frechet derivative*, a natural extension of the familiar notion of differentiability of functions $\mathbf{f} : \mathbb{R}^N \to \mathbb{R}^N$, is defined as follows.

Definition 5.5.4. Let $\mathcal{X}$ and $\mathcal{Y}$ be Banach spaces. The mapping $F : \mathcal{X} \to \mathcal{Y}$ is *(Frechet) differentiable* at $x_0 \in \mathcal{X}$ if there exists a linear operator $F'(x_0) : \mathcal{X} \to \mathcal{Y}$ such that

$$\underbrace{\overbrace{F(x_0 + \triangle x)}^{\text{in } \mathcal{X}} - F(x_0)}_{\text{in } \mathcal{Y}} = \underbrace{F'(x_0)}_{\text{mapping}} \overbrace{(\triangle x)}^{\text{applied to } \triangle x} + \underbrace{\varepsilon(x_0, \triangle x)}_{\text{Error term in } \mathcal{Y}},$$

where $\lim\limits_{\|\triangle x\|_{\mathcal{X}} \to 0} \frac{\|\varepsilon(x_0, \triangle x)\|_{\mathcal{Y}}}{\|\triangle x\|_{\mathcal{X}}} = 0$.

Remarks.
1. <u>A useful estimate</u>: Let $\eta > 0$. There exists $\delta > 0$ such that

$$0 < \|\triangle x\|_{\mathcal{X}} < \delta \implies \|\varepsilon(x_0, \triangle x)\|_{\mathcal{Y}} < \eta \|\triangle x\|_{\mathcal{X}}$$

(Why?) and so,

$$\|(F(x_0 + \triangle x) - F(x_0)) - F'(x_0)(\triangle x)\|_{\mathcal{Y}} < \eta \|\triangle x\|_{\mathcal{X}}. \tag{5.90}$$

2. We shall use the notation $\frac{\partial f}{\partial x}(x_0)$ interchangeably with $F'(x_0)$. When it is not confusing to do so, we omit the prefix "Frechet" when referring to this type of derivative.

Proposition 5.5.5. Properties of $F'(x_0)$
Let $\mathcal{X}, \mathcal{Y}, \mathcal{Z}$ be Banach spaces and $F : \mathcal{X} \to \mathcal{Y}$ and $G : \mathcal{Y} \to \mathcal{Z}$ given mappings.
i.) *If $F'(x_0)$ exists, then it is unique.*
ii.) **(Chain Rule)** *If G is differentiable at x_0 and F is differentiable at $G(x_0)$, then $F \circ G$ is differentiable at x_0 and*

$$(F \circ G)'(x_0) = F'(G(x_0))G'(x_0).$$

iii.) *If F is strongly continuous, then $F'(x_0) \in \mathbb{B}(\mathcal{X}, \mathcal{Y})$.*

As such, $\frac{\partial H}{\partial z}$ and $\frac{\partial^2 H}{\partial z^2}$ are meaningful and are used in place of ∇H and $\mathscr{J}(h)$, respectively, in (4.65). We require that H be differentiable in t and twice continuously Frechet differentiable in z.

Now, in order to be able to extend (4.65) to the Hilbert space setting, we must make certain that all the terms are meaningful and belong to $\mathscr{H}$. Observe that $\forall t \in [0,T]$, $f(t) \in \mathscr{H}$ and $g(t)d\mathbf{W}(t) \in \mathscr{H}$. In order for the terms in (4.65) involving $\frac{\partial H}{\partial z}$ to be meaningful, we must impose the condition that

$$\frac{\partial H}{\partial z}(t,z)h \in \mathscr{K}, \forall t \in [0,T], h \in \mathscr{H}. \tag{5.91}$$

Finally, the term involving $\frac{\partial^2 H}{\partial z^2}$ remains. Looking back at how this term arose in (4.65), it is reasonable to interpret the expression $\frac{1}{2}\text{trace}\left(\mathscr{J}(h)\mathbf{G}(t)\mathbf{G}^T(t)\right)$ in the present setting as

$$\frac{1}{2}\sum_{k=1}^{m}\frac{\partial^2 H}{\partial z^2}(t,z)\langle g(\mathbf{e}_k), g(\mathbf{e}_k)\rangle_{\mathscr{H}} = \frac{1}{2}\frac{\partial^2 H}{\partial z^2}(t,z)\|g\|^2_{\mathscr{B}_0}, \tag{5.92}$$

where $\{\mathbf{e}_k | k = 1, \ldots, m\}$ is an orthonormal basis for $\mathbb{R}^m$. (Why?) As such, because $\langle \cdot, \cdot \rangle_{\mathscr{H}}$ is a real number, the summand in (5.92) belongs to $\mathscr{K}$, and hence the entire finite sum in (5.92) belong to $\mathscr{K}$ (because $\mathscr{K}$ is a linear space).

The above discussion can be formally summarized as the following proposition.

Proposition 5.5.6. *Let $\mathscr{H}$ be a Hilbert space and $H : [0,T] \times \mathscr{H} \to \mathscr{K}$ a continuous mapping such that H is differentiable in t, twice continuously Frechet differentiable in z, and satisfies (5.91). If X is an Itó process, then*

$$d(H(t,X(t))) = \left[\frac{\partial H}{\partial t}(t,X(t)) + \frac{\partial H}{\partial z}(t,X(t))f(t) + \frac{1}{2}\frac{\partial^2 H}{\partial z^2}(t,X(t))\|g\|^2_{\mathscr{B}_0}\right]dt$$

$$+ \frac{\partial H}{\partial z}(t,X(t))(g(t)d\mathbf{W}(t)). \tag{5.93}$$

Remark. In applications, it is typical to take $\mathscr{K}$ to be $\mathbb{R}$, $\mathbb{R}^N$, or $\mathscr{H}$ itself.

5.6 The Cauchy Problem — Formulation

Assume (S.A.1). We shall consider SEEs of the abstract form

$$\begin{cases} dX(t;\omega) = AX(t;\omega)dt + \sum_{k=1}^{m}B_kX(t;\omega)dW_k(t), 0 < t < T, \omega \in \Omega, \\ X(0;\omega) = X_0(\omega), \omega \in \Omega, \end{cases} \tag{5.94}$$

in a separable Hilbert space $\mathscr{H}$, where $X : [0,T] \times \Omega \longrightarrow \mathscr{H}$, $A : \text{dom}(A) \subset \mathscr{H} \longrightarrow \mathscr{H}$ is a linear (possibly unbounded) operator, $B_k : \mathscr{H} \to \mathscr{B}_0(\mathbb{R}^m, \mathscr{H})$ $(k = 1, \ldots, m)$

are given mappings, $\{W_k : k = 1, \ldots, m\}$ are independent real-valued Wiener processes, and $X_0 : \Omega \to \mathscr{H}$.

Exercise 5.6.1. Explain carefully why $\int_0^t \sum_{k=1}^m B_k X(s; \omega) dW_k(s)$ is well-defined.

We assume the following:
(H5.1) $A : \text{dom}(A) \subset \mathscr{H} \longrightarrow \mathscr{H}$ generates a C_0-semigroup $\{e^{At} : t \geq 0\}$ on $\mathscr{H}$.
(H5.2) $\mathbf{W}(t) = \langle W_1(t), \ldots, W_m(t) \rangle^T$, $0 \leq t \leq T$, is an m-dimensional Wiener process.
(H5.3) $X_0 \in \mathfrak{L}^2(\Omega; \mathscr{H})$ is an $\mathscr{F}_0$-measurable random variable independent of the Wiener process $\mathbf{W}$.
(H5.4) $B_k : \mathscr{H} \to \mathscr{B}_0(\mathbb{R}^m, \mathscr{H})$, $k = 1, \ldots, m$, are bounded linear operators.
(H5.5) $\forall t \in [0, T]$, $\sigma(X_0, \mathbf{W}(s) | 0 \leq s \leq t) \subset \mathscr{F}_t$.

The Principle of Uniform Boundedness guarantees that

$$M_A = \max_{0 \leq t \leq T} \left\| e^{At} \right\|_{\mathbb{B}(\mathscr{H})} < \infty. \tag{5.95}$$

Consider the integrated form of (5.94):

$$X(t; \omega) = X_0(\omega) + \int_0^t AX(s; \omega) ds + \sum_{k=1}^m \int_0^t B_k X(s; \omega) dW_k(s), \tag{5.96}$$

$\forall 0 \leq t \leq T$, $\omega \in \Omega$. The notion of a strong solution of (5.94) closely resembles Def. 4.5.1. Precisely, we have

Definition 5.6.1. A stochastic process $X : [0, T] \times \Omega \longrightarrow \mathscr{H}$ is a *strong solution* of (5.94) on $[0, T]$ if
i.) $X \in \mathbb{C}\left([0, T]; \mathfrak{L}^2(\Omega; \mathscr{H})\right)$,
ii.) $X(t; \omega) \in \text{dom}(A)$, $\forall t \in [0, T]$, a.s. $[\mathscr{P}]$,
iii.) $\int_0^T \|AX(t; \omega)\|_{\mathscr{H}} dt < \infty$ a.s. $[\mathscr{P}]$,
iv.) $X(t; \cdot)$ satisfies (5.96), $\forall 0 \leq t \leq T$, a.s. $[\mathscr{P}]$.

The one notable difference between Def. 4.5.1 and Def. 5.6.1 is condition (ii). This is indicative of the infinite-dimensional nature of the problem.

Recall that the existence of a classical solution in the deterministic setting was assured only under rather restrictive conditions involving A, e^{At}, and X_0 in order for the terms in the integrated version of the Cauchy problem to be defined. The same is true for the existence of strong solutions in the sense of Def. 5.6.1. As such, we introduce the following more practical notion of a *mild solution*:

Definition 5.6.2. A stochastic process $X : [0, T] \times \Omega \longrightarrow \mathscr{H}$ is a *mild solution* of (5.94) on $[0, T]$ if
i.) $X \in \mathbb{C}\left([0, T]; \mathfrak{L}^2(\Omega; \mathscr{H})\right)$,
ii.) $X(t; \omega) = e^{At} X_0(\omega) + \sum_{k=1}^m \int_0^t e^{A(t-s)} B_k X(s; \omega) dW_k(s)$, $\forall 0 \leq t \leq T$, a.s. $[\mathscr{P}]$.

The formula in Def. 5.6.2(ii) is often referred to as the *variation of parameters formula*. Itó's formula can be used to prove that a strong solution of (5.94) is also a mild solution of (5.94). (See **[203, 204]**.)

5.7   The Basic Theory

The existence and uniqueness of a mild solution of (5.94) is a straightforward consequence of the Contraction Mapping Principle.

Theorem 5.7.1. *If* **(H5.1)** *through* **(H5.5)** *hold, then (5.94) has a unique mild solution on* $[0,T]$.

Proof. Define $\Phi : \mathbb{C}\left([0,T]; \mathfrak{L}^2\left(\Omega; \mathscr{H}\right)\right) \to \mathbb{C}\left([0,T]; \mathfrak{L}^2\left(\Omega; \mathscr{H}\right)\right)$ by

$$(\Phi X)(t;\omega) = e^{At}X_0(\omega) + \sum_{k=1}^{m}\int_0^t e^{A(t-s)}B_k X(s;\omega)dW_k(s). \tag{5.97}$$

First, observe that for any $X \in \mathbb{C}\left([0,T]; \mathfrak{L}^2\left(\Omega; \mathscr{H}\right)\right)$,

$$E\left\|X(t;\cdot)\right\|_{\mathscr{H}}^2 \leq 2\left[E\left\|e^{At}X_0(\cdot)\right\|_{\mathscr{H}}^2 + E\left\|\sum_{k=1}^{m}\int_0^t e^{A(t-s)}B_k X(s;\cdot)dW_k(s)\right\|_{\mathscr{H}}^2\right]$$

$$\leq 2M_A^2\left[\|X_0\|_{\mathfrak{L}^2}^2 + m\sum_{k=1}^{m}E\left\|\int_0^t B_k X(s;\cdot)dW_k(s)\right\|_{\mathscr{H}}^2\right]$$

$$\leq 2M_A^2\left[\|X_0\|_{\mathfrak{L}^2}^2 + m\zeta_g(t)\sum_{k=1}^{m}\int_0^t E\left\|B_k X(s;\cdot)\right\|_{\mathscr{B}_0}^2 ds\right] \tag{5.98}$$

$$\leq 2M_A^2\|X_0\|_{\mathfrak{L}^2}^2 + 2M_A^2 m\zeta_g(t)\sum_{k=1}^{m}\|B_k\|_{\mathscr{B}_0}^2\int_0^t E\left\|X(s;\cdot)\right\|_{\mathscr{H}}^2 ds.$$

Applying Gronwall's Lemma then yields

$$E\left\|X(t;\cdot)\right\|_{\mathscr{H}}^2 \leq 2M_A^2\|X_0\|_{\mathfrak{L}^2}^2\, e^{2M_A^2 m\zeta_g(t)\sum_{k=1}^m\|B_k\|_{\mathscr{B}_0}^2 t}, \tag{5.99}$$

$\forall 0 \leq t \leq T$. Hence, $\|X\|_{\mathbb{C}}^2 < \infty$, so that Φ is a well-defined mapping.
Next, let $X,Y \in \mathbb{C}\left([0,T]; \mathfrak{L}^2\left(\Omega; \mathscr{H}\right)\right)$. Arguing as in (5.98) yields

$$E\left\|(\Phi X)(t;\cdot) - (\Phi Y)(t;\cdot)\right\|_{\mathscr{H}}^2 \leq \left(2M_A^2 m\zeta_g(t)\sum_{k=1}^{m}\|B_k\|_{\mathscr{B}_0}^2\right) \times \tag{5.100}$$

$$\int_0^t E\left\|X(s;\cdot) - Y(s;\cdot)\right\|_{\mathscr{H}}^2 ds.$$

Let $\xi = 2M_A^2 m\zeta_g(t)\sum_{k=1}^{m}\|B_k\|_{\mathscr{B}_0}^2$. Iterating (5.100) shows that

$$E\left\|(\Phi^n X)(t;\cdot) - (\Phi^n Y)(t;\cdot)\right\|_{\mathscr{H}}^2 \leq \frac{(\xi T)^n}{n!}\|X - Y\|_{\mathbb{C}}^2, \tag{5.101}$$

and so, taking the supremum over $(0,T)$ yields

$$\|\Phi^n X - \Phi^n Y\|_{\mathbb{C}}^2 \leq \frac{(\xi T)^n}{n!} \|X - Y\|_{\mathbb{C}}^2. \qquad (5.102)$$

There exists $n_0 \in \mathbb{N}$ such that $\frac{(\xi T)^{n_0}}{n_0!} < 1$. (Why?) Hence, Φ^{n_0} is a strict contraction. As such, we conclude that Φ has a unique fixed point that coincides with a mild solution of (5.94). $\qquad \square$

Exercise 5.7.1. Consider the Cauchy problem

$$\begin{cases} dX(t;\omega) = \left(A + \sum_{k=1}^n C_k\right) X(t;\omega)dt + \sum_{k=1}^m B_k X(t;\omega)dW_k(t), \\ X(0;\omega) = X_0(\omega), \end{cases} \qquad (5.103)$$

where $0 < t < T, \omega \in \Omega$; $A, B_k, W_k(t)$ $(k = 1,\ldots,m)$ and X_0 satisfy **(H5.1)** through **(H5.5)**; and $C_k \in \mathbb{B}(\mathscr{H})$ $(k = 1,\ldots,n)$. Prove that (5.103) has a unique mild solution on $[0,T]$.

Exercise 5.7.2. Verify (5.101).

Exercise 5.7.3. Consider the advection model

$$\begin{cases} \partial c(z,t;\omega) = V\frac{\partial}{\partial z}c(z,t;\omega)\partial t + c(z,t;\omega)dW(t), \ z > 0, t > 0, \omega \in \Omega, \\ c(z,0;\omega) = c_0(z;\omega), \ z > 0, \omega \in \Omega, \\ c(0,t;\omega) = 0, \ t > 0, \omega \in \Omega, \end{cases} \qquad (5.104)$$

where $W(t)$ is a one-dimensional Wiener process.
i.) Prove that (5.104) has a unique mild solution on $[0,T]$.
ii.) Use the variation of parameters formula for the solution to estimate $E\,|c(z,t;\cdot)|^p$, where $p > 2$.

Exercise 5.7.4. Consider the two-dimensional stochastic heat conduction IBVP

$$\begin{cases} \partial z(x,y,t;\omega) = & k\triangle z(x,y,t;\omega)\partial t + \alpha_1 z(x,y,t;\omega)dW_1(t) \\ & +\alpha_2 z(x,y,t;\omega)dW_2(t), \\ z(x,y,0;\omega) = & z_0(x,y;\omega), \\ z(x,0,t;\omega) = & 0 = z(x,b,t;\omega), \\ z(0,y,t;\omega) = & 0 = z(a,y,t;\omega), \end{cases} \qquad (5.105)$$

where $0 < x < a, 0 < y < b, t > 0$, and $\omega \in \Omega$. Prove that (5.105) has a unique mild solution on $[0,T]$.

In general, a mild solution of a Cauchy problem is not automatically a strong solution. For instance, the existence of an event $\mathscr{D} \subset \Omega$ with $\mathscr{P}(\mathscr{D}) > 0$ for which $\forall \omega \in \mathscr{D}, X_0(\omega) \in \mathscr{H} \setminus \text{dom}(A)$ is enough to prevent a mild solution from reaching strong solution status in many cases. (Compare this to the deterministic setting!) But, we can always approximate a mild solution by a sequence of strong solutions of a well-chosen sequence of Cauchy problems. Indeed, we can use the resolvent of A to form an abstract Cauchy problem to which a strong solution exists. Specifically, $\forall \lambda \in \rho(A)$, consider the Cauchy problem

$$\begin{cases} dX_\lambda(t;\omega) = AX_\lambda(t;\omega)dt + \sum_{k=1}^m R_\lambda(A)B_kX_\lambda(t;\omega)dW_k(t), \\ X_\lambda(0;\omega) = R_\lambda(A)X_0(\omega), \end{cases} \quad (5.106)$$

where $0 < t < T$, $\omega \in \Omega$ and $R_\lambda(A)$ is the resolvent operator (cf. Def. 5.2.7). We can argue as in Thrm. 5.7.1 to show that (5.106) has a unique mild solution $X_\lambda : [0,T] \rightarrow \mathscr{H}$. (Show this!) We further assert the following.

Proposition 5.7.2. *The mild solution X_λ of (5.106) is also a strong solution of (5.106).*

Proof. The mild solution of (5.106) is given by

$$X_\lambda(t;\omega) = e^{At}\left(R_\lambda(A)X_0(\omega)\right) + \sum_{k=1}^m \int_0^t e^{A(t-s)}R_\lambda(A)B_kX_\lambda(s;\omega)dW_k(s). \quad (5.107)$$

It is immediate that $X_\lambda(t;\omega) \in \mathrm{dom}(A)$, $\forall 0 \leq t \leq T$, a.s. $[\mathscr{P}]$, because

$$R_\lambda(A)h \in \mathrm{dom}(A), \forall h \in \mathscr{H}.$$

(Why?) Moreover, the requirement that $X_\lambda \in \mathbb{C}\left([0,T]; \mathfrak{L}^2(\Omega; \mathscr{H})\right)$ is common to the definitions of both a mild solution and a strong solution. We must argue that

$$\int_0^T \|AX_\lambda(t;\omega)\|_{\mathscr{H}} dt < \infty, \text{ a.s. } [\mathscr{P}], \quad (5.108)$$

and

$$X_\lambda(t;\omega) = R_\lambda(A)X_0(\omega) + \int_0^t AX_\lambda(s;\omega)ds + \sum_{k=1}^m \int_0^t R_\lambda(A)B_kX_\lambda(s;\omega)dW_k(s), \quad (5.109)$$

$\forall 0 \leq t \leq T$, a.s. $[\mathscr{P}]$. We proceed by using the semigroup properties (cf. Thrm. 5.2.3) and the stochastic Fubini theorem in the calculation of $\int_0^t AX_\lambda(s;\omega)ds$. To this end, applying A to both sides of (5.107) and then integrating over $(0,t)$ yields

$$\int_0^t AX_\lambda(s;\omega)ds = \int_0^t Ae^{As}\left(R_\lambda(A)X_0(\omega)\right)ds$$

$$+ \sum_{k=1}^m \int_0^t \int_0^s Ae^{A(s-u)}R_\lambda(A)B_kX_\lambda(u;\omega)dW_k(u)ds$$

$$= I_1 + I_2. \quad (5.110)$$

Observe that

$$I_1 = A \int_0^t e^{As} \left(R_\lambda(A) X_0(\omega) \right) ds = e^{At} \left(R_\lambda(A) X_0(\omega) \right) - R_\lambda(A) X_0(\omega) \quad (5.111)$$

(Why?) and

$$I_2 = \sum_{k=1}^m \int_0^t \int_u^s e^{A(s-u)} A \left[R_\lambda(A) B_k X_\lambda(u;\omega) \right] ds dW_k(u) \qquad (5.112)$$

$$= \sum_{k=1}^m \int_0^t \left[e^{A(t-u)} R_\lambda(A) B_k X_\lambda(u;\omega) - R_\lambda(A) B_k X_\lambda(u;\omega) \right] dW_k(u),$$

a.s. $[\mathscr{P}]$. Substituting (5.111) and (5.112) into (5.110) yields

$$\int_0^t A X_\lambda(s;\omega) ds = \left[e^{At} \left(R_\lambda(A) X_0(\omega) \right) + \sum_{k=1}^m \int_0^t e^{A(t-u)} R_\lambda(A) B_k X_\lambda(u;\omega) dW_k(u) \right]$$

$$- \left[R_\lambda(A) X_0(\omega) + \sum_{k=1}^m \int_0^t R_\lambda(A) B_k X_\lambda(u;\omega) dW_k(u) \right] \qquad (5.113)$$

$$= X_\lambda(t;\omega) - \left[R_\lambda(A) X_0(\omega) + \sum_{k=1}^m \int_0^t R_\lambda(A) B_k X_\lambda(u;\omega) dW_k(u) \right]$$

a.s. $[\mathscr{P}]$. So, moving the quantity enclosed within brackets on the right-hand side of (5.113) to the left-hand side, we see that $X_\lambda(t;\omega)$ satisfies (5.109) a.s. $[\mathscr{P}]$, as needed. Moreover, condition (5.108) has been shown to be satisfied along the way. (Tell why.) This completes the proof. ☐

Next, we assert that the sequence of strong solutions $\{X_\lambda(t;\omega) \,|\, 0 \le t \le T, \omega \in \Omega\}$ approximates the mild solution $\{X(t;\omega) \,|\, 0 \le t \le T, \omega \in \Omega\}$ of (5.94).

Proposition 5.7.3. $\lim_{\lambda \to \infty} \|X_\lambda - X\|_{\mathbb{C}} = 0$, *where* X_λ *is the strong solution of (5.106) and X is the mild solution of (5.94).*

The proof of Prop. 5.7.3 is routine and relies on the contractive nature of the resolvent, strong continuity of the semigroup, and LDC. A more general result is proven in Chapter 7 to which Prop. 5.7.3 is an immediate corollary. As such, the proof is momentarily left to you to complete as an exercise.

Exercise 5.7.5. Prove Prop. 5.7.3.

The following result closely resembles those in the previous two settings with the caveat that we are now working in a Hilbert space and are using semigroup properties. Otherwise, the approach is essentially the same. The details are left as an exercise.

Consider (5.94) and the related Cauchy problem

$$\begin{cases} dY(t;\omega) = AY(t;\omega)dt + \sum_{k=1}^{m} B_k Y(t;\omega)dW_k(t), 0 < t < T, \omega \in \Omega, \\ Y(0;\omega) = Y_0(\omega), \omega \in \Omega, \end{cases} \quad (5.114)$$

both under hypotheses (H5.1) through (H5.5).

Exercise 5.7.6. Formulate a continuous dependence estimate for $E \|X(t;\cdot) - Y(t;\cdot)\|_{\mathscr{H}}^2$ in terms of $\|X_0 - Y_0\|_{\mathcal{L}^2(\Omega;\mathscr{H})}^2$.

Consider the deterministic IVP (5.14), where A satisfies (H5.1), and $\forall 0 < \varepsilon < 1$, consider the IVP

$$\begin{cases} dY_\varepsilon(t;\omega) = A_\varepsilon Y\varepsilon(t;\omega)dt + \sum_{k=1}^{m} (B_k)_\varepsilon Y_\varepsilon(t;\omega)dW_k(t), \\ Y_\varepsilon(0;\omega) = u_0, \omega \in \Omega, \end{cases} \quad (5.115)$$

where $0 < t < T$, $\omega \in \Omega$, and $u_0 \in \mathcal{L}^2(\Omega;\mathscr{H})$ is a constant random variable (that is, a fixed element of $\mathscr{H}$). Assume the following hold $\forall 0 < \varepsilon < 1$:

(H5.6) $A_\varepsilon : \mathrm{dom}(A_\varepsilon) \subset \mathscr{H} \longrightarrow \mathscr{H}$ generates a C_0-semigroup $\{e^{A_\varepsilon t} : 0 \le t \le T\}$ on $\mathscr{H}$,

$$\left\| e^{A_\varepsilon t} - e^{At} \right\|_{\mathbb{B}(\mathscr{H})} \longrightarrow 0 \text{ as } \varepsilon \to 0^+ \text{ uniformly in } t \in [0,T],$$

and

$$\sup_{0<\varepsilon<1} \left(\sup_{0 \le t \le T} \left\| e^{A_\varepsilon t} \right\|_{\mathbb{B}(\mathscr{H})} \right) \le M_A, \quad (5.116)$$

where M_A is defined in (5.95).

(H5.7) For every $k = 1, \ldots, m$, $(B_k)_\varepsilon : \mathscr{H} \to \mathscr{B}_0(\mathbb{R}^m, \mathscr{H})$ is a bounded linear operator for which

$$\|(B_k)_\varepsilon\|_{\mathbb{B}(\mathscr{H}, \mathscr{B}_0(\mathbb{R}^m, \mathscr{H}))} \longrightarrow 0 \text{ as } \varepsilon \to 0^+ \quad (5.117)$$

and

$$\sup_{0<\varepsilon<1} \|(B_k)_\varepsilon\|_{\mathbb{B}(\mathscr{H}, \mathscr{B}_0(\mathbb{R}^m, \mathscr{H}))} \le \|B_k\|_{\mathbb{B}(\mathscr{H}, \mathscr{B}_0(\mathbb{R}^m, \mathscr{H}))}. \quad (5.118)$$

Then, we have

Proposition 5.7.4. *Let Y_ε and u be the unique mild solutions of (5.115) and (5.14), respectively. For every $p > 2$, there exist $\zeta > 0$ and $\psi : I \subset [0,1] \to (0,\infty)$ for which $\lim_{\varepsilon \to 0^+} \psi(\varepsilon) = 0$ and for sufficiently small $\varepsilon > 0$,*

$$E \|Y_\varepsilon(t;\cdot) - u(t)\|_{\mathscr{H}}^p \le \psi(\varepsilon)\zeta, \forall 0 \le t \le T. \quad (5.119)$$

Proof. Observe that $\forall 0 \le t \le T$ and $0 < \varepsilon < 1$,

$$E \|Y_\varepsilon(t;\cdot) - u(t)\|_{\mathcal{H}}^p = E \left\| \left(e^{A_\varepsilon t} - e^{At}\right) u_0 + \sum_{k=1}^m \int_0^t e^{A_\varepsilon(t-s)} (B_k)_\varepsilon Y_\varepsilon(s;\cdot) dW_k(s) \right\|_{\mathcal{H}}^p$$

$$\le p \left[E \left\| \left(e^{A_\varepsilon t} - e^{At}\right) u_0 \right\|_{\mathbb{B}(\mathcal{H})}^p u_0^p + \right. \tag{5.120}$$

$$\left. m \sum_{k=1}^m E \left\| \int_0^t e^{A_\varepsilon(t-s)} (B_k)_\varepsilon Y_\varepsilon(s;\cdot) dW_k(s) \right\|_{\mathcal{H}}^p \right]$$

$$= I_3 + I_4.$$

By **(H5.5)**, $I_3 \to 0$ uniformly for $t \in [0,T]$ as $\varepsilon \to 0^+$ (Why?). So, for sufficiently small $\varepsilon > 0$, there exist $K_1 > 0$ and $\psi_1 : I \subset [0,1] \to (0,\infty)$ for which $\lim_{\varepsilon \to 0^+} \psi_1(\varepsilon) = 0$ such that

$$I_3 \le K_1 \psi_1(\varepsilon), \ \forall 0 \le t \le T. \tag{5.121}$$

Next, Prop. 5.5.3 guarantees that

$$I_4 = pm \sum_{k=1}^m E \left\| \int_0^t \left(e^{A_\varepsilon(t-s)} (B_k)_\varepsilon Y_\varepsilon(s;\cdot) - e^{A_\varepsilon(t-s)} (B_k)_\varepsilon u(s) \right. \right.$$

$$\left. \left. + e^{A_\varepsilon(t-s)} (B_k)_\varepsilon u(s) \right) dW_k(s) \right\|_{\mathcal{H}}^p$$

$$\le p^2 m \zeta(p) \sum_{k=1}^m \left[\int_0^t E \left\| e^{A_\varepsilon(t-s)} (B_k)_\varepsilon (Y_\varepsilon(s;\cdot) - u(s)) \right\|_{\mathcal{B}_0}^p ds \right.$$

$$\left. + \int_0^t E \left\| e^{A_\varepsilon(t-s)} (B_k)_\varepsilon u(s) \right\|_{\mathcal{B}_0}^p ds \right] \tag{5.122}$$

$$\le p^2 m \zeta(p) M_A^p \sum_{k=1}^m \left[\|B_k\|_{\mathbb{B}}^p \int_0^t E \|Y_\varepsilon(s;\cdot) - u(s)\|_{\mathcal{H}}^p ds \right.$$

$$\left. + \int_0^t \|(B_k)_\varepsilon\|_{\mathbb{B}}^p E \|u(s)\|_{\mathcal{H}}^p ds \right].$$

Because u is a mild solution of (5.14), we know that

$$\sup_{0 \le s \le T} E \|u(s)\|_{\mathcal{H}}^p \le M^\star. \tag{5.123}$$

As such, by **(H5.7)**, $\exists K_2 > 0$ and $\psi_2 : I \subset [0,1] \to (0,\infty)$ for which $\lim_{\varepsilon \to 0^+} \psi_2(\varepsilon) = 0$ such that

$$\int_0^t \|(B_k)_\varepsilon\|_{\mathbb{B}}^p E \|u(s)\|_{\mathcal{H}}^p ds \le M^\star \int_0^t \|(B_k)_\varepsilon\|_{\mathbb{B}}^p ds \le K_2 \psi_2(\varepsilon), \tag{5.124}$$

$\forall 0 \le t \le T$. (Tell why carefully.) Combining these estimates yields

$$E \|Y_\varepsilon(t;\cdot) - u(t)\|_{\mathcal{H}}^p \le \zeta_1 \psi(\varepsilon) + \zeta_2 \int_0^t E \|Y_\varepsilon(s;\cdot) - u(s)\|_{\mathcal{H}}^p ds, \tag{5.125}$$

where $\zeta_1, \zeta_2 > 0$ and $\psi : I \subset [0,1] \to (0,\infty)$ is such that $\lim_{\varepsilon \to 0^+} \psi(\varepsilon) = 0$. Applying Gronwall's Lemma then yields

$$E \, \|Y_\varepsilon(t;\cdot) - u(t)\|_{\mathscr{H}}^p \leq \zeta_1 e^{\zeta_2 t} \psi(\varepsilon) \leq \zeta \psi(\varepsilon), \tag{5.126}$$

where $\zeta = \zeta_1 e^{\zeta_2 T}$. This completes the proof. $\qquad\qquad \square$

Exercise 5.7.7. Formulate results in the spirit of Prop. 5.7.4 directly for the IBVPs discussed in Section 5.4.

5.8 Looking Ahead

By way of preparation for Chapter 6, we consider a more complex version of IBVP (4.16).

Suppose that we now incorporate a term $D(t)dt$ into IVP (2.4) that describes the time variability of the drug dosage from the GI tract viewpoint. Because the rate at which the quantity y changes is directly affected by variations in the dosage over time, it is reasonable to add this forcing term to the right-hand side of the differential equation describing dy (Why?) As such, the resulting IVP is

$$\begin{cases} dy(t;\omega) = & -a_1 y(t;\omega)dt - a_2 y(t;\omega)dW_1(t) + D(t)dt, t > 0, \omega \in \Omega, \\ dz(t;\omega) = & (a_1 y(t;\omega) - b_1 z(t;\omega)) \, dt + a_2 y(t;\omega)dW_1(t) \\ & -b_2 z(t;\omega)dW_2(t), t > 0, \omega \in \Omega, \\ y(0;\omega) = & y_0(\omega), \ z(0;\omega) = 0, \omega \in \Omega. \end{cases} \tag{5.127}$$

Note that $a, b > 0$ and $D(\cdot)$ should decrease to zero as $t \to \infty$. (Why?) Because y_0 is the initial full dosage, it should coincide with $D(0)$. System (5.127) in matrix form is

$$\begin{cases} d\begin{bmatrix} y(t;\omega) \\ z(t;\omega) \end{bmatrix} = \begin{bmatrix} -a_1 & 0 \\ a_1 & -b_1 \end{bmatrix} \begin{bmatrix} y(t;\omega) \\ z(t;\omega) \end{bmatrix} dt + \begin{bmatrix} -a_2 & 0 \\ a_2 & 0 \end{bmatrix} \begin{bmatrix} y(t;\omega) \\ z(t;\omega) \end{bmatrix} dW_1(t) \\ \qquad + \begin{bmatrix} 0 & 0 \\ 0 & -b_2 \end{bmatrix} \begin{bmatrix} y(t;\omega) \\ z(t;\omega) \end{bmatrix} dW_2(t) + \begin{bmatrix} D(t) \\ 0 \end{bmatrix} dt, t > 0, \omega \in \Omega, \\ \begin{bmatrix} y(0;\omega) \\ z(0;\omega) \end{bmatrix} = \begin{bmatrix} y_0(\omega) \\ 0 \end{bmatrix}, \ \omega \in \Omega. \end{cases}$$

$$\tag{5.128}$$

Of particular interest is the effect that the presence of $D(t)$ has on the behavior of the solution of (5.127), in comparison to the solution of (4.16).

5.9   Guidance for Selected Exercises

5.9.1   Level 1: A Nudge in a Right Direction

5.1.1. (i) Verify the conditions of Def. 1.7.3.

(ii) Use Def. 5.1.1(ii) with standard norm properties.

5.1.2. For each $m \in \mathbb{N}$, find a ...

5.1.3. (i) Show $\exists m > 0$ such that

$$\sup_{\|x\|_C = 1} \left\| \int_a^b g(\cdot, s) x(\cdot)\, ds \right\|_{C([a,b];\mathbb{R})} \leq m \|x\|_{C([a,b];\mathbb{R})} .$$

(ii) Use the $\mathbb{L}^2$-norm in the above inequality.

5.1.4. In general, this operator is not linear (Why?), but it is when $g(x, y, z(y)) = \bar{g}(x, y) z(y)$. (Why?) Prove this operator is bounded in a manner similar to the one used in Exer. 5.1.3(i).

5.1.5. (i) Use $g_n(x) = x^n$, where $n \in \mathbb{N}$.

(ii) Carefully use the definition of norm.

5.1.6. Apply the hint for Exer. 5.1.5(i), where $n \geq 2$. (How?)

5.1.7. (i) Note that $\|\mathscr{F} x_n - \mathscr{F} x\|_{\mathscr{Y}} = \|\mathscr{F}(x_n - x)\|_{\mathscr{Y}}$. (Now what?)

5.1.8. Linearity follows easily because $\mathscr{F}$ and $\mathscr{G}$ are both linear. Use the first of two versions of Def. 5.1.1(ii) with Exer. 5.1.1(ii).

5.1.9. Adapt the hint provided for Exer. 5.1.7 to this situation.

5.1.10. (i) ($\Longrightarrow$) $\mathscr{F}$ must be one-to-one, so $\mathscr{F} x = \mathscr{F} y$ implies what?

($\Longleftarrow$) $\mathscr{F} x = \mathscr{F} y \Longrightarrow \mathscr{F}(x - y) = 0$. (So what?)

(ii) Let $\alpha, \beta \in \mathbb{R}$ and choose $x = \mathscr{F} z_1$ and $y = \mathscr{F} z_2$ in the expression $\mathscr{F}^{-1}(\alpha x + \beta y)$.

5.1.11. Prove that $\{S_n\}$ is a Cauchy sequence in $\mathscr{X}$.

5.1.12. For any $\alpha \neq 0$, find an operator $\mathscr{B}_\alpha$ such that $\mathscr{B}_\alpha (\alpha \mathscr{A}) = (\alpha \mathscr{A}) \mathscr{B}_\alpha = I$.

5.1.13. Verify the properties in Def. 1.7.3 directly to prove that $\|\cdot\|_{\mathscr{X} \times \mathscr{Y}}$ is a norm. Completeness of $\mathscr{X} \times \mathscr{Y}$ follows from the completeness of $\mathscr{X}$ and $\mathscr{Y}$. (How?)

5.1.14. Adapt the argument used in the example directly following Def. 5.1.14.

5.1.15. Determine the graph of $\mathscr{B}^{-1}$.

5.1.16. If $\mathscr{X}_1 \subset \mathscr{X}_2 \subset \mathscr{X}$ and all are equipped with the same norm $\|\cdot\|_{\mathscr{X}}$, then $\mathrm{cl}_{\mathscr{X}}(\mathscr{X}_1) = \mathscr{X}$. (So what?)

5.1.17. Yes, because every $\mathbb{L}^2$-function has a unique Fourier representation. (So what?)

5.3.1. Use linearity for both parts.

5.4.4. i.) The process is practically the same, but be careful with the details.

ii.) Integrate by parts to compute b_m.

iii.) The only change from the previous example occurs in how dom(A) is defined.

5.4.6. i.) The main change is that the cosine terms remain instead of the sine terms. There might be one extra Fourier coefficient also, so be careful.

ii.) Everything remains the same except for one change in dom(A). What is it?

5.4.7. Let $z = z(x_1, x_2, \ldots, x_n, t)$. The IBVP becomes

$$\begin{cases} \frac{\partial z}{\partial t} = k \sum_{i=1}^{n} \frac{\partial^2 z}{\partial x_i^2}, \ (x_1, x_2, \ldots, x_n) \in [0, a_1] \times \ldots \times [0, a_n], t > 0, \\ \frac{\partial z}{\partial x_i}(x_1, \ldots, x_{i-1}, 0, x_{i+1}, \ldots, x_n, t) = \frac{\partial z}{\partial x_i}(x_1, \ldots, x_{i-1}, a_i, x_{i+1}, \ldots, x_n, t) = 0, \end{cases}$$
(5.129)

where $i = 1, \ldots, n$, $t > 0$. In order to find the solution of (5.129), suitably modify the steps of the separation of variables approach. In fact, how many different separations of variable does this entail?

5.4.8. You would need to apply B^{-1} on both sides. But, what must be true in order to justify such action?

5.7.1. $A + \sum_{k=1}^{n} C_k$ generates a C_0-semigroup on $\mathscr{H}$. (Why?) How do you handle the Wiener process term?

5.7.6. Subtract the integrated versions of (5.114).

5.9.2 Level 2: An Additional Thrust in a Right Direction

5.1.1. For both parts, apply standard norm properties in $\mathscr{X}$ or $\mathscr{Y}$, whichever is appropriate, together with properties of sup and inf.

5.1.2. $\ldots x \in \mathscr{X}$ such that $\|\mathscr{F}(x)\|_{\mathscr{Y}} > m \|x\|_{\mathscr{X}}$.

5.1.3. **(i)** Show that $\|\mathscr{F}\|_{\mathbb{B}(\mathbb{C})} \leq M_g (b - a)$, where

$$M_g = \sup \{|g(t, s)| : (t, s) \in [a, b] \times [a, b]\}.$$

(ii) $\int_a^b \left| \int_a^b g(t, s) x(s) \, ds \right|^2 dt \leq \int_a^b \left(\int_a^b g^2(t, s) ds \right)^{1/2} \|x\|_{\mathbb{L}^2} dt$. (So what?)

5.1.4. Show that $\|\mathscr{F}\|_{\mathbb{B}(\mathbb{C})} \leq m_g (b - a)$. (So what?)

5.1.5. **(i)** $g_n'(x) = nx^{n-1}$, where $n \in \mathbb{N}$, and $\|g_n'\|_{C([0,a];\mathbb{R})} = na^{n-1}$. (So what?)

(ii) $\|\mathscr{F}(g)\|_{\mathbb{L}^2}^2 = \int_0^a |g'(x)|^2 dx \leq a \|g\|_{C^1([0,a];\mathbb{R})}$

5.1.6. For any $n \geq 2$, compute $\|g''\|_{C([0,a];\mathbb{R})}$. What can you conclude?

5.1.7. Use Exer. 5.1.1(ii).

5.1.8. Show that $\|\mathscr{F}\mathscr{G}x\|_{\mathscr{X}} \leq \|\mathscr{F}\|_{\mathbb{B}(\mathscr{X})} \|\mathscr{G}\|_{\mathbb{B}(\mathscr{X})} \|x\|_{\mathscr{X}}, \forall x \in \mathscr{X}$.

5.1.9. $\|\mathscr{F}_n(x) - \mathscr{F}(x)\|_{\mathscr{Y}} = \|(\mathscr{F}_n - \mathscr{F})(x)\|_{\mathscr{Y}} \leq \|\mathscr{F}_n - \mathscr{F}\|_{\mathbb{B}(\mathscr{X})} \|x\|_{\mathscr{X}}, \forall x \in \mathscr{X}$.

5.1.10. **(i)** ($\Longrightarrow$) $x = 0$. (Why?) and ($\Longleftarrow$) $x - y = 0$. (Why?)

(ii) Simplify to get $\alpha \mathscr{F}^{-1} x + \beta \mathscr{F}^{-1} y$.

5.1.11. Observe that $\forall p \in \mathbb{N}, \|S_{n+p} - S_n\|_{\mathscr{X}} \leq \sum_{k=n+1}^{n+p} \|x_k\|_{\mathscr{X}}$. (So what?)

5.1.12. Use $\mathscr{B}_\alpha = \frac{1}{\alpha} \mathscr{A}^{-1}$.

5.1.13. Start with a Cauchy sequence $\{(x_n, y_n)\}$ in $\mathscr{X} \times \mathscr{Y}$ and use the completeness of $\mathscr{X}$ and $\mathscr{Y}$ independently to produce the natural candidate $(x, y) \in \mathscr{X} \times \mathscr{Y}$ to which the sequence converges.

5.1.14. Adapt the argument used in the example directly following Def. 5.1.14.

5.1.15. graph $(\mathscr{B}^{-1}) = \{(\mathscr{B}x, x) \mid x \in \text{dom}(\mathscr{B})\}$ is closed. (Why?)

5.1.16. It must be the case that $\text{cl}_{\mathscr{X}}(\mathscr{X}_2) = \mathscr{X}$. (Why?)

5.1.17. You can show that the set used in Prop. 5.1.17 is dense in $\mathbb{L}^2(a, b; \mathbb{R})$.

5.4.4. iii.) Incorporate the BCs into the definition of dom(A). Is the resulting set contained within $\mathscr{X}$?

5.4.6. i.) In applying the BCs, you will get cosines instead of sines. Also, because $\cos(0) = 1$, you get one additional term in the solution formula.

ii.) dom(A) $= \left\{ z \in \mathbb{H}^2 \left((0,a) \times (0,b) ; \mathbb{R} \right) \mid \frac{\partial z}{\partial x}, \frac{\partial z}{\partial y} \big|_{\partial R} = 0 \right\}$. (The condition on the partials amounts to saying that the outward normal to the boundary is zero.)

5.4.7. If you follow the steps of the solution in Volume 1 , make note of the following changes:

(3.50) becomes

$$
\begin{cases}
(X_i)_{m_i}(x_i) = \cos\left(\frac{m_i \pi_i}{a_i} x_i \right), \ 0 < x_i < a_i, \ m_i \in \mathbb{N}, \ i = 1, 2, \ldots \\
T_{m_1 \ldots m_n}(t) = Ce^{-kt \sum_{i=1}^{n} \left(\frac{m_i \pi_i}{a_i} \right)^2}, t > 0.
\end{cases}
$$

(3.51) becomes

$$
z_{m_1 \ldots m_n}(x_1, \ldots, x_n, t) = b_{m_1 \ldots m_n} T_{m_1 \ldots m_n}(t) \prod_{i=1}^{n} (X_i)_{m_i}(x_i).
$$

(3.52) becomes

$$
z(x_1, \ldots, x_n, t) = \sum_{m_1=1}^{\infty} \cdots \sum_{m_n=1}^{\infty} z_{m_1 \ldots m_n}(x_1, \ldots, x_n, t),
$$

where

$$
b_{m_1 \ldots m_n} = \frac{2^n}{a_1 \cdots a_n} \int_0^{a_n} \cdots \int_0^{a_1} z_0(w_1, \ldots, w_n, t) \prod_{i=1}^{n} \cos\left(\frac{m_i \pi_i}{a_i} w_i \right) dw_1 \cdots dw_n.
$$

Finally, use $\mathscr{X} = \mathbb{L}^2 \left((0, a_1) \times (0, a_n) \right)$ along with the natural modifications to A, dom(A), etc. in order to write the IBVP in the desired abstract form.

5.4.8. The domains of A and B must be "compatible." Additional restrictions must be imposed on the operator $B^{-1}A$ in order for the theory that we are about to develop to be applicable. (See Chapter 9.)

Chapter 6

Nonhomogenous Linear Stochastic Evolution Equations

Overview

More often than not, significant external forces, which are also subject to noise, impact the evolution of the process. An accurate mathematical model must account for this. How does one incorporate such external forces into the IVPs and IBVPs and the subsequent abstract formulation of them? What effect does this have on the solution in the sense of existence, continuous dependence, long-term behavior, etc.? We focus on these questions in this chapter.

6.1 Finite-Dimensional Setting

Assume **(S.A.1)** throughout this chapter.

6.1.1 Motivation by Models

Model II.2 Pharmacokinetics with Time-Varying Drug Dosage
Suppose that we now incorporate a term $D(t)dt$ into IVP (2.4) describing the time variability of the drug dosage from the GI tract viewpoint. Because the rate at which the quantity y changes is directly affected by variations in the dosage over time, it is reasonable to add this forcing term to the right-hand side of the differential equation describing dy (Why?) As such, the resulting IVP is

$$
\begin{cases}
dy(t;\omega) = & -a_1 y(t;\omega)dt - a_2 y(t;\omega)dW_1(t) + D(t)dt, \\
dz(t;\omega) = & (a_1 y(t;\omega) - b_1 z(t;\omega))\,dt + a_2 y(t;\omega)dW_1(t) \\
& -b_2 z(t;\omega)dW_2(t), \\
y(0;\omega) = & y_0(\omega),\ z(0;\omega) = 0,
\end{cases}
\tag{6.1}
$$

where $t > 0$, $\omega \in \Omega$. Note that $a, b > 0$ and $D(\cdot)$ should decrease to zero as $t \to \infty$. Because y_0 is the initial full dosage, it should coincide with $D(0)$. The matrix form of (6.1) is

$$
\begin{cases}
d\begin{bmatrix} y(t;\omega) \\ z(t;\omega) \end{bmatrix} = \begin{bmatrix} -a_1 & 0 \\ a_1 & -b_1 \end{bmatrix} \begin{bmatrix} y(t;\omega) \\ z(t;\omega) \end{bmatrix} dt + \begin{bmatrix} -a_2 & 0 \\ a_2 & 0 \end{bmatrix} \begin{bmatrix} y(t;\omega) \\ z(t;\omega) \end{bmatrix} dW_1(t) \\
\qquad + \begin{bmatrix} 0 & 0 \\ 0 & -b_2 \end{bmatrix} \begin{bmatrix} y(t;\omega) \\ z(t;\omega) \end{bmatrix} dW_2(t) + \begin{bmatrix} D(t) \\ 0 \end{bmatrix} dt, \, t > 0, \, \omega \in \Omega, \\
\begin{bmatrix} y(0;\omega) \\ z(0;\omega) \end{bmatrix} = \begin{bmatrix} y_0(\omega) \\ 0 \end{bmatrix}, \, \omega \in \Omega.
\end{cases}
$$

(6.2)

Of particular interest is the effect that the presence of $D(t)$ has on the behavior of the solution of (6.1), in comparison to the solution of (4.16).

Exercise 6.1.1. Suppose that $a_2 = b_2 = 0$ and that noise enters into the model only through the initial data. Formulate a random version of this model (like (4.2), for instance) and apply the variation of parameters method to determine the solution.

Model III.2 Spring-Mass System with External Force

Suppose that an external force described by the function $f : [0,\infty) \to \mathbb{R}$ acts on a spring-mass system and that noise is incorporated into the model through the initial data, resulting in the following generalization of (4.19):

$$
\begin{cases}
\frac{d^2 x}{dt^2}(t;\omega) + \alpha^2 x(t;\omega) = f(t), \, t > 0, \, \omega \in \Omega, \\
x(0;\omega) = x_0(\omega), \, \frac{dx}{dt}(0;\omega) = x_1(\omega), \, \omega \in \Omega.
\end{cases}
$$

(6.3)

The random IVP (6.3) can be written in the following equivalent matrix form:

$$
\begin{cases}
\frac{d}{dt}\begin{bmatrix} y(t;\omega) \\ z(t;\omega) \end{bmatrix} = \begin{bmatrix} 0 & 1 \\ -\alpha^2 & 0 \end{bmatrix} \begin{bmatrix} y(t;\omega) \\ z(t;\omega) \end{bmatrix} + \begin{bmatrix} 0 \\ f(t) \end{bmatrix}, \, t > 0, \, \omega \in \Omega, \\
\begin{bmatrix} y(0;\omega) \\ z(0;\omega) \end{bmatrix} = \begin{bmatrix} x_0(\omega) \\ x_1(\omega) \end{bmatrix}, \, \omega \in \Omega.
\end{cases}
$$

(6.4)

Assuming that the function f is sufficiently smooth, the variation of parameters method can be used to show that the solution of (6.3) is

$$
x(t;\omega) = x_0(\omega)\cos(\alpha t) + \frac{x_1(\omega)}{\alpha}\sin(\alpha t) + \int_0^t \frac{1}{\alpha} f(s)\sin(\alpha(t-s))\,ds, \quad (6.5)
$$

$\forall t \geq 0$, a.s. $[\mathscr{P}]$.

Exercise 6.1.2. Suppose that $f(t) = \beta \sin(\gamma t)$, where $\beta, \gamma, t > 0$.
i.) Simplify (6.5) for this particular choice of $f(t)$.
ii.) For a fixed $\omega \in \Omega$, does there exist $\lim_{t \to \infty} x(t;\omega)$?

Exercise 6.1.3. Incorporate forcing terms into IVP (4.28) describing the motion of two attached springs. Reformulate the resulting random IVP in matrix form.

Now, we investigate what happens when noise is incorporated into such a model via a white noise process. Suppose that the the spring-mass system described by (4.19) is affected by two external forces, each of which is driven by a white noise process, say

$$f_1(t) + \overline{f_1}(t)\frac{dW_1(t)}{dt}, \quad f_2(t) + \overline{f_2}(t)\frac{dW_2(t)}{dt}. \tag{6.6}$$

Assume that $\{W_1(t) : 0 \le t \le T\}$ and $\{W_2(t) : 0 \le t \le T\}$ are independent. Incorporating these into the right-hand side of (4.19) (and relabeling $W(t)$ by $W_0(t)$) results in the SDE

$$\frac{d^2x}{dt^2}(t;\omega) + (\overline{\alpha_1})^2 x(t;\omega) = -(\overline{\alpha_2})^2 x(t;\omega)\frac{dW_0(t)}{dt} + f_1(t) + f_2(t)$$

$$+\overline{f_1}(t)\frac{dW_1(t)}{dt} + \overline{f_2}(t)\frac{dW_2(t)}{dt}, t > 0, \, \omega \in \Omega. \tag{6.7}$$

In order to reformulate (6.7) in matrix form, we use the change of variable

$$\begin{cases} y = y(t;\omega) \equiv x \\ z = z(t;\omega) \equiv \frac{dx}{dt} \end{cases} \tag{6.8}$$

so that

$$\begin{cases} \frac{dy}{dt} = \frac{dx}{dt} & = z, \\ \frac{dz}{dt} = \frac{d^2x}{dt^2} & = -(\overline{\alpha_1})^2 \underbrace{x(t;\omega)}_{=y} - (\overline{\alpha_2})^2 \underbrace{x(t;\omega)}_{=y} \frac{dW_0(t)}{dt} \\ & \quad + f_1(t) + f_2(t) + \overline{f_1}(t)\frac{dW_1(t)}{dt} + \overline{f_2}(t)\frac{dW_2(t)}{dt}. \end{cases}$$

The equivalent differential form of (6.8) is

$$\begin{cases} dy = zdt, \\ dz = -(\overline{\alpha_1})^2 dt + (f_1(t) + f_2(t)) dt + \\ \quad + \left(-(\overline{\alpha_2})^2 ydW_0(t) + \overline{f_1}(t)dW_1(t) + \overline{f_2}(t)dW_2(t)\right). \end{cases} \tag{6.9}$$

(Why?) Now, look back at (4.26) and (4.27). The zero place-holders were used to assist us in correctly identifying the matrices to use in order to reformulate the system (4.25) as the matrix equation (4.27). We do the same thing here, albeit using many more such place-holders, to obtain:

$$\begin{cases} dy = & (0y + z + 0) \, dt + (0y + 0z + 0) dW_0 \\ & + (0y + 0z + 0) dW_1 + (0y + 0z + 0) dW_2, \\ dz = & \left(-(\overline{\alpha_1})^2 y + 0z + (f_1(t) + f_2(t))\right) dt + \left(-(\overline{\alpha_2})^2 y + 0z + 0\right) dW_0 \\ & + \left(0y + 0z + \overline{f_1}(t)\right) dW_1 + \left(0y + 0z + \overline{f_2}(t)\right) dW_2, \\ y(0;\omega) = & x_0(\omega), z(0;\omega) = x_1(\omega). \end{cases}$$

$$\tag{6.10}$$

Subsequently, (6.10) can be written in the following matrix form:

$$
\begin{cases}
\frac{d}{dt}\begin{bmatrix} y(t;\omega) \\ z(t;\omega) \end{bmatrix} = \begin{bmatrix} 0 & 1 \\ -(\overline{\alpha_1})^2 & 0 \end{bmatrix}\begin{bmatrix} y(t;\omega) \\ z(t;\omega) \end{bmatrix}dt + \begin{bmatrix} 0 \\ f_1(t)+f_2(t) \end{bmatrix}dt \\
\qquad + \left(\begin{bmatrix} 0 & 1 \\ -(\overline{\alpha_2})^2 & 0 \end{bmatrix}\begin{bmatrix} y(t;\omega) \\ z(t;\omega) \end{bmatrix} + \begin{bmatrix} 0 \\ 0 \end{bmatrix} \right)dW_0(t) \\
\qquad + \left(\begin{bmatrix} 0 & 0 \\ 0 & 0 \end{bmatrix}\begin{bmatrix} y(t;\omega) \\ z(t;\omega) \end{bmatrix} + \begin{bmatrix} 0 \\ \overline{f_1}(t) \end{bmatrix} \right)dW_1(t) \qquad (6.11) \\
\qquad + \left(\begin{bmatrix} 0 & 0 \\ 0 & 0 \end{bmatrix}\begin{bmatrix} y(t;\omega) \\ z(t;\omega) \end{bmatrix} + \begin{bmatrix} 0 \\ \overline{f_2}(t) \end{bmatrix} \right)dW_2(t), \\
\begin{bmatrix} y(0;\omega) \\ z(0;\omega) \end{bmatrix} = \begin{bmatrix} x_0(\omega) \\ x_1(\omega) \end{bmatrix}.
\end{cases}
$$

Exercise 6.1.4. Conjecture the form of the abstract nonhomogenous stochastic IVP in $\mathbb{R}^N$ that subsumes all the IVPs discussed above as a special case.

Common Theme: As in previous chapters, we encountered two types of stochastic IVPs. One, when the initial data are random variables and no other noise is introduced into the model, the resulting stochastic IVP is of the form

$$
\begin{cases}
\mathbf{U}'(t;\omega) = \mathbf{A}\mathbf{U}(t;\omega) + \mathbf{F}(t), t > 0, \omega \in \Omega, \\
\mathbf{U}(0;\omega) = \mathbf{U}_0(\omega), \omega \in \Omega,
\end{cases} \qquad (6.12)
$$

where $\mathbf{U} : [0,\infty) \times \Omega \to \mathbb{R}^N$ is the solution stochastic process, $\mathbf{A}$ is an $N \times N$ constant matrix, $\mathbf{U}_0 \in \mathcal{L}^2(\Omega;\mathbb{R}^N)$ is the vector containing the initial conditions, and $\mathbf{F} : [0,T] \to \mathbb{R}^N$ is the forcing term. Assuming that $\mathbf{F}$ is a continuous mapping, (6.12) can be solved explicitly, $\forall \omega \in \Omega$, as follows:

$$
\begin{aligned}
\mathbf{U}'(s;\omega) - \mathbf{A}\mathbf{U}(s;\omega) &= \mathbf{F}(s) \\
e^{-\mathbf{A}s}\left(\mathbf{U}'(s;\omega) - \mathbf{A}\mathbf{U}(s;\omega)\right) &= e^{-\mathbf{A}s}\mathbf{F}(s) \\
\frac{d}{ds}\left(e^{-\mathbf{A}s}\mathbf{U}(s;\omega)\right) &= e^{-\mathbf{A}s}\mathbf{F}(s) \qquad (6.13) \\
\int_0^t \frac{d}{ds}\left(e^{-\mathbf{A}s}\mathbf{U}(s;\omega)\right)ds &= \int_0^t e^{-\mathbf{A}s}\mathbf{F}(s)ds \\
e^{-\mathbf{A}t}\mathbf{U}(t;\omega) - e^{-\mathbf{A}(0)}\mathbf{U}(0;\omega) &= \int_0^t e^{-\mathbf{A}s}\mathbf{F}(s)ds \\
e^{-\mathbf{A}t}\mathbf{U}(t;\omega) &= \mathbf{U}_0(\omega) + \int_0^t e^{-\mathbf{A}s}\mathbf{F}(s)ds
\end{aligned}
$$

$$\mathbf{U}(t;\omega) = e^{\mathbf{A}t}\mathbf{U}_0(\omega) + e^{\mathbf{A}t}\int_0^t e^{-\mathbf{A}s}\mathbf{F}(s)ds$$

$$\mathbf{U}(t;\omega) = e^{\mathbf{A}t}\mathbf{U}_0(\omega) + \int_0^t e^{\mathbf{A}(t-s)}\mathbf{F}(s)ds.$$

As such, the deterministic theory established in Volume 1 can be used to study (6.12) and the related random IVPs.

Incorporating noise into the model through parameter estimation and time-dependent forcing terms via Wiener processes results in a stochastic IVP in $\mathbb{R}^N$ of the abstract form

$$\begin{cases} d\mathbf{U}(t;\omega) = [\mathbf{A}\mathbf{U}(t;\omega) + \mathbf{F}(t)]\,dt + \sum_{k=1}^m [\mathbf{B}_k\mathbf{U}(t;\omega) + \mathbf{g}_k(t)]\,dW_k(t), \\ \mathbf{U}(0;\omega) = \mathbf{U}_0(\omega), \end{cases} \tag{6.14}$$

where $t > 0$, $\omega \in \Omega$. Here, $\mathbf{U} : [0,\infty) \times \Omega \to \mathbb{R}^N$ is the solution stochastic process; $\mathbf{A}, \mathbf{B}_1, \ldots, \mathbf{B}_m$ are an $N \times N$ constant matrices; $\{W_i(t) : 0 \le t \le T\}$ $(i = 1,\ldots,m)$ are independent $\mathbb{R}$-valued Wiener processes; $\mathbf{U}_0 \in \mathfrak{L}^2\left(\Omega; \mathbb{R}^N\right)$ is an $\mathscr{F}_0$-measurable random variable independent of $\{W_i(t) : 0 \le t \le T\}$ $(i = 1,\ldots,m)$; and $\mathbf{F} : [0,T] \to \mathbb{R}^N$ and $\mathbf{g}_k : [0,T] \to \mathbb{R}^N$ $(k = 1,\ldots,m)$ are given mappings.

The hope is that there is a variation of parameters formula similar to (6.13) for a mild solution of (6.14). As it turns out, as long as the matrices $\mathbf{A}, \mathbf{B}_1, \ldots, \mathbf{B}_m$ all mutually commute, such a formula can be derived, albeit it is not nearly as nice as (6.13) due to the presence of the so-called *multiplicative noise* terms $\mathbf{B}_k\mathbf{U}(t;\omega)dW_k(t)$ terms in (6.14). However, we shall see that when $\mathbf{B}_k = \mathbf{0}$, $\forall k \in \{1,\ldots,m\}$, the strong solution of (6.14) *can* be represented using a very natural variation of parameters formula that will serve as the basis of our study in the remainder of the text.

We shall establish the theory first for the one-dimensional case, and then comment on the extension of the theory for SDEs in $\mathbb{R}^N$.

6.2 Nonhomogenous Linear SDEs in $\mathbb{R}$

6.2.1 The Cauchy Problem — Existence/Uniqueness Theory

We focus in this section on linear SDEs in $\mathbb{R}$ that are now equipped with a time-dependent forcing term, namely

$$\begin{cases} dX(t;\omega) = [aX(t;\omega) + f(t)]\,dt + \sum_{k=1}^m [c_kX(t;\omega) + g_k(t)]\,dW_k(t), \\ X(0;\omega) = X_0(\omega), \end{cases} \tag{6.15}$$

where $0 < t < T$, $\omega \in \Omega$; $a, c_k(k = 1,\ldots,m)$ are real constants; and $f : [0,T] \to \mathbb{R}$ and $g_k : [0,T] \to \mathbb{R}$ $(k = 1,\ldots,m)$ are mappings.

We assume **(H3.1)** through **(H3.3)** (suitably modified for (6.15)). But, what must we impose on the mappings f and g_k? Consider the integrated form of (6.15):

$$X(t;\omega) = X_0(\omega) + \int_0^t [aX(s;\omega) + f(s)]\,ds + \sum_{k=1}^m \int_0^t [c_k X(s;\omega) + g_k(s)]\,dW_k(s).$$

(6.16)

First, if we are to reasonably expect to prove the existence of a strong solution, it is clear that the integrals $\int_0^t f(s)\,ds$ and $\int_0^t g_k(s)\,dW_k(s)$ $(k = 1,\dots,m)$ must be defined. (Why?) Moreover, looking at the proof of Thrm. 3.4.1 with the foresight of mimicking the approach closely, we see that we will need to also compute the quantities

$$\int_0^t E\,|f(s)|^2\,ds \text{ and } \int_0^t E\,|g_k(s)|\,ds\ (k = 1,\dots,m),$$

and ultimately we will want the sample paths of the solution processes to be continuous a.s. $[\mathscr{P}]$. All of these conditions are satisfied if we assume the following:

(H6.1) $f : [0,T] \to \mathbb{R}$ and $g_k : [0,T] \to \mathbb{R}$ $(k = 1,\dots,m)$ are continuous.

(Why?) This condition is actually stronger than necessary, but it suffices for our purposes. We seek a strong solution of (6.15) in the following sense.

Definition 6.2.1. A stochastic process $X : [0,T] \times \Omega \longrightarrow \mathbb{R}$ is a *strong solution* of (6.15) if $X \in \mathbb{C}\left([0,T]; \mathscr{L}^2\left(\Omega;\mathbb{R}\right)\right)$ and $X(t)$ satisfies (6.16), $\forall 0 \le t \le T$, a.s. $[\mathscr{P}]$.

Our first main result is an extension of Thrm. 3.4.1 that takes into account time-dependent forcing terms.

Theorem 6.2.2. *If (H3.1) through (H3.3) and (H6.1) are satisfied, then (6.15) has a unique strong solution on* $[0,T]$.

Outline of Proof. We follow the proof of Thrm. 3.4.1 and make appropriate modifications. Consider the sequence

$$\begin{cases} X_0(t;\omega) &= X_0(\omega), \\ X_n(t;\omega) &= X_0(\omega) + \int_0^t [aX_{n-1}(s;\omega) + f(s)]\,ds \\ & \quad + \sum_{k=1}^m \int_0^t [c_k X_{n-1}(s;\omega) + g_k(s)]\,dW_k(s), \end{cases}$$

(6.17)

where $0 \le t \le T$, $\omega \in \Omega$, $n \in \mathbb{N}$. The uniqueness portion of the proof is identical to the argument used for Thrm. 3.4.1 because upon subtracting two strong solutions given by (6.16), the terms containing f and g_k cancel immediately (due to their lack of dependence on $X(t;\omega)$). Consequently, we still have the estimate (3.117) and the same reasoning works thereafter. (Convince yourself.)

As for the existence portion of the proof, the main steps remain the same, but the estimates now involve f and g_k. We only provide an outline of the proof and comment on where the changes occur.

Exercise 6.2.1. Try formulating the existence proof BEFORE proceeding!

<u>Claim 1</u>: X_n defined in (6.17) is a well-defined martingale, $\forall n \geq 0$.
Proof. Because deterministic continuous functions are automatically $\mathscr{F}_t$-adapted, the fact that X_n is a real-valued martingale, $\forall n \in \mathbb{N}$, follows as before.$\diamond$

<u>Claim 2</u>: The sequence $\{X_n\}$ is a uniformly bounded subset of $\mathbb{C}\left([0,T];\mathfrak{L}^2(\Omega;\mathbb{R})\right)$.
Proof. Let $0 \leq t \leq T$ and $n \in \mathbb{N}$. We must verify (3.87), but the constants $\zeta_1, \zeta_2 > 0$ now also depend on f and g_k. Indeed, observe that (3.91) is now replaced by

$$
E\left|X_n(t;\cdot)\right|^2 = E\left[\underbrace{X_0(\cdot) + \int_0^t aX_{n-1}(s;\cdot)ds + \sum_{k=1}^m \int_0^t c_k X_{n-1}(s;\cdot)dW_k(s)}_{\text{Old terms}}\right.
$$

$$
\left.\left.+\underbrace{\int_0^t f(s)ds + \sum_{k=1}^m \int_0^t g_k(s)dW_k(s)}_{\text{New terms}}\right]\right|^2 \tag{6.18}
$$

$$
\leq \underbrace{5\|X_0\|_{\mathfrak{L}^2}^2 + 5T\left(a^2 + 8m\sum_{k=1}^m c_k^2\right)\int_0^t E\left|X_{n-1}(s;\cdot)\right|^2 ds}_{\text{Previous estimate}}
$$

$$
+5E\left(\int_0^t f(s)ds\right)^2 + 5m\sum_{k=1}^m E\left|\int_0^t g_k(s)dW_k(s)\right|^2.
$$

Using Lemma 3.2.7 with (**H6.1**) yields

$$
E\left(\int_0^t f(s)ds\right)^2 \leq T\|f\|_{\mathbb{C}([0,T];\mathbb{R})}^2 \tag{6.19}
$$

$$
E\left|\int_0^t g_k(s)dW_k(s)\right|^2 \leq 4T\int_0^T E\left|g_k(s)\right|^2 ds \leq 4T^2\|g_k\|_{\mathbb{C}([0,T];\mathbb{R})}^2. \tag{6.20}
$$

Using (6.19) and (6.20) in (6.18) and simplifying reveals that we can use the following constants $\zeta_1, \zeta_2 > 0$ to verify (3.87):

$$
\zeta_1 = 5\|X_0\|_{\mathfrak{L}^2}^2 + T\|f\|_{\mathbb{C}([0,T];\mathbb{R})}^2 + 20mT^2\sum_{k=1}^m \|g_k\|_{\mathbb{C}([0,T];\mathbb{R})}^2
$$

$$
\zeta_2 = 5T\left(a^2 + 8m\sum_{k=1}^m c_k^2\right)
$$

The rest of the argument remains the same. $\diamond$

<u>Claim 3</u>: There exists $\zeta > 0$ (independent of n) such that $\forall n \geq 0$ and $\forall 0 \leq t \leq T$,

$$E\,|X_{n+1}(t;\cdot) - X_n(t;\cdot)|^2 \leq \frac{(\zeta t)^{n+1}}{(n+1)!}. \qquad (6.21)$$

Proof. The only significant change occurs in the estimate of $E\,|X_1(t;\cdot) - X_0(t;\cdot)|^2$. Using (6.17), we proceed as in Claim 2 to see that

$$E\,|X_1(t;\cdot) - X_0(t;\cdot)|^2 \leq 4\left[a^2 + 8m\sum_{k=1}^{m} c_k^2 + \|f\|_{C([0,T];\mathbb{R})}^2 + \right.$$

$$\left. 4mT\sum_{k=1}^{m}\|g_k\|_{C([0,T];\mathbb{R})}^2\right]\|X_0\|_{\mathfrak{L}^2}^2\, t \qquad (6.22)$$

$$= \frac{(\zeta^\star t)^{0+1}}{(0+1)!}.$$

(Tell why.) The induction step is identical to the one used in the homogenous argument because the terms involving f and g_k cancel. (Convince yourself.) $\Diamond$

<u>Claims 4, 5, and 6</u>: The reasoning is identical to what was used in the proof of Thrm. 3.4.1, using the modified estimates established above. $\Diamond$

This completes the outline of the proof of Thrm. 6.2.2. $\square$

Exercise 6.2.2.
i.) Write up a polished proof of Thrm. 6.2.2.
ii.) Can hypothesis (**H6.1**) be weakened and still produce the same result?

Now that we know that a strong solution exists, can we find a representation formula for it? We have asked this question in earlier chapters and each time Itó's formula was the tool that enables us to employ a typical deterministic approach (e.g., separation of variables) to derive a representation formula for the SDE. The same is true here. For simplicity, we shall consider (6.15) with only one noise source, namely

$$\begin{cases} dX(t;\omega) = [aX(t;\omega) + f(t)]\,dt + [cX(t;\omega) + g(t)]\,dW(t), \\ X(0;\omega) = X_0(\omega),\ \omega \in \Omega \end{cases} \qquad (6.23)$$

where $t > 0$, $\omega \in \Omega$. Assume (**H3.1**) through (**H3.3**) and (**H6.1**), appropriately modified. We follow the standard variation of parameters technique (cf. (6.13)), but adapted to the stochastic setting using Itó's formula, as explained in **[142]**.

<u>Step 1</u>: Solve the following IVP that corresponds to the homogenous part of (6.23):

$$\begin{cases} dX_h(t;\omega) = aX_h(t;\omega)dt + cX_h(t;\omega)dW(t),\ t > 0,\ \omega \in \Omega, \\ X_h(0;\omega) = 1,\ \omega \in \Omega. \end{cases} \qquad (6.24)$$

(Note: For simplicity, we assume that the initial condition in (6.24) is identically 1. This is typical, even in the deterministic setting, in the sense that the initial condition is applied at the very end of the procedure in order to identify the arbitrary constant arising in the computations. Using more than one such constant arising from several anti-differentiations simply creates unnecessary notational complexity.)

We have from (3.126) that

$$X_h(t;\omega) = e^{at}\left[e^{-\frac{1}{2}c^2 t + cW(t)}\right]. \tag{6.25}$$

Step 2: Assume that the strong solution of (6.24) is given by

$$X(t;\omega) = k(t;\omega)X_h(t;\omega), \tag{6.26}$$

where $k(t;\omega)$ plays the role of the "varying parameter." If $k(t;\omega)$ were a constant random variable, then (6.26) would not be a strong solution of (6.24). (Why?) As such, we must identify $k(t;\omega)$ such that (6.26) satisfies (6.23). To this end, observe that (6.26) is equivalent to

$$k(t;\omega) = X(t;\omega)X_h^{-1}(t;\omega) = X(t;\omega)e^{-at}\left[e^{\frac{1}{2}c^2 t - cW(t)}\right]. \tag{6.27}$$

(Why?) Thinking ahead, if we could compute $d\left[k(s;\omega)\right]$ using Itó's formula, then we could integrate the result over $(0,t)$ and subsequently multiply both sides by $X_h(t;\omega)$ (given by (6.25)) to arrive at the desired representation formula. We are in luck because both $X(t;\omega)$ and $X_h^{-1}(t;\omega)$ are Itó processes involving the same Wiener process (Why?). As such, we can apply the integration by parts formula (cf. Exer. 3.2.13) to obtain

$$d\left[k(s;\omega)\right] = d\left[X(s;\omega)X_h^{-1}(s;\omega)\right] \tag{6.28}$$
$$= X(s;\omega)d\left[X_h^{-1}(s;\omega)\right] + X_h^{-1}(s;\omega)d[X(s;\omega)] + d[X(s;\omega)]d\left[X_h^{-1}(s;\omega)\right].$$

Using (6.24) with Itó's formula yields

$$d\left[X_h^{-1}(s;\omega)\right] = -X_h^{-1}(s;\omega)\left(a - c^2\right)ds - cX_h^{-1}(s;\omega)dW(s). \tag{6.29}$$

Substituting (6.24) and (6.29) into (6.28) yields, after simplification,

$$d\left[X(s;\omega)X_h^{-1}(s;\omega)\right] = X(s;\omega)\left[-X_h^{-1}(s;\omega)\left(a - c^2\right)ds - cX_h^{-1}(s;\omega)dW(s)\right]$$
$$+ X_h^{-1}(s;\omega)\left[[aX(s;\omega) + f(s)]ds + [cX(s;\omega) + g(s)]dW(s)\right]$$
$$+ c\left(cX(s;\omega) + g(s)\right)X_h^{-1}(s;\omega)ds \tag{6.30}$$
$$= X_h^{-1}(s;\omega)\left[(f(s) - cg(s))ds + g(s)dW(s)\right].$$

Integrating (6.30) over $(0,t)$ implies

$$X(t;\omega)X_h^{-1}(t;\omega) - \underbrace{X(0;\omega)X_h^{-1}(0;\omega)}_{=X_0(\omega)} = \int_0^t X_h^{-1}(s;\omega)\left(f(s) - cg(s)\right)ds$$

$$+ \int_0^t X_h^{-1}(s;\omega)g(s)dW(s). \tag{6.31}$$

Using (6.25) and the formula for $X_h^{-1}(s;\omega)$ (as extracted from (6.27)) further yields the following representation formula:

$$X(t;\omega) = X_0(\omega)e^{at}\left[e^{-\frac{1}{2}c^2t+cW(t)}\right]$$
$$+ \int_0^t e^{a(t-s)}\left[e^{-\frac{1}{2}c^2(t-s)+c(W(t)-W(s))}\right](f(s)-cg(s))\,ds$$
$$+ \int_0^t e^{-as}\left[e^{\frac{1}{2}c^2s-cW(s)}\right]g(s)dW(s). \tag{6.32}$$

Exercise 6.2.3. Verify (6.29).

Exercise 6.2.4. Derive a formula in the spirit of (6.32) directly for (6.15).

The special case when $c_k = 0, \forall k \in \{1,\ldots,m\}$ in (6.15) (i.e., the case of *additive noise*) and the corresponding representation formula will be of particular interest in the upcoming chapters. Specifically, the strong solution of the stochastic IVP

$$\begin{cases} dX(t;\omega) = [aX(t;\omega)+f(t)]\,dt + \sum_{k=1}^m g_k(t)dW_k(t), t > 0, \omega \in \Omega, \\ X(0;\omega) = X_0(\omega), \omega \in \Omega, \end{cases} \tag{6.33}$$

is given by the variation of parameters formula

$$X(t;\omega) = e^{at}X_0(\omega) + \int_0^t e^{a(t-s)}f(s)ds + \int_0^t e^{a(t-s)}\left[\sum_{k=1}^m g_k(s)\right]dW_k(s). \tag{6.34}$$

(Tell why.) When $g_k \equiv 0, \forall k \in \{1,\ldots,m\}$ in (6.33), the formula (6.34) reduces to the variation of parameters formula for (6.12).

Remark. Observe that (6.34) is a stochastic integral equation related to (6.33), but it is different from the directly integrated form of (6.16). Presently, formula (6.34) coincides with the strong solution of (6.33). However, we will see that this is not necessarily true in the Hilbert space setting. As such, we introduce the following weaker notion of a solution of (6.33):

Definition 6.2.3. A stochastic process $X : [0,T] \times \Omega \longrightarrow \mathbb{R}$ is a *mild solution* of (6.33) if $X \in \mathcal{C}\left([0,T]; \mathcal{L}^2(\Omega;\mathbb{R})\right)$ and $X(t;\omega)$ satisfies (6.34), $\forall 0 \le t \le T$, a.s. $[\mathscr{P}]$.

The distinction between mild and strong solutions of a given stochastic Cauchy problem begins to play a crucial role in Chapter 7 where we focus solely on mild solutions due to the complexity of the forcing term.

6.2.2   Continuous Dependence Estimates

Consider (6.15) and the related Cauchy problem

$$\begin{cases} dY(t;\omega) = [aY(t;\omega)+f^\star(t)]\,dt + \sum_{k=1}^m \left[c_k Y(t;\omega)+g_k^\star(t)\right]dW_k(t), \\ Y(0;\omega) = Y_0(\omega), \end{cases} \tag{6.35}$$

where $0 < t < T$, $\omega \in \Omega$, both under hypotheses **(H3.1)** through **(H3.3)** and **(H6.1)**. We seek an estimate for the quantity $E|X(t;\cdot) - Y(t;\cdot)|^2$ in terms of the data and parameters, namely $a, c_k, X_0, Y_0, f, f^\star, g_k, g_k^\star$, and T. We shall assume that the corresponding parameters and initial data for (6.15) and (6.35) are "close" in the following sense:

(H6.2) There exists $\delta_1 > 0$ such that $\|X_0 - Y_0\|^2_{\mathscr{L}^2(\Omega;\mathbb{R})} < \delta_1$.

(H6.3) There exists $\delta_2 > 0$ such that $\|f - f^\star\|^2_{\mathbb{C}([0,T];\mathbb{R})} < \delta_2$.

(H6.4) For every $\forall k \in \{1, \ldots, m\}$, $\exists \delta_{k+2} > 0$ such that $\|g_k - g_k^\star\|^2_{\mathbb{C}([0,T];\mathbb{R})} < \delta_{k+2}$.

The continuous dependence result is as follows.

Proposition 6.2.4. *If **(H3.1)** through **(H3.3)** and **(H6.1)** - **(H6.4)** hold, then $\forall t \in [0,T]$,*

$$E|X(t;\cdot) - Y(t;\cdot)|^2 \leq 5\left(\delta_1 + T^2\left(\delta_2 + m\sum_{k=1}^m \delta_{k+2}\right)\right)e^{T\left(a^2 + m\sum_{k=1}^m c_k^2\right)t}. \quad (6.36)$$

Proof. Let $0 \leq t \leq T$. Subtracting the integrated forms of (6.15) and (6.35) yields

$$E|X(t;\cdot) - Y(t;\cdot)|^2 \leq E\left|(X_0(\cdot) - Y_0(\cdot)) + \int_0^t a(X(s;\cdot) - Y(s;\cdot))\,ds\right.$$

$$+ \sum_{k=1}^m \int_0^t c_k(X(s;\cdot) - Y(s;\cdot))\,dW_k(s)$$

$$\left. + \int_0^t (f(s) - f^\star(s))\,ds + \sum_{k=1}^m \int_0^t (g_k(s) - g_k^\star(s))\,dW_k(s)\right|^2$$

$$\leq 5\left[\|X_0 - Y_0\|^2_{\mathscr{L}^2(\Omega;\mathbb{R})} + 5T^2\|f - f^\star\|^2_{\mathbb{C}([0,T];\mathbb{R})}\right.$$

$$+ \left(a^2 + m\sum_{k=1}^m c_k^2\right)T\int_0^t E|X(s;\cdot) - Y(s;\cdot)|^2\,ds$$

$$\left. + 5mT^2\sum_{k=1}^m \|g_k - g_k^\star\|^2_{\mathbb{C}([0,T];\mathbb{R})}\right] \quad (6.37)$$

$$\leq 5\left(\delta_1 + T^2\left(\delta_2 + m\sum_{k=1}^m \delta_{k+2}\right)\right)$$

$$+ T\left(a^2 + m\sum_{k=1}^m c_k^2\right)\int_0^t E|X(s;\cdot) - Y(s;\cdot)|^2\,ds.$$

An application of Gronwall's Lemma in (6.37) immediately yields (6.36). (Tell how.)

$\square$

Exercise 6.2.5. Let $0 < \varepsilon < 1$. Identify a natural way in which to restrict the δ_j values in Prop. 6.2.4 (in the sense of requiring $\delta_j < C_j(\varepsilon)$, for appropriately chosen

$C_j(\varepsilon)$, $j = 1,\ldots,m+2$) to ensure that $\|X - Y\|_{\mathbb{C}([0,T];\mathbb{R})} < \varepsilon$.

Exercise 6.2.6. Suppose that the $\mathbb{C}([0,T];\mathbb{R})$-norm in **(H6.3)** and **(H6.4)** is replaced by the $\mathbb{L}^2(0,T;\mathbb{R})$-norm.
i.) How does this change affect the estimate (6.36)?
ii.) Which estimate is better? Why?

Now, consider the special case of (6.15) and (6.35) in which $c_k = 0$, $\forall k \in \{1,\ldots,m\}$. We can establish a continuous dependence estimate for $E\,|X(t;\cdot) - Y(t;\cdot)|^2$ by appealing directly to the variation of parameters (6.34) instead of using the integrated form of the Cauchy problem. The estimate thusly obtained will not be the same as those established above, but it will involve the same parameters and initial data.

Exercise 6.2.7.
i.) Assume **(H6.2)** through **(H6.4)**. Subtract the variation of parameters formulae for the mild solutions of (6.15) and (6.35) and derive an estimate for $E\,|X(t;\cdot) - Y(t;\cdot)|^2$. How does this estimate compare to the one provided by (6.36)? Which seems less restrictive? Why?
ii.) Redo part (i), but this time replace the $\mathbb{C}([0,T];\mathbb{R})$-norm by the $\mathbb{L}^2(0,T;\mathbb{R})$-norm in **(H6.3)** and **(H6.4)**. Compare the result to the estimate obtained in (i).

Exercise 6.2.8. Assume that $X_0, Y_0 \in \mathscr{L}^p(\Omega;\mathbb{R})$, for $p > 2$, and replace the power on all norms in **(H6.2)** through **(H6.4)** by p. Formulate an estimate for $E\,|X(t;\cdot) - Y(t;\cdot)|^p$.

Exercise 6.2.9. Suppose that a in (6.35) is replaced by $a^\star$ and assume:

(H6.5) There exists $\delta_0 > 0$ such that $|a - a^\star| < \delta_0$.

i.) Assume that **(H6.2)** through **(H6.4)** hold. Formulate an estimate for $E\,|X(t;\cdot) - Y(t;\cdot)|^2$. What differences do you encounter in comparison to Prop. 6.2.4?
ii.) Redo (i) assuming that the $\mathbb{C}([0,T];\mathbb{R})$-norm is replaced by the $\mathbb{L}^2(0,T;\mathbb{R})$-norm in **(H6.3)** and **(H6.4)**.
iii.) Assuming the modification introduced in Exer. 6.2.8, redo (i) and (ii) to establish analogous estimates for $E\,|X(t;\cdot) - Y(t;\cdot)|^p$, for $p > 2$.

6.2.3 Statistical Properties of the Solution

Mean and Variance

Let $0 \le t \le T$ and consider the strong solution $X(t;\omega)$ of (6.15). We seek formulae for $\mu_X(t)$ and $Var_X(t)$ in the spirit of those derived in Section 3.6.1. As motivation, we begin with the simpler Cauchy problem

$$\begin{cases} dX(t;\omega) = [aX(t;\omega) + f(t)]dt + cX(t;\omega)dW(t), t > 0, \omega \in \Omega, \\ X(0;\omega) = X_0(\omega), \omega \in \Omega. \end{cases} \tag{6.38}$$

Taking expectations on both sides of the integrated form of (6.38) yields

$$\mu_X(t) = E[X_0(\cdot)] + E\left[\int_0^t [aX(s;\cdot) + f(s)]ds\right] + E\left[\int_0^t cX(s;\cdot)dW(s)\right]$$

$$= \mu_X(0) + a\int_0^t [\mu_X(s) + f(s)]ds. \tag{6.39}$$

(Why?) Observe that $t \mapsto \mu_X(t)$ satisfies the deterministic IVP

$$\begin{cases} \frac{d\mu_X(t)}{dt} = a\mu_X(t) + f(t), \quad 0 \le t \le T, \\ \mu_X(0) = E[X_0(\cdot)], \end{cases} \tag{6.40}$$

the unique solution of which is

$$\mu_X(t) = e^{at}E[X_0(\cdot)] + \int_0^t e^{a(t-s)}f(s)ds. \tag{6.41}$$

Exercise 6.2.10. Verify that if X is the strong solution of (6.15), then $t \mapsto \mu_X(t)$ is *still* given by (6.41).

Exercise 6.2.11. Consider (6.15) with $c_k = 0, \forall k \in \{1,\dots,m\}$. Repeat Exer. 6.2.10, but this time directly using the variation of parameters formula (6.34) for the solution. Which approach seems easier?

Next, we shall compute $Var_X(t)$ for (6.38). As in (3.139), we need an expression for $E[X^2(t;\cdot)]$.

Exercise 6.2.12.
i.) Show that

$$d(X^2(t;\cdot)) = [(2a + c^2)X^2(t;\cdot) + 2X(t;\cdot)f(t)]dt + 2cX^2(t;\cdot)dW(t). \tag{6.42}$$

ii.) Deduce from (i) that

$$X^2(t;\cdot) = X_0^2(\cdot) + (2a + c^2)\int_0^t X^2(s;\cdot)ds \tag{6.43}$$

$$+ \int_0^t 2X(s;\cdot)f(s)ds + 2c\int_0^t X^2(s;\cdot)dW(s).$$

Taking the expectation on both sides of (6.43) yields

$$E[X^2(t;\cdot)] = E[X_0^2(\cdot)] + (2a + c^2)\int_0^t E[X^2(s;\cdot)]ds + \int_0^t 2\mu_X(s)f(s)ds. \tag{6.44}$$

(Why?) So, the deterministic function $t \mapsto E\left[X^2(t;\cdot)\right]$ satisfies the IVP

$$\begin{cases} \frac{d\left(E\left[X^2(t;\cdot)\right]\right)}{dt} = \left(2a+c^2\right) E\left[X^2(t;\cdot)\right] + 2\mu_X(t)f(t), & 0 \le t \le T, \\ E\left[X^2(0;\cdot)\right] = E\left[X_0^2(\cdot)\right]. \end{cases} \tag{6.45}$$

Hence, we conclude that

$$E\left[X^2(t;\cdot)\right] = E\left[X_0^2(\cdot)\right] e^{\left(2a+c^2\right)t} + \int_0^t e^{\left(2a+c^2\right)(t-s)} 2\mu_X(s)f(s)ds. \tag{6.46}$$

Substituting (6.41) and (6.46) into (3.139) yields the desired formula.

Exercise 6.2.13. Derive a formula for $Var_X(t)$ for (6.33).

Exercise 6.2.14. Derive a formula for $Cov_X(t,s)$ for the strong solution of (6.15) with $c_k = 0$, $\forall k \in \{1,\ldots,m\}$.

Exercise 6.2.15. Let $\{f_n\}$ be a sequence in $\mathbb{C}\left([0,T];\mathbb{R}\right)$ for which $f_n \longrightarrow f$ uniformly on $[0,T]$. For each $n \in \mathbb{N}$, consider the IVP

$$\begin{cases} dX_n(t;\omega) = \left[aX_n(t;\omega) + f_n(t)\right]dt + cX_n(t;\omega)dW(t), t > 0, \omega \in \Omega, \\ X_n(0;\omega) = X_0(\omega), \omega \in \Omega. \end{cases} \tag{6.47}$$

i.) Compute $\lim_{n\to\infty} \mu_{X_n}(t)$ and $\lim_{n\to\infty} Var_{X_n}(t)$.
ii.) How are the quantities in (i) related to $\mu_X(t)$ and $Var_X(t)$, where X is the strong solution of (6.38)?

Moment Estimates

Different approaches can be used to establish p^{th} moment estimates of (6.15), depending on which terms are present and the level of detail required of the constants arising in the estimate. For instance, consider (6.15) with $c_k = 0$, $\forall k \in \{1,\ldots,m\}$. In such case, the strong solution of (6.15) is given by (6.34). We derive a pth moment estimate directly, as follows:

$$\begin{aligned} E\left|X(t;\cdot)\right|^p &= E\left|e^{at}X_0(\cdot) + \int_0^t e^{a(t-s)}f(s)ds + \sum_{k=1}^m \int_0^t e^{a(t-s)}g_k(s)dW_k(s)\right|^p \\ &\le 3^{p-1}\left[e^{apt}E\left|X_0(\cdot)\right|^p + E\left|\int_0^t e^{a(t-s)}f(s)ds\right|^p \right. \\ &\quad \left. + m^{p-1}\sum_{k=1}^m E\left|\int_0^t e^{a(t-s)}g_k(s)dW_k(s)\right|^p\right] \\ &= 3^{p-1}\left[I_1 + I_2 + I_3\right]. \end{aligned} \tag{6.48}$$

Using Lemma 3.2.7 yields

$$I_2 \leq t^{\frac{p}{q}} \int_0^t e^{ap(t-s)} f^p(s)\, ds, \tag{6.49}$$

$$I_3 \leq m^{p-1} \left(\frac{tp(p-1)}{2} \right)^{\frac{p}{2}} \int_0^t e^{ap(t-s)} g_k^p(s)\, ds. \tag{6.50}$$

Substituting (6.49) and (6.50) into (6.48) yields the following p^{th} moment estimate, upon simplification:

$$E\,|X(t;\cdot)|^p \leq 3^{p-1} e^{apt} \left[\|X_0\|_{\mathscr{L}^p(\Omega;\mathscr{H})}^p + t^p \left(\int_0^t e^{-aps} \left[f^p(s) + g_k^p(s) \right] ds \right) \times \right.$$
$$\left. \left(t^{\frac{1}{q}} + m^{p-1} t^{\frac{1}{2}} m^{p-1} \left(\frac{p(p-1)}{2} \right)^{\frac{p}{2}} \right) \right]. \tag{6.51}$$

However, when the noise includes a multiplicative part, the representation formula is not nearly as nice and so, we apply an alternate approach. For simplicity, consider (6.23).

Proposition 6.2.5. *Assume that $\forall p \in \mathbb{N}$, $E\,|X_0|^p < \infty$, and let X be the strong solution of (6.23). Then, $\forall p \in \mathbb{N}$, there exist $\alpha_p, \beta_p > 0$ such that*

$$E\left[X^p(t;\cdot) \right] \leq \alpha_p \left(1 + E\,|X_0(\cdot)|^p \right) e^{\beta_p t}, \quad 0 \leq t \leq T. \tag{6.52}$$

Proof. We use strong induction on p.
Let $p = 1$. Observe that by (6.41), the following holds, $\forall 0 \leq t \leq T$:

$$E\left[X(t;\cdot) \right] = E\,|X_0(\cdot)|\, e^{at} + e^{at} \int_0^t e^{-as} f(s)\, ds$$
$$\leq e^{|a|t} \left(E\,|X_0(\cdot)| + \left\| e^{-a(\cdot)} f(\cdot) \right\|_{\mathbb{L}^1(0,T;\mathbb{R})} \right) \tag{6.53}$$
$$\leq \max\left\{ 1, \left\| e^{-a(\cdot)} f(\cdot) \right\|_{\mathbb{L}^1(0,T;\mathbb{R})} \right\} \left[1 + E\,|X_0(\cdot)| \right] e^{|a|t}.$$

Choosing $\alpha_1 = \max\left\{ 1, \left\| e^{-a(\cdot)} f(\cdot) \right\|_{\mathbb{L}^1(0,T;\mathbb{R})} \right\}$ and $\beta_1 = |a|$ is sufficient to establish the base case. (Why?)

Next, let $p_0 \in \mathbb{N}$ and assume that $\forall p \in \mathbb{N}$ for which $p \leq p_0$, there exist $\alpha_p, \beta_p > 0$ such that (6.52) holds. We shall show that there exist $\alpha_{p+1}, \beta_{p+1} > 0$ such that

$$E\left[X^{p_0+1}(t;\cdot) \right] \leq \alpha_{p_0+1} \left(1 + E\,|X_0(\cdot)|^{p_0+1} \right) e^{\beta_{p_0+1} t}, \quad 0 \leq t \leq T. \tag{6.54}$$

An application of Itó's formula with $h(s,x) = x^{p_0+1}$ yields the following formula, for $0 \le s \le T$:

$$
\begin{aligned}
d\left[X^{p_0+1}(s;\cdot)\right] &= \left[a(p_0+1) + c^2 p_0(p_0+1)\right] X^{p_0+1}(s;\cdot) ds \\
&\quad + \left[(p_0+1)f(s) + 2cp_0(p_0+1)g(s)\right] X^{p_0}(s;\cdot) ds \\
&\quad + p_0(p_0+1)g^2(s) X^{p_0-1}(s;\cdot) ds \\
&\quad + (p_0+1)(cX(s;\cdot) + g(s)) X^{p_0}(s;\cdot) dW(s).
\end{aligned}
\tag{6.55}
$$

(Why?) Integrating (6.55) over $(0,t)$ and subsequently taking expectations yields

$$
\begin{aligned}
E\left[X^{p_0+1}(t;\cdot)\right] &\le E\left[X_0^{p_0+1}(\cdot)\right] + \left[a(p_0+1) + c^2 p_0(p_0+1)\right] \int_0^t E\left[X^{p_0+1}(s;\cdot)\right] ds \\
&\quad + \int_0^t \left[(p_0+1)f(s) + 2cp_0(p_0+1)g(s)\right] E\left[X^{p_0}(s;\cdot)\right] ds \\
&\quad + p_0(p_0+1) \int_0^t g^2(s) E\left[X^{p_0-1}(s;\cdot)\right] ds.
\end{aligned}
\tag{6.56}
$$

The strong induction hypothesis implies, in particular, that

$$
E\left[X^{p_0}(s;\cdot)\right] \le \alpha_{p_0}\left(1 + E\,|X_0(\cdot)|^{p_0}\right) e^{\beta_{p_0}s}, \quad 0 \le s \le T,
\tag{6.57}
$$

$$
E\left[X^{p_0-1}(s;\cdot)\right] \le \alpha_{p_0-1}\left(1 + E\,|X_0(\cdot)|^{p_0-1}\right) e^{\beta_{p_0-1}s}, \quad 0 \le s \le T,
\tag{6.58}
$$

for some positive constants $\alpha_{p_0}, \beta_{p_0}, \alpha_{p_0-1}, \beta_{p_0-1} > 0$. Substituting (6.57) and (6.58) into (6.56) and simplifying leads to an inequality of the form

$$
E\left[X^{p_0+1}(t;\cdot)\right] \le \left(E\left|X_0^{p_0+1}(\cdot)\right| + \zeta_1\right) + \zeta_2 \int_0^t E\left[X^{p_0+1}(s;\cdot)\right] ds
\tag{6.59}
$$

for some positive constants ζ_1 and ζ_2. (Verify this!) An application of Gronwall's Lemma in (6.59) then yields

$$
E\left[X^{p_0+1}(t;\cdot)\right] \le \left(E\left|X_0^{p_0+1}(\cdot)\right| + \zeta_1\right) e^{\zeta_2 t}, \quad 0 \le t \le T.
\tag{6.60}
$$

Now, arguing as in (6.53) yields the desired result. (Tell how.) $\qquad\square$

Continuity in the pth Moment

The proof of the analog of Prop. 3.6.2 for (6.15) follows the same reasoning, except that now the continuity of f and g is used to deduce that $\forall 0 \le \tau' \le \tau \le T$,

$$
\int_{\tau'}^{\tau} |f(s)|^p\, ds \le \|f\|_{\mathbb{C}([0,T];\mathbb{R})}^p \left(\tau - \tau'\right),
\tag{6.61}
$$

$$
\int_{\tau'}^{\tau} |g_k(s)|^p\, ds \le \|g_k\|_{\mathbb{C}([0,T];\mathbb{R})}^p \left(\tau - \tau'\right), \quad k = 1,\ldots,m.
\tag{6.62}
$$

These estimates are used at the key moment to establish (3.150) for (6.38).

Exercise 6.2.16. State and prove the analog of Prop. 3.6.2 for (6.15).

In the absence of multiplicative noise (that is, when $c_k = 0$ in (6.15)), the variation of parameters formula for the strong solution of (6.15) can be used directly to establish a similar continuity result. Doing so would require us to consider the following expression, for $0 \leq \tau' \leq \tau \leq T$:

$$E\left|X(\tau;\cdot) - X(\tau';\cdot)\right|^p = E\left|\left(e^{a\tau} - e^{a\tau'}\right)X_0(\cdot)\right.$$

$$+ \left(\int_0^\tau e^{a(\tau-s)}f(s)ds - \int_0^{\tau'} e^{a(\tau'-s)}f(s)ds\right)$$

$$+ \left.\left(\int_0^\tau e^{a(\tau-s)}g(s)dW(s) - \int_0^{\tau'} e^{a(\tau'-s)}g(s)dW(s)\right)\right|^p$$

$$\leq 3^{p-1}\left[E\left|\left(e^{a\tau} - e^{a\tau'}\right)X_0\right|^p \right. \tag{6.63}$$

$$+E\left|\int_0^\tau e^{a(\tau-s)}f(s)ds - \int_0^{\tau'} e^{a(\tau'-s)}f(s)ds\right|^p$$

$$+E\left.\left|\int_0^\tau e^{a(\tau-s)}g(s)dW(s) - \int_0^{\tau'} e^{a(\tau'-s)}g(s)dW(s)\right|^p\right]$$

$$= 3^{p-1}\left(I_1 + I_2 + I_3\right).$$

We must show that each of $I_1, I_2, I_3 \longrightarrow 0$ as $|\tau - \tau'| \to 0$. Using the fact that

$$e^{a(\tau-\tau')} = \sum_{n=0}^\infty \frac{(a(\tau-\tau'))^n}{n!}, \tag{6.64}$$

we see that

$$I_1 = E\left|\left(e^{a\tau} - e^{a\tau'}\right)X_0(\cdot)\right|^p = \left|\left(e^{a\tau} - e^{a\tau'}\right)\right|^p E|X_0(\cdot)|^p \tag{6.65}$$

$$= \left|e^{a\tau'}\left(e^{a(\tau-\tau')} - 1\right)\right|^p E|X_0(\cdot)|^p = e^{pa\tau'}\left|\sum_{n=1}^\infty \frac{(a(\tau-\tau'))^n}{n!}\right|^p E|X_0(\cdot)|^p$$

$$\leq e^{p|a|\tau'}\underbrace{\left|\sum_{n=1}^\infty \frac{(a(\tau-\tau'))^{n-1}}{n!}\right|}_{\leq e^{|a|T}}^p E|X_0(\cdot)|^p \leq e^{(p+1)|a|T}\left|\tau - \tau'\right|.$$

(Why?) As such, we conclude that $I_1 \longrightarrow 0$ as $|\tau - \tau'| \to 0$.

Next, observe that

$$I_2 = E \left| \int_0^{\tau'} \left(e^{a\tau} - e^{a\tau'} \right) e^{-as} f(s) ds - \int_{\tau'}^{\tau} e^{a\tau} e^{-as} f(s) ds \right|^p \tag{6.66}$$

$$\leq 2^{p-1} \left| e^{a\tau'} \left(e^{a(\tau-\tau')} - 1 \right) \right|^p \left\| e^{-a(\cdot)} f(\cdot) \right\|_{\mathbb{L}^1(0,T;\mathbb{R})}^p + e^{p|a|T} \left(\sup_{0 \leq s \leq T} \left| e^{-as} f(s) \right| \right)^p.$$

Both terms on the right-hand side of (6.66) approach 0 as $|\tau - \tau'| \to 0$. (Why?) Hence, $I_2 \longrightarrow 0$ as $|\tau - \tau'| \to 0$. In a similar fashion, it follows that $I_3 \longrightarrow 0$ as $|\tau - \tau'| \to 0$. (Tell how.)

Exercise 6.2.17. Write up a polished proof of the above continuity result.

Exercise 6.2.18. Establish such a continuity result for (6.15).

Other Statistical Properties

Consider (6.33) and assume that X_0 is a Gaussian random variable. We claim that the strong solution of (6.33) is a Gaussian process. This can be argued in two natural ways using different properties of Gaussian processes. These two approaches are outlined below.

Approach 1: The following argument follows the reasoning used in **[212]**. Consider the successive approximations for (6.33) given by

$$\begin{cases} X_0(t;\omega) & = X_0(\omega), \\ X_n(t;\omega) & = X_0(\omega) + \int_0^t \left(aX_{n-1}(s;\omega) + f(s) \right) ds + \sum_{k=1}^m \int_0^t g_k(s) dW_k(s), \, n \in \mathbb{N}. \end{cases} \tag{6.67}$$

First, argue inductively that $X_n(t;\cdot)$ is a Gaussian random variable, $\forall n \in \mathbb{N}$, $0 \leq t \leq T$. For $n = 1$, observe that $\forall 0 \leq t \leq T$,

$$\int_0^t \left(aX_0(s;\cdot) + f(s) \right) ds = atX_0(\cdot) + \|f\|_{\mathbb{L}^1(0,t;\mathbb{R})}. \tag{6.68}$$

The right-hand side of (6.68) is a linear transformation of a Gaussian random variable and thus, is itself a Gaussian random variable. (Why?) Also, because the functions g_k $(k = 1, \ldots, m)$ are deterministic, it follows that $\int_0^t g_k(s) dW_k(s)$ is Gaussian (Why?) and hence, $\sum_{k=1}^m \int_0^t g_k(s) dW_k(s)$ is Gaussian. (Why?) Thus, the right-hand side of the formula for $X_1(t;\cdot)$, as given by (6.67), is a linear transformation of a Gaussian random variable and thus, is itself Gaussian.

Next, assume that $X_n(t;\cdot)$ is a Gaussian random variable, $\forall 0 \leq t \leq T$. Argue as above to show that $X_{n+1}(t;\cdot)$, as given by (6.67), is a linear transformation of a Gaussian random variable. This completes the induction proof.

Because $X(t;\cdot)$ is the $\mathscr{L}^2$-limit of a sequence of Gaussian random variables $\{X_n(t;\cdot)\}$, we conclude that $X(t;\cdot)$ is itself Gaussian. (Why?) This completes the proof. $\Diamond$

<u>Approach 2</u>: The following argument follows the reasoning in **[20]**. The variation of parameters formula for $X(t;\cdot)$ is given by (6.34). Observe that $\forall 0 \le t \le T$, $e^{at}X_0$ is a scalar multiple of a Gaussian random variable and so, is itself Gaussian. Also, the second term in (6.34) is deterministic (Why?), and each of the m terms of which the third quantity is comprised in (6.34) is Gaussian. (Why?) Thus, the sum is itself a Gaussian random variable. (Why?) As such, $X(t;\cdot)$ is a linear transformation of a Gaussian random variable (Why?) and hence, must be Gaussian. This completes the proof. $\Diamond$

Exercise 6.2.19. Fill in the details in the above proofs.

Remark. It can be shown that if at least one of the $c_k \ne 0$, then the solution of (6.33) is not Gaussian. (See **[20]**.)

Exercise 6.2.20. Argue as in Section 3.6.5 that the solution of (6.33) is a Markov process.

Exercise 6.2.21. Assume that X_0 is a Gaussian random variable and consider (6.23) with $c = 0$. We have shown that $X(t;\cdot)$ is a Gaussian process and so its density is characterized by $\mu_X(t)$ and $Cov_X(t,s)$. Let $0 \le t \le T$ and $h \in \mathbb{R}$ be such that $0 \le t+h \le T$. Show that

$$\mu_X(t+h) = \mu_X(t),$$
$$Cov_X(t+h,s+h) = Cov_X(t,s).$$

(That is, $X(t;\cdot)$ has stationary increments and so its statistical properties do not change over time.)

6.2.4   Convergence Results

Convergence results enable us to effectively approximate IVPs containing complicated parameters and expressions that would otherwise be difficult to analyze directly. Ultimately, we want a convergence scheme for (6.15) that allows the parameters and data (namely a, c_k, f, and g_k) to be approximated in a single convergence scheme. For instructional purposes, we begin with a simpler problem.
    For every $n \in \mathbb{N}$, consider the IVPs

$$\begin{cases} dX_n(t;\omega) = a_n X_n(t;\omega)dt + \beta_n dW(t), 0 \le t \le T, \omega \in \Omega, \\ X_n(0;\omega) = (X_0)_n(\omega), \omega \in \Omega, \end{cases} \tag{6.69}$$

and

$$\begin{cases} dX(t;\omega) = aX(t;\omega)dt + \beta dW(t), 0 \le t \le T, \omega \in \Omega, \\ X(0;\omega) = X_0(\omega), \omega \in \Omega. \end{cases} \tag{6.70}$$

Assume **(H3.6)** and that $a, \beta, \{a_n\}$ and $\{\beta_n\}$ are such that

(H6.6) $\lim\limits_{n\to\infty} |a_n - a| = 0$.
(H6.7) $\lim\limits_{n\to\infty} |\beta_n - \beta| = 0$.

Proposition 6.2.6. *If (H3.2), (H3.3), (H6.6), and (H6.7) hold, then* $\lim\limits_{n\to\infty} \|X_n - X\|_{\mathbb{C}} = 0$.

Outline of Proof. We provide two different approaches.
Approach 1: The proof of Prop. 3.7.1 can be easily adapted to prove this result. Indeed, subtracting the integrated versions of (6.69) and (6.70) reveals that we need to estimate the quantities

$$E \left| \int_0^t (a_n X_n(s;\cdot) - aX(s;\cdot)) \, ds \right|^2 \tag{6.71}$$

and

$$E \left| \int_0^t (\beta_n - \beta) \, dW(s) \right|^2. \tag{6.72}$$

We treat (6.71) exactly as in the proof of Prop. 3.7.1, with the slight modification that the ε-tolerance is chosen so that final quantity corresponding to (3.149) works out nicely. Also, applying Lemma 3.2.7 in (6.72) yields

$$E \left| \int_0^t (\beta_n - \beta) \, dW(s) \right|^2 \leq T^2 |\beta_n - \beta|^2, \tag{6.73}$$

and **(H6.7)** enables us to control the right-hand side of (6.73) in terms of ε, for sufficiently large n. (How?)

Exercise 6.2.22. Write up a polished proof of Prop. 6.2.6.

Approach 2: We use the variation of parameters formulae for the solutions of (6.69) and (6.70). Let $\varepsilon > 0$. The continuity of $h(t) = e^t$, together with **(H6.6)**, guarantees the existence of $M_0 > 0$ such that

$$\sup_{n\in\mathbb{N}} \left(\sup_{0\leq t\leq T} |e^{a_n t}|^2 \right) \leq M_0. \tag{6.74}$$

Hypothesis **(H3.6)** implies that $\exists N_1 \in \mathbb{N}$ such that

$$n \geq N_1 \implies \|(X_0)_n - X_0\|_{\mathcal{L}_0^2(\Omega;\mathbb{R})}^2 < \frac{\varepsilon^2}{12 M_0}. \tag{6.75}$$

Computations similar to those used in (6.65) show that

$$|e^{a_n t} - e^{at}|^2 \leq \left[|e^{a_n t}| t e^{|a - a_n|t} \right]^2 |a - a_n|^2. \tag{6.76}$$

(Tell how carefully.) From **(H6.6)**, we know that $\exists M_1 > 0$ such that

$$\sup_{n\in\mathbb{N}} \left(\sup_{0\le t\le T} \left[e^{|a-a_n|t} \right]^2 \right) \le M_1. \tag{6.77}$$

Moreover, $\exists N_2 \in \mathbb{N}$ such that

$$n \ge N_2 \Longrightarrow |a - a_n|^2 < \frac{\varepsilon^2}{12M_0 T^2 M_1 \left(\|X_0\|^2_{\mathscr{L}^2_0(\Omega;\mathbb{R})} + 1 \right)}. \tag{6.78}$$

Finally, **(H6.7)** implies that $\exists N_3 \in \mathbb{N}$ such that

$$n \ge N_3 \Longrightarrow |\beta - \beta_n|^2 < \frac{\varepsilon^2}{6T^2}. \tag{6.79}$$

Let $N = \max\{N_1, N_2, N_3\}$. Then, (6.75), (6.78), and (6.79) hold, $\forall n \ge N$. Observe that $\forall 0 \le t \le T$,

$$
\begin{aligned}
E\left|X_n(t;\cdot) - X(t;\cdot)\right|^2 &\le 2E\left| e^{a_n t}(X_0)_n(\cdot) - e^{at}X_0(\cdot) \right|^2 + 2E\left| \int_0^t (\beta_n - \beta)\,dW(s) \right|^2 \\
&= 2E\left| e^{a_n t}(X_0)_n(\cdot) - e^{a_n t}X_0(\cdot) + e^{a_n t}X_0(\cdot) - e^{at}X_0(\cdot) \right|^2 \\
&\quad + 2E\left| \int_0^t (\beta_n - \beta)\,dW(s) \right|^2 \\
&\le 4\left| e^{a_n t} \right|^2 \|(X_0)_n - X_0\|^2_{\mathscr{L}^2_0(\Omega;\mathbb{R})} \\
&\quad + 4\left| e^{a_n t} - e^{at} \right|^2 \|X_0\|^2_{\mathscr{L}^2_0(\Omega;\mathbb{R})} + 2\left| \beta - \beta_n \right|^2 T^2.
\end{aligned} \tag{6.80}
$$

Using (6.75) through (6.79) in (6.80) shows that, $\forall 0 \le t \le T$,

$$n \ge N \Longrightarrow E\left|X_n(t;\cdot) - X(t;\cdot)\right|^2 < \varepsilon^2. \tag{6.81}$$

Taking the supremum over $(0,T)$ in (6.81), followed by taking square roots, yields

$$n \ge N \Longrightarrow \|X_n - X\|_{\mathbb{C}\left([0,T];\mathscr{L}^2(\Omega;\mathbb{R})\right)} < \varepsilon,$$

as desired. This completes the proof. $\square$

Exercise 6.2.23. Verify (6.76).

Exercise 6.2.24.
i.) Compute $\mu_{X_n}(t)$ and $Var_{X_n}(t)$ for (6.69).
ii.) Compute the limit as $n \to \infty$ of each of the quantities in (i).

Exercise 6.2.25. Formulate and prove an analog of Prop. 6.2.6 for IVPs formed by replacing $\beta_n dW(s)$ and $\beta dW(s)$ in (6.69) and (6.70), respectively, by

$\sum_{k=1}^{m}(\beta_k)_n dW_k(s)$ and $\sum_{k=1}^{m}\beta_k dW_k(s)$.

Exercise 6.2.26. Formulate and prove a version of Prop. 6.2.6 for the IVPs

$$\begin{cases} dX_n(t;\omega) = a_n X_n(t;\omega)dt + g_n(t)dW(t), 0 \le t \le T, \omega \in \Omega, \\ X_n(0;\omega) = (X_0)_n(\omega), \omega \in \Omega \end{cases} \tag{6.82}$$

and

$$\begin{cases} dX(t;\omega) = aX(t;\omega)dt + g(t)dW(t), 0 \le t \le T, \omega \in \Omega, \\ X(0;\omega) = X_0(\omega), \omega \in \Omega. \end{cases} \tag{6.83}$$

Make certain to impose appropriate hypotheses on g_n and g.

6.2.5   Approximation by a Deterministic IVP

We have encountered this notion for linear homogenous SDEs with multiplicative noise. To begin, we consider some simple extensions. Consider the deterministic IVP

$$\begin{cases} dY(t) = aY(t)dt, 0 < t < T, \\ Y(0) = X_0 \end{cases} \tag{6.84}$$

and $\forall 0 < \varepsilon < 1$, the stochastic IVP

$$\begin{cases} dX_\varepsilon(t;\omega) = a_\varepsilon X_\varepsilon(t;\omega)dt + \varepsilon dW(t), 0 < t < T, \omega \in \Omega, \\ X_\varepsilon(0;\omega) = X_0, \omega \in \Omega, \end{cases} \tag{6.85}$$

where X_0 is a constant random variable. Assume that **(H3.7)** holds.

For any $p > 2$, we estimate $E|X_\varepsilon(t;\cdot) - Y(t)|^p$ using the variation of parameters formulae for X_ε and Y. Observe that

$$E|X_\varepsilon(t;\cdot) - X(t)|^p = E\left|e^{a_\varepsilon t}X_0(\cdot) - e^{at}X_0(\cdot) + \int_0^t e^{a_\varepsilon(t-s)}\varepsilon dW(s)\right|^p \tag{6.86}$$

$$\le \left[\left|e^{a_\varepsilon t} - e^{at}\right|^p\|X_0\|_{\mathscr{L}^p}^p + \varepsilon^p E\left|\int_0^t e^{a_\varepsilon(t-s)}dW(s)\right|^p\right].$$

Arguing as in Exer. 6.2.23, we can show that

$$\lim_{\varepsilon \to 0^+}\left(\sup_{0 \le t \le T}\left|e^{a_\varepsilon t} - e^{at}\right|\right) = 0. \tag{6.87}$$

Also, **(H3.7)** implies that

$$\sup_{0<\varepsilon<1}|a_\varepsilon| \le M. \tag{6.88}$$

Hence, applying Lemma 3.2.7 to the integral term in (6.86) yields

$$\sup_{0<\varepsilon<1}\left(\sup_{0\le t\le T}\left(E\left|\int_0^t e^{a_\varepsilon(t-s)}dW(s)\right|^p\right)\right)\le \tag{6.89}$$

$$\sup_{0<\varepsilon<1}\left(\sup_{0\le t\le T}\left(\frac{tp(p-1)}{2}\right)^{\frac{p}{2}}\int_0^t e^{pa_\varepsilon(t-s)}ds\right)\le$$

$$\sup_{0<\varepsilon<1}\left(\sup_{0\le t\le T}\left(\frac{tp(p-1)}{2}\right)^{\frac{p}{2}}te^{p|a_\varepsilon|t}\right)<\infty.$$

In view of (6.87) and (6.89), we know that the right-hand side of (6.86) goes to zero as $\varepsilon\to 0^+$. As such, $\exists\zeta>0$ and a function $\psi:I\subset[0,1]\longrightarrow(0,\infty)$ for which $\lim_{\varepsilon\to 0^+}\psi(\varepsilon)=0$ and

$$E\left|X_\varepsilon(t;\cdot)-Y(t)\right|^p\le\zeta\psi(\varepsilon),\ 0\le t\le T. \tag{6.90}$$

Exercise 6.2.27. Replace $\varepsilon dW(t)$ in (6.85) by $g_\varepsilon(t)dW(t)$, where $g_\varepsilon:[0,T]\to\mathbb{R}$ and assume that

(H6.8) $\lim_{\varepsilon\to 0^+}\left(\sup_{0\le t\le T}|g_\varepsilon(t)|\right)=0.$

Rework the above argument to verify that (6.90) still holds.

Exercise 6.2.28. Now, use the insight gained by investigating the previous two scenarios to formulate and prove an approximation result in the spirit of the above discussion for the more general IVP (6.15).

6.3 Nonhomogenous Linear SDEs in $\mathbb{R}^N$

We now consider the most general form of a nonhomogenous linear SDE in $\mathbb{R}^N$ given by

$$\begin{cases} d\mathbf{X}(t;\omega)=[\mathbf{A}\mathbf{X}(t;\omega)+\mathbf{F}(t)]dt+\sum_{k=1}^m[\mathbf{B}_k\mathbf{X}(t;\omega)+\mathbf{g}_k(t)]dW_k(t), \\ \mathbf{X}(0;\omega)=\mathbf{X}_0(\omega), \end{cases} \tag{6.91}$$

where $0<t<T$, $\omega\in\Omega$; $\mathbf{X}:[0,\infty)\times\Omega\to\mathbb{R}^N$ is the solution stochastic process; $\mathbf{A},\mathbf{B}_1,\ldots,\mathbf{B}_m$ are an $N\times N$ constant matrices; $\{W_i(t):0\le t\le T\}$, $i=1,\ldots,m$ are independent $\mathbb{R}$-valued Wiener processes; $\mathbf{X}_0\in\mathfrak{L}^2\left(\Omega;\mathbb{R}^N\right)$ is an $\mathscr{F}_0$-measurable random variable independent of $\{W_i(t):0\le t\le T\}$, $i=1,\ldots,m$; and $\mathbf{F}:[0,T]\to\mathbb{R}^N$ and $\mathbf{g}_k:[0,T]\to\mathbb{R}^N$ ($k=1,\ldots,m$) are given mappings.

Assume that **(H4.1)** through **(H4.3)** hold. In analogy with (6.15) through (6.16), the integrated version of (6.91) is

$$\mathbf{X}(t;\omega) = \mathbf{X}_0(\omega) + \int_0^t [\mathbf{AX}(s;\omega) + \mathbf{F}(s)]\,ds + \sum_{k=1}^m \int_0^t [\mathbf{B}_k\mathbf{X}(s;\omega) + \mathbf{g}_k(s)]\,dW_k(s),$$

$$(6.92)$$

where $0 < t < T$, $\omega \in \Omega$. As in the one-dimensional setting discussed in Section 6.2, it suffices to assume

(H6.9) F, $\mathbf{g}_k(k = 1,\ldots,m)$ are continuous mappings.

The changes to the proofs of the results in Chapter 3 that were implemented in order to incorporate a time-dependent forcing term into the theory established in Section 6.2 are essentially the same modifications necessary when establishing the analogous theory for (6.91). The details should be checked carefully, but we leave this as an instructional exercise for you to complete prior to moving onward. We shall focus on the specific case of additive noise only, namely

$$\begin{cases} dX_1(t;\omega) = & (a_{11}X_1(t;\omega) + \ldots + a_{1N}X_N(t;\omega))\,dt \\ & + g_{11}(t)dW_1(t) + \ldots + g_{1m}(t)dW_m(t) + f_1(t)dt \\ \vdots \\ dX_N(t;\omega) = & (a_{N1}X_1(t;\omega) + \ldots + a_{NN}X_N(t;\omega))\,dt \\ & + g_{N1}(t)dW_1(t) + \ldots + g_{Nm}(t)dW_m(t) + f_N(t)dt \\ X_1(0;\omega) = & (X_0)_1(\omega) \ldots, X_N(0;\omega) = (X_0)_N(\omega), \end{cases}$$

$$(6.93)$$

where $0 < t < T$, $\omega \in \Omega$. System (6.93) can be written in the equivalent matrix form

$$\begin{cases} d\mathbf{X}(t;\omega) = [\mathbf{AX}(t;\omega) + \mathbf{F}(t)]\,dt + \mathbf{g}(t)d\mathbf{W}(t), t > 0, \omega \in \Omega, \\ \mathbf{X}(0;\omega) = \mathbf{X}_0(\omega), \omega \in \Omega, \end{cases}$$

$$(6.94)$$

where $\mathbf{X}(t;\cdot)$ and $\mathbf{X}_0(\cdot)$ are as in (4.67), $\mathbf{A}$ is given by (4.68), $\mathbf{W}(t) = \langle W_1(t),\ldots,W_m(t)\rangle$ is an m-dimensional Wiener process, and $\mathbf{g} : [0,T] \rightarrow \mathscr{B}_0(\mathbb{R}^m, \mathbb{R}^N)$ is given by

$$\mathbf{g}(t) = [g_{ij}(t)], \ 1 \le i, j \le N. \tag{6.95}$$

The theory for (6.91) applies to (6.94). (Why?) So, in particular, the existence and uniqueness of a strong solution of (6.94) on $[0,T]$ is guaranteed. Below, we derive the variation of parameters formula for (6.94).

Proposition 6.3.1. *The unique strong solution* $\mathbf{X} : [0,T] \times \Omega \rightarrow \mathbb{R}^N$ *of (6.94) is given by*

$$\mathbf{X}(t;\omega) = e^{\mathbf{A}t}\mathbf{X}_0(\omega) + \int_0^t e^{\mathbf{A}(t-s)}\mathbf{F}(s)\,ds + \int_0^t e^{\mathbf{A}(t-s)}\mathbf{g}(s)d\mathbf{W}(s). \tag{6.96}$$

Proof. Define the stochastic process $\mathbf{Y} : [0, T] \times \Omega \to \mathbb{R}^N$ by

$$\mathbf{Y}(t; \omega) = \mathbf{X}_0(\omega) + \int_0^t e^{-\mathbf{A}s} \mathbf{F}(s) ds + \int_0^t e^{-\mathbf{A}s} \mathbf{g}(s) d\mathbf{W}(s). \tag{6.97}$$

(Why is $\mathbf{Y}$ well-defined?) Observe that

$$d(\mathbf{Y}(t; \omega)) = e^{-\mathbf{A}t} \mathbf{F}(t) dt + e^{-\mathbf{A}t} \mathbf{g}(t) d\mathbf{W}(t). \tag{6.98}$$

The stochastic process $\mathbf{Z} : [0, T] \times \Omega \to \mathbb{R}^N$ given by

$$\mathbf{Z}(t; \omega) = e^{\mathbf{A}t} \mathbf{Y}(t; \omega) \tag{6.99}$$

is equal to the right-hand side of (6.96). As such, it suffices to show that its stochastic differential is given by (6.94). (Why?) We apply Itó's formula with $\mathbf{H}(t, \mathbf{z}) = e^{\mathbf{A}t} \mathbf{z}$. The properties of the matrix exponential (cf. Prop. 4.2.4) imply that

$$\frac{\partial \mathbf{H}}{\partial t} = \mathbf{A}e^{\mathbf{A}t} \mathbf{z}, \quad \frac{\partial \mathbf{H}}{\partial \mathbf{z}} = e^{\mathbf{A}t}, \quad \frac{\partial^2 \mathbf{H}}{\partial \mathbf{z}^2} = 0. \tag{6.100}$$

Taking $\mathbf{z} = \mathbf{Y}(t; \omega)$ in our application of Itó's formula yields

$$
\begin{aligned}
d(\mathbf{H}(t, \mathbf{z})) &= d\left(e^{\mathbf{A}t} \mathbf{Y}(t; \omega)\right) = d(\mathbf{Z}(t; \omega)) \\
&= \left(\mathbf{A}e^{\mathbf{A}t}\right) \mathbf{Y}(t; \omega) dt + e^{\mathbf{A}t} d(\mathbf{Y}(t; \omega)) \\
&= \left(\mathbf{A}e^{\mathbf{A}t}\right) \mathbf{Y}(t; \omega) dt + e^{\mathbf{A}t} \left[e^{-\mathbf{A}t} \mathbf{F}(t) dt + e^{-\mathbf{A}t} \mathbf{g}(t) d\mathbf{W}(t)\right] \\
&= \mathbf{A}\left(e^{\mathbf{A}t} \mathbf{Y}(t; \omega)\right) dt + \mathbf{F}(t) dt + \mathbf{g}(t) d\mathbf{W}(t) \\
&= [\mathbf{A}\mathbf{Z}(t; \omega) + \mathbf{F}(t)] dt + \mathbf{g}(t) d\mathbf{W}(t).
\end{aligned} \tag{6.101}
$$

Also,

$$\mathbf{Z}(0; \omega) = e^{\mathbf{A}(0)} \mathbf{Y}(0; \omega) = \mathbf{I}(\mathbf{X}_0(\omega)) = \mathbf{X}_0(\omega). \tag{6.102}$$

Therefore, by uniqueness, we conclude that $\mathbf{Z}(t; \omega) = \mathbf{X}(t; \omega)$, $\forall t \in [, 0, T]$, a.s. $[\mathscr{P}]$. This completes the proof. $\qquad \square$

BIG Exercise 6.3.1. Let $\mathbf{X}$ be the strong solution of (6.94) given by (6.96).
i.) Show that

$$\mu_{\mathbf{X}}(t) = e^{\mathbf{A}t} E[\mathbf{X}_0(\cdot)] + \int_0^t e^{\mathbf{A}(t-s)} \mathbf{F}(s) ds. \tag{6.103}$$

ii.) Let $p > 2$. Derive an estimate for $E \|\mathbf{X}(t; \cdot)\|_{\mathbb{R}^N}^p$.
iii.) Let $p > 2$. Prove that $E \|\mathbf{X}(\tau; \cdot) - \mathbf{X}(\tau'; \cdot)\|_{\mathbb{R}^N}^p \to 0$ as $|\tau - \tau'| \to 0$.
iv.) Prove that if $\mathbf{X}_0$ is Gaussian, then $\{\mathbf{X}(t; \cdot) : 0 \le t \le T\}$ is a Gaussian process.
v.) Formulate and prove a version of Prop. 6.2.6 for (6.94).
vi.) Because the strong solution of (6.94) exists on $[0, T]$, $\forall T > 0$, we can consider the solution on $[0, \infty)$. How does this solution behave as $t \to \infty$? This is a loaded

question, but as in the deterministic case, it can be characterized in some situations. For instance, show that if the eigenvalues of $\mathbf{A}$ all have negative real parts, then $\lim_{t \to \infty} E \|\mathbf{X}(t;\cdot)\|^2_{\mathbb{R}^N} = 0$.

Exercise 6.3.2. (The Ornstein-Uhlenbeck Process Revisited)
i.) Reformulate (3.188) and (3.189) in matrix form and derive a formula for its strong solution.
ii.) Let $0 < \varepsilon < 1$ and suppose that β is replaced by β_ε in (3.188), where $\beta_\varepsilon \to 0$ as $\varepsilon \to 0^+$. Denote the strong solution of the resulting IVP by $\begin{bmatrix} v_\varepsilon \\ s_\varepsilon \end{bmatrix}$. Prove that $\exists \mathbf{y}^\star \in \mathbb{C}\left([0,T];\mathcal{L}^2\left(\Omega;\mathbb{R}^2\right)\right)$ such that

$$\lim_{\varepsilon \to 0^+} \left\| \begin{bmatrix} v_\varepsilon \\ s_\varepsilon \end{bmatrix} - \mathbf{y}^\star \right\|_{\mathbb{C}} = 0.$$

To what deterministic IVP is $\mathbf{y}^\star$ a solution?

6.4 Abstract Nonhomogenous Linear SEEs

External forces naturally arise in IBVPs as well. The hope is that we can reformulate such IBVPs abstractly as stochastic evolution equations similar to (6.14), but in a separable Hilbert space $\mathscr{H}$, and then adapt the theory developed in Sections 6.2 and 6.3 to handle this more general form.

6.4.1 Motivation by Models

Model V.2 Heat Conduction with External Source
Suppose that a heat source is positioned in proximity to one edge of a rectangular slab of material for which we are monitoring the temperature over time. Assuming that its intensity increases with time, a possible generalization of IBVP (3.47) describing this scenario is given by

$$\begin{cases} \partial z(x,y,t;\omega) & = k\triangle z(x,y,t;\omega)\partial t + \left(t^2 + x + 2y\right)\partial t \\ & \quad + \left(\alpha_1 z(x,y,t;\omega) + g_1(t)\right)dW_1(t) + \left(\alpha_2 z(x,y,t;\omega) + g_2(t)\right)dW_2(t) \\ z(x,y,0;\omega) & = \sin 2x + \cos 2y, \ 0 < x < a, 0 < y < b, \\ z(x,0,t;\omega) & = 0 = z(x,b,t;\omega), \ 0 < x < a, t > 0, \\ z(0,y,t;\omega) & = 0 = z(a,y,t;\omega), \ 0 < y < b, t > 0, \end{cases}$$

$$(6.104)$$

where $0 < x < a, 0 < y < b, t > 0, \omega \in \Omega$; $z(x,y,t;\omega)$ represents the temperature at the point (x,y) on the plate at time t; $\alpha_1, \alpha_2 > 0$; and $W_1(t)$ and $W_2(t)$ are independent one-dimensional Wiener processes.

Model VI.2 Sobolev Equation with Forcing Term
External forces naturally arise when modeling fluid flow through fissured rocks. A simplified one-dimensional version of an IBVP arising in the modeling of such a scenario is given by

$$\begin{cases} \partial \left(z(x,t;\omega) - \frac{\partial^2}{\partial x^2} z(x,t;\omega) \right) = \left(\frac{\partial^2}{\partial x^2} z(x,t;\omega) + 1 + x \right) \partial t + (z(x,t;\omega) + g(t)) dW(t) \\ z(x,0;\omega) = 1 + x^3, \ 0 < x < \pi, \omega \in \Omega, \\ z(0,t;\omega) = z(\pi,t;\omega) = 0, \ t > 0, \omega \in \Omega, \end{cases}$$

(6.105)

where $0 < x < \pi, t > 0, \omega \in \Omega$.

Exercise 6.4.1. Try to express (6.104) and (6.105) as abstract stochastic evolution equations similar to (6.14), but now in a separable Hilbert space.

6.4.2 The Cauchy Problem

Equipping the IBVPs discussed in Chapter 5 with time-dependent forcing terms gives rise to the following abstract linear nonhomogenous SEE of the same form as (6.14), but now taking values in a separable Hilbert space:

$$\begin{cases} dX(t;\omega) = (AX(t;\omega) + f(t)) dt + \sum_{k=1}^{m} (B_k X(t;\omega) + g_k(t)) dW_k(t), t > 0, \ \omega \in \Omega, \\ X(0;\omega) = X_0(\omega), \omega \in \Omega, \end{cases}$$

(6.106)

in a separable Hilbert space $\mathscr{H}$, where $f : [0,T] \to \mathscr{H}$ and $g_k : [0,T] \to \mathscr{B}_0(\mathbb{R}^m; \mathscr{H})$ $(k = 1,\ldots,m)$. We assume **(H5.1)** - **(H5.4)**, as well as

(H6.10) $f : [0,T] \to \mathscr{H}$ and $g_k : [0,T] \to \mathscr{B}_0(\mathbb{R}^m; \mathscr{H})$ $(k = 1,\ldots,m)$ are continuous.

Our main focus is to establish the existence and uniqueness of a mild solution of (6.106) in the following sense.

Definition 6.4.1. A stochastic process $X : [0,T] \times \Omega \to \mathscr{H}$ is a *mild solution* of (6.106) on $[0,T]$ if
i.) $X \in \mathbb{C}\left([0,T]; \mathcal{L}^2(\Omega; \mathscr{H})\right)$,
ii.) $X(t;\omega) = e^{At} X_0(\omega) + \int_0^t e^{A(t-s)} f(s) ds + \sum_{k=1}^{m} \int_0^t e^{A(t-s)} (B_k X(s;\omega) + g_k(s)) dW_k(s)$, $\forall 0 \le t \le T$, a.s. $[\mathscr{P}]$.

The following result can be proven in a nearly identical fashion to the proof of Thrm. 5.7.1 using a fixed-point argument.

Theorem 6.4.2. *If* **(H5.1)** *through* **(H5.4)** *and* **(H6.10)** *hold, then (6.106) has a unique mild solution on* $[0,T]$.

Exercise 6.4.2. Prove Thrm. 6.4.2.

Exercise 6.4.3. Use the variation of parameters formula to establish an estimate for $E \|X(t;\cdot)\|_{\mathcal{H}}^p$, for $p > 2$, where X is the mild solution of (6.106).

Exercise 6.4.4. Show that the mild solution of (6.106) can be approximated by a sequence of strong solutions, as in Prop. 5.7.2.

Exercise 6.4.5. Formulate and prove an analog of Prop. 5.7.4 for (6.106).

Exercise 6.4.6. Consider the following modification of the advection equation (5.26):

$$
\begin{cases}
\partial c(z,t;\omega) &= \left(V \frac{\partial}{\partial z} c(z,t) + \arctan\left(1 + \sqrt{1+t^2}\right)\right) \partial t \\
&\quad + \left(e^t + \frac{1}{2}c(z,t;\omega)\right) dW(t), z > 0, t > 0, \omega \in \Omega, \\
c(z,0;\omega) &= 1 + 2z, z > 0, \omega \in \Omega, \\
c(0,t;\omega) &= 0, t > 0, \omega \in \Omega,
\end{cases} \tag{6.107}
$$

where W is a one-dimensional Wiener process.
i.) Reformulate (6.107) as the abstract stochastic evolution equation (6.106) in an appropriate Hilbert space.
ii.) Argue that (6.107) has a unique mild solution. Find an explicit formula for this solution.

6.5   Introducing Some New Models

Model VII.1 Classical Wave Equations
The evolution over time of the vertical displacement of a vibrating string of length L subject to small vibrations can be described by the so-called *wave equation*. Precisely, suppose that the deflection of the string at position x along the string at time t for a given $\omega \in \Omega$ is given by $z(x,t;\omega)$. An argument based on elementary physical principles (see **[111, 336]**) yields the following random IBVP:

$$
\begin{cases}
\frac{\partial^2}{\partial t^2} z(x,t;\omega) + c^2 \frac{\partial^2}{\partial x^2} z(x,t;\omega) = 0, \ 0 < x < L, t > 0, \omega \in \Omega, \\
z(x,0;\omega) = z_0(x;\omega), \ \frac{\partial z}{\partial t}(x,0;\omega) = z_1(x;\omega), \ 0 < x < L, \omega \in \Omega, \\
z(0,t;\omega) = z(L,t;\omega) = 0, t > 0, \omega \in \Omega,
\end{cases} \tag{6.108}
$$

where $z_0, z_1 \in \mathcal{L}^2(\Omega;\mathbb{R})$.

The separation of variables technique can be used to show that the solution $z(x,t;\omega)$ of (6.108) is given by

$$
\sum_{n=1}^{\infty} 2\left[\langle z_0(\cdot;\omega), e_n(\cdot)\rangle_{\mathbb{L}^2} \cos(\lambda_n ct) + \frac{1}{L\lambda_n}\langle z_1(\cdot;\omega), e_n(\cdot)\rangle_{\mathbb{L}^2} \sin(\lambda_n ct)\right] \cdot \sin(\lambda_n x),
$$

$$
\tag{6.109}
$$

where $e_n(\cdot) = \sin(\lambda_n \cdot)$ and $\lambda_n = \sqrt{\frac{n\pi}{L}}, \forall n \in \mathbb{N}$.

Now, we argue that (6.108) can be written as the abstract stochastic evolution equation (6.106) by suitably choosing the state space $\mathscr{H}$ and the operator A. We do this by adapting the approach used when studying the spring-mass system. Indeed, applying the change of variable

$$v_1 = z, \qquad v_2 = \frac{\partial z}{\partial t}$$

$$\frac{\partial v_1}{\partial t} = v_2, \ \frac{\partial v_2}{\partial t} = -c^2 \frac{\partial^2 v_1}{\partial x^2} \qquad (6.110)$$

enables us to express (6.108) as the equivalent system

$$\begin{cases} \frac{\partial}{\partial t} \begin{bmatrix} v_1 \\ v_2 \end{bmatrix}(x,t;\omega) = \begin{bmatrix} 0 & I \\ -c^2 \frac{\partial^2}{\partial x^2} & 0 \end{bmatrix} \begin{bmatrix} v_1 \\ v_2 \end{bmatrix}(x,t;\omega), \ 0 < x < L, t > 0, \omega \in \Omega, \\ \begin{bmatrix} v_1 \\ v_2 \end{bmatrix}(x,0;\omega) = \begin{bmatrix} z_0 \\ z_1 \end{bmatrix}(x,0;\omega), \ 0 < x < L, \omega \in \Omega, \\ v_1(0,t;\omega) = v_1(L,t;\omega) = 0, \ t > 0, \omega \in \Omega. \end{cases}$$

$$(6.111)$$

This time, the state space must be a product space $\mathscr{H}_1 \times \mathscr{H}_2$ because the unknown is a vector consisting of two components. Arguing as in Volume 1, we use the space

$$\mathscr{H} = \mathbb{H}_0^1(0,L;\mathbb{R}) \times \mathbb{L}^2(0,L;\mathbb{R}) \qquad (6.112)$$

$$\left\langle \begin{bmatrix} v_1 \\ v_2 \end{bmatrix}, \begin{bmatrix} v_1^\star \\ v_2^\star \end{bmatrix} \right\rangle_{\mathscr{H}} \equiv \int_0^L \left[\frac{\partial v_1}{\partial x} \frac{\partial v_1^\star}{\partial x} + v_2 v_2^\star \right] dx.$$

It can be shown that $\mathscr{H}$ is a Hilbert space with norm

$$\left\| \begin{bmatrix} h \\ k \end{bmatrix} \right\|_{\mathscr{H}} = \|h\|_{\mathbb{H}_0^1} + \|k\|_{\mathbb{L}^2}, \qquad (6.113)$$

where $\|h\|_{\mathbb{H}_0^1}^2 \equiv \left\| \frac{\partial h}{\partial x} \right\|_{\mathbb{L}^2}^2 + \|h\|_{\mathbb{L}^2}^2$.

Now, define the operator $A : \text{dom}(A) \subset \mathscr{H} \to \mathscr{H}$ by

$$A \begin{bmatrix} v_1 \\ v_2 \end{bmatrix} = \begin{bmatrix} 0 & I \\ -c^2 \frac{\partial^2}{\partial x^2} & 0 \end{bmatrix} \begin{bmatrix} v_1 \\ v_2 \end{bmatrix} = \begin{bmatrix} v_1 \\ -c^2 \frac{\partial^2 v_2}{\partial x^2} \end{bmatrix} \qquad (6.114)$$

$$\text{dom}(A) = \left(\mathbb{H}^2(0,L;\mathbb{R}) \cap \mathbb{H}_0^1(0,L;\mathbb{R}) \right) \times \mathbb{H}_0^1(0,L;\mathbb{R}).$$

Theorem 5.2.13 can be used to show that A generates a C_0-semigroup on $\mathscr{H}$. (Actually, A generates a *group* on $\mathscr{H}$ in the sense that e^{At} is also defined $\forall t < 0$.) Further, if (6.111) is viewed as the abstract stochastic evolution equation (5.94) using the above identifications, then Thrm. 5.7.1 ensures that $\forall u_0 = \begin{bmatrix} z_0 \\ z_1 \end{bmatrix} \in \mathcal{L}^2(\Omega;\mathscr{H})$ and (6.111) has a unique mild solution given by

$$\begin{bmatrix} z(t;\omega) \\ \frac{\partial z}{\partial t}(t;\omega) \end{bmatrix} = e^{At} \begin{bmatrix} z_0(\omega) \\ z_1(\omega) \end{bmatrix}. \qquad (6.115)$$

Generalizing the above discussion from a one-dimensional spatial domain to a bounded domain $\mathscr{D} \subset \mathbb{R}^N$ with smooth boundary $\partial \mathscr{D}$ is not difficult. Indeed, the resulting IBVP (6.108) is

$$
\begin{cases}
\frac{\partial^2}{\partial t^2} z(x,t;\omega) + c^2 \triangle z(x,t;\omega) = 0, \ x \in \mathscr{D}, t > 0, \omega \in \Omega, \\
z(x,0;\omega) = z_0(x;\omega), \ \frac{\partial z}{\partial t}(x,0;\omega) = z_1(x;\omega), \ x \in \mathscr{D}, \omega \in \Omega, \\
z(x,t;\omega) = 0, \ x \in \partial \mathscr{D}, t > 0, \omega \in \Omega.
\end{cases}
\tag{6.116}
$$

Transforming (6.116) into a system comparable to (6.111) amounts to using the more general matrix operator $\begin{bmatrix} 0 & I \\ -c^2 \triangle & 0 \end{bmatrix}$ and subsequently replacing every occurrence of the interval $(0,L)$ by $\mathscr{D}$. The resulting function spaces are Hilbert spaces. Of course, the detail-checking becomes more involved; specifically, showing that the new operator A generates a C_0-semigroup on $\mathbb{H}_0^1(\mathscr{D};\mathbb{R}) \times \mathbb{L}^2(\mathscr{D};\mathbb{R})$ relies partly on the Lax-Milgram Theorem and the theory of elliptic PDEs (see [296, 407]). But, the process closely resembles the one used in the one-dimensional setting. We summarize these observations below. (See [147, 148, 149] for details.)

Proposition 6.5.1. *Let $\mathscr{D} \subset \mathbb{R}^N$ be a bounded domain with smooth boundary $\partial \mathscr{D}$ and $\mathscr{H} = \mathbb{H}_0^1(\mathscr{D};\mathbb{R}) \times \mathbb{L}^2(\mathscr{D};\mathbb{R})$. The operator $A : \mathrm{dom}(A) \subset \mathscr{H} \to \mathscr{H}$ defined by*

$$
A = \begin{bmatrix} 0 & I \\ -c^2 \triangle & 0 \end{bmatrix}
\tag{6.117}
$$

$$
\mathrm{dom}(A) = \left(\mathbb{H}^2(\mathscr{D};\mathbb{R}) \cap \mathbb{H}_0^1(\mathscr{D};\mathbb{R}) \right) \times \mathbb{H}_0^1(\mathscr{D};\mathbb{R})
$$

generates a C_0-semigroup on $\mathscr{H}$.

Next, incorporating *viscous damping* into the model (as we did for the spring-mass system model) leads to the following variant of (6.108):

$$
\begin{cases}
\frac{\partial^2}{\partial t^2} z(x,t;\omega) + \alpha \frac{\partial}{\partial t} z(x,t;\omega) + c^2 \frac{\partial^2}{\partial x^2} z(x,t;\omega) = 0, \\
z(x,0;\omega) = z_0(x;\omega), \ \frac{\partial z}{\partial t}(x,0;\omega) = z_1(x;\omega), \\
z(0,t;\omega) = z(L,t;\omega) = 0, t > 0, \omega \in \Omega,
\end{cases}
\tag{6.118}
$$

where $0 < x < L, t > 0, \omega \in \Omega$, and $\alpha > 0$ is the damping coefficient.

Exercise 6.5.1. Assume that z_0 and z_1 belong to $\mathfrak{L}^2(\Omega;\mathbb{R})$.
i.) Determine the solution of (6.118) using the separation of variables technique.
ii.) Rewrite (6.118) as a system in $\mathscr{H}$ (defined in (6.112)) and show that the new operator A is m-accretive on $\mathscr{H}$.

Next, we incorporate randomness into the model via a white noise process $\frac{dW}{dt}$.

We consider the following nonhomogenous IBVP:

$$\begin{cases} \partial\left(\frac{\partial}{\partial t}z(x,t;\omega) + \alpha z(x,t;\omega)\right) + \left(c^2\frac{\partial^2}{\partial x^2}z(x,t;\omega)\right)\partial t = F(x,t)\partial t \\ \quad + G(x,t)dW(t)\ 0 < x < L,\, t > 0,\, \omega \in \Omega, \\ z(x,0;\omega) = z_0(x;\omega),\ \frac{\partial z}{\partial t}(x,0;\omega) = z_1(x;\omega),\ 0 < x < L,\, \omega \in \Omega, \\ z(0,t;\omega) = z(L,t;\omega) = 0,\, t > 0,\, \omega \in \Omega, \end{cases} \tag{6.119}$$

where $F : [0,T] \times [0,L] \to \mathbb{R}$ and $G : [0,T] \times [0,L] \to \mathbb{R}$ are continuous mappings and α, c^2, z_0 and z_1 are as above. We use the following similar change of variable:

$$\begin{aligned} v_1 &= z, \\ v_2 &= \frac{\partial z}{\partial t}, \\ \frac{\partial v_1}{\partial t} &= v_2, \\ \frac{\partial v_2}{\partial t} &= -\alpha v_2 - c^2\frac{\partial^2 v_1}{\partial x^2} + F + G\frac{dW}{dt}. \end{aligned} \tag{6.120}$$

Converting the change of variable equations to differential form yields

$$dv_1 = v_2 dt, \tag{6.121}$$

$$dv_2 = \left(-\alpha v_2 - c^2\frac{\partial^2 v_1}{\partial x^2}\right)dt + F dt + G dW.$$

Hence, we have

$$\begin{cases} \frac{\partial}{\partial t}\begin{bmatrix} v_1 \\ v_2 \end{bmatrix}(x,t;\omega) = \begin{bmatrix} 0 & I \\ -c^2\frac{\partial^2}{\partial x^2} & 0 \end{bmatrix}\begin{bmatrix} v_1 \\ v_2 \end{bmatrix}(x,t;\omega)dt \\ \quad + \begin{bmatrix} 0 \\ F(x,t) \end{bmatrix}dt + \begin{bmatrix} 0 \\ G(x,t) \end{bmatrix}dW(t), \\ \begin{bmatrix} v_1 \\ v_2 \end{bmatrix}(x,0;\omega) = \begin{bmatrix} z_0 \\ z_1 \end{bmatrix}(x,0;\omega),\ 0 < x < L, \omega \in \Omega, \\ v_1(0,t;\omega) = v_1(L,t;\omega) = 0,\ t > 0, \omega \in \Omega, \end{cases} \tag{6.122}$$

where $0 < x < L, t > 0, \omega \in \Omega$. Using the space $\mathscr{H}$ defined in (6.112), defining $(A, \text{dom}(A))$ as in Exer. 6.5.1, and defining the mappings $f : [0,T] \to \mathscr{H}$ and $g : [0,T] \to \mathscr{B}_0(\mathbb{R};\mathscr{H})$ by

$$f(t)(\cdot) = \begin{bmatrix} 0 \\ F(\cdot,t) \end{bmatrix},\quad g(t)(\cdot) = \begin{bmatrix} 0 \\ G(\cdot,t) \end{bmatrix} \tag{6.123}$$

enables us to express (6.122) abstractly as (6.106) (with $m = 1$ and $B_1 = 0$) in $\mathscr{H}$.

Exercise 6.5.2.
i.) Argue that (6.122) has a unique mild solution on $[0,T]$.
ii.) Let $0 < \varepsilon < 1$ and consider the IVP obtained by replacing $G : [0,T] \times [0,L] \to \mathbb{R}$ in (6.122) by a continuous mapping $G_\varepsilon : [0,T] \times [0,L] \to \mathbb{R}$ where $G_\varepsilon \to 0$ uniformly on $[0,T] \times [0,L]$ as $\varepsilon \to 0^+$. Denote the mild solution of this IVP by $\begin{bmatrix} v_1 \\ v_2 \end{bmatrix}_\varepsilon$. Prove

$$\exists \begin{bmatrix} v^\star \\ w^\star \end{bmatrix} \in \mathbb{C}\left([0,T];\mathscr{L}^2\left(\Omega;\mathscr{H}\right)\right) \text{ such that}$$

$$\lim_{\varepsilon \to 0^+} \left\| \begin{bmatrix} v_1 \\ v_2 \end{bmatrix}_\varepsilon - \begin{bmatrix} v^\star \\ w^\star \end{bmatrix} \right\|_{\mathbb{C}} = 0.$$

Model VIII.1 Advection 2 — Gas Flow in a Large Container

A stochastic version of the linearized system governing the flow of gas in a large container (as discussed in **[168]**) is given by

$$\begin{cases} \partial v(x,t;\omega) + c^2 \frac{\partial p}{\partial x}(x,t;\omega)\partial t = \left(2t\sin^3\left(x^2+1\right)\right)\partial t + t^2 dW(t), \\ \partial p(x,t;\omega) + c^2 \frac{\partial v}{\partial x}(x,t;\omega)\partial t = \left(-t\cos\left(x^2+1\right)\right)\partial t + 2t^2 dW(t), \\ p(x,0;\omega) = h_1(x;\omega),\ v(x,0;\omega) = h_2(x;\omega),\ 0 < x < \infty,\ \omega \in \Omega, \\ p(0,t;\omega) = v(0,t;\omega) = 0,\ 0 < t < T,\ \omega \in \Omega, \end{cases} \quad (6.124)$$

where $0 < x < \infty, 0 < t < T, \omega \in \Omega$, v is the velocity of the gas, p is the variation in density, $\{W(t) : 0 \le t \le T\}$ is a one-dimensional Wiener process, and $h_1, h_2 : [0,\infty) \times \Omega \to \mathbb{R}$ are $\mathscr{F}_0$-measurable random variables independent of $W(t)$ with finite second moments. This IBVP can be expressed equivalently as

$$\begin{cases} \partial \begin{bmatrix} v \\ p \end{bmatrix}(x,t;\omega) = \begin{bmatrix} 0 & -c^2 \frac{\partial}{\partial x} \\ -\frac{\partial}{\partial x} & 0 \end{bmatrix} \begin{bmatrix} v \\ p \end{bmatrix}(x,t;\omega)\partial t + \begin{bmatrix} 2t\sin^3\left(x^2+1\right) \\ -t\cos\left(x^2+1\right) \end{bmatrix}\partial t \\ \qquad + \begin{bmatrix} t^2 \\ 2t^2 \end{bmatrix} dW(t),\ 0 < x < \infty, 0 < t < T,\ \omega \in \Omega, \\ \begin{bmatrix} v \\ p \end{bmatrix}(x,0;\omega) = \begin{bmatrix} h_1(x;\omega) \\ h_2(x;\omega) \end{bmatrix},\ 0 < x < \infty,\ \omega \in \Omega. \end{cases} \quad (6.125)$$

The structure of this IBVP resembles a cross between the wave equation (due to the 2×2 matrix of operators involved) and a diffusion equation (due to the presence of the first-order time derivative). A combination of the approaches used when reformulating these IBVPs abstractly can be implemented here. Precisely, to view (6.125) as the abstract stochastic evolution equation (6.106), consider the Hilbert space

$$\mathscr{H} = \mathbb{L}^2\left(0,\infty;\mathbb{R}\right) \times \mathbb{L}^2\left(0,\infty;\mathbb{R}\right) \quad (6.126)$$

equipped with the inner product

$$\left\langle \begin{bmatrix} v_1 \\ v_2 \end{bmatrix}, \begin{bmatrix} v_1^\star \\ v_2^\star \end{bmatrix} \right\rangle_{\mathscr{H}} \equiv \int_0^\infty [v_1 v_1^\star + v_2 v_2^\star]\, dx. \quad (6.127)$$

Define the operator $A : \text{dom}(A) \subset \mathscr{H} \to \mathscr{H}$ by

$$A \begin{bmatrix} v \\ p \end{bmatrix} = \begin{bmatrix} 0 & -c^2 \frac{\partial}{\partial x} \\ -\frac{\partial}{\partial x} & 0 \end{bmatrix} \begin{bmatrix} v \\ p \end{bmatrix},$$

$$\text{dom}(A) = \mathbb{H}^1(0, \infty; \mathbb{R}) \times \mathbb{H}^1(0, \infty; \mathbb{R}); \tag{6.128}$$

the mappings $f : [0, T] \to \mathscr{H}$ and $g : [0, T] \to \mathscr{B}_0(\mathbb{R}; \mathscr{H})$ by

$$f(t) = \begin{bmatrix} 2t \sin^3 (\cdot^2 + 1) \\ -t \cos (\cdot^2 + 1) \end{bmatrix}, \tag{6.129}$$

$$g(t) = \begin{bmatrix} t^2 \\ 2t^2 \end{bmatrix}; \tag{6.130}$$

and the IC by

$$X_0(\omega) = \begin{bmatrix} h_1(\cdot; \omega) \\ h_2(\cdot; \omega) \end{bmatrix}. \tag{6.131}$$

It can be shown that $(A, \text{dom}(A))$ is an m-accretive operator that generates a C_0-semigroup on $\mathscr{H}$.

Exercise 6.5.3. Explain why (6.125) has a unique mild solution on $[0, T]$.

6.6 Looking Ahead

External forces acting on a system are often state dependent. For instance, if the forcing term represents temperature regulation of a material, then it necessarily takes into account the temperature of the material at various times t and makes appropriate adjustments. This is easily illustrated by the following adaption of the forced heat equation (4.20) discussed in Model V.2:

$$\begin{cases} \partial z(x, y, t; \omega) = \left(k \triangle z(x, y, t; \omega) + \alpha e^{-\frac{\beta}{z(x,y,t;\omega)}} \right) \partial t + \\ \quad (t^2 + 2t) \sin z(x, y, t; \omega) dW(t) \\ z(x, y, 0; \omega) = (\sin 2x + \cos 2y) h(\omega), \ 0 < x < a, 0 < y < b, \omega \in \Omega, \\ z(x, 0, t; \omega) = 0 = z(x, b, t; \omega), \ 0 < x < a, t > 0, \omega \in \Omega, \\ z(0, y, t; \omega) = 0 = z(a, y, t; \omega), \ 0 < y < b, t > 0, \omega \in \Omega, \end{cases} \tag{6.132}$$

where $0 < x < a, 0 < y < b, t > 0, \omega \in \Omega$; $z(x, y, t)$ represents the temperature at the point (x, y) on the plate at time $t > 0$ for $\omega \in \Omega$, $\omega \mapsto h(\omega)$ is a uniform random variable on $(0, a) \times (0, b)$; and $W(t)$ is a one-dimensional Wiener process.

Exercise 6.6.1.
i.) Reformulate (6.132) as an abstract stochastic evolution equation. Indicate any new complications or changes that arise.
ii.) Conjecture a representation formula for a mild solution of the abstract stochastic evolution equation formulated in (i).

Consider the abstract stochastic evolution equation formulated in Exer. 6.6.1(i) and look back at the results developed in this chapter. What new obstacles are present that might complicate the extension of the theory needed to study this stochastic evolution equation?

6.7   Guidance for Selected Exercises

6.7.1   Level 1: A Nudge in a Right Direction

6.2.2. How about using $f, g_k \in \mathbb{L}^2(0, T : \mathbb{R})$? The question is whether or not the solution process can still have continuous sample paths in such case.
6.2.3. Note that $X_h^{-1}(t; \omega) = \frac{1}{X_h(t;\omega)}$. Use $h(t,x) = \frac{1}{x}$ with Itó's formula. (How?)
6.2.6. (i) A factor of T in the final constant (obtained upon application of Gronwall's Lemma) will be subsumed into the $\mathbb{L}^2$-norm.
6.2.8. The constants change slightly, but the approach is the same.
6.2.10. Note that $E\left[\sum_{k=1}^{m} \int_0^t [c_k X(s; \omega) + g_k(s)] dW_k(s)\right] = 0$.
6.2.12. Argue as in Chapter 5.
6.2.23. Observe that

$$
\begin{aligned}
\left| e^{a_n t} - e^{at} \right| &= \left| e^{a_n t} \left[1 - \sum_{n=0}^{\infty} \frac{((a-a_n)t)^n}{n!} \right] \right| \\
&\leq \left| e^{a_n t} \right| |a - a_n| \sum_{n=1}^{\infty} \frac{|a-a_n|^{n-1} t^n}{n!} \\
&\leq \left[\left| e^{a_n t} \right| t e^{|a-a_n|t} \right] |a - a_n|.
\end{aligned}
$$

6.2.26. Assume that g_n, g are continuous and $g_n \to g$ uniformly on $[0, T]$ as $n \to \infty$. (What else do you need to impose here?)
6.2.27. The change occurs in (6.89). (How?)
6.2.28. You will need to be particularly careful when using LDC and especially when dealing with the term

$$
\int_0^t \left(e^{a_\varepsilon(t-s)} f_\varepsilon(s) - e^{a(t-s)} f(s) \right) ds.
$$

6.4.1. For (6.104), let $\mathscr{H} = \mathbb{L}^2(0, a; \mathbb{R}) \times \mathbb{L}^2(0, b; \mathbb{R})$ and identify $f : [0, \infty) \to \mathscr{X}$ by $f(t)(x, y) = t^2 + x + 2y$. For a fixed $t^* \in [0, \infty)$, observe that $\int_0^a \int_0^b f(t^*)(x, y) dy dx <$

∞. (Why?) For (6.105), let $\mathcal{H} = \mathbb{L}^2(0,\pi;\mathbb{R})$ and use the natural choice for f as for (6.104). Finally, check in both cases that $u_0 \in \mathcal{L}^2(\Omega;\mathcal{H})$.

6.4.6. i.) Define $f : [0,T] \to \mathcal{X}$ by $f(t) = \arctan\left(1+\sqrt{1+t^2}\right)$. (Now what?)

ii.) We need only to check the regularity of f. (Why?)

iii.) Consider the FTC. (So what?)

6.5.1. i.) The ODE for $T(t)$ will be different, but the general approach is the same.

6.7.2 Level 2: An Additional Thrust in a Right Direction

6.2.3. Observe that

$$d\left[X_h^{-1}(s;\omega)\right] = 0dt - \frac{1}{X_h^2(s;\omega)}d\left[X_h(s;\omega)\right] + \frac{1}{2}\cdot\frac{2}{X_h^3(s;\omega)}\left(d\left[X_h(s;\omega)\right]\right)^2$$

$$= -\frac{1}{X_h^2(s;\omega)}\left[aX_h(s;\omega)dt + cX_h(s;\omega)dW(s)\right] + \frac{c^2}{X_h(s;\omega)}ds$$

$$= -\frac{1}{X_h(s;\omega)}\left[a - c^2\right]ds - \frac{c}{X_h(s;\omega)}dW(s)$$

$$= -X_h^{-1}(s;\omega)\left[a - c^2\right]ds - cX_h^{-1}(s;\omega)dW(s).$$

6.2.26. We will need to use LDC on the difference of integral terms involving g_n and g. Specifically, observe that

$$E\left|\int_0^t (g_n(s) - g(s))\,dW(s)\right|^2 \leq tE\int_0^t |g_n(s) - g(s)|^2\,ds.$$

Because $g_n \to g$ uniformly on $[0,T]$ as $n \to \infty$, LDC applies to conclude that

$$\lim_{n\to\infty} tE\int_0^t |g_n(s) - g(s)|^2\,ds = tE\int_0^t \lim_{n\to\infty} |g_n(s) - g(s)|^2\,ds,$$

and the integrand can be controlled. The remaining results are routine.

6.2.27. Observe that

$$E\left|\int_0^t e^{a_\varepsilon(t-s)}g_\varepsilon(s)dW(s)\right|^p \leq \zeta(p,t)\int_0^t \left|e^{a_\varepsilon(t-s)}\right|^p |g_\varepsilon(s)|^p\,ds.$$

Here, $g_\varepsilon \to 0$ uniformly as $\varepsilon \to 0^+$ and

$$\sup_{0<\varepsilon<1}\left(\sup_{0\leq t\leq T}\left|e^{a_\varepsilon(t-s)}\right|^p\right) < \infty.$$

(So what?)

6.4.6. ii.) Use the variation of parameters formula with the representation formula for e^{At}.

iii.) It should be an integral operator of the form $\int_0^x g(z)dz$. (Why?)

6.5.1. i.) The solution is

$$X(x,t;\omega) = e^{-\frac{\alpha t}{2}} \sum_{n=0}^{\infty} [a_n \cos(\mu_n t) + b_n \sin(\mu_n t)] \sin(\lambda_n x),$$

where $a_n = \frac{2}{L} \int_0^L z_0(x;\omega) \sin(\lambda_n x)\,dx$, $b_n = \frac{2}{L} \int_0^L z_1(x;\omega) \sin(\lambda_n x)\,dx + \frac{\alpha}{2} a_n$, $\lambda_n = \frac{n\pi}{L}$ and $\mu_n = 4c^2 \lambda_n^2 - \alpha^2 > 0$.

Chapter 7

Semi-Linear Stochastic Evolution Equations

Overview

We develop an extension of the theory established in Chapter 6 that enables us to formally study mathematical models whose forcing terms are also state dependent. The general nature of such perturbations creates various complications, even when trying to establish the existence of a mild solution. The focus of the current chapter is to precisely describe how these complications can be overcome.

7.1 Motivation by Models

We briefly consider variants of several models explored in earlier chapters and introduce two new applications. Our investigation of them will progress gradually throughout the chapter.

7.1.1 Some Models Revisited

Model II.3 Pharmacokinetics with Concentration-Dependent Dosage
The rate of absorption of a drug into the bloodstream is affected by the dosage $D(t)$ at any time $t > 0$. In turn, the dosage might depend on the concentration of the drug in the GI tract and/or bloodstream, which in turn varies with time. Abstractly, this leads to the following *semi-linear* version of (6.1) in Model II.2:

$$\begin{cases} dy(t;\omega) = \left(-ay(t;\omega) + D(t,y(t;\omega),z(t;\omega))\right)dt \\ \quad + G(t,y(t;\omega),z(t;\omega))dW(t), \\ dz(t;\omega) = \left(ay(t;\omega) - bz(t;\omega) + \overline{D}(t,y(t;\omega),z(t;\omega))\right)dt \\ \quad + \overline{G}(t,y(t;\omega),z(t;\omega))dW(t), \\ y(0;\omega) = y_0(\omega),\ z(0;\omega) = 0, \end{cases} \qquad (7.1)$$

where $0 < t < T$, $\omega \in \Omega$, and the terms $\overline{D}$ and $\overline{G}$ have been added in the second equation as a result of the state dependence within the term D in the first equation. Though somewhat more general, the basic form of (7.1) is not too dissimilar from (6.1).

Exercise 7.1.1. Rewrite (7.1) in vector form.

Model VII.2 Semi-Linear Wave Equations
Dispersion can be incorporated into the classical wave equation by altering the form of the forcing term. Precisely, consider the following extension of IBVP (6.119), where $\alpha, \beta > 0$:

$$
\begin{cases}
\partial \left(\frac{\partial}{\partial t} z(x,t;\omega) + \alpha z(x,t;\omega) \right) + \left(c^2 \frac{\partial^2}{\partial x^2} z(x,t;\omega) \right) \partial t = z(x,t;\omega) \partial t \\
\quad + \sum_{k=1}^{m} g_k(x,t) dW_k(t) \ 0 < x < L, \ 0 \le t \le T, \ \omega \in \Omega, \\
z(x,0;\omega) = z_0(x;\omega), \ \frac{\partial z}{\partial t}(x,0;\omega) = z_1(x;\omega), \ 0 < x < L, \ \omega \in \Omega, \\
z(0,t;\omega) = z(L,t;\omega) = 0, \ 0 \le t \le T, \ \omega \in \Omega.
\end{cases}
\tag{7.2}
$$

Exercise 7.1.2. Reformulate (7.2) as an abstract stochastic evolution equation in the space defined in (6.112). How does the resulting form compare to (6.122)?

Similarly, a system of weakly coupled damped wave equations with nonlinear dispersion can be described by

$$
\begin{cases}
\partial \left(\frac{\partial}{\partial t} z(x,t;\omega) + \alpha z(x,t;\omega) \right) + \left(c^2 \frac{\partial^2}{\partial x^2} z(x,t;\omega) \right) \partial t = (f_1(z) + g_1(w)) \partial t \\
\quad + \left(\overline{f_1}(z) + \overline{g_1}(w) \right) dW(t), \ 0 < x < L, \ 0 \le t \le T, \ \omega \in \Omega, \\
\partial \left(\frac{\partial}{\partial t} w(x,t;\omega) + \alpha w(x,t;\omega) \right) + \left(c^2 \frac{\partial^2}{\partial x^2} w(x,t;\omega) \right) \partial t = (f_2(z) + g_2(w)) \partial t \\
\quad + \left(\overline{f_2}(z) + \overline{g_2}(w) \right) dW(t), \ 0 < x < L, \ 0 \le t \le T, \ \omega \in \Omega, \\
z(x,0;\omega) = z_0(x;\omega), \ \frac{\partial z}{\partial t}(x,0;\omega) = z_1(x;\omega), \ 0 < x < L, \ \omega \in \Omega, \\
w(x,0;\omega) = w_0(x;\omega), \ \frac{\partial w}{\partial t}(x,0;\omega) = w_1(x;\omega), \ 0 < x < L, \ \omega \in \Omega, \\
\frac{\partial z}{\partial x}(0,t;\omega) = \frac{\partial z}{\partial x}(L,t;\omega) = \frac{\partial w}{\partial x}(0,t;\omega) = \frac{\partial w}{\partial x}(L,t;\omega) = 0, \ 0 \le t \le T, \ \omega \in \Omega.
\end{cases}
\tag{7.3}
$$

Exercise 7.1.3. Reformulate (7.3) as an abstract stochastic evolution equation in the space $\mathscr{H} = \left(\mathbb{H}_0^1(0,L;\mathbb{R}) \times \mathbb{L}^2(0,L;\mathbb{R}) \right)^2$. What hurdles do you encounter?

Model V.3 Heat Conduction with State-Dependent Heat Source
The heat production source can change with both time and the temperature of the material being heated. For instance, it is reasonable for certain chemical reactions to exhibit an exponentially decaying heat source. The following IBVP for heat conduction through a metal sheet takes this into account and constitutes a natural generalization

of (6.104):

$$
\begin{cases}
\partial z(x,y,t;\omega) = \left(k\triangle z(x,y,t;\omega) + \alpha e^{-\frac{\beta}{z(x,y,t;\omega)}} \right) \partial t + g(t)dW_1(t) \\
+ g(t,z(x,y,t;\omega))dW_2(t)\, 0 < x < a,\, 0 < y < b,\, 0 \leq t \leq T,\, \omega \in \Omega, \\
z(x,y,0;\omega) = z_0(x,y;\omega),\, 0 < x < a,\, 0 < y < b,\, \omega \in \Omega, \\
\frac{\partial z}{\partial y}(x,0,t;\omega) = 0 = \frac{\partial z}{\partial y}(x,b,t;\omega),\, 0 < x < a,\, 0 \leq t \leq T,\, \omega \in \Omega, \\
\frac{\partial z}{\partial x}(0,y,t;\omega) = 0 = \frac{\partial z}{\partial x}(a,y,t;\omega),\, 0 < y < b,\, 0 \leq t \leq T,\, \omega \in \Omega.
\end{cases}
\tag{7.4}
$$

Exercise 7.1.4. Reformulate (7.4) as an abstract stochastic evolution equation.

7.1.2 Introducing Two New Models

Model IX.1 Neural Networks

Hopfield initiated the study of neural networks in 1982. Applications range from the modeling of physiological functions to using artificial neural networks to perform parallel computations. (See **[136, 194, 245, 339, 392, 410]**.) A heuristic development of a basic neural network is as follows.

Suppose that we begin with M neurons interconnected via a network of synapses. Every neuron receives an input signal from the other $M - 1$ neurons, each with a varying degree of strength. The neuron acts on the total signal and produces an output that is subsequently broken down and emitted as inputs into the other $M - 1$ neurons comprising the network. To form the mathematical model, label the neurons as $1 \leq i \leq M$, and for each i and time $t \geq 0$, let

$x_i(t) =$	voltage of input from i at time t
$\eta_{ij}(t) =$	strength of signal that j contributes to i at time t
$\sum_{j=1}^{M} \eta_{ij}(t)x_j(t) =$	total input signal from the networked neurons
$g\left(\sum_{j=1}^{M} \eta_{ij}(t)x_j(t) \right) =$	output signal at time t from i due to internal activity

Momentarily, we assume for simplicity that the output signal can be expressed as

$$
g\left(\sum_{j=1}^{M} \eta_{ij}(t)x_j(t) \right) = \sum_{j=1}^{M} \eta_{ij}(t)g_j(x_j(t))
$$

so that it is clear how the neuron acts on each input signal individually. (We acknowledge that this is somewhat unrealistic because it assumes that we can distinguish among the M inputs.) These quantities evolve over time. As such, the rates at which the voltages of these M neurons change is governed by a system of ODEs formulated under the assumption that the rate of voltage change for each neuron is proportional to its present voltage and its output signal. Assuming that noise is incorporated into the system via the coefficients of $x_i(t;\omega)$ ($i = 1,\ldots,M$), we obtain the following stochastic system

$$
\begin{cases}
dx_1(t;\omega) = \left(a_1 x_1(t;\omega) + \sum_{j=1}^{M} \eta_{1j}(t) g_j(x_j(t;\omega))\right) dt + \overline{a_1} x_1(t;\omega) dW(t), \\
\;\;\vdots \\
dx_M(t;\omega) = \left(a_M x_M(t;\omega) + \sum_{j=1}^{M} \eta_{Mj}(t) g_j(x_j(t;\omega))\right) dt + \overline{a_M} x_M(t;\omega) dW(t), \\
x_i(0;\omega) = x_{i,0}(\omega),\ i = 1,\ldots,M,
\end{cases}
$$

$$(7.5)$$

where $0 \le t \le T$, $\omega \in \Omega$, and $W(t)$ is a one-dimensional Wiener process.

Exercise 7.1.5. Reformulate (7.5) as an abstract stochastic evolution equation.

Model X.1 Spatial Pattern Formation

Diffusion, without the intervening effects of kinetics, disperses a pattern. Chemicals react and diffuse in different ways, thereby resulting in a distribution of varying concentrations that can be viewed as a distinct spatial pattern. Morphogenesis is the development of pattern and form in a living thing, and arises naturally in ecology by way of describing migratory patterns, the formation of animal coatings (e.g., dispersion and pattern of spots on a leopard), butterfly wing patterns, etc. There are differing viewpoints as to how, biologically, such patterns are formed. In some manner, though, the reaction-diffusion equations enter into the mathematical modeling of this phenomenon. In 1952, Turing asserted that diffusion need not lead to a uniformly distributed concentration, but rather certain perturbations could redirect its action to form patterns. (See [120, 298, 309, 321, 354, 413].) We discuss a two-dimensional version of his model below.

Let $\mathscr{D}$ be a bounded domain in $\mathbb{R}^3$ with smooth boundary $\partial\mathscr{D}$ and suppose that there are N chemicals interacting within this domain. For each $\mathbf{x} = (x,y,z) \in \mathscr{D}, t \ge 0$, $1 \le i \le N$, and $\omega \in \Omega$, let

$C_i = C_i(\mathbf{x},t;\omega) =$	Concentration of chemical i at $(\mathbf{x},t;\omega)$
$\alpha_i =$	Diffusion coefficient for chemical i
$f_i(t,C_1,\ldots,C_N) =$	Reaction among N chemicals affecting how C_i changes

The dynamics of these N chemicals within $\mathscr{D}$ can be described by the following system of diffusion equations:

$$
\begin{cases}
\partial C_1(\mathbf{x},t;\omega) &= (\alpha_1 \triangle C_1 + f_1(t,C_1,\ldots,C_N))\,\partial t + \overline{f_1}(t,C_1,\ldots,C_N)\,dW(t), \\
\;\;\vdots \\
\partial C_N(\mathbf{x},t;\omega) &= (\alpha_N \triangle C_N + f_N(t,C_1,\ldots,C_N))\,\partial t + \overline{f_N}(t,C_1,\ldots,C_N)\,dW(t), \\
C_i(\mathbf{x},0;\omega) &= C_i^\star(\mathbf{x};\omega),\ \mathbf{x} \in \mathscr{D},\ \omega \in \Omega,\ 1 \le i \le N, \\
C_i(\mathbf{x},t;\omega) &= 0,\ \mathbf{x} \in \partial\mathscr{D},\ 0 \le t \le T,\ \omega \in \Omega,\ 1 \le i \le N,
\end{cases}
$$

$$(7.6)$$

where $\mathbf{x} \in \mathscr{D}, t > 0, \omega \in \Omega$, and $W(t)$ is a one-dimensional Wiener process. We can express the equation portion of (7.6) in matrix form as

$$
\partial \underbrace{\begin{bmatrix} C_1 \\ \vdots \\ C_N \end{bmatrix}}_{=\mathbf{C}} = \underbrace{\begin{bmatrix} \alpha_1 \triangle & 0 & \cdots & 0 \\ 0 & \ddots & & \vdots \\ \vdots & & \ddots & 0 \\ 0 & \cdots & 0 & \alpha_N \triangle \end{bmatrix}}_{=A} \underbrace{\begin{bmatrix} C_1 \\ \vdots \\ C_N \end{bmatrix}}_{=\mathbf{C}} \partial t
$$

$$
+ \underbrace{\begin{bmatrix} f_1(t, C_1, \ldots, C_N) \\ \vdots \\ f_N(t, C_1, \ldots, C_N) \end{bmatrix}}_{=\mathbf{f}(t,\mathbf{C}) \text{ Kinetics}} \partial t + \underbrace{\begin{bmatrix} \overline{f}_1(t, C_1, \ldots, C_N) \\ \vdots \\ \overline{f}_N(t, C_1, \ldots, C_N) \end{bmatrix}}_{=\mathbf{\overline{f}}(t,\mathbf{C}) \text{ Kinetic Noise}} dW(t) \tag{7.7}
$$

or more succinctly as

$$
\partial \mathbf{C}(\mathbf{x},t;\omega) = (A\mathbf{C}(\mathbf{x},t;\omega) + \mathbf{f}(t,\mathbf{C}(\mathbf{x},t;\omega))) \partial t + \mathbf{\overline{f}}(t,\mathbf{C}(\mathbf{x},t;\omega)) dW(t). \tag{7.8}
$$

For simplicity, we consider the case in which there are only two chemicals (i.e., $N = 2$ above) interacting in $\mathscr{D}$.

Exercise 7.1.6. On what Hilbert space would it be natural to reformulate (7.7) abstractly when $N = 2$?

Different kinetic terms f_i have been derived theoretically and experimentally by researchers in the field. One classical model of activator-inhibitor type [120] describes the kinetics by

$$
\begin{cases} f_1(C_1, C_2) = \beta_1 - \beta_2 C_1 + \beta_3 C_1^2 C_2, \\ f_2(C_1, C_2) = \beta_4 - \beta_3 C_1^2 C_2, \end{cases} \tag{7.9}
$$

where β_i $(i = 1, 2, 3, 4)$ are the rate constants. Of course, noise can be incorporated into the model naturally through any of the rate constants β_i $(i = 1, 2, 3, 4)$.

Several examples illustrating how imposing different conditions on β_i guarantees diffusive instability leading to pattern formation are discussed in the references cited within [120, 309].

Common Theme: All of the IBVPs considered in this section can be reformulated as the abstract stochastic evolution equation

$$
\begin{cases} dX(t;\omega) = (AX(t;\omega) + f(t,X(t;\omega))) dt + g(t,X(t;\omega)) dW(t), 0 < t < T, \omega \in \Omega, \\ X(0;\omega) = X_0(\omega), \omega \in \Omega, \end{cases}
$$

$$
\tag{7.10}
$$

in an appropriate separable Hilbert space $\mathscr{H}$. Mere symbolic identification suggests that a variation of parameters formula for a mild solution of (7.10) might be given by

$$X(t;\omega) = e^{At}X_0(\omega) + \int_0^t e^{A(t-s)}f(s,X(s;\omega))ds + \int_0^t e^{A(t-s)}g(s,X(s;\omega))dW(s),$$

$$(7.11)$$

where $0 \le t \le T$ and $\omega \in \Omega$. This is intuitive, but the dependence of the forcing term on the state $X(s;\omega)$ creates a self-referential situation in (7.11) that was not present before. Somehow, we need a technique that allows us to temporarily suspend this interdependence in order to make use of the theory in Chapter 6.

7.2　Some Essential Preliminary Considerations

We shall take a slight departure from our usual tack in that we will not develop the entire theory first for the finite-dimensional case and then for the case of a general Banach space. Rather, we shall explore certain special cases as the need arises to spark our intuition as to what a concept or result "ought to be." Of course, upon completion of the development of our theory, we will effortlessly recover the results for finite-dimensional SDEs as a special case, at times under weaker hypotheses. In the latter case, we will explore various improvements of the theory by critically analyzing the proof in order to identify where, and how, the hypotheses can be weakened.

Consider the abstract IVP (7.10). Suppose that $X : [0,\infty) \times \Omega \to \mathscr{H}$ satisfies (7.10) in a separable Hilbert space $\mathscr{H}$, assuming whatever level of regularity seems necessary to render (7.10) meaningful. The new struggle we face is the self-referential nature of (7.11). In essence, this is simply another equation to solved, albeit one of a different type. We have simply managed to replace the solvability of (7.10) by the solvability of (7.11), which is hopefully easier.

When does (7.11) have a mild solution? The equation (7.11) is more complicated than the nonhomogenous IVP studied in Chapter 6 because the forcing term f now changes according to a second variable. "Solvability" then naturally boils down to the behavior of the mappings

$$t \mapsto \int_0^t e^{A(t-s)}f(s,X(s;\omega))ds,$$

$$t \mapsto \int_0^t e^{A(t-s)}g(s,X(s;\omega))dW(s).$$

The dependence of the forcing term f on the state process $X(s;\omega)$ opens the door to possibilities that did not arise in Chapter 6. Indeed, we encounter new issues, some of which are explored below.

Consider the random stochastic evolution equation

$$\begin{cases} X'(t;\omega) = AX(t;\omega) + f(t,X(t;\omega), 0 < t < T, \omega \in \Omega, \\ X(0;\omega) = X_0(\omega), \omega \in \Omega. \end{cases} \tag{7.12}$$

Exercise 7.2.1. For simplicity, consider (7.12) with $A = 0$ and $\mathscr{H} = \mathbb{R}$. As in **[285]**, let $E \in \mathscr{F}_0$ and define the random variable $\omega \mapsto X_0(\omega) = \chi_E(\omega)$.
i.) Show that $\forall \omega \in \Omega$, the following IVP has more than one solution:

$$\begin{cases} x'(t;\omega) = 3[x(t;\omega)]^{5/8}, t > 0, \omega \in \Omega, \\ x(0;\omega) = \chi_E(\omega), \omega \in \Omega. \end{cases} \tag{7.13}$$

(In fact, this IVP has infinitely many solutions!)
ii.) Consider the IVP

$$\begin{cases} x'(t;\omega) = (1+2x(t;\omega))^4, t > 0, \omega \in \Omega, \\ x(0;\omega) = x_0(\omega), \omega \in \Omega. \end{cases} \tag{7.14}$$

a.) Show that $\forall \omega \in \Omega$, a solution of (7.14) is given by

$$x(t;\omega) = \frac{1}{2}\left[-1 + \left((2x_0(\omega)+1)^{-3} - 6t\right)^{-1/3}\right].$$

b.) Let $T = \frac{1}{6}(2x_0(\omega)+1)^{-3}$. If $x_0(\omega) > -\frac{1}{2}$, show that $\lim_{t \to T^-} |x(t;\omega)| = \infty$. As such, (7.14) does not have a continuous solution on $[0,\infty)$ and, in fact, cannot be extended past $t = T$.

For every $\omega \in \Omega$, the right-hand sides $f(t,x(t;\omega))$ of (7.13) and (7.14) are continuous functions of both variables. This was certainly sufficient to guarantee the existence and uniqueness of a mild solution of (6.106) on $[0,\infty)$ when $f(t,x(t;\omega)) = f(t)$. However, this is false for (7.13), and in a big way for (7.14). The situation is even more bleak when noise is incorporated into the equation via a white noise process. How can we overcome these issues in order to formulate a theory analogous to Chapter 6 for (7.10). We shall explore this question and along the way develop various strategies of attack that will be used throughout the remainder of the text.

7.3 Growth Conditions

For this section, $\mathscr{X}$ and $\mathscr{Y}$ are general Banach spaces.

Exercise 7.2.1 revealed that mere continuity of f in both variables is insufficient to guarantee the uniqueness of a mild solution of (7.10) on $[0,\infty)$, even when $\mathscr{H} = \mathbb{R}$. Even worse, such continuity does not even guarantee the existence of a mild solution.

(See **[329]** for an example.) As such, it is sensible to ask what conditions would ensure the existence (and possibly uniqueness) of a mild solution of (5.10) on at least some interval $[0, T_0]$. There is a plentiful supply of such conditions that can be imposed on f which further control its "growth." We introduce several common ones in this section and investigate how they are interrelated.

A close investigation of (7.13) reveals that the curve corresponding to the forcing term was sufficiently steep in a vicinity of $(0,0)$ as to enable us to construct a sequence of chord lines, all passing through $(0,0)$, whose slopes became infinitely large.

Exercise 7.3.1. Show that the sequence of chord line slopes for $f(x) = 3x^{5/8}$ connecting $(0,0)$ to $\left(\frac{1}{n}, f\left(\frac{1}{n}\right)\right)$ approaches infinity as $n \to \infty$.

As such, close to the initial starting point, the behavior of f changes very quickly, and this in turn affects the behavior of x' in a short interval of time. Moreover, this worsens the closer you get to the origin. Thus, if we were to try to generate the solution path on a given time interval $[0, T]$ numerically, refining the partition of $[0, T]$ (in order to increase the number of time points used to construct the approximate solution) would subsequently result in a sequence of paths that does not approach a single recognizable continuous curve.

The presence of the cusp in the graph is the troublemaker! Can we somehow control the chord line slopes without demanding that f be differentiable (because this would exclude functions like the absolute value)? The search for such control over chord line slopes prompts us to make the following definition.

Definition 7.3.1. A function $f : \mathscr{X} \to \mathscr{Y}$ is *globally Lipschitz on* $\mathscr{D} \subset \mathscr{X}$ if $\exists M_f > 0$ such that

$$\|f(x) - f(y)\|_{\mathscr{Y}} \leq M_f \|x - y\|_{\mathscr{X}}, \ \forall x, y \in \mathscr{D}. \tag{7.15}$$

(M_f is called a *Lipschitz constant* for f.)

This definition is easily adapted to functions of more than one independent variable, but we must carefully indicate to which of the independent variables we intend the condition to apply. Functions of the form $f : [0, T] \times \mathscr{X} \to \mathscr{Y}$ commonly arise in practice. We introduce the following modification of Def. 7.3.1 as it applies to such functions.

Definition 7.3.2. A function $f : [0, T] \times \mathscr{X} \to \mathscr{Y}$ is *globally Lipschitz on* $\mathscr{D} \subset \mathscr{X}$ *(uniformly in t)* if $\exists M_g > 0$ (independent of t) such that

$$\|g(t, x) - g(t, y)\|_{\mathscr{Y}} \leq M_g \|x - y\|_{\mathscr{X}}, \ \forall t \in [0, T] \text{ and } x, y \in \mathscr{D}. \tag{7.16}$$

Exercise 7.3.2. Interpret (7.16) geometrically. For simplicity, assume that $\mathscr{X} = \mathscr{Y} = \mathbb{R}$. How does this interpretation change if M_g depends on t.

The space $\mathscr{X}$ could be a product space $\mathscr{X}_1 \times \cdots \times \mathscr{X}_n$. Assuming that it is equipped with the usual norm, (7.16) becomes

$$\|g(t,x_1,\ldots,x_n) - g(t,y_1,\ldots,y_n)\|_{\mathscr{Y}} \leq M_g \sum_{i=1}^{n} \|x_i - y_i\|_{\mathscr{X}}, \qquad (7.17)$$

$\forall t \in [0,T]$ and $(x_1,\ldots,x_n),(y_1,\ldots,y_n) \in \mathscr{D} \subset \mathscr{X}_1 \times \cdots \times \mathscr{X}_n$.

Exercise 7.3.3. If $f : I \subset \mathbb{R} \to \mathbb{R}$ has a bounded derivative on I, prove that f is globally Lipschitz on I.

Exercise 7.3.4. Let $f \in \mathbf{L}^1 (a,b;\mathbb{R})$ and define $g : [a,b] \to \mathbb{R}$ by $g(x) = \int_a^x f(z)dz$. Is g globally Lipschitz on (a,b)? If not, try to identify the least amount of additional regularity that could be imposed on f to ensure that g is globally Lipschitz.

Exercise 7.3.5. Suppose $f : [0,T] \times \mathbb{R} \to \mathbb{R}$ is continuous and $v \in \mathbb{C}^1 ((0,\infty);(0,\infty))$. Define $h : [0,T] \times (0,\infty) \to \mathbb{R}$ by $h(t,x) = \int_0^{v(x)} f(t,z)dz$.
i.) Is h globally Lipschitz on $(0,\infty)$ (uniformly in t)? If not, try to identify the least amount of additional regularity that could be imposed on f and v to ensure that h is globally Lipschitz.
ii.) Let $k \in \mathbb{C}^1 (\mathbb{R};\mathbb{R})$ and define $h_k : [0,\infty) \times \mathbb{R} \to \mathbb{R}$ by $h_k(t,x) = \int_0^{v(x)} f(t,k(z))dz$. Is h_k globally Lipschitz on $[0,\infty)$ (uniformly in t)?

Exercise 7.3.6. Let $f : [0,T] \times \mathbb{R} \times \mathbb{R} \to \mathbb{R}$ and $g : \mathbb{R} \to \mathbb{R}$ be given mappings. Define $j : [0,T] \times \mathbb{R} \to \mathbb{R}$ by $j(t,x) = f\left(t,x,\int_0^x g(z)dz\right)$. Provide sufficient conditions on f and g that ensure that j is globally Lipschitz on $\mathbb{R}$ (uniformly in t).

Exercise 7.3.7. Must a finite linear combination of functions $f_i : \mathscr{D} \subset \mathscr{X} \to \mathscr{Y}$, $1 \leq i \leq n$, that are globally Lipschitz on $\mathscr{D}$ also be globally Lipschitz on $\mathscr{D}$?

Of course, imposing such a Lipschitz condition on a function f over an *entire space* like $[0,T] \times \mathbb{R}$ is still restrictive because the same Lipschitz constant is used throughout the space, which essentially demands that f grow no faster than a linear function on this space. This eliminates many functions from consideration, including relatively tame examples such as $f(t,x) = e^x$ or $f(t,x) = tx^2$. (Why?) Perhaps we can weaken the condition slightly so that rather than on the whole space, we can demand that the function be Lipschitz on any closed ball contained within in, with the caveat that the Lipschitz constant can change from ball to ball. This suggests the following localized version of Def. 7.3.1.

Definition 7.3.3. A function $f : \mathscr{X} \to \mathscr{Y}$ is *locally Lipschitz on $\mathscr{X}$* if $\forall x_0 \in \mathscr{X}$ and $\varepsilon > 0$, $\exists$ a constant $M_{(x_0,\varepsilon)} > 0$ (depending on x_0 and ε) such that

$$\|f(x) - f(y)\|_{\mathscr{Y}} \leq M_{(x_0,\varepsilon)} \|x - y\|_{\mathscr{X}}, \ \forall x,y \in \mathfrak{B}_{\mathscr{X}}(x_0;\varepsilon). \qquad (7.18)$$

Exercise 7.3.8. Formulate local versions of Def. 7.3.2 and (7.17) in the spirit of Def. 7.3.3.

Exercise 7.3.9. Must all continuous real-valued functions $f : \mathbb{R} \to \mathbb{R}$ be locally Lipschitz on $\mathbb{R}$? Explain.

The inequalities (7.16) and (7.18) used in Def. 7.3.1 and Def. 7.3.3, respectively, can be generalized in various ways, two of which are

$$\|f(t,x) - f(t,y)\|_{\mathscr{Y}} \leq k(t) \|x - y\|_{\mathscr{X}}, \ \forall t \in [0,T] \text{ and } x, y \in \mathscr{X}, \quad (7.19)$$

$$\|f(t,x) - f(t,y)\|_{\mathscr{Y}} \leq k(t) \|x - y\|_{\mathscr{X}}^{p}, \ \forall t \in [0,T] \text{ and } x, y \in \mathscr{X}, \quad (7.20)$$

where $p > 1$ and k typically belongs to either $\mathbb{L}^1(0,T;\mathbb{R})$ or $\mathbb{C}([0,T];(0,\infty))$.

Exercise 7.3.10. If $f : [0,T] \times \mathscr{X} \to \mathscr{Y}$ satisfies Def. 7.3.2, must it satisfy either (7.19) or (7.20) for some $k \in \mathbb{C}([0,T];\mathbb{R})$? What if $k \in \mathbb{L}^1(0,T;\mathbb{R})$? How about the converse implications?

Exercise 7.3.11. If $f : \mathscr{X} \to \mathscr{Y}$ satisfies Def. 7.3.3, must it satisfy either (7.19) or (7.20) if $k \in \mathbb{C}([0,T];\mathbb{R})$? What if $k \in \mathbb{L}^1(0,T;\mathbb{R})$? How about the converse implications?

In addition to a Lipschitz-type condition, we will need to control the growth of a single term $\|f(t,x)\|_{\mathscr{Y}}$ rather than the norm of a difference of functional values, in the following sense.

Definition 7.3.4. A function $f : [0,T] \times \mathscr{X} \to \mathscr{Y}$ has *sublinear growth (uniformly in t)* if $\exists M_1 > 0$ such that

$$\|f(t,x)\|_{\mathscr{Y}} \leq M_1 [\|x\|_{\mathscr{X}} + 1], \ \forall t \in [0,T] \text{ and } x \in \mathscr{X}. \quad (7.21)$$

More generally, (7.21) can be replaced by one of the following:

$$\|f(t,x)\|_{\mathscr{Y}} \leq M_1(t) [\|x\|_{\mathscr{X}} + 1], \ \forall t \in [0,T] \text{ and } x \in \mathscr{X}, \quad (7.22)$$

$$\|f(t,x)\|_{\mathscr{Y}} \leq M_1(t) [\|x\|_{\mathscr{X}}^{p} + 1], \ \forall t \in [0,T] \text{ and } x \in \mathscr{X}, \quad (7.23)$$

where $p > 1$ and $M_1(\cdot)$ is typically assumed to belong to either $\mathbb{L}^q(0,T;\mathbb{R})$ (where $\frac{1}{p} + \frac{1}{q} = 1$) or $\mathbb{C}([0,T];\mathbb{R})$.

Exercise 7.3.12. Define $\mathbf{f} : \mathbb{R}^N \to \mathbb{R}^N$ by $\mathbf{f}(\mathbf{x}) = \mathbf{Ax} + \mathbf{B}$, where $\mathbf{A} \in \mathbb{M}^N(\mathbb{R})$ and $\mathbf{B} \in \mathbb{R}^N$. Is $\mathbf{f}$ locally Lipschitz?

Exercise 7.3.13. Suppose that $f : [0,T] \times \mathscr{X} \to \mathscr{Y}$ has sublinear growth in the sense of one of (7.21), (7.22), or (7.23), where $M_1 \in \mathbb{C}([0,T];\mathbb{R})$. Let $\mathscr{D} \subset \mathscr{X}$ be a bounded set.
i.) Show that the image $f([0,T] \times \mathscr{D})$ is a bounded subset of $\mathscr{Y}$.
ii.) Must the image $f([0,T] \times \mathscr{D})$ also be precompact in $\mathscr{Y}$? Explain.

The growth conditions discussed in this section, when coupled with the correct technique, can be used to formulate a rich existence theory for (7.10).

7.4 The Cauchy Problem

Unless otherwise specified, we impose the following standing assumption:
(H$_A$) $A : \text{dom}(A) \subset \mathscr{X} \to \mathscr{X}$ generates a C_0-semigroup $\{e^{At} : t \geq 0\}$ on $\mathscr{X}$ for which $\exists M^\star > 0$ and $\alpha \in \mathbb{R}$ such that

$$\left\| e^{tA} \right\|_{\mathbb{B}(\mathscr{X})} \leq M^\star e^{\alpha t}, \ \forall t \geq 0. \tag{7.24}$$

In particular, $\forall T > 0$, the Principle of Uniform Boundedness ensures that

$$\overline{M_A} = \sup_{0 \leq t \leq T} \left\| e^{At} \right\|_{\mathbb{B}(\mathscr{X})} < \infty. \tag{7.25}$$

7.4.1 Problem Formulation

Assume **(S.A.1)**. Consider the abstract semi-linear SEE

$$\begin{cases} dX(t;\omega) = (AX(t;\omega) + f(t,X(t;\omega)))\, dt + g(t,X(t;\omega))dW(t), \\ X(0;\omega) = X_0(\omega), \omega \in \Omega \end{cases} \tag{7.26}$$

where $0 < t < T$, $\omega \in \Omega$ in a separable Hilbert space $\mathscr{H}$. Here, $f : [0,T] \times \mathscr{H} \to \mathscr{H}$ and $g : [0,T] \times \mathscr{H} \to \mathscr{B}_0(\mathbb{R}^m; \mathscr{H})$ are given mappings. We assume **(H$_A$)**, **(H5.1)** through **(H5.3)**, and **(H5.5)**, as well as

(H7.1) $f : [0,T] \times \mathscr{H} \to \mathscr{H}$ is an $\mathscr{F}_t$-adapted, progressively measurable mapping such that $\exists M_f, \overline{M_f} > 0$ for which

$$\|f(t,x) - f(t,y)\|_{\mathscr{H}} \leq M_f \|x - y\|_{\mathscr{H}}$$
$$\|f(t,x)\|_{\mathscr{H}} \leq \overline{M_f}[1 + \|x\|_{\mathscr{H}}],$$

$\forall x, y \in \mathscr{H}$, uniformly in $t \in [0,T]$;
(H7.2) $g : [0,T] \times \mathscr{H} \to \mathscr{B}_0(\mathbb{R}^m; \mathscr{H})$ is an $\mathscr{F}_t$-adapted, progressively measurable mapping such that $\exists M_g, \overline{M_g} > 0$ for which

$$\|g(t,x) - g(t,y)\|_{\mathscr{B}_0(\mathbb{R}^m;\mathscr{H})} \leq M_g \|x - y\|_{\mathscr{H}}$$
$$\|g(t,x)\|_{\mathscr{B}_0(\mathbb{R}^m;\mathscr{H})} \leq \overline{M_g}[1 + \|x\|_{\mathscr{H}}],$$

$\forall x, y \in \mathscr{H}$, uniformly in $t \in [0,T]$.

Our interest lies primarily with mild solutions of (7.26) in the following sense.

Definition 7.4.1. A stochastic process $X : [0,T_0] \subset [0,T] \times \Omega \to \mathscr{H}$ is a *mild solution* of (7.26) on $[0,T_0]$ if
i.) $X \in \mathbb{C}\left([0,T_0]; \mathscr{L}^2(\Omega; \mathscr{H})\right)$,
ii.) $X(t;\omega) = e^{At}X_0(\omega) + \int_0^t e^{A(t-s)} f(s,X(s;\omega))ds + \int_0^t e^{A(t-s)}g(s,X(s;\omega))dW(s)$, $\forall 0 \leq t \leq T_0$, a.s. $[\mathscr{P}]$.

Remark. We defined the notion of a mild solution of (7.26) on a subinterval $[0, T_0] \subset [0, T]$ because, in general, global existence is not guaranteed a priori (cf. Exer. 7.2.1).

7.4.2 Existence and Uniqueness Results

The main theorem is as follows.

Theorem 7.4.2. *If* **(H$_A$)**, **(H5.1)** *through* **(H5.3)**, **(H7.1)**, *and* **(H7.2)** *are satisfied, then (7.26) has a unique mild solution on* $[0, T]$.

The deterministic case of such Cauchy problems is discussed in **[315, 417]**. We shall present two methods of attack used to prove this theorem. Both approaches rely on the following estimates in the spirit of Lemma 3.2.7 and Prop. 5.5.3:

$$E \left\| \int_0^t e^{A(t-s)} f(s, X(s; \cdot)) ds \right\|_{\mathscr{H}}^p \leq M_A^p T^{\frac{p}{q}} \int_0^t E \|f(s, X(s; \cdot))\|_{\mathscr{H}}^p ds \quad (7.27)$$

$$E \left\| \int_0^t e^{A(t-s)} g(s, X(s; \cdot)) dW(s) \right\|_{\mathscr{H}}^p \leq M_A^p \zeta_g(t, p) \quad (7.28)$$

$$\times \int_0^t E \|g(s, ; X(s; \cdot))\|_{\mathscr{B}_0}^p ds,$$

where $p > 2$ and $\zeta_g(t, p)$ is the constant defined in Prop. 5.5.3. These estimates are established using an approach similar to the one used to establish the estimates in Section 5.5. (Try it!)

Approach 1: A Typical Contraction Argument
With the Fixed-Point Approach in mind, we use (7.11) to define the solution map
$\Phi : \mathbb{C}\left([0, T_0]; \mathcal{L}^2(\Omega; \mathscr{H})\right) \to \mathbb{C}\left([0, T_0]; \mathcal{L}^2(\Omega; \mathscr{H})\right)$ by

$$(\Phi X)(t; \omega) = e^{At} X_0(\omega) + \int_0^t e^{A(t-s)} f(s, X(s; \omega)) ds + \int_0^t e^{A(t-s)} g(s, X(s; \omega)) dW(s). \quad (7.29)$$

Exercise 7.4.1. Assume **(H7.1)** and **(H7.2)**.
i.) Why is the right-hand side of (7.29) $\mathscr{F}_t$-measurable?
ii.) Let $X \in \mathbb{C} = \mathbb{C}\left([0, T_0]; \mathcal{L}^2(\Omega; \mathscr{H})\right)$ and take $\{X_n\} \subset \mathbb{C}\left([0, T_0]; \mathcal{L}^2(\Omega; \mathscr{H})\right)$ such that $X_n \to X$ in $\mathbb{C}$. Show that $\Phi(X_n) \to \Phi(X)$ in $\mathbb{C}$ as $n \to \infty$.
iii.) Show that $\forall 0 \leq t \leq T_0$,

$$E \|\Phi(X)(t) - \Phi(Y)(t)\|_{\mathscr{H}}^2 \leq 4M_A^2 T \left(TM_f^2 + \zeta_g(t, p)M_g^2\right) \|X - Y\|_{\mathbb{C}}^2. \quad (7.30)$$

Note that Φ is not automatically a contraction, in general. What condition is sufficient to impose in (7.30) in order to render it one? The moment Φ is a contraction, we know from Thrm. 1.11.2 that Φ has a unique fixed point that satisfies (7.29) and, hence, is the mild solution of (7.26) we seek. (Explain why.)
iv.) Alternatively, we can avoid restricting the size of $4M_A^2 T \left(TM_f^2 + \zeta_g(t, p)M_g^2\right)$ by

successively iterating Φ. Precisely, show that $\forall n \in \mathbb{N}$,

$$\|\Phi^n(X) - \Phi^n(Y)\|_{\mathbb{C}}^2 \leq \frac{\left(4M_A^2 T \left(TM_f^2 + \zeta_g(t,p)M_g^2\right)\right)^n}{n!} \|X - Y\|_{\mathbb{C}}^2. \tag{7.31}$$

Because $\exists n_0 \in \mathbb{N}$ such that

$$\frac{\left(4M_A^2 T \left(TM_f^2 + \zeta_g(t,p)M_g^2\right)\right)^{n_0}}{n_0!} < 1,$$

we can conclude that Φ^{n_0} is a contraction and so, by Cor. 1.11.3, Φ itself has a unique fixed point that coincides with the mild solution of (7.26) we seek.

Upon completion of Exer. 7.4.1, we can conclude that (7.26) has a unique mild solution on $[0,T]$.

Approach 2: A Typical Convergence Argument
Another standard approach is the technique used to prove Thrm. 3.4.1. There, we defined a recursive sequence of so-called *successive approximations* and proved that it converged (in an appropriate sense) to a stochastic process. Ultimately, it was shown that the uniform limit of this sequence was the strong solution of the IVP under consideration. We now use a similar approach in that we define

$$\begin{cases} X_0(t;\omega) = e^{At}X_0(\omega), \\ X_{n+1}(t;\omega) = e^{At}X_0(\omega) + \int_0^t e^{A(t-s)}f(s,X_n(s;\omega))ds + \int_0^t e^{A(t-s)}g(s,X_n(s;\omega))dW(s), \end{cases} \tag{7.32}$$

where $0 \leq t \leq T$, $\omega \in \Omega$, and follow the same steps used in the proof of Thrm. 3.4.1.

Exercise 7.4.2. Write up a detailed proof of Thrm. 7.4.2 using successive approximations.

Exercise 7.4.3. Consider the Cauchy problem

$$\begin{cases} dX(t;\omega) = (A + \sum_{k=1}^n C_k)X(t;\omega)dt + f(t,X(t;\omega))dt + g(t,X(t;\omega))dW(t), \\ X(0;\omega) = X_0(\omega), \omega \in \Omega, \end{cases} \tag{7.33}$$

where $0 < t < T$, $\omega \in \Omega$; A, B_k, $\mathbf{W}(t)$ and X_0 satisfy **(H5.1)** through **(H5.5)**; and $C_k \in \mathbb{B}(\mathscr{H})$ $(k = 1, \ldots, n)$. Prove that (7.33) has a unique mild solution on $[0,T]$.

Exercise 7.4.4. Assume $(\mathbf{H_A})$ and that $\exists k \in \mathbb{C}([0,T];(0,\infty))$ for which $f : [0,T] \times \mathscr{H} \to \mathscr{H}$ satisfies (7.19). Prove that (7.26) has a unique mild solution on $[0,T]$.

Exercise 7.4.5. Assume $(\mathbf{H_A})$ and that $\exists M \in \mathbb{C}([0,T];(0,\infty))$ for which $f : [0,T] \times \mathscr{H} \to \mathscr{H}$ satisfies

$$\int_0^t \|f(s,x) - f(s,y)\|_{\mathscr{H}}^2 ds \leq M(t) \|x - y\|_{\mathscr{H}}^2, \quad \forall 0 \leq t \leq T, x,y \in \mathscr{H}. \tag{7.34}$$

Must (7.26) have a unique mild solution on $[0,T]$?

Exercise 7.4.6. Assume $(\mathbf{H_A})$, $f : [0,T] \times \mathcal{H} \to \mathcal{H}$ is globally Lipschitz (uniformly in t), and $a \in \mathbb{C}\left([0,T]; \left[0,\frac{1}{2}\right]\right)$. Consider the IVP

$$\begin{cases} dX(t;\omega) = \left(AX(t;\omega) + f\left(t,X(t;\omega), \int_0^t a(s)X(s;\omega)ds\right)\right)dt + g(t,X(t;\omega))d\mathbf{W}(t), \\ X(0;\omega) = X_0(\omega), \omega \in \Omega, \end{cases}$$

(7.35)

where $0 < t < T, \omega \in \Omega$.

i.) Determine an expression for the solution map Φ for (7.35) in the spirit of (7.29).

ii.) Show that there exist positive constants C_1 and C_2 such that $\forall x, y \in \mathcal{H}$,

$$E\left\|(\Phi x)(t;\cdot) - (\Phi y)(t;\cdot)\right\|_{\mathcal{H}}^2 \leq C_1 \int_0^t E\left\|x(s;\cdot) - y(s;\cdot)\right\|_{\mathcal{H}}^2 ds$$

$$+ C_2 \int_0^t \int_0^s E\left\|x(\tau;\cdot) - y(\tau;\cdot)\right\|_{\mathcal{H}}^2 d\tau ds.$$

iii.) Can Prop. 1.10.4 be used to argue that $\exists N \in \mathbb{N}$ such that Φ^N is a strict contraction? Clearly indicate any additional restrictions that can be imposed on the data $(X_0, M_f, \text{etc.})$ in order to make this possible.

7.4.3   Continuous Dependence Estimates

Consider (7.26), together with the related Cauchy problem

$$\begin{cases} dY(t;\omega) = (AY(t;\omega) + f(t,Y(t;\omega)))dt + g(t,Y(t;\omega))d\mathbf{W}(t), 0 < t < T, \omega \in \Omega, \\ Y(0;\omega) = Y_0(\omega), \omega \in \Omega, \end{cases}$$

(7.36)

both under hypotheses $(\mathbf{H_A})$, $(\mathbf{H5.1})$ through $(\mathbf{H5.3})$, $(\mathbf{H7.1})$, and $(\mathbf{H7.2})$.

Proposition 7.4.3. *If* $(\mathbf{H_A})$, $(\mathbf{H5.1})$ *through* $(\mathbf{H5.3})$, $(\mathbf{H7.1})$, *and* $(\mathbf{H7.2})$ *hold, then*

$$E\|X(t;\cdot) - Y(t;\cdot)\|_{\mathcal{H}}^2 \leq 4M_A^2 e^{4M_A^2\left(TM_f^2 + M_g^2\right)t} \|X_0 - Y_0\|_{\mathfrak{L}^2(\Omega;\mathcal{H})}^2,$$

(7.37)

$\forall 0 \leq t \leq T$, *where X and Y are the mild solutions of (7.26) and (7.37), respectively.*

Proof. Begin by subtracting the variation of parameters formulae for the mild solutions of (7.26) and (7.37), then taking the $\mathcal{H}$-norm of both sides, and then taking the expectation. Applying the triangle law and estimates (7.27) and (7.28) subsequently yields

$$
\begin{aligned}
E\left\|X(t;\cdot)-Y(t;\cdot)\right\|_{\mathscr{H}}^2 \le 4\Big[& E\left\|e^{At}\left(X_0(\cdot)-Y_0(\cdot)\right)\right\|_{\mathscr{H}}^2 \\
&+E\left\|\int_0^t e^{A(t-s)}\left(f(s,X(s;\cdot))-f(s,Y(s;\cdot))\right)ds\right\|_{\mathscr{H}}^2 \\
&+E\left\|\int_0^t e^{A(t-s)}\left(g(s,X(s;\cdot))-g(s,Y(s;\cdot))\right)dW(s)\right\|_{\mathscr{H}}^2\Big]
\end{aligned}
$$

$$
\begin{aligned}
\le 4M_A^2\Big[& \|X_0-Y_0\|_{\mathfrak{L}^2(\Omega;\mathscr{H})}^2 \\
&+T\int_0^t E\left\|f(s,X(s;\cdot))-f(s,Y(s;\cdot))\right\|_{\mathscr{H}}^2 ds \\
&+\int_0^t E\left\|g(s,X(s;\cdot))-g(s,Y(s;\cdot))\right\|_{\mathscr{B}_0(\mathbb{R}^m,\mathscr{H})}^2 ds\Big]
\end{aligned}
\tag{7.38}
$$

$$
\begin{aligned}
\le \; & 4M_A^2\|X_0-Y_0\|_{\mathfrak{L}^2(\Omega;\mathscr{H})}^2 \\
&+4M_A^2\left(TM_f^2+M_g^2\right)\int_0^t E\left\|X(s;\cdot)-Y(s;\cdot)\right\|_{\mathscr{H}}^2 ds.
\end{aligned}
$$

An application of Gronwall's Lemma yields (7.37). $\qquad\square$

Exercise 7.4.7. Establish an estimate similar to (7.37) for $E\left\|X(t;\cdot)-Y(t;\cdot)\right\|_{\mathscr{H}}^p$ in terms of $\|X_0-Y_0\|_{\mathfrak{L}^p(\Omega;\mathscr{H})}^2$, for $p>2$.

Exercise 7.4.8. If the Lipschitz condition imposed in **(H7.1)** and **(H7.2)** is replaced by (7.19), how does (7.37) change?

Next, consider (7.26) and the related Cauchy problem

$$
\begin{cases}
dY(t;\omega)=\left(\widehat{A}Y(t;\omega)+\widehat{f}(t,Y(t;\omega))\right)dt+\widehat{g}(t,Y(t;\omega))d\mathbf{W}(t), \; 0<t<T, \omega\in\Omega, \\
Y(0;\omega)=Y_0(\omega), \omega\in\Omega,
\end{cases}
\tag{7.39}
$$

both under hypotheses **(H5.1)** through **(H5.3)**, **(H7.1)**, and **(H7.2)**, and assume that $\widehat{A}:\mathrm{dom}\left(\widehat{A}\right)\subset\mathscr{H}\to\mathscr{H}$ satisfies $(\mathbf{H}_{\widehat{A}})$.

Proposition 7.4.4. *Assume that* $(\mathbf{H_A})$, *(H5.1) through (H5.3), (H7.1), and (H7.2) hold for* $A,\widehat{A},f,\widehat{f},g,$ *and* $\widehat{g}$*, as well as*

(H7.3) *There exists* $\varepsilon_1>0$ *such that* $\sup_{0\le t\le T}\left\|f(t,x)-\widehat{f}(t,x)\right\|_{\mathscr{H}}^2<\varepsilon_1, \forall x\in\mathscr{H}$.

(H7.4) *There exists* $\varepsilon_2>0$ *such that* $\sup_{0\le t\le T}\|g(t,x)-\widehat{g}(t,x)\|_{\mathscr{B}_0}^2<\varepsilon_2, \forall x\in\mathscr{H}$.

(H7.5) *There exists* $\varepsilon_3>0$ *such that* $\|X_0-Y_0\|_{\mathfrak{L}^2(\Omega;\mathscr{H})}^2<\varepsilon_3$.

(H7.6) *There exists* $\varepsilon_4>0$ *such that* $\sup_{0\le t\le T}\left\|e^{At}-e^{\widehat{A}t}\right\|_{\mathbb{B}(\mathscr{H})}^2<\varepsilon_4$.

Then, $\exists \xi_1, \xi_2 > 0$ *such that*

$$E \|X(t;\cdot) - Y(t;\cdot)\|_{\mathscr{H}}^2 \le \xi_1 e^{\xi_2 t}, \forall 0 \le t \le T. \tag{7.40}$$

Proof. Begin by subtracting the variation of parameters formulae for the mild solutions of (7.26) and (7.39). Then, take the $\mathscr{H}$-norm of both sides followed by taking the expectation. Applying the triangle law and estimates (7.27) and (7.28) subsequently yields

$$E \|X(t;\cdot) - Y(t;\cdot)\|_{\mathscr{H}}^2 \le$$
$$8 \Bigg[E \left\| e^{At} X_0(\cdot) - e^{\widehat{A}t} Y_0(\cdot) \right\|_{\mathscr{H}}^2$$
$$+ E \left\| \int_0^t e^{A(t-s)} f(s, X(s;\cdot)) ds - \int_0^t e^{\widehat{A}(t-s)} \widehat{f}(s, Y(s;\cdot)) ds \right\|_{\mathscr{H}}^2 \tag{7.41}$$
$$+ E \left\| \int_0^t e^{A(t-s)} g(s, X(s;\cdot)) dW(s) - \int_0^t e^{\widehat{A}(t-s)} \widehat{g}(s, Y(s;\cdot)) dW(s) \right\|_{\mathscr{H}}^2 \Bigg] =$$
$$I_1 + I_2 + I_3.$$

We estimate $I_1, I_2,$ and I_3 individually. To this end, observe that

$$I_1 \le 2E \left[\left\| e^{At} (X_0(\cdot) - Y_0(\cdot)) \right\|_{\mathscr{H}}^2 + \left\| \left(e^{At} - e^{\widehat{A}t} \right) Y_0(\cdot) \right\|_{\mathscr{H}}^2 \right]$$
$$< 2M_A^2 \varepsilon_4 + 2\varepsilon_3 \|Y_0\|_{\mathfrak{L}^2(\Omega;\mathscr{H})}^2. \tag{7.42}$$

(Why?) Next,

$$I_2 \le 8E \Bigg[\left\| \int_0^t e^{A(t-s)} (f(s, X(s;\cdot)) - f(s, Y(s;\cdot))) ds \right\|_{\mathscr{H}}^2$$
$$+ \left\| \int_0^t \left(e^{A(t-s)} f(s, Y(s;\cdot)) - e^{\widehat{A}(t-s)} f(s, Y(s;\cdot)) \right) ds \right\|_{\mathscr{H}}^2 \tag{7.43}$$
$$+ \left\| \int_0^t e^{\widehat{A}(t-s)} \left(f(s, Y(s;\cdot)) - \widehat{f}(s, Y(s;\cdot)) \right) ds \right\|_{\mathscr{H}}^2 \Bigg]$$
$$\le 8E [I_4 + I_5 + I_6].$$

Observe that

$$E [I_4] \le T M_A^2 M_f^2 \int_0^t E \|X(s;\cdot) - Y(s;\cdot)\|_{\mathscr{H}}^2 ds. \tag{7.44}$$

Because the pth moment of $Y(s;\cdot)$ is bounded, $\forall s \in [0,T]$, we have

$$
\begin{aligned}
E[I_5] &\leq T\varepsilon_3 \int_0^t E \|f(s,Y(s;\cdot))\|_{\mathscr{H}}^2 \, ds \\
&\leq 2T\varepsilon_3 \left(\overline{M_f}\right)^2 \int_0^t \underbrace{\left[1 + E\|Y(s;\cdot)\|_{\mathscr{H}}^2\right]}_{\leq C} ds \\
&\leq 2T^2\varepsilon_3 \left(\overline{M_f}\right)^2 C.
\end{aligned}
\tag{7.45}
$$

Further,

$$
E[I_6] \leq TM_A^2 \int_0^t E \left\|f(s,Y(s;\cdot)) - \widehat{f}(s,Y(s;\cdot))\right\|_{\mathscr{H}}^2 ds < T^2 M_A^2 \varepsilon_1. \tag{7.46}
$$

Substituting (7.44) through (7.46) into (7.43) yields an estimate of the form

$$
I_2 \leq \gamma_1 + \gamma_2 \int_0^t E\|X(s;\cdot) - Y(s;\cdot)\|_{\mathscr{H}}^2 ds, \tag{7.47}
$$

where $\gamma_1, \gamma_2 > 0$.

The same approach is used to estimate I_3, only now we use Prop. 5.5.3 to estimate the Itó integrals. Otherwise, the computations are very similar and lead to an estimate of the form

$$
I_3 \leq \gamma_3 + \gamma_4 \int_0^t E\|X(s;\cdot) - Y(s;\cdot)\|_{\mathscr{H}}^2 ds, \tag{7.48}
$$

where $\gamma_3, \gamma_4 > 0$. (Show this!) As such, using (7.42), (7.47), and (7.48) in (7.41) yields the estimate

$$
E\|X(t;\cdot) - Y(t;\cdot)\|_{\mathscr{H}}^2 \leq \xi_1 + \xi_2 \int_0^t E\|X(s;\cdot) - Y(s;\cdot)\|_{\mathscr{H}}^2 ds, \tag{7.49}
$$

so that an application of Gronwall's Lemma finally leads to (7.40). $\qquad\square$

Exercise 7.4.9. Interpret the meaning of Prop. 7.4.4.

Exercise 7.4.10.
i.) Modify the hypotheses of Prop. 7.4.4 to obtain an estimate for $E\|X(t;\cdot) - Y(t;\cdot)\|_{\mathscr{H}}^p$, where $p > 2$.
ii.) Interpret (i) for the specific IBVPs introduced in Section 7.1.

Exercise 7.4.11. Consider the following system of abstract semi-linear stochastic evolution equations:

$$
\begin{cases}
dX(t;\omega) = (A_1 X(t;\omega) + f(t)X(t;\omega))\,dt + \sum_{k=1}^m h_k(t)X^2(t;\omega))dW_k(t), \\
dY(t;\omega) = (A_2 Y(t;\omega) + g(t)Y(t;\omega))\,dt + \sum_{k=1}^m \widehat{h}_k(t)Y^4(t;\omega))dW_k(t), \\
X(0;\omega) = X_0(\omega),\ Y(0;\omega) = Y_0(\omega),
\end{cases}
\tag{7.50}
$$

where $0 < t < T$, $\omega \in \Omega$. Here, $A_1 : \text{dom}(A_1) \subset \mathcal{H}_1 \to \mathcal{H}_1$ and $A_2 : \text{dom}(A_2) \subset \mathcal{H}_2 \to \mathcal{H}_2$ satisfy **(H5.1)**, where $\mathcal{H}_1$ and $\mathcal{H}_2$ are separable Hilbert spaces; $f : [0,T] \to \mathcal{H}_1$, $g : [0,T] \to \mathcal{H}_2$, and $h_k : [0,T] \to \mathbb{R}$ and $\hat{h}_k : [0,T] \to \mathbb{R}$ $(k = 1,\dots,m)$ are continuous, measurable mappings; $W_1(t),\dots,W_m(t)$ are independent one-dimensional Wiener processes; and X_0, Y_0 satisfy **(H5.3)**.

i.) Reformulate (7.50) as an abstract semi-linear SEE on the Hilbert space $\mathcal{H}_1 \times \mathcal{H}_2$.

ii.) Prove that (7.50) has a unique mild solution on $[0,T]$.

iii.) Assuming that $\mathcal{H}_1 = \mathbb{R}^n$ and $\mathcal{H}_2 = \mathbb{R}^l$ and the operators A_1 and A_2 are diagonal matrices, determine a representation formula for the mild solution of (7.50).

iv.) Formulate continuous dependence results in the spirit of Prop. 7.4.3 and Prop. 7.4.4 specifically for (7.50).

Exercise 7.4.12. Consider the following abstract semi-linear SEE:

$$\begin{cases} dX(t;\omega) = (AX(t;\omega) + h_1(t)f(X(t;\omega)))\,dt + h_2(t)d\mathbf{W}(t), \\ X(0;\omega) = X_0(\omega), \omega \in \Omega, \end{cases} \quad (7.51)$$

where $0 < t < T$, $\omega \in \Omega$ and $h_i : [0,T] \to \mathbb{R}$ $(i = 1,2)$ are continuous mappings. Formulate an existence and uniqueness theorem for (7.51).

7.4.4 p^{th} Moment Continuity

We begin with the following lemma, which can be verified using computations similar to those used to prove Prop. 7.4.3 and Prop. 7.4.4. (Tell how.)

Lemma 7.4.5. *Assume that* **(H$_A$)**, **(H5.1)**, **(H5.2)**, **(H5.5)**, **(H7.1)**, *and* **(H7.2)** *hold, and that* $X_0 \in \mathcal{L}^p(\Omega;\mathcal{H})$, *where* $p > 2$. *Then,* $\exists \xi_1, \xi_2 > 0$ *such that*

$$E\|X(t;\cdot)\|_{\mathcal{H}}^p \leq \xi_1 e^{\xi_2 t}, \forall 0 \leq t \leq T. \quad (7.52)$$

Consider the following continuity result.

Proposition 7.4.6. *Assume that* **(H$_A$)**, **(H5.1)**, **(H5.2)**, **(H5.5)**, **(H7.1)**, *and* **(H7.2)** *hold, and that* $X_0 \in \mathcal{L}^p(\Omega;\mathcal{H})$, *where* $p > 2$. *Then,*

$$E\|X(\tau;\cdot) - X(\tau';\cdot)\|_{\mathcal{H}}^p \to 0 \text{ as } |\tau - \tau'| \to 0. \quad (7.53)$$

Proof. Let $t_0 \in [0,T]$ and $\varepsilon > 0$. We must produce $\delta > 0$ for which

$$t \in [0,T] \text{ and } 0 < |t - t_0| < \delta \implies E\|X(t;\cdot) - X(t_0;\cdot)\|_{\mathcal{H}}^p < \varepsilon. \quad (7.54)$$

Observe that

$$E \|X(t;\cdot) - X(t_0;\cdot)\|_{\mathscr{H}}^p \leq$$
$$3^p \Big[E \left\| \left(e^{At} - e^{At_0} \right) X_0(\cdot) \right\|_{\mathscr{H}}^p$$
$$+ E \left\| \int_0^t e^{A(t-s)} f(s, X(s;\cdot)) ds - \int_0^{t_0} e^{A(t_0-s)} f(s, X(s;\cdot)) ds \right\|_{\mathscr{H}}^p \qquad (7.55)$$
$$+ E \left\| \int_0^t e^{A(t-s)} g(s, X(s;\cdot)) dW(s) - \int_0^{t_0} e^{A(t_0-s)} g(s, X(s;\cdot)) dW(s) \right\|_{\mathscr{H}}^p \Big] =$$
$$3^p [I_1 + I_2 + I_3].$$

We shall show that each of I_1, I_2, I_3 can be bounded above by $\frac{\varepsilon}{3^p}$ for values of t sufficiently near t_0. Assume, for the moment, that $0 < t_0 \leq t < T$. Observe that

$$I_1 = E \left\| e^{At_0} \left(e^{A(t-t_0)} - I \right) X_0(\cdot) \right\|_{\mathscr{H}}^p \leq M_A^p \left\| e^{A(t-t_0)} - I \right\|_{\mathbb{B}(\mathscr{H})}^p E \|X_0(\cdot)\|_{\mathscr{H}}^p. \quad (7.56)$$

The strong continuity of $\{e^{At}\}$ guarantees the existence of $\delta_1 > 0$ such that

$$t \in [0,T] \text{ and } 0 < |t - t_0| < \delta_1 \implies \left\| e^{A(t-t_0)} - I \right\|_{\mathbb{B}(\mathscr{H})}^p < \qquad (7.57)$$

$$\min \left\{ \frac{\varepsilon}{3^p \left(3M_A^p \left(E \|X_0(\cdot)\|_{\mathscr{H}}^p + 1 \right) \right)}, \frac{\varepsilon}{6 \cdot 3^p \left[(M_A \overline{M_f})^p t_0^{\frac{p}{q}+1} C + 1 \right]} \right\}.$$

Next, observe that

$$I_2 \leq 2^p \Bigg[\left\| \int_{t_0}^t e^{A(t-s)} f(s, X(s;\cdot)) ds \right\|_{\mathscr{H}}^p$$
$$+ E \left\| \int_0^{t_0} \left(e^{A(t-t_0)} - I \right) e^{A(t_0-s)} f(s, X(s;\cdot)) ds \right\|_{\mathscr{H}}^p \Bigg]$$
$$= 2^p [I_4 + I_5]. \qquad (7.58)$$

Using standard computations involving the triangle and Hölder inequalities, **(H7.1)**, and Lemma 7.4.5 yields

$$I_4 \leq E \left[\int_{t_0}^t \left\| e^{A(t-s)} \right\|_{\mathbb{B}(\mathscr{H})} \|f(s, X(s;\cdot))\|_{\mathscr{H}} ds \right]^p$$
$$\leq M_A^p (t - t_0)^{\frac{p}{q}} \int_{t_0}^t E \|f(s, X(s;\cdot))\|_{\mathscr{H}}^p ds \qquad (7.59)$$
$$\leq 2^p M_A^p (t - t_0)^{\frac{p}{q}} (\overline{M_f})^p \int_{t_0}^t \underbrace{\left[1 + E \|X(s;\cdot)\|_{\mathscr{H}}^p \right]}_{\leq C} ds$$
$$\leq \left(2M_A \overline{M_f} \right)^p (t - t_0)^{\frac{p}{q}+1} C.$$

There exists $\delta_2 > 0$ such that

$$t \in [0,T] \text{ and } 0 < |t - t_0| < \delta_2 \implies (2M_A \overline{M_f})^p (t - t_0)^{\frac{p}{q}+1} C < \frac{\varepsilon}{6 \cdot 3^p}. \qquad (7.60)$$

Similarly,

$$
\begin{aligned}
I_5 &\leq \left[\int_0^{t_0} \left\| e^{A(t-t_0)} - I \right\|_{\mathbb{B}(\mathscr{H})} \left\| e^{A(t_0-s)} \right\|_{\mathbb{B}(\mathscr{H})} \| f(s, X(s; \cdot)) \|_{\mathscr{H}} \, ds \right]^p \\
&\leq M_A^p \left\| e^{A(t-t_0)} - I \right\|_{\mathbb{B}(\mathscr{H})}^p t_0^{\frac{p}{q}} \int_0^{t_0} E \, \| f(s, X(s; \cdot)) \|_{\mathscr{H}}^p \, ds \qquad (7.61) \\
&\leq M_A^p \left\| e^{A(t-t_0)} - I \right\|_{\mathbb{B}(\mathscr{H})}^p t_0^{\frac{p}{q}} (\overline{M_f})^p \underbrace{\int_0^{t_0} \left[1 + E \, \| X(s; \cdot) \|_{\mathscr{H}}^p \right] ds}_{\leq C} \\
&\leq (M_A \overline{M_f})^p \left\| e^{A(t-t_0)} - I \right\|_{\mathbb{B}(\mathscr{H})}^p t_0^{\frac{p}{q}+1} C
\end{aligned}
$$

and the right-hand side of (7.61) is less than $\frac{\varepsilon}{6}$ provided that $t \in [0,T]$ and $0 < |t - t_0| < \delta_1$ (by (7.57)).

The computations are very similar to those above when estimating I_3 because all of the integral properties used also hold for the Itó integral. It can be shown that there exists $\delta_3 > 0$ such that

$$t \in [0,T] \text{ and } 0 < |t - t_0| < \delta_3 \implies I_3 < \frac{\varepsilon}{3 \cdot 3^p}. \qquad (7.62)$$

(Show the details!)

Let $\delta = \min \{\delta_1, \delta_2, \delta_3\}$. Based on our initial restriction that $0 < t_0 \leq t < T$, combining all of the above estimates leads to the conclusion that

$$t \in [0,T] \text{ and } 0 < t - t_0 < \delta \implies E \, \| X(t; \cdot) - X(t_0; \cdot) \|_{\mathscr{H}}^p < \varepsilon. \qquad (7.63)$$

The case when $0 < t < t_0 < T$ is left as an exercise. $\square$

Exercise 7.4.13. Provide a detailed argument of the case when $0 < t < t_0 < T$ in the proof of Prop. 7.4.6.

Exercise 7.4.14. Provide an alternate proof of Prop. 7.4.6 using the sequential criterion. Specifically, let $t_0 \in [0,T]$ and show that for any $\{t_n\} \subset \mathbb{R}$ for which $t_n \to t_0$, $E \, \| X(t_n; \cdot) - X(t_0; \cdot) \|_{\mathscr{H}}^p \longrightarrow 0$.

7.4.5 Convergence of Yosida Approximations

For every $n \in \mathbb{N}$, consider the following Yosida approximation of (7.26):

$$
\begin{cases}
dX_n(t; \omega) = (AX_n(t; \omega) + nR_n(A) f(t, X_n(t; \omega))) \, dt \\
\qquad + nR_n(A) g(t, X_n(t; \omega)) dW(t), \ 0 < t < T, \ \omega \in \Omega, \\
X_n(0; \omega) = nR_n(A) X_0(\omega), \ \omega \in \Omega,
\end{cases} \qquad (7.64)
$$

where $nR_n(A) = (I - nA)^{-1}$ is the resolvent of A. Under assumptions $(\mathbf{H_A})$, $(\mathbf{H5.1})$, $(\mathbf{H5.2})$, $(\mathbf{H5.5})$, $(\mathbf{H7.1})$, and $(\mathbf{H7.2})$, we know that (7.64) has a unique mild solution $X_n \in \mathbb{C}\left([0,T]; \mathcal{L}^2(\Omega; \mathcal{H})\right)$ given by the variation of parameters formula

$$X_n(t; \omega) = e^{At}\left(nR_n(A)X_0(\omega)\right) + \int_0^t e^{A(t-s)}nR_n(A)f(s, X_n(s; \omega))ds$$

$$+ \int_0^t e^{A(t-s)}nR_n(A)g(s, X_n(s; \omega))dW(s). \tag{7.65}$$

In fact, because $X_n(t; \omega) \in \mathrm{dom}(A)$, $\forall t \in [0,T]$, a.s. $[\mathscr{P}]$, it can be shown that X_n is actually a strong solution of (7.64) in the following sense. (See [203, 204].)

Definition 7.4.7. A continuous stochastic process $X_n : [0,T] \times \Omega \to \mathcal{H}$ is a *strong solution* of (7.64) if
i.) $X_n(t; \cdot)$ is progressively measurable and $\mathscr{F}_t$-adapted, $\forall t \in [0,T]$,
ii.) $X_n(t; \omega) \in \mathrm{dom}(A)$, $\forall t \in [0,T]$, a.s. $[\mathscr{P}]$,
iii.) $\int_0^T \|AX_n(s; \omega)\|_{\mathcal{H}} ds < \infty$, a.s. $[\mathscr{P}]$,
iv.) $X_n(t; \omega)$ is given by the formula

$$X_n(t; \omega) = nR_n(A)X_0(\omega) + \int_0^t AX_n(s; \omega)ds + \int_0^t nR_n(A)f(s, X_n(s; \omega))ds$$

$$+ \int_0^t nR_n(A)g(s, X_n(s; \omega))dW(s), \forall t \in [0,T], \text{ a.s. }[\mathscr{P}].$$

We have the following convergence result.

Proposition 7.4.8. *Assume* $(\mathbf{H_A})$, $(\mathbf{H5.1})$, $(\mathbf{H5.2})$, $(\mathbf{H5.5})$, $(\mathbf{H7.1})$, *and* $(\mathbf{H7.2})$. *For every* $n \in \mathbb{N}$, *let* X_n *be the strong solution of (7.64) and let* X *be the mild solution of (7.26). Then,*

$$\lim_{n \to \infty} \|X_n - X\|_{\mathbb{C}\left([0,T]; \mathcal{L}^2(\Omega; \mathcal{H})\right)} = 0. \tag{7.66}$$

Proof. We begin by estimating each term in the expression $E\|X_n(t; \cdot) - X(t; \cdot)\|_{\mathcal{H}}^2$. To this end, observe that subtracting the variation of parameters formulae of the solutions X_n and X, and applying the usual triangle inequality yields

$$E\|X_n(t; \cdot) - X(t; \cdot)\|_{\mathcal{H}}^2 \le$$

$$4\left[E\left\|e^{At}\left(nR_n(A) - I\right)X_0(\cdot)\right\|_{\mathcal{H}}^2\right.$$

$$+E\left\|\int_0^t e^{A(t-s)}\left(nR_n(A)f(s, X_n(s; \cdot)) - f(s, X(s; \cdot))\right)ds\right\|_{\mathcal{H}}^2 \tag{7.67}$$

$$+E\left.\left\|\int_0^t e^{A(t-s)}\left(nR_n(A)g(s, X_n(s; \cdot)) - g(s, X(s; \cdot))\right)dW(s)\right\|_{\mathcal{H}}^2\right] =$$

$$4\left[I_1 + I_2 + I_3\right].$$

We will show that each of I_1, I_2, I_3 can be bounded above by $C\|nR_n(A) - I\|_{\mathbb{B}(\mathcal{H})}^2$, for some constant C independent of n.

Observe that

$$I_1 \leq M_A^2 \|X_0\|_{\mathfrak{L}^2}^2 \|nR_n(A) - I\|_{\mathbb{B}(\mathscr{H})}^2 \tag{7.68}$$

and

$$I_2 = E \left\| \int_0^t e^{A(t-s)} \left[nR_n(A)f(s, X_n(s; \cdot)) - f(s, X_n(s; \cdot)) \right. \right.$$
$$\left. \left. + f(s, X_n(s; \cdot)) - f(s, X(s; \cdot)) \right] ds \right\|_{\mathscr{H}}^2 \tag{7.69}$$
$$\leq 2TM_A^2 \int_0^t E \|(nR_n(A) - I)f(s, X_n(s; \cdot))\|_{\mathscr{H}}^2 ds$$
$$+ 2TM_A^2 M_f^2 \int_0^t E \|X_n(s; \cdot) - X(s; \cdot)\|_{\mathscr{H}}^2 ds.$$

We estimate the term $\int_0^t E \|(nR_n(A) - I)f(s, X_n(s; \cdot))\|_{\mathscr{H}}^2 ds$ in (7.69), as follows:

$$\int_0^t E \|(nR_n(A) - I)f(s, X_n(s; \cdot))\|_{\mathscr{H}}^2 ds =$$

$$\int_0^t E \|(nR_n(A) - I)\left(f(s, X_n(s; \cdot)) - f(s, X(s; \cdot)) + f(s, X(s; \cdot))\right)\|_{\mathscr{H}}^2 ds$$

$$2\|nR_n(A) - I\|_{\mathbb{B}(\mathscr{H})}^2 M_f^2 \int_0^t E \|X_n(s; \cdot) - X(s; \cdot)\|_{\mathscr{H}}^2 ds$$

$$+ 2\|nR_n(A) - I\|_{\mathbb{B}(\mathscr{H})}^2 \int_0^t E \|f(s, X(s; \cdot))\|_{\mathscr{H}}^2 ds \leq \tag{7.70}$$

$$2\|nR_n(A) - I\|_{\mathbb{B}(\mathscr{H})}^2 M_f^2 \int_0^t E \|X_n(s; \cdot) - X(s; \cdot)\|_{\mathscr{H}}^2 ds$$

$$+ 4\|nR_n(A) - I\|_{\mathbb{B}(\mathscr{H})}^2 \left(\overline{M_f}\right)^2 \int_0^t \underbrace{\left[1 + E \|X(s; \cdot)\|_{\mathscr{H}}^2\right]}_{\leq C} ds.$$

As such, we have

$$I_2 \leq \left(2TM_A^2 M_f^2\right)\left(1 + 2\|nR_n(A) - I\|_{\mathbb{B}(\mathscr{H})}^2\right) \int_0^t E \|X_n(s; \cdot) - X(s; \cdot)\|_{\mathscr{H}}^2 ds$$
$$+ 8T^2 M_A^2 \left(\overline{M_f}\right)^2 C \|nR_n(A) - I\|_{\mathbb{B}(\mathscr{H})}^2. \tag{7.71}$$

A similar argument can be used to show that

$$I_3 \leq \left(2\zeta_g(T, p)M_A^2 M_g^2 + 4\zeta_g(T, p)M_A^2 M_g^2 \|nR_n(A) - I\|_{\mathbb{B}(\mathscr{H})}^2\right) \times$$
$$\int_0^t E \|X_n(s; \cdot) - X(s; \cdot)\|_{\mathscr{H}}^2 ds + 2\zeta_g(T, p)M_A^2 \left(\overline{M_g}\right)^2 C \|nR_n(A) - I\|_{\mathbb{B}(\mathscr{H})}^2.$$

(Show the details!) Using (7.68) through (7.71) in (7.67) yields the following inequality

$$E \|X_n(t; \cdot) - X(t; \cdot)\|_{\mathscr{H}}^2 \leq \xi_1 \|nR_n(A) - I\|_{\mathbb{B}(\mathscr{H})}^2 + \xi_2 \int_0^t E \|X_n(s; \cdot) - X(s; \cdot)\|_{\mathscr{H}}^2 ds, \tag{7.72}$$

$\forall 0 \leq t \leq T$, where ξ_1 and ξ_2 are positive constants independent of n. Applying Gronwall's Lemma in (7.72) subsequently yields

$$E \left\| X_n(t;\cdot) - X(t;\cdot) \right\|_{\mathscr{H}}^2 \leq \xi_1 \left\| nR_n(A) - I \right\|_{\mathbb{B}(\mathscr{H})}^2 e^{\xi_2 t}, \forall 0 \leq t \leq T. \tag{7.73}$$

Note that

$$\xi_2 = \xi_3 \left\| nR_n(A) - I \right\|_{\mathbb{B}(\mathscr{H})}^2 + \xi_4, \tag{7.74}$$

where ξ_4 is a positive constant independent of n. As such, (7.73) becomes

$$E \left\| X_n(t;\cdot) - X(t;\cdot) \right\|_{\mathscr{H}}^2 \leq \xi_1 e^{\xi_4 t} \left\| nR_n(A) - I \right\|_{\mathbb{B}(\mathscr{H})}^2 e^{\xi_3 \left\| nR_n(A) - I \right\|_{\mathbb{B}(\mathscr{H})}^2 T}, \forall 0 \leq t \leq T. \tag{7.75}$$

Taking the supremum over $[0,T]$ on both sides of (7.75) then yields

$$\left\| X_n - X \right\|_{\mathbb{C}\left([0,T];\mathcal{L}^2(\Omega;\mathscr{H})\right)}^2 \leq \xi_1 e^{\xi_4 T} \left\| nR_n(A) - I \right\|_{\mathbb{B}(\mathscr{H})}^2 e^{\xi_3 \left\| nR_n(A) - I \right\|_{\mathbb{B}(\mathscr{H})}^2 T}. \tag{7.76}$$

Let $\varepsilon > 0$. The strong convergence of $\left\| nR_n(A) - I \right\|_{\mathbb{B}(\mathscr{H})}^2 \to 0$ as $n \to \infty$ implies that $\exists M > 0$ such that

$$e^{\xi_3 \left\| nR_n(A) - I \right\|_{\mathbb{B}(\mathscr{H})}^2 T} \leq M, \forall n \in \mathbb{N}. \tag{7.77}$$

Moreover, $\exists N \in \mathbb{N}$ such that

$$n \geq N \implies \left\| nR_n(A) - I \right\|_{\mathbb{B}(\mathscr{H})}^2 < \frac{\varepsilon^2}{\left(M\xi_1 e^{\xi_4 T} \right)}. \tag{7.78}$$

Consequently,

$$n \geq N \implies \left\| X_n - X \right\|_{\mathbb{C}\left([0,T];\mathcal{L}^2(\Omega;\mathscr{H})\right)}^2 < \varepsilon^2$$
$$\implies \left\| X_n - X \right\|_{\mathbb{C}\left([0,T];\mathcal{L}^2(\Omega;\mathscr{H})\right)} < \varepsilon. \tag{7.79}$$

This completes the proof. □

7.4.6 Convergence of Induced Probability Measures

This short section is devoted to a discussion of a result requiring somewhat more advanced probability theory and can be viewed as a convergence result in the spirit of the Central Limit Theorem. The Central Limit Theorem is concerned with the weak convergence of a sequence of random variables (that is, convergence of the distributions of this sequence of random variables). In the present setting, we consider a sequence of probability measures generated by a sequence of stochastic processes and investigate its convergence. The difference is that we now study the convergence of the distributions of a sequence of stochastic processes, not just random variables. Such approximation issues in stochastic analysis, and specifically those arising in the study of SEEs, pervade the field because they allow us to get close to an object of interest with a sequence of nicer objects whose properties and behavior are more readily able to be described. Doing so usually tells us something about the behavior of the more complicated object. We will not discuss the full power of the weak convergence of probability measures here, but rather refer you to **[49, 51]** for further study.

We begin with some terminology and a general strategy.

Definition 7.4.9. A *probability measure* $\mathscr{P}$ *induced by an* $\mathscr{H}$-*valued random variable X*, denoted $\mathscr{P}_X$, is defined by

$$\mathscr{P}_X = \mathscr{P} \circ X^{-1} : \mathfrak{B}(\mathscr{H}) \rightarrow [0, 1], \tag{7.80}$$

where $\mathfrak{B}(\mathscr{H})$ is the collection of Borel sets on $\mathscr{H}$. The collection of all such measures is denoted by $\mathfrak{P}(\mathscr{H})$.

The following are some properties of probability measures.

Definition 7.4.10. Let $\{\mathscr{P}_n\} \subset \mathfrak{P}(\mathscr{H})$. We say that $\{\mathscr{P}_n\}$ is
i.) *weakly convergent* to $\mathscr{P} \in \mathfrak{P}(\mathscr{H})$ if $\int_\Omega f d\mathscr{P}_n \longrightarrow \int_\Omega f d\mathscr{P}$, for every bounded, continuous function $f : \mathscr{H} \rightarrow \mathbb{R}$. In such case, we write $\mathscr{P}_n \rightarrow \mathscr{P}$ weakly.
ii.) *tight* if $\forall \varepsilon > 0$, there exists a compact set K_ε such that $\mathscr{P}_n(K_\varepsilon) \geq 1 - \varepsilon$, $\forall n \in \mathbb{N}$.
iii.) *relatively compact* if every subsequence $\{\mathscr{P}_{n_k}\} \subset \{\mathscr{P}_n\}$, in turn, has a subsequence $\left\{\mathscr{P}_{(n_k)_j}\right\}$ such that $\mathscr{P}_{(n_k)_j} \rightarrow \mathscr{Q}$ weakly, for some $\mathscr{Q} \in \mathfrak{P}(\mathscr{H})$. (That is, every subsequence has a weakly convergent subsequence in $\mathfrak{P}(\mathscr{H})$.)

The notions of tightness and relative compactness of a family of measures are closely related, as Prokorov **[246]** proved.

Theorem 7.4.11. (Prokorov's Theorem)
A sequence $\{\mathscr{P}_n\} \subset \mathfrak{P}(\mathscr{H})$ *is tight iff* $\{\mathscr{P}_n\}$ *is relatively compact.*

As such, the Arzela–Ascoli Theorem (cf. Thrm. 1.11.7) can be used to establish tightness. However, this is just one step in establishing weak convergence. We also need the following.

Definition 7.4.12. Let $\mathscr{P} \in \mathfrak{P}(\mathscr{H})$, $k \in \mathbb{N}$, and $0 \leq t_1 < t_2 < \ldots < t_k \leq T$. Define the mapping $\pi_{t_1,\ldots,t_k} : \mathbb{C}([0, T]; \mathscr{H}) \rightarrow \underbrace{\mathscr{H} \times \ldots \times \mathscr{H}}_{k \text{ times}}$ by

$$\pi_{t_1,\ldots,t_k}(X) = \langle X(t_1), \ldots, X(t_k) \rangle.$$

The probability measure induced by the collection of functions $\left\{\pi_{t_1,\ldots,t_k}\right\}$ is the *finite-dimensional joint distribution of* $\mathscr{P}$.

The following result is proven in **[246]**.

Proposition 7.4.13. *If a sequence* $\{X_n\}$ *of* $\mathscr{H}$-*valued random variables converges weakly to an* $\mathscr{H}$-*valued random variable X in* $\mathfrak{L}^2(\Omega; \mathscr{H})$, *then the sequence of finite dimensional joint distributions corresponding to* $\{\mathscr{P}_{X_n}\}$ *converges weakly to the finite dimensional joint distribution of* $\mathscr{P}_X$.

The next theorem (proved in **[51]**), in conjunction with Prop. 7.4.13, is the main tool in the proof of the next convergence result.

Theorem 7.4.14. *Let $\{\mathscr{P}_n\} \subset \mathfrak{P}(\mathscr{H})$. If the sequence of finite dimensional joint distributions corresponding to $\{\mathscr{P}_n\}$ converges weakly to the finite dimensional joint distribution of $\mathscr{P}$, and $\{\mathscr{P}_n\}$ is relatively compact, then $\mathscr{P}_n \to \mathscr{P}$ weakly.*

Our main convergence result is:

Proposition 7.4.15. *Assume that $E \|X_0\|_{\mathscr{H}}^4 < \infty$ and that (HA), (H5.1), (H5.2), (H5.5), (H7.1), and (H7.2) hold. For each $n \in \mathbb{N}$, let X_n be the strong solution of (7.64) and X the mild solution of (7.26). Then, $\mathscr{P}_{X_n} \to \mathscr{P}_X$ weakly as $n \to \infty$.*

Proof. We shall argue that $\{\mathscr{P}_{X_n}\}$ is relatively compact using the Arzela–Ascoli Theorem. We break this proof into three subclaims.
<u>Claim 1</u>: $\{X_n\}$ is a uniformly bounded subset of $\mathbb{C}([0,T];\mathfrak{L}^2(\Omega;\mathscr{H}))$; that is, $\exists M^\star > 0$ such that

$$\sup_{n\in\mathbb{N}} \left(\sup_{0 \leq t \leq T} E \|X_n(t;\cdot)\|_{\mathscr{H}}^2 \right) \leq M^\star. \tag{7.81}$$

Proof: We estimate each term on the right-hand side of

$$E \|X_n(t;\cdot)\|_{\mathscr{H}}^2 \leq 4 \left[E \left\| e^{At} nR_n(A)X_0(\cdot) \right\|_{\mathscr{H}}^2 + E \left\| \int_0^t e^{A(t-s)} nR_n(A)f(s,X_n(s;\cdot)) \right\|_{\mathscr{H}}^2 \right.$$
$$\left. + E \left\| \int_0^t e^{A(t-s)} nR_n(A)g(s,X_n(s;\cdot))dW(s) \right\|_{\mathscr{H}}^2 \right] \tag{7.82}$$
$$= 4 [I_1 + I_2 + I_3].$$

Because the operator $nR_n(A)$ is contractive, $\forall n \in \mathbb{N}$, it follows that

$$\|nR_n(A)\|_{\mathbb{B}(\mathscr{H})} \leq 1, \forall n \in \mathbb{N}. \tag{7.83}$$

Using this fact with routine computations yields the following estimates:

$$I_1 \leq M_A^2 \|X_0\|_{\mathfrak{L}^2}^2, \tag{7.84}$$

$$I_2 \leq 2T^2 M_A^2 + 2t M_A^2 \int_0^t E \|X_n(s;\cdot)\|_{\mathscr{H}}^2 ds, \tag{7.85}$$

$$I_3 \leq 2\zeta_g(T,2) T M_A^2 + 2\zeta_g(T,2) M_A^2 \int_0^t E \|X_n(s;\cdot)\|_{\mathscr{H}}^2 ds. \tag{7.86}$$

(Show the details!) Using (7.84) through (7.86) in (7.82) yields the inequality

$$E \|X_n(t;\cdot)\|_{\mathscr{H}}^2 \leq M_A^2 \left(\|X_0\|_{\mathfrak{L}^2}^2 + 2T^2 + 2\zeta_g(T,2) \right) \tag{7.87}$$
$$+ 2(T + \zeta_g(T,2)) M_A^2 \int_0^t E \|X_n(s;\cdot)\|_{\mathscr{H}}^2 ds.$$

Applying Gronwall's Lemma in (7.87) yields

$$E \|X_n(t;\cdot)\|^2_{\mathscr{H}} \le M_A^2 \left(\|X_0\|^2_{\mathscr{L}^2} + 2T^2 + 2\zeta_g(T,2) \right) e^{2\left(T + \zeta_g(T,2)\right)M_A^2}, \forall n \in \mathbb{N}. \quad (7.88)$$

Taking the supremum over $n \in \mathbb{N}$ and $t \in [0,T]$ establishes (7.81).

<u>Claim 2</u>: $\{X_n\}$ is equicontinuous.
Proof: We shall show that $\forall n \in \mathbb{N}$ and $0 \le s \le t \le T$, $E \|X_n(t;\cdot) - X_n(s;\cdot)\|^2_{\mathscr{H}} \longrightarrow 0$ as $t - s \to 0$, independently of n. As usual, subtract the variation of parameters formulae to arrive at

$$E \|X_n(t;\cdot) - X_n(s;\cdot)\|^4_{\mathscr{H}} \le$$

$$16 \left[E \left\| \left(e^{At} - e^{As} \right) nR_n(A)X_0(\cdot) \right\|^4_{\mathscr{H}} + \right.$$

$$E \left\| \int_0^t e^{A(t-\tau)} nR_n(A)f(\tau, X_n(\tau;\cdot))d\tau - \int_0^s e^{A(s-\tau)} nR_n(A)f(\tau, X_n(\tau;\cdot))d\tau \right\|^4_{\mathscr{H}} =$$

$$E \left\| \int_0^t e^{A(t-\tau)} nR_n(A)g(\tau, X_n(\tau;\cdot))dW(\tau) - \right.$$

$$\left. \int_0^s e^{A(s-\tau)} nR_n(A)g(\tau, X_n(\tau;\cdot))dW(\tau) \right\|^4_{\mathscr{H}} \right] =$$

$$I_4 + I_5 + I_6. \quad (7.89)$$

Using Thrm. 5.2.3(vii), together with the boundedness of $\{AnR_n(A) : n \in \mathbb{N}\}$, we see that

$$I_4 \le E \left[\left(\int_s^t \left\| e^{A\tau} AnR_n(A)X_0(\cdot) \right\|_{\mathscr{H}} d\tau \right)^4 \right] \quad (7.90)$$

$$\le M_A^4 \left(M^{\star\star} \right)^4 E \|X_0(\cdot)\|^4_{\mathscr{H}} (t-s)^4.$$

Next, observe that

$$I_5 \le E \left[\int_0^s \left\| \left(e^{A(t-\tau)} - e^{A(s-\tau)} \right) nR_n(A)f(\tau, X_n(\tau;\cdot)) \right\|_{\mathscr{H}} d\tau \right.$$

$$\left. + \int_s^t \left\| e^{A(t-\tau)} nR_n(A)f(\tau, X_n(\tau;\cdot)) \right\|_{\mathscr{H}} d\tau \right]^4$$

$$\le E \left[\int_0^s \int_{s-\tau}^{t-\tau} \left\| e^{Au} nR_n(A)f(u, X_n(u;\cdot)) \right\|_{\mathscr{H}} du \, d\tau \right. \quad (7.91)$$

$$\left. + M_A \|nR_n(A)\|_{\mathbb{B}(\mathscr{H})} \overline{M_f} \int_s^t \left(1 + \|X_n(\tau;\cdot)\|_{\mathscr{H}} \right) d\tau \right]^4$$

$$\le \left[\overline{M_f} M_A M^{\star\star} T \int_{s-\tau}^{t-\tau} \left(1 + E \|X_n(u;\cdot)\|_{\mathscr{H}} \right) du \right.$$

$$\left. + \overline{M_f} M_A \int_s^t \left(1 + E \|X_n(\tau;\cdot)\|_{\mathscr{H}} \right) d\tau \right]^4.$$

Applying Hölder's inequality yields

$$E\,\|X_n(u;\cdot)\|_{\mathcal{H}} \le \underbrace{\left(E\,\|X_n(u;\cdot)\|_{\mathcal{H}}^2\right)^{\frac{1}{2}}}_{\le M^\star,\,\forall n\in\mathbb{N}}\underbrace{(E(1))^{\frac{1}{2}}}_{=1}. \tag{7.92}$$

Using (7.92) in (7.91) then yields

$$I_5 \le \left[\overline{M_f}M_A\,(1+M^\star)\,(M^{\star\star}T+1)\right]^4(t-s)^4. \tag{7.93}$$

Similarly, it can be shown that

$$I_6 \le C^\star(t-s)^4, \tag{7.94}$$

for some constant $C^\star > 0$ independent of n. (Show the details!)

Consequently, we deduce from (7.89) that

$$E\,\|X_n(t;\cdot)-X_n(s;\cdot)\|_{\mathcal{H}}^4 \le C\,(t-s)^4, \tag{7.95}$$

where $C > 0$ is a constant independent of n. As such, we conclude that

$$E\,\|X_n(t;\cdot)-X_n(s;\cdot)\|_{\mathcal{H}}^4 \longrightarrow 0 \text{ as } (t-s) \longrightarrow 0. \tag{7.96}$$

This proves Claim 2.

<u>Claim 3:</u> $\mathscr{P}_{X_n} \to \mathscr{P}_X$ weakly as $n \to \infty$.

Proof: In light of Claims 1 and 2, the Arzela–Ascoli Theorem guarantees that the family $\{X_n(t) : t \in [0,T]\}$ is relatively compact in $\mathcal{H}$. So, $\left\{\mathscr{P}\circ X_n^{-1} = \mathscr{P}_{X_n} : n \in \mathbb{N}\right\}$ is relatively compact in $\mathbb{C}\left([0,T];\mathscr{L}^2(\Omega;\mathcal{H})\right)$ and is therefore tight (by Thrm. 7.4.11). Hence, by Prop. 7.4.13, the sequence of finite-dimensional joint distributions of $\mathscr{P}_{X_n}$ converges weakly to the finite-dimensional joint distribution of $\mathscr{P}_X$. Thus, by Thrm. 7.4.14, we conclude that $\mathscr{P}_{X_n} \to \mathscr{P}_X$ weakly as $n \to \infty$.

This completes the proof. $\qquad\qquad\qquad\qquad\qquad\qquad\qquad\qquad\qquad\qquad\square$

Exercise 7.4.15. Fill in the missing details in the proof of Prop. 7.4.15.

7.4.7 Zeroth-Order Approximation

Let $0 < \varepsilon < 1$. Consider the stochastic IVP

$$\begin{cases} dX_\varepsilon(t;\omega) = (A\varepsilon X_\varepsilon(t;\omega)+f_\varepsilon(t,X_\varepsilon(t;\omega)))\,dt+g_\varepsilon(t,X_\varepsilon(t;\omega))d\mathbf{W}(t), \\ X_\varepsilon(0;\omega) = X_0, \end{cases} \tag{7.97}$$

where $0 < t < T$, $\omega \in \Omega$, together with the deterministic IVP

$$\begin{cases} dX(t) = (AX(t)+f(t,X(t))\,dt, 0 < t < T, \\ X(0) = X_0. \end{cases} \tag{7.98}$$

Assume that X_0 is a fixed element of $\mathcal{H}$ (so that it is deterministic), as well as the following.

(H7.7) For every $0 < \varepsilon < 1$, $A_\varepsilon : \mathrm{dom}(A_\varepsilon) \subset \mathcal{H} \to \mathcal{H}$ generates a C_0-semigroup $\left\{ e^{A_\varepsilon t} : t \geq 0 \right\}$ on $\mathcal{H}$, $A : \mathrm{dom}(A) \subset \mathcal{H} \to \mathcal{H}$ generates a C_0-semigroup $\left\{ e^{At} : t \geq 0 \right\}$ on $\mathcal{H}$, and $\left\| e^{A_\varepsilon t} - e^{At} \right\|_{\mathbb{B}(\mathcal{H})} \longrightarrow 0$ as $\varepsilon \to 0^+$ uniformly in $t \in [0,T]$. Moreover,

$$\sup_{0 < \varepsilon < 1} \left(\sup_{0 \leq t \leq T} \left\| e^{A_\varepsilon t} \right\|_{\mathbb{B}(\mathcal{H})} \right) \leq M_A e^{\alpha T}. \tag{7.99}$$

(H7.8) For every $0 < \varepsilon < 1$, $f_\varepsilon : [0,T] \times \mathcal{H} \to \mathcal{H}$ is an $\mathscr{F}_t$-adapted, progressively measurable mapping such that

$$\| f_\varepsilon(t,x) - f_\varepsilon(t,y) \|_{\mathcal{H}} \leq M_f \| x - y \|_{\mathcal{H}}$$
$$\| f_\varepsilon(t,x) \|_{\mathcal{H}} \leq \overline{M_f} \left[1 + \| x \|_{\mathcal{H}} \right],$$

$\forall x, y \in \mathcal{H}$, uniformly in $t \in [0,T]$, where the growth constants M_f and $\overline{M_f}$ are the same $\forall 0 < \varepsilon < 1$ and actually coincide with the growth constants used for the mapping $f : [0,T] \times \mathcal{H} \to \mathcal{H}$ appearing in (7.98) under **(H7.1)**. Moreover,

$$\| f_\varepsilon(t,x) - f(t,x) \|_{\mathcal{H}} \longrightarrow 0 \text{ as } \varepsilon \to 0^+, \forall x, y \in \mathcal{H}, \text{ uniformly in } t \in [0,T]. \tag{7.100}$$

(H7.9) For every $0 < \varepsilon < 1$, $g_\varepsilon : [0,T] \times \mathcal{H} \to \mathscr{B}_0 \left(\mathbb{R}^m, \mathcal{H} \right)$ is an $\mathscr{F}_t$-adapted, progressively measurable mapping such that

$$\| g_\varepsilon(t,x) - g_\varepsilon(t,y) \|_{\mathscr{B}_0} \leq M_g \| x - y \|_{\mathcal{H}}$$
$$\| g_\varepsilon(t,x) \|_{\mathcal{H}} \leq \overline{M_g} \left[1 + \| x \|_{\mathcal{H}} \right],$$

$\forall x, y \in \mathcal{H}$, uniformly in $t \in [0,T]$, where the growth constants M_g and $\overline{M_g}$ are the same $\forall 0 < \varepsilon < 1$. Moreover,

$$\| g_\varepsilon(t,x) \|_{\mathscr{B}_0} \longrightarrow 0 \text{ as } \varepsilon \to 0^+, \forall x, y \in \mathcal{H}, \text{ uniformly in } t \in [0,T]. \tag{7.101}$$

We have the following result.

Proposition 7.4.16. *Assume that $p > 2$ and let X_ε and X be the mild solutions of (7.97) and (7.98), respectively. Then, $\exists \zeta > 0$ and a function $\psi : I \subset [0,1] \longrightarrow (0,\infty)$ for which $\lim_{\varepsilon \to 0^+} \psi(\varepsilon) = 0$ and*

$$E \| X_\varepsilon(t;\cdot) - X(t) \|_{\mathcal{H}}^p \leq \zeta \psi(\varepsilon), \tag{7.102}$$

$\forall 0 \leq t \leq T$ *and $\varepsilon > 0$ sufficiently small. (That is, $\lim_{\varepsilon \to 0^+} \| X_\varepsilon - X \|_{\mathbb{C}([0,T]; \mathcal{L}^p(\Omega; \mathcal{H}))} = 0$.)*

Proof. Subtract the variation of parameters formulae for the mild solutions of (7.97) and (7.98) and estimate to obtain

$$
E \|X_\varepsilon(t;\cdot) - X(t)\|_{\mathcal{H}}^p \leq p \left[\|e^{A_\varepsilon t} - e^{At}\|_{\mathbb{B}(\mathcal{H})}^p \underbrace{E \|X_0\|_{\mathcal{H}}^p}_{=X_0^p} \right.
$$

$$
+ T^{\frac{p}{q}} \int_0^t E \left\| e^{A_\varepsilon(t-s)} f_\varepsilon(s, X_\varepsilon(s;\cdot)) - e^{A(t-s)} f(s, X(s)) \right\|_{\mathcal{H}}^p ds
$$

$$
\left. + \varsigma_g(T, p) \int_0^t E \left\| e^{A_\varepsilon(t-s)} g_\varepsilon(s, X_\varepsilon(s;\cdot)) \right\|_{\mathcal{B}_0}^p ds \right] \qquad (7.103)
$$

$$
= p [I_1 + I_2 + I_3].
$$

From **(H7.7)**, $I_1 \longrightarrow 0$ uniformly in $t \in [0, T]$ as $\varepsilon \to 0^+$. So, for sufficiently small $\varepsilon > 0$, $\exists K_1 > 0$ and a function $\psi_1 : I \subset [0, 1] \longrightarrow (0, \infty)$ for which $\lim_{\varepsilon \to 0^+} \psi_1(\varepsilon) = 0$ and

$$
I_1 \leq K_1 \psi_1(\varepsilon), \ \forall t \in [0, T]. \qquad (7.104)
$$

Next, observe that

$$
I_2 = T^{\frac{p}{q}} \int_0^t E \left\| e^{A_\varepsilon(t-s)} f_\varepsilon(s, X_\varepsilon(s;\cdot)) - e^{A_\varepsilon(t-s)} f_\varepsilon(s, X(s)) \right.
$$

$$
+ e^{A_\varepsilon(t-s)} f_\varepsilon(s, X(s)) - e^{A(t-s)} f_\varepsilon(s, X(s))
$$

$$
\left. + e^{A(t-s)} f_\varepsilon(s, X(s)) - e^{A(t-s)} f(s, X(s)) \right\|_{\mathcal{H}}^p ds \qquad (7.105)
$$

$$
\leq 3^p T^{\frac{p}{q}} \left[(M_A e^{\alpha T} M_f)^p \int_0^t E \|X_\varepsilon(s;\cdot) - X(s)\|_{\mathcal{H}}^p ds \right.
$$

$$
+ \int_0^t E \left\| \left(e^{A_\varepsilon(t-s)} - e^{A(t-s)} \right) f_\varepsilon(s, X(s)) \right\|_{\mathcal{H}}^p ds
$$

$$
\left. + (M_A e^{\alpha T})^p \int_0^t E \|f_\varepsilon(s, X(s)) - f(s, X(s))\|_{\mathcal{H}}^p ds \right].
$$

The first integral term on the right-hand side of (7.105) is fine as is, but we must further estimate the second and third integral terms. To this end, observe that

$$
E \|f_\varepsilon(s, X(s))\|_{\mathcal{H}}^p \leq 2^p \left[E \|f_\varepsilon(s, X(s)) - f(s, X(s))\|_{\mathcal{H}}^p + E \|f(s, X(s))\|_{\mathcal{H}}^p \right] \ (7.106)
$$

$$
\leq 2^p \left[E \|f_\varepsilon(s, X(s)) - f(s, X(s))\|_{\mathcal{H}}^p + 2^p \left[1 + E \|X(s)\|_{\mathcal{H}}^p \right] \right].
$$

By **(H7.8)**, for sufficiently small ε, $\exists K_2 > 0$ and a function $\psi_2 : I \subset [0, 1] \longrightarrow (0, \infty)$ for which $\lim_{\varepsilon \to 0^+} \psi_2(\varepsilon) = 0$ and

$$
E \|f_\varepsilon(s, X(s)) - f(s, X(s))\|_{\mathcal{H}}^p \leq K_2 \psi_2(\varepsilon), \ \forall s \in [0, T]. \qquad (7.107)
$$

Taking the supremum over $[0, T]$ in (7.106) and using (7.107) yields

$$\sup_{0 \leq s \leq T} E \, \|f_\varepsilon(s, X(s))\|_{\mathcal{H}}^p \leq 2^p \left[K_2 \psi_2(\varepsilon) + 2^p \underbrace{\left[1 + E \, \|X(s)\|_{\mathcal{H}}^p\right]}_{\leq C} \right]. \quad (7.108)$$

Using these observations, we conclude that for sufficiently small $\varepsilon > 0$, $\exists K_3 > 0$ and a function $\psi_3 : I \subset [0,1] \longrightarrow (0, \infty)$ for which $\lim_{\varepsilon \to 0^+} \psi_3(\varepsilon) = 0$ and

$$\int_0^t E \, \left\| \left(e^{A_\varepsilon(t-s)} - e^{A(t-s)} \right) f_\varepsilon(s, X(s)) \right\|_{\mathcal{H}}^p ds \leq K_3 \psi_3(\varepsilon), \, \forall t \in [0, T]. \quad (7.109)$$

(Tell why carefully.) As for the third integral term in (7.105), **(H8.7)** immediately implies that $\exists K_4 > 0$ and a function $\psi_4 : I \subset [0,1] \longrightarrow (0, \infty)$ for which $\lim_{\varepsilon \to 0^+} \psi_4(\varepsilon) = 0$ and

$$\left(M_A e^{\alpha T}\right)^p \int_0^t E \, \|f_\varepsilon(s, X(s)) - f(s, X(s))\|_{\mathcal{H}}^p ds \leq K_4 \psi_4(\varepsilon), \, \forall t \in [0, T]. \quad (7.110)$$

Finally, similar computations can be used to show that $\exists K_5 > 0$ and a function $\psi_5 : I \subset [0,1] \longrightarrow (0, \infty)$ for which $\lim_{\varepsilon \to 0^+} \psi_5(\varepsilon) = 0$ and

$$I_3 \leq K_5 \psi_5(\varepsilon), \, \forall t \in [0, T]. \quad (7.111)$$

(Show the details!) Combining all of these estimates in (7.103) yields the inequality

$$E \, \|X_\varepsilon(t; \cdot) - X(t)\|_{\mathcal{H}}^p \leq \xi_1 \psi(\varepsilon) + \xi_2 \int_0^t E \, \|X_\varepsilon(s; \cdot) - X(s)\|_{\mathcal{H}}^p ds, \quad (7.112)$$

where $\xi_1, \xi_2 > 0$ and $\lim_{\varepsilon \to 0^+} \psi(\varepsilon) = 0$. Hence, applying Gronwall's Lemma in (7.112) yields

$$E \, \|X_\varepsilon(t; \cdot) - X(t)\|_{\mathcal{H}}^p \leq \xi_1 e^{\xi_2 t} \psi(\varepsilon), \, \forall t \in [0, T], \quad (7.113)$$

as desired. $\square$

Remarks.
1. The results that we have established in Section 7.4 assuming that the forcing terms were globally Lipschitz continuous in their second variable can be strengthened in the sense that the growth condition can be weakened to a local Lipschitz one (in the sense of Def. 7.3.3). The arguments are more delicate just as in the deterministic setting. (See **[105]** for details.)

2. Questions of stability in its various forms become more delicate when the forcing terms are state dependent. Some references for stability of finite-dimensional SDEs include **[285, 414]** and **[89, 267, 285]** for the infinite-dimensional case.

7.5 Models Revisited

We now apply the results established in Section 7.4 to draw some conclusions about each of the models discussed in Section 7.1, including some new variants.

Model II.4 Semi-linear Pharmacokinetics
Consider IVP (7.1) and complete the following exercises.

Exercise 7.5.1.
i.) Assume that D and $\overline{D}$ are globally Lipschitz. Argue that (7.1) has a unique mild solution $U(t;\omega) = \begin{bmatrix} y(t;\omega) \\ z(t;\omega) \end{bmatrix}$.
ii.) Suppose D depends on an additional real parameter μ, which could be viewed as an environmental quantity that impacts concentration level. Formally, we replace $D(t,y(t;\omega),z(t;\omega))$ in (7.1) by the mapping $\widehat{D}: [0,\infty) \times \mathbb{R} \times \mathbb{R} \times \mathbb{R} \to \mathbb{R}$ given by $\widehat{D}(t,y(t;\omega),z(t;\omega),\mu)$. Assume that $\forall \mu_0$, there exists a positive constant $M(\mu_0)$ such that

$$\left| \widehat{D}(t,x,y,\mu_0) - \widehat{D}(t,\widehat{x},\widehat{y},\mu_0) \right| \le M(\mu_0) \left[|x - \widehat{x}| + |y - \widehat{y}| \right], \qquad (7.114)$$

$\forall x,\widehat{x},y,\widehat{y} \in \mathbb{R}$ and $t > 0$.
 a.) For every $\mu_0 \in \mathbb{R}$, show that (7.1) has a unique mild solution U_{μ_0} on $[0,T]$, for any $T > 0$. (As such, the mild solution exists on $[0,\infty)$.)
 b.) Compute $\lim_{t \to \infty} U_{\mu_0}(t;\omega)$ in the $\mathscr{L}^2$-sense assuming that $a < b$ in (7.1).
 c.) Assume that $\mu \mapsto M(\mu)$ is a continuous mapping such that $\lim_{\mu \to 0} M(\mu) = 0$.
Establish a continuous dependence estimate for (7.1) with $\widehat{D}$ in place of D.
 d.) Show that $\mu \mapsto U_\mu(\cdot)$ is continuous. Interpret the result.

Exercise 7.5.2. For every $n \in \mathbb{N}$, consider the IVP

$$\begin{cases} dy_n(t;\omega) = \left((\alpha + \frac{1}{n}) y_n(t;\omega) + \sin(2t) f(y_n(t;\omega)) g(z_n(t;\omega)) \right) dt \\ \qquad + \sum_{k=1}^m \cos(2t) h_k(y_n(t;\omega)) j_k(z_n(t;\omega)) dW(t), 0 < t < T, \omega \in \Omega, \\ dz_n(t;\omega) = \left((\beta - \frac{1}{n}) z_n(t;\omega) + \sin(2t) \widehat{f}(y_n(t;\omega)) \widehat{g}(z_n(t;\omega)) \right) dt \\ \qquad + \sum_{k=1}^m \cos(2t) \widehat{h}_k(y_n(t;\omega)) \widehat{j}_k(z_n(t;\omega)) dW(t), 0 < t < T, \omega \in \Omega, \\ y_n(0;\omega) = y_0(\omega) + \frac{1}{n^2}, z_n(0;\omega) = z_0(\omega) + \frac{1}{n^2}, \omega \in \Omega, \end{cases} \qquad (7.115)$$

where $W(t)$ is a one-dimensional Wiener process. Assume that $f,\widehat{f},g,\widehat{g},h_k,j_k,\widehat{h}_k,\widehat{j}_k$ $(k = 1,\dots,m) \in C(\mathbb{R};\mathbb{R})$ and $\alpha,\beta < 0$.
i.) Prove that $\forall T > 0$, (7.115) has a unique mild solution on $[0,T]$. Determine an explicit formula for this solution.

ii.) Let $\varepsilon > 0$. Determine a value of $N \in \mathbb{N}$ for which

$$n, m \geq N \implies \left\| \begin{bmatrix} y_n \\ z_n \end{bmatrix} - \begin{bmatrix} y_m \\ z_m \end{bmatrix} \right\|_{\mathbb{C}\left([0,T];\mathcal{L}^2\left(\Omega;\mathbb{R}^2\right)\right)} < \varepsilon. \qquad (7.116)$$

What does this tell you?

iii.) Does $\exists \begin{bmatrix} y^\star \\ z^\star \end{bmatrix} \in \mathbb{C}\left([0,T];\mathcal{L}^2\left(\Omega;\mathbb{R}^2\right)\right)$ such that

$$\lim_{n \to \infty} \left\| \begin{bmatrix} y_n \\ z_n \end{bmatrix} - \begin{bmatrix} y^\star \\ z^\star \end{bmatrix} \right\|_{\mathbb{C}\left([0,T];\mathcal{L}^2\left(\Omega;\mathbb{R}^2\right)\right)} = 0? \qquad (7.117)$$

To which related IVP is $\begin{bmatrix} y^\star \\ z^\star \end{bmatrix}$ a mild solution?

Next, $\forall 0 < \varepsilon < 1$, consider the stochastic IVP

$$\begin{cases} dy_\varepsilon(t;\omega) &= \left(-y_\varepsilon(t;\omega) + t \cdot \frac{\varepsilon}{\varepsilon+1}\left(\alpha y_\varepsilon(t;\omega) + \beta z_\varepsilon(t;\omega)\right)\right) dt + \varepsilon e^{-\varepsilon t} dW_1(t) \\ &\quad + \sqrt{\varepsilon} \sin\left(\frac{1}{2} y_\varepsilon(t;\omega)\right) dW_2(t) + \sqrt{\varepsilon} \cos^2\left(y_\varepsilon^2(t;\omega) + z_\varepsilon(t;\omega)\right) dW_3(t), \\ dz_\varepsilon(t;\omega) &= \left(y_\varepsilon(t;\omega) - 2z_\varepsilon(t;\omega) + 2t\left(\frac{\varepsilon^2}{3(\varepsilon^2+4)}\right)\left(\overline{\alpha} y_\varepsilon(t;\omega) + \overline{\beta} z_\varepsilon(t;\omega)\right)\right) dt \\ &\quad + \varepsilon e^{-\frac{1}{2}t+\varepsilon} dW_1(t) + \varepsilon \cos\left(y_\varepsilon(t;\omega)\right) dW_2(t) \\ &\quad + \sqrt{\varepsilon} \sin^4\left(y_\varepsilon(t;\omega) + z_\varepsilon(t;\omega)\right) dW_3(t), \\ y_\varepsilon(0;\omega) &= y_0, \\ z_\varepsilon(0;\omega) &= z_0, \end{cases}$$
$$(7.118)$$

where $0 < t < T$, $\omega \in \Omega$; $y_0, z_0 \in \mathbb{R}$; and $W_1(t), W_2(t)$, and $W_3(t)$ are independent one-dimensional Wiener processes. Also, consider the deterministic IVP

$$\begin{cases} dy(t) = \left(-y(t) + t\left(\alpha y(t) + \beta z(t)\right)\right) dt, 0 < t < T, \\ dz(t) = y(t) - 2z(t) + \frac{2}{3}t\left(\overline{\alpha} y(t) + \overline{\beta} z(t)\right) dt, 0 < t < T, \\ y(0) = y_0, \\ z(0) = z_0. \end{cases} \qquad (7.119)$$

Exercise 7.5.3.
i.) Argue that (7.118) has a unique mild solution on $[0,T]$.
ii.) Reformulate (7.118) as an abstract SEE in $\mathbb{R}^2$, and reformulate (7.119) as an abstract deterministic evolution equation in $\mathbb{R}^2$.
iii.) Carefully verify that the conditions of Prop. 7.4.16 are satisfied.
iv.) Conclude that

$$\lim_{\varepsilon \to 0^+} \left\| \begin{bmatrix} y_\varepsilon \\ z_\varepsilon \end{bmatrix} - \begin{bmatrix} y \\ z \end{bmatrix} \right\|_{\mathbb{C}\left([0,T];\mathcal{L}^2\left(\Omega;\mathbb{R}^2\right)\right)} = 0.$$

Model III.3 Semi-linear Spring-Mass Systems

We can account for damping and friction in the model by incorporating the term $\beta x'(t)$ into the left-hand side of the ODE in IVP (4.3). Specifically, consider the IVP

$$\begin{cases} (dx'(t;\omega) + \beta x(t;\omega)) + \eta^2 x(t;\omega)dt + \widehat{\eta}^2 x(t;\omega)dW(t) = 0, \\ x(0;\omega) = x_0(\omega), \ x'(0;\omega) = x_1(\omega), \ \omega \in \Omega \end{cases} \tag{7.120}$$

where $0 < t < T$, $\omega \in \Omega$. (Technically, $\beta = \frac{\overline{\beta}}{m}$, where $\overline{\beta}$ is the damping constant and $\widehat{\eta}^2$ was introduced into the model via a white noise process arising in the measurement of η^2.)

Exercise 7.5.4.
i.) Reformulate (7.120) as an abstract SEE in $\mathbb{R}^2$.
ii.) Argue that (7.120) has a unique mild solution on $[0,T]$.
iii.) Determine an explicit formula for the mild solution of (7.120).

Next, equip (7.120) with an external force driven by two time-periodic functions to obtain the following generalization of (7.120):

$$\begin{cases} (dx'(t;\omega) + \beta x(t;\omega)) + \eta^2 x(t;\omega)dt + \widehat{\eta}^2 x(t;\omega)dW(t) = \\ (\alpha_1 \sin(\gamma t) + \alpha_2 \sin(\Gamma t))dt + \widehat{\alpha_1} \sin(\gamma t)dW_1(t) + \widehat{\alpha_2} \sin(\Gamma t)dW_2(t), \\ x(0;\omega) = x_0(\omega), \ x'(0;\omega) = x_1(\omega), \ \omega \in \Omega, \end{cases} \tag{7.121}$$

where $0 < t < T$, $\omega \in \Omega$, and $\gamma, \Gamma, \alpha_1, \alpha_2, \widehat{\alpha_1}, \widehat{\alpha_2} > 0$. (Here, randomness has been introduced into the model via white noise processes arising in the measurement of α_1 and α_2.)

Exercise 7.5.5.
i.) Prove that (7.121) has a unique mild solution on $[0,T]$, $\forall T > 0$. Determine an explicit formula for this solution.
ii.) Establish a pth moment estimate for $x(t;\cdot)$, for $p > 2$.

Suppose that in (7.121), α_1 and α_2 are replaced by $\overline{\alpha_1}(\varepsilon)$ and $\overline{\alpha_2}(\varepsilon)$, respectively, and that both $\overline{\alpha_1}(\varepsilon)$ and $\overline{\alpha_2}(\varepsilon)$ approach zero as $\varepsilon \to 0^+$. It follows from Exer. 7.5.5 that $\forall 0 < \varepsilon < 1$, the newly formed IVP has a unique mild solution $x_\varepsilon(t;\omega)$ on $[0,T]$.

Exercise 7.5.6. Does $\exists y \in \mathbb{C}([0,T];\mathfrak{L}^2(\Omega;\mathbb{R}))$ for which

$$\lim_{\varepsilon \to 0^+} \|x_\varepsilon - y\|_{\mathbb{C}([0,T];\mathfrak{L}^2(\Omega;\mathbb{R}))} = 0?$$

To what IVP is y a mild solution?

Exercise 7.5.7.
i.) Replace the right-hand side of (7.121) by the more general function $g(x) + h(t)$. What condition(s) can be imposed on g and h to ensure the existence of a mild solution of (7.121) on $[0,T]$, $\forall T > 0$? Under such conditions, show that this existence of

mild solutions of both (7.120) and (7.121) can be recovered as special cases.
ii.) Replace the right-hand side of (7.121) by $g_1(x) + g_1(x') + h(t)$ and repeat (i).

The motion of an oscillator in a double-well potential described by

$$\Gamma(x) = -\frac{1}{2}\left(\alpha_1^2 + (\widehat{\alpha_1})^2 \frac{dW_1(t)}{dt}\right)x^2 + \frac{1}{4}\left(\alpha_2^2 + (\widehat{\alpha_2})^2 \frac{dW_2(t)}{dt}\right)x^4, \quad (7.122)$$

where $\alpha_1, \alpha_2, \widehat{\alpha_1}$, and $\widehat{\alpha_2}$ are positive constants and $W_1(t)$ and $W_2(t)$ are independent Wiener processes, can be modeled by the following IVP:

$$\begin{cases} (dx'(t;\omega) + \beta x(t;\omega)) = \left(-\frac{1}{2}\alpha_1^2 x^2(t;\omega) + \frac{1}{4}\alpha_2^2 x^4(t;\omega)\right)dt \\ \quad + \left(-\frac{1}{2}(\widehat{\alpha_1})^2 x^2(t;\omega)dW_1(t) + \frac{1}{4}(\widehat{\alpha_2})^2 x^4(t;\omega)dW_2(t)\right), \\ x(0;\omega) = x_0(\omega),\ x'(0;\omega) = x_1(\omega),\ \omega \in \Omega. \end{cases} \quad (7.123)$$

where $0 < t < T,\ \omega \in \Omega$.

Exercise 7.5.8. Must (7.123) have a unique mild solution on $[0,T]$? Prove your assertion.

Model VII.3 Semi-linear Wave Equations
We have accounted for dispersion in a model of elementary waves and have studied a system of coupled waves. The following is an extension of those IBVPs.

Exercise 7.5.9. Consider the following generalization of (7.3), where $z = z(x,t;\omega)$ and $w = w(x,t;\omega)$:

$$\begin{cases} \partial\left(\frac{\partial}{\partial t}z(x,t;\omega) + \alpha z(x,t;\omega)\right) + \left(c^2 \frac{\partial^2}{\partial x^2}z(x,t;\omega)\right)\partial t = f(t,x,z,w,z_t,w_t)\partial t \\ \quad + g(t,x,z,w,z_t,w_t)dW(t),\ 0 < x < L,\ 0 \le t \le T,\ \omega \in \Omega, \\ \partial\left(\frac{\partial}{\partial t}w(x,t;\omega) + \alpha w(x,t;\omega)\right) + \left(c^2 \frac{\partial^2}{\partial x^2}w(x,t;\omega)\right)\partial t = \widehat{f}(t,x,z,w,z_t,w_t)\partial t \\ \quad + g(t,x,z,w,z_t,w_t)dW(t),\ 0 < x < L,\ 0 \le t \le T,\ \omega \in \Omega, \\ z(x,0;\omega) = z_0(x;\omega),\ \frac{\partial z}{\partial t}(x,0;\omega) = z_1(x;\omega),\ 0 < x < L,\ \omega \in \Omega, \\ w(x,0;\omega) = w_0(x;\omega),\ \frac{\partial w}{\partial t}(x,0;\omega) = w_1(x;\omega),\ 0 < x < L,\ \omega \in \Omega, \\ \frac{\partial z}{\partial x}(0,t;\omega) = \frac{\partial z}{\partial x}(L,t;\omega) = \frac{\partial w}{\partial x}(0,t;\omega) = \frac{\partial w}{\partial x}(L,t;\omega) = 0,\ 0 \le t \le T,\ \omega \in \Omega. \end{cases} \quad (7.124)$$

i.) Reformulate (7.124) as an abstract SEE in an appropriate space.
ii.) Impose conditions on the forcing terms that ensure that (7.124) has a unique mild solution on $[0,T]$. Prove your assertion.
iii.) Formulate an IBVP in which the one-dimensional domain is replaced by a two-dimensional rectangular domain $\mathscr{R}$ in (7.124). Repeat (i) and (ii).
iv.) Formulate an IBVP similar to (7.124) for a system of N such coupled waves on a two-dimensional rectangular domain $\mathscr{R}$ described by $z_i = z_i(x,y,t;\omega),\ i = 1,\dots,N$. Repeat (i) and (ii).

Model V.4 Diffusion Revisited
The semi-linear modifications of the type suggested below can be applied to all diffusion models discussed thus far.

Exeicse 7.5.10. Consider IBVP (7.4).
i.) Prove that (7.4) has a unique mild solution on $[0,T]$, $\forall T > 0$.
ii.) Suppose that $z_0(x,y) = ax + by$, where $a, b > 0$, and consider the related IBVP

$$
\begin{cases}
\partial \widehat{z}(x,y,t;\omega) = \left(\widehat{k} \triangle \widehat{z}(x,y,t;\omega) + \alpha e^{-\frac{\beta}{\widehat{z}(x,y,t;\omega)}} \right) \partial t + \widehat{g}(t) dW_1(t) \\
\quad + \overline{g}(t, z(x,y,t;\omega)) dW_2(t)\, 0 < x < a,\, 0 < y < b,\, 0 \le t \le T,\, \omega \in \Omega, \\
z(x,y,0;\omega) = \widehat{a}x + \widehat{b}y,\, 0 < x < a,\, 0 < y < b,\, \omega \in \Omega, \\
\frac{\partial z}{\partial y}(x,0,t;\omega) = 0 = \frac{\partial z}{\partial y}(x,b,t;\omega),\, 0 < x < a,\, 0 \le t \le T,\, \omega \in \Omega, \\
\frac{\partial z}{\partial x}(0,y,t;\omega) = 0 = \frac{\partial z}{\partial x}(a,y,t;\omega),\, 0 < y < b,\, 0 \le t \le T,\, \omega \in \Omega.
\end{cases}
\tag{7.125}
$$

Assume that $\exists \delta_1, \delta_2, \delta_3$, and $\delta_4 > 0$ such that the following hold:

$$
\left| k - \widehat{k} \right| < \delta_1,
\tag{7.126}
$$

$$
\sup_{0 \le t \le T} \left\| g(t) - \widehat{g}(t) \right\|_{\mathscr{B}_0\left(\mathbb{R}^2; \mathbb{L}^2(0,a) \times (0,b); \mathbb{R} \right)} < \delta_2,
\tag{7.127}
$$

$$
\left| a - \widehat{a} \right| < \delta_3,
\tag{7.128}
$$

$$
\left| b - \widehat{b} \right| < \delta_4.
\tag{7.129}
$$

a.) Reformulate (7.125) as an abstract SEE in the same space $\mathscr{H}$ as (7.4). Determine an explicit formula for the mild solution.
b.) Derive an estimate for $E \left\| z(\cdot,\cdot,t;\cdot) - \widehat{z}(\cdot,\cdot,t;\cdot) \right\|_{\mathscr{H}}^2$.

Next, consider the IBVP

$$
\begin{cases}
\partial z(x,t;\omega) + \left(\alpha \frac{\partial^2 z}{\partial x^2}(x,t;\omega) + \beta z(x,t;\omega) \right) \partial t = f(t, z(x,t;\omega)) \partial t \\
\quad + \left(\gamma_1 t^2 + \gamma_2 t + \gamma_3 \right) z(x,t;\omega) dW(t),\, 0 < x < 2\pi,\, 0 < t < T,\, \omega \in \Omega, \\
\frac{\partial z}{\partial x}(0,t;\omega) = \frac{\partial z}{\partial x}(2\pi,t;\omega) = 0,\, 0 < t < T,\, \omega \in \Omega, \\
z(x,0;\omega) = \cos(2x),\, 0 < x < 2\pi,\, \omega \in \Omega,
\end{cases}
\tag{7.130}
$$

where $\gamma_i > 0$ $(i = 1,2,3)$, $W(t)$ is a one-dimensional Wiener process, and $f : [0,T] \times \mathbb{R} \to \mathbb{R}$ is defined by

$$
f(t,w) = \begin{cases} |\cos(t)| \sin(w),\, w > 0 \text{ and } 0 \le t \le T, \\ 0,\, w \le 0 \text{ and } 0 \le t \le T. \end{cases}
\tag{7.131}
$$

Exercise 7.5.11.
i.) Prove that (7.130) has a unique mild solution on $[0,T]$.

ii.) More generally, suppose that $f : [0, \infty) \times \mathbb{R} \to \mathbb{R}$ is defined by

$$f(t, w) = \begin{cases} g(t)h(w), & w > 0 \text{ and } 0 \le t \le T, \\ 0, & w \le 0 \text{ and } 0 \le t \le T, \end{cases} \tag{7.132}$$

where $g : [0, \infty) \to \mathbb{R}$ is continuous and bounded and $h : \mathbb{R} \to \mathbb{R}$ is globally Lipschitz. Must (7.130) have a mild solution on $[0, T]$?

iii.) Let $\mathscr{D}$ denote the open unit disc in $\mathbb{R}^2$ and consider the IBVP

$$\begin{cases} \partial z + \left(\alpha \left(\frac{\partial^2 z}{\partial x^2} + \frac{\partial^2 z}{\partial y^2} \right) + \beta z \right) \partial t = f(t, x, y, z) \, \partial t \\ \qquad + \left(\gamma_1 t^2 + \gamma_2 t + \gamma_3 \right) z(x, t; \omega) dW(t), \ (x, y) \in \mathscr{D}, 0 < t < T, \ \omega \in \Omega, \\ \frac{\partial z}{\partial \mathbf{n}}(x, y, t; \omega) = 0, \ (x, y) \in \mathscr{D}, 0 < t < T, \ \omega \in \Omega, \\ z(x, y, 0; \omega) = z_0(x, y; \omega), \ (x, y) \in \partial \mathscr{D} = \{ (x, y) : x^2 + y^2 = 1 \}, \ \omega \in \Omega, \end{cases} \tag{7.133}$$

where $z = z(x, y, t; \omega)$, $\frac{\partial}{\partial \mathbf{n}}$ is the outward unit normal to $\partial \mathscr{D}$, and $f : [0, T] \times \mathscr{D} \times \mathbb{R} \to \mathbb{R}$ is given by

$$f(t, x, y, z) = \begin{cases} \left| \cos(t) \cos^3(x) \cos^5(y) \right| \sin(z), & 0 \le t \le T, (x, y) \in \mathscr{D}, z > 0, \\ 0, & 0 \le t \le T, (x, y) \in \mathscr{D}, z \le 0. \end{cases}$$

Show that (7.133) has a unique mild solution on $[0, T]$.

iv.) Formulate a continuous dependence result for (7.133).

Model IX.2 Neural Networks Continued
We now formally study IVP (7.5).

Exercise 7.5.12. Assume that $\forall 1 \le j \le M$, $g_j : [0, \infty) \to \mathbb{R}$ is globally Lipschitz and that $\forall 1 \le i, j \le M$, $\omega_{ij}(t)$ is continuous. (In fact, $\text{rng}(\omega_{ij}) \subset [-1, 1]$ because ω_{ij} represents a proportion in the positive and negative directions.)

i.) Show that (7.5) has a unique mild solution on $[0, T]$.

ii.) Formulate a continuous dependence result for (7.5).

Exercise 7.5.13. For every $\varepsilon > 0$, consider the IVP (7.5) in which ω_{ij} is replaced by an approximation $(\omega_{ij})_\varepsilon$. Assume that $\forall 1 \le j \le M$, $g_j : [0, \infty) \to \mathbb{R}$ is globally Lipschitz. Denote the corresponding unique mild solution of (7.5) on $[0, T]$ by $\mathbf{U}_\varepsilon$. If $\forall 1 \le i, j \le M$, $\lim_{\varepsilon \to 0^+} (\omega_{ij})_\varepsilon = \omega_{ij}$, show that $\forall T > 0$, $\exists \mathbf{U} \in \mathbb{C}\left([0, T]; \mathscr{L}^2\left(\Omega; \mathbb{R}^M\right)\right)$ such that $\lim_{\varepsilon \to 0^+} \|\mathbf{U}_\varepsilon - \mathbf{U}\|_{\mathbb{C}} = 0$.

Model X.2 Spatial Pattern Formation Revisited
Consider the IBVP (7.6) for $N = 2$ in which f_1 and f_2 are given by (7.9), and $\overline{f_1}$ and $\overline{f_2}$ are given by the same formulae with β_i replaced by $\overline{\beta_i}$. This IBVP can be reformulated abstractly as (7.10) on the space $\mathscr{H} = \mathbb{L}^2(\mathscr{D}) \times \mathbb{L}^2(\mathscr{D})$. (Why?)

Let $\mathbf{C} = (C_1, C_2)$. Motivated by (7.9), define the mappings $F : \mathscr{H} \to \mathscr{H}$ and $G : \mathscr{H} \to \mathscr{B}_0(\mathbb{R}, \mathscr{H})$ by

$$F(\mathbf{C})(\mathbf{x}, t; \omega) = \begin{bmatrix} \beta_1 - \beta_2 C_1(\mathbf{x}, t; \omega) + \beta_3 C_1^2(\mathbf{x}, t; \omega) C_2(\mathbf{x}, t; \omega) \\ \beta_4 - \beta_3 C_1^2(\mathbf{x}, t; \omega) C_2(\mathbf{x}, t; \omega) \end{bmatrix}, \quad (7.134)$$

$$G(\mathbf{C})(\mathbf{x}, t; \omega) = \begin{bmatrix} \overline{\beta_1} - \overline{\beta_2} C_1(\mathbf{x}, t; \omega) + \overline{\beta_3} C_1^2(\mathbf{x}, t; \omega) C_2(\mathbf{x}, t; \omega) \\ \overline{\beta_4} - \overline{\beta_3} C_1^2(\mathbf{x}, t; \omega) C_2(\mathbf{x}, t; \omega) \end{bmatrix}. \quad (7.135)$$

For brevity, we shall write $F(\mathbf{C})$ and $G(\mathbf{C})$ in place of $F(\mathbf{C})(\mathbf{x}, t; \omega)$ and $G(\mathbf{C})(\mathbf{x}, t; \omega)$, respectively.

Exercise 7.5.14.
i.) For each $R > 0$, show that

$$\|F(\mathbf{w}) - F(\mathbf{z})\|_{\mathscr{H}} \le 2R^2 \|\mathbf{w} - \mathbf{z}\|_{\mathscr{H}}, \quad \forall \mathbf{w}, \mathbf{z} \in \mathfrak{B}_{\mathscr{H}}(0; R)$$

$$\|F(\mathbf{w}) - F(\mathbf{z})\|_{\mathscr{B}_0(\mathbb{R}; \mathscr{H})} \le 2R^2 \|\mathbf{w} - \mathbf{z}\|_{\mathscr{H}}, \quad \forall \mathbf{w}, \mathbf{z} \in \mathfrak{B}_{\mathscr{H}}(0; R).$$

ii.) Deduce that (7.6) has a unique mild solution on $[0, T]$ and determine an explicit variation of parameters formula for it. To what space must the solution belong?

Without significantly increasing the complexity of the IBVP, we can replace the first components in (7.134) and (7.135) by the following more general forms of an activator inhibitor:

$$f_1(C_1, C_2) = \beta_1 - \beta_2 C_1 + \mu \frac{|g_1(C_1, C_2)|}{|1 + g_2(C_1, C_2)|} \quad (7.136)$$

$$\overline{f_1}(C_1, C_2) = \overline{\beta_1} - \overline{\beta_2} C_1 + \overline{\mu} \frac{|\overline{g_1}(C_1, C_2)|}{|1 + \overline{g_2}(C_1, C_2)|}, \quad (7.137)$$

where $g_1, \overline{g_1} : \mathbb{R} \times \mathbb{R} \to \mathbb{R}$ are globally Lipschitz, $g_2, \overline{g_2} : \mathbb{R} \times \mathbb{R} \to (0, \infty)$ are continuous, and $\mu, \overline{\mu} \in \mathbb{R}$. Such functions are used to describe the kinetics in various biological models. For instance, in the so-called *Thomas model*, the functions

$$\begin{cases} g_1(C_1, C_2) = v_1 C_1 C_2 \\ g_2(C_1, C_2) = v_2 C_1 + v_3 C_1^2, \end{cases} \quad (7.138)$$

are used, where $v_i > 0 \, (i = 1, 2, 3)$, while different functions appear in the *Gray-Scott model* (see [**120**]). (Similar functions can be used for $\overline{g_1}$ and $\overline{g_2}$.)

Exercise 7.5.15. Consider the IBVP (7.6) for $N = 2$ in which the forcing terms F and G are now taken to be

$$F(\mathbf{C})(\mathbf{x}, t; \omega) = \begin{bmatrix} \beta_1 - \beta_2 C_1(\mathbf{x}, t; \omega) + \mu \frac{|g_1(C_1(\mathbf{x}, t; \omega), C_2(\mathbf{x}, t; \omega))|}{|1 + g_2(C_1(\mathbf{x}, t; \omega), C_2(\mathbf{x}, t; \omega))|} \\ \beta_4 - \beta_3 C_2(\mathbf{x}, t; \omega) \end{bmatrix}, \quad (7.139)$$

$$G(\mathbf{C})(\mathbf{x}, t; \omega) = \begin{bmatrix} \overline{\beta_1} - \overline{\beta_2} C_1(\mathbf{x}, t; \omega) + \overline{\mu} \frac{|\overline{g_1}(C_1(\mathbf{x}, t; \omega), C_2(\mathbf{x}, t; \omega))|}{|1 + \overline{g_2}(C_1(\mathbf{x}, t; \omega), C_2(\mathbf{x}, t; \omega))|} \\ \overline{\beta_4} - \overline{\beta_3} C_2(\mathbf{x}, t; \omega) \end{bmatrix}. \quad (7.140)$$

Show that (7.6) has a unique mild solution on $[0, T]$.

The dynamics of (7.6) depending on the rate constants β_i are interesting to study. As alluded to in **[120]**, certain conditions on β_i lead to the generation of a discernible pattern, while others produce apparently random behavior.

Model XI.1 Diffusion Revisited – Effects of Random Motility on a Bacteria Population

Lauffenburger, in 1981, suggested a model for the effects of random motility on a bacterial population that consume a diffusible substrate. (See **[254]**.) A three-dimensional stochastic version of this model in a bounded region $\mathscr{D} \subset \mathbb{R}^3$ with smooth boundary $\partial \mathscr{D}$ is described by the following IBVP:

$$\begin{cases} \partial B = (\alpha \triangle B + (G(S) - K_D) B(x,y,z,t;\omega)) \, \partial t \\ \quad + G^\star(S) B(x,y,z,t;\omega) dW_1(t), \ (x,y,z) \in \mathscr{D}, \ 0 < t < T, \ \omega \in \Omega, \\ \partial S = \left(\beta \triangle S - \frac{G(S)}{M} B(x,y,z,t;\omega) \right) \partial t \\ \quad - \frac{G^\star(S)}{M} B(x,y,z,t;\omega) dW_2(t), \ (x,y,z) \in \mathscr{D}, \ 0 < t < T, \ \omega \in \Omega, \\ B(x,y,z,t;\omega) = S(x,y,z,t;\omega) = 0, \ (x,y,z) \in \partial \mathscr{D}, \ 0 < t < T, \ \omega \in \Omega, \\ B(x,y,z,0;\omega) = B_0(x,y,z,t;\omega), \ (x,y,z) \in \mathscr{D}, \ \omega \in \Omega, \\ S(x,y,z,0;\omega) = S_0(x,y,z,t), \ (x,y,z) \in \mathscr{D}, \ \omega \in \Omega. \end{cases} \tag{7.141}$$

Here,

$B = B(x,y,z,t;\omega) = $ density of bacteria population at position $(x,y,z) \in \Omega$ at time t and $\omega \in \Omega$,

$S = S(x,y,z,t;\omega) = $ substrate concentration at position $(x,y,z) \in \Omega$ at time t and $\omega \in \Omega$,

$M = $ mass of bacteria per unit mass of nutrient,

$K_D = $ bacterial death rate in the absence of the substrate,

$G(S), G^\star(S) = $ substrate-dependent growth rates,

$W_1(t), W_2(t)$ are independent one-dimensional Wiener processes,

$B_0, S_0 \in \mathfrak{L}^2 \left(\Omega; \mathbb{R}^3 \right)$ are $\mathscr{F}_0$-measurable random variables independent of $W_1(t)$ and $W_2(t)$.

We can rewrite (7.141) equivalently as

$$\begin{cases} \partial \begin{bmatrix} B \\ S \end{bmatrix} = \begin{bmatrix} \alpha \triangle & -K_D \\ 0 & \beta \triangle \end{bmatrix} \begin{bmatrix} B \\ S \end{bmatrix} \partial t + \begin{bmatrix} G(S)B \\ -\frac{G(S)}{M} B \end{bmatrix} \partial t \\ \quad + \begin{bmatrix} G^\star(S)B \\ 0 \end{bmatrix} dW_1(t) + \begin{bmatrix} 0 \\ -\frac{G^\star(S)}{M} B \end{bmatrix} dW_2(t), \\ \begin{bmatrix} B \\ S \end{bmatrix} (x,y,z,0;\omega) = \begin{bmatrix} B_0 \\ S_0 \end{bmatrix} (x,y,z;\omega), \ (x,y,z) \in \mathscr{D}, \ \omega \in \Omega, \\ \begin{bmatrix} B \\ S \end{bmatrix} (x,y,z,t) = 0, \ (x,y,z) \in \partial \mathscr{D}, \ 0 < t < T, \ \omega \in \Omega, \end{cases} \tag{7.142}$$

where $(x,y,z) \in \mathcal{D}, 0 < t < T$, and $\omega \in \Omega$. In order to reformulate (7.142) as an abstract stochastic evolution equation in the space $\mathcal{H} = \mathbb{L}^2(\mathcal{D}) \times \mathbb{L}^2(\mathcal{D})$, we identify the unknown as $X = \begin{bmatrix} B \\ S \end{bmatrix}$ and the initial condition as $X_0 = \begin{bmatrix} B_0 \\ S_0 \end{bmatrix}$, and define the forcing terms $F : [0,T] \times \mathcal{H} \to \mathcal{H}$ and $G : [0,T] \times \mathcal{H} \to \mathcal{B}_0\left(\mathbb{R}^2, \mathcal{H}\right)$ by

$$F\left(t, \begin{bmatrix} B \\ S \end{bmatrix}\right) = \begin{bmatrix} G(S)B \\ -\frac{G(S)}{M}B \end{bmatrix} \tag{7.143}$$

$$G\left(t, \begin{bmatrix} B \\ S \end{bmatrix}\right) = \begin{bmatrix} G^\star(S)B & 0 \\ 0 & -\frac{G^\star(S)}{M}B \end{bmatrix}. \tag{7.144}$$

Also, define the operator $A : \mathrm{dom}(A) \subset \mathcal{H} \to \mathcal{H}$ by

$$A = \begin{bmatrix} \alpha\triangle & 0 \\ 0 & \beta\triangle \end{bmatrix}, \quad \mathrm{dom}(A) = \mathrm{dom}(\triangle) \times \mathrm{dom}(\triangle), \tag{7.145}$$

and $B : \mathrm{dom}(B) \subset \mathcal{H} \to \mathcal{H}$ by

$$B = \begin{bmatrix} 0 & -K_D \\ 0 & 0 \end{bmatrix}, \quad \mathrm{dom}(B) = \mathcal{H}. \tag{7.146}$$

It can be shown that A generates a C_0-semigroup on $\mathcal{H}$ and that $B \in \mathbb{B}(\mathcal{H})$. (Why?)

Exercise 7.5.16. Assume that $G : [0,\infty) \to [0,\infty)$ is continuous.
i.) Show that (7.142) has a unique mild solution on $[0,T]$.
ii.) Formulate a continuous dependence result directly for (7.142).

Model XII.1 Chemotaxis
Substances, organisms, and collections of animals (such as bacteria and schools of fish) tend to travel in the direction dictated by a specific gradient (chemically driven or otherwise) that coincides with the direction of maximum increase in food supply, optimal temperature, pheromone concentration, etc. The equation governing the population concentration of such substances or organisms cannot be simply diffusion based due to the existence of such directive gradients. However, the concentration of the attractant is subject to diffusion. As such, the description of this scenario requires two equations. Below, we study a one-dimensional version in which randomness is incorporated into the model only through the initial conditions. To this end, let
$p(x,t;\omega) = $ density of the population being attracted (e.g., by a pheromone),
$c(x,t;\omega) = $ concentration level of the pheromone.
Consider the following IBVP:

$$\begin{cases} \frac{\partial p}{\partial t}(x,t;\omega) = \alpha_p \frac{\partial^2 p}{\partial x^2}(x,t;\omega) - \beta\frac{\partial}{\partial x}\left(p(x,t;\omega)\frac{\partial c}{\partial x}(x,t;\omega)\right), \\ \frac{\partial c}{\partial t}(x,t;\omega) = \alpha_c \frac{\partial^2 c}{\partial x^2}(x,t;\omega) + f(t,x,c(x,t;\omega)), \\ p(x,0;\omega) = p_0(x;\omega), \ c(x,0;\omega) = c_0(x;\omega), \\ \frac{\partial p}{\partial x}(0,t;\omega) = \frac{\partial p}{\partial x}(L,t;\omega) = \frac{\partial c}{\partial x}(0,t;\omega) = \frac{\partial c}{\partial x}(L,t;\omega) = 0, \end{cases} \tag{7.147}$$

where $0 < x < L$, $0 \le t \le T$, $\omega \in \Omega$ and $p_0, c_0 : [0,L] \times \Omega \to \mathbb{R}$ are $\mathscr{F}_0$-measurable random variables with finite second moments.

Remarks.
1. The diffusivity constants α_p, α_c are assumed to be positive, and the parameter $\beta > 0$ is a measure of attractivity of c and generally depends on c.
2. The second equation in (7.147) governs the signal emitted by the pheromone. Usually, it also involves a perturbation involving $p(x,t;\omega)$, the presence of which renders the two equations more strongly coupled. For simplicity, we assume no such dependence upon p here, but refer you to [189, 190, 312, 406] for other analyses in this case.
3. The forcing term f describes the kinetics of the process.

How do we attack this problem? Without using any context-specific knowledge, the space $\mathbb{L}^2(0,L)$ would seem to be a reasonable space to use for both p and c when reformulating the problem as an abstract SEE. (Why?) Further, the independence of the second equation of p suggests that a viable approach might be to first solve the IBVP consisting of the second equation in (7.147), together with its IC and BC, and then substitute the result into the first equation in (7.147) and solve the resulting IBVP. Seems reasonable, right? Well, note that even if we have $c(x,t;\omega)$, we must be able to prove the existence of a mild solution of an IBVP of the form

$$\begin{cases} \frac{\partial p}{\partial t}(x,t;\omega) = \alpha_p \frac{\partial^2 p}{\partial x^2}(x,t;\omega) + a(x,t)\frac{\partial p}{\partial x}(x,t;\omega) + b(x,t)p(x,t;\omega), \\ p(x,0;\omega) = p_0(x;\omega), 0 < x < L, \omega \in \Omega, \\ \frac{\partial p}{\partial x}(0,t;\omega) = \frac{\partial p}{\partial x}(L,t;\omega) = 0, 0 < t < T, \omega \in \Omega, \end{cases} \tag{7.148}$$

where $0 < x < L$, $0 < t < T$, $\omega \in \Omega$, which can be reformulated abstractly as

$$\begin{cases} X'(t;\omega) = AX(t;\omega), 0 < t < T, \omega \in \Omega, \\ X(0;\omega) = X_0(\omega), \omega \in \Omega, \end{cases} \tag{7.149}$$

in a separable Hilbert space $\mathscr{H}$, where the operator $A : \mathrm{dom}(A) \subset \mathscr{H} \to \mathscr{H}$ is given by

$$A = c_1 \frac{\partial^2}{\partial x^2} + \underbrace{c_2(x)\frac{\partial}{\partial x} + c_3(x)}_{\text{Call this } \mathscr{M}(x)} = c_1 \frac{\partial^2}{\partial x^2} + \mathscr{M}(x). \tag{7.150}$$

It can be shown that the operator $\mathscr{M} : \mathrm{dom}(A) \subset \mathscr{H} \to \mathscr{H}$ is A-bounded whenever c_2, c_3 are continuous functions (see [149]). Hence, we can conclude that A satisfies $(\mathbf{H_A})$, so that (7.148) has a unique mild solution on $[0,T]$. (Why?)

7.6 Theory for Non-Lipschitz-Type Forcing Terms

The results of the previous section were formulated under the assumption that the forcing term satisfied a Lipschitz-type growth condition on the space in which the mild solution of (7.10) was sought. This assumption was central to the development of the theory. But, forcing terms are not always that nicely behaved. For instance, consider a function $f : [0,T] \times (0,\infty) \to (0,\infty)$ satisfying

$$|f(t,x) - f(t,y)| \leq t|x-y|^4, \quad \forall x,y \in (0,\infty), t \in [0,T], \tag{7.151}$$

$$|f(t,x)| \leq p(t)|x|\ln\left(\frac{1}{|x|}\right), \quad \forall x \in (0,\infty), t \in [0,T], \tag{7.152}$$

where $p \in \mathbb{L}^1(0,\infty;(0,\infty))$.

Exercise 7.6.1. Show that f is not locally Lipschitz on its domain.

In light of Exer. 7.6.1, the theory developed in Section 7.4 does not apply to (7.10) when equipped with such a forcing term. In order to handle situations of this type, the following notion of convexity is introduced.

Definition 7.6.1. Let $\mathscr{D}$ be an open subset of $\mathbb{R}$. A function $g : \mathscr{D} \subset \mathbb{R} \to \mathbb{R}$ is
i.) *convex* if

$$g(\alpha x + (1-\alpha)y) \leq \alpha g(x) + (1-\alpha)g(y), \quad \forall 0 \leq \alpha \leq 1, x,y \in \mathscr{D}. \tag{7.153}$$

ii.) *concave* if $-g$ is convex.

Exercise 7.6.2. It can be shown that a convex function $f : \mathscr{D} \subset \mathbb{R} \to \mathbb{R}$ is locally Lipschitz, but not conversely. (See **[348]**.) Show that the following are convex.
i.) $f_1 : \mathbb{R} \to (0,\infty)$ defined by $f_1(x) = |x|^p$, where $p \geq 1$.
ii.) $f_2 : (0,\infty) \to (0,\infty)$ defined by $f_2(x) = x\log(x)$.

Exercise 7.6.3. It can be shown that if $f : \mathbb{R} \to \mathbb{R}$ is convex and $g : \mathbb{R} \to \mathbb{R}$ is continuous, then $f \circ g$ is convex. Use this fact to show that the function $h : \mathbb{R} \to (0,\infty)$ defined by $h(x) = \left|\sin(2x) - x\cos\left(x^2\right)\right|^{5/3}$ is convex.

The remainder of this section is devoted to studying (7.10) equipped with forcing terms $f : [0,T] \times \mathscr{H} \to \mathscr{H}$ and $g : [0,T] \times \mathscr{H} \to \mathscr{B}_0(\mathbb{R}^m, \mathscr{H})$ that satisfy the following hypotheses:

(H7.10) There exists $K_1 : [0,T] \times [0,\infty) \to [0,\infty)$ such that
    i.) $K_1(\cdot,\cdots)$ is continuous in both variables,
    ii.) $K_1(t,\cdots)$ is nondecreasing and concave, $\forall t \in [0,T]$,
    iii.) $E\|f(t,x)\|_{\mathscr{H}}^2 + E\|g(t,x)\|_{\mathscr{B}_0}^2 \leq K_1\left(t, E\|x\|_{\mathscr{H}}^2\right), \forall t \in [0,T], x \in \mathscr{H}.$
(H7.11) There exists $K_2 : [0,T] \times [0,\infty) \to [0,\infty)$ such that

i.) $K_2(\cdot,\cdot\cdot)$ is continuous in both variables,

ii.) $K_2(t,\cdot\cdot)$ is nondecreasing, concave, and $K_2(t,0) = 0$, $\forall t \in [0,T]$,

iii.) $\forall t \in [0,T]$, $x,y \in \mathcal{H}$,

$$E\,\|f(t,x) - f(t,y)\|_{\mathcal{H}}^2 + E\,\|g(t,x) - g(t,y)\|_{\mathcal{B}_0}^2 \leq K_2\left(t, E\,\|x-y\|_{\mathcal{H}}^2\right).$$

(H7.12) Any function $w : [0,T] \to [0,\infty)$ that is continuous, $w(0) = 0$, and satisfies

$$w(t) \leq 2\left(\overline{M_A}\right)^2 (T + \zeta_g(T,2)) \int_0^t K_2(s,w(s))\,ds, \quad \forall 0 \leq t \leq T^\star \leq T, \qquad (7.154)$$

must be identically 0 on $[0,T^\star]$.

Remarks. Some standard examples of functions K_1 satisfying **(H7.10)** are

1. $K_1(t,z) = az$. (In this case, a is actually a Lipschitz constant.)

2. $K_1(t,z) = p(t)\phi(z)$, where $p \in \mathbb{L}^1(0,\infty;(0,\infty))$ and $\phi : (0,\infty) \to (0,\infty)$ is a continuous, nondecreasing, concave function such that $\phi(0^+) = 0$ and $\int_0^\varepsilon \frac{1}{\phi(r)}dr = \infty$. Some typical choices for ϕ are

$$\phi(z) = z\ln\left(\frac{1}{z}\right) \qquad (7.155)$$

$$\phi(z) = z\ln\left(\frac{1}{z}\right)\ln\left(\ln\left(\frac{1}{z}\right)\right). \qquad (7.156)$$

(See **[122]**.)

Our approach begins with the usual sequence of successive approximations, but this time showing that $\{X_n\}$ converges to a unique function X which is a mild solution of (7.10) is more involved. This is equivalent to showing that $\{X_n\}$ is a Cauchy sequence. Taking a step back, if we could argue that

$$\|X_n - X_m\|_{\mathbb{C}([0,\tilde{T}];\mathcal{L}^2(\Omega;\mathcal{H}))} \leq C \int_0^t K^\star\left(\|X_n(s) - X_m(s)\|_{\mathcal{H}}\right)ds, \qquad (7.157)$$

$\forall 0 \leq t \leq \tilde{T} \leq T$, for some $K^\star$ for which the right-hand side of (7.157) goes to zero as $n,m \to \infty$, then we would conclude that $\{X_n\}$ is a Cauchy sequence in the space $\mathbb{C}\left([0,\tilde{T}];\mathcal{L}^2(\Omega;\mathcal{H})\right)$. This constitutes the foundation of our strategy. This technique has been used frequently in the study of ODEs (see **[93]**) and has recently been applied in the study of evolution equations (see **[36, 122, 277, 377, 411]**).

We shall make use of the following lemma established in **[93]**.

Lemma 7.6.2. *For every $\beta_1, \beta_2 > 0$, $\exists 0 < T_1 \leq T$ such that the equation*

$$z(t) = \beta_1 + \beta_2 \int_0^t K_1(s,z(s))ds \qquad (7.158)$$

has a continuous local solution $z : [0,T_1) \to [0,\infty)$.

Our main result is

Theorem 7.6.3. *If* $(\mathbf{H_A})$, $(\mathbf{H7.10})$ *through* $(\mathbf{H7.12})$ *hold, then* $\exists 0 < T^\star \leq T$ *such that* (7.10) *has a unique mild solution* $u \in \mathbb{C}\left([0, T^\star]; \mathfrak{L}^2\left(\Omega; \mathcal{H}\right)\right)$.

Proof. We use the successive approximations defined by

$$\begin{cases} X_0(t; \omega) & = e^{At} X_0(\omega), \\ X_n(t; \omega) & = e^{At} X_0(\omega) + \int_0^t e^{A(t-s)} f(s, X_{n-1}(s; \omega)) ds \\ & \quad + \int_0^t e^{A(t-s)} g(s, X_{n-1}(s; \omega)) dW(s), \end{cases} \tag{7.159}$$

where $0 < t < T, \omega \in \Omega$, but under different assumptions. For any $\beta_1 > 3\left(\overline{M_A}\right)^2 \|X_0\|_{\mathfrak{L}^2}^2$, Lemma 7.6.2 guarantees the existence of $0 < T_1 \leq T$ for which the equation

$$z(t) = \beta_1 + 3\left(\overline{M_A}\right)^2 (T + \zeta_g(T, 2)) \int_0^t K_1(s, z(s)) ds \tag{7.160}$$

has a unique solution $z : [0, T_1) \to [0, \infty)$. We divide the proof into several claims.

<u>Claim 1</u>: For each $n \in \mathbb{N}$, $E \|X_n(t; \cdot)\|_{\mathcal{H}}^2 \leq z(t)$, $\forall 0 \leq t < T_1 \leq T$.
Proof: By induction on n. For $n = 1$, observe that $\forall 0 \leq t < T_1$,

$$\begin{aligned} E \|X_1(t; \cdot)\|_{\mathcal{H}}^2 &\leq 3\left[\left(\overline{M_A}\right)^2 \|X_0\|_{\mathfrak{L}^2}^2 + T\left(\overline{M_A}\right)^2 \int_0^t E \|f(s, X_0(\cdot))\|_{\mathcal{H}}^2 ds \right. \\ & \qquad \left. \left(\overline{M_A}\right)^2 \zeta_g(T, 2) \int_0^t E \|g(s, X_0(\cdot))\|_{\mathscr{B}_0}^2 ds \right] \\ &= \xi_1^\star + \xi_2^\star \int_0^t \left(E \|f(s, X_0(\cdot))\|_{\mathcal{H}}^2 + E \|g(s, X_0(\cdot))\|_{\mathscr{B}_0}^2\right) ds \\ &\leq \xi_1^\star + \xi_2^\star \int_0^t K_1\left(s, E \|X_0(\cdot)\|_{\mathcal{H}}^2\right) ds \\ &\leq \xi_1^\star + \xi_2^\star \int_0^t K_1\left(s, \left(\overline{M_A}\right)^2 \|X_0\|_{\mathfrak{L}^2}^2\right) ds \\ &\leq \xi_1^\star + \xi_2^\star \int_0^t K_1(s, z(s)) ds \\ &= z(t) + (\xi_1^\star - \beta_1) \\ &\leq z(t), \end{aligned} \tag{7.161}$$

where

$$\begin{aligned} \xi_1^\star &= 3\left(\overline{M_A}\right)^2 \|X_0\|_{\mathfrak{L}^2}^2, \\ \xi_2^\star &= 3\left(\overline{M_A}\right)^2 (T + \zeta_g(T, 2)). \end{aligned}$$

Now, assume that $E\left\|X_n(t;\cdot)\right\|_{\mathcal{H}}^2 \le z(t)$, $\forall 0 \le t < T_1$, and observe that

$$E\left\|X_{n+1}(t;\cdot)\right\|_{\mathcal{H}}^2 \le 3\left(\overline{M_A}\right)^2 \|X_0\|_{\mathfrak{L}^2}^2 + 3\left(\overline{M_A}\right)^2 (T + \zeta_g(T,2))$$

$$\times \int_0^t \left(E\left\|f(s,X_n(s;\cdot))\right\|_{\mathcal{H}}^2 + E\left\|g(s,X_n(s;\cdot))\right\|_{\mathcal{B}_0}^2\right)ds$$

$$\le \xi_1^\star + \xi_2^\star \int_0^t K_1\left(s, E\left\|X_n(s;\cdot)\right\|_{\mathcal{H}}^2\right)ds$$

$$\le \xi_1^\star + \xi_2^\star \int_0^t K_1\left(s, z(s)\right)ds$$

$$= z(t).$$

This proves the claim. $\Diamond$

<u>Claim 2</u>: For every $\delta_0 > 0$, $\exists 0 < T_2 \le T_1$ such that $\forall n \in \mathbb{N}$,

$$E\left\|X_n(t;\cdot) - e^{At}X_0(\cdot)\right\|_{\mathcal{H}}^2 \le \delta_0, \ \forall 0 \le t < T_2 \le T_1 \le T. \tag{7.162}$$

Proof: Let $\delta_0 > 0$ be fixed. We argue inductively. For $n = 1$, observe that

$$E\left\|X_1(t;\cdot) - e^{At}X_0(\cdot)\right\|_{\mathcal{H}}^2 \le 2\left[T\left(\overline{M_A}\right)^2 \int_0^t E\left\|f(s,X_0(\cdot))\right\|_{\mathcal{H}}^2 ds\right.$$

$$\left. + \left(\overline{M_A}\right)^2 \zeta_g(T,2) \int_0^t E\left\|g(s,X_0(\cdot))\right\|_{\mathcal{B}_0}^2 ds\right]$$

$$= 2\left(\overline{M_A}\right)^2 (T + \zeta_g(T,2)) \times$$

$$\int_0^t \left(E\left\|f(s,X_0(\cdot))\right\|_{\mathcal{H}}^2 + E\left\|g(s,X_0(\cdot))\right\|_{\mathcal{B}_0}^2\right)ds$$

$$\le \xi_2^\star \int_0^t K_1\left(s, \|X_0\|_{\mathfrak{L}^2}^2\right)ds \tag{7.163}$$

$$\le \xi_2^\star \int_0^t K_1\left(s, z(s)\right)ds.$$

The continuity of z and K_1 (and hence the absolute continuity property of the integral) guarantees the existence of $0 < T_2 \le T_1$ such that

$$\xi_2^\star \int_0^t K_1\left(s, z(s)\right)ds \le \delta_0, \ \forall 0 \le t < T_2 \le T_1 \le T. \tag{7.164}$$

(Why?) As such, (7.162) holds for $n = 1$.

Next, assume that (7.162) holds for n and observe that $\forall 0 \le t < T_2 \le T_1 \le T$,

$$E\left\|X_{n+1}(t;\cdot) - e^{At}X_0(\cdot)\right\|_{\mathcal{H}}^2 \le \xi_2^\star \int_0^t K_1\left(s, E\left\|X_n(s;\cdot)\right\|_{\mathcal{H}}^2\right)ds$$

$$\le \xi_2^\star \int_0^t K_1\left(s, z(s)\right)ds$$

$$\le \delta_0.$$

This proves the Claim. ◊

Claim 3: For every $n, m \in \mathbb{N}$,

$$E \|X_{n+m}(t;\cdot) - X_n(t;\cdot)\|_{\mathcal{H}}^2 \leq \xi_3^\star \int_0^t K_2(s, 4\delta_0)\, ds, \ \forall 0 \leq t < T_2, \qquad (7.165)$$

where

$$\xi_3^\star = 2 \left(\overline{M_A}\right)^2 (T + \zeta_g(T, 2)).$$

Proof: Let $n, m \in \mathbb{N}$. The monotonicity of K_2 implies that $\forall 0 \leq t < T_2 \leq T_1 \leq T$,

$$E \|X_{n+m}(t;\cdot) - X_n(t;\cdot)\|_{\mathcal{H}}^2 \leq$$

$$2 \left[E \left\| \int_0^t e^{A(t-s)} [f(s, X_{n+m-1}(s;\cdot)) - f(s, X_{n-1}(s;\cdot))]\, ds \right\|_{\mathcal{H}}^2 \right.$$

$$\left. + E \left\| \int_0^t e^{A(t-s)} [g(s, X_{n+m-1}(s;\cdot)) - g(s, X_{n-1}(s;\cdot))]\, dW(s) \right\|_{\mathcal{H}}^2 \right] \leq$$

$$\xi_3^\star \int_0^t K_2 \left(s, E \|X_{n+m-1}(s;\cdot) - X_{n-1}(s;\cdot)\|_{\mathcal{H}}^2 \right) ds \leq$$

$$\xi_3^\star \int_0^t K_2 \left(s, 2E \|X_{n+m-1}(s;\cdot) - e^{As}X_0(\cdot)\|_{\mathcal{H}}^2 \right. \qquad (7.166)$$

$$\left. + 2E \|e^{As}X_0(\cdot) - X_{n-1}(s;\cdot)\|_{\mathcal{H}}^2 \right) ds \leq$$

$$\xi_3^\star \int_0^t K_2(s, 4\delta_0)\, ds.$$

This proves the Claim. ◊

Now, define $\gamma_n : [0, T_2] \to (0, \infty)$ and $\theta_{m,n} : [0, T_2] \to (0, \infty)$ by

$$\gamma_1(t) = \xi_3^\star \int_0^t K_2(s, 4\delta_0)\, ds,$$

$$\gamma_n(t) = \xi_3^\star \int_0^t K_2(s, \gamma_{n-1}(s))\, ds, \ n \geq 2, \qquad (7.167)$$

$$\theta_{m,n}(t) = E \|X_{n+m}(t;\cdot) - X_n(t;\cdot)\|_{\mathcal{X}}, \ n, m \in \mathbb{N}. \qquad (7.168)$$

The continuity of K_2 (by **(H7.11)(ii)**) (and hence the absolute continuity of the integral) guarantees the existence of $0 < T_3 \leq T_2$ such that

$$\gamma_1(t) \leq 4\delta_0, \ \forall 0 \leq t < T_3. \qquad (7.169)$$

Claim 4: For every $n \geq 2$,

$$\gamma_n(t) \leq \gamma_{n-1}(t) \leq \ldots \leq \gamma_1(t), \ \forall 0 \leq t < T_3. \qquad (7.170)$$

Proof: By induction on n. For $n = 2$, we use (7.167) and (7.169) to conclude that $\forall 0 \le t < T_3$,

$$\gamma_2(t) = \xi_3^\star \int_0^t K_2(s, \gamma_1(s)) \, ds \le \xi_3^\star \int_0^t K_2(s, 4\delta_0) \, ds = \gamma_1(t).$$

Next, assuming that (7.170) holds for n, it follows immediately that $\forall 0 \le t < T_3$,

$$\gamma_{n+1}(t) = \xi_3^\star \int_0^t K_2(s, \gamma_n(s)) \, ds \le \xi_3^\star \int_0^t K_2(s, \gamma_{n-1}(s)) \, ds = \gamma_n(t).$$

This proves the claim. $\Diamond$

<u>Claim 5:</u> For every $n, m \in \mathbb{N}$,

$$\theta_{m,n}(t) \le \gamma_n(t), \ \forall 0 \le t < T_3. \tag{7.171}$$

Proof: By induction on n. For $n = 1$, note that $\forall m \in \mathbb{N}$ and $0 \le t < T_3$, Claim 3 implies that

$$\theta_{m,1}(t) = E \left\| X_{m+1}(t; \cdot) - X_1(t; \cdot) \right\|_{\mathscr{H}}^2 \le \xi_3^\star \int_0^t K_2(s, 4\delta_0) \, ds = \gamma_1(t).$$

Next, assume that (7.171) holds for a fixed n, uniformly $\forall m \in \mathbb{N}$. Observe that $\forall m \in \mathbb{N}$ and $0 \le t < T_3$,

$$
\begin{aligned}
\theta_{m,(n+1)}(t) &= E \left\| X_{n+1+m}(t; \cdot) - X_{n+1}(t; \cdot) \right\|_{\mathscr{H}}^2 \\
&\le \xi_3^\star \int_0^t K_2 \left(s, E \left\| X_{n+m}(s; \cdot) - X_n(s; \cdot) \right\|_{\mathscr{H}}^2 \right) ds \\
&= \xi_3^\star \int_0^t K_2(s, \theta_{m,n}(s)) \, ds \\
&\le \xi_3^\star \int_0^t K_2(s, \gamma_n(s)) \, ds \\
&= \gamma_{n+1}(t).
\end{aligned}
$$

This proves the claim. $\Diamond$

Observe that Claim 4 implies that

$$\begin{cases} \text{For every } t \in [0, T_3), \ \{\gamma_n(t)\} \text{ is decreasing in } n. \\ \text{For every } n_0 \in \mathbb{N}, \ \{\gamma_{n_0}(t)\} \text{ is increasing in } t. \end{cases} \tag{7.172}$$

<u>Claim 6:</u> $\exists X \in \mathbb{C}\left([0, T_3]; \mathcal{L}^2(\Omega; \mathscr{H})\right)$ such that $\lim_{n \to \infty} \left\| X_n - X \right\|_{\mathbb{C}([0,T_3]; \mathcal{L}^2(\Omega; \mathscr{H}))} = 0$.

Proof: Define $\gamma : [0, T_3] \to \mathbb{R}$ by

$$\gamma(t) = \lim_{n \to \infty} \gamma_n(t) = \inf_{n \in \mathbb{N}} \gamma_n(t). \tag{7.173}$$

The well-definedness of $\gamma(\cdot)$ follows from (7.172). It is easy to see that γ is non-negative and $\gamma(0) = 0$. It turns out that γ is also continuous. Indeed, because $K_2(s, \gamma_n(s)) \leq K_2(s, \gamma_1(s))$, $\forall n \in \mathbb{N}$ and $\forall s \in [0, T_3]$, we can use LDC to obtain the following string of equalities:

$$\gamma(t) = \lim_{n \to \infty} \gamma_{n+1}(t)$$

$$= \lim_{n \to \infty} \xi_3^\star \int_0^t K_2(s, \gamma_n(s))\, ds$$

$$= \xi_3^\star \int_0^t \lim_{n \to \infty} K_2(s, \gamma_n(s))\, ds \qquad (7.174)$$

$$= \xi_3^\star \int_0^t K_2\left(s, \lim_{n \to \infty} \gamma_n(s)\right) ds$$

$$= \xi_3^\star \int_0^t K_2(s, \gamma(s))\, ds.$$

The continuity (in the second variable) of K_2 was used in going from line three to four in (7.174). Because the right-hand side of (7.174) is a continuous function of t, we conclude that γ is continuous. Moreover, γ satisfies

$$\gamma(t) \leq \xi_3^\star \int_0^t K_2(s, \gamma(s))\, ds, \quad \forall 0 \leq t < T_3.$$

Hence, **(H7.12)** implies that

$$\gamma(t) = 0, \quad \forall 0 \leq t < T_3. \qquad (7.175)$$

Further, using Claim 5 and (7.172) yields

$$\|X_{n+m} - X_n\|_{\mathbb{C}\left([0,T_3];\mathcal{L}^2(\Omega;\mathcal{H})\right)} = \sup_{0 \leq t < T_3} \theta_{m,n}(t) \leq \sup_{0 \leq t < T_3} \gamma_n(t) \leq \gamma_n(T_3). \quad (7.176)$$

The right-side of (7.175) goes to zero as $n \to \infty$ by (7.173) and (7.174). Consequently, $\{X_n\}$ is a Cauchy sequence in $\mathbb{C}\left([0, T_3]; \mathcal{L}^2(\Omega; \mathcal{H})\right)$ and hence, convergent to some $X \in \mathbb{C}\left([0, T_3]; \mathcal{L}^2(\Omega; \mathcal{H})\right)$. This proves the claim. $\Diamond$

<u>Claim 7</u>: The function u from Claim 6 is a mild solution of (7.10) on $[0, T_3]$.
Proof: Define $Y : [0, T_3] \times \Omega \longrightarrow \mathcal{H}$ by

$$Y(t; \omega) = e^{At} X_0(\omega) + \int_0^t e^{A(t-s)} f(s, X(s; \omega))\, ds + \int_0^t e^{A(t-s)} g(s, X(s; \omega))\, dW(s).$$

We must argue that the $\lim_{n\to\infty} \|X_n - Y\|_{\mathbb{C}([0,T_3];\mathcal{L}^2(\Omega;\mathcal{H}))} = 0$. To this end, observe that

$$
\begin{aligned}
E \|X_n(t;\cdot) - Y(t;\cdot)\|_{\mathcal{H}}^2 = E &\left\| \int_0^t e^{A(t-s)} \left[f(s, X_{n-1}(s;\cdot)) - f(s, X(s;\cdot)) \right] ds \right\|_{\mathcal{H}}^2 \\
&+ E \left\| \int_0^t e^{A(t-s)} \left[g(s, X_{n-1}(s;\cdot)) - g(s, X(s;\cdot)) \right] dW(s) \right\|_{\mathcal{H}}^2 \\
\leq &\ \xi_3^\star \int_0^t K_2 \left(s, E \|X_{n-1}(s;\cdot) - X(s;\cdot)\|_{\mathcal{H}}^2 \right) ds \\
\leq &\ \xi_3^\star \int_0^t K_2 \left(s, \|X_{n-1} - X\|_{\mathbb{C}([0,T_3];\mathcal{L}^2(\Omega;\mathcal{H}))}^2 \right) ds
\end{aligned}
$$

and so

$$
\|X_n - Y\|_{\mathbb{C}([0,T_3];\mathcal{L}^2(\Omega;\mathcal{H}))} \leq \xi_3^\star \int_0^T K_2 \left(s, \|X_{n-1} - X\|_{\mathbb{C}([0,T_3];\mathcal{L}^2(\Omega;\mathcal{H}))}^2 \right) ds.
$$

Taking the limit as $n \to \infty$ now yields (due to Claim 6, the continuity of K_2, and LDC)

$$
0 \leq \lim_{n\to\infty} \|X_n - Y\|_{\mathbb{C}([0,T_3];\mathcal{L}^2(\Omega;\mathcal{H}))} \leq \xi_3^\star \int_0^T K_2(s, 0) ds = 0,
$$

as needed. This proves the claim. $\Diamond$

<u>Claim 8:</u> The mild solution of (7.10) on $[0, T_3]$ from Claim 7 is unique.
Proof: Suppose that $x, y \in \mathbb{C}\left([0, T_3]; \mathcal{L}^2(\Omega; \mathcal{H})\right)$ are two mild solutions of (7.10). Subtracting their representation formulae and using **(H7.10)(iii)** and **(H7.11)(iii)** yields

$$
E \|x(t;\cdot) - y(t;\cdot)\|_{\mathcal{H}}^2 \leq \xi_3^\star \int_0^t K_2 \left(s, E \|x(s;\cdot) - y(s;\cdot)\|_{\mathcal{H}}^2 \right) ds, \ \forall 0 \leq t < T_3.
$$

Thus, by **(H7.12)**, we conclude that

$$
\sup_{0 \leq t \leq T} E \|x(t;\cdot) - y(t;\cdot)\|_{\mathcal{H}}^2 = 0.
$$

This proves the claim. $\Diamond$

The proof of the theorem is complete. $\qquad\qquad\qquad\qquad\qquad\qquad\Box$

Remarks.
1. While we have consistently defined the notion of a mild solution on a closed interval, the definition remains the same if the endpoint(s) are excluded (as in Claim 7 in the above proof). That said, we do not continue this line of reasoning to define a mild solution on a union of disjoint intervals.
2. If $\{e^{At} : t \geq 0\}$ is contractive, then $\sup \left\{ \|e^{At}\|_{\mathbb{B}(\mathcal{H})} \leq 1 : 0 \leq t \leq T \right\} \leq 1$. In such

case, the calculations in the proof of Thrm. 7.6.3 simplify slightly. (Tell how.)

3. More general versions of the models considered in Section 7.5 can be easily formed under conditions (**H7.10**) through (**H7.12**). The results are of interest because uniqueness is still guaranteed, but it is more difficult to work with the hypotheses. Indeed, consider the following open exercise and pay particular attention to the difficulties you encounter when attempting it.

Open Exercise 7.6.4. Assume that (**H$_A$**) and (**H7.10**) through (**H7.12**) hold.

i.) Establish a continuous dependence result for (7.10) .

ii.) Determine a pth moment estimate for the mild solution of (7.10) under these conditions.

iii.) Formulate and prove a zeroth-order approximation result in the spirit of Prop. 7.4.16.

7.7 Looking Ahead

The form of the forcing term arising in the mathematical modeling of applications can be too complicated to conveniently treat as a special case of the semi-linear term $f(t,X)$. For instance, consider the following two IBVPs arising in the modeling of pharmacokinetics and wave phenomena in which the forcing term describes a build-up over time of toxin in the GI tract (in the case of pharmacokinetics) and an external force (in the case of the wave equation):

Pharmacokinetics:

$$
\begin{cases}
dy(t;\omega) = & \left(-ay(t;\omega) + \int_0^t k(s)D(s,y(s;\omega),z(s;\omega))ds\right) dt \\
& +G(t,y(t;\omega),z(t;\omega))dW(t), \\
dz(t;\omega) = & \left(ay(t;\omega) - bz(t;\omega) + \overline{D}(s,y(s;\omega),z(s;\omega))\right) dt \\
& +\overline{G}(t,y(t;\omega),z(t;\omega))dW(t), \\
y(0;\omega) = & y_0(\omega),\ z(0;\omega) = 0,
\end{cases}
\tag{7.177}
$$

where $0 < t < T$, $\omega \in \Omega$; $k \in \mathbb{L}^2(0,T;(0,\infty))$ is a positive, bounded function; and $D, \overline{D}$ are continuous in their first variables and globally Lipschitz in their second and third variables (uniformly in t).

Coupled Wave Equations:

Consider the IBVP

$$
\begin{cases}
\partial \left(\frac{\partial}{\partial t} z(x,t;\omega) + \alpha z(x,t;\omega) \right) + \left(c^2 \frac{\partial^2}{\partial x^2} z(x,t;\omega) \right) \partial t = \\
\left(\int_0^t a_1(t-s) f_1(s,x,z,w) ds \right) \partial t + \left(\int_0^t \widehat{a_1}(t-s) \widehat{f_1}(s,x,z,w) ds \right) dW(t), \\
\partial \left(\frac{\partial}{\partial t} w(x,t;\omega) + \alpha w(x,t;\omega) \right) + \left(c^2 \frac{\partial^2}{\partial x^2} w(x,t;\omega) \right) \partial t = \\
\left(\int_0^t a_2(t-s) f_2(s,x,z,w) ds \right) \partial t + \left(\int_0^t \widehat{a_2}(t-s) \widehat{f_2}(s,x,z,w) ds \right) dW(t), \\
z(x,0;\omega) = z_0(x;\omega), \ \frac{\partial z}{\partial t}(x,0;\omega) = z_1(x;\omega), \ 0 < x < L, \ \omega \in \Omega, \\
w(x,0;\omega) = w_0(x;\omega), \ \frac{\partial w}{\partial t}(x,0;\omega) = w_1(x;\omega), \ 0 < x < L, \ \omega \in \Omega, \\
\frac{\partial z}{\partial x}(0,t;\omega) = \frac{\partial z}{\partial x}(L,t;\omega) = \frac{\partial w}{\partial x}(0,t;\omega) = \frac{\partial w}{\partial x}(L,t;\omega) = 0, \ 0 \le t \le T, \ \omega \in \Omega,
\end{cases}
\tag{7.178}
$$

where $0 < x < L$, $0 \le t \le T$, $\omega \in \Omega$ and $z = z(x,t;\omega)$, $w = w(x,t;\omega)$.

A key observation to make is that when formulating these IBVPs as abstract stochastic evolution equations, the forcing terms in both of them is operator-like and, in some sense, can be viewed more generally as a mapping between function spaces. Specifically, for (7.178), consider the mapping $\mathfrak{F} : \mathbb{C}\left([0,T]; \mathscr{L}^2\left(\Omega; \mathbb{R}^2\right)\right) \to \mathbb{C}\left([0,T]; \mathscr{L}^2\left(\Omega; \mathbb{R}^2\right)\right)$ defined by

$$
\mathfrak{F}\left(\begin{bmatrix} y \\ z \end{bmatrix}\right)(t;\cdot) = \begin{bmatrix} \int_0^t k(s) D(s, y(s;\cdot), z(s;\cdot)) ds \\ \overline{D}(t, y(t;\cdot), z(t;\cdot)) \end{bmatrix}.
\tag{7.179}
$$

Next, define $\mathfrak{H} : \left(\mathbb{C}\left([0,T]; \mathscr{L}^2\left(\Omega; L^2(0,\pi)\right)\right)\right)^2 \to \left(\mathbb{L}^2\left(0,T; \mathscr{L}^2\left(\Omega; L^2(0,\pi)\right)\right)\right)^2$ by

$$
\mathfrak{H}\left(\begin{bmatrix} z \\ w \end{bmatrix}\right)(t,\cdot) = \begin{bmatrix} \int_0^t a_1(t-s) f_1(s,\cdot,z(s,\cdot), w(s,\cdot)) ds \\ \int_0^t a_2(t-s) f_2(s,\cdot,z(s,\cdot), w(s,\cdot)) ds \end{bmatrix}.
\tag{7.180}
$$

Exercise 7.7.1. Under what conditions on a_i and f_i can you conclude that $\mathfrak{H}$ is globally Lipschitz?

Loosely speaking, we are considering a modified version of (7.10) in which the semi-linear forcing term $f(t, X(t;\cdot))$ is replaced by a more general form $\mathfrak{F}(X)(t)$. Can the theory developed for (7.10) be adapted to study an abstract evolution equation of the form

$$
\begin{cases}
dX(t;\omega) = (AX(t;\omega) + \mathfrak{F}(X)(t;\omega)) dt + \mathfrak{G}(X)(t;\omega) dW(t), \\
X(0;\omega) = X_0(\omega), \ \omega \in \Omega
\end{cases}
\tag{7.181}
$$

where $0 < t < T$, $\omega \in \Omega$, and if so, how?

7.8 Guidance for Selected Exercises

7.8.1 Level 1: A Nudge in a Right Direction

7.1.5. This is a system of M ODEs. (Now what?)

7.1.6. Think about the classical heat equation.

7.3.1 Evaluate the expression $\frac{f(\frac{1}{n})-f(0)}{\frac{1}{n}-0}$ and then compute its limit as $n \to \infty$.

7.3.2. The rate of growth of the graphs of all cross-sections in the t-direction is globally bounded above over the entire space $\mathscr{D}$.

7.3.3. First think about how you would argue this for real-valued functions $f : \mathbb{R} \to \mathbb{R}$. Then, generalize.

7.3.4. Consider $\int_y^x |f(z)|\,dz$.

7.3.5. i.) Must there exist positive constants M_1, M_2 for which

$$\int_{v(y)}^{v(x)} |f(t,z)|\,dz \leq M_1\,|v(x)-v(y)| \leq M_2\,|x-y|,$$

$\forall x,y \in (0,\infty)$ and $0 \leq t \leq T$? (Now what?)

ii.) Is this any different from (i)? Explain.

7.3.6. It is natural to begin by requiring f to be globally Lipschitz in its second and third variables (uniformly in t). Under this assumption, estimate

$$\left| f\left(t,x,\int_0^x g(z)dz\right) - f\left(t,y,\int_0^y g(z)dz\right) \right|.$$

7.3.7. Estimate $\|\sum_{i=1}^n a_i f_i(x) - \sum_{i=1}^n a_i f_i(y)\|_{\mathscr{X}}$, for $x,y \in \mathscr{D}$, and $\{a_i | i = 1,\ldots,n\} \subset \mathbb{R}$.

7.3.8. Local version: Can you simply require the condition to hold uniformly in t?

7.3.9. Can such a function exhibit cusp-like behavior at any point?

7.3.10. Do constant functions belong to $\mathbb{C}([0,T];\mathbb{R})$? How about $\mathbb{L}^1(0,T;\mathbb{R})$? Conversely, consider the function $f : [0,T] \times \mathbb{R} \to \mathbb{R}$ defined by $f(t,x) = x^2$. Does it satisfy either of the two forms of the condition?

7.3.11. Proceed as in Exer. 7.3.10.

7.3.12. Estimate the quantity $\|(\mathbf{A}x+\mathbf{B}) - (\mathbf{A}y+\mathbf{B})\|_{\mathbb{R}^N}$.

7.4.1. (i) Using **(H7.1)** and **(H7.2)** with the strong continuity of the semigroup, we conclude that both integrands are $\mathscr{F}_t$-measurable mappings with finite p^{th} moments.

(ii) Observe that

$$E\,\|\Phi(X_n)(t) - \Phi(X)(t)\|_{\mathscr{H}}^2 \leq$$

$$4\left[\left\|\int_0^t e^{A(t-s)}\left(f(s,X_n(s;\omega)) - f(s,X(s;\omega))\right)ds\right\|_{\mathscr{H}}^2\right.$$

$$\left.+ \left\|\int_0^t e^{A(t-s)}\left(g(s,X_n(s;\omega)) - g(s,X(s;\omega))\right)dW(s)\right\|_{\mathscr{H}}^2\right] =$$

$$I_1 + I_2.$$

Now, estimate both I_1 and I_2 using LDC. (How?)

7.4.3. This is similar to Exer. 5.7.1.

7.4.4. Note that $\|k\|_C < \infty$. (So what?)

7.4.6. i.) The only change to the expression occurs in the forcing term.

    ii.) Use norm and integral properties with the global Lipschitz condition on f.

    iii.) Compute $\|\Phi^n(x)(t;\omega) - \Phi^n(y)(t;\omega)\|_{\mathscr{X}}$, $\forall n \in \mathbb{N}$. Does an explicit formula emerge?

7.4.9. We can choose $\varepsilon_1, \varepsilon_2, \varepsilon_3$, and ε_4 arbitrarily small in order to get the corresponding solutions close in the $\mathfrak{L}^2$-sense.

7.5.11. i.) Formulate abstractly in $\mathbb{C}([0, 2\pi]; \mathbb{R})$. Argue that f is globally Lipschitz on this space (uniformly in t).

    ii.) Is f is globally Lipschitz on this space (uniformly in t)?

    iii.) Identify the forcing term abstractly and prove that it is globally Lipschitz.

    iv.) Use the variation of parameters formula and proceed as before, taking into account this particular forcing term.

7.5.13. i.) and ii.) Subtract the variation of parameters formulae. Use the continuity of the integral to compute the limit.

7.5.14. Argue that $\forall \mathbf{w}, \mathbf{z} \in \mathscr{B}_{\mathscr{H}}(0; R)$,

$$\|F(\mathbf{w}) - F(\mathbf{z})\|_{\mathscr{H}} \leq \beta_2 \|w_1 - z_1\|_{\mathbb{L}^2} + \|w_1^2\|_{\mathbb{L}^2} \|w_2 - z_2\|_{\mathbb{L}^2}$$
$$+ \|z_2\|_{\mathbb{L}^2} \|w_1 + z_1\|_{\mathbb{L}^2} \|w_2 - z_2\|_{\mathbb{L}^2}.$$

(Now what?)

7.5.15. Argue that F is locally Lipschitz using standard estimates.

7.6.1. Consider f on the the interval $(0, \varepsilon)$, for arbitrarily small ε and consider appropriate secant lines.

7.6.2. i.) Use properties of absolute value, including Minkowski's inequality.

    ii.) Use the fact that $\log(\cdot)$ is increasing and convex.

7.7.1. Set up the left-hand side of the inequality in the definition of *globally Lipschitz* for (i) and *sublinear* for (ii) for such a mapping.

7.8.2   Level 2: An Additional Thrust in a Right Direction

7.1.5. Use $\mathbf{A}$ with diagonal entries $a_1, \ldots, a_M$ and define $\mathbf{F} : [0, T] \times \mathbb{R}^M \to \mathbb{R}^M$ by

$$\mathbf{F}\left(t, \begin{bmatrix} x_1 \\ \vdots \\ x_M \end{bmatrix}\right) = \begin{bmatrix} \sum_{j=1}^M \omega_{1j}(t) g_j(x_j(t; \omega)) \\ \vdots \\ \sum_{j=1}^M \omega_{Mj}(t) g_j(x_j(t; \omega)) \end{bmatrix}.$$

(Now what?)

7.1.6. Try $\mathscr{H} = \mathbb{L}^2(\Omega) \times \mathbb{L}^2(\Omega)$. (Why?) What is the inner product on this space?

7.3.1. The difference quotient simplifies to $nf\left(\frac{1}{n}\right) = 3n^{3/8}$.

7.3.2. For simplicity, visualize the surface in $\mathbb{R}^3$ and interpret this as meaning that we cannot form a sequence of tangent planes to the surface that become asymptotically closer to a plane perpendicular to the xy-plane. Intuitively, there are bounds on the slopes of each of the three cross-sections of the set of all tangent planes, and so the plane whose slope is comprised of these bounds would serve as a boundary plane controlling the steepness of the surface throughout the domain. Of course, if the function is time dependent, these bounds can change.

7.3.3. Not in general, but if the derivative is bounded, then yes.

7.3.4. Is $f(x) = \frac{1}{\sqrt{x}}$ in $\mathbb{L}^1(0, 1; \mathbb{R})$? If so, this suggests an additional condition to impose on f in order to ensure that g is globally Lipschitz.

7.3.5. i.) Neither of these constants M_1, M_2 must exist, in general. The function f must be bounded and v would need to satisfy what condition?

ii.) We can handle this as in (i). Alternatively, is it sufficient to assume that

$$|f(t, k(z))| \le M |k(z)|,$$

where $k(\cdot)$ is Lipschitz?

7.3.6. Now use Exer. 7.3.4. (How?) Identify a possible Lipschitz constant.

7.3.7. Yes. Use the triangle inequality and choose $M = \max\left\{ |a_i M_{f_i}| : i = 1, \ldots, n \right\}$ as the Lipschitz constant.

7.3.8. For the second part, the space is $\mathscr{X}_1 \times \ldots \times \mathscr{X}_n$. Require the conditiont to hold uniformly in t.

7.3.9. Consider $f(x) = x^{2/3}$. Show that the condition fails on $(-\varepsilon, \varepsilon)$, $\forall \varepsilon > 0$. (So what?)

7.3.10. The definition implies (5.59) and (5.60). The converse implications do not hold.

7.3.12. This mapping is globally Lipschitz with $M_f = \|\mathbf{A}\|_{\mathbb{M}^N}$.

7.4.1. ii.) Use the Lipschitz condition in both integrals, together with (7.27) and (7.28), to show that both I_1 and I_2 are dominated by expressions of the form $C\|X_n - X\|_{\mathbb{C}}$. (Tell how.)

7.5.11. i.) Is f continuously differentiable? Does the IC belong to dom(A)?

ii.) If the solution exists on $[0, T]$, $\forall T > 0$, then yes.

iii.) It is easier to check if f is continuously differentiable.

7.5.13. ii.) Observe that

$$\|\mathbf{U}_\varepsilon(t; \omega) - \mathbf{U}(t; \omega)\|_{\mathbb{R}^M} \le$$

$$\int_0^t \left\|e^{\mathbf{A}(t-s)}\right\|_{\mathbb{M}^M(\mathbb{R})} \times \left\|\sum_{j=1}^M \left[(\omega_{1j})_\varepsilon(t) g_j\left(x_j^\varepsilon(t; \omega)\right) - \omega_{1j}(t) g_j\left(x_j(t; \omega)\right)\right]\right.$$

$$\left. + \ldots + \sum_{j=1}^M \left[(\omega_{Mj})_\varepsilon(t) g_j\left(x_j^\varepsilon(t; \omega)\right) - \omega_{Mj}(t) g_j\left(x_j(t; \omega)\right)\right]\right\|_{\mathbb{R}^M} ds,$$

where $\mathbf{U}_\varepsilon(t; \omega) = \begin{bmatrix} x_1^\varepsilon(t; \omega) \\ \vdots \\ x_M^\varepsilon(t; \omega) \end{bmatrix}$. (Now what?)

7.5.14. i.) Use the fact that $\mathbf{w}, \mathbf{z} \in \mathscr{B}_{\mathscr{H}}(0; R)$ to conclude.

ii.) The solution must be at least continuous. (So what?)

7.6.1. Can either hold uniformly $\forall t \in (0, \infty)$?

7.7.1. i.) Try assuming that $a_i \in \mathbb{L}^p$, for appropriate values of p, and f_i globally Lipschitz.

ii.) Weaken the Lipschitz growth assumption on f_i to sublinear growth. Does this work? Does the value of p change from (i)?

Chapter 8

Functional Stochastic Evolution Equations

Overview

Incorporating more realistic assumptions into the formulation of a mathematical model leads to increased complexity in the resulting IBVP, which in turn must be accounted for in its formulation as an abstract SEE. The new forcing terms arising in these IBVPs can often be viewed as mappings between function spaces. This chapter focuses on the study of SEEs equipped with such forcing terms.

8.1 Motivation by Models

We begin by revisiting two IBVPs that are now equipped with more complex forcing terms.

Model II.5 Pharmacokinetics with an Accumulation Effect

Consider the IVP

$$\begin{cases} dy(t;\omega) = & \left(-ay(t;\omega) + \int_0^t k(s)D(s,y(s;\omega),z(s;\omega))ds\right)dt \\ & +G(t,y(t;\omega),z(t;\omega))dW(t), \\ dz(t;\omega) = & \left(ay(t;\omega) - bz(t;\omega) + \overline{D}(s,y(s;\omega),z(s;\omega))\right)dt \\ & +\overline{G}(t,y(t;\omega),z(t;\omega))dW(t), \\ y(0;\omega) = & y_0(\omega),\ z(0;\omega) = 0, \end{cases} \tag{8.1}$$

where $0 < t < T$, $\omega \in \Omega$; $k \in \mathbb{L}^2(0,T;(0,\infty))$ is bounded; $W(t)$ is a one-dimensional Wiener process; and $D, \overline{D}, G,$ and $\overline{G}$ are globally Lipschitz in their second and third variables. Also, $\mathfrak{F}_1, \mathfrak{G}_1 : \mathbb{C}\left([0,T];\mathcal{L}^2\left(\Omega;\mathbb{R}^2\right)\right) \to \mathbb{C}\left([0,T];\mathcal{L}^2\left(\Omega;\mathbb{R}^2\right)\right)$ defined by

$$\mathfrak{F}_1\left(\begin{bmatrix} y \\ z \end{bmatrix}\right)(t;\cdot) = \begin{bmatrix} \int_0^t k(s)D(s,y(s;\cdot),z(s;\cdot))ds \\ \overline{D}(t,y(t;\cdot),z(t;\cdot)) \end{bmatrix} \tag{8.2}$$

$$\mathfrak{G}_1\left(\begin{bmatrix} y \\ z \end{bmatrix}\right)(t;\cdot) = \begin{bmatrix} G(t,y(t;\cdot),z(t;\cdot)) \\ \overline{G}(t,y(t;\cdot),z(t;\cdot)) \end{bmatrix} \tag{8.3}$$

are globally Lipschitz. (Show this!) These mappings effectively describe the forcing terms of this system when (8.1) is reformulated as an abstract stochastic evolution equation.

Model V.5 Two-Dimensional Diffusion Equation Revisited
Consider the following more general two-dimensional version of IBVP (7.4):

$$\begin{cases} \partial z(x,y,t;\omega) = \left(\frac{\partial^2}{\partial x^2}z(x,y,t;\omega) + \frac{\partial^2}{\partial y^2}z(x,y,t;\omega)\right)\partial t + j(t)dW(t) \\ \quad + \left(\int_0^t a(t,s)g\left(s,z(x,y,s;\omega), \int_0^s h(s,\tau,z(x,y,\tau;\omega))d\tau\right)ds\right)\partial t, \\ z(x,y,0;\omega) = z_0(x,y;\omega),\ 0<x<a, 0<y<b, \omega\in\Omega, \\ \frac{\partial z}{\partial y}(x,0,t;\omega) = 0 = \frac{\partial z}{\partial y}(x,b,t;\omega),\ 0<x<a, 0\le t\le T, \omega\in\Omega, \\ \frac{\partial z}{\partial x}(0,y,t;\omega) = 0 = \frac{\partial z}{\partial x}(a,y,t;\omega),\ 0<y<b, 0\le t\le T, \omega\in\Omega, \end{cases} \tag{8.4}$$

where $0<x<a, 0<y<b, 0\le t\le T, \omega\in\Omega$; $z : [0,a]\times[0,b]\times[0,T]\times\Omega\to\mathbb{R}, j : [0,T]\to\mathbb{R}, g : [0,T]\times\mathbb{R}\times\mathbb{R}\to\mathbb{R}, h : [0,T]\times[0,T]\times\mathbb{R}\to\mathbb{R}$ are mappings; and $W(t)$ is a one-dimensional Wiener process. We can again view the forcing term in a somewhat more general manner as the mapping

$$\mathfrak{F}_2 : \mathbb{C}\left([0,T];\mathcal{L}^2\left(\Omega;\mathbb{L}^2\left((0,a)\times(0,b)\right)\right)\right) \to \mathbb{C}\left([0,T];\mathcal{L}^2\left(\Omega;\mathbb{L}^2\left((0,a)\times(0,b)\right)\right)\right)$$

defined by

$$\mathfrak{F}_2(z)(\cdot,\cdots,t;\omega) = \int_0^t a(t,s)g\left(s,z(\cdot,\cdots,s;\omega), \int_0^s h(s,\tau,z(\cdot,\cdots,\tau;\omega))d\tau\right)ds. \tag{8.5}$$

Loosely speaking, for a given time t, the forcing term in (8.4) ought to be captured by $\mathfrak{F}_2(z)(\cdot,\cdots,t;\omega)$ when viewing the IBVP abstractly. The big difference here is that $\mathfrak{F}_2$ is itself a mapping between function spaces (similar to the solution map Φ in (7.29)), while in the previous chapter the domain and range spaces of the forcing terms were, in effect, one level lower in that they included terms that involved only $[0,T]$ and $\mathscr{H}$.

We can further improve the above diffusion model by accounting for advection. Doing so results in the alternate, somewhat more complex, version of IBVP (8.4):

$$\begin{cases} \partial z(x,y,t;\omega) = \left(k\left(\frac{\partial^2 z}{\partial x^2} + \frac{\partial^2 z}{\partial y^2}\right) + \alpha_1\frac{\partial z}{\partial x} + \alpha_2\frac{\partial z}{\partial y}\right)\partial t + \mathfrak{F}_2(z)(\cdot,\cdots,t;\omega) \\ \quad + \left(\int_0^t b(s)h(s,x,y,z(x,y,s;\omega))ds\right)dW(t), \\ z(x,y,0;\omega) = z_0(x,y;\omega),\ 0<x<a, 0<y<b, \omega\in\Omega, \\ \frac{\partial z}{\partial y}(x,0,t;\omega) = 0 = \frac{\partial z}{\partial y}(x,b,t;\omega),\ 0<x<a, 0\le t\le T, \omega\in\Omega, \\ \frac{\partial z}{\partial x}(0,y,t;\omega) = 0 = \frac{\partial z}{\partial x}(a,y,t;\omega),\ 0<y<b, 0\le t\le T, \omega\in\Omega, \end{cases} \tag{8.6}$$

where $0 < x < a, 0 < y < b, \omega \in \Omega, 0 \leq t \leq T; z = z(x,y,t;\omega), h : [0,T] \times [0,a] \times [0,b] \times \mathbb{R} \rightarrow \mathbb{R}$ are mappings; $k > 0$; and $W(t)$ is a one-dimensional Wiener process. The following symbolic substitutions are often used to simplify the notation:

$$\triangle z = \frac{\partial^2 z}{\partial x^2} + \frac{\partial^2 z}{\partial y^2}, \tag{8.7}$$

$$\overrightarrow{\alpha} \cdot \nabla z = \alpha_1 \frac{\partial z}{\partial x} + \alpha_2 \frac{\partial z}{\partial y}. \tag{8.8}$$

Substituting (8.7) and (8.8) into (8.6) yields the following equivalent, more succinct form of IBVP (8.6):

$$\begin{cases} \partial z(x,y,t;\omega) = & \left(k\triangle z + \overrightarrow{\alpha} \cdot \nabla z\right) \partial t + \mathfrak{F}_2(z)(\cdot,\cdot\cdot,t;\omega) \\ & + \left(\int_0^t b(s)h(s,x,y,z(x,y,s;\omega))\,ds\right) dW(t), \\ z(x,y,0;\omega) = & z_0(x,y;\omega), 0 < x < a, 0 < y < b, \omega \in \Omega, \\ \frac{\partial z}{\partial y}(x,0,t;\omega) = & 0 = \frac{\partial z}{\partial y}(x,b,t;\omega), 0 < x < a, 0 < t < T, \omega \in \Omega, \\ \frac{\partial z}{\partial x}(0,y,t;\omega) = & 0 = \frac{\partial z}{\partial x}(a,y,t;\omega), 0 < y < b, 0 < t < T, \omega \in \Omega, \end{cases} \tag{8.9}$$

where $0 < x < a, 0 < y < b, \omega \in \Omega, 0 < t < T, z = z(x,y,t;\omega)$, and $W(t)$ is a one-dimensional Wiener process. The additional wrinkle we encounter when considering (8.9) is that the operator $k\triangle z$, which is known to generate a compact C_0-semigroup on $\mathbb{L}^2((0,a) \times (0,b))$ (see **[118, 329]**), is now perturbed by the first-order term $\overrightarrow{\alpha} \cdot \nabla z$. The question we need to address is whether or not the operator $k\triangle z + \overrightarrow{\alpha} \cdot \nabla z$ still generates a C_0-semigroup on $\mathbb{L}^2((0,a) \times (0,b))$. If not, an alternative approach would be to determine if the mapping $\overline{\mathfrak{F}_2}(z) = \mathfrak{F}_2(z) + \overrightarrow{\alpha} \cdot \nabla z$ can be defined on convenient function spaces in a manner for which $\overline{\mathfrak{F}_2}$ satisfies appropriate growth conditions, and we can define a solution map to which a fixed point will be a mild solution of (8.9) we seek. We shall revisit this problem later in the chapter.

The following model is an extension of (8.9) that arises when studying pollution **[218]**, chemotactic phenomena **[189, 190, 254, 312, 406]**, population ecology **[119, 177, 370, 380]**, and tumor growth **[171]**.

Model XIII.1 A Reactive-Convective-Diffusive Model

A consequence of not accounting for diffusion in a pollution model is that a pollution "cloud" would simply travel in a single direction at a constant speed and never disperse. A similar interpretation can be made if instead of a pollutant we consider a migratory herd or a traveling pathogen. We now consider a more general model that provides a more realistic description of such phenomena.

Introducing new elements into a local environment (such as a new species, pollution, etc., the descriptions of which are subject to noise) can affect the evolution of populations present within the environment. We begin by considering a one-dimensional diffusive model involving two populations into which a convective term is introduced to account for the directed flow or spread of a new element affecting the system. Suppose that the region to which the population is confined is represented as the interval $[0,a]$ and denote the population of the ith entity at position x in $[0,a]$ at time $t \geq 0$ by $P_i = P_i(x,t;\omega)$ $(i = 1,2)$. A description of this phenomenon is given by the following IBVP:

$$
\begin{cases}
\partial P_1 = \left(\alpha_{P_1}\dfrac{\partial^2 P_1}{\partial x^2} - \dfrac{\partial}{\partial x}(\beta_{P_1}P_1) + \sum_{i=1}^{N_1}\int_0^t k_1(t,s)g_{1,i}(s,P_1,P_2)\,ds\right)\partial t \\
\quad + h_1(t)dW(t),\ 0 < x < a,\ 0 \leq t \leq T,\ \omega \in \Omega, \\
\partial P_2 = \left(\alpha_{P_2}\dfrac{\partial^2 P_2}{\partial x^2} - \dfrac{\partial}{\partial x}(\beta_{P_2}P_2) + \sum_{j=1}^{N_2}\int_0^t k_2(t,s)g_{2,j}(s,P_1,P_2)\,ds\right)\partial t \\
\quad + h_2(t)dW(t),\ 0 < x < a,\ 0 \leq t \leq T,\ \omega \in \Omega, \\
P_1(x,0;\omega) = P_1^0(x,0;\omega),\ P_2(x,0;\omega) = P_2^0(x,0;\omega),\ 0 < x < a,\ \omega \in \Omega, \\
\dfrac{\partial P_1}{\partial x}(0,t;\omega) = \dfrac{\partial P_1}{\partial x}(a,t;\omega) = \dfrac{\partial P_2}{\partial x}(0,t;\omega) = \dfrac{\partial P_2}{\partial x}(a,t;\omega) = 0,\ , 0 \leq t \leq T,\ \omega \in \Omega,
\end{cases}
\tag{8.10}
$$

where $\alpha_{P_i}, \beta_{P_i}$ $(i = 1,2)$ are real constants; $k_i : [0,T] \times [0,T] \to \mathbb{R}$ $(i = 1,2)$ and $h_i : [0,T] \to \mathbb{R}$ $(i = 1,2)$ are sufficiently smooth functions; $g_{1,i} : [0,T] \times \mathbb{R} \times \mathbb{R} \to \mathbb{R}$ $(i = 1,\ldots,N_1)$ and $g_{2,j} : [0,T] \times \mathbb{R} \times \mathbb{R} \to \mathbb{R}$ $(j = 1,\ldots,N_2)$ are globally Lipschitz in their second and third variables; and $W(t)$ is a one-dimensional Wiener process. We can reformulate (8.10) as the system

$$
\begin{cases}
\partial \begin{bmatrix} P_1 \\ P_2 \end{bmatrix} = \begin{bmatrix} \alpha_{P_1}\dfrac{\partial^2}{\partial x^2} - \beta_{P_1}\dfrac{\partial}{\partial x} & 0 \\ 0 & \alpha_{P_2}\dfrac{\partial^2}{\partial x^2} - \beta_{P_2}\dfrac{\partial}{\partial x} \end{bmatrix} \begin{bmatrix} P_1 \\ P_2 \end{bmatrix} \partial t + \begin{bmatrix} h_1(t) \\ h_2(t) \end{bmatrix} dW(t) \\
\quad + \begin{bmatrix} \sum_{i=1}^{N_1}\int_0^t k_1(t,s)g_{1,i}(s,P_1,P_2)\,ds \\ \sum_{j=1}^{N_2}\int_0^t k_2(t,s)g_{2,j}(s,P_1,P_2)\,ds \end{bmatrix} \partial t,\ 0 < x < a,\ 0 \leq t \leq T,\ \omega \in \Omega, \\
\begin{bmatrix} P_1 \\ P_2 \end{bmatrix}(x,0;\omega) = \begin{bmatrix} P_1^0(x,0;\omega) \\ P_2^0(x,0;\omega) \end{bmatrix},\ 0 < x < a,\ \omega \in \Omega, \\
\dfrac{\partial}{\partial x}\begin{bmatrix} P_1 \\ P_2 \end{bmatrix}(0,t;\omega) = \dfrac{\partial}{\partial x}\begin{bmatrix} P_1 \\ P_2 \end{bmatrix}(a,t;\omega) = 0,\ 0 \leq t \leq T,\ \omega \in \Omega.
\end{cases}
\tag{8.11}
$$

As with the other IBVPs, the forcing term can be described by the mapping

$$
\mathfrak{F}_3 : \mathbb{C}\left([0,T];\mathscr{L}^2\left(\Omega;\left(\mathbb{L}^2(0,a)\right)^2\right)\right) \to \mathbb{C}\left([0,T];\mathscr{L}^2\left(\Omega;\left(\mathbb{L}^2(0,a)\right)^2\right)\right)
$$

defined by

$$
\mathfrak{F}\left(\begin{bmatrix} P_1 \\ P_2 \end{bmatrix}\right)(t;\omega) = \begin{bmatrix} \sum_{i=1}^{N_1}\int_0^t k_1(t,s)g_{1,i}(s,P_1(\cdot,s;\omega),P_2(\cdot,s;\omega))\,ds \\ \sum_{j=1}^{N_2}\int_0^t k_2(t,s)g_{2,j}(s,P_1(\cdot,s;\omega),P_2(\cdot,s;\omega))\,ds \end{bmatrix}.
\tag{8.12}
$$

Exercise 8.1.1. Formulate an extension of this model to the case in which the region Ω to which the population is restricted is a bounded subset of $\mathbb{R}^3$ with smooth boundary $\partial\Omega$.

Remarks.
1. We encounter the same perturbation issue of the operator $\triangle$ as in (8.9).

2. The functions $g_{1,i}$, $g_{2,j}$ often have logistic growth.

3. Interesting questions regarding the dynamics of (8.11) (such as when a population will approach extinction or when the collection of all populations settles toward an equilibrium) are of particular interest and are explored in various references (see [381]).

4. A related model of genetically engineered microbes is studied in [258]. This has particular utility in the agricultural community. With this context in mind, IBVP (8.11) can be improved by further subdividing the population into regions with which the microbes interact, such as top soil, deep soil, surface water, plants, etc. Each of these has a different mechanism of dispersal that should be accounted for in the model.

5. A related IBVP arises in the modeling of semiconductor technology in [145], although the underlying physical setting in this case requires that the IBVP be formulated in a function space consisting of more regular functions.

6. Other interpretations of IBVP (8.11) include an epidemiological model of the elementary spread of a viral infection through a community [134], or as a description of the damage incurred by a plague of locusts on grasslands [347].

Common Theme: The challenge with which we are now presented is handling a forcing term whose description involves the use of operators. We refer to such a mapping (from one function space into another) generically as a *functional*. Specifically, when reformulating the above IBVPs as abstract stochastic evolution equations, how *exactly* do we define the mappings $\mathfrak{F}(X)$ and $\mathfrak{G}(X)$ below?

$$\begin{cases} dX(t;\omega) = (AX(t;\omega) + \mathfrak{F}(X)(t;\omega))\,dt + \mathfrak{G}(X)(t;\omega)\,dW(t), \\ X(0;\omega) = X_0(\omega), \omega \in \Omega. \end{cases} \tag{8.13}$$

where $0 < t < T$, $\omega \in \Omega$. Ideally, we need a term that fills the above gap in a manner for which the fixed-point approach can be effectively used to study (8.13).

8.2 Functionals

Up to now, the forcing terms we have encountered in the IBVPs were, at worst, of the form $f : [0, T] \times \mathscr{H} \to \mathscr{H}$. The existence proof for the corresponding evolution equation involved the use of a mapping between function spaces when defining the solution map. We now wish to incorporate such a mapping as a term into the evolution equation itself and investigate its properties. The purpose of this section is to expose you to different types of mappings arising in the mathematical modeling of various phenomena and to subsequently study their growth properties in preparation for the study of related IBVPs. The basic properties (e.g., Lipschitz growth conditions, sublinear growth conditions, etc.) are the same as those used in Chapter 7, but the concomitant calculations are more intricate. At the moment, we shall illustrate these computations outside the context of concrete models to help you to develop a working knowledge of them.

Example 8.2.1. Consider $\mathfrak{F}_1 : \mathbb{C}\left([0,T]; \mathcal{L}^2\left(\Omega; \mathbb{R}^2\right)\right) \to \mathbb{C}\left([0,T]; \mathcal{L}^2\left(\Omega; \mathbb{R}^2\right)\right)$ defined by (8.2). Assume the following:

(H8.1) $D : [0,T] \times \mathbb{R} \times \mathbb{R} \to \mathbb{R}$ is continuous in the first variable and $\exists M_D, \widehat{M_D} > 0$ such that

$$|D(t,y_1,z_1) - D(t,y_2,z_2)| \le M_D [|y_1 - y_2| + |z_1 - z_2|], \tag{8.14}$$

$\forall y_i, z_i \in \mathbb{R}\, (i = 1, 2)$, uniformly in $t \in [0,T]$, and

$$|D(t,y,z)| \le \widehat{M_D} [1 + |y| + |z|], \tag{8.15}$$

$\forall y, z \in \mathbb{R}$, uniformly in $t \in [0,T]$.

(H8.2) $\overline{D} : [0,T] \times \mathbb{R} \times \mathbb{R} \to \mathbb{R}$ is continuous in the first variable and $\exists M_{\overline{D}}, \widehat{M_{\overline{D}}} > 0$ such that

$$\left|\overline{D}(t,y_1,z_1) - \overline{D}(t,y_2,z_2)\right| \le M_{\overline{D}} [|y_1 - y_2| + |z_1 - z_2|], \tag{8.16}$$

$\forall y_i, z_i \in \mathbb{R}\, (i = 1, 2)$, uniformly in $t \in [0,T]$, and

$$\left|\overline{D}(t,y,z)\right| \le \widehat{M_{\overline{D}}} [1 + |y| + |z|], \tag{8.17}$$

$\forall y, z \in \mathbb{R}$, uniformly in $t \in [0,T]$.

(H8.3) $k : [0,T] \to [0,\infty)$ is continuous with $M_k = \sup\{k(s) : 0 \le s \le T\}$.

Claim 1: There exist $M_{\mathfrak{F}_1}, \overline{M_{\mathfrak{F}_1}} > 0$ such that

$$\left\| \mathfrak{F}_1\left(\begin{bmatrix} y_1 \\ z_1 \end{bmatrix}\right) - \mathfrak{F}_1\left(\begin{bmatrix} y_2 \\ z_2 \end{bmatrix}\right) \right\|_{\mathbb{C}} \le M_{\mathfrak{F}_1} \left\| \begin{bmatrix} y_1 \\ z_1 \end{bmatrix} - \begin{bmatrix} y_2 \\ z_2 \end{bmatrix} \right\|_{\mathbb{C}}, \tag{8.18}$$

$$\left\| \mathfrak{F}_1\left(\begin{bmatrix} y \\ z \end{bmatrix}\right) \right\|_{\mathbb{C}} \le \overline{M_{\mathfrak{F}_1}} \left[1 + \left\| \begin{bmatrix} y \\ z \end{bmatrix} \right\|_{\mathbb{C}} \right], \tag{8.19}$$

$\forall \begin{bmatrix} y_i \\ z_i \end{bmatrix}, \begin{bmatrix} y \\ z \end{bmatrix} \in \mathbb{C}, (i = 1, 2).$

Proof: Let $\begin{bmatrix} y_1 \\ z_1 \end{bmatrix}, \begin{bmatrix} y_2 \\ z_2 \end{bmatrix} \in \mathbb{C}$. Using the $\mathbb{R}^2$-norm given by (1.46) and estimating yields

$$\left\| \begin{bmatrix} \int_0^t k(s)\,(D(s,y_1(s;\cdot),z_1(s;\cdot)) - D(s,y_2(s;\cdot),z_2(s;\cdot)))\,ds \\ \overline{D}(t,y_1(t;\cdot),z_1(t;\cdot)) - \overline{D}(t,y_2(t;\cdot),z_2(t;\cdot)) \end{bmatrix} \right\|_{\mathbb{R}^2}^2 \le$$

$$\max\left\{1, \|k\|_{\mathbb{L}^1(0,T;\mathbb{R})}\right\} \times \sup_{0 \le t \le T}\{|D(s,y_1(s;\cdot),z_1(s;\cdot)) - D(s,y_2(s;\cdot),z_2(s;\cdot))|\tag{8.20}$$

$$\left|\overline{D}(t,y_1(t;\cdot),z_1(t;\cdot)) - \overline{D}(t,y_2(t;\cdot),z_2(t;\cdot))\right|\}.$$

(Why?) The continuity of the semigroup guarantees that the supremum on the right-hand side of (8.20) is actually attained at some $t_0 \in [0,T]$. Thus, subsequently using **(H8.1)** and **(H8.2)** yields the estimate

$$\left\| \begin{bmatrix} \int_0^t k(s) \left(D(s,y_1(s;\cdot),z_1(s;\cdot)) - D(s,y_2(s;\cdot),z_2(s;\cdot)) \right) ds \\ \overline{D}(t,y_1(t;\cdot),z_1(t;\cdot)) - \overline{D}(t,y_2(t;\cdot),z_2(t;\cdot)) \end{bmatrix} \right\|_{\mathbb{R}^2}^2 \leq$$

$$\max\left\{1, \|k\|_{\mathbb{L}^1(0,T;\mathbb{R}}\right\} \times \max\{M_D,M_{\overline{D}}\} \left(|y_1(t_0;\cdot) - y_2(t_0;\cdot)| + |z_1(t_0;\cdot) - z_2(t_0;\cdot)| \right).$$

Furthermore,

$$|y_1(t_0;\cdot) - y_2(t_0;\cdot)| + |z_1(t_0;\cdot) - z_2(t_0;\cdot)| \leq$$
$$2\max\left\{|y_1(t_0;\cdot) - y_2(t_0;\cdot)|, |z_1(t_0;\cdot) - z_2(t_0;\cdot)|\right\} \leq \qquad (8.21)$$
$$2\left\| \begin{bmatrix} y_1(t_0;\cdot) \\ z_1(t_0;\cdot) \end{bmatrix} - \begin{bmatrix} y_2(t_0;\cdot) \\ z_2(t_0;\cdot) \end{bmatrix} \right\|_{\mathbb{R}^2}.$$

Using (8.20) and (8.21) then yields

$$\left\| \mathfrak{F}_1\left(\begin{bmatrix} y_1 \\ z_1 \end{bmatrix} \right) - \mathfrak{F}_1\left(\begin{bmatrix} y_2 \\ z_2 \end{bmatrix} \right) \right\|_{\mathbb{C}} =$$

$$\left(\sup_{0 \leq t \leq T} E \left\| \begin{bmatrix} \int_0^t k(s) \left(D(s,y_1(s;\cdot),z_1(s;\cdot)) - D(s,y_2(s;\cdot),z_2(s;\cdot)) \right) ds \\ \overline{D}(t,y_1(t;\cdot),z_1(t;\cdot)) - \overline{D}(t,y_2(t;\cdot),z_2(t;\cdot)) \end{bmatrix} \right\|_{\mathbb{R}^2}^2 \right)^{\frac{1}{2}} \leq$$

$$\left(\mathscr{M} \sup_{0 \leq t_0 \leq T} E \left\| \begin{bmatrix} y_1(t_0;\cdot) \\ z_1(t_0;\cdot) \end{bmatrix} - \begin{bmatrix} y_2(t_0;\cdot) \\ z_2(t_0;\cdot) \end{bmatrix} \right\|_{\mathbb{R}^2}^2 \right)^{\frac{1}{2}} =$$

$$\sqrt{\mathscr{M}} \left\| \begin{bmatrix} y_1 \\ z_1 \end{bmatrix} - \begin{bmatrix} y_2 \\ z_2 \end{bmatrix} \right\|_{\mathbb{C}}, (8.22)$$

where $\mathscr{M} = 2\max\left\{1, \|k\|_{\mathbb{L}^1(0,T;\mathbb{R}}\right\} \times \max\{M_D,M_{\overline{D}}\}$. Choosing

$$M_{\mathfrak{F}_1} = \sqrt{2\max\left\{1, \|k\|_{\mathbb{L}^1(0,T;\mathbb{R}}\right\} \times \max\{M_D,M_{\overline{D}}\}}$$

proves that (8.18) holds.

A similar argument can be used to verify (8.19). (Try it!) This proves the claim. □

Exercise 8.2.1.
i.) Verify (8.19).
ii.) Can you use the $\mathbb{R}^2$-norm defined by (1.45) more readily? Explain.

Exercise 8.2.2. Suppose that the space $\mathbb{C}\left([0,T];\mathcal{L}^2\left(\Omega;\mathbb{R}^2\right)\right)$ in the definition of the functional $\mathfrak{F}_1$ given in (8.2) is replaced by the space $\mathbb{C}\left([0,T];\mathcal{L}^r\left(\Omega;\mathbb{R}^2\right)\right)$, where $r > 2$. How do the computations, and in particular the growth constants, change?

Exercise 8.2.3. Formulate hypotheses in the spirit of those used in Claim 1 under which the functional $\mathfrak{G}_1 : \mathbb{C}\left([0,T]; \mathcal{L}^2\left(\Omega; \mathbb{R}^2\right)\right) \to \mathbb{C}\left([0,T]; \mathcal{L}^2\left(\Omega; \mathbb{R}^2\right)\right)$ defined by (8.3) satisfies (8.18) and (8.19). Prove this result.

Next, we can view the functional defined in (8.2) as one with an enlarged range space, namely as a mapping

$$\mathfrak{F}_1 : \mathbb{C}\left([0,T]; \mathcal{L}^2\left(\Omega; \mathbb{R}^2\right)\right) \to \mathrm{L}^2\left(0,T; \mathcal{L}^2\left(\Omega; \mathbb{R}^2\right)\right).$$

This change affects how the initial norm on the right-hand side of the computations in (8.20) is applied. Precisely, under **(H8.1)** through **(H8.3)**, we have

$$\left\| \mathfrak{F}_1\left(\begin{bmatrix} y_1 \\ z_1 \end{bmatrix}\right) - \mathfrak{F}_1\left(\begin{bmatrix} y_2 \\ z_2 \end{bmatrix}\right) \right\|_{\mathrm{L}^2} =$$

$$\left(\int_0^T \left\| \mathfrak{F}_1\left(\begin{bmatrix} y_1 \\ z_1 \end{bmatrix}\right)(t;\cdot) - \mathfrak{F}_1\left(\begin{bmatrix} y_2 \\ z_2 \end{bmatrix}\right)(t;\cdot) \right\|_{\mathcal{L}^2(\Omega;\mathbb{R}^2)}^2 dt \right)^{\frac{1}{2}} = \quad (8.23)$$

$$\left(\int_0^T E\left\| \mathfrak{F}_1\left(\begin{bmatrix} y_1 \\ z_1 \end{bmatrix}\right)(t;\cdot) - \mathfrak{F}_1\left(\begin{bmatrix} y_2 \\ z_2 \end{bmatrix}\right)(t;\cdot) \right\|_{\mathbb{R}^2}^2 dt \right)^{\frac{1}{2}} =$$

$$\left(\int_0^T E\left\| \begin{bmatrix} \int_0^t k(s)\left(D(s,y_1(s;\cdot),z_1(s;\cdot)) - D(s,y_2(s;\cdot),z_2(s;\cdot))\right)ds \\ \overline{D}(t,y_1(t;\cdot),z_1(t;\cdot)) - \overline{D}(t,y_2(t;\cdot),z_2(t;\cdot)) \end{bmatrix} \right\|_{\mathbb{R}^2}^2 dt \right)^{\frac{1}{2}}.$$

The computation would now proceed in a standard way, except that instead of applying the sup over $(0,T)$, we would integrate over this interval.

Exercise 8.2.4.
i.) Continue the computation in (8.23). What is the new Lipschitz constant?
ii.) Establish (8.19) for this variant of $\mathfrak{F}_1$.
iii.) Redo (i) and (ii), but now replace $\mathcal{L}^2\left(\Omega; \mathbb{R}^2\right)$ by $\mathcal{L}^r\left(\Omega; \mathbb{R}^2\right)$, where $r > 2$, in the space $\mathbb{C}\left([0,T]; \mathcal{L}^2\left(\Omega; \mathbb{R}^2\right)\right)$. How does this change the computations?

Example 8.2.2. Let $n \in \mathbb{N}$, $0 < t_1 < t_2 < \ldots < t_n < T$ be fixed times, and $\mathscr{D} = \{\mathbf{z} = (u,v) \in \mathbb{R}^2 \,|\, u^2 + v^2 \leq 1\}$. We make the following assumptions:
(H8.4) $\alpha \in \mathrm{L}^2\left(0,T; \mathbb{R}\right)$,
(H8.5) $f : [0,T] \times \mathbb{R} \to \mathbb{R}$ is continuous in the first variable and globally Lipschitz in the second variable (uniformly in t) with Lipschitz constant M_f,
(H8.6) $\beta_i \in \mathbb{C}\left(\mathscr{D}; \mathbb{R}\right)$, $i = 1, \ldots, n$.

Define the functional $\mathfrak{H}_1 : \mathbb{C}\left([0,T]; \mathcal{L}^2\left(\Omega; \mathrm{L}^2\left(\mathscr{D}\right)\right)\right) \to \mathcal{L}^2\left(\Omega; \mathrm{L}^2\left(\mathscr{D}\right)\right)$ by

$$\mathfrak{H}_1(x)(\mathbf{z};\cdot) = \sum_{i=1}^n \beta_i(\mathbf{z})x(t_i)(\mathbf{z};\cdot) + \int_0^T \alpha(s)f\left(s,x(s)(\mathbf{z};\cdot)\right)ds. \quad (8.24)$$

<u>Claim 2</u>: *If* **(H8.4)** *through* **(H8.6)** *hold, then* $\mathfrak{H}_1$ *is globally Lipschitz. For such a mapping, this means that* $\exists M_{\mathfrak{H}_1} > 0$ *such that* $\forall x, y \in \mathbb{C}\left([0,T]; \mathcal{L}^2\left(\Omega; L^2(\mathcal{D})\right)\right)$,

$$\|\mathfrak{H}_1(x) - \mathfrak{H}_1(y)\|_{\mathcal{L}^2(\Omega; L^2(\mathcal{D}))} \leq M_{\mathfrak{H}_1} \|x - y\|_{\mathbb{C}([0,T]; \mathcal{L}^2(\Omega; L^2(\mathcal{D})))}. \tag{8.25}$$

Proof: By definition,

$$\|h\|_{\mathcal{L}^2(\Omega; L^2(\mathcal{D}))} = \left(E\|h(\cdot)\|^2_{L^2(\mathcal{D})}\right)^{\frac{1}{2}} = \left(E\int_{\mathcal{D}} |h(\mathbf{z}; \cdot)|^2 \, d\mathbf{z}\right)^{\frac{1}{2}}, \tag{8.26}$$

$$\|g\|_{\mathbb{C}([0,T]; \mathcal{L}^2(\Omega; L^2(\mathcal{D})))} = \left(\sup_{0 \leq t \leq T} E\|g(t)(\cdot; \cdots)\|^2_{L^2(\mathcal{D})}\right)^{\frac{1}{2}}$$

$$= \left(\sup_{0 \leq t \leq T} E\int_{\mathcal{D}} |g(t)(\mathbf{z}; \cdots)|^2 \, d\mathbf{z}\right)^{\frac{1}{2}}. \tag{8.27}$$

Observe that $\forall x, y \in \mathbb{C}\left([0,T]; \mathcal{L}^2\left(\Omega; L^2(\mathcal{D})\right)\right)$,

$$\|\mathfrak{H}_1(x) - \mathfrak{H}_1(y)\|_{\mathcal{L}^2(\Omega; L^2(\mathcal{D}))} =$$

$$\left(E\left[\int_{\mathcal{D}} \left| \sum_{i=1}^{n} \beta_i(\mathbf{z}) \left(x(t_i)(\mathbf{z}; \cdot) - y(t_i)(\mathbf{z}; \cdot)\right)\right.\right.\right. \tag{8.28}$$

$$\left.\left.\left. + \int_0^T \alpha(s)\left(f(s, x(s)(\mathbf{z}; \cdot)) - f(s, y(s)(\mathbf{z}; \cdot))\right) ds\right|^2 \, d\mathbf{z}\right]\right)^{\frac{1}{2}}.$$

We estimate each of the two main terms on the right-hand side of (8.28) below. Successive applications of Minkowski's inequality on the right-hand side of (8.28) yields

$$\|\mathfrak{H}_1(x) - \mathfrak{H}_1(y)\|_{\mathcal{L}^2(\Omega; L^2(\mathcal{D}))} \leq$$

$$2^n \int_{\mathcal{D}} \sum_{i=1}^{n} |\beta_i(\mathbf{z})\left(x(t_i)(\mathbf{z}; \cdot) - y(t_i)(\mathbf{z}; \cdot)\right)|^2 \, d\mathbf{z} \tag{8.29}$$

$$+2 \int_{\mathcal{D}} \left(\int_0^T |\alpha(s)\left(f(s, x(s)(\mathbf{z}; \cdot)) - f(s, y(s)(\mathbf{z}; \cdot))\right)| ds\right)^2 \, d\mathbf{z} =$$

$$I_1 + I_2.$$

In order to estimate I_1, note that **(H8.6)** implies

$$\max_{1 \leq i \leq n} \|\beta_i\|_{\mathbb{C}(\mathcal{D}; \mathbb{R})} = M_\beta < \infty. \tag{8.30}$$

Using (8.30), then taking the expectation, followed by the supremum over $(0, T)$, and then taking the square root yields

$$I_1 \leq 2^{\frac{n}{2}} M_\beta^{\frac{1}{2}} \left(\sup_{0 \leq t \leq T} E\left[\sum_{i=1}^{n} \int_{\mathcal{D}} |x(t_i)(\mathbf{z}; \cdot) - y(t_i)(\mathbf{z}; \cdot)|^2 \, d\mathbf{z}\right]\right)^{\frac{1}{2}}. \tag{8.31}$$

Because

$$|x(t_i)(\mathbf{z};\cdot) - y(t_i)(\mathbf{z};\cdot)| \leq \sup_{0 \leq t \leq T} |x(t)(\mathbf{z};\cdot) - y(t)(\mathbf{z};\cdot)|, \qquad (8.32)$$

it follows that

$$\sum_{i=1}^{n} \int_{\mathscr{D}} |x(t_i)(\mathbf{z};\cdot) - y(t_i)(\mathbf{z};\cdot)|^2 \, d\mathbf{z} \leq n \sup_{0 \leq t \leq T} \int_{\mathscr{D}} |x(t)(\mathbf{z};\cdot) - y(t)(\mathbf{z};\cdot)| \, d\mathbf{z}. \quad (8.33)$$

Using (8.33) in (8.31) then yields the estimate

$$I_1 \leq \left(n 2^n M_\beta\right)^{\frac{1}{2}} \|x - y\|_{C\left([0,T];\mathscr{L}^2\left(\Omega;L^2(\mathscr{D})\right)\right)}. \qquad (8.34)$$

Next, we estimate I_2. Applying Holder's inequality and then rearranging terms yields

$$I_2 \leq \sqrt{2} \|\alpha\|_{L^2(0,T;\mathbb{R})} \left(E\left[\int_{\mathscr{D}} \int_0^T |f(s,x(s)(\mathbf{z};\cdot)) - f(s,y(s)(\mathbf{z};\cdot))|^2 \, ds d\mathbf{z}\right]\right)^{\frac{1}{2}}.$$
$$(8.35)$$

Applying **(H8.5)** and Fubini's Theorem in (8.35) subsequently yields

$$\begin{aligned}
I_2 &\leq \sqrt{2} \|\alpha\|_{L^2(0,T;\mathbb{R})} M_f \left(\int_0^T \int_{\mathscr{D}} E |x(s)(\mathbf{z};\cdot) - y(s)(\mathbf{z};\cdot)|^2 \, d\mathbf{z} ds\right)^{\frac{1}{2}} \\
&= \sqrt{2} \|\alpha\|_{L^2(0,T;\mathbb{R})} M_f \left(\int_0^T E \|x(s)(\cdot) - y(s)(\cdot)\|_{L^2(\mathscr{D})}^2\right)^{\frac{1}{2}} \qquad (8.36) \\
&\leq \sqrt{2} \|\alpha\|_{L^2(0,T;\mathbb{R})} M_f \left(\int_0^T \sup_{0 \leq s \leq T} E \|x(s)(\cdot) - y(s)(\cdot)\|_{L^2(\mathscr{D})}^2\right)^{\frac{1}{2}} \\
&= \sqrt{2} \|\alpha\|_{L^2(0,T;\mathbb{R})} M_f \left(\int_0^T \|x - y\|_{C\left([0,T];\mathscr{L}^2\left(\Omega;L^2(\mathscr{D})\right)\right)}^2\right)^{\frac{1}{2}} \\
&= \sqrt{2T} \|\alpha\|_{L^2(0,T;\mathbb{R})} M_f \|x - y\|_{C\left([0,T];\mathscr{L}^2\left(\Omega;L^2(\mathscr{D})\right)\right)}.
\end{aligned}$$

The claim then follows by identifying the Lipschitz constant as

$$M_{\mathfrak{H}_1} = \left(n 2^n M_\beta\right)^{\frac{1}{2}} + \sqrt{2T} \|\alpha\|_{L^2(0,T;\mathbb{R})} M_f.$$

This completes the proof of Claim 2. □

Example 8.2.3. Assume that
(H8.7) $a : [0,T] \times [0,T] \to (0,\infty)$ is such that $\sup_{0 \leq t \leq T} \|a(t,\cdot)\|_{L^2(0,T)}^2 < \infty$.
(H8.8) Let $\mathscr{U} = \left\{(t,s) \in \mathbb{R}^2 \,|\, 0 \leq s \leq t \leq T\right\}$ and assume that $k : \mathscr{U} \times \mathbb{R} \to \mathbb{R}$ is continuous in all variables and $\exists M_k > 0$ such that

$$|k(t,s,x_1) - k(t,s,x_2)| \leq M_k |x_1 - x_2|, \forall (t,s) \in \mathscr{U}, x_1, x_2 \in \mathbb{R}. \qquad (8.37)$$

(H8.9) $g : [0,T] \times \mathbb{R} \times \mathbb{R} \to \mathbb{R}$ is continuous in all three variables and satisfies
    (a) $g(\cdot,0,0) \in \mathbb{L}^2(0,T)$, and
    (b) there exists $M_g > 0$ such that

$$|g(t,x_1,y_1) - g(t,x_2,y_2)| \le M_g [|x_1 - x_2| + |y_1 - y_2|], \tag{8.38}$$

$\forall (x_1,y_1),(x_2,y_2) \in \mathbb{R} \times \mathbb{R}$, uniformly in $t \in [0,T]$.

Let $\mathscr{D}$ be a bounded set in $\mathbb{R}^N$. Define the functional $\mathfrak{H}_2 : \mathbb{C}\left([0,T];\mathscr{L}^2\left(\Omega;\mathbb{L}^2(\mathscr{D})\right)\right) \to \mathbb{L}^2 = \mathbb{L}^2\left(0,T;\mathscr{L}^2\left(\Omega;\mathbb{L}^2(\mathscr{D})\right)\right)$ by

$$\mathfrak{H}_2(x)(t)(\mathbf{z};\cdot) = \int_0^t a(t,s)g\left(s,x(s)(\mathbf{z};\cdot),\int_0^s k(s,\tau,x(\tau)(\mathbf{z};\cdot))d\tau\right) ds. \tag{8.39}$$

<u>Claim 3:</u> *$\mathfrak{H}_2$ is globally Lipschitz.*
Proof: It is equivalent (and more convenient) to produce a Lipschitz constant $M_{\mathfrak{H}_4}$ for which

$$\|\mathfrak{H}_2(x) - \mathfrak{H}_2(y)\|_{\mathbb{L}^2}^2 \le M_{\mathfrak{H}_2} \|x - y\|_{\mathbb{C}}^2, \forall x,y \in \mathbb{C}\left([0,T];\mathscr{L}^2\left(\Omega;\mathbb{L}^2(\mathscr{D})\right)\right).$$

Using Holder's inequality and properties of the integral yields

$$\|\mathfrak{H}_2(x) - \mathfrak{H}_2(y)\|_{\mathbb{L}^2}^2 =$$

$$\int_0^T \|\mathfrak{H}_2(x)(t) - \mathfrak{H}_2(y)(t)\|_{\mathbb{L}^2(\Omega;\mathbb{L}^2(\mathscr{D}))}^2 =$$

$$\int_0^T E |\mathfrak{H}_2(x)(t)(\cdot) - \mathfrak{H}_2(y)(t)(\cdot)|_{\mathbb{L}^2(\mathscr{D})}^2 dt =$$

$$\int_0^T \int_{\mathscr{D}} E |\mathfrak{H}_2(x)(t)(\mathbf{z};\cdot) - \mathfrak{H}_2(y)(t)(\mathbf{z};\cdot)|^2 d\mathbf{z}dt \le$$

$$\int_0^T \int_{\mathscr{D}} E \left[\left(\int_0^t |a(t,s)|^2 ds \right) \left(\int_0^t \left(g\left(s,x(s)(\mathbf{z};\cdot),\int_0^s k(s,\tau,x(\tau)(\mathbf{z};\cdot))d\tau\right) \right.\right.\right.$$
$$\left.\left.\left. -g\left(s,y(s)(\mathbf{z};\cdot),\int_0^s k(s,\tau,y(\tau)(\mathbf{z};\cdot))d\tau\right) \right)^2 ds \right] d\mathbf{z}dt \le$$

$$\int_0^T \sup_{0 \le t \le T} \|a(t,\cdot)\|_{\mathbb{L}^2(0,T)}^2 \int_{\mathscr{D}} E \left[\left(\int_0^t \left(g\left(s,x(s)(\mathbf{z};\cdot),\int_0^s k(s,\tau,x(\tau)(\mathbf{z};\cdot))d\tau\right) \right.\right.\right.$$
$$\left.\left.\left. -g\left(s,y(s)(\mathbf{z};\cdot),\int_0^s k(s,\tau,y(\tau)(\mathbf{z};\cdot))d\tau\right) \right)^2 ds \right] d\mathbf{z}dt.$$

Now, apply (8.38), and then (8.37). Then what? How do you proceed? $\square$

Exercise 8.2.5. Complete the proof of Claim 3.

Exercise 8.2.6. If **(H8.7)** and **(H8.8)** are replaced by appropriate sublinear growth conditions (as in (8.15)), must there exist positive constants $M_{\mathfrak{H}_2}^1, M_{\mathfrak{H}_2}^2$ such that

$$\|\mathfrak{H}_2(x)\|_{\mathbb{L}^2} \leq \overline{M_{\mathfrak{H}_2}} \left[1 + \|x\|_{\mathbb{C}}\right], \; \forall x \in \mathbb{C}\left([0,T]; \mathcal{L}^2\left(\Omega; \mathbb{L}^2(\mathscr{D})\right)\right)? \qquad (8.40)$$

If so, prove it; otherwise, explain why.

Exercise 8.2.7. Restrict the range of $\mathfrak{H}_2$ to be the subspace $\mathbb{C}\left([0,T]; \mathcal{L}^2\left(\Omega; \mathbb{L}^2(\mathscr{D})\right)\right)$ of $\mathbb{L}^2\left(0,T; \mathcal{L}^2\left(\Omega; \mathbb{L}^2(\mathscr{D})\right)\right)$. Do the conclusions of Claim 3 and Exer. 8.2.6 still hold? If so, how do the growth constants $M_{\mathfrak{H}_2}, \overline{M_{\mathfrak{H}_2}}$ change?

Example 8.2.4. Assume $k_1, k_2 \in \mathbb{L}^2\left(0,T; [0,\infty)\right)$ and $f_i : [0,T] \times \mathscr{H} \times \mathscr{H} \to \mathscr{H}$ $(i = 1,2)$ is continuous in the first variable and globally Lipschitz on $\mathscr{H} \times \mathscr{H}$ uniformly in $t \in [0,T]$. Define $\mathfrak{H}_3 : \left(\mathbb{C}\left([0,T]; \mathcal{L}^2\left(\Omega; \mathscr{H}\right)\right)\right)^2 \to \left(\mathbb{C}\left([0,T]; \mathcal{L}^2\left(\Omega; \mathscr{H}\right)\right)\right)^2$ by

$$\mathfrak{H}_3 \begin{pmatrix} u \\ v \end{pmatrix}(t; \omega) = \begin{pmatrix} \int_0^t k_1(s) f_1(s, u(s; \omega), v(s; \omega)) ds \\ \int_0^t k_2(s) f_2(s, u(s; \omega), v(s; \omega)) ds \end{pmatrix}. \qquad (8.41)$$

Exercise 8.2.8. Prove that $\mathfrak{H}_3$ is globally Lipschitz.

8.3   The Cauchy Problem

A comparison between the examples explored in Section 8.2 and the right-hand sides of the IVPs and IBVPs described in Section 8.1 suggests that the mappings $\mathfrak{F}(X)$ and $\mathfrak{G}(X)$ in (8.13) can be viewed as functionals defined on appropriate spaces. As motivation for our discussion of the abstract theory, consider the following exercise.

Exercise 8.3.1. Show that each of the IVPs and IBVPs in Section 8.1 can be reformulated as an abstract stochastic evolution equation of the form (8.13) in an appropriate separable Hilbert space $\mathscr{H}$. In each case, clearly identify the space $\mathscr{H}$, the operator $A : \text{dom}(A) \subset \mathscr{X} \to \mathscr{X}$, the functionals $\mathfrak{F} : \mathscr{Y}_1 \to \mathscr{Y}_2$ and $\mathfrak{G} : \mathscr{Y}_3 \to \mathscr{Y}_4$ (including the spaces $\mathscr{Y}_i$ $(i = 1,2,3,4)$), and the initial condition X_0.

8.3.1   Problem Formulation

Assume **(S.A.1)**. We begin by considering the following abstract functional SEE in a separable Hilbert space $\mathscr{H}$:

$$\begin{cases} dX(t; \omega) = (AX(t; \omega) + \mathfrak{F}(X)(t; \omega)) dt + g(t)) d\mathbf{W}(t), \\ X(0; \omega) = X_0(\omega), \omega \in \Omega, \end{cases} \qquad (8.42)$$

where $0 < t < T$, $\omega \in \Omega$, $\mathfrak{F} : \mathbb{C}\left([0,T]; \mathcal{L}^2\left(\Omega; \mathscr{H}\right)\right) \to \mathbb{C}\left([0,T]; \mathcal{L}^2\left(\Omega; \mathscr{H}\right)\right)$ is a given functional and $g : [0,T] \to \mathscr{B}_0\left(\mathbb{R}^m; \mathscr{H}\right)$ is a given mapping. We assume $(\mathbf{H_A})$,

(H5.1) through **(H5.3)**, and **(H5.5)**, as well as

(H8.10) There exist $M_{\mathfrak{F}}, \overline{M_{\mathfrak{F}}} > 0$ such that

$$\|\mathfrak{F}(X) - \mathfrak{F}(Y)\|_C \leq M_{\mathfrak{F}} \|X - Y\|_C,$$
$$\|\mathfrak{F}(X)\|_C \leq \overline{M_{\mathfrak{F}}} [1 + \|X\|_C],$$

$\forall X, Y \in \mathbb{C}\left([0,T]; \mathfrak{L}^2(\Omega; \mathscr{H})\right)$.

(H8.11) $g : [0,T] \to \mathscr{B}_0(\mathbb{R}^m; \mathscr{H})$ belongs to $\mathbb{C}\left([0,T]; \mathscr{B}_0(\mathbb{R}^m; \mathscr{H})\right)$ and is $\mathscr{F}_t$-measurable.

Remark. The $\mathscr{F}_t$-measurability of $\mathscr{F}$ is built into the space $\mathbb{C}\left([0,T]; \mathfrak{L}^2(\Omega; \mathscr{H})\right)$. (Why?)

Although (8.42) is distinct from (7.26), the difference is not so great as to require us to formulate a drastically different notion of a mild solution for (8.42). Indeed, it is reasonable to slightly tweak Def. 7.4.1 to obtain the following definition.

Definition 8.3.1. A stochastic process $X : [0,T] \times \Omega \to \mathscr{H}$ is a *mild solution* of (8.13) on $[0,T]$ if

i.) $X \in \mathbb{C}\left([0,T]; \mathfrak{L}^2(\Omega; \mathscr{H})\right)$,

ii.) $X(t; \omega) = e^{At} X_0(\omega) + \int_0^t e^{A(t-s)} \mathfrak{F}(X)(s; \omega) ds + \int_0^t e^{A(t-s)} g(s) d\mathbf{W}(s), \forall 0 \leq t \leq T_0$, a.s. $[\mathscr{P}]$.

The remainder of this chapter is devoted to forming an extension of the basic results from Chapter 7 to the present setting. Consider (8.13), where $\mathfrak{F}$ is a prescribed functional. The nature of how $\mathfrak{F}$ is defined suggests that the fixed-point theorems introduced in Section 1.11 might be useful.

8.3.2 Existence Results

We begin with the following theorem.

Theorem 8.3.2. *Assume* **(H$_A$)**, **(H5.1)** *through* **(H5.3)**, **(H5.5)**, **(H8.10)**, *and* **(H8.11)**. *If* $T M_A M_{\mathfrak{F}} < 1$, *then* (8.42) *has a unique mild solution on* $[0,T]$.

Proof. Define the solution map $\Phi : \mathbb{C}\left([0,T]; \mathfrak{L}^2(\Omega; \mathscr{H})\right) \to \mathbb{C}\left([0,T]; \mathfrak{L}^2(\Omega; \mathscr{H})\right)$ by

$$(\Phi X)(t; \cdot) = e^{At} X_0(\cdot) + \int_0^t e^{A(t-s)} \mathfrak{F}(X)(s; \cdot) ds + \int_0^t e^{A(t-s)} g(s) d\mathbf{W}(s). \quad (8.43)$$

In order to show that Φ is well-defined, we need to show that $(\Phi X)(t; \cdot) \in \mathfrak{L}^2(\Omega; \mathscr{H}), \forall t \in [0,T]$. First, the $\mathscr{F}_t$-measurability is assured because of the measurability assumptions in **(H5.3)** and **(H8.11)** and the fact that $(\mathfrak{F}X)(s; \cdot) \in \mathfrak{L}^2(\Omega; \mathscr{H})$ by the very way $\mathfrak{F}$ is defined. (Why?)

Next, we verify the $\mathcal{L}^2$-continuity of Φ. Let $\{X_n\}$ belong to $\mathbb{C}\left([0,T];\mathcal{L}^2\left(\Omega;\mathcal{H}\right)\right)$ such that $X_n \longrightarrow X$ in $\mathbb{C}\left([0,T];\mathcal{L}^2\left(\Omega;\mathcal{H}\right)\right)$ as $n \to \infty$. We must show that

$$\sup_{0 \le t \le T} E\left\|\Phi\left(X_n\right)\left(t;\cdot\right) - \Phi\left(X\right)\left(t;\cdot\right)\right\|_{\mathcal{H}}^2 \longrightarrow 0 \text{ as } n \to \infty. \tag{8.44}$$

To this end, observe that

$$E\left\|\Phi\left(X_n\right)\left(t;\cdot\right) - \Phi\left(X\right)\left(t;\cdot\right)\right\|_{\mathcal{H}}^2 \le E\left\|\int_0^t e^{A(t-s)}\left(\mathfrak{F}(X_n)(s;\cdot) - \mathfrak{F}(X)(s;\cdot)\right)ds\right\|_{\mathcal{H}}^2$$

$$\le TM_A^2 \int_0^t E\left\|\mathfrak{F}(X_n)(s;\cdot) - \mathfrak{F}(X)(s;\cdot)\right\|_{\mathcal{H}}^2 ds$$

$$\le TM_A^2 \int_0^t \sup_{0 \le s \le T} E\left\|\mathfrak{F}(X_n)(s;\cdot) - \mathfrak{F}(X)(s;\cdot)\right\|_{\mathcal{H}}^2 ds$$

$$\le T^2 M_A^2 \left\|\mathfrak{F}(X_n) - \mathfrak{F}(X)\right\|_{\mathbb{C}}^2 \tag{8.45}$$

$$\le T^2 M_A^2 M_{\mathfrak{F}}^2 \left\|X_n - X\right\|_{\mathbb{C}}^2.$$

Taking the supremum over $[0,T]$ and then applying the square root on both sides of (8.45) yields

$$\left\|\Phi(X_n) - \Phi(X)\right\|_{\mathbb{C}} \le TM_A M_{\mathfrak{F}} \left\|X_n - X\right\|_{\mathbb{C}}. \tag{8.46}$$

Let $\varepsilon > 0$. There exists $N \in \mathbb{N}$ such that

$$n \ge N \implies \left\|X_n - X\right\|_{\mathbb{C}} < \frac{\varepsilon}{TM_A M_{\mathfrak{F}} + 1}. \tag{8.47}$$

As such,

$$n \ge N \implies \left\|\Phi(X_n) - \Phi(X)\right\|_{\mathbb{C}} \le TM_A M_{\mathfrak{F}} \left\|X_n - X\right\|_{\mathbb{C}} < \varepsilon. \tag{8.48}$$

So, Φ is $\mathcal{L}^2$-continuous.

Finally, we must argue that Φ is a contraction. Let $X,Y \in \mathbb{C}\left([0,T];\mathcal{L}^2\left(\Omega;\mathcal{H}\right)\right)$. Arguing as in (8.45) yields

$$E\left\|\Phi\left(X\right)\left(t;\cdot\right) - \Phi\left(Y\right)\left(t;\cdot\right)\right\|_{\mathcal{H}}^2 \le E\left\|\int_0^t e^{A(t-s)}\left(\mathfrak{F}(X)(s;\cdot) - \mathfrak{F}(Y)(s;\cdot)\right)ds\right\|_{\mathcal{H}}^2$$

$$\le T^2 M_A^2 M_{\mathfrak{F}}^2 \left\|X - Y\right\|_{\mathbb{C}}^2. \tag{8.49}$$

Taking the supremum over $[0,T]$, applying the square root on both sides of (8.49), and using the assumption that $TM_A M_{\mathfrak{F}} < 1$, yields

$$\left\|\Phi(X) - \Phi(Y)\right\|_{\mathbb{C}} \le TM_A M_{\mathfrak{F}} \left\|X - Y\right\|_{\mathbb{C}} < \left\|X - Y\right\|_{\mathbb{C}}. \tag{8.50}$$

Hence, Φ has a unique fixed point that coincides with the mild solution we seek. $\square$

Remark. Note that in the proof of the above theorem, we cannot simply iterate the solution map as we did in the proof of Thrm. 7.4.2(i) because the growth condition imposed on $\mathfrak{F}$ is expressed as an inequality involving a norm taken on the entire space

$\mathbb{C}\left([0,T];\mathscr{L}^2\left(\Omega;\mathscr{H}\right)\right)$ rather than simply in $\mathscr{H}$. (Try doing so to see what happens.)

If $g(t) \equiv 0$ and X_0 is a nonrandom initial condition, then (8.42) would be a deterministic IVP. When studying such an IVP in Volume 1, we imposed precisely the <u>same</u> growth restriction on the data in order to obtain a unique mild solution. This is interesting because incorporating noise into that IVP did not require us to impose any additional restriction on the data to establish the analogous result in the stochastic setting. However, this is no longer the case if the noise term is of a more general form. For instance, consider the following IVP:

$$\begin{cases} dX(t;\omega) = (AX(t;\omega) + \mathfrak{F}(X)(t;\omega))dt + g(t,X(t;\omega)))d\mathbf{W}(t), \\ X(0;\omega) = X_0(\omega), \omega \in \Omega, \end{cases} \tag{8.51}$$

where $0 < t < T$, $\omega \in \Omega$ in a separable Hilbert space $\mathscr{H}$. Assume $(\mathbf{H_A})$, $(\mathbf{H5.1})$ through $(\mathbf{H5.3})$, $(\mathbf{H5.5})$, $(\mathbf{H8.10})$, and $(\mathbf{H7.2})$ (instead of $(\mathbf{H8.11})$).

Exercise 8.3.2.
i.) Define the solution map for (8.51).
ii.) Formulate and prove a result analogous to Thrm. 8.3.2 for (8.51). How does the data restriction change?
iii.) Recover Thrm. 7.4.2 as a corollary of the result established in (ii).
iv.) Formulate a more general result than (ii) by replacing $\mathbb{C}\left([0,T];\mathscr{L}^2\left(\Omega;\mathscr{H}\right)\right)$ by $\mathbb{C}\left([0,T];\mathscr{L}^r\left(\Omega;\mathscr{H}\right)\right)$ (where $r > 2$). Pay particular attention to if, and how, the data restriction changes.

Consider the following more general version of (8.51):

$$\begin{cases} dX(t;\omega) = (AX(t;\omega) + BX(t;\omega) + \mathfrak{F}(X)(t;\omega))dt + g(t,X(t;\omega)))d\mathbf{W}(t), \\ X(0;\omega) = X_0(\omega), \end{cases}$$

$$\tag{8.52}$$

where $0 < t < T$, $\omega \in \Omega$ and $B : \mathscr{H} \to \mathscr{H}$ is a bounded linear operator.

Exercise 8.3.3. Formulate and prove a result analogous to Thrm. 8.3.2 for (8.52), assuming that the functional $\mathfrak{F}$ is defined on (and into) the space $\mathbb{C}\left([0,T];\mathscr{L}^r\left(\Omega;\mathscr{H}\right)\right)$ (where $r > 2$).

We can consider a slightly more general functional $\mathfrak{F}$ used in all of the above IVPs by enlarging its range space to $\mathbb{L}^p\left(0,T;\mathscr{L}^2\left(\Omega;\mathscr{H}\right)\right)$, where $p > 2$. By definition, if $f \in \mathbb{L}^p\left(0,T;\mathscr{L}^2\left(\Omega;\mathscr{H}\right)\right)$, then

$$\|f\|_{\mathbb{L}^p} \equiv \left(\int_0^T \left(E\|f(s;\cdot)\|_{\mathscr{H}}^2\right)^{\frac{p}{2}} ds\right)^{\frac{1}{p}} < \infty. \tag{8.53}$$

Specifically, replace $(\mathbf{H8.10})$ by

(H8.12) The functional $\mathfrak{F} : \mathbb{C}\left([0,T];\mathcal{L}^2\left(\Omega;\mathscr{H}\right)\right) \to \mathbb{L}^p\left(0,T;\mathcal{L}^2\left(\Omega;\mathscr{H}\right)\right)$ is such that there exist $M_{\mathfrak{F}}, \overline{M_{\mathfrak{F}}} > 0$ for which

$$\|\mathfrak{F}(X) - \mathfrak{F}(Y)\|_{\mathbb{L}^p} \le M_{\mathfrak{F}} \|X - Y\|_{\mathbb{C}},$$
$$\|\mathfrak{F}(X)\|_{\mathbb{L}^p} \le \overline{M_{\mathfrak{F}}} [1 + \|X\|_{\mathbb{C}}],$$

$\forall X, Y \in \mathbb{C}\left([0,T];\mathcal{L}^2\left(\Omega;\mathscr{H}\right)\right).$

We have the following proposition.

Proposition 8.3.3. *Assume* **(H$_A$)**, **(H5.1)** *through* **(H5.3)**, **(H5.5)**, **(H8.11)**, *and* **(H8.12)**. *If* $T^{1-\frac{1}{p}} M_A M_{\mathfrak{F}} < 1$, *then (8.42) has a unique mild solution on* $[0,T]$.

Proof. The proof is virtually the same as the proof of Thrm. 8.3.2, with the exception that the actual estimates are different due to the change of space. Specifically, let $X, Y \in \mathbb{C}\left([0,T];\mathcal{L}^2\left(\Omega;\mathscr{H}\right)\right)$. Arguing as in (8.45) and using Holder's inequality yields

$$E\|\Phi(X)(t;\cdot) - \Phi(Y)(t;\cdot)\|_{\mathscr{H}}^2 \le$$
$$E\left\|\int_0^t e^{A(t-s)}\left(\mathfrak{F}(X)(s;\cdot) - \mathfrak{F}(Y)(s;\cdot)\right)ds\right\|_{\mathscr{H}}^2 \le$$
$$TM_A^2 \int_0^t E\|\mathfrak{F}(X)(s;\cdot) - \mathfrak{F}(Y)(s;\cdot)\|_{\mathscr{H}}^2 ds \le$$
$$TM_A^2\left[\left(\int_0^t \left(E\|\mathfrak{F}(X)(s;\cdot) - \mathfrak{F}(Y)(s;\cdot)\|_{\mathscr{H}}^2\right)^{\frac{p}{2}} ds\right)^{\frac{2}{p}} \cdot \left(\int_0^t 1 ds\right)^{\frac{p-2}{p}}\right] \le$$
$$TM_A^2 T^{\frac{p-2}{p}}\left(\int_0^t \|\mathfrak{F}(X)(s;\cdot) - \mathfrak{F}(Y)(s;\cdot)\|_{\mathcal{L}^2(\Omega;\mathscr{H})}^p ds\right)^{\frac{2}{p}} = (8.54)$$
$$M_A^2 T^{2-\frac{2}{p}} \|\mathfrak{F}(X) - \mathfrak{F}(Y)\|_{\mathbb{L}^p}^2 \le$$
$$M_{\mathfrak{F}}^2 M_A^2 T^{2-\frac{2}{p}} \|X - Y\|_{\mathbb{C}}^2.$$

Taking the supremum over $[0,T]$, applying the square root on both sides of (8.54), and using the assumption that $T^{1-\frac{1}{p}} M_A M_{\mathfrak{F}} < 1$, yields

$$\|(X) - (Y)\|_{\mathbb{C}} \le T^{1-\frac{1}{p}} M_A M_{\mathfrak{F}} \|X - Y\|_{\mathbb{C}} < \|X - Y\|_{\mathbb{C}}.$$

Hence, Φ has a unique fixed point that coincides with the mild solution we seek. $\square$

Exercise 8.3.4. Formulate and prove results analogous to Prop. 8.3.3 for (8.51) and (8.52).

Exercise 8.3.5. Let $r, p > 2$.
i.) What is the norm for the space $\mathbb{L}^p\left(0,T;\mathcal{L}^r\left(\Omega;\mathscr{H}\right)\right)$?

ii.) Formulate and prove an extension of Prop. 8.3.3 in which the functional $\mathfrak{F}$ is now viewed as a mapping from $\mathbb{C}\left([0,T]; \mathscr{L}^r\left(\Omega; \mathscr{H}\right)\right)$ into $\mathbb{L}^p\left(0,T; \mathscr{L}^r\left(\Omega; \mathscr{H}\right)\right)$.

As seen in Sections 8.1 and 8.2, several different functionals can be defined and appear as forcing terms on the right-hand sides of partial differential equations in models of various phenomena. We consider some abstract SEEs whose right-hand sides are now some of these specific functionals and show how to transform each of them into a functional SEE.

Consider the IVP

$$\begin{cases} dX(t;\omega) = \left(AX(t;\omega) + \int_0^t k(s)f(s,X(s;\omega))ds\right)dt + g(t,X(t;\omega)))d\mathbf{W}(t), \\ X(0;\omega) = X_0(\omega), \omega \in \Omega \end{cases}$$

$$(8.55)$$

in a separable Hilbert space $\mathscr{H}$, where $0 < t < T$, $\omega \in \Omega$. Here, we assume **(H$_A$)**, **(H5.1)** through **(H5.3)**, **(H5.5)**, **(H7.1)**, **(H7.2)**, and

(H8.13) $k \in \mathbb{C}\left([0,T]; [0,\infty)\right)$.

If we can prove that the mapping $t \mapsto \int_0^t k(s)f(s,X(s;\omega))ds$ can viewed as a functional (defined on an appropriate space), then the existence and uniqueness of a mild solution of (8.55) on $[0,T]$ can be deduced from the above results. Certainly, it is sensible to take the domain space of such a functional to be $\mathbb{C}\left([0,T]; \mathscr{L}^2\left(\Omega; \mathscr{H}\right)\right)$ because that is precisely the space to which we want the mild solution to belong. But, how about the range space? Ideally, we would use the most general range space possible under which we can still conclude the existence and uniqueness of a mild solution of (8.55). The choice of this space is directly connected to the growth conditions imposed on f and k. (Why?) This is explored in the following exercise.

Exercise 8.3.6.
i.) a.) Impose conditions on f and k that ensure that the functional $\mathfrak{F}$ given by

$$\mathfrak{F}(X)(t;\omega) = \int_0^t k(s)f(s,X(s;\omega))ds \qquad (8.56)$$

satisfies **(H8.10)**.
b.) Now, applying Thrm. 8.3.2, under what restriction on the data is it guaranteed that (8.55) has a unique mild solution on $[0,T]$?
ii.) Redo (i), but now assuming that $\mathfrak{F}$ satsifies **(H8.12)**.

Next, consider the slightly more general IVP

$$\begin{cases} dX(t;\omega) = \left(AX(t;\omega) + \int_0^t k_1(s)f(s,X(s;\omega)), \int_0^s h(s,X(s;\omega)))ds\right)dt \\ \qquad + \int_0^t k_2(s)g(s,X(s;\omega)))d\mathbf{W}(t), \ 0 < t < T, \ \omega \in \Omega, \\ X(0;\omega) = X_0(\omega), \omega \in \Omega \end{cases}$$

$$(8.57)$$

in a separable Hilbert space $\mathscr{H}$. We assume $(\mathbf{H_A})$, $(\mathbf{H5.1})$ through $(\mathbf{H5.3})$, and $(\mathbf{H5.5})$, as usual, and $k_i : [0,T] \to [0,\infty)$ $(i = 1,2)$, $f : [0,T] \times \mathscr{H} \times \mathscr{H} \to \mathscr{H}$, $g : [0,T] \times \mathscr{H} \to \mathscr{H}$, and $g : [0,T] \times \mathscr{H} \to \mathscr{B}_0(\mathbb{R}^m; \mathscr{H})$ are given mappings.

Exercise 8.3.7. Repeat Exer. 8.3.6 for (8.57). How do you now handle the more general noise term?

The more general noise term appearing on the right-hand side of (8.57) suggests that it might be useful to directly study the more general functional SEE given by

$$\begin{cases} dX(t;\omega) = (AX(t;\omega) + \mathfrak{F}(X)(t;\omega))dt + \mathfrak{G}(X)(t;\omega)d\mathbf{W}(t), \\ X(0;\omega) = X_0(\omega), \omega \in \Omega \end{cases} \tag{8.58}$$

where $0 < t < T$, $\omega \in \Omega$ in a separable Hilbert space $\mathscr{H}$. Assume that $(\mathbf{H_A})$, $(\mathbf{H5.1})$ through $(\mathbf{H5.3})$, and $(\mathbf{H5.5})$ hold. Let

$$\mathfrak{F} : \mathbb{C}\left([0,T]; \mathscr{L}^2\left(\Omega; \mathscr{H}\right)\right) \to \mathbb{C}\left([0,T]; \mathscr{L}^2\left(\Omega; \mathscr{H}\right)\right),$$
$$\mathfrak{G} : \mathbb{C}\left([0,T]; \mathscr{L}^2\left(\Omega; \mathscr{H}\right)\right) \to \mathbb{C}\left([0,T]; \mathscr{L}^2\left(\Omega; \mathscr{B}_0\left(\mathbb{R}^m; \mathscr{H}\right)\right)\right)$$

be given functionals.

Exercise 8.3.8. Explain why the choice of range space for such a functional $\mathscr{G}$ is sensible.

We can use a modified version of the argument used to prove Thrm. 8.3.2 to establish an analogous result for (8.58), but the solution map $\Phi : \mathbb{C}\left([0,T]; \mathscr{L}^2\left(\Omega; \mathscr{H}\right)\right) \to \mathbb{C}\left([0,T]; \mathscr{L}^2\left(\Omega; \mathscr{H}\right)\right)$ is now defined by

$$(\Phi X)(t;\cdot) = e^{At}X_0(\cdot) + \int_0^t e^{A(t-s)}\mathfrak{F}(X)(s;\cdot)ds$$
$$+ \int_0^t e^{A(t-s)}\mathfrak{G}(X)(s;\cdot)d\mathbf{W}(s). \tag{8.59}$$

Complete the following exercise to establish the above results for (8.59).

Exercise 8.3.9.
i.) For any $t \in [0,T]$, prove that $(\Phi X)(t;\cdot)$ is $\mathscr{F}_t$-measurable.
ii.) Determine the hypothesis corresponding to $(\mathbf{H8.10})$ for the functional $\mathfrak{G}$.
iii.) Let $r \geq 2$. Establish an upper bound (in terms of $\|x\|_{\mathbb{C}}$) for the term

$$E\left\|\int_0^t e^{A(t-s)}\mathfrak{G}(X)(s;\cdot)d\mathbf{W}(s)\right\|_{\mathscr{H}}^r.$$

iv.) Assuming $(\mathbf{H_A})$, $(\mathbf{H5.1})$ through $(\mathbf{H5.3})$, $(\mathbf{H5.5})$, $(\mathbf{H8.10})$, and the hypothesis in (ii), determine a restriction on the data under which an argument similar to the one used to prove Thrm. 8.3.2 can be used to prove (8.58) has a unique mild solution on $[0,T]$. Then, prove the result.

v.) Recover the result established for (8.57) as a corollary of (iv).

vi.) Redo (i) through (v), now replacing $\mathcal{L}^2(\Omega;\mathcal{H})$ in the spaces by $\mathcal{L}^r(\Omega;\mathcal{H})$, $r > 2$.

Exercise 8.3.10. Redo Exer. 8.3.9 for the IVP

$$\begin{cases} dX(t;\omega) = (AX(t;\omega) + BX(t;\omega) + \mathfrak{F}(X)(t;\omega))dt + \mathfrak{G}(X)(t;\omega)d\mathbf{W}(t), \\ X(0;\omega) = X_0(\omega), \omega \in \Omega \end{cases}$$

(8.60)

in a separable Hilbert space $\mathcal{H}$, where $0 < t < T$, $\omega \in \Omega$ and $B : \mathcal{H} \to \mathcal{H}$ is a bounded linear operator.

Exercise 8.3.11. Consider the following system of functional SEEs:

$$\begin{cases} dX(t;\omega) = (A_1X(t;\omega) + \mathfrak{F}_1(X)(t;\omega))dt + \mathfrak{G}_1(X)(t;\omega)d\mathbf{W}(t), \\ dY(t;\omega) = (A_2Y(t;\omega) + \mathfrak{F}_2(Y)(t;\omega))dt + \mathfrak{G}_2(Y)(t;\omega)d\mathbf{W}(t), \\ X(0;\omega) = X_0(\omega), \omega \in \Omega, \\ Y(0;\omega) = Y_0(\omega), \omega \in \Omega \end{cases}$$

(8.61)

in a separable Hilbert space $\mathcal{H}_1 \times \mathcal{H}_2$ under **(H5.2)** and **(H5.5)**, where $0 < t < T$, $\omega \in \Omega$; $A_i : \text{dom}(A_i) \subset \mathcal{H}_i \to \mathcal{H}_i$ $(i = 1,2)$ satisfies **(H$_A$)**;

$$\mathfrak{F}_i : \mathbb{C}\left([0,T];\mathcal{L}^2(\Omega;\mathcal{H}_i)\right) \to \mathbb{C}\left([0,T];\mathcal{L}^2(\Omega;\mathcal{H}_i)\right),$$

$$\mathfrak{G}_i : \mathbb{C}\left([0,T];\mathcal{L}^2(\Omega;\mathcal{H}_i)\right) \to \mathbb{C}\left([0,T];\mathcal{L}^2(\Omega;\mathscr{B}_0(\mathbb{R}^m;\mathcal{H}_i))\right)\ (i=1,2)$$

are functionals that satisfy global Lipschitz conditions as in **(H8.10)**; and $X_0 \in \mathcal{L}^2(\Omega;\mathcal{H}_1)$ and $Y_0 \in \mathcal{L}^2(\Omega;\mathcal{H}_2)$ are $\mathscr{F}_0$-measurable random variables independent of the Wiener processes.

i.) Reformulate (8.61) as the abstract functional SEE (8.58) on the space $\mathcal{H}_1 \times \mathcal{H}_2$.

ii.) Determine a data restriction under which (8.61) has a unique mild solution on $[0,T]$.

8.3.3 Convergence Results

Now, we consider various convergence results for some of the IVPs introduced in Section 8.3.2, including continuous dependence on initial data.

Consider (8.55) and the following related IVP

$$\begin{cases} dY(t;\omega) = \left(AY(t;\omega) + \int_0^t k(s)\widehat{f}(s,Y(s;\omega))ds\right)dt + \widehat{g}(t,Y(t;\omega)))d\mathbf{W}(t), \\ Y(0;\omega) = Y_0(\omega), \omega \in \Omega \end{cases}$$

(8.62)

in a separable Hilbert space $\mathcal{H}$, where $0 < t < T$, $\omega \in \Omega$, both under assumptions **(H$_A$)**, **(H5.1)** through **(H5.3)**, **(H5.5)**, **(H7.1)** through **(H7.6)**, and **(H8.13)**.

Exercise 8.3.12.

i.) Estimate $E \|X(t;\cdot) - Y(t;\cdot)\|_{\mathscr{H}}^2$ in terms of $\delta_i (i = 1,2,3,4)$ and $\|k\|_{\mathbb{C}}$.

ii.) Redo (i), now replacing $\mathscr{L}^2(\Omega;\mathscr{H})$ in the spaces by $\mathscr{L}^r(\Omega;\mathscr{H})$, $r > 2$.

Similar results can be established for each of the stochastic IVPs considered in Section 8.3.2. Next, we present several different convergence results.

Let $\varepsilon > 0$. Consider the stochastic IVP

$$\begin{cases} dX_\varepsilon(t;\omega) = (AX_\varepsilon(t;\omega) + \mathfrak{F}(X_\varepsilon)(t;\omega))\, dt + \mathfrak{G}(X_\varepsilon)(t;\omega)d\mathbf{W}(t), \\ X_\varepsilon(0;\omega) = X_0(\omega) + \varepsilon, \omega \in \Omega \end{cases} \tag{8.63}$$

in a separable Hilbert space $\mathscr{H}$, where $0 < t < T$, $\omega \in \Omega$. Assume that $(\mathbf{H_A})$, $(\mathbf{H5.1})$ through $(\mathbf{H5.3})$, $(\mathbf{H5.5})$, $(\mathbf{H8.10})$ hold, as well as

$(\mathbf{H8.14})$ $\mathfrak{G} : \mathbb{C}\left([0,T];\mathscr{L}^2(\Omega;\mathscr{H})\right) \to \mathbb{C}\left([0,T];\mathscr{L}^2(\Omega;\mathscr{B}_0(\mathbb{R}^m;\mathscr{H}))\right)$ is such that there exist $M_\mathfrak{G}, \overline{M_\mathfrak{G}} > 0$ such that $\forall X, Y \in \mathbb{C}\left([0,T];\mathscr{L}^2(\Omega;\mathscr{H})\right)$,

$$\|\mathfrak{G}(X) - \mathfrak{G}(Y)\|_{\mathbb{C}} \le M_\mathfrak{G} \|X - Y\|_{\mathbb{C}},$$
$$\|\mathfrak{G}(X)\|_{\mathbb{C}} \le \overline{M_\mathfrak{G}} [1 + \|X\|_{\mathbb{C}}].$$

Observe that $(X_0(\cdot) + \varepsilon) \in \mathscr{L}^2(\Omega;\mathscr{H})$. (Why?) As such, Exer. 8.3.9 ensures that (8.63) has a unique mild solution $X_\varepsilon(t;\cdot)$ on $[0,T]$, provided the data restriction imposed in Exer 8.3.9 is satisfied. This mild solution is given by

$$X_\varepsilon(t;\cdot) = e^{At}(X_0(\cdot) + \varepsilon) + \int_0^t e^{A(t-s)}\mathfrak{F}(X_\varepsilon)(s;\cdot)ds$$

$$+ \int_0^t e^{A(t-s)}\mathfrak{G}(X_\varepsilon)(s;\cdot)d\mathbf{W}(s). \tag{8.64}$$

Exercise 8.3.13.

i.) Determine a data restrction under which it can be shown that $\|X_\varepsilon - X\|_{\mathbb{C}} \longrightarrow 0$ as $\varepsilon \to 0^+$, where X is the mild solution of (8.58) (considered under the same hypotheses as (8.63)). Prove the result.

ii.) Suppose that the functional $\mathfrak{F}$ now satisfies $(\mathbf{H8.12})$. Can you establish a result analogous to (i)? If so, how does the data restriction change?

For the duration of our discussion of the next main result, $X(t;\cdot)$ denotes the mild solution of (8.58) as guaranteed to exist by Exer. 8.3.9. For every $n \in \mathbb{N}$, consider the IVP

$$\begin{cases} dX_n(t;\omega) = (A_n X_n(t;\omega) + \mathfrak{F}_n(X_n)(t;\omega))\, dt + \mathfrak{G}_n(X_n)(t;\omega)d\mathbf{W}(t), \\ X_n(0;\omega) = (X_0)_n(\omega), \omega \in \Omega \end{cases} \tag{8.65}$$

in a separable Hilbert space $\mathscr{H}$, where $0 < t < T$, $\omega \in \Omega$. We impose the following assumptions:

(H8.15) $A : \mathrm{dom}(A) \subset \mathscr{H} \longrightarrow \mathscr{H}$ satisfies **($\mathbf{H_A}$)**. For every $n \in \mathbb{N}$, $A_n : \mathrm{dom}(A_n) \subset \mathscr{H} \longrightarrow \mathscr{H}$ generates a C_0-semigroup $\{e^{A_n t} : t \geq 0\}$ on $\mathscr{H}$, and
i.) $\mathrm{dom}(A_n) = \mathrm{dom}(A)$,
ii.) $\sup \left\{ \left\| e^{A_n t} \right\|_{\mathbb{B}(\mathscr{H})} : 0 \leq t \leq T, n \in \mathbb{N} \right\} \leq M_A$, where M_A is the same bound appearing in **($\mathbf{H_A}$)**.
(H8.16) $\forall n \in \mathbb{N}$, $\mathfrak{F}, \mathfrak{F}_n : \mathbb{C}\left([0,T]; \mathcal{L}^2\left(\Omega; \mathscr{H}\right)\right) \rightarrow \mathbb{C}\left([0,T]; \mathcal{L}^2\left(\Omega; \mathscr{H}\right)\right)$ satisfy **(H8.10)** such that the growth constants satisfy $M_{\mathfrak{F}_n} = M_{\mathfrak{F}}$ and $\overline{M_{\mathfrak{F}_n}} = \overline{M_{\mathfrak{F}}}$,
(H8.17) $\forall n \in \mathbb{N}$, $\mathfrak{G}, \mathfrak{G}_n : \mathbb{C}\left([0,T]; \mathcal{L}^2\left(\Omega; \mathscr{H}\right)\right) \rightarrow \mathbb{C}\left([0,T]; \mathcal{L}^2\left(\Omega; \mathscr{B}_0\left(\mathbb{R}^m; \mathscr{H}\right)\right)\right)$ satisfy **(H8.14)** such that the growth constants satisfy $M_{\mathfrak{G}_n} = M_{\mathfrak{G}}$ and $\overline{M_{\mathfrak{G}_n}} = \overline{M_{\mathfrak{G}}}$.
(H8.18) $X_0, (X_0)_n \in \mathcal{L}^2\left(\Omega; \mathscr{H}\right)$.
(H8.19) $\lim\limits_{n \to \infty} \sup \left\{ \left\| e^{A_n t} - e^{At} \right\|_{\mathbb{B}(\mathscr{H})} : 0 \leq t \leq T \right\} = 0$.
(H8.20) $\lim\limits_{n \to \infty} \left\| (X_0)_n - X_0 \right\|_{\mathcal{L}^2(\Omega; \mathscr{H})} = 0$.
(H8.21) $\lim\limits_{n \to \infty} \left\| \mathfrak{F}_n(X) - \mathfrak{F}(X) \right\|_{\mathbb{C}\left([0,T]; \mathcal{L}^2(\Omega; \mathscr{H})\right)} = 0$, $\forall X \in \mathbb{C}$.
(H8.22) $\lim\limits_{n \to \infty} \left\| \mathfrak{G}_n(X) - \mathfrak{G}(X) \right\|_{\mathbb{C}\left([0,T]; \mathcal{L}^2(\Omega; \mathscr{B}_0(\mathbb{R}^m; \mathscr{H}))\right)} = 0$, $\forall X \in \mathbb{C}$.

Exercise 8.3.14. Verify under the above assumptions that (8.65) has a unique mild solution X_n on $[0, T]$.

Before establishing our first main convergence result, we need a lemma. Specifically, $\forall n \in \mathbb{N}$, consider the following IVP related to (8.65):

$$\begin{cases} dY_n(t; \omega) = \left(A_n Y_n(t; \omega) + \mathfrak{F}_n(X)(t; \omega)\right) dt + \mathfrak{G}_n(X)(t; \omega) d\mathbf{W}(t), \\ Y_n(0; \omega) = X_0(\omega), \omega \in \Omega \end{cases} \tag{8.66}$$

in a separable Hilbert space $\mathscr{H}$, where $0 < t < T$, $\omega \in \Omega$, under the above assumptions.

Exercise 8.3.15. Prove that $\forall n \in \mathbb{N}$, (8.66) has a unique mild solution X_n on $[0, T]$.

Lemma 8.3.4. *If* **(H8.15)** *through* **(H8.22)** *hold, then* $\lim\limits_{n \to \infty} \left\| Y_n - X \right\|_{\mathbb{C}\left([0,T]; \mathcal{L}^2(\Omega; \mathscr{H})\right)} = 0$.

Proof. Let $\varepsilon > 0$. There exist $N_1, N_2, N_3 \in \mathbb{N}$ such that

$$n \geq N_1 \implies \sup_{0 \leq t \leq T} \left\| e^{A_n t} - e^{At} \right\|_{\mathbb{B}(\mathcal{H})}^2 <$$

$$\left(\frac{\varepsilon}{6 \max \left\{ \left(\|X_0\|_{\mathfrak{L}^2}^2 + 1, 2T^2 \left(1 + \|x\|_{\mathbb{C}}^2\right), 2T \zeta_{\mathfrak{G}}(T,2) \left(1 + \|x\|_{\mathbb{C}}^2\right) \right) \right\}} \right)^2$$

$$n \geq N_2 \implies \|\mathfrak{F}_n(X) - \mathfrak{F}(X)\|_{\mathbb{C}\left([0,T]; \mathfrak{L}^2(\Omega; \mathcal{H})\right)} < \left(\frac{\varepsilon}{6 M_A^2 T^2} \right)^2 \quad (8.67)$$

$$n \geq N_3 \implies \|\mathfrak{G}_n(X) - \mathfrak{G}(X)\|_{\mathbb{C}\left([0,T]; \mathfrak{L}^2(\Omega; \mathcal{B}_0(\mathbb{R}^m; \mathcal{H}))\right)} < \left(\frac{\varepsilon}{6 M_A^2 T \zeta_{\mathfrak{G}}(T,2)} \right)^2.$$

Observe that

$$\sup_{0 \leq s \leq t \leq T} \left\| e^{A_n(t-s)} - e^{A(t-s)} \right\|_{\mathbb{B}(\mathcal{H})}^2 = \sup_{0 \leq t \leq T} \left\| e^{A_n t} - e^{At} \right\|_{\mathbb{B}(\mathcal{H})}^2. \quad (8.68)$$

(Why?) Let $N = \max \{N_1, N_2, N_3\}$. Then, subtract the variation of parameters formulae for Y_n and X, and estimate this difference to obtain

$$n \geq N \implies E \|Y_n(t; \cdot) - X(t; \cdot)\|_{\mathcal{H}}^2 < \left(\frac{5\varepsilon}{6} \right)^2, \forall t \in [0,T]. \quad (8.69)$$

Taking the supremum over $(0,T)$ followed by the square root in (8.69) then yields

$$n \geq N \implies \|Y_n - X\|_{\mathbb{C}\left([0,T]; \mathfrak{L}^2(\Omega; \mathcal{H})\right)} < \frac{5\varepsilon}{6} < \varepsilon.$$

This proves the lemma. $\qquad\qquad\qquad\qquad\qquad\qquad\qquad\qquad\qquad\qquad\qquad\square$

Exercise 8.3.16. Carefull verify (8.69).

We use Lemma 8.3.4 as a stepping stone to prove the following convergence result.

Proposition 8.3.5. *If* $1 - 8M_A^2 T \left(TM_{\mathfrak{F}}^2 + \zeta_{\mathfrak{G}}(T,2) M_{\mathfrak{G}}^2 \right) > 0$, *then*

$$\lim_{n \to \infty} \|X_n - X\|_{\mathbb{C}\left([0,T]; \mathfrak{L}^2(\Omega; \mathcal{H})\right)} = 0.$$

Proof. Let $n \in \mathbb{N}$ and Y_n be the unique mild solution of (8.66). Observe that

$$E \|X_n(t; \cdot) - X(t; \cdot)\|_{\mathcal{H}}^2 \leq 4 \left[E \|X_n(t; \cdot) - Y_n(t; \cdot)\|_{\mathcal{H}}^2 + E \|Y_n(t; \cdot) - X(t; \cdot)\|_{\mathcal{H}}^2 \right]. \quad (8.70)$$

Estimating the first term on the right-hand side of (8.70) yields

$$E \|X_n(t; \cdot) - X(t; \cdot)\|_{\mathcal{H}}^2 \leq 8M_A^2 \|(X_0)_n - X_0\|_{\mathfrak{L}^2(\Omega; \mathcal{H})}^2 +$$
$$8M_A^2 T \left(TM_{\mathfrak{F}}^2 + \zeta_{\mathfrak{G}}(T,2) M_{\mathfrak{G}}^2 \right) \|X_n - X\|_{\mathbb{C}}^2. \quad (8.71)$$

Substituting (8.71) into (8.70) and subsequently taking the supremum over $[0,T]$ on both sides and rearranging terms leads to

$$K_1 \|X_n - X\|_{\mathbb{C}}^2 \leq 32M_A^2 \|(X_0)_n - X_0\|_{\mathfrak{L}^2(\Omega;\mathscr{H})}^2 + 4\|Y_n - X\|_{\mathbb{C}}^2, \qquad (8.72)$$

where $K_1 = 1 - 8M_A^2T\left(TM_{\mathfrak{F}}^2 + \zeta_{\mathfrak{G}}(T,2)M_{\mathfrak{G}}^2\right)$. Because $K_1 > 0$, by assumption, we can divide both sides of (8.72) by K_1 to obtain

$$\|X_n - X\|_{\mathbb{C}}^2 \leq \frac{32M_A^2}{K_1} \|(X_0)_n - X_0\|_{\mathfrak{L}^2(\Omega;\mathscr{H})}^2 + \frac{4}{K_1}\|Y_n - X\|_{\mathbb{C}}^2. \qquad (8.73)$$

Let $\varepsilon > 0$. Hypothesis **(H8.20)** implies that $\exists N_1 \in \mathbb{N}$ such that

$$n \geq N_1 \implies \|(X_0)_n - X_0\|_{\mathfrak{L}^2(\Omega;\mathscr{H})}^2 < \left(\frac{K_1}{32M_A^2}\right)\left(\frac{\varepsilon}{\sqrt{2}}\right)^2, \qquad (8.74)$$

and Lemma 8.3.4 implies that $\exists N_2 \in \mathbb{N}$ such that

$$n \geq N_2 \implies \|Y_n - X\|_{\mathbb{C}}^2 < \frac{K_1\varepsilon^2}{8}. \qquad (8.75)$$

Let $N = \max\{N_1, N_2\}$. Using (8.74) and (8.75) in (8.73) then yields

$$n \geq N \implies \|X_n - X\|_{\mathbb{C}([0,T];\mathfrak{L}^2(\Omega;\mathscr{H}))} < \varepsilon.$$

This completes the proof. $\qquad\qquad\qquad\qquad\qquad\qquad\qquad\qquad\qquad\qquad\qquad$ □

Exercise 8.3.17. Carefully verify (8.71).

Exercise 8.3.18. Suppose that the functionals $\mathfrak{F}, \mathfrak{F}_n$ now have the enlarged range space $\mathbb{L}^p\left(0,T;\mathfrak{L}^2(\Omega;\mathscr{H})\right)$. Under suitably modified hypotheses to account for this change of space, establish results analogous to Lemma 8.3.4 and Prop. 8.3.5.

For the same reasons as those mentioned in Chapter 7, (8.58) need not have a strong solution, in general. But, we can approximate the mild solution of (8.58) that is known to exist (under the hypotheses stated earlier) by a sequence of strong solutions of IVPs formed using the Yosida approximations. Indeed, $\forall n \in \mathbb{N}$, consider the IVP

$$\begin{cases} dZ_n(t;\omega) = (AZ_n(t;\omega) + nR_n(A)\mathfrak{F}(Z_n)(t;\omega))\,dt + nR_n(A)\mathfrak{G}(Z_n)(t;\omega)d\mathbf{W}(t), \\ Z_n(0;\omega) = nR_n(A)X_0(\omega), \omega \in \Omega \end{cases}$$

$$(8.76)$$

in a separable Hilbert space $\mathscr{H}$, where $0 < t < T$, $\omega \in \Omega$. Under the assumptions that guarantee (8.58) has a unique mild solution on $[0,T]$, we can show that (8.76) has a unique mild solution Z_n on $[0,T]$. In fact, because $Z_n(t;\omega) \in \text{dom}(A)$, $\forall t \in [0,T]$ a.s. $[\mathscr{P}]$, it can be shown that Z_n is actually a strong solution of (8.76) in the sense of Def. 7.4.7 with (iv) appropriately modified.

Exercise 8.3.19. Let X be the mild solution of (8.58).

i.) Determine a data restriction that, if imposed, enables us to argue (as in Prop. 7.4.8) that

$$\lim_{n \to \infty} \|Z_n - X\|_{C([0,T];\mathcal{L}^2(\Omega;\mathcal{H}))} = 0.$$

ii.) Verify that the sequence of induced probability measures $\mathscr{P}_{Z_n}$ converges weakly to $\mathscr{P}_X$.

8.3.4 Zeroth-Order Approximation

Let $0 < \varepsilon < 1$. Consider the stochastic IVP

$$\begin{cases} dX_\varepsilon(t;\omega) = (A\varepsilon X_\varepsilon(t;\omega) + \mathfrak{F}_\varepsilon(X_\varepsilon)(t;\omega))\,dt + g_\varepsilon(t, X_\varepsilon(t;\omega))d\mathbf{W}(t), \\ X_\varepsilon(0;\omega) = X_0, \omega \in \Omega \end{cases} \tag{8.77}$$

in a separable Hilbert space $\mathcal{H}$, where $0 < t < T$, $\omega \in \Omega$, together with the deterministic IVP

$$\begin{cases} dX(t) = (AX(t) + \mathfrak{F}(X)(t))\,dt, 0 < t < T, \\ X(0) = X_0. \end{cases} \tag{8.78}$$

Assume that X_0 is a fixed element of $\mathcal{H}$ (so that it is deterministic), A satisfies $(\mathbf{H}_A)$, and $\mathfrak{F} : C([0,T];\mathcal{H}) \to \mathbb{L}^p(0,T;\mathcal{H})$ $(p > 2)$ satisfies $(\mathbf{H8.10})$, appropriately adapted to such a functional, as well as the following:

$(\mathbf{H8.23})$ For every $0 < \varepsilon < 1$, $A_\varepsilon : \text{dom}(A_\varepsilon) \subset \mathcal{H} \to \mathcal{H}$ generates a C_0-semigroup $\{e^{A_\varepsilon t} : t \ge 0\}$ on $\mathcal{H}$, $A : \text{dom}(A) \subset \mathcal{H} \to \mathcal{H}$ generates a C_0-semigroup $\{e^{At} : t \ge 0\}$ on $\mathcal{H}$, and $\|e^{A_\varepsilon t} - e^{At}\|_{\mathbb{B}(\mathcal{H})} \longrightarrow 0$ as $\varepsilon \to 0^+$ uniformly in $t \in [0,T]$. Moreover, assume that

$$\sup_{0<\varepsilon<1}\left(\sup_{0 \le t \le T} \|e^{A_\varepsilon t}\|_{\mathbb{B}(\mathcal{H})}\right) \le M_A e^{\alpha T}. \tag{8.79}$$

$(\mathbf{H8.24})$ For every $0 < \varepsilon < 1$, $\mathfrak{F}, \mathfrak{F}_\varepsilon : C([0,T];\mathcal{L}^2(\Omega;\mathcal{H})) \to \mathbb{L}^p(0,T;\mathcal{L}^2(\Omega;\mathcal{H}))$ satisfy $(\mathbf{H8.10})$ such that the growth constants satisfy $M_{\mathfrak{F}_\varepsilon} = M_{\mathfrak{F}}$ and $\overline{M_{\mathfrak{F}_\varepsilon}} = \overline{M_{\mathfrak{F}}}$, $\forall 0 < \varepsilon < 1$, and

$$\|\mathfrak{F}_\varepsilon(Z) - \mathfrak{F}(Z)\|_{\mathbb{L}^p} \longrightarrow 0 \text{ as } \varepsilon \to 0^+, \forall Z \in C.$$

$(\mathbf{H8.25})$ For every $0 < \varepsilon < 1$, $g_\varepsilon : [0,T] \times \mathcal{H} \to \mathcal{B}_0(\mathbb{R}^m, \mathcal{H})$ satisfies $(\mathbf{H7.9})$.

We have the following result.

Proposition 8.3.6. *Let X_ε and X be the mild solutions of (8.77) and (8.78), respectively. If*

$$1 - \left(T^{2-\frac{2}{p}}(M_A M_{\mathfrak{F}})^2 + 4\zeta_g(T,2)(M^\star M_g)^2 T\right) > 0,$$

then $\exists \zeta > 0$ *and a function* $\psi : I \subset [0,1] \longrightarrow (0,\infty)$ *for which* $\lim_{\varepsilon \to 0^+} \psi(\varepsilon) = 0$ *and*

$$E \left\| X_\varepsilon(t;\cdot) - X(t) \right\|_{\mathscr{H}}^2 \leq \zeta \psi(\varepsilon), \tag{8.80}$$

$\forall 0 \leq t \leq T$ *and* $\varepsilon > 0$ *sufficiently small. (That is,* $\lim_{\varepsilon \to 0^+} \|X_\varepsilon - X\|_{\mathbb{C}([0,T];\mathfrak{L}^p(\Omega;\mathscr{H}))} = 0.)$

Proof. The proof closely follows the argument of Prop. 7.4.16. First, observe that

$$E \left\| X_\varepsilon(t;\cdot) - X(t) \right\|_{\mathscr{H}}^2 \leq 8 \left[\left\| e^{A_\varepsilon t} - e^{At} \right\|_{\mathbb{B}(\mathscr{H})}^2 \|X_0\|_{\mathfrak{L}^2}^2 + \right.$$

$$+ E \left\| \int_0^t \left(e^{A_\varepsilon(t-s)} \mathfrak{F}_\varepsilon(X_\varepsilon)(s;\cdot) - e^{A(t-s)} \mathfrak{F}(X)(s;\cdot) \right) ds \right\|_{\mathscr{H}}^2$$

$$+ E \left\| \int_0^t e^{A_\varepsilon(t-s)} g_\varepsilon(s, X_\varepsilon(s;\cdot)) d\mathbf{W}(s) \right\|_{\mathscr{H}}^2 \right]. \tag{8.81}$$

It follows from the proof of Prop. 7.4.16 that $\left\| e^{A_\varepsilon t} - e^{At} \right\|_{\mathbb{B}(\mathscr{H})}^2 \|X_0\|_{\mathfrak{L}^2}^2 \longrightarrow 0$ uniformly in $t \in [0,T]$ as $\varepsilon \to 0^+$ by **(H8.23)**. So, for sufficiently small ε, $\exists K_1 > 0$ and a function $\psi_1 : I \subset [0,1] \longrightarrow (0,\infty)$ for which $\lim_{\varepsilon \to 0^+} \psi_1(\varepsilon) = 0$ and

$$\left\| e^{A_\varepsilon t} - e^{At} \right\|_{\mathbb{B}(\mathscr{H})}^2 \|X_0\|_{\mathfrak{L}^2}^2 \leq K_1 \psi_1(\varepsilon), \ \forall t \in [0,T]. \tag{8.82}$$

Also, it can be shown that for sufficiently small ε, $\exists K_2, M^\star > 0$ and a function $\psi_2 : I \subset [0,1] \longrightarrow (0,\infty)$ for which $\lim_{\varepsilon \to 0^+} \psi_2(\varepsilon) = 0$ and

$$E \left\| \int_0^t e^{A_\varepsilon(t-s)} g_\varepsilon(s, X_\varepsilon(s;\cdot)) d\mathbf{W}(s) \right\|_{\mathscr{H}}^2 \leq$$

$$4\zeta_g(T,2)(M^\star M_g)^2 \int_0^t E \left\| X_\varepsilon(s;\cdot) - X(s) \right\|_{\mathscr{H}}^2 ds + K_2 \psi_2(\varepsilon) \leq \tag{8.83}$$

$$4\zeta_g(T,2)(M^\star M_g)^2 T \left\| X_\varepsilon - X \right\|_{\mathbb{C}} + K_2 \psi_2(\varepsilon),$$

$\forall 0 \leq t \leq T$. (Why?)

It remains to estimate the second term on the right-hand side of (8.81). Observe that

$$E \left\| \int_0^t \left(e^{A\varepsilon(t-s)} \mathfrak{F}_\varepsilon (X_\varepsilon)(s;\cdot) - e^{A(t-s)} \mathfrak{F}(X)(s;\cdot) \right) ds \right\|_{\mathscr{H}}^2 \le$$

$$8 \left[E \left\| \int_0^t \left(e^{A\varepsilon(t-s)} \mathfrak{F}_\varepsilon (X_\varepsilon)(s;\cdot) - e^{A\varepsilon(t-s)} \mathfrak{F}_\varepsilon (X)(s;\cdot) \right) ds \right\|_{\mathscr{H}}^2 \right.$$

$$+ E \left\| \int_0^t \left(e^{A\varepsilon(t-s)} \mathfrak{F}_\varepsilon (X)(s;\cdot) - e^{A(t-s)} \mathfrak{F}_\varepsilon (X)(s;\cdot) \right) ds \right\|_{\mathscr{H}}^2$$

$$\left. E \left\| \int_0^t \left(e^{A(t-s)} \mathfrak{F}_\varepsilon (X)(s;\cdot) - e^{A(t-s)} \mathfrak{F}(X)(s;\cdot) \right) ds \right\|_{\mathscr{H}}^2 \right] \le \qquad (8.84)$$

$$8 \left[T M_A^2 \int_0^t E \left\| \mathfrak{F}_\varepsilon (X_\varepsilon)(s;\cdot) - \mathfrak{F}_\varepsilon (X)(s;\cdot) \right\|_{\mathscr{H}}^2 ds \right.$$

$$+ T \sup_{0 \le t \le T} \left\| e^{A\varepsilon t} - e^{At} \right\|_{\mathbb{B}(\mathscr{H})}^2 \int_0^t E \left\| \mathfrak{F}_\varepsilon (X)(s;\cdot) \right\|_{\mathscr{H}}^2 ds$$

$$\left. + T M_A^2 \int_0^t E \left\| \mathfrak{F}_\varepsilon (X)(s;\cdot) - \mathfrak{F}(X)(s;\cdot) \right\|_{\mathscr{H}}^2 ds \right] =$$

$$8 \left[I_1 + I_2 + I_3 \right].$$

We estimate each of I_1, I_2, I_3 in turn. Applying Holder's inequality and using **(H8.24)** yields

$$I_1 \le T^{2-\frac{2}{p}} M_A^2 M_{\mathfrak{F}}^2 \left\| X_\varepsilon - X \right\|_{\mathbb{C}}^2 . \qquad (8.85)$$

(Tell how.) In order for $I_2 \to 0$ as $\varepsilon \to 0^+$, we must show that

$$\sup_{0<\varepsilon<1} \left(\sup_{0 \le t \le T} \int_0^t E \left\| \mathfrak{F}_\varepsilon (X)(s;\cdot) \right\|_{\mathscr{H}}^2 ds \right) < \infty . \qquad (8.86)$$

Applying Holder's inequality, **(H8.24)**, and the fact that $\left\| X \right\|_{\mathbb{C}}^2 < \infty$ yields

$$\int_0^t E \left\| \mathfrak{F}_\varepsilon (X)(s;\cdot) \right\|_{\mathscr{H}}^2 ds \le T^{\frac{p-2}{p}} \left\| \mathfrak{F}_\varepsilon (X) \right\|_{\mathbb{L}^p}^2$$

$$\le 2 T^{\frac{p-2}{p}} \left(\overline{M_{\mathfrak{F}}} \right)^2 \left[1 + \left\| X \right\|_{\mathbb{C}}^2 \right]$$

$$< \infty .$$

Hence, **(H8.23)** guarantees that for sufficiently small ε, $\exists K_3 > 0$ and a function $\psi_3 : I \subset [0,1] \longrightarrow (0,\infty)$ for which $\lim_{\varepsilon \to 0^+} \psi_3(\varepsilon) = 0$ and

$$T \left(\sup_{0 \le t \le T} \left\| e^{A\varepsilon t} - e^{At} \right\|_{\mathbb{B}(\mathscr{H})}^2 \int_0^t E \left\| \mathfrak{F}_\varepsilon (X)(s;\cdot) \right\|_{\mathscr{H}}^2 ds \right) \le K_3 \psi_3(\varepsilon). \qquad (8.87)$$

Finally, applying Holder's inequality yields

$$T M_A^2 \int_0^t E \left\| \mathfrak{F}_\varepsilon (X)(s;\cdot) - \mathfrak{F}(X)(s;\cdot) \right\|_{\mathscr{H}}^2 ds \le T M_A^2 T^{\frac{p-2}{p}} \left\| \mathfrak{F}_\varepsilon (X) - \mathfrak{F}(X) \right\|_{\mathbb{L}^p}^2 ,$$

$$(8.88)$$

and so, **(H8.24)** ensures that for sufficiently small ε, $\exists K_4 > 0$ and a function $\psi_4 : I \subset [0,1] \longrightarrow (0,\infty)$ for which $\lim_{\varepsilon \to 0^+} \psi_4(\varepsilon) = 0$ and

$$TM_A^2 \int_0^t E \left\| \mathfrak{F}_\varepsilon(X)(s;\cdot) - \mathfrak{F}(X)(s;\cdot) \right\|_{\mathscr{H}}^2 ds \leq K_4 \psi_4(\varepsilon). \tag{8.89}$$

Using all of the above estimates in (8.81) leads to

$$E \left\| X_\varepsilon(t;\cdot) - X(t) \right\|_{\mathscr{H}}^2 \leq \xi_1 \left\| X_\varepsilon - X \right\|_{\mathbb{C}}^2 + \xi_2 \Psi(\varepsilon), \tag{8.90}$$

where

$$\begin{aligned}
\xi_1 &= T^{2-\frac{2}{p}} (M_A M_{\mathfrak{F}})^2 + 4\zeta_g(T,2)(M^\star M_g)^2 T, \\
\xi_2 &= K_1 + K_2 + K_3 + K_4, \\
\Psi(\varepsilon) &= \psi_1(\varepsilon) + \psi_2(\varepsilon) + \psi_3(\varepsilon) + \psi_4(\varepsilon).
\end{aligned}$$

Taking the supremum over $[0,T]$ and subsequently subtracting $\xi_1 \left\| X_\varepsilon - X \right\|_{\mathbb{C}}^2$ from both sides of (8.90) yields

$$(1 - \xi_1) \left\| X_\varepsilon - X \right\|_{\mathbb{C}}^2 \leq \xi_2 \Psi(\varepsilon).$$

Because $1 - \xi_1 > 0$ by assumption, we have

$$\left\| X_\varepsilon - X \right\|_{\mathbb{C}}^2 \leq \frac{\xi_2}{1 - \xi_1} \Psi(\varepsilon). \tag{8.91}$$

Because the right-hand side of (8.91) goes to zero as $\varepsilon \to 0^+$, the proposition is proved. $\qquad \square$

Exercise 8.3.20. Formulate and prove a result analogous to Prop. 8.3.6 for (8.55) directly.

8.4 Models — New and Old

We illustrate how the theory developed in this chapter can be applied to several different models.

Model VII.4 Functional Stochastic Wave Equations
Consider the following stochastic IBVP:

$$\begin{cases}
\partial \left(\frac{\partial z}{\partial t} + \alpha z \right) + c^2 \frac{\partial^2 z}{\partial x^2} \partial t = \left(\int_0^t a(t,s) f \left(s, z, \frac{\partial z}{\partial s} \right) ds \right) \partial t \\
\quad + \left(\int_0^t b(t,s) f \left(s, z, \frac{\partial z}{\partial s} \right) ds \right) dW(t) \, 0 < x < L, \, 0 \leq t \leq T, \, \omega \in \Omega, \\
z(x,0;\omega) = z_0(x;\omega), \, \frac{\partial z}{\partial t}(x,0;\omega) = z_1(x;\omega), \, 0 < x < L, \, \omega \in \Omega, \\
z(0,t;\omega) = z(L,t;\omega) = 0, \, 0 \leq t \leq T, \, \omega \in \Omega,
\end{cases} \tag{8.92}$$

where $z = z(x,t;\omega)$, $f,g : [0,T] \times \mathbb{R} \times \mathbb{R} \to \mathbb{R}$, $a \in \mathbb{L}^2((0,T) \times (0,T))$, $W(t)$ is a one-dimensional Wiener process, and z_0, z_1 are $\mathscr{F}_0$-measurable random variables independent of $W(t)$. Related problems are investigated in **[54, 81, 82, 114, 291, 314, 325]**.

Exercise 8.4.1.
i.) Reformulate (8.92) abstractly as (8.13) in an appropriate space.
ii.) Assume that

(H8.26) $f : [0,T] \times \mathbb{R} \times \mathbb{R} \to \mathbb{R}$ is continuous in all three variables and is such that there exist positive constants m_1, m_2 for which

$$|f(t,x,y)| \le m_1 [1 + |x| + |y|],$$
$$|f(t,x,y) - f(t,\bar{x},\bar{y})| \le m_2 [|x - \bar{x}| + |y - \bar{y}|],$$

$\forall x, y, \bar{x}, \bar{y} \in \mathbb{R}$, uniformly in $t \in [0,T]$.

(H8.27) $g : [0,T] \times \mathbb{R} \times \mathbb{R} \to \mathbb{R}$ is continuous in all three variables and is such that there exist positive constants m_3, m_4 for which

$$|g(t,x,y)| \le m_3 [1 + |x| + |y|],$$
$$|g(t,x,y) - f(t,\bar{x},\bar{y})| \le m_4 [|x - \bar{x}| + |y - \bar{y}|],$$

$\forall x, y, \bar{x}, \bar{y} \in \mathbb{R}$, uniformly in $t \in [0,T]$.

Show that if **(H8.26)** and **(H8.27)** hold, then (8.92) has a unique mild solution on $[0,T]$.

Exercise 8.4.2. For every $\varepsilon > 0$, consider the IBVP obtained by replacing the expression $\left(\int_0^t b(t,s) f\left(s,z,\frac{\partial z}{\partial s}\right) ds \right) dW(t)$ in (8.92) by $\left(\int_0^t g_\varepsilon(s) ds \right) dW(t)$, where $g_\varepsilon : [0,T] \to \mathbb{R}$ is continuous and

$$\lim_{\varepsilon \to 0^+} \left(\sup \left\{ \int_0^t g_\varepsilon(s) ds : 0 \le t \le T \right\} \right) = 0.$$

i.) Verify that the resulting IBVP has a unique mild solution z_ε on $[0,T]$.
ii.) Formulate and prove a zeroth approximation result in the spirit of Prop. 8.3.6.
iii.) Redo (i), but now assuming that α in (8.92) is also replaced by α_ε, where $\lim_{\varepsilon \to 0^+} \alpha_\varepsilon = 0$. To what IBVP is the limit function a mild solution?

Exercise 8.4.3. Consider the following IBVP:

$$\begin{cases} \partial \left(\frac{\partial z}{\partial t} + \alpha z \right) + c^2 \frac{\partial^2 z}{\partial x^2} \partial t = \left(\int_0^t a_1(t-s) g_1(s,w,z) ds \right) \partial t \\ \quad + \left(\int_0^t \widehat{a_1}(t-s) \widehat{g_1}(s,w,z) ds \right) dW(t), \, 0 < x < L, \, 0 \le t \le T, \, \omega \in \Omega, \\ \partial \left(\frac{\partial w}{\partial t} + \alpha w \right) + c^2 \frac{\partial^2 w}{\partial x^2} \partial t = \left(\int_0^t a_2(t-s) g_2(s,w,z) ds \right) \partial t \\ \quad + \left(\int_0^t \widehat{a_2}(t-s) \widehat{g_2}(s,w,z) ds \right) dW(t), \, 0 < x < L, \, 0 \le t \le T, \, \omega \in \Omega, \\ z(x,0;\omega) = z_0(x;\omega), \, \frac{\partial z}{\partial t}(x,0;\omega) = z_1(x;\omega), \, 0 < x < L, \, \omega \in \Omega, \\ w(x,0;\omega) = w_0(x;\omega), \, \frac{\partial w}{\partial t}(x,0;\omega) = w_1(x;\omega), \, 0 < x < L, \, \omega \in \Omega, \\ \frac{\partial z}{\partial x}(0,t;\omega) = \frac{\partial z}{\partial x}(L,t;\omega) = \frac{\partial w}{\partial x}(0,t;\omega) = \frac{\partial w}{\partial x}(L,t;\omega) = 0, \, 0 \le t \le T, \, \omega \in \Omega, \end{cases}$$
$$(8.93)$$

where $z = z(x,t;\omega)$, $w = w(x,t;\omega)$, $g_i, \widehat{g_i} : [0,T] \times \mathbb{R} \times \mathbb{R} \to \mathbb{R}$ $(i = 1,2)$, and $a_1, a_2, \widehat{a_1}, \widehat{a_2} \in \mathbb{L}^2((0,T))$.

i.) Reformulate (8.93) abstractly as (8.13) in an appropriate space.

ii.) Impose conditions on $g_i, \widehat{g_i}, a_i, \widehat{a_i}$ $(i = 1,2)$ that ensure (8.93) has a unique mild solution on $[0,T]$.

Model V.6 A Stochastic Functional Diffusion-Advection Equation

Let $n \in \mathbb{N}$ and $0 < t_1 < t_2 < \ldots < t_n < T$ be fixed times. Consider the following IBVP governing a diffusive-advective process with accumulative external force:

$$\begin{cases} \partial z + \left(\alpha^2 \frac{\partial^2 z}{\partial x^2} + \gamma \frac{\partial z}{\partial x} \right) \partial t = \left(\sum_{i=1}^n \beta_i(x) z(x,t_i) + \int_0^T \zeta(s) f(s,z) ds \right) \partial t \\ \quad + \left(\int_0^t \frac{g(s)}{1+|g(s)|} ds \right) dW(t), \, 0 < x < L, \, 0 \le t \le T, \, \omega \in \Omega, \\ \frac{\partial z}{\partial x}(0,t;\omega) = \frac{\partial z}{\partial x}(L,t;\omega) = 0, \, 0 \le t \le T, \, \omega \in \Omega, \\ z(x,0;\omega) = z_0(x;\omega), \, 0 < x < L, \, \omega \in \Omega, \end{cases}$$
$$(8.94)$$

where $z = z(x,t;\omega)$, $g : [0,T] \to \mathbb{R}$, and $W(t)$ is a one-dimensional Wiener process. Such an IBVP arises, for instance, when describing atmospheric diffusion properties. (See [335].)

Assume **(H8.4)** through **(H8.6)** (with $\mathscr{D}$ replaced by the interval $[0,L]$). When reformulating (8.94) abstractly as (8.13), we would like to identify the operator A as $\left(\alpha^2 \frac{\partial^2}{\partial x^2} + \gamma \frac{\partial}{\partial x} \right)$. We know that the operator $\alpha^2 \frac{\partial^2}{\partial x^2}$ generates a C_0-semigroup on $\mathbb{L}^2(0,L)$, but does adding the operator $\gamma \frac{\partial}{\partial x}$ prevent the sum operator from also generating a C_0-semigroup on $\mathbb{L}^2(0,L)$? We encountered a similar notion when the perturbing operator is bounded. Note that this is not true in the present scenario. (Why?) Even so, as long as the perturbation does not overpower $\alpha^2 \frac{\partial^2}{\partial x^2}$, we are okay. The answer to this question is given by the following theorem. (See [86, 149] for a proof.)

Theorem 8.4.1. *Assume that $A : \mathrm{dom}(A) \subset \mathscr{X} \to \mathscr{X}$ generates a contractive C_0-semigroup on $\mathscr{X}$ and that $B : \mathrm{dom}(B) \subset \mathscr{X} \to \mathscr{X}$ is a dissipative operator for which $\mathrm{dom}(A) \subset \mathrm{dom}(B)$. If there exist constants $0 \le \delta_1 < 1$ and $\delta_2 \ge 0$ for which*

$$\|Bg\|_{\mathscr{X}} \le \delta_1 \|Ag\|_{\mathscr{X}} + \delta_2 \|g\|_{\mathscr{X}}, \forall g \in \mathrm{dom}(A), \qquad (8.95)$$

then the operator $A + B : \mathrm{dom}(A) \subset \mathscr{X} \to \mathscr{X}$ generates a contractive C_0-semigroup on $\mathscr{X}$.

Remark. If (8.95) holds, we say that B is *A-bounded.*

It can be shown that there exist $0 \leq \delta_1 < 1$ and $\delta_2 \geq 0$ for which

$$\left\| \frac{\partial h}{\partial x} \right\|_{\mathbb{L}^2(0,a)} \leq \delta_1 \left\| -\frac{\partial^2 h}{\partial x^2} \right\|_{\mathbb{L}^2(0,a)} + \delta_2 \|h\|_{\mathbb{L}^2(0,a)}, \forall h \in \mathbb{L}^2(0,a). \tag{8.96}$$

(See **[228]**.) In light of (8.96) and Thrm. 8.4.1, we can conclude that the operator $C : \mathrm{dom}(C) \subset \mathbb{L}^2(0,a) \to \mathbb{L}^2(0,a)$ defined by

$$Cu = \alpha^2 \frac{d^2 u}{dx^2} + \gamma \frac{du}{dx} \tag{8.97}$$

$$\mathrm{dom}(C) = \left\{ u \in \mathbb{L}^2(0,a) \Big| \exists \frac{du}{dx}, \frac{d^2 u}{dx^2}, \frac{d^2 f}{dx^2} \in \mathbb{L}^2(0,a) \wedge u(0) = u(a) = 0 \right\}$$

generates a C_0-semigroup on $\mathbb{L}^2(0,a)$. If the BCs are of Neumann or convective type (as defined in (5.46) and (5.47)), then the domain can be modified in order to draw the same conclusion. Use this fact to complete the following exercise.

Exercise 8.4.4.
i.) Reformulate (8.94) abstractly as (8.13) in $\mathscr{H} = \mathbb{L}^2(0,a)$.
ii.) Impose conditions to ensure that (8.94) has a unique mild solution on $[0,T]$.

Exercise 8.4.5. Consider the following IBVP with logistic forcing:

$$\begin{cases} \partial P_1 = \left(\alpha_{P_1} \frac{\partial^2 P_1}{\partial x^2} + \frac{\partial}{\partial x}(\beta_{P_1} P_1) + \frac{\gamma_1 g_1(P_1,P_2)}{1+\gamma_1|g_1(P_1,P_2)|} \right) \partial t \\ \quad + \frac{\widehat{\gamma}_1 \widehat{g}_1(P_1,P_2)}{1+\widehat{\gamma}_1|\widehat{g}_1(P_1,P_2)|} dW_1(t) + \varepsilon dW_2(t) \, 0 < x < a, \, 0 \leq t \leq T, \, \omega \in \Omega, \\ \partial P_2 = \left(\alpha_{P_2} \frac{\partial^2 P_2}{\partial x^2} + \frac{\partial}{\partial x}(\beta_{P_2} P_2) + \frac{\gamma_2 g_2(P_1,P_2)}{1+\gamma_2|g_2(P_1,P_2)|} \right) \partial t \\ \quad + \frac{\widehat{\gamma}_2 \widehat{g}_2(P_1,P_2)}{1+\widehat{\gamma}_2|\widehat{g}_2(P_1,P_2)|} dW_3(t) + \varepsilon dW_4(t), \, 0 < x < a, \, 0 \leq t \leq T, \, \omega \in \Omega, \\ P_1(x,0;\omega) = P_1^0(x,0;\omega), \, P_2(x,0;\omega) = P_2^0(x,0;\omega), \, 0 < x < a, \, \omega \in \Omega, \\ P_1(0,t;\omega) = P_1(a,t;\omega) = \frac{\partial P_2}{\partial x}(0,t;\omega) = \frac{\partial P_2}{\partial x}(a,t;\omega) = 0, \, 0 \leq t \leq T, \, \omega \in \Omega, \end{cases} \tag{8.98}$$

where $P_i = P_i(x,t;\omega)$, $\varepsilon > 0$, $g_i, \widehat{g}_i : \mathbb{R} \times \mathbb{R} \to \mathbb{R}$, $\alpha_{P_i}, \beta_{P_i}$ $(i=1,2)$ are real constants, $\gamma_i, \widehat{\gamma}_i$ $(i=1,2)$ are positive constants, and $W_i(t)$ $(i=1,2,3,4)$ are independent one-dimensional Wiener processes.
i.) Reformulate (8.98) abstractly as (8.13) in an appropriate space.
ii.) Impose conditions to ensure that (8.98) has a unique mild solution on $[0,T]$.

Model III.4 Spring-Mass Systems with Logistic Forcing

Logistic forcing can be incorporated into the model of a spring-mass system. For instance, consider the following variant of (7.120):

$$\begin{cases} (dx'(t;\omega) + \beta x(t;\omega)) + \eta^2 x(t;\omega)dt = \left(\frac{\alpha g(x)}{1+g(x)}\right) dt \\ + \left(\frac{\widehat{\alpha}\widehat{g}(x)}{1+\widehat{g}(x)}\right) dW(t) = 0, \ 0 < t < T, \ \omega \in \Omega, \\ x(0;\omega) = x_0(\omega), \ x'(0;\omega) = x_1(\omega), \ \omega \in \Omega, \end{cases} \tag{8.99}$$

where $g, \widehat{g} : \mathbb{R} \to (0,\infty)$ and $\alpha, \widehat{\alpha} > 0$.

Exercise 8.4.6.
i.) Reformulate (8.99) abstractly as (8.13) in an appropriate space.
ii.) If $g, \widehat{g} \in \mathbb{C}(\mathbb{R};(0,\infty))$, must (8.99) have a unique mild solution on $[0,T]$?
iii.) Let $h : \mathbb{R} \times \mathbb{R} \to (0,\infty)$ and suppose that $\frac{\alpha g(x)}{1+g(x)}$ is replaced by the more general logistic term $\frac{\alpha h(x,x')}{1+h(x,x')}$, and that $\frac{\widehat{\alpha}\widehat{g}(x)}{1+\widehat{g}(x)}$ is replaced by $\frac{\alpha \widehat{h}(x,x')}{1+\widehat{h}(x,x')}$. Impose sufficient conditions on h and $\widehat{h}$ ensuring that this modified version of IVP (8.99) has a unique mild solution on $[0,T]$.

Exercise 8.4.7. Consider the following coupled system of two springs with accumulative forcing terms where all constants are the same as in (2.9):

$$\begin{cases} d\left(m_A x'_A(t;\omega)\right) + ((k_A + k_B)x_A(t;\omega) + k_A x_B(t;\omega))dt = \\ \left(\int_0^t a_1(t-s)g_1(s,x_A(s;\omega),x_B(s;\omega))ds\right)dt + \alpha_1 x_A(t)dW_1(t) + \beta_1 x_B(t)dW_2(t), \\ d\left(m_B x'_B(t;\omega)\right) - (k_B x_A(t;\omega) - k_B x_B(t;\omega))dt = \\ \left(\int_0^t a_2(t-s)g_2(s,x_A(s;\omega),x_B(s;\omega))ds\right)dt + \widehat{\alpha}_1 x_A(t)dW_1(t) + \widehat{\beta}_1 x_B(t)dW_2(t), \\ x_A(0;\omega) = x_{0,A}(\omega), \ x'_A(0;\omega) = x_{1,A}(\omega), \\ x_B(0;\omega) = x_{0,B}(\omega), \ x'_B(0;\omega) = x_{1,B}(\omega), \end{cases} \tag{8.100}$$

where $0 < t < T$, $\omega \in \Omega$; $\alpha_1, \widehat{\alpha}_1, \beta_1, \widehat{\beta}_1 > 0$; $W_1(t)$ and $W_2(t)$ are independent one-dimensional Wiener processes with finite second moments; and $x_{0,A}, x_{1,A}, x_{0,B}, x_{1,B}$ are $\mathscr{F}_0$-measurable random variables independent of $W_1(t)$ and $W_2(t)$.
i.) Reformulate (8.100) abstractly as (8.13) in an appropriate space.
ii.) Formulate and prove an existence-uniqueness result for (8.100).

Model XII.2 Stochastic Pollution Model

A more elaborate stochastic model of pollution is investigated in **[218]**. We consider a version of this model with different perturbative effects. Let $\mathscr{D}$ be a bounded region in $\mathbb{R}^N$ with smooth boundary $\partial\mathscr{D}$ and let $z(\mathbf{x},t;\omega)$ denote the pollution concentration at position $\mathbf{x} \in \Omega$ at time $0 < t < T$ for $\omega \in \Omega$. Consider the following version of (8.9):

$$
\begin{cases}
dz = \left(k\triangle z + \overrightarrow{\alpha} \cdot \nabla z + \int_0^t a(t-s)g_1(s,z)ds\right) \partial t + \left(\int_0^t \beta g_2(s,z)ds\right) dW(t), \\
z(\mathbf{x},0;\omega) = z_0(\mathbf{x};\omega),\ \mathbf{x} \in \mathscr{D},\ \omega \in \Omega, \\
\frac{\partial z}{\partial \mathbf{n}}(\mathbf{x},t;\omega) = 0,\ \mathbf{x} \in \partial\mathscr{D},\ 0 < t < T,\ \omega \in \Omega,
\end{cases}
$$

(8.101)

where $\mathbf{x} \in \mathscr{D}$, $0 < t < T$, $\omega \in \Omega$, $z = z(\mathbf{x},t;\omega)$, $\beta > 0$, $W(t)$ is a one-dimensional Wiener process, $\frac{\partial z}{\partial \mathbf{n}}$ is the outward unit normal vector to $\partial\mathscr{D}$, and z_0 is an $\mathscr{F}_0$-measurable random variable independent of $W(t)$ with finite second moment. Here, k is the dispersion coefficient, $\overrightarrow{\alpha} \cdot \nabla$ represents the water/air velocity (assumed, for simplicity, to be independent of spatial and temporal variables), and N is the dimension of the region under consideration. For instance, pollution in a river could be modeled by (8.101) with $N = 1$, while describing the concentration of pollution across the surface of infected algae throughout a bay would require that we use $N = 2$. The forcing term in (8.101) can be interpreted as an accumulation of concentration.

Exercise 8.4.8.
i.) Reformulate (8.101) abstractly as (8.13) in an appropriate space.
ii.) Formulate and prove an existence-uniqueness result for (8.101).

Exercise 8.4.9. Formulate a continuous dependence result, with respect to $\overrightarrow{\alpha}$ and β only, for (8.101). What happens to the mild solution of (8.101), as guaranteed to exist by Exer. 8.4.8 (ii), as $\left\|\overrightarrow{\alpha}\right\|_{\mathbb{R}^N} \longrightarrow 0$ and $\beta \to 0$?

We now introduce several new models.

Model XIV.1 Epidemiological Models
Diffusive phenomena occur in a wide variety of settings ranging from the spreading of ideas and rumors in a social setting **[91]**, to the spread of a virus through a region **[134]** or even a worm through the Internet, to the effects of predation on rain forests **[43]** and wetlands. The study of *epidemiology* is concerned with developing models that describe the evolution of such spreading. We consider such models below.

Many variables affect population density (e.g., demographics, environment, geographic elements, etc.). We begin with a version of a classical two-dimensional model explored in **[120]**. Let $\mathscr{D} = [0,L_1] \times [0,L_2]$ represent the spatial region of interest and $t > 0$. We consider two interrelated populations, given by

$P_H = P_H(x,y,t;\omega) = $ Host population at position $(x,y) \in \mathscr{D}$ at time $t > 0$ and $\omega \in \Omega$,
$P_V = P_V(x,y,t;\omega) = $ Viral population at position $(x,y) \in \mathscr{D}$ at time $t > 0$ and $\omega \in \Omega$.

Assume that:
(a) Both P_H and P_V are subject to diffusion (with diffusion constants α_H and α_V).
(b) The birth rate of the host population is the positive constant β_H.

(c) The rate at which the virus becomes inviable is the positive constant γ_V.
(d) The virus is transmitted via human interaction.

We account for noise in the measurement of the rates β_H and γ_V so that

$$\beta_H = \beta_{1,H} + \beta_{2,H}\frac{dW_0(t)}{dt},$$

$$\gamma_V = \gamma_{1,V} + \gamma_{2,V}\frac{dW_1(t)}{dt},$$

where $W_0(t)$ and $W_1(t)$ are independent one-dimensional Wiener processes. We account for nonlinearity in the transmission dynamics via the forcing term g and arrive at the IBVP

$$\begin{cases} \partial P_H = (\alpha_H \triangle P_H + \beta_{1,H}P_H - g\,(t,x,y,P_H,P_V)\,P_H P_V)\,\partial t + \beta_{2,H}dW_0(t), \\ \partial P_V = (\alpha_V \triangle P_V - \gamma_{1,V}P_V + g\,(t,x,y,P_H,P_V)\,P_H P_V)\,\partial t + \gamma_{2,V}dW_1(t), \\ P_H(x,y,0;\omega) = P_H^0(x,y;\omega),\ P_V(x,y,0;\omega) = P_V^0(x,y;\omega),\ (x,y) \in \mathscr{D}, \omega \in \Omega, \\ \frac{\partial P_H}{\partial \mathbf{n}}(x,y,t;\omega) = \frac{\partial P_V}{\partial \mathbf{n}}(x,y,t;\omega) = 0,\ (x,y) \in \partial\mathscr{D},\ 0 < t < T,\ \omega \in \Omega, \end{cases} \tag{8.102}$$

where $(x,y) \in \mathscr{D}, 0 < t < T, \omega \in \Omega$. Observe that (8.102) is equivalent to

$$\begin{cases} \partial\begin{bmatrix} P_H \\ P_V \end{bmatrix} = \begin{bmatrix} \alpha_H\triangle + \beta_{1,H}I & 0 \\ 0 & \alpha_V\triangle - \gamma_{1,V}I \end{bmatrix}\begin{bmatrix} P_H \\ P_V \end{bmatrix} + \begin{bmatrix} -g\,(t,x,y,P_H,P_V)\,P_H P_V \\ g\,(t,x,y,P_H,P_V)\,P_H P_V \end{bmatrix} \\ \quad + \begin{bmatrix} \beta_{2,H} & 0 \\ 0 & \gamma_{2,V} \end{bmatrix} d\begin{bmatrix} W_0(t) \\ W_1(t) \end{bmatrix},\ (x,y) \in \mathscr{D}, 0 < t < T, \omega \in \Omega, \\ \begin{bmatrix} P_H \\ P_V \end{bmatrix}(x,y,0;\omega) = \begin{bmatrix} P_H^0 \\ P_V^0 \end{bmatrix}(x,y;\omega),\ (x,y) \in \mathscr{D}, \omega \in \Omega, \\ \frac{\partial}{\partial \mathbf{n}}\begin{bmatrix} P_H \\ P_V \end{bmatrix}(x,y,t;\omega) = 0,\ (x,y) \in \partial\mathscr{D}, 0 < t < T, \omega \in \Omega, \end{cases}$$

$$\tag{8.103}$$

where $g : [0,\infty) \times \mathscr{D} \times [0,\infty) \times [0,\infty) \to [0,\infty)$ is a continuous mapping.

Exercise 8.4.10.
i.) Reformulate (8.103) abstractly as (6.31) in an appropriate space.
ii.) Formulate and prove an existence-uniqueness result for (8.103).
iii.) Formulate a zeroth-order approximation result directly for (8.103).

Exercise 8.4.11. Assume that

$$g(t,x,y,P_H,P_V) = \int_0^t \frac{\eta P_H(x,y,s;\omega)P_V(x,y,s;\omega)}{(1+P_H(x,y,s;\omega)+P_V(x,y,s;\omega))^2}ds, \tag{8.104}$$

so that the rate of infection/transmission increases over time due to the fact that more people are being infected. Does the IBVP (8.103), where g is defined as in (8.104), have a unique mild solution on $[0,T]$, or must additional restrictions be imposed on the data in order to draw this conclusion?

Next, we incorporate a more general type of dispersion into (8.103) by adding appropriate integral terms. Doing so yields the modified forcing term

$$\begin{bmatrix} -g\left(t,x,y,P_H,P_V\right)P_HP_V + \int_\Omega a_1(x,y,w,z)P_H(w,z,t;\omega)dwdz \\ g\left(t,x,y,P_H,P_V\right)P_HP_V + \int_\Omega a_2(x,y,w,z)P_V(w,z,t;\omega)dwdz \end{bmatrix}, \tag{8.105}$$

where $a_i : \mathscr{D} \times \mathscr{D} \to (0,\infty)$ $(i=1,2)$ are continuous mappings and g is globally Lipschitz in its last four variables.

Exercise 8.4.12. Show that (8.105) can be expressed as a functional

$$\mathfrak{F} : \mathbb{C}\left([0,T];\mathscr{L}^2\left(\Omega;\left(\mathbb{L}^2(\mathscr{D})\right)^2\right)\right) \to \mathbb{L}^2\left(0,T;\mathscr{L}^2\left(\Omega;\left(\mathbb{L}^2(\mathscr{D})\right)^2\right)\right).$$

Is $\mathfrak{F}$ Lipschitz?

The complexity of the model increases if there exist N different strains of the virus, each of which attacks the population separately and is governed by its own dispersal and infection rates. Assuming no interaction among strains leads to the following system of $N+1$ equations:

$$\begin{cases} \partial P_H = \left(\alpha_H \triangle P_H + \beta_{1,H} P_H - \sum_{i=1}^N g_i\left(t,x,y,P_H,P_{V_1},\ldots,P_{V_N}\right)P_H P_{V_i}\right)\partial t \\ \quad + \beta_{2,H} dW_0(t), \ (x,y) \in \mathscr{D}, 0 < t < T, \ \omega \in \Omega, \\ \partial P_{V_1} = \left(\alpha_{V_1} \triangle P_{V_1} - \sum_{i=1}^N \gamma_{1,V_i} P_{V_i} + \sum_{i=1}^N g_i\left(t,x,y,P_H,P_{V_1},\ldots,P_{V_N}\right)P_H P_{V_i}\right)\partial t \\ \quad + \sum_{i=1}^N \gamma_{2,V_i} dW_i(t), \ (x,y) \in \mathscr{D}, 0 < t < T, \ \omega \in \Omega, \\ \vdots \\ \partial P_{V_N} = \left(\alpha_{V_N} \triangle P_{V_N} - \sum_{i=1}^N \gamma_{1,V_i} P_{V_i} + \sum_{i=1}^N g_i\left(t,x,y,P_H,P_{V_1},\ldots,P_{V_N}\right)P_H P_{V_i}\right)\partial t \\ \quad + \sum_{i=1}^N \gamma_{2,V_i} dW_i(t), \ (x,y) \in \mathscr{D}, 0 < t < T, \ \omega \in \Omega, \\ P_H(x,y,0;\omega) = P_H^0(x,y;\omega), \ P_{V_i}(x,y,0;\omega) = P_{V_i}^0(x,y;\omega), \ (x,y) \in \mathscr{D}, i = 1,\ldots,N, \\ \frac{\partial P_H}{\partial \mathbf{n}}(x,y,t;\omega) = \frac{\partial P_{V_i}}{\partial \mathbf{n}}(x,y,t;\omega) = 0, \ (x,y) \in \partial\mathscr{D}, 0 < t < T, i = 1,\ldots,N, \end{cases} \tag{8.106}$$

where $P_H = P_H(x,y,t;\omega)$, $P_{V_i} = P_{V_i}(x,y,t;\omega)$, all constants are positive, and $W_i(t)$ $(i=1,\ldots,N)$ are independent one-dimensional Wiener processes.

Exercise 8.4.13.
i.) Reformulate (8.106) abstractly as (8.13) in an appropriate space.
ii.) Formulate and prove an existence-uniqueness result for (8.106).

Exercise 8.4.14. Establish a continuous dependence result for (8.106) in terms of the dispersal and rate constants.

We could further subdivide the host population into subclasses based on susceptibility, age, and other factors to generate similar systems, albeit involving more equations and more complex nonlinearities.

Model XV.1 Aeroelasticity – A Linear Approximation

Airplane wings are designed to bend and flap in a controlled manner during flight. Helicopter rotor blades undergo vibrations whose dynamics depend on the material with which the blades are composed, aerodynamic forces, etc. An important part of ensuring successful flight is controlling the vertical displacement of the wing/rotor blades in order to prevent *flutter*, an increasingly rapid and potentially destructive and uncontrollable vibration.

We consider a model introduced by Dowell in 1975 (see [141]) describing the deflection of a rectangular panel devoid of two-dimensional effects, meaning that we track only the cross-section of an edge of the panel. We further assume that the edges are supported so that there is no movement along them. If this panel is part of an aircraft moving through an airstream, it makes sense that it will deform as forces due to wind act on it; otherwise, it would snap. (If you have ever glanced out the window during a plane ride, you have undoubtedly noticed a small bounce in the wing.) We also expect that the panel has a natural state to which it reverts in the absence of external forces, because otherwise the aircraft would permanently deform as a result of even the most insignificant of forces acting on it. Detailed discussions can be found in [34, 52, 141, 146, 289].

We model the panel as a one-dimensional segment, say $[0,a]$, because all cross-sections of the panel are assumed to be identical. (Accounting for twisting of the panel would lead to a more complicated nonlinear model, mentioned later in Chapter 9.)

Let $w = w(z,t)$ represent the panel deflection at (z,t), where $0 \le z \le a, t > 0$. Dowell considered the following second-order IBVP that serves as a linearized approximation of panel flutter:

$$
\begin{cases}
\frac{\partial^2 w}{\partial t^2} + \beta_1 \frac{\partial w}{\partial t} + \beta_2 \frac{\partial}{\partial t}\left(\frac{\partial^4 w}{\partial z^4}\right) + \frac{\partial^4 w}{\partial z^4} - \beta_3 \frac{\partial^2 w}{\partial z^2} + \beta_4 \frac{\partial w}{\partial z} = 0, \\
w(0,t) = w(a,t) = 0, t > 0, \\
\frac{\partial^2 w}{\partial z^2}(0,t) + \beta_2 \frac{\partial}{\partial t}\left(\frac{\partial^2 w}{\partial z^2}\right)(0,t) = 0, t > 0, \\
\frac{\partial^2 w}{\partial z^2}(a,t) + \beta_2 \frac{\partial}{\partial t}\left(\frac{\partial^2 w}{\partial z^2}\right)(a,t) = 0, t > 0, \\
w(z,0) = w_0(z), \frac{\partial w}{\partial t}(z,0) = w_1(z), 0 \le z \le a.
\end{cases}
\tag{8.107}
$$

The physical parameters $\beta_1, \beta_2, \beta_3, \beta_4$ are assumed to be positive constants and represent the measures of viscoelastic structural damping, aerodynamic pressure, inplane tensile load, and aerodynamic damping, respectively.

Now, noise can be introduced into this model in various ways. For instance, assuming that the viscoelastic structural damping parameter β_1 is replaced by

$$
\widehat{\beta}_1 + \widehat{\beta}_2 \frac{dW}{dt}
$$

in the deterministic IBVP (8.107) yields the stochastic IBVP

$$
\begin{cases}
\partial\left(\frac{\partial w}{\partial t} + \beta_1 w + \beta_2\left(\frac{\partial^4 w}{\partial z^4}\right)\right) + \left(\frac{\partial^4 w}{\partial z^4} - \beta_3\frac{\partial^2 w}{\partial z^2} + \beta_4\frac{\partial w}{\partial z}\right)\partial t + \widehat{\beta_2}dW(t) = 0, \\
w(0,t;\omega) = w(a,t;\omega) = 0, t > 0, \omega \in \Omega, \\
\frac{\partial^2 w}{\partial z^2}(0,t;\omega) + \beta_2\frac{\partial}{\partial t}\left(\frac{\partial^2 w}{\partial z^2}\right)(0,t;\omega) = 0, t > 0, \omega \in \Omega, \\
\frac{\partial^2 w}{\partial z^2}(a,t;\omega) + \beta_2\frac{\partial}{\partial t}\left(\frac{\partial^2 w}{\partial z^2}\right)(a,t;\omega) = 0, t > 0, \omega \in \Omega, \\
w(z,0;\omega) = w_0(z;\omega), \ \frac{\partial w}{\partial t}(z,0;\omega) = w_1(z;\omega), 0 \le z \le a, \omega \in \Omega.
\end{cases}
$$

(8.108)

We want to reformulate (8.108) as an abstract stochastic evolution equation. But, which of the forms studied thus far is most appropriate? The equation involves several more terms than previous models, and it is not initially clear which terms should be used to form the operator A and which should be subsumed into the forcing term. We outline one possible reformulation of (8.108), the deterministic version of which is developed in [289], below.

First, the presence of a second-order time derivative suggests that viewing (8.108) as a system of two equations, as we did in the study of the classical wave equation, might be a prudent first step. To this end, we use the Hilbert space $\mathscr{H} = \mathbb{H}^2(0,a) \times \mathbb{L}^2(0,a)$ equipped with the following inner product and norm:

$$
\left\langle \begin{bmatrix} f \\ \frac{\partial f}{\partial t} \end{bmatrix}, \begin{bmatrix} g \\ \frac{\partial g}{\partial t} \end{bmatrix} \right\rangle_{\mathscr{H}} \equiv \left\langle \frac{\partial^2 f}{\partial z^2}, \frac{\partial^2 g}{\partial z^2} \right\rangle_{\mathbb{L}^2(0,a)} + \left\langle \frac{\partial f}{\partial t}, \frac{\partial g}{\partial t} \right\rangle_{\mathbb{L}^2(0,a)}
$$

(8.109)

$$
\left\| \begin{bmatrix} f \\ \frac{\partial f}{\partial t} \end{bmatrix} \right\|_{\mathscr{H}}^2 \equiv \left\| \frac{\partial^2 f}{\partial z^2} \right\|_{\mathbb{L}^2(0,a)}^2 + \left\| \frac{\partial f}{\partial t} \right\|_{\mathbb{L}^2(0,a)}^2
$$

$$
= \int_0^a \left[\left| \frac{\partial^2 f}{\partial z^2}(y,\cdot) \right|^2 + \left| \frac{\partial f}{\partial t}(y,\cdot) \right|^2 \right] dy.
$$

(8.110)

The stochastic PDE in (8.108) can be written equivalently as

$$
\partial \begin{bmatrix} w \\ \frac{\partial w}{\partial t} \end{bmatrix} = \begin{bmatrix} 0 & I \\ -\frac{\partial^4}{\partial z^4} & -\beta_2\frac{\partial^4}{\partial z^4} \end{bmatrix} \begin{bmatrix} w \\ \frac{\partial w}{\partial t} \end{bmatrix} \partial t
$$

$$
+ \begin{bmatrix} 0 & 0 \\ \beta_3\frac{\partial^2}{\partial z^2} - \beta_4\frac{\partial}{\partial z} & -\beta_1 I \end{bmatrix} \begin{bmatrix} w \\ \frac{\partial w}{\partial t} \end{bmatrix} \partial t + \begin{bmatrix} \widehat{\beta_2} \\ 0 \end{bmatrix} dW(t).
$$

(8.111)

Symbolically, it is reasonable to identify the unknown $U(t) = \begin{bmatrix} w \\ \frac{\partial w}{\partial t} \end{bmatrix}$ and the operators $A : \mathrm{dom}(A) \subset \mathscr{H} \to \mathscr{H}$ and $B : \mathrm{dom}(B) \subset \mathscr{H} \to \mathscr{H}$ by

$$
A \begin{bmatrix} w \\ \frac{\partial w}{\partial t} \end{bmatrix} = \begin{bmatrix} 0 & I \\ -\frac{\partial^4}{\partial z^4} & -\beta_2\frac{\partial^4}{\partial z^4} \end{bmatrix} \begin{bmatrix} w \\ \frac{\partial w}{\partial t} \end{bmatrix},
$$

(8.112)

$$
B \begin{bmatrix} w \\ \frac{\partial w}{\partial t} \end{bmatrix} = \begin{bmatrix} 0 & 0 \\ \beta_3\frac{\partial^2}{\partial z^2} - \beta_4\frac{\partial}{\partial z} & -\beta_1 I \end{bmatrix} \begin{bmatrix} w \\ \frac{\partial w}{\partial t} \end{bmatrix}.
$$

(8.113)

These identfications render (8.108) as an abstract <u>homogenous</u> stochastic evolution equation similar to (5.94)!

Exercise 8.4.15. Keeping in mind the space and the BCs, determine $\text{dom}(A)$ and $\text{dom}(B)$. Why are the operators A and B well-defined?

It can be shown that $(A, \text{dom}(A))$ is a dissipative operator and $(B, \text{dom}(B))$ is a bounded linear operator. As such, $A + B$ generates a C_0-semigroup on $\mathscr{H}$. (Why?)

Exercise 8.4.16.
i.) Prove that (8.108) has a unique mild solution on $[0, T]$.
ii.) Determine $\mathbf{y} \in \mathbb{C}\left([0, T]; \mathcal{L}^2\left(\Omega; \mathscr{H}\right)\right)$ such that

$$\lim_{\beta_2 \to 0} \left\| \mathbf{y} - \begin{bmatrix} w \\ \frac{\partial w}{\partial t} \end{bmatrix} \right\|_{\mathbb{C}} = 0.$$

As with the other models explored in this chapter, we can account for an accumulation effect of external forces by adding the following forcing term on the right-hand side of the equation in (8.108):

$$g\left(t, z, \int_0^t a_1(t, s) w(s, z) ds, \int_0^t a_2(t, s) \frac{\partial w}{\partial s}(s, z) ds\right) dt \qquad (8.114)$$

A similar term can be used to replace $\widehat{\beta}_2 dW(t)$.

Exercise 8.4.17. Consider the IBVP obtained by adding the forcing term (8.114) to the right-hand side of (8.108) and keeping all other conditions the same.
i.) Reformulate (8.108) abstractly as (8.13) in an appropriate space.
ii.) Formulate and prove an existence-uniqueness result for (8.108).

Model XVI.1 Transverse Vibrations of Extensible Beams
We now consider a model similar to the one used to describe waves in a vibrating elastic string, but now we account for a different kind of vibration, namely transverse vibrations of a uniform bar (see [**32, 35, 135, 271, 327**]).

Let $w = w(z, t; \omega)$ represent the deflection of the beam at position $0 \le z \le a$ at time $t > 0$ for $\omega \in \Omega$. The most rudimentary equation, coupled with its initial profile, involved in describing such a phenomenon is given by

$$\begin{cases} \partial\left(\frac{\partial w}{\partial t}\right) + \alpha \frac{\partial^4 w}{\partial z^4} \partial t + \varepsilon dW(t) = 0, \ 0 < z < a, t > 0, \omega \in \Omega, \\ w(z, 0; \omega) = w_0(z; \omega), \ \frac{\partial w}{\partial t}(z, 0; \omega) = w_1(z; \omega), \ 0 < z < a, \omega \in \Omega, \end{cases} \qquad (8.115)$$

where $\alpha > 0$ describes the bending stiffness, $\varepsilon > 0$, and $W(t)$ is a one-dimensional Wiener process. A complete description of this phenomenon requires that we prescribe what happens at both ends of the rod. This can be done in many natural ways, two of which are described below:

1. Clamped at $z = a$:

$$\frac{\partial w}{\partial z}(a,t;\omega) = w(a,t;\omega) = 0, t > 0, \omega \in \Omega; \qquad (8.116)$$

2. Hinged at $z = a$:

$$\frac{\partial^2 w}{\partial z^2}(a,t;\omega) = w(a,t;\omega) = 0, t > 0, \omega \in \Omega. \qquad (8.117)$$

Exercise 8.4.18. Consider (8.115) equipped with the following four BCs:

$$\frac{\partial w}{\partial z}(0,t;\omega) = \frac{\partial^2 w}{\partial z^2}(0,t;\omega) = \frac{\partial w}{\partial z}(a,t;\omega) = \frac{\partial^2 w}{\partial z^2}(a,t;\omega) = 0, \qquad (8.118)$$

where $t > 0, \omega \in \Omega$.
i.) Assume $\varepsilon = 0$ in (8.115) and solve (8.115) coupled with (8.118) using the separation of variables method.
ii.) Using the form of the solution obtained in (i), conjecture the form of the semigroup.

As in our discussion of the wave equation, we can view (8.115) abstractly by making the following change of variable:

$$v_1 = w, \; v_2 = \frac{\partial w}{\partial t}, \; \frac{\partial v_1}{\partial t} = v_2, \; \frac{\partial v_2}{\partial t} = -\alpha \frac{\partial^4 v_1}{\partial z^4}. \qquad (8.119)$$

Then, (8.115) becomes

$$\begin{cases} \partial \begin{bmatrix} v_1 \\ v_2 \end{bmatrix}(z,t;\omega) = \begin{bmatrix} 0 & I \\ -\alpha \frac{\partial^4}{\partial z^4} & 0 \end{bmatrix} \begin{bmatrix} v_1 \\ v_2 \end{bmatrix}(z,t;\omega)\partial t + \begin{bmatrix} \varepsilon \\ 0 \end{bmatrix} dW(t), \\[2mm] \begin{bmatrix} v_1 \\ v_2 \end{bmatrix}(z,0;\omega) = \begin{bmatrix} w_0 \\ w_1 \end{bmatrix}(z,0;\omega), \; 0 < z < a, \; \omega \in \Omega, \\[2mm] \frac{\partial w}{\partial z}(0,t;\omega) = \frac{\partial^2 w}{\partial z^2}(0,t;\omega) = \frac{\partial w}{\partial z}(a,t;\omega) = \frac{\partial^2 w}{\partial z^2}(a,t;\omega) = 0, t > 0, \omega \in \Omega, \end{cases}$$
$$(8.120)$$

where $0 < z < a, t > 0, \omega \in \Omega$. Guided by our discussion of the wave equation, we shall reformulate (8.120) as an abstract stochastic evolution equation in the Hilbert space

$$\mathscr{H} = \mathrm{dom}\left(-\frac{d^2}{dz^2}\right) \times \mathbb{L}^2(0,a) \qquad (8.121)$$

$$\left\langle \begin{bmatrix} v_1 \\ v_2 \end{bmatrix}, \begin{bmatrix} v_1^\star \\ v_2^\star \end{bmatrix} \right\rangle_{\mathscr{H}} \equiv \int_0^a \left[\frac{\partial^2 v_1}{\partial z^2} \frac{\partial^2 v_1^\star}{\partial z^2} + v_2 v_2^\star \right] dz \qquad (8.122)$$

by defining the operator $A : \text{dom}(A) \subset \mathscr{H} \to \mathscr{H}$ by

$$A \begin{bmatrix} v_1 \\ v_2 \end{bmatrix} = \begin{bmatrix} 0 & I \\ -\alpha \frac{\partial^4}{\partial z^4} & 0 \end{bmatrix} \begin{bmatrix} v_1 \\ v_2 \end{bmatrix} = \begin{bmatrix} v_1 \\ -\alpha \frac{\partial^4 v_2}{\partial z^4} \end{bmatrix}, \qquad (8.123)$$

$$\text{dom}(A) = \text{dom}\left(-\frac{d^4}{dz^4}\right) \times \text{dom}\left(-\frac{d^2}{dz^2}\right),$$

where

$$\text{dom}\left(-\frac{d^4}{dz^4}\right) = \left\{ w \in \mathbb{L}^2(0,a) \,\Big|\, w, \frac{\partial w}{\partial z}, \frac{\partial^2 w}{\partial z^2}, \frac{\partial^3 w}{\partial z^3} \text{ are AC, } \frac{\partial^4 w}{\partial z^4} \in \mathbb{L}^2(0,a), \wedge \right.$$

$$\left. \frac{\partial w}{\partial z}(0,\cdot) = \frac{\partial^2 w}{\partial z^2}(0,\cdot) = \frac{\partial w}{\partial z}(a,\cdot) = \frac{\partial^2 w}{\partial z^2}(a,\cdot) = 0 \right\}. \qquad (8.124)$$

and

$$\text{dom}\left(-\frac{d^2}{dz^2}\right) = \left\{ w \in \mathbb{L}^2(0,a) \,\Big|\, w, \frac{\partial w}{\partial z} \text{ are AC, } \frac{\partial^2 w}{\partial z^2} \in \mathbb{L}^2(0,a), \wedge \right.$$

$$\left. \frac{\partial w}{\partial z}(0,\cdot) = \frac{\partial^2 w}{\partial z^2}(0,\cdot) = \frac{\partial w}{\partial z}(a,\cdot) = \frac{\partial^2 w}{\partial z^2}(a,\cdot) = 0 \right\}. \qquad (8.125)$$

It can be shown that $(\mathscr{H}, \langle \cdot, \cdot \rangle_{\mathscr{H}})$ is a Hilbert space (see **[398]**) and that A generates a C_0-semigroup on $\mathscr{H}$ (see Volume 1). As such, the IBVP (8.120) can be reformulated as the abstract stochastic evolution equation (5.94). As such, we know that the IVP (8.115) coupled with (8.118) has a unique mild solution on $[0,T]$.

Exercise 8.4.19. Let $\beta > 0$. Prove that the following IBVP has a unique classical solution:

$$\begin{cases} \partial\left(\frac{\partial w}{\partial t}\right) + \left(\alpha \frac{\partial^4 w}{\partial z^4} + \beta w\right)\partial t + \varepsilon dW(t) = 0, \ 0 < z < a, t > 0, \omega \in \Omega, \\ w(z,0;\omega) = w_0(z;\omega), \ \frac{\partial w}{\partial t}(z,0;\omega) = w_1(z;\omega), \ 0 \leq z \leq a, \omega \in \Omega, \\ \frac{\partial^2 w}{\partial z^2}(0,t;\omega) = w(0,t;\omega) = 0 = \frac{\partial^2 w}{\partial z^2}(a,t;\omega) = w(a,t;\omega), t > 0, \omega \in \Omega. \end{cases}$$
$$(8.126)$$

Next, we incorporate additional physical terms into the IBVP to improve the model. For instance, consider the following more general IBVP:

$$\begin{cases} \partial\left(\frac{\partial w}{\partial t}\right) + \left(\alpha \frac{\partial^4 w}{\partial z^4} + \beta w\right)\partial t = g\left(t,z,w(z,t)\right)\partial t \\ \quad + \sum_{k=1}^{m} \overline{g_k}\left(t,z,w(z,t)\right) dW_k(t), \ 0 < z < a, t > 0, \omega \in \Omega, \\ w(z,0;\omega) = w_0(z;\omega), \ \frac{\partial w}{\partial t}(z,0;\omega) = w_1(z;\omega), \ 0 \leq z \leq a, \omega \in \Omega, \\ \frac{\partial^2 w}{\partial z^2}(0,t;\omega) = w(0,t;\omega) = 0 = \frac{\partial^2 w}{\partial z^2}(a,t;\omega) = w(a,t;\omega), t > 0, \omega \in \Omega, \end{cases}$$
$$(8.127)$$

where $g : [0,T] \times [0,a] \times \mathbb{R} \to \mathbb{R}$ is continuous on $[0,T] \times [0,a]$ and globally Lipschitz in the third variable (uniformly in (t,z)) with Lipschitz constant M_g. We impose the same conditions on g_k ($k = 1, \ldots, m$) and label the Lipschitz constants as M_{g_k}. We can reformulate (8.127) abstractly as (7.26), in which the forcing term $f : [0, T] \times \mathcal{H} \to \mathcal{H}$ (where $\mathcal{H}$ is given by (8.121)) is defined by

$$f\left(t, \begin{pmatrix} v_1 \\ v_2 \end{pmatrix}\right)(\cdot) = \begin{bmatrix} 0 \\ g(t, \cdot, v_1) \end{bmatrix}. \tag{8.128}$$

To see that f is globally Lipschitz on $\mathcal{H}$, observe that

$$\left\| f\left(t, \begin{pmatrix} v_1 \\ v_2 \end{pmatrix}\right) - f\left(t, \begin{pmatrix} v_1^\star \\ v_2^\star \end{pmatrix}\right) \right\|_{\mathcal{H}}^2 = \left\| \begin{bmatrix} 0 \\ g(t, \cdot, v_1) \end{bmatrix} - \begin{bmatrix} 0 \\ g(t, \cdot, v_1^\star) \end{bmatrix} \right\|_{\mathcal{H}}^2$$

$$= \int_0^a |g(t, v_1(t, z)) - g(t, v_1^\star(t, z))|^2 \, dz$$

$$\leq M_g^2 \int_0^a |v_1(t, z) - v_1^\star(t, z)|^2 \, dz \tag{8.129}$$

$$\leq M_g^2 \left\| \begin{bmatrix} v_1 \\ v_2 \end{bmatrix} - \begin{bmatrix} v_1^\star \\ v_2^\star \end{bmatrix} \right\|_{\mathcal{H}}^2.$$

Exercise 8.4.20. Carefully explain how the noise term is handled in (8.127). Specifically, reformulate (8.127) abstractly as (7.26) and then invoke Thrm. 7.4.2 to conclude that (8.127) has a unique mild solution on $[0, a]$.

Exercise 8.4.21. Let $\gamma > 0$. Modify the above reasoning to deduce that the following IBVP has a unique mild solution on $[0, a]$:

$$\begin{cases} \partial \left(\frac{\partial w}{\partial t}\right) + \left(\alpha \frac{\partial^4 w}{\partial z^4} + \beta w - \gamma \frac{\partial^2 w}{\partial z^2}\right) \partial t + g(t) dW(t) = 0, \ 0 < z < a, t > 0, \omega \in \Omega, \\ w(z, 0; \omega) = w_0(z; \omega), \ \frac{\partial w}{\partial t}(z, 0; \omega) = w_1(z; \omega), \ 0 \leq z \leq a, \omega \in \Omega, \\ \frac{\partial^2 w}{\partial z^2}(0, t; \omega) = w(0, t; \omega) = 0 = \frac{\partial^2 w}{\partial z^2}(a, t; \omega) = w(a, t; \omega), t > 0, \omega \in \Omega, \end{cases} \tag{8.130}$$

where $\alpha, \beta, \gamma > 0$, and $g : [0, T] \to \mathbb{R}$ is continuous.

Exercise 8.4.22.
i.) Let $\delta > 0$. Show that the following IBVP has a unique mild solution on $[0, a]$:

$$\begin{cases} \partial \left(\frac{\partial w}{\partial t} + \delta w\right) + \left(\alpha \frac{\partial^4 w}{\partial z^4} + \beta w - \gamma \frac{\partial^2 w}{\partial z^2}\right) \partial t + g(t) dW(t) = 0, \\ w(z, 0; \omega) = w_0(z; \omega), \ \frac{\partial w}{\partial t}(z, 0; \omega) = w_1(z; \omega), \ 0 \leq z \leq a, \omega \in \Omega, \\ \frac{\partial^2 w}{\partial z^2}(0, t; \omega) = w(0, t; \omega) = 0 = \frac{\partial^2 w}{\partial z^2}(a, t; \omega) = w(a, t; \omega), t > 0, \omega \in \Omega, \end{cases} \tag{8.131}$$

where $0 < z < a, t > 0, \omega \in \Omega$,

ii.) More generally, suppose that the term $\partial(\delta w)$ is replaced by $\partial \left(\left(\frac{\partial w}{\partial t}\right)^2 \right)$. Argue that the same conclusion as in (i) holds.
iii.) Argue that if $J : \mathbb{R} \to \mathbb{R}$ is an increasing globally Lipschitz function, then the

IBVP (8.131) with $\partial(\delta w)$ replaced by $\partial\left(J\left(\frac{\partial w}{\partial t}\right)\right)$ has a unique mild solution on $[0,a]$.

Going one step further, we incorporate a more general forcing term into the above IBVPs to obtain

$$
\begin{cases}
\partial\left(\frac{\partial w}{\partial t}\right) + \left(\alpha\frac{\partial^4 w}{\partial z^4} + \beta w - \gamma\frac{\partial^2 w}{\partial z^2}\right)\partial t = \left(\int_0^t a(t,s)f\left(s,z,w,\frac{\partial^2 w}{\partial z^2},\frac{\partial w}{\partial s}\right)ds\right)\partial t \\
\quad + \left(\int_0^t b(t,s)g\left(s,z,w,\frac{\partial^2 w}{\partial z^2},\frac{\partial w}{\partial s}\right)ds\right)d\mathbf{W}(t),\ 0 < z < a, t > 0,\ \omega \in \Omega, \\
w(z,0;\omega) = w_0(z;\omega),\ \frac{\partial w}{\partial t}(z,0;\omega) = w_1(z;\omega),\ 0 \le z \le a,\ \omega \in \Omega, \\
\frac{\partial^2 w}{\partial z^2}(0,t;\omega) = w(0,t;\omega) = 0 = \frac{\partial^2 w}{\partial z^2}(a,t;\omega) = w(a,t;\omega),\ t > 0,\ \omega \in \Omega,
\end{cases}
$$
$$(8.132)$$

where $w = w(z,t;\omega)$, $f,g : [0,T] \times [0,a] \times \mathbb{R} \times \mathbb{R} \times \mathbb{R} \to \mathbb{R}$ are continuous on $[0,T] \times [0,a]$ and globally Lipschitz in the last three variables (uniformly in (t,z)), and $a,b : [0,T] \times [0,T] \to \mathbb{R}$ are continuous. Of course, this is only one of many different types of possible forcing terms. Observe that (8.132) is a particular case of the abstract functional evolution equation

$$
\begin{cases}
dX(t;\omega) + (AX(t;\omega) + BX(t;\omega))dt = \mathfrak{F}(X)(t;\omega)dt \\
\quad + \mathfrak{G}(X)(t;\omega)d\mathbf{W}(t), \\
X(0;\omega) = X_0(\omega),\ \omega \in \Omega,
\end{cases}
$$
$$(8.133)$$

in a separable Hilbert space $\mathscr{H}$, where $0 < t < T, \omega \in \Omega$; $U(\cdot) = \begin{bmatrix} v_1(\cdot) \\ v_2(\cdot) \end{bmatrix}$, $A : \mathrm{dom}(A) \subset \mathscr{H} \to \mathscr{H}$ and $B : \mathrm{dom}(B) \subset \mathscr{H} \to \mathscr{H}$ are operators satisfying the usual assumptions;

$$\mathfrak{F} : \mathbb{C}\left([0,T];\mathscr{L}^2\left(\Omega;\mathscr{H}\right)\right) \to \mathbb{C}\left([0,T];\mathscr{L}^2\left(\Omega;\mathscr{H}\right)\right)$$

$$\mathfrak{G} : \mathbb{C}\left([0,T];\mathscr{L}^2\left(\Omega;\mathscr{H}\right)\right) \to \mathbb{C}\left([0,T];\mathscr{L}^2\left(\Omega;\mathscr{B}_0\left(\mathbb{R}^m;\mathscr{H}\right)\right)\right)$$

are Lipschitz functionals; and $\mathbf{W}(t)$ is an m-dimensional Wiener process.

Exercise 8.4.23. Prove that (8.133) has a unique mild solution on $[0,T]$.

Remark. The study of fiber dynamics (or elastodynamics) is related to that of beam dynamics, but for a material that has different inherent characteristics. The goal is to model the motion of long flexible fibers in a moving "fluid," such as an airstream or liquid. There have been numerous articles written on this subject (see [191, 286]). The following simple stochastic linearized model of the horizontal component of the fiber is very similar to the beam models discussed above:

$$
\begin{cases}
\partial\left(\frac{\partial w}{\partial t}\right) + \left(\alpha\frac{\partial^4 w}{\partial z^4} - \beta\frac{\partial}{\partial z}\left(c(z)\frac{\partial w}{\partial z}\right)\right)\partial t = F(z,t)\partial t + G(z,t)d\mathbf{W}(t), \\
w(z,0;\omega) = w_0(z;\omega),\ \frac{\partial w}{\partial t}(z,0;\omega) = w_1(z;\omega),\ 0 < z < a,\ \omega \in \Omega, \\
\frac{\partial^2 w}{\partial z^2}(0,t;\omega) = w(0,t;\omega) = 0 = \frac{\partial^2 w}{\partial z^2}(a,t;\omega) = w(a,t;\omega),\ 0 < t < T,\ \omega \in \Omega,
\end{cases}
$$
$$(8.134)$$

where $0 < z < a$, $0 < t < T$, $\omega \in \Omega$, and $w = w(z,t;\omega)$. The external forcing term can be due to aerodynamic drag.

Model XVII.1 Some Important Equations from Mathematical Physics
Partial differential equations arise in the study of physical phenomena, many of which can be studied under the parlance of the theory developed in this text. We provide a brief encounter with five such equations. More general perturbations of them can be studied using the techniques developed thus far. Indeed, the versions mentioned below can all be reformulated abstractly as (8.13). We refer you to **[12, 21, 56, 85, 103, 117, 313, 346, 381, 420]** for detailed analyses of these, and related, equations of mathematical physics.

1. Burger's Equation
The following equation is an elementary quasilinear diffusion equation arising in the mathematical modeling of fluid dynamics, magneto-hydrodynamics, and traffic flow:

$$\begin{cases} \partial(x,t;\omega) + u(x,t;\omega)\frac{\partial u}{\partial x}(x,t;\omega)\partial t - \gamma\frac{\partial^2 u}{\partial x^2}(x,t;\omega)\partial t + g(t)dW(t) = 0, \\ u(x,0;\omega) = \cos(2x)\chi_E(\omega), \, 0 < x < L, \omega \in \Omega, \\ u(0,t;\omega) = u(L,t;\omega) = 0, t > 0, \omega \in \Omega, \end{cases}$$

$$(8.135)$$

where $0 < x < L$, $t > 0$, $\omega \in \Omega$, $\gamma > 0$, E is a given event, $g : [0,T] \to \mathbb{R}$, and $W(t)$ is a one-dimensional Wiener process.

2. Schrodinger Equations
These complex-valued PDEs arise in the study of quantum physics, specifically in the modeling of the dynamics of free particles in a bounded region $\mathscr{D}$ with smooth boundary $\partial\mathscr{D}$. The most basic form of the stochastic version of these equations is given by

$$\begin{cases} \partial\psi(\mathbf{x},t;\omega) - i\triangle\psi(\mathbf{x},t;\omega)\partial t = 0, \, \mathbf{x} \in \mathscr{D}, t > 0, \omega \in \Omega, \\ \psi(\mathbf{x},t;\omega) = 0, \mathbf{x} \in \partial\mathscr{D}, t > 0, \omega \in \Omega, \\ \psi(\mathbf{x},0;\omega) = \Psi_0(\mathbf{x};\omega), \mathbf{x} \in \mathscr{D}, \omega \in \Omega. \end{cases}$$

$$(8.136)$$

Here, ψ is a complex-valued function expressed as $\psi = \psi_1 + i\psi_2$, where ψ_1 and ψ_2 are real-valued functions. Using the fact that

$$a + bi = c + di \text{ iff } a = c \text{ and } b = d,$$

we can formulate (8.136) as a system of two real-valued stochastic PDEs. (Tell how.) We can apply that same process to the following perturbed version of (8.136):

$$\begin{cases} \partial\psi(\mathbf{x},t;\omega) - i\triangle\psi(\mathbf{x},t;\omega)\partial t = \left(\int_0^t a(t-s)\psi(\mathbf{x},s;\omega)ds\right)\partial t \\ \left(\int_0^t b(t-s)\psi(\mathbf{x},s;\omega)ds\right)dW(t), \mathbf{x} \in \mathscr{D}, t > 0, \omega \in \Omega, \\ \psi(\mathbf{x},t;\omega) = 0, \mathbf{x} \in \partial\mathscr{D}, t > 0, \omega \in \Omega, \\ \psi(\mathbf{x},0;\omega) = \Psi_0(\mathbf{x};\omega), \mathbf{x} \in \mathscr{D}, \omega \in \Omega, \end{cases}$$

$$(8.137)$$

where $a, b : [0, T] \rightarrow [0, \infty)$ are continuous functions.

Exercise 8.4.24.
i.) Prove that (8.137) has a unique mild solution on $[0, T]$, for any $T > 0$.
ii.) Establish a continuous dependence result for (8.137).
iii.) Suppose the right-hand side of (8.137) is replaced by

$$f(\mathbf{x}, t, \psi(\mathbf{x}, t; \omega)) dt + g(\mathbf{x}, t, \psi(\mathbf{x}, t; \omega)) dW(t),$$

where f, g satisfies the non-Lipschitz conditions. Apply the theory established in Section 7.6 to argue that (8.137) has a unique mild solution on $[0, T]$ under these weaker conditions.

3. Sine–Gordon Equation
This second-order PDE arises in the theory of semiconductors, lasers, and particle physics. Its most basic form in bounded region $\mathscr{D}$ with smooth boundary $\partial \mathscr{D}$ is given by

$$\begin{cases} \frac{\partial^2 u}{\partial t^2}(\mathbf{x}, t) + \alpha \frac{\partial u}{\partial t}(\mathbf{x}, t) - \beta \triangle u(\mathbf{x}, t) + \gamma \sin(u(\mathbf{x}, t)) = 0, \ \mathbf{x} \in \mathscr{D}, t > 0, \\ u(\mathbf{x}, t) = 0, \ \mathbf{x} \in \partial \mathscr{D}, t > 0, \\ u(\mathbf{x}, 0) = u_0(\mathbf{x}), \frac{\partial u}{\partial t}(\mathbf{x}, 0) = u_1(\mathbf{x}), \ \mathbf{x} \in \mathscr{D}. \end{cases} \tag{8.138}$$

We can introduce noise through any of the parameters. In doing so, we can more generally view (8.138) as a special case of the semi-linear stochastic wave equation

$$\begin{cases} \partial \left(\frac{\partial u}{\partial t}(\mathbf{x}, t; \omega) \right) + \left(\alpha \frac{\partial u}{\partial t}(\mathbf{x}, t; \omega) - \beta \triangle u(\mathbf{x}, t; \omega) \right) \partial t = F(\mathbf{x}, t, u(\mathbf{x}, t; \omega)) \partial t \\ + F(\mathbf{x}, t, u(\mathbf{x}, t; \omega)) dW(t), \ \mathbf{x} \in \mathscr{D}, t > 0, \omega \in \Omega, \\ u(\mathbf{x}, t; \omega) = 0, \ \mathbf{x} \in \partial \mathscr{D}, t > 0, \omega \in \Omega, \\ u(\mathbf{x}, 0; \omega) = u_0(\mathbf{x}; \omega), \frac{\partial u}{\partial t}(\mathbf{x}, 0; \omega) = u_1(\mathbf{x}; \omega), \ \mathbf{x} \in \mathscr{D}, \omega \in \Omega. \end{cases} \tag{8.139}$$

The IBVP (8.139) can be expressed as an equivalent system via the following substitution:

$$v = \frac{\partial u}{\partial t}, \ \partial v = (-\alpha v + \beta \triangle u - F(u)) dt - G(u) dW(t). \tag{8.140}$$

Indeed, using (8.140) then enables us to rewrite (8.139) as

$$\begin{cases} \partial \begin{bmatrix} u \\ v \end{bmatrix} = \begin{bmatrix} v \\ -\alpha v + \beta \triangle u - F(u) \end{bmatrix} \partial t + \begin{bmatrix} 0 \\ G(u) \end{bmatrix} dW(t) \\ = \begin{bmatrix} 0 & I \\ \beta \triangle & -\alpha I \end{bmatrix} \begin{bmatrix} u \\ v \end{bmatrix} \partial t + \begin{bmatrix} 0 \\ F(u) \end{bmatrix} \partial t + \begin{bmatrix} 0 \\ G(u) \end{bmatrix} dW(t) \\ \begin{bmatrix} u \\ v \end{bmatrix} (0; \omega) = \begin{bmatrix} u_0(\omega) \\ u_1(\omega) \end{bmatrix} \end{cases} \tag{8.141}$$

in the space $\mathbb{L}^2(\mathscr{D}) \times \mathbb{L}^2(\mathscr{D})$, where $t > 0, \omega \in \Omega$.

Exercise 8.4.25. Prove that if F and G are globally Lipschitz, then (8.141), and hence (8.139), has a unique mild solution on $[0, T]$, for any $T > 0$.

4. Klein–Gordon Equation
This second-order complex-valued PDE arises in quantum theory and the study of nonlinear dispersion. Its most basic form is given by

$$\begin{cases} \frac{\partial^2 u}{\partial t^2}(\mathbf{x},t) - \alpha \triangle u(\mathbf{x},t) + \beta u(\mathbf{x},t) = 0, \ \mathbf{x} \in \mathscr{D}, t > 0, \\ u(\mathbf{x},t) = 0, \ \mathbf{x} \in \partial \mathscr{D}, t > 0, \\ u(\mathbf{x},0) = v_0(\mathbf{x}), \ \frac{\partial u}{\partial t}(\mathbf{x},0) = v_1(\mathbf{x}), \ \mathbf{x} \in \mathscr{D}, \end{cases} \quad (8.142)$$

where $u = u_1 + iu_2$. The term βu is usually taken to be the derivative of the potential function and so it generally takes on a more complicated form. Introducing noise through the parameter β via a white noise process yields the following stochastic variant of (8.142):

$$\begin{cases} \partial \left(\frac{\partial u}{\partial t}(\mathbf{x},t;\omega) \right) - \alpha \triangle u(\mathbf{x},t;\omega)\partial t = \beta_1 u(\mathbf{x},t;\omega)\partial t + \beta_2 u(\mathbf{x},t;\omega)dW(t), \\ u(\mathbf{x},t;\omega) = 0, \ \mathbf{x} \in \partial \mathscr{D}, t > 0, \ \omega \in \Omega, \\ u(\mathbf{x},0;\omega) = v_0(\mathbf{x};\omega), \ \frac{\partial u}{\partial t}(\mathbf{x},0;\omega) = v_1(\mathbf{x};\omega), \ \mathbf{x} \in \mathscr{D}, \omega \in \Omega, \end{cases}$$

$$(8.143)$$

where $\mathbf{x} \in \mathscr{D}, t > 0, \omega \in \Omega$, and $W(t)$ is a one-dimensional Wiener process.

Exercise 8.4.26.
i.) Express the second-order IBVP (8.143) as a system of two real-valued PDEs, and rewrite the ICs and BCs in a suitable manner. Then, prove that (8.143) has a unique mild solution on $[0, T]$, $\forall T > 0$.
ii.) Equip (8.143) with a forcing term

$$F(\mathbf{x},t,u(\mathbf{x},t;\omega), \frac{\partial u}{\partial t}(\mathbf{x},t;\omega))\partial t + G(\mathbf{x},t,u(\mathbf{x},t;\omega), \frac{\partial u}{\partial t}(\mathbf{x},t;\omega))dW(t).$$

Formulate existence-uniqueness results for the newly formed IBVP under global Lipschitz growth conditions.

5. Cahn–Hillard Equation
This PDE arises in the study of pattern formation in materials undergoing phase transitions, especially in alloys and glass. A stochastic version of the model is given by

$$\begin{cases} \partial(\mathbf{x},t;\omega) - \triangle(\alpha \triangle u(\mathbf{x},t;\omega))\partial t + \triangle \left(\beta u(\mathbf{x},t;\omega) + \gamma u^3(\mathbf{x},t;\omega) \right) \partial t = \\ \quad + \sum_{k=1}^m g_k(t,u(\mathbf{x},t;\omega))dW_k(t), \ \mathbf{x} \in \mathscr{D}, t > 0, \omega \in \Omega, \\ \frac{\partial u}{\partial \mathbf{n}}(\mathbf{x},t;\omega) = 0, \ \mathbf{x} \in \partial \mathscr{D}, t > 0, \omega \in \Omega, \\ u(\mathbf{x},0;\omega) = u_0(\mathbf{x};\omega), \ \mathbf{x} \in \mathscr{D}, \omega \in \Omega, \end{cases}$$

$$(8.144)$$

where $\alpha, \beta, \gamma > 0$ and $W_k(t)$ $(k = 1,\ldots,m)$ are independent one-dimensional Wiener processes.

Exercise 8.4.27.
i.) Use the techniques from the discussion of the beam equation to reformulate (8.144) as an abstract stochastic evolution equation.
ii.) Prove the existence and uniqueness of a unique mild solution on $[0,T]$, $\forall T > 0$.

8.5   Looking Ahead

The IBVPs considered thus far have all been reformulated as abstract stochastic evolution equations in which one could solve for $dX(t;\omega)$ explicitly. But, this is not the case for all IBVPs arising in practice. Indeed, consider, for instance, the following IBVP arising in the study of soil mechanics:

$$\begin{cases} \partial \left(z - \frac{\partial^2 z}{\partial x^2}\right) + \frac{\partial^2 z}{\partial x^2} \partial t = \left(\int_0^t h(s)f(s,z)ds\right) \partial t + \left(\int_0^t k(s)g(s,z)ds\right) dW(t), \\ z(x,0;\omega) = z_0(x;\omega), \ 0 < x < \pi, \ \omega \in \Omega, \\ z(0,t;\omega) = z(\pi,t;\omega) = 0, \ 0 < t < T, \omega \in \Omega, \end{cases}$$

(8.145)

where $0 < x < \pi, 0 < t < T, \omega \in \Omega$, $z = z(x,t)$, $h,k : [0,T] \to \mathbb{R}$ are continuous functions, and $f,g : [0,T] \times \mathbb{R} \to \mathbb{R}$ are given mappings. In order to reformulate (8.145) as an abstract stochastic evolution equation, let $\mathscr{H} = \mathbb{L}^2(0,\pi;\mathbb{R})$ and define the operators $A : \text{dom}(A) \subset \mathscr{H} \to \mathscr{H}$ and $B : \text{dom}(B) \subset \mathscr{H} \to \mathscr{H}$ as in (5.62). Also, define the functionals $\mathfrak{F} : \mathbb{C}\left([0,T];\mathscr{L}^2(\Omega;\mathscr{H})\right) \to \mathbb{C}\left([0,T];\mathscr{L}^2(\Omega;\mathscr{H})\right)$ and $\mathfrak{G} : \mathbb{C}\left([0,T];\mathscr{L}^2(\Omega;\mathscr{H})\right) \to \mathbb{C}\left([0,T];\mathscr{L}^2(\Omega;\mathscr{B}_0(\mathbb{R};\mathscr{H}))\right)$ by

$$\mathfrak{F}(z)(\cdot)(t;\omega) = \int_0^t h(s)f(s,z(\cdot,s;\omega))ds,$$

$$\mathfrak{G}(z)(\cdot)(t;\omega) = \int_0^t k(s)g(s,z(\cdot,s;\omega))ds.$$

These identifications enable us to reformulate (8.145) as the abstract stochastic evolution equation

$$\begin{cases} d(BX)(t;\omega) = (AX(t;\omega) + \mathfrak{F}(X)(t;\omega))dt + \mathfrak{G}(X)(t;\omega)dW(t), \\ X(0;\omega) = X_0(\omega), \omega \in \Omega. \end{cases}$$

(8.146)

where $0 < t < T, \omega \in \Omega$. Ideally, we would like to further express (8.146) in the form (8.13). The only way, symbolically, to do so is to use the formal substitution $v(t;\omega) = BX(t;\omega)$ in (8.146). While making this substitution does create an isolated term $dv(t;\omega)$, it comes at the expense of requiring that the operator B be at least invertible. (Why?)

Exercise 8.5.1. What other technical complications arise as a result of making this substitution? Specifically, what compatibility requirements must exist between the operators A and B?

Despite the apparent shortcomings of the suggested substitution, our discussion in the next chapter reveals that it constitutes a viable approach.

8.6   Guidance for Selected Exercises

8.6.1   Level 1: A Nudge in a Right Direction

8.2.7. Yes. Use the sup norm in the last step of the calculation instead of integrating from 0 to T. (How does the estimate change as a result?)

8.2.8. Let $\begin{pmatrix} x \\ y \end{pmatrix} \in \left(\mathbb{C}\left([0,T]; \mathcal{L}^2(\Omega; \mathscr{H})\right) \right)^2$. Note that

$$\left\| \begin{pmatrix} x \\ y \end{pmatrix} \right\|_{\mathbb{C}^2} = \sup_{0 \le t \le T} \left(\|x(t)\|_{\mathcal{L}^2(\Omega;\mathscr{H})} + \|y(t)\|_{\mathcal{L}^2(\Omega;\mathscr{H})} \right).$$

Apply this with calculations analogous to those used to verify Claims 2 and 3.

8.3.2. ii.) You must now also estimate the term

$$E \left\| \int_0^t e^{A(t-s)} \left(g(s, X(s; \cdot)) - g(s, Y(s; \cdot)) \right) d\mathbf{W}(s) \right\|_{\mathscr{H}}^2 .$$

How does this affect the data restriction?

8.3.6. i.)(a) Impose a Lipschitz condition on f and assume $k \in \mathbb{C}([0,T]; \mathbb{R})$.

8.3.10. $A + B$ generates a C_0-semigroup on $\mathscr{H}$. (So what?)

8.3.11. $\begin{bmatrix} A_1 & 0 \\ 0 & A_2 \end{bmatrix}$ generates a C_0-semigroup on $\mathscr{H}_1 \times \mathscr{H}_2$.

8.4.15. The easier one to identify is

$$\text{dom}(B) = \left\{ \begin{pmatrix} w \\ \frac{\partial w}{\partial t} \end{pmatrix} \in \mathbb{H}^2(0,a) \times \mathbb{L}^2(0,a) \,\middle|\, w(0,t) = w(a,t) = 0 \right\}.$$

Explain why this makes sense.

8.4.18. i.) Assume $w(z,t) = Z(z)T(t)$. Substituting this into the PDE yields

$$\frac{Z^{(4)}(z)}{Z(z)} = -\frac{T''(t)}{\alpha T(t)} = \lambda.$$

Take into account the BCs and solve the resulting BVP for Z, which only has nontrivial solutions when $\lambda > 0$. (Why?) For convenience, write say $\lambda = c^2$ and continue...

ii.) Identify this in a manner similar to the example in the discussion of the classical stochastic wave equation.

8.4.24. i.) Reformulate the IBVP as a system of PDEs:

$$
\begin{cases}
\frac{\partial \psi_1}{\partial t}(\mathbf{x},t;\omega) + \triangle \psi_2(\mathbf{x},t;\omega) = \int_0^t a(t-s)\psi_1(\mathbf{x},s;\omega)ds \\
+ \left(\int_0^t b(t-s)\psi_1(\mathbf{x},s;\omega)ds \right) dW(t) \; \mathbf{x} \in \mathscr{D}, t > 0, \omega \in \Omega, \\
\frac{\partial \psi_2}{\partial t}(\mathbf{x},t;\omega) - \triangle \psi_1(\mathbf{x},t;\omega) = \int_0^t a(t-s)\psi_2(\mathbf{x},s;\omega)ds \\
+ \left(\int_0^t b(t-s)\psi_2(\mathbf{x},s;\omega)ds \right) dW(t) \; \mathbf{x} \in \mathscr{D}, t > 0, \omega \in \Omega, \\
\psi_1(\mathbf{x},t;\omega) = \psi_2(\mathbf{x},t;\omega) = 0, \mathbf{x} \in \partial\mathscr{D}, t > 0, \omega \in \Omega, \\
\psi_1(\mathbf{x},0;\omega) = \Psi_0^1(\mathbf{x};\omega), \; \psi_2(\mathbf{x},0;\omega) = \Psi_0^2(\mathbf{x};\omega), \mathbf{x} \in \mathscr{D}, \omega \in \Omega.
\end{cases}
\tag{8.147}
$$

(Now what?)

ii.) This is a special case of the typical abstract semi-linear stochastic evolution equation studied in Chapter 7. So, the continuous dependence estimate has already been established. Interpret the abstract result using the particular functions in (8.147).

8.4.25. If we can guarantee that the operator $A : \text{dom}(A) \subset \left(\mathbb{L}^2(\Omega)\right)^2 \to \left(\mathbb{L}^2(\Omega)\right)^2$ defined by

$$
A \begin{bmatrix} u \\ v \end{bmatrix} = \begin{bmatrix} 0 & I \\ \beta\triangle & -\alpha I \end{bmatrix} \begin{bmatrix} u \\ v \end{bmatrix}
$$

generates a C_0-semigroup on $\left(\mathbb{L}^2(\mathscr{D})\right)^2$, then the result will follow from the theory established in Chapter 7 because the mapping $(t,u) \mapsto \begin{bmatrix} 0 \\ F(u) \end{bmatrix}$ is globally Lipschitz.

8.4.26. i.) Let $u = u_1 + iu_2$. Then, (8.143) can be viewed as the system

$$
\begin{cases}
\frac{\partial^2 u_1}{\partial t^2}(\mathbf{x},t;\omega) - \alpha\triangle u_1(\mathbf{x},t;\omega) + \beta_1 u_1(\mathbf{x},t;\omega) \\
+ \beta_2 u(\mathbf{x},t;\omega)dW(t) = 0, \mathbf{x} \in \mathscr{D}, t > 0, \omega \in \Omega, \\
\frac{\partial^2 u_2}{\partial t^2}(\mathbf{x},t;\omega) - \alpha\triangle u_2(\mathbf{x},t;\omega) + \beta_1 u_2(\mathbf{x},t;\omega) \\
+ \beta_2 u(\mathbf{x},t;\omega)dW(t) = 0, \mathbf{x} \in \mathscr{D}, t > 0, \omega \in \Omega, \\
u_1(\mathbf{x},t;\omega) = u_2(\mathbf{x},t;\omega) = 0, \mathbf{x} \in \partial\mathscr{D}, t > 0, \\
u_1(\mathbf{x},0;\omega) = v_0^1(\mathbf{x};\omega), u_2(\mathbf{x},0;\omega) = v_0^2(\mathbf{x};\omega), \mathbf{x} \in \mathscr{D}, \\
\frac{\partial u_1}{\partial t}(\mathbf{x},0;\omega) = v_1^1(\mathbf{x};\omega), \frac{\partial u_2}{\partial t}(\mathbf{x},0;\omega) = v_1^2(\mathbf{x};\omega), \mathbf{x} \in \mathscr{D}.
\end{cases}
\tag{8.148}
$$

ii.) Once you have incorporated the forcing term into (8.148), the abstract form of the resulting system is the standard semi-linear abstract SEE from Chapter 7.

8.6.2 Level 2: An Additional Thrust in a Right Direction

8.2.7. The constants involve multiples of T because we are now taking sups and hence can invoke the linearity of the integral.

8.2.8. Observe that

$$\left\| \mathfrak{H}_6 \begin{pmatrix} u_1 \\ v_1 \end{pmatrix} - \mathfrak{H}_6 \begin{pmatrix} u_2 \\ v_2 \end{pmatrix} \right\|_{\mathbb{C}} \leq \left(M_{f_1} \|k_1\|_{\mathbb{L}^1} + M_{f_2} \|k_2\|_{\mathbb{L}^1} \right) \cdot$$

$$[\|u_1 - u_2\|_{\mathbb{C}} + \|v_1 - v_2\|_{\mathbb{C}}].$$

(So what?)

8.3.2. (ii) The new data restriction is $2M_A^2 \left(T^2 M_{\mathfrak{F}}^2 + \zeta_g(T,2) M_g^2 T \right) < 1$.

8.4.15. The easier one to identify is

$$\text{dom}(A) = \left\{ \begin{pmatrix} w \\ \frac{\partial w}{\partial t} \end{pmatrix} \in \mathbb{H}^2(0,a) \times \mathbb{L}^2(0,a) | w \text{ satisfies } (\mathbf{i}) - (\mathbf{vi}) \right\},$$

where the conditions (i) through (vi) are as follows:

(i) $w(0,t) = w(a,t) = 0$, **(ii)** $\frac{\partial w}{\partial t}(0,t) = \frac{\partial w}{\partial t}(a,t) = 0$, **(iii)** $\frac{\partial^2 w}{\partial x^2}(0,t) + \beta_2 \frac{\partial^2 w}{\partial x^2}(0,t) = 0$, **(iv)** $\frac{\partial^2 w}{\partial x^2}(a,t) + \beta_2 \frac{\partial^2 w}{\partial x^2}(a,t) = 0$, **(v)** $w + \beta_2 \frac{\partial w}{\partial t} \in \mathbb{H}^4(0,a)$, and **(vi)** $\frac{\partial w}{\partial t} \in \mathbb{H}^2(0,a)$.

8.4.24. i.) This IBVP is of the form (7.10) in $\mathscr{H} = \left(\mathbb{L}^2(\Omega) \right)^2$, where

$$A \begin{bmatrix} \psi_1 \\ \psi_2 \end{bmatrix} = \begin{bmatrix} 0 & -\triangle \\ \triangle & 0 \end{bmatrix} \begin{bmatrix} \psi_1 \\ \psi_2 \end{bmatrix}.$$

Does this operator generate a C_0-semigroup on $\mathscr{H}$?

8.4.25. Refer to our discussion of the classical wave equation and the beam equation.

Chapter 9

Sobolev-Type Stochastic Evolution Equations

Overview

We consider a special class of stochastic evolution equations in which the time derivative of the unknown is defined implicitly in the equation, but for which we can still generate a variation of parameters formula for a mild solution. Such equations arise in a vast assortment of fields, including soil mechanics, thermodynamics, civil engineering, and the dynamics of non-Newtonian fluids.

9.1 Motivation by Models

Assume throughout this chapter that **(S.A.1)** holds.

The intention of the following discussion is to focus only on the forms of the equations arising in many different models rather than to provide a rigorous derivation of them. As such, references are provided throughout the section to facilitate further study of the underlying detail in the development of these models.

Model XVIII.1 Soil Mechanics and Clay Consolidation

The erosion of beaches and grasslands is an ongoing environmental concern for various species of wildlife and human development. Avalanches occur due to the movement and changing of soil. Understanding such phenomena has important environmental ramifications. *Hypoplasticity* is an area of study that examines the behavior of granular solids, such as soil, sand, and clay. The IBVPs that arise in the modeling of this phenomena are complicated, mainly due to the presence of phase changes that the material undergoes. Refer to **[167, 237, 307, 378, 403, 408]** for further study.

We examine a stochastic version of a particular form of a system of equations discussed in **[362, 363, 364, 367]** relating the fluid pressure and structural displacement, ignoring the physical meaning of the constants involved. We begin with the deterministic IBVP and then incorporate noise through certain parameters.

Let $\mathscr{D} \subset \mathbb{R}^3$ be a bounded domain with smooth boundary $\partial \mathscr{D}$. The fluid pressure is denoted by $p(\mathbf{x},t)$ and the (three-dimensional) structural displacement by $\mathbf{w}(\mathbf{x},t)$ at position $\mathbf{x} = (x,y,z)$ in the soil at time t. Consider the following IBVP:

$$\begin{cases} -\beta_1 \nabla(\nabla \cdot \mathbf{w}(\mathbf{x},t)) - \beta_2 \triangle \mathbf{w}(\mathbf{x},t) + \beta_3 \nabla p(\mathbf{x},t) = \mathbf{f}(\mathbf{x},t), \ \mathbf{x} \in \mathscr{D}, t > 0, \\ \frac{\partial}{\partial t}(\beta_4 p(\mathbf{x},t) + \beta_3 \nabla \cdot \mathbf{w}(\mathbf{x},t)) - \nabla \cdot \beta_5 \nabla p(\mathbf{x},t) = h(\mathbf{x},t), \ \mathbf{x} \in \mathscr{D}, t > 0, \\ p(\mathbf{x},0) = p_0(\mathbf{x}), \ w(\mathbf{x},0) = w_0(\mathbf{x}), \ \mathbf{x} \in \mathscr{D}, \\ \frac{\partial}{\partial \mathbf{n}} p(\mathbf{x},t) = \frac{\partial}{\partial \mathbf{n}} \mathbf{w}(\mathbf{x},t) = 0, \ x \in \partial \mathscr{D}, t > 0. \end{cases} \quad (9.1)$$

It can be shown that the PDE obtained by solving the first PDE in (9.1) for $\mathbf{w}(\mathbf{x},t)$ and then substituting this into the second PDE can be formulated as an abstract evolution equation in the space $(\mathbb{L}^2(\mathscr{D}))^3$ of the form

$$\begin{cases} \frac{d}{dt}(\alpha p(t) - \nabla \cdot v^{-1}(\nabla p(t))) + \triangle p(t) = h(t), \ t > 0, \\ p(0) = p_0, \end{cases} \quad (9.2)$$

where

$$v(\mathbf{w}) = -\beta_1 \nabla(\nabla \cdot \mathbf{w})) - \beta_2 \triangle \mathbf{w}. \quad (9.3)$$

An overly simplified, yet comprehensible, one-dimensional version of this evolution equation is given by

$$\begin{cases} \frac{\partial}{\partial t}\left(z(x,t) - \frac{\partial^2}{\partial x^2}z(x,t)\right) + \alpha\frac{\partial^2}{\partial x^2}z(x,t) = f(x,t), \ 0 < x < a, t > 0, \\ z(x,0) = z_0(x), \ 0 < x < a, \\ \frac{\partial z}{\partial x}(0,t) = \frac{\partial z}{\partial x}(a,t) = 0, \ t > 0, \end{cases} \quad (9.4)$$

where $\alpha > 0$. The following more general version of this model in $\mathscr{D} \subset \mathbb{R}^3$ is of neutral-type (discussed more extensively in Section 7.2):

$$\begin{cases} \frac{\partial}{\partial t}(z(\mathbf{x},t) - \triangle z(\mathbf{x},t)) + \alpha\triangle z(\mathbf{x},t) = f(\mathbf{x},t), \ \mathbf{x} \in \mathscr{D}, t > 0, \\ z(\mathbf{x},0) = z_0(\mathbf{x}), \ \mathbf{x} \in \mathscr{D}, \\ \frac{\partial z}{\partial \mathbf{n}}(\mathbf{x},t) = 0, \ \mathbf{x} \in \partial \mathscr{D}, t > 0. \end{cases} \quad (9.5)$$

A detailed discussion of (9.5), including some subtle complications, is given in **[26]**.

It is conceivable that the external forcing term $f(\mathbf{x},t)$ is itself subject to randomness for different reasons and thus, we can formulate a stochastic version of (9.5) by incorporating a white noise process in the forcing term to obtain

$$\begin{cases} \partial(z(\mathbf{x},t;\omega) - \triangle z(\mathbf{x},t;\omega)) + \alpha\triangle z(\mathbf{x},t;\omega)\partial t = f(\mathbf{x},t)\partial t + g(\mathbf{x},t)d\mathbf{W}(t), \\ z(\mathbf{x},0;\omega) = z_0(\mathbf{x};\omega), \ \mathbf{x} \in \mathscr{D}, \omega \in \Omega, \\ \frac{\partial z}{\partial \mathbf{n}}(\mathbf{x},t;\omega) = 0, \ \mathbf{x} \in \partial \mathscr{D}, t > 0, \omega \in \Omega, \end{cases}$$
$$\quad (9.6)$$

where $\mathbf{x} \in \mathscr{D}, t > 0, \omega \in \Omega$, and $\mathbf{W}(t)$ is an m-dimensional Wiener process. A different stochastic variant of (9.5) in a two-dimensional domain with a more general functional forcing term, related to those discussed in [30, 31, 33, 220, 261, 386], is as follows:

$$
\begin{cases}
\partial \left(z(x,y,t;\omega) - \frac{\partial^2}{\partial x^2} z(x,y,t;\omega) - \frac{\partial^2}{\partial y^2} z(x,y,t;\omega) \right) \\
+ \left(\frac{\partial^2}{\partial x^2} z(x,y,t;\omega) + \frac{\partial^2}{\partial y^2} z(x,y,t;\omega) \right) \partial t \\
= \left(\int_0^{a_2} \int_0^{a_1} k(t,x,y) f(t,z(x,y,t;\omega)) \, dxdy \right) \partial t + \left(\int_0^t k(s)g(s,z(x,y,s;\omega)ds \right) dW(t), \\
z(x,y,0;\omega) = z_0(x,y;\omega), \ 0 < x < a_1, 0 < y < a_2, \omega \in \Omega, \\
z(0,y,t;\omega) = z(a_1,y,t;\omega) = 0, \ 0 < x < a_1, 0 < t < T, \omega \in \Omega, \\
z(x,0,t;\omega) = z(x,a_2,t;\omega) = 0, \ 0 < y < a_2, 0 < t < T, \omega \in \Omega,
\end{cases}
$$
(9.7)

where $0 < x < a_1, 0 < y < a_2, 0 < t < T, \omega \in \Omega$.

Exercise 9.1.1. Try to reformulate (9.7) as an abstract stochastic evolution equation.

Model XIX.1 Seepage of Fluid Through Fissured Rocks

Try to visualize a sizeable stack of rocks separated by a network of mini-cracks or fissures. Liquid flows along the arteries of this network, but also through tiny pores in the rocks themselves. The modeling of this situation in a bounded domain $\mathscr{D} \subset \mathbb{R}^3$ involves PDEs governing the pressure of the liquid in the fissures. A simplified version of one such stochastic IBVP is as follows:

$$
\begin{cases}
\partial \left(p(x,y,z,t;\omega) - \alpha \left(\triangle p(x,y,z,t;\omega) \right) \right) = \beta \triangle p(x,y,z,t;\omega) \partial t, \\
+ \sum_{k=1}^m g_k(x,y,z,t;\omega) dW_k(t), \ (x,y,z) \in \mathscr{D}, t > 0, \omega \in \Omega, \\
p(x,y,z,0;\omega) = p_0(x,y,z;\omega), \ (x,y,z) \in \mathscr{D}, \omega \in \Omega, \\
\frac{\partial p}{\partial \mathbf{n}}(x,y,z,,t;\omega) = 0, \ (x,y,z) \in \partial \mathscr{D}, t > 0, \omega \in \Omega,
\end{cases}
$$
(9.8)

where $W_k(t) \, (k = 1, \ldots, m)$ are independent one-dimensional Wiener processes. The parameters α and β are dependent on the characteristics of the rocks (e.g., porosity and permeability). A detailed discussion of such models can be found in [39].

Exercise 9.1.2. Thinking ahead, if it can be shown that $\forall \alpha > 0$, (9.8) has a unique mild solution p_α, must $\exists p \in \mathbb{C}\left([0,T]; \mathscr{L}^2(\Omega; \mathscr{H})\right)$ for which $\lim_{\alpha \to 0^+} p_\alpha = p$? To what stochastic IBVP is p a mild solution?

Model XX.1 Second-Order Fluids

Non-Newtonian fluids are fluids characterized by a variable viscosity (e.g., some oils and grease, shampoo, blood, and polymer melts). The dynamics of such fluids have been investigated extensively (see [124, 202, 385]). Assuming a unidirectional, nonsteady flow, a model of the velocity field $w(x,t;\omega)$ for the flow over a wall can be characterized by

$$\begin{cases} \partial w(x,t;\omega) = \left(\alpha \frac{\partial^2 w}{\partial x^2}(x,t;\omega) + \beta \frac{\partial^3 w}{\partial x^2 \partial t}(x,t;\omega) \right) \partial t + g(x,t,w(x,t;\omega))dW(t), \\ w(0,t;\omega) = 0, \, t > 0, \omega \in \Omega, \\ w(x,0;\omega) = w_0(x;\omega), \, x > 0, \omega \in \Omega, \end{cases}$$

$$(9.9)$$

where $x > 0, t > 0, \omega \in \Omega$, and $W(t)$ is a one-dimensional Wiener process.

Model VII.5 Wave Equations of Sobolev-Type

Consider our earlier discussion of classical wave equations. The various characteristics that we incorporated into the model (e.g., dissipation, diffusion, advection, etc.) manifested as distinct differential terms being incorporated into the PDE portion of the IBVP. In some cases, depending on the term being added, the resulting PDE is of Sobolev-type. For instance, consider the following IBVP:

$$\begin{cases} \partial \left(\frac{\partial z}{\partial t}(x,t;\omega) + \frac{\partial^2 z}{\partial x^2}(x,t;\omega) \right) + \frac{\partial^2 z}{\partial x^2}(x,t;\omega)\partial t = g(x,t)\partial t + \overline{g}(x,t)dW(t), \\ z(x,0;\omega) = z_0(x;\omega), \, \frac{\partial z}{\partial t}(x,0;\omega) = z_1(x;\omega), \, 0 < x < L, \omega \in \Omega, \\ z(0,t;\omega) = z(L,t;\omega) = 0, \, t > 0, \omega \in \Omega, \end{cases}$$

$$(9.10)$$

where $0 < x < L, t > 0, \omega \in \Omega$.

Exercise 9.1.3. How would you convert this second-order IBVP into an abstract stochastic evolution equation?

Remark. There are other applications in which such equations arise. We refer you to **[216]** for a discussion of such a model in thermodynamics and **[3]** for one arising in civil engineering.

9.2   The Abstract Framework

The IBVPs in Section 9.1, when equipped with forcing terms of the semi-linear variety, can be reformulated as an abstract stochastic evolution equation of the form

$$\begin{cases} d(BX)(t;\omega) = (AX(t;\omega) + f(t,X(t;\omega)))dt + g(t,X(t;\omega))d\mathbf{W}(t), \\ X(0;\omega) = X_0(\omega), \omega \in \Omega \end{cases}$$

$$(9.11)$$

in a separable Hilbert space $\mathscr{H}$, where $0 < t < T, \omega \in \Omega$. Here, $X : [0,T] \times \Omega \to \text{dom}(B) \subset \mathscr{H}$, $A : \text{dom}(A) \subset \mathscr{H} \to \mathscr{H}$, $B : \text{dom}(B) \subset \mathscr{H} \to \mathscr{H}$, $f : [0,T] \times \mathscr{H} \to \mathscr{H}$, $g : [0,T] \times \mathscr{H} \to \mathscr{B}_0(\mathbb{R}^m; \mathscr{H})$, $\mathbf{W}(t)$ is an m-dimensional Wiener process, and $X_0 \in \mathcal{L}^2(\Omega; \mathscr{H})$ is independent of $\mathbf{W}(t)$. Such a stochastic evolution equation is said to be of *Sobolev-type*. The main difference from (7.26), of course, is the presence of the operator B.

What can we do to transform (9.11) into an equivalent stochastic evolution equation of the form (7.26) so that the approach used in Chapter 7 is applicable? Different applications require that different choices for the operators A and B be used in the abstract formulation of the problem, and these choices naturally lead to different relationships between A and B. As such, we can be assured that a single approach will not handle all possibilities. We shall focus in this section on one particular scenario guided by a change of variable suggested in Section 8.5. This is the approach adopted in **[68, 363, 386]**. Doing so requires that we impose certain assumptions on the operators A and B, the first one of which is

(H9.1) $A : \text{dom}(A) \subset \mathscr{H} \to \mathscr{H}$ and $B : \text{dom}(B) \subset \mathscr{H} \to \mathscr{H}$ are linear operators.

Of course, additional restrictions must be imposed in order to express (9.11) as (7.26). To this end, a natural approach is to define the new function $v : [0,T] \times \Omega \to \text{rng}(B)$ by

$$v(t;\omega) = BX(t;\omega), \ 0 \leq t \leq T, \omega \in \Omega. \tag{9.12}$$

Keep in mind that the goal of the substitution is to produce a stochastic evolution equation equivalent to (9.11), but for which the time-derivative term is not obstructed by an operator. If we substitute (9.12) into (9.11), we need to replace each occurrence of $X(t;\omega)$ by an equivalent term involving $v(t;\omega)$. As such, we would like to further say that (9.12) is equivalent to

$$B^{-1}v(t;\omega) = X(t;\omega), \ 0 \leq t \leq T, \omega \in \Omega. \tag{9.13}$$

This leads to the second assumption:

(H9.2) $B : \text{dom}(B) \subset \mathscr{H} \to \mathscr{H}$ is invertible.

Now, substituting (9.13) into (9.11) yields

$$\begin{cases} dv(t;\omega) = \left(AB^{-1}v(t;\omega) + f(t,B^{-1}v(t;\omega))\right) dt + g(t,B^{-1}v(t;\omega))d\mathbf{W}(t), \\ B^{-1}v(0;\omega) = X_0(\omega), \omega \in \Omega, \end{cases}$$

$$\tag{9.14}$$

where $0 < t < T$, $\omega \in \Omega$. Since $B^{-1}v(t;\omega) \in \text{dom}(B)$, $\forall 0 \leq t \leq T$, $\omega \in \Omega$, (Why?) we need to further impose the following two assumptions:

(H9.3) $X_0 \in \mathcal{L}^2\left(\Omega;\text{dom}(B)\right)$,
(H9.4) $\text{dom}(B) \subset \text{dom}(A)$.

It is now meaningful to rewrite (9.14) as

$$\begin{cases} dv(t;\omega) = \left(AB^{-1}v(t;\omega) + f(t,B^{-1}v(t;\omega))\right)dt + g(t,B^{-1}v(t;\omega))d\mathbf{W}(t), \\ v(0;\omega) = BX_0(\omega), \omega \in \Omega, \end{cases}$$

$$(9.15)$$

where $0 < t < T$, $\omega \in \Omega$. In order to apply the existence results from Chapter 7 and 8, $AB^{-1} : \text{dom}(AB^{-1}) \subset \mathscr{H} \to \mathscr{H}$ must generate a C_0-semigroup $\left\{ e^{(AB^{-1})t} : t \geq 0 \right\}$ on $\mathscr{H}$. The assumptions imposed up to now merely guarantee that AB^{-1} is a well-defined, linear operator on $\mathscr{H}$. But, if $AB^{-1} \in \mathbb{B}(\mathscr{H})$, then it must generate a C_0-semigroup on $\mathscr{H}$. So, the question is whether or not we can impose natural assumptions on A and/or B in order to guarantee this. After all, individually, each of them can be an unbounded operator. The following proposition, proven in Volume 1, answers this question.

Proposition 9.2.1. *Assume that **(H9.1)** through **(H9.4)** hold, as well as*
(H9.5) *A and B are closed operators, and*
(H9.6) $B^{-1} : \text{rng}(B) \subset \mathscr{H} \to \mathscr{H}$ *is a compact operator.*
Then, $AB^{-1} \in \mathbb{B}(\mathscr{H})$.

Summarizing, we have shown that (9.11) is equivalent to (9.15) via the substitution (9.12) provided that **(H9.1)** through **(H9.6)** hold. Moreover, under these assumptions, it follows that $AB^{-1} : \text{dom}(AB^{-1}) \subset \mathscr{H} \to \mathscr{H}$ generates a C_0-semigroup $\left\{ e^{(AB^{-1})t} : t \geq 0 \right\}$ on $\mathscr{H}$.

Of course, this is very convenient from a theoretical perspective, but is it applicable in the various applied settings introduced in Section 9.1? Thankfully, yes. It has been shown that the models in Section 9.1 can be treated within this framework. For instance, consider the following exercise.

Exercise 9.2.1. Show directly that IBVP (9.4) can be transformed into a first-order semi-linear abstract stochastic evolution equation by verifying the hypotheses formulated in the above discussion.

9.3   Semi-Linear Sobolev Stochastic Equations

Imposing the assumptions **(H9.1)** through **(H9.6)** essentially renders (9.11) as an abstract stochastic evolution equation of the form (7.26). As such, it should not be surprising that the results and proofs of Chapter 7 carry over with minimal changes.

Exercise 9.3.1 Before proceeding, try to formulate the definitions of mild and strong solutions for (9.11), as well as the theoretical results by suitably modifying the theory developed in Chapter 7. Pay particular attention to the spaces to which various terms must belong.

The definition of a mild solution for (9.15) is a simple extension of Def. 7.4.1. Indeed, we have

Definition 9.3.1. A stochastic process $v : [0,T] \times \Omega \to \mathcal{H}$ is a *mild solution* of (9.15) on $[0,T]$ if

i.) $v \in \mathbb{C}\left([0,T];\mathcal{L}^2\left(\Omega;\mathrm{rng}(B)\right)\right)$,

ii.) $v(t;\omega) = e^{\left(AB^{-1}\right)t}BX_0(\omega) + \int_0^t e^{\left(AB^{-1}\right)(t-s)}f(s,B^{-1}v(s;\omega))ds$
$+ \int_0^t e^{\left(AB^{-1}\right)(t-s)}g(s,B^{-1}v(s;\omega))dW(s)$, a.s. $[\mathscr{P}]$, $\forall 0 \le t \le T$.

We would like to say that v is a mild solution of (9.15) if and only if $X(t;\omega) = B^{-1}v(t;\omega)$ is a mild solution of (9.11). For if this were the case, then substituting (9.12) into Def. 9.3.1(ii) would yield

$$BX(t;\omega) = e^{\left(AB^{-1}\right)t}BX_0(\omega) + \int_0^t e^{\left(AB^{-1}\right)(t-s)}f\left(s,\underbrace{B^{-1}\left(BX(s;\omega)\right)}_{=X(s;\omega)}\right)ds$$

$$+ \int_0^t e^{\left(AB^{-1}\right)(t-s)}g\left(s,\underbrace{B^{-1}\left(BX(s;\omega)\right)}_{=X(s;\omega)}\right)dW(s)$$

so that

$$X(t;\omega) = B^{-1}e^{\left(AB^{-1}\right)t}BX_0(\omega) + \int_0^t B^{-1}e^{\left(AB^{-1}\right)(t-s)}f\left(s,\underbrace{B^{-1}\left(BX(s;\omega)\right)}_{=X(s;\omega)}\right)ds$$

$$+ \int_0^t B^{-1}e^{\left(AB^{-1}\right)(t-s)}g\left(s,\underbrace{B^{-1}\left(BX(s;\omega)\right)}_{=X(s;\omega)}\right)dW(s). \tag{9.16}$$

Exercise 9.3.2. Explain why B^{-1} can be brought inside the integral sign in (9.16).

This suggests that the following definition of a mild solution of (9.11) is practical.

Definition 9.3.2. A stochastic process $X : [0,T] \times \Omega \to \mathcal{H}$ is a *mild solution* of (9.11) on $[0,T]$ if

i.) $X \in \mathbb{C}\left([0,T];\mathcal{L}^2\left(\Omega;\mathcal{H}\right)\right)$,

ii.) $X(t;\omega)$ satisfies (9.16) a.s. $[\mathscr{P}]$, $\forall 0 \le t \le T$.

Consequently, we define $\Phi : \mathbb{C}\left([0,T];\mathcal{L}^2\left(\Omega;\mathcal{H}\right)\right) \to \mathbb{C}\left([0,T];\mathcal{L}^2\left(\Omega;\mathcal{H}\right)\right)$ by

$$(\Phi X)(t;\omega) = B^{-1}e^{\left(AB^{-1}\right)t}BX_0(\omega) + \int_0^t B^{-1}e^{\left(AB^{-1}\right)(t-s)}f(s,X(s;\omega))ds$$

$$+ \int_0^t B^{-1}e^{\left(AB^{-1}\right)(t-s)}g(s,X(s;\omega))dW(s). \tag{9.17}$$

Remark. It is very tempting to first reduce $B^{-1}e^{(AB^{-1})t}BX_0(\omega)$ to $e^{(AB^{-1})t}B^{-1}BX_0(\omega)$ and then to $e^{(AB^{-1})t}X_0(\omega)$. However, recall that operators do not commute in general (even in the finite-dimensional case). Specifically, $B^{-1}e^{(AB^{-1})t} \neq e^{(AB^{-1})t}B^{-1}$. So, we must leave (9.17) as is. Nevertheless, the compactness of B^{-1} enables us to establish a reasonable estimate on this term. Indeed, $\forall x \in \text{dom}(B)$,

$$
\begin{aligned}
\left\| B^{-1}e^{(AB^{-1})t}Bx \right\|_{\mathscr{H}} &\leq \left\| B^{-1} \right\|_{\mathbb{B}(\mathscr{H})} \left\| e^{(AB^{-1})t}Bx \right\|_{\mathscr{H}} \\
&\leq \left\| B^{-1} \right\|_{\mathbb{B}(\mathscr{H})} \left\| e^{(AB^{-1})t} \right\|_{\mathbb{B}(\mathscr{H})} \left\| Bx \right\|_{\mathscr{H}} \\
&\leq \left\| B^{-1} \right\|_{\mathbb{B}(\mathscr{H})} (M_{AB^{-1}}) \left\| Bx \right\|_{\mathscr{H}},
\end{aligned}
$$

where $M_{AB^{-1}}$ is defined as in (5.95).

The formulation and proofs of the standard results from Chapter 7 carry over to the present setting without issue. For instance, the following is the analog of Thrm. 7.4.2. The proof of the existence portion follows from a straightforward application of the Contraction Mapping Theorem, while the continuous dependence portion is a direct consequence of Gronwall's Lemma.

Proposition 9.3.3. *Assume that (H5.1), (H5.2), (H5.5), (H7.1), (H7.2), and (H9.1) through (H9.6) hold. Then, (9.11) has a unique mild solution on $[0,T]$. Moreover, if $X_0, Y_0 \in \mathcal{L}^2(\Omega; \text{dom}(B))$ and X and Y are the corresponding mild solutions of (9.11), then $\forall 0 \leq t \leq T$,*

$$
E \left\| X(t;\cdot) - Y(t;\cdot) \right\|_{\mathscr{H}}^2 \leq \xi \left\| X_0 - Y_0 \right\|_{\mathcal{L}^2(\Omega;\text{dom}(B))}^2, \tag{9.18}
$$

where ξ is a positive constant involving

$$
M_A, \left\| B^{-1} \right\|_{\mathbb{B}(\mathscr{H})}, (M_{AB^{-1}}), \left\| B \right\|_{\mathbb{B}(\mathscr{H})}, M_f, M_g, \zeta_g(T,2), T.
$$

Proof. The proof is nearly identical to the argument used in Thrm. 7.4.2 and Prop. 7.4.3. Indeed, note that the presence of B^{-1} and B in the definition of the solution map Φ does not, in any way, obstruct the $\mathscr{F}_t$-measurability of Φ. (Why?) Also, $\forall x, y \in \mathbb{C}([0,T]; \mathcal{L}^2(\Omega; \mathscr{H}))$ and $p > 1$,

$$
E \left\| \int_0^t B^{-1}e^{(AB^{-1})(t-s)}\left[f(s,X(s;\cdot)) - f(s,Y(s;\cdot)) \right] ds \right\|_{\mathscr{H}}^p \leq
$$
$$
\left\| B^{-1} \right\|_{\mathbb{B}(\mathscr{H})}^p (M_{AB^{-1}})^p T^{\frac{p}{q}} M_f^p \int_0^t E \left\| X(s;\cdot) - Y(s;\cdot) \right\|_{\mathscr{H}}^p ds. \tag{9.19}
$$

$$
E \left\| \int_0^t B^{-1}e^{(AB^{-1})(t-s)}\left[g(s,X(s;\cdot)) - g(s,Y(s;\cdot)) \right] dW(s) \right\|_{\mathscr{H}}^p \leq
$$
$$
\left\| B^{-1} \right\|_{\mathbb{B}(\mathscr{H})}^p (M_{AB^{-1}})^p M_g^p \zeta_g(T,2) \int_0^t E \left\| X(s;\cdot) - Y(s;\cdot) \right\|_{\mathscr{H}}^p ds. \tag{9.20}
$$

As such, the mean square continuity of Φ is established as in the proof of Thrm. 7.4.2, yielding

$$\|\Phi(X) - \Phi(Y)\|_{\mathbb{C}}^2 \leq \underbrace{4 \left\|B^{-1}\right\|_{\mathbb{B}(\mathscr{H})}^2 (M_{AB^{-1}})^2 T \left(TM_f^2 + \zeta_g(T,2)M_g^2\right) \|X - Y\|_{\mathbb{C}}^2,}_{=\xi}$$

so that by iteration, it follows that $\forall n \in \mathbb{N}$,

$$\|\Phi^n(X) - \Phi^n(Y)\|_{\mathbb{C}}^2 \leq \frac{\xi^n}{n!} \|X - Y\|_{\mathbb{C}}^2.$$

Hence, $\exists n_0 \in \mathbb{N}$ such that Φ^{n_0} is a strict contraction and so, Φ has a unique fixed point that coincides with the mild solution of (9.11) that we seek.

The verification of (9.18) follows as in the proof of Prop. 7.4.3. (Try it.) □

Exercise 9.3.3.
i.) Identify conditions guaranteeing that IBVP (9.4) has a unique mild solution on $[0,T]$.
ii.) Repeat (i) for IBVP (9.7).

Given that estimates (9.19) and (9.20) and the definition of the solution map Φ bear such close resemblance to those established in Chapter 7 for (7.26), a moment's thought suggests that results for semi-linear Sobolev SEEs of the form (9.11) concerning continuous dependence, p^{th} moment continuity, convergence schemes, and approximation should be established in the same manner as in Chapter 7, with the main changes occurring in the actual numerical estimates obtained. As such, we do not provide a detailed presentation of those results. Rather, you are encouraged to complete the following exercises.

Exercise 9.3.4. (Continuous Dependence)
Consider (9.11), together with the IVP

$$\begin{cases} d(BY)(t;\omega) = \left(AY(t;\omega) + \widehat{f}(t,Y(t;\omega))\right) dt + \widehat{g}(t,Y(t;\omega))d\mathbf{W}(t), \\ Y(0;\omega) = Y_0(\omega), \omega \in \Omega \end{cases} \tag{9.21}$$

in a separable Hilbert space $\mathscr{H}$, where $0 < t < T$, $\omega \in \Omega$, and $\widehat{f}$ and $\widehat{g}$ satisfy **(H7.1)** and **(H7.2)**.
i.) Establish an estimate for $E \|X(t;\cdot) - Y(t;\cdot)\|_{\mathscr{H}}^2$.
ii.) A more general continuous dependence result can be formulated by further replacing the operators A and B in (9.21) by appropriate operators $\widehat{A}$ and $\widehat{B}$, respectively. Where does the difficulty arise in doing this? How can it be overcome?

Exercise 9.3.5. (Yosida Approximations)
i.) Set up the sequence of Yosida approximations (like (7.64)) for **(9.11)**.
ii.) Argue that each IVP in the sequence defined in (i) has a strong solution on $[0,T]$.

iii.) Determine the variation of parameters formula for the mild solution X_n of the IVPs in the sequence defined in (i).

iv.) Prove that $\lim_{n\to\infty} \|X_n - X\|_{C([0,T];\mathscr{L}^2(\Omega;\mathscr{H}))} = 0$, where X is the mild solution of (9.11).

v.) Deduce the weak convergence of the sequence of induced probability measures $\mathscr{P}_{X_n}$.

Exercise 9.3.6. (Non-Lipschitz Conditions)

Formulate existence results for (9.11) analogous to the results developed in Section 7.6. Carefully indicate the necessary modifications to the hypotheses in the statements, as well as to the proofs of the results. Pay particular attention to how the computations change due to the presence of B^{-1} and B.

Exercise 9.3.7. (Zeroth-Order Approximation)

For each $0 < \varepsilon < 1$, consider the IVP

$$\begin{cases} \partial\left(z_\varepsilon(\mathbf{x},t;\omega) - \triangle z_\varepsilon(\mathbf{x},t;\omega)\right) + \alpha\triangle z_\varepsilon(\mathbf{x},t;\omega))\partial t = \\ f(\mathbf{x},t)\partial t + g_\varepsilon(z_\varepsilon(\mathbf{x},t;\omega)))d\mathbf{W}(t),\ \mathbf{x}\in\mathscr{D}, t > 0,\ \omega\in\Omega, \\ z_\varepsilon(\mathbf{x},0;\omega) = z_0(\mathbf{x};\omega),\ \mathbf{x}\in\mathscr{D},\ \omega\in\Omega, \\ \frac{\partial z_\varepsilon}{\partial\mathbf{n}}(\mathbf{x},t;\omega) = 0,\ \mathbf{x}\in\partial\mathscr{D}, t > 0,\ \omega\in\Omega, \end{cases} \tag{9.22}$$

where $\mathscr{D}$ is a bounded domain in $\mathbb{R}^3$, $\alpha > 0$, $z : \mathscr{D}\times[0,T]\times\Omega\to\mathbb{R}$, and $\mathbf{W}(t)$ is an m-dimensional Wiener process. Assume that

(H9.7) $f : \mathscr{D}\times[0,T]\to\mathbb{R}$ is an $\mathscr{F}_t$-adapted, continuous mapping.
(H9.8) For each $0 < \varepsilon < 1$, $g_\varepsilon : \mathbb{R}\to\mathbb{R}$ is globally Lipschitz (with the same Lipschitz constant $M_{g_\varepsilon} = M$, $\forall 0 < \varepsilon < 1$) and

$$\|g_\varepsilon(y)\|_{\mathscr{B}_0} \longrightarrow 0\ \text{as}\ \varepsilon\to 0^+, \forall y\in\mathbb{R},\ \text{uniformly in}\ t\in[0,T].$$

The question is whether or not there exist $z^\star : \mathscr{D}\times[0,T]\to\mathbb{R}$ such that

$$\lim_{\varepsilon\to 0^+} \|z_\varepsilon - z^\star\|_{C([0,T];\mathscr{L}^2(\Omega;\mathbb{L}^2(\mathscr{D})))} = 0.$$

To answer this question, proceed as follows:
i.) Reformulate (9.22) as the abstract Sobolev SEE (9.11).
ii.) Formulate and prove a result in the spirit of Prop. 7.4.16.
iii.) Use (ii) to answer the question posed above.

9.4 Functional Sobolev SEEs

We now turn our attention to the more general functional Sobolev-type stochastic evolution equation

$$\begin{cases} d(BX)(t;\omega) = (AX(t;\omega) + \mathfrak{F}(X)(t;\omega))\,dt + \mathfrak{G}(X)(t;\omega)d\mathbf{W}(t), \\ X(0;\omega) = X_0(\omega), \omega \in \Omega \end{cases} \tag{9.23}$$

in a separable Hilbert space $\mathscr{H}$, where $0 < t < T$, $\omega \in \Omega$, under the usual hypotheses $(\mathbf{H_A})$, $(\mathbf{H5.1})$ through $(\mathbf{H5.3})$ and $(\mathbf{H5.5})$, $(\mathbf{H9.1})$ through $(\mathbf{H9.6})$, where

$$\mathfrak{F} : \mathbb{C}\left([0,T];\mathfrak{L}^2\left(\Omega;\mathscr{H}\right)\right) \to \mathbb{C}\left([0,T];\mathfrak{L}^2\left(\Omega;\mathscr{H}\right)\right),$$

$$\mathfrak{G} : \mathbb{C}\left([0,T];\mathfrak{L}^2\left(\Omega;\mathscr{H}\right)\right) \to \mathbb{C}\left([0,T];\mathfrak{L}^2\left(\Omega;\mathscr{B}_0\left(\mathbb{R}^m;\mathscr{H}\right)\right)\right).$$

We say that $X : [0,T] \times \Omega \to \mathscr{H}$ is a *mild solution* of (9.23) if u satisfies Def. 9.3.2 with $f(t,X(t;\omega))$ and $g(t,X(t;\omega))$ replaced by $\mathfrak{F}(X)(t;\omega)$ and $\mathfrak{G}(X)(t;\omega)$, respectively.

Guided by the approach used in Section 8.3, let $v \in \mathbb{C}\left([0,T];\mathfrak{L}^2\left(\Omega;\mathscr{H}\right)\right)$ and define the solution map $\Phi : \mathbb{C}\left([0,T];\mathfrak{L}^2\left(\Omega;\mathscr{H}\right)\right) \to \mathbb{C}\left([0,T];\mathfrak{L}^2\left(\Omega;\mathscr{H}\right)\right)$ by $\Phi(v) = X_v$, where X_v is the unique mild solution of the IVP

$$\begin{cases} d(BX_v)(t;\omega) = (AX_v(t;\omega) + \mathfrak{F}(v)(t;\omega))\,dt + \mathfrak{G}(v)(t;\omega)d\mathbf{W}(t), \\ X_v(0;\omega) = X_0(\omega), \omega \in \Omega. \end{cases} \tag{9.24}$$

where $0 < t < T$, $\omega \in \Omega$. Define the mappings $f : [0,T] \times \mathscr{H} \to \mathscr{H}$ and $g : [0,T] \times \mathscr{H} \to \mathscr{B}_0(\mathbb{R}^m,\mathscr{H})$ by $f(t,X_v(t;\omega)) = \mathfrak{F}(v)(t;\omega)$ and $g(t,X_v(t;\omega)) = \mathfrak{G}(v)(t;\omega)$, respectively. As long as $\mathfrak{F}$ and $\mathfrak{G}$ are continuous, we can invoke Prop. 9.3.3 to conclude that Φ is well-defined. (Why?)

We apply the theory developed in Chapter 8 to establish results for (9.23). Particular versions of (9.23) corresponding to specific choices for the functionals $\mathfrak{F}$ and $\mathfrak{G}$ have been studied in the literature under various assumptions (see, for instance, [229]). The theory outlined below encompasses many of these results as special cases. We begin with the following extension of Prop. 9.3.3.

Proposition 9.4.1. *Assume that* $(\mathbf{H_A})$, $(\mathbf{H5.1})$ *through* $(\mathbf{H5.3})$ *and* $(\mathbf{H5.5})$, $(\mathbf{H8.10})$, $(\mathbf{H8.14})$, *and* $(\mathbf{H9.1})$ *through* $(\mathbf{H9.6})$ *hold. Then,* (9.23) *has a unique mild solution on* $[0,T]$ *provided that*

$$2\left\|B^{-1}\right\|_{\mathbb{B}(\mathscr{H})}^2 (M_{AB^{-1}})^2 \left[M_{\mathfrak{F}}^2 T^2 + \zeta_{\mathfrak{G}}(T,2)TM_{\mathfrak{G}}^2\right] < 1.$$

Proof. Let $v \in \mathbb{C}\left([0,T];\mathfrak{L}^2\left(\Omega;\mathscr{H}\right)\right)$. Then, $\mathfrak{F}(v) \in \mathbb{C}\left([0,T];\mathfrak{L}^2\left(\Omega;\mathscr{H}\right)\right)$ and $\mathfrak{G}(v) \in \mathbb{C}\left([0,T];\mathfrak{L}^2\left(\Omega;\mathscr{B}_0\left(\mathbb{R}^m;\mathscr{H}\right)\right)\right)$, so that Prop. 9.3.3 ensures that (9.24) has a unique mild solution on $[0, T]$. (Tell why carefully.) Because

$$(\Phi v)(t; \omega) = B^{-1} e^{(AB^{-1})t} BX_0(\omega) + \int_0^t B^{-1} e^{(AB^{-1})(t-s)} \mathfrak{F}(v)(s; \omega) ds$$

$$+ \int_0^t B^{-1} e^{(AB^{-1})(t-s)} \mathfrak{G}(v)(s; \omega) dW(s), \tag{9.25}$$

we know that $\forall v, w \in \mathbb{C}\left([0, T]; \mathcal{L}^2(\Omega; \mathcal{H})\right)$ and $0 < t < T$,

$$E \left\| (\Phi v)(t; \cdot) - (\Phi w)(t; \cdot) \right\|_{\mathcal{H}}^2 \le$$

$$2E \left\| \int_0^t B^{-1} e^{(AB^{-1})(t-s)} \left[\mathfrak{F}(v)(s; \cdot) - \mathfrak{F}(w)(s; \cdot)\right] ds \right\|_{\mathcal{H}}^2$$

$$+2E \left\| \int_0^t B^{-1} e^{(AB^{-1})(t-s)} \left[\mathfrak{G}(v)(s; \cdot) - \mathfrak{G}(w)(s; \cdot)\right] dW(s) \right\|_{\mathcal{H}}^2 \le$$

$$2 \left\|B^{-1}\right\|_{\mathbb{B}(\mathcal{H})}^2 (M_{AB^{-1}})^2 \int_0^t \left[T \left\|\mathfrak{F}(v) - \mathfrak{F}(w)\right\|_{\mathbb{C}}^2 + \right.$$

$$\left. + \zeta_{\mathfrak{G}}(T, 2) \left\|\mathfrak{G}(v) - \mathfrak{G}(w)\right\|_{\mathbb{C}}^2 \right] ds \le \tag{9.26}$$

$$2 \left\|B^{-1}\right\|_{\mathbb{B}(\mathcal{H})}^2 (M_{AB^{-1}})^2 \left[M_{\mathfrak{F}}^2 T^2 + \zeta_{\mathfrak{G}}(T, 2) T M_{\mathfrak{G}}^2\right] \left\|v - w\right\|_{\mathbb{C}}^2.$$

Taking the supremum over $[0, T]$, followed by taking the square root in (9.26), subsequently yields

$$\left\|(\Phi v) - (\Phi w)\right\|_{\mathbb{C}} \le \sqrt{2 \left\|B^{-1}\right\|_{\mathbb{B}(\mathcal{H})}^2 (M_{AB^{-1}})^2 \left[M_{\mathfrak{F}}^2 T^2 + \zeta_{\mathfrak{G}}(T, 2) T M_{\mathfrak{G}}^2\right]} \left\|v - w\right\|_{\mathbb{C}}$$

$$< \left\|v - w\right\|_{\mathbb{C}}.$$

Thus, Φ is a contraction and thus, has a unique fixed-point by the Contraction Mapping Theorem. This fixed-point coincides with a mild solution of (9.23). (Why?) This completes the proof. $\square$

A similar result formulated under a slightly different data restriction can be established if the range space of $\mathfrak{F}$ is enlarged to $\mathbb{L}^1(0, T; \mathcal{X})$ and the semigroup is required to be contractive. Consider the following exercise.

Exercise 9.4.1. Assume that $\mathfrak{F} : \mathbb{C}\left([0, T]; \mathcal{L}^2(\Omega; \mathcal{H})\right) \to \mathbb{L}^2\left(0, T; \mathcal{L}^2(\Omega; \mathcal{H})\right)$ satisfies **(H8.12)** in Prop. 9.4.1 and that A generates a contractive C_0-semigroup on $\mathcal{H}$. Show that (9.23) has a unique mild solution on $[0, T]$ provided that

$$2 \left\|B^{-1}\right\|_{\mathbb{B}(\mathcal{H})}^2 (M_{AB^{-1}})^2 \left[M_{\mathfrak{F}}^2 T^{2 - \frac{2}{p}} + \zeta_{\mathfrak{G}}(T, 2) T M_{\mathfrak{G}}^2\right] < 1.$$

Corollary 9.4.2. *Consider the evolution equation*

$$
\begin{cases}
d(BX)(t;\omega) = \left(AX(t;\omega) + f\left(t,X(t;\omega),\int_0^t a_1(t-s)h_1(s,X(s;\omega))ds\right)\right)dt \\
+g\left(t,X(t;\omega),\int_0^t a_2(t-s)h_2(s,X(s;\omega))ds\right)d\mathbf{W}(t),\ 0<t<T,\ \omega\in\Omega, \\
X(0;\omega) = X_0(\omega),\ \omega\in\Omega,
\end{cases}
$$

(9.27)

where $f : [0,T] \times \mathcal{H} \times \mathcal{H} \to \mathcal{H}$, $g : [0,T] \times \mathcal{H} \times \mathcal{H} \to \mathcal{B}_0(\mathbb{R}^m;\mathcal{H})$ *and* $h_i :$ $[0,T] \times \mathcal{H} \to \mathcal{H}$ $(i=1,2)$ *are continuous in the first variable and globally Lipschitz in the remaining variables (uniformly in t), and* $a_i \in C(\mathbb{R};[0,\infty))$ $(i=1,2)$. *Then,* (9.27) *has a unique mild solution on* $[0,T]$.

Exercise 9.4.2.
i.) Prove Cor. 9.4.2 directly without first reformulating (9.27) as (9.23).
ii.) Alternatively, reformulate (9.27) as (9.23) in an appropriate space and recover the result directly from Prop. 9.4.1.

Exercise 9.4.3. Formulate and prove an existence-uniqueness result for IBVP (9.7).

Consider the stochastic evolution equation

$$
\begin{cases}
d(BX)(t;\omega) = (AX(t;\omega) + CX(t;\omega) + \mathfrak{F}(X)(t;\omega))dt + \mathfrak{G}(X)(t;\omega)d\mathbf{W}(t), \\
X(0;\omega) = X_0(\omega),\ \omega\in\Omega
\end{cases}
$$

(9.28)

in a separable Hilbert space $\mathcal{H}$, where $0<t<T$, $\omega\in\Omega$, under the same hypotheses imposed on (9.23), where $C \in \mathbb{B}(\mathcal{H})$.

Exercise 9.4.4.
i.) Explain why the operator $(A+C)B^{-1}$ generates a C_0-semigroup on $\mathcal{H}$.
ii.) Assuming the hypotheses of Prop. 9.4.1, show that (9.28) has a unique mild solution on $[0,T]$
iii.) Redo (ii), now assuming that $\mathfrak{F} : C\left([0,T];\mathscr{L}^2(\Omega;\mathcal{H})\right) \to L^2\left(0,T;\mathscr{L}^2(\Omega;\mathcal{H})\right)$ satisfies **(H8.12)**.

As an application of (9.28), consider the following system of Sobolev PDEs governing the behavior of $z = z(x,t;\omega)$ and $w = w(x,t;\omega)$ for $0<x<a$, $0<t<T$:

$$
\begin{cases}
\partial\left(z - \frac{\partial^2 z}{\partial x^2}\right) + \left(\alpha_1\frac{\partial^2 z}{\partial x^2} + \beta_1 z\right)\partial t = \left(h_1(z,w)\int_0^t a_1(t-s)f_1(s,x,z,w)ds\right)\partial t \\
+\sum_{k=1}^m g_k(t)dW_k(t),\ 0<x<a,\ 0<t<T,\ \omega\in\Omega, \\
\partial\left(w - \frac{\partial^2 w}{\partial x^2}\right) + \left(\alpha_2\frac{\partial^2 w}{\partial x^2} + \beta_2 w\right)\partial t = \left(h_2(z,w)\int_0^t a_2(t-s)f_2(s,x,z,w)ds\right)\partial t \\
+\sum_{k=1}^m \widehat{g}_k(t)dW_k(t),\ 0<x<a,\ 0<t<T,\ \omega\in\Omega, \\
z(x,0;\omega) = z_0(x;\omega),\ w(x,0;\omega) = w_0(x;\omega),\ 0<x<a,\ \omega\in\Omega, \\
\frac{\partial z}{\partial x}(0,t;\omega) = \frac{\partial z}{\partial x}(a,t;\omega) = 0 = \frac{\partial w}{\partial x}(0,t;\omega) = \frac{\partial w}{\partial x}(a,t;\omega),\ 0<t<T,\ \omega\in\Omega,
\end{cases}
$$

(9.29)

where α_i, β_i $(i = 1, 2)$ are real constants, $f_i : [0, T] \times [0, a] \times \mathbb{R}^2 \to \mathbb{R}$, $h_i : \mathbb{R}^2 \to \mathbb{R}$, $g_k : [0, T] \to \mathscr{B}_0 \left(\mathbb{R}; \left(L^2(0, a) \right)^2 \right)$ $(k = 1, \ldots, m)$, $a_i \in \mathbb{C} \left([0, T]; (0, \infty) \right)$ $(i = 1, 2)$, and $W_k(t)$ $(k = 1, \ldots, m)$ are independent one-dimensional Wiener processes.

Exercise 9.4.5.
i.) Reformulate (9.29) abstractly as (9.28) in an appropriate space.
ii.) Verify that (**H9.1**) through (**H9.6**) are satisfied.
iii.) Impose appropriate growth and/or regularity restrictions on a_i, f_i, h_i $(i = 1, 2)$, g_k $(k = 1, \ldots, m)$ that ensure (9.29) has a unique mild solution on $[0, T]$.

9.5   Guidance for Selected Exercises

9.5.1   Level 1: A Nudge in a Right Direction

9.1.1. Define $A : \text{dom}(A) \subset L^2(\Omega) \to L^2(\Omega)$ and $B : \text{dom}(B) \subset L^2(\Omega) \to L^2(\Omega)$ by

$$A[z] = \left(\frac{\partial^2}{\partial x^2} + \frac{\partial^2}{\partial y^2} \right) [z],$$
$$B[z] = (I - A)[z].$$

Identify $\text{dom}(A)$ and $\text{dom}(B)$. How do you handle the forcing term?
9.1.2. To what operator must $I - \alpha \frac{\partial}{\partial t}$ converge as $\alpha \to 0^+$?
9.1.3. Does a change of variable similar to the one used to convert a classical wave equation into a system of first-order PDEs work?
9.2.1. Define $A : \text{dom}(A) \subset L^2(0, a) \to L^2(0, a)$ and $B : \text{dom}(B) \subset L^2(0, a) \to L^2(0, a)$ by

$$A[u] = \alpha \frac{\partial^2}{\partial x^2}[u],$$
$$B[u] = (I - \frac{1}{\alpha} A)[u],$$

where $u(t) = z(\cdot, t)$. Then, loosely speaking, (9.4) can be reformulated as (9.9) in $\mathscr{X} = L^2(0, a)$ with $f(t, u(t)) = f(t)$. Now, verify (**H9.1**) through (**H9.6**).
9.3.2. Because B^{-1} is a linear operator, it commutes with certain limit operators and behaves well when applied to inputs in the forms of finite sums. (So what?)

9.3.4. ii.) The operator $\widehat{A} \left(\widehat{B} \right)^{-1}$ must generate a C_0-semigroup so that the same approach outlined in this chapter can be used. So, (**H9.1**) through (**H9.6**) must be appropriately adapted to ensure this happens.
9.4.2. i.) Modify the solution map appropriately and then argue it is a strict contraction in the usual manner. Of course, the estimate involves more terms and this affects

the data restriction needed to ensure that the solution map is a contraction.

9.4.3. Technically, this follows from the corollary, if you define f appropriately.

9.4.5. i.) Use $\mathscr{H} = \left(\mathbb{L}^2(0,a)\right)^2$. The functional $\mathfrak{F}$ will consist of two components, while each of the operators A, B, and C are 2×2 matrices. (Now what?)

ii.) See Exer. 7.1.4.

iii.) Be careful here. The functional must be globally Lipschitz, but the initial thought of simply imposing Lipschitz conditions on all of the functions involved might need to be reconsidered.

iv.) Check to see whether the functional satisfies **(H6.16)**. If so, then what?

9.5.2 Level 2: An Additional Thrust in a Right Direction

9.1.1. Make certain to incorporate the BCs into $\text{dom}(A)$. Define the functional $\mathfrak{F}$:
$\mathbb{C}\left([0,T]; \mathscr{L}^2(\Omega; \mathbb{L}^2(\mathscr{D}))\right) \to \mathbb{C}\left([0,T]; \mathscr{L}^2(\Omega; \mathbb{L}^2(\mathscr{D}))\right)$ by

$$\mathfrak{F}(z)(t) = \int_0^{a_2} \int_0^{a_1} k(t,x,y) f\left(t, z(x,y,t)\right) dx dy.$$

Is $\mathfrak{F}$ well-defined? What form is the abstract stochastic evolution equation?

9.1.2. It can be shown that the limit operator is I. As such, p is a mild solution to a classical diffusion equation.

9.1.3. It is tempting to try to use a change of variable similar to the one used to convert the wave equation to a system of first-order PDEs. However, mimicking that approach results in having to make the following identification:

$$A \begin{bmatrix} z \\ \frac{\partial z}{\partial t} \end{bmatrix} = \begin{bmatrix} 0 & I \\ \frac{\partial^2}{\partial x^2} & \frac{\partial^2}{\partial x^2} \end{bmatrix} \begin{bmatrix} z \\ \frac{\partial z}{\partial t} \end{bmatrix},$$

which does not generate a C_0-semigroup on $\mathscr{H}$. As such, we can at best reformulate the given IBVP as the abstract SEE

$$\begin{cases} d\left(u'(t;\omega) + (Au)(t;\omega)\right) + Au(t;\omega)dt = f(t,u(t;\omega))dt \\ + g(t,u(t;\omega))dW(t), \ 0 < t < T, \ \omega \in \Omega, \\ u(0;\omega) = u_0(\omega), \ u'(0;\omega) = u_1(\omega), \ \omega \in \Omega. \end{cases} \quad (9.30)$$

It remains to be seen if such a second-order evolution equation can be transformed into a system of first-order PDEs that can be handled using the theory developed thus far.

9.1.4. Here are the highlights:

Define $\text{dom}(A) = \text{dom}(B)$ to be

$$\left\{ f \in \mathbb{L}^2(0,a) \mid f, \frac{df}{dx} \text{ are AC}, \frac{d^2 f}{dx^2} \in \mathbb{L}^2(0,a), \wedge f(0) = f(a) = 0 \right\}.$$

Then, **(H9.4)** holds. Moreover, A and B can be expressed by

$$Az = \sum_{n=1}^{\infty} n^2 \langle z, e_n \rangle_{\mathbb{L}^2} e_n,$$

$$Bz = \sum_{n=1}^{\infty} \left(1 + n^2\right) \langle z, e_n \rangle_{\mathbb{L}^2} e_n,$$

where $e_n = \sqrt{\frac{2}{a}} \sin\left(\frac{2n\pi x}{a}\right)$, $n \in \mathbb{N}$. The linearity of A implies the linearity of B, so that **(H9.1)** holds. Also, **(H9.5)** holds by an earlier exercise, as does the fact that B^{-1} is closed. You can check directly that B is invertible with

$$B^{-1}z = \sum_{n=1}^{\infty} \frac{1}{1+n^2} \langle z, e_n \rangle_{\mathbb{L}^2} e_n,$$

so that **(H9.2)** holds. (Do so!) The fact that **(H9.3)** holds is an assumption that must be imposed on z_0. Finally, try to verify **(H9.6)** directly using the definition.

9.3.2. This implies that $B^{-1}\left(\sum_{i=1}^{n} H\left(x_i^*\right) \triangle x_i\right) = \sum_{i=1}^{n} B^{-1}\left(H\left(x_i^*\right) \triangle x_i\right)$, for all partitions of $(0, t)$. Hence, taking $\lim_{n \to \infty}$ of both sides yields equal results. By definition of the integral, this suggests that B^{-1} commutes with the integral operator.

9.4.2. Proving that the forcing term is globally Lipschitz is the crucial part of this approach. When viewing it as a functional, make certain to adapt $\mathfrak{F}_2$ so that it maps into the smaller space $\mathbb{C}\left([0,T]; \mathfrak{L}^2(\Omega; \mathscr{H})\right)$.

9.4.3. Alternatively, define the functional

$$\mathfrak{F}: \mathbb{C}\left([0,T]; \mathfrak{L}^2(\Omega; \mathscr{H})\right) \to \mathbb{C}\left([0,T]; \mathfrak{L}^2(\Omega; \mathscr{H})\right),$$

where $\mathscr{H} = \mathbb{L}^2\left((0,a_1) \times (0,a_2)\right)$, as in Exer. 9.1.1. Assuming that k is continuous is sufficient (but might be overly strong) and that f is globally Lipschitz are sufficient.

9.4.4. ii.) What is true about the operator $A + C$? So what?

iii.) The only modification is that $(A + C)B^{-1}$ is the generator of interest, so we change the notation in the constant $\overline{M}_{AB^{-1}}$ in the main existence theorem accordingly to $\overline{M}_{(A+C)B^{-1}}$.

9.4.5. i.) Symbolically,

$$A\begin{bmatrix} z \\ w \end{bmatrix} = \begin{bmatrix} \alpha_1 \frac{\partial^2}{\partial x^2} & 0 \\ 0 & \alpha_2 \frac{\partial^2}{\partial x^2} \end{bmatrix}\begin{bmatrix} z \\ w \end{bmatrix},$$

$$B\begin{bmatrix} z \\ w \end{bmatrix} = \begin{bmatrix} I - \frac{\partial^2}{\partial x^2} & 0 \\ 0 & I - \frac{\partial^2}{\partial x^2} \end{bmatrix}\begin{bmatrix} z \\ w \end{bmatrix},$$

$$C\begin{bmatrix} z \\ w \end{bmatrix} = \begin{bmatrix} \beta_1 I & 0 \\ 0 & \beta_2 I \end{bmatrix}\begin{bmatrix} z \\ w \end{bmatrix}.$$

Make certain to identify their domains and incorporate the BCs appropriately. Define the functional $\mathfrak{F}$ by

$$\mathfrak{F}\begin{bmatrix} z \\ w \end{bmatrix}(t) = \begin{bmatrix} h_1(z,w) \int_0^t a_1(t-s) f_1(s,x,z,w) ds \\ h_2(z,w) \int_0^t a_2(t-s) f_2(s,x,z,w) ds \end{bmatrix}.$$

iii.) Suppose that h_1 and f_1 are globally Lipschitz. How would you argue that

$$\left| \mathfrak{F} \begin{bmatrix} z \\ w \end{bmatrix} (t) - \mathfrak{F} \begin{bmatrix} \bar{z} \\ \bar{w} \end{bmatrix} (t) \right| \le M \left[|z - \bar{z}| + |w - \bar{w}| \right]?$$

The triangle inequality can help only so much. Alternatively, you might consider assuming that h_i continuous and globally bounded and that f_i are globally Lipschitz. Can these be weakened?

iv.) Yes, but prove it.

Chapter 10

Beyond Volume 2

Overview

The material developed in this volume has provided an introductory look at stochastic evolution equations designed to acquaint you with some essential and foundational notions and techniques, as well as a wealth of applications to which they apply. But, this is just the beginning. Several more chapters could have been included in this text that explored other interesting classes of abstract stochastic evolution equations, similarly rooted in concrete applications, using a similar approach. And, volumes more could be written for the study of abstract stochastic evolution equations for which a nice variation of parameters formula is no longer available. As encouragment for you to continue the current line of study, we provide very short encounters with different, yet related, classes of equations below, each of which can be studied using a similar approach to the one developed in this text.

10.1   Fully Nonlinear SEEs

Using a modest dose of linear semigroup theory and a bit of probability theory, we were able to develop a rather rich existence theory that formed a theoretical basis for a formal mathematical study of vastly different phenomena subject to noise. We introduced a significant amount of complexity into the IBVPs by way of perturbations and complex forcing terms. But, as rich as the theory is, it all hinges on the crucial assumption that the operator $A : \text{dom}(A) \subset \mathscr{X} \to \mathscr{X}$ is linear and generates a C_0-semigroup on $\mathscr{H}$. The problem is that these two assumptions do not hold for many phenomena, including the various improvements on the models discussed throughout the text. As such, the question is whether or not we can somehow argue analogously as we did when generalizing the setting of Chapter 4 to Chapter 5 to develop a theory in the so-called *nonlinear* case. The answer is a tentative yes, but the extension from the linear to the nonlinear setting takes place on a much grander scale than the generalization of the finite-dimensional to the infinite-dimensional setting in the linear case, and requires a considerably higher degree of sophistication. We refer you to the following references to assist you in launching a formal study of such equations: **[2,**

5, 6, 9, 13, 14, 40, 42, 61, 65, 98, 100, 132, 140, 150, 169, 198, 210, 226, 233, 239, 240, 253, 260, 290, 304, 305, 328, 334, 344, 358, 359, 361, 365, 366, 375, 384, 393, 395, 401, 402, 412, 415, 419].

Model XXI.1 Nonlinear Diffusive Phenomena
We have encountered numerous phenomena whose mathematical description involved a diffusion term of the general form $\alpha \triangle$, where α is a positive constant. At its root, this operator arises from the premise that diffusion is governed by Fick's law (see [336]), which yields a natural, albeit linear, description of dispersion. But, more complicated phenomena are governed by more complex laws often resulting in nonlinear diffusivity. Indeed, it is often necessary to replace $\alpha \triangle u$ by a more general operator of the form $\triangle f(u)$, where $f : \mathbb{R} \to \mathbb{R}$ is a continuous, increasing function for which $f(0) = 0$. For instance, such a term with $f(u) = u |u|^{m-1}$, where $m > 1$, occurs in an IBVP arising in the study of porous media. Other variants arise in models in the context of differential geometry with Ricci flow (see [393]), a nonlinear model of brain tumor growth (see [171]), porous media (see [100]), and other diffusive processes (see [239, 240, 266, 389, 402, 412]).

The most rudimentary IBVP involving this nonlinear diffusion operator is

$$
\begin{cases}
\partial u(x,t;\omega) = \triangle f(u(x,t;\omega))\partial t + g(u(x,t;\omega)dW(t), x \in \mathscr{D}, t > 0, \omega \in \Omega, \\
u(x,t;\omega) = 0, x \in \partial \mathscr{D}, t > 0, \omega \in \Omega, \\
u(x,0;\omega) = u_0(x;\omega), x \in \mathscr{D}, \omega \in \Omega.
\end{cases}
$$
(10.1)

Given that the IBVP is homogenous, it is natural to reformulate (10.1) as an abstract stochastic evolution equation of a form similar to (5.63). This requires that we identify the operator $\mathscr{A} : \text{dom}(\mathscr{A}) \subset \mathscr{H} \to \mathscr{H}$ (for an appropriate space $\mathscr{H}$) as

$$\mathscr{A}u = \triangle f(u),$$

which is not linear unless $f(u) = au + b$, the typical *linear* diffusion operator. As such, the theory in Chapter 3 and all subsequent results are inapplicable to (10.1).

Model XXII.1 Hydrology and Groundwater Flow
A stochastic PDE governing one-dimensional lateral groundwater flow, referred to as the *Boussinesq equation* (discussed in [42, 213, 361]) is given by

$$
\partial h(x,t;\omega) = \frac{1}{S} \frac{\partial}{\partial x} \left(Kh(x,t;\omega) \frac{\partial h}{\partial x}(x,t;\omega) \right) \partial t,
$$
$$
+ f(t, h(x,t;\omega))dW(t), 0 \le x \le L, t > 0, \omega \in \Omega, \quad (10.2)
$$

where the aquifer is modeled as the interval $[0,L]$, h is the hydraulic head, K is the hydraulic conductivity, and S is the specific yield. Assuming that there is no replenishment of water via rainfall by seepage through the soil surrounding the aquifer, (10.2) can be coupled with the following BCs

$$
h(0,t;\omega) = M(t;\omega), \frac{\partial h}{\partial x}(L,t;\omega) = 0, \omega \in \Omega, \quad (10.3)
$$

so that the aquifer is replenished at the end $x = 0$ and experiences no change at $x = L$.

Model VII.6 Nonlinear Waves

We have accounted for linear dissipation in the models of wave phenomena via the inclusion of the term $\alpha \frac{\partial z}{\partial t}$, where $\alpha > 0$, in the stochastic PDE, where z represents the unknown displacement function. What if the dissipation is governed by a *nonlinear* function of $\frac{\partial z}{\partial t}$? For instance, consider the IBVP

$$
\begin{cases}
\partial \left(\frac{\partial z}{\partial t} + \alpha z \right) + \left(\beta \left(\frac{\partial z}{\partial t} \right)^3 - c^2 \frac{\partial^2 z}{\partial x^2} \right) \partial t = \left(\int_0^t a(t-s) \sin(z(x,s;\omega)) ds \right) \partial t + \\
\quad f(t, z(x,t;\omega)) dW(t), \ 0 < x < L, t > 0, \omega \in \Omega, \\
z(x,0;\omega) = z_0(x;\omega), \ \frac{\partial z}{\partial t}(x,0;\omega) = z_1(x;\omega), \ 0 < x < L, \omega \in \Omega, \\
z(0,t;\omega) = z(L,t;\omega) = 0, t > 0, \omega \in \Omega,
\end{cases}
$$

$$(10.4)$$

where $z_0 \in \mathscr{L}^2 \left(\Omega; \mathbb{H}^2(0,L) \cap \mathbb{H}_0^1(0,L) \right)$, $z_1 \in \mathscr{L}^2 \left(\Omega; \mathbb{H}_0^1(0,L) \right)$, and $a : [0,T] \to \mathbb{R}$ is continuous (see **[81, 82, 169, 260, 331]**). The second-order PDE portion of (10.4) can be written as the following equivalent system of first-order PDEs:

$$
\frac{\partial}{\partial t} \begin{bmatrix} v_1 \\ v_2 \end{bmatrix} (x,t) = \begin{bmatrix} 0 & I \\ -c^2 \frac{\partial^2}{\partial x^2} & \alpha + \beta (v_2)^2 \end{bmatrix} \begin{bmatrix} v_1 \\ v_2 \end{bmatrix} (x,t) \qquad (10.5)
$$
$$
+ \begin{bmatrix} 0 \\ \int_0^t a(t-s) \sin(v_1) ds \end{bmatrix} \partial t + \begin{bmatrix} 0 \\ f(t,v_1) \end{bmatrix} dW(t),
$$

where $0 < x < L, t > 0, \omega \in \Omega$. It is natural to view (10.5) abstractly as (8.51) in $\mathscr{X} = \mathbb{H}_0^1(0,L) \times \mathbb{H}^0(0,L)$. Doing so requires that we define $\mathscr{A} : \mathrm{dom}(\mathscr{A}) \subset \mathscr{X} \to \mathscr{X}$ by

$$
\mathscr{A} \begin{bmatrix} v_1 \\ v_2 \end{bmatrix} = \begin{bmatrix} 0 & I \\ -c^2 \frac{\partial^2}{\partial x^2} & \alpha + \beta (v_2)^2 \end{bmatrix} \begin{bmatrix} v_1 \\ v_2 \end{bmatrix} = \begin{bmatrix} v_2 \\ -c^2 \frac{\partial^2 v_1}{\partial x^2} + \alpha v_2 + \beta (v_2)^3 \end{bmatrix}
$$
$$
\mathrm{dom}(\mathscr{A}) = \left(\mathbb{H}^2(0,L) \cap \mathbb{H}_0^1(0,L) \right) \times \mathbb{H}_0^1(0,L). \qquad (10.6)
$$

Model XXVII.1 Nonlinear Beams

The model of the deflection of beams can also be improved by accounting for nonlinear dissipation. The following generalization of IBVP (8.115) is one such improvement:

$$
\begin{cases}
\partial \left(\frac{\partial w}{\partial t} \right) + \left(\alpha \frac{\partial^4 w}{\partial z^4} + \beta w + \left(\frac{\partial w}{\partial t} \right)^{2m+1} \right) \partial t = \left(\int_0^t a(t,s) f \left(s,z,w, \frac{\partial^2 w}{\partial z^2}, \frac{\partial w}{\partial s} \right) ds \right) \partial t, \\
+ f \left(t, w, \frac{\partial w}{\partial t} \right) 0 < z < a, 0 < t < T, \omega \in \Omega, \\
w(z,0;\omega) = w_0(z;\omega), \ \frac{\partial w}{\partial t}(z,0;\omega) = w_1(z;\omega), \ 0 < z < a, \omega \in \Omega, \\
\frac{\partial^2 w}{\partial z^2}(0,t;\omega) = w(0,t;\omega) = 0 = \frac{\partial^2 w}{\partial z^2}(a,t;\omega) = w(a,t;\omega), t > 0, \omega \in \Omega,
\end{cases}
$$

$$(10.7)$$

where $m \in \mathbb{N}$. Note the similarity to the nonlinear wave equation (see [398]).

There are many other interesting nonlinear models, some of which are explored in [240, 366]. Certainly, the nonlinearity of the operator $\mathscr{A}$ is an obstacle, but this is just the beginning.

Loosely speaking, all of the above IBVPs can be reformulated as an abstract stochastic evolution equation of the form

$$\begin{cases} du(t;\omega) = \mathscr{A}u(t;\omega)dt + \mathfrak{F}(u)(t;\omega)dt + \mathscr{G}(u)(t;\omega)dW(t), \\ u(0;\omega) = u_0(\omega), \omega \in \Omega \end{cases} \tag{10.8}$$

where $0 < t < T, \omega \in \Omega$ in a Banach space $\mathscr{X}$ as in earlier chapters. As such, it might be tempting to apply the theory already established without hesitation. That would be fine, <u>provided that</u> the new operators $\mathscr{A} : \text{dom}(\mathscr{A}) \subset \mathscr{X} \to \mathscr{X}$ satisfied the necessary hypotheses. However, we are immediately faced with the fact that each of these operators is NON-linear, which throws a huge wrench into the works. Indeed, the assumption of linearity has crept into all aspects of our development in both subtle and very apparent ways.

Research on nonlinear abstract SEEs is not as extensive as in the linear case, and there remain several open problems and directions for research. Refer to the references cited at the beginning of this section for some foundational work in the deterministic and stochastic settings.

10.2   Time-Dependent SEEs

Model IX.3 Time-Dependent Neural Networks
The following is a time-dependent version of the IVP (7.5) describing an elementary neural network:

$$\begin{cases} dx_1(t;\omega) = \left(a_1(t)x_1(t;\omega) + \sum_{j=1}^{M} \eta_{1j}(t)g_j(x_j(t;\omega))\right) dt + \overline{a_1}x_1(t;\omega)dW(t), \\ \vdots \\ dx_M(t;\omega) = \left(a_M(t)x_M(t;\omega) + \sum_{j=1}^{M} \eta_{Mj}(t)g_j(x_j(t;\omega))\right) dt + \overline{a_M}x_M(t;\omega)dW(t), \\ x_i(0;\omega) = x_{i,0}(\omega), i = 1,\ldots,M, . \end{cases}$$

$$\tag{10.9}$$

where $0 \leq t \leq T, \omega \in \Omega$; $\alpha_i \in \mathbb{L}^2(0,T;\mathbb{R})$, $(i = 1,\ldots,M)$; and $W(t)$ is a one-dimensional Wiener process.

Model V.7 Time-Dependent Functional Diffusion-Advection Equation
Let $n \in \mathbb{N}$ and $0 < t_1 < t_2 < \ldots < t_n < T$ be fixed times. Consider the following generalization of IBVP (8.94) governing a diffusive-advective process with accumulative

external force and more general time-dependent diffusion:

$$
\begin{cases}
\partial z + \left(\alpha^2 \frac{\partial}{\partial x} \left(a(t,x) \frac{\partial z}{\partial x} \right) + \gamma b(t,x) \frac{\partial z}{\partial x} \right) \partial t = \\
\left(\sum_{i=1}^n \beta_i(x) z(x,t_i) + \int_0^T \zeta(s) f(s,z) \, ds \right) \partial t + \left(\int_0^t \frac{g(s)}{1+|g(s)|} \, ds \right) dW(t), \\
\frac{\partial z}{\partial x}(0,t;\omega) = \frac{\partial z}{\partial x}(L,t;\omega) = 0, \, 0 \le t \le T, \, \omega \in \Omega, \\
z(x,0;\omega) = z_0(x;\omega), \, 0 < x < L, \, \omega \in \Omega,
\end{cases}
\tag{10.10}
$$

where $0 < x < L$, $0 \le t \le T$, $\omega \in \Omega$, and a, b are sufficiently smooth functions.

We can consider similar generalizations of all models developed in the text. The main change is that the operator identified by A in each case is now time dependent. As such, we can naively reformulate (10.9) and (10.10) as the abstract time-dependent stochastic evolution equation

$$
\begin{cases}
dX(t;\omega) = (A(t)X(t;\omega) + \mathfrak{F}(X)(t;\omega)) \, dt + \mathfrak{G}(X)(t;\omega) d\mathbf{W}(t), \\
X(0;\omega) = X_0(\omega), \, \omega \in \Omega,
\end{cases}
\tag{10.11}
$$

where $0 \le t \le T$, $\omega \in \Omega$. The theory is understandably more complicated from the very beginning due to the time dependence of the operator A. However, it is possible to develop a theory in both the linear and nonlinear settings that resembles our development when the operator is not time dependent. Indeed, knowing nothing else and ignoring all technical details, we might expect each member of the family of operators $\{A(t)|t \ge 0\}$ to generate a semigroup. Of course, there is no reason to expect the semigroup to remain the same for each value of t. Rather, we have the following loose association:

$$
A \mapsto \{e^{As}|s \ge 0\},
\tag{10.12}
$$

$$
A(t) \mapsto \{U(t,s)|0 \le s \le t < \infty\}.
\tag{10.13}
$$

For each fixed t, the hope is that the family of operators in (10.13) somehow resembles a semigroup as in (10.12). Indeed, as developed in [328], this interpretation can be made formal, and in the linear case, the variation of parameters formula for a mild solution for (10.11) is given by

$$
X(t;\omega) = U(t,0)X_0(\omega) + \int_0^t U(t,s)\mathfrak{F}(X)(s;\omega) ds
$$

$$
+ \int_0^t U(t,s)\mathfrak{G}(X)(s;\omega)dW(s).
\tag{10.14}
$$

See [10, 264, 280, 328, 389] for some foundational work for deterministic time-dependent evolution equations.

10.3 Quasi-Linear SEEs

Further generalizing the time-dependent case, we can incorporate state dependence
into the operators $A(t)$ in (10.11). Such operators arise naturally when reformulating the IBVPs arising in environmental science and mathematical physics abstractly.
Some standard references include **[19, 26, 28, 113, 227, 242, 288, 306, 376, 400]**.
For instance, consider the following examples.

Model XII.3 Quasi-Linear Pollution Model
Let $\mathscr{D}$ be a bounded region in $\mathbb{R}^N$ with smooth boundary $\partial\mathscr{D}$, and let $w(\mathbf{x},t;\omega)$
denote the pollution concentration at position $\mathbf{x} \in \mathscr{D}$ and time $t > 0$. Consider the
following generalization of (8.101) that now accounts for a more elaborate wind
trajectory:

$$\begin{cases} \partial z = \left(k\triangle z + \sum_{i=1}^N a_i(t,\mathbf{x},z(\mathbf{x},t))\frac{\partial z}{\partial x_i} + \int_0^t a(t-s)g_1(s,z)ds\right)\partial t \\ + \left(\int_0^t \beta g_2(s,z)ds\right)dW(t), \\ z(\mathbf{x},0;\omega) = z_0(\mathbf{x};\omega),\ \mathbf{x} \in \mathscr{D},\ \omega \in \Omega, \\ \frac{\partial z}{\partial \mathbf{n}}(\mathbf{x},t;\omega) = 0,\ \mathbf{x} \in \partial\mathscr{D}, 0 < t < T, \omega \in \Omega, \end{cases} \tag{10.15}$$

where $\mathbf{x} \in \mathscr{D}, 0 < t < T, \omega \in \Omega$, $z = z(\mathbf{x},t;\omega)$, $\beta > 0$, $W(t)$ is a one-dimensional
Wiener process, $\frac{\partial z}{\partial \mathbf{n}}$ is the outward unit normal vector to $\partial\mathscr{D}$, z_0 is an $\mathscr{F}_0$-
measurable random variable independent of $W(t)$ with finite second moment, and
$a_i (i = 1,\ldots,N)$ are sufficiently smooth.

Model VII.8 Quasi-Linear Wave Equations
Certain wave phenomena are too complicated to be described using the classical
wave equation. Some specific areas in which more general wave equations naturally arise include fluid dynamics (via the Navier-Stokes equations), magneto-
hydrodynamics, and forestation. (See **[227, 306]** for details.) An example of a typical
quasi-linear wave equation is

$$\partial\left(\frac{\partial z}{\partial t}\right) - \alpha(z)\frac{\partial^2 z}{\partial x^2}\partial t = f\left(t,z,\frac{\partial z}{\partial t},\frac{\partial z}{\partial x}\right)\partial t + g\left(t,z,\frac{\partial z}{\partial x}\right)dW(t), \tag{10.16}$$

where $\alpha(\cdot)$ is sufficiently smooth.
  The equation portion of IBVPs (10.15) and (10.16) can be viewed as an abstract
quasi-linear stochastic evolution equation of the form

$$dX(t;\omega) = (A(t,X)X(t;\omega) + \mathfrak{F}(X)(t;\omega))dt + \mathfrak{G}(X)(t;\omega)d\mathbf{W}(t). \tag{10.17}$$

The theory is considerably more complicated to develop for (10.17) and largely remains open.

10.4 McKean-Vlasov SEEs

As explained in [157, 158], in chromatography, one analyzes a mixture of L different species. An inert fluid pushes the mixture through a long column containing an adsorbent medium. The particles of each species have different mobilities and affinities with the adsorbent material and so, take different amounts of time to pass through the column. There is competition among molecules both for access to the adsorbent medium and for space to diffuse. There is a nonlinear effect in that each molecules does not encounter individual molecules, but reacts to their distributions. Incorporating such dependence into the mathematical model of phenomena subject to noise results in a so-called *McKean-Vlasov* equation. Such equations have been studied extensively in both the finite and infinite-dimensional settings; see [4, 16, 50, 57, 58, 79, 84, 87, 88, 99, 109, 110, 151, 157, 158, 217, 230, 247, 248, 249, 275, 276, 299, 310, 356, 374, 394].

A typical abstract McKean-Vlasov SEE, as studied in [4], is given by

$$\begin{cases} dX(t;\omega) = (AX(t;\omega) + f(t,X(t;\omega),\mu(t)))\,dt + g(t,X(t;\omega))))dW(t), \\ X(0;\omega) = X_0(\omega), \omega \in \Omega \end{cases} \quad (10.18)$$

in a separable Hilbert space $\mathscr{H}$, where $0 < t < T$, $\omega \in \Omega$, and $t \mapsto \mu(t)$ is the probability law of $\{X(t;\omega) : 0 < t < T, \omega \in \Omega\}$. Under the same hypotheses imposed in Chapter 7, the same approach can essentially be used to study (10.18), with extra care taken due to the presence of the probability law in the forcing term. There remain many interesting open problems for such classes of equations; refer to the references cited at the beginning of this section for some background reading.

10.5 Even More Classes of SEEs

We have only scratched the surface of the theory of stochastic evolution equations. As we bring our discussion to a close, we mention a few last directions of interest in the research realm that you might find interesting to explore.

For some applications, especially in the mathematical modeling of communication networks and some diffusive processes, the noise term can be modeled much more accurately by replacing the standard Brownian motion by a so-called *fractional* Brownian motion (or, fBm, for short). The underlying theory of fBM is rather technical and involves hefty use of Malliavin calculus. Some references to get you started in this direction are [15, 112, 130, 165, 184, 262, 281, 317, 326, 405].

While some of external forces can have an impact on a system that is realized immediately (e.g., a sharp blow or electrical surge), the effects of the others are noticeable only after a certain time delay. Incorporating this fact into a mathematical model results in a so-called *delay evolution equation*. The deterministic case has been discussed in Volume 1. For some related work, some of which is in the stochastic setting, consult the following references: **[41, 44, 45, 60, 69, 71, 76, 101, 106, 156, 162, 174, 175, 179, 181, 186, 187, 192, 208, 209, 211, 224, 231, 259, 263, 269, 282, 283, 316, 349, 379]**.

The Sobolev-type SEEs investigated in Chapter 9 are one particular example of so-called *implicit SEEs*. Another common class of equations that arises in oscillation theory consists of *neutral SEEs*. Such equations with delay, especially of the state-dependent and infinite varieties, are currently under investigation and there are several open questions regarding them. Consult the following references: **[18, 23, 25, 27, 29, 63, 115, 129, 152, 238, 274, 294, 409]**.

We encountered specific second-order SEEs that were able to be effectively reformulated as a first-order SEE in a suitable product space. An alternative approach to study such SEEs is to use the theory of cosine operators. For a thorough discussion of the background of this approach, as well as some work in this direction in both the deterministic and stochastic settings, consult the following references: **[53, 55, 131, 180, 182, 183, 185, 188, 293, 308, 342, 343, 360, 387, 388]**.

Backward SEEs are of particular utility in the theory of optimal control, and can also be studied using the techniques discussed in this text. Some good references to get started in a study of such equations are **[24, 166, 176, 199, 200, 201, 272, 278, 279, 284, 322, 323, 324, 332, 369, 382]**.

Volterra integral and integro-differential evolution equations are of interest in the mathematical modeling of different phenomena, including viscoelasticity. Some references in this direction include **[7, 8, 74, 75, 95, 104, 153, 164, 223, 229, 232, 340, 341]**.

Finally, while we discussed more than twenty different models in this text, there are many, many more that can be studied abstractly using the techniques developed in this text. Among these are population models **[77]**, forced elongation **[96, 172, 292, 330]**, sensory system **[207]**, geotropic movement **[214]**, diffusion of hadrons **[335]**, and RC circuits **[345]**.

Good luck to you on your continued journey in this rich and interesting field of mathematics!

Bibliography

[1] R.A. Adams and J.J.F. Fournier. *Sobolev Spaces*. Academic Press, Amsterdam, 1975.

[2] S. Agarwal and D. Bahuguna. Existence and uniqueness of strong solutions to nonlinear nonlocal functional differential equations. *Electronic Journal of Differential Equations*, 2004(52):1–9, 2004.

[3] N.U. Ahmed. *Dynamic Systems and Control with Applications*. World Scientific, Singapore, 2006.

[4] N.U. Ahmed and X. Ding. A semilinear McKean-Vlasov stochastic evolution equation in Hilbert space. *Stochastic Processes and their Applications*, 60(1):65–85, 1995.

[5] S. Aizawa. A semigroup treatment of the Hamilton-Jacobí equation in several space variables. *Hiroshima Mathematics Journal*, 6:15–30, 1976.

[6] S. Aizicovici and Y. Gao. Functional differential equations with nonlocal initial conditions. *Journal of Applied Mathematics and Stochastic Analysis*, 10(2):145–156, 1997.

[7] S. Aizicovici and K.B. Hannsgen. Local existence for abstract semilinear Volterra integrodifferential equations. *Journal of Integral Equations and Applications*, 5(3):299–313, 1993.

[8] S. Aizicovici and M.A. McKibben. Semilinear Volterra integrodifferential equations with nonlocal initial conditions. *Abstract and Applied Analysis*, 4(2):127–139, 1999.

[9] S. Aizicovici and M.A. McKibben. Existence results for a class of abstract nonlinear nonlocal Cauchy problems. *Nonlinear Analysis*, 39(5):649–668, 2000.

[10] A.R. Al-Hussein. Time-dependent backward stochastic evolution equations. *Bulletin of the Malaysian Mathematical Sciences Society*, 30(2):159–183, 2007.

[11] E. Allen. *Modeling with Ito Stochastic Differential Equations*. Springer, Berlin, 2007.

[12] J.M. Alonso, J. Mawhin, and R. Ortega. Bounded solutions of second order semilinear evolution equations and applications to the telegraph equation. *Journal de Mathématiques Pures et Appliquées*, 78(1):49–63, 1999.

[13] H. Amann and G. Metzen. *Ordinary Differential Equations: An Introduction to Nonlinear Analysis.* Walter de Gruyter, Berlin, 1990.

[14] D. Andrade. A note on solvability of the nonlinear abstract viscoelastic problem in Banach spaces. *Journal of Partial Differential Equations*, 12:337–344, 1999.

[15] V.V. Anh and W. Grecksch. A fractional stochastic evolution equation driven by fractional Brownian motion. *Monte Carlo Methods and Applications*, 9(3):189–199, 2003.

[16] F. Antonelli and A. Kohatsu-Higa. Rate of convergence of a particle method to the solution of the McKean-Vlasov equation. *Annals of Applied Probability*, 12(2):423–476, 2002.

[17] T.M. Apostol. *Mathematical Analysis.* Addison-Wesley, Reading, MA, 1974.

[18] O. Arino, R. Benkhalti, and K. Ezzinbi. Existence results for initial value problems for neutral functional differential equations. *Journal of Differential Equations*, 138(1):188–193, 1997.

[19] L. Arlotti and J. Banasiak. Strictly substochastic semigroups with application to conservative and shattering solutions to fragmentation equations with mass loss. *Journal of Mathematical Analysis and Applications*, 293(2):693–720, 2004.

[20] L. Arnold. *Stochastic Differential Equations: Theory and Applications.* Wiley Interscience, New York, 1974.

[21] A. Ashyralyev and A. Sirma. Nonlocal boundary value problems for the Schrodinger equation. *Computers and Mathematics with Applications*, 55(3):392–407, 2008.

[22] K.J. Astrom. *Introduction to Stochastic Control Theory.* Academic Press, New York, 1970.

[23] S. Baghli and M. Benchohra. Perturbed functional and neutral functional evolution equations with infinite delay in Fréchet spaces. *Electronic Journal of Differential Equations*, 2008(69):1–19, 2008.

[24] K. Bahlali. Existence and uniqueness of solutions for BSDEs with locally Lipschitz coefficient. *Electronic Communications in Probability*, 7:169–179, 2002.

[25] D. Bahuguna and S. Agarwal. Approximations of solutions to neutral functional differential equations with nonlocal history conditions. *Journal of Mathematical Analysis and Applications*, 317(2):583–602, 2006.

[26] D. Bahuguna and R. Shukla. Approximations of solutions to nonlinear Sobolev type evolution equations. *Electronic Journal of Differential Equations*, 2003(31):1–16, 2003.

[27] D.D. Bainov and D.P. Mishev. *Oscillation Theory for Neutral Differential Equations with Delay*. Institute of Physics Publishing, Bristol, England, 1991.

[28] K. Balachandran and M. Chandrasekaran. Nonlocal Cauchy problem for quasilinear integrodifferential equation in Banach spaces. *Dynamic Systems and Applications*, 8:35–44, 1999.

[29] K. Balachandran, D.G. Park, and S.M. Anthoni. Existence of solutions of abstract nonlinear second-order neutral functional integrodifferential equations. *Computers and Mathematics with Applications*, 46(8-9):1313–1324, 2003.

[30] K. Balachandran and J.Y. Park. Nonlocal Cauchy problem for Sobolev type functional integrodifferential equation. *Bulletin of the Korean Mathematical Society*, 39(4):561–570, 2002.

[31] K. Balachandran, J.Y. Park, and M. Chandrasekaran. Nonlocal Cauchy problem for delay integrodifferential equations of Sobolev type in Banach spaces. *Applied Mathematics Letters*, 15(7):845–854, 2002.

[32] K. Balachandran, J.Y. Park, and I.H. Jung. Existence of solutions of nonlinear extensible beam equations. *Mathematical and Computer Modelling*, 36(7-8):747–754, 2002.

[33] K. Balachandran and K. Uchiyama. Existence of solutions of nonlinear integrodifferential equations of Sobolev type with nonlocal condition in Banach spaces. *Proceedings of Indian Academy of Mathematical Sciences*, 110(2):225–232, 2000.

[34] A.V. Balakrishnan. On the (Non-numeric) Mathematical Foundations of Linear Aeroelasticity. In *Fourth International Conference on Nonlinear Problems in Aviation and Aerospace*, pages 179–194.

[35] J.M. Ball. Initial-boundary value problems for an extensible beam. *Journal of Mathematical Analysis and Applications*, 42(1):61–90, 1973.

[36] D. Barbu. Local and global existence for mild solutions of stochastic differential equations. *Portugaliae Mathematica*, 55(4):411–424, 1998.

[37] V. Barbu. *Nonlinear Semigroups and Differential Equations in Banach Spaces*. Editura Academiei Bucharest-Noordhoff, Leyden, 1976.

[38] V. Barbu. *Analysis and Control of Nonlinear Infinite Dimensional Systems*, volume 190 of *Mathematics in Science and Engineering*. Academic Press, San Diego, CA, 1993.

[39] G.I. Barenblatt, Y.P. Zheltov, and I.N. Kochina. Basic concepts in the theory of seepage of homogeneous liquids in fissured rocks (strata). *PMM, Journal of Applied Mathematics and Mechanics*, 24:1286–1303, 1961.

[40] E. Barone and A. Belleni-Morante. A nonlinear initial-value problem arising from kinetic theory of vehicular traffic. *Transport Theory and Statistical Physics*, 7(1):61–79, 1978.

[41] A. Batkai and S. Piazzera. *Semigroups for Delay Equations.* AK Peters, Ltd., Wellesley, MA, 2005.

[42] J. Bear. *Hydraulics of Groundwater.* McGraw-Hill, New York, 1979.

[43] C. Bellehumeur, P. Legendre, and D. Marcotte. Variance and spatial scales in a tropical rain forest: Changing the size of sampling units. *Plant Ecology,* 130(1):89–98, 1997.

[44] A. Bellen and N. Guglielmi. Solving neutral delay differential equations with state-dependent delays. *Journal of Computational and Applied Mathematics,* 229(2):350–362, 2009.

[45] A. Bellen, N. Guglielmi, and A.E. Ruehli. Methods for linear systems of circuit delay differential equationsof neutral type. *IEEE Transactions on Circuits and Systems I: Fundamental Theory and Applications,* 46(1):212–215, 1999.

[46] A. Belleni-Morante. *A Concise Guide to Ssemigroups and Evolution Equations,* volume 19 of *Advances in Mathematics for Applied Sciences.* World Scientific, Singapore, 1994.

[47] A. Belleni-Morante and A.C. McBride. *Applied Nonlinear Semigroups: An Introduction.* John Wiley & Sons Inc., New York, 1998.

[48] H.C. Berg. Chemotaxis in bacteria. *Annual Review of Biophysics and Bioengineering,* 4(1):119–136, 1975.

[49] H. Bergstrom. *Weak Convergence of Measures.* Academic Press, New York, 1982.

[50] A.G. Bhatt, G. Kallianpur, R.L. Karandikar, and J. Xiong. On interacting systems of Hilbert-space-valued diffusions. *Applied Mathematics and Optimization,* 37(2):151–188, 1998.

[51] P. Billingsley. *Weak Convergence of Measures: Applications in Probability.* Society for Industrial and Applied Mathematics, Bristol, England, 1971.

[52] R.L. Bisplinghoff, H. Ashley, and R.L. Halfman. *Aeroelasticity.* Dover Publications, Mineola, NY, 1996.

[53] J. Bochenek. An abstract nonlinear second order differential equation. *Annales Polonici Mathematici,* 54:155–166, 1991.

[54] J. Bochenek. Second order semilinear volterra integro-differential equation in banach space. *Annales Polonici Mathematici,* 57:231–241, 1992.

[55] J. Bochenek. Existence of the fundamental solution of a second order evolution equation. *Annales Polonici Mathematici,* 66:15–35, 1997.

[56] A. Bonfoh and A. Miranville. On Cahn-Hilliard-Gurtin equations. *Nonlinear Analysis,* 47(5):3455–3466, 2001.

[57] M. Bossy. Some stochastic particle methods for nonlinear parabolic PDEs. In *ESAIM: Proceedings*, volume 15, pages 18–57, 2005.

[58] M. Bossy and D. Talay. A stochastic particle method for the McKean-Vlasov and the Burgers equation. *Mathematics of Computation*, 66(217):157–192, 1997.

[59] M. Boudart. *Kinetics of Chemical Processes*. Prentice-Hall, Engelwood Cliffs, NJ, 1968.

[60] H. Bouzahir. Semigroup approach to semilinear partial functional differential equations with infinite delay. *Journal of Inequalities and Applications*, vol. 2007:ArticleID49125, 13 pages, 2007.

[61] M. Brandau. Stochastic differential equations with nonlinear semigroups. *ZAMM-Journal of Applied Mathematics and Mechanics/Zeitschrift fur Angewandte Mathematik und Mechanik*, 82(11-12):737–743, 2002.

[62] F. Brauer and J.A. Nohel. *The Qualitative Theory of Ordinary Differential Equations: An Introduction*. Dover Publications, Mineola, NY, 1989.

[63] R. Brayton. Nonlinear oscillations in a distributed network. *Quarterly of Applied Mathematics*, 24:289–301, 1967.

[64] L. Breiman. *Probability*. SIAM, Philadelphia, PA, 1992.

[65] H. Brézis. *Operateurs Maximaux Monotones*. North-Holland, Amsterdam, 1973.

[66] H. Brézis and F. Browder. Partial differential equations in the 20th century. *Advances in Mathematics*, 135(1):76–144, 1998.

[67] H. Brézis, P.G. Ciarlet, and J.L. Lions. *Analyse Fonctionnelle: Théorie et Applications*. Masson, Paris, 1983.

[68] H. Brill. A semilinear Sobolev evolution equation in a Banach space. *Journal of Differential Equations*, 24(3):412–425, 1977.

[69] T.A. Burton. *Stability by Fixed Point Theory for Functional Differential Equations*. Dover Publications, Mineola, NY, 2006.

[70] T.A. Burton and C. Kirk. A fixed point theorem of Krasnoselskii-Schaefer type. *Mathematische Nachrichten*, 189(1):23–31, 1998.

[71] S.N. Busenberg and C.C. Travis. On the use of reducible-functional-differential equations in biological models. *Journal of Mathematical Analysis and Applications*, 89:46–66, 1982.

[72] C. Capellos and B.H.J. Bielski. *Kinetic Systems*. John Wiley and Sons, New York, 1972.

[73] M. Capiński and E. Kopp. *Measure, Integral and Probability*. Springer, Berlin, 2nd edition, 2004.

[74] R.W. Carr and K.B. Hannsgen. A nonhomogeneous integrodifferential equation in Hilbert space. *SIAM Journal of Mathematical Analysis*, 10:961–984, 1979.

[75] R.W. Carr and K.B. Hannsgen. Resolvent formulas for a Volterra equation in Hilbert space. *SIAM Journal of Mathematical Analysis*, 13:459–483, 1982.

[76] A. Casal and A. Somolinos. Forced oscillations for the sunflower equation, Entrainment. *Nonlinear Analysis: Theory, Methods & Applications*, 6(4):397–414, 1982.

[77] H. Caswell. *Matrix Population Models*. Sinauer Associates, Sunderland, MA, 2001.

[78] T. Cazenave, A. Haraux, and Y. Martel. *An Introduction to Semilinear Evolution Equations*. Clarendon Press, Oxford, 1998.

[79] T. Chan. Dynamics of the McKean-Vlasov equation. *The Annals of Probability*, 22(1):431–441, 1994.

[80] F.R. Chang. *Stochastic Optimization in Continuous Time*. Cambridge University Press, Cambridge, UK, 2004.

[81] G. Chen. Control and stabilization for the wave equation in a bounded domain. *SIAM Journal of Control and Optimization*, 17:66–81, 1979.

[82] G. Chen. Control and stabilization for the wave equation, part II. *SIAM Journal of Control and Optimization*, 19:114–122, 1981.

[83] G. Chen, G. Chen, and S.H. Hsu. *Linear Stochastic Control Systems*. CRC Press, Boca Raton, FL, 1995.

[84] P.J. Chen and M.E. Gurtin. On a theory of heat conduction involving two temperatures. *Zeitschrift für Angewandte Mathematik und Physik (ZAMP)*, 19(4):614–627, 1968.

[85] V.V. Chepyzhov and M.I. Vishik. *Attractors for Equations of Mathematical Physics*. American Mathematical Society, Providence, RI, 2002.

[86] P.R. Chernoff. Perturbations of dissipative operators with relative bound one. *Proceedings of the American Mathematical Society*, 33(1):72–74, 1972.

[87] T.S. Chiang. McKean-Vlasov equations with discontinuous coefficients. *Soochow Journal of Mathematics*, 20(4):507–526, 1994.

[88] T.S. Chiang, G. Kallianpur, and P. Sundar. Propagation of chaos and the McKean-Vlasov equation in duals of nuclear spaces. *Applied Mathematics and Optimization*, 24(1):55–83, 1991.

[89] P.L. Chow. *Stochastic Partial Differential Equations*. Chapman & Hall/CRC Press, Boca Raton, FL, 2007.

[90] P. Clement. *One-Parameter Semigroups*. Elsevier Science Ltd., 1987.

[91] L. Cobb. Stochastic differential equations for the social sciences. *Mathematical Frontiers of the Social and Policy Sciences, Westview Press, Boulder, CO*, pages 37–68, 1981.

[92] E.A. Coddington and R. Carlson. *Linear Ordinary Differential Equations.* Society for Industrial and Applied Mathematics, Philadelphia, PA, 1997.

[93] E.A. Coddington and N. Levinson. *Theory of Ordinary Differential Equations.* Tata McGraw-Hill, New York, 1972.

[94] C. Corduneanu. *Principles of Differential and Integral equations.* Chelsea Publishing Company, New York, 1977.

[95] C. Corduneanu. *Integral Equations and Applications.* Cambridge University Press, Cambridge, UK, 1991.

[96] R.G. Cox. The motion of long slender bodies in a viscous fluid. Part 1. General theory. *Journal of Fluid Mechanics*, 44(part 3):790–810, 1970.

[97] H. Cramér and M.R. Leadbetter. *Stationary and Related Stochastic Processes: Sample Function Properties and Their Applications.* John Wiley & Sons, New York, 1967.

[98] M.G. Crandall and T.M. Liggett. Generation of semi-groups of nonlinear transformations on general Banach spaces. *American Journal of Mathematics*, 93(2):265–298, 1971.

[99] D. Crisan and J. Xiong. Approximate McKean-Vlasov representations for a class of SPDEs. *Arxiv preprint math/0510668*, 2005.

[100] Z. Cui and Z. Yang. Roles of weight functions to a nonlinear porous medium equation with nonlocal source and nonlocal boundary condition. *Journal of Mathematical Analysis and Applications*, 342:559–570, 2007.

[101] R.V. Culshaw and S. Ruan. A delay-differential equation model of HIV infection of CD4+ T-cells. *Mathematical Biosciences*, 165(1):27–39, 2000.

[102] R.F. Curtain and H.J. Zwart. *An Introduction to Infinite-Dimensional Linear Systems Theory.* Springer, Berlin, 1995.

[103] M. Cyrot. Ginzburg-Landau theory for superconductors. *Rep. Progress Physics*, 36(2):103–158, 1973.

[104] G. Da Prato and M. Iannelli. Linear integro-differential equations in Banach spaces. *Rendiconti del Seminario Matematico dell Universita di Padova*, 62:207–219, 1980.

[105] G. Da Prato and J. Zabczyk. *Stochastic Equations in Infinite Dimensions.* Cambridge University Press, Cambridge, UK, 1992.

[106] J.P. Dauer and K. Balachandran. Existence of solutions of nonlinear neutral integrodifferential equations in Banach spaces. *Journal of Mathematical Analysis and Applications*, 251(1):93–105, 2000.

[107] R. Dautray and J.L. Lions. *Evolution Problems I, Volume 5 of Mathematical analysis and numerical methods for science and technology.* Springer, Berlin, 1992.

[108] E.B. Davies. *One-Parameter Semigroups.* Academic Press, New York, 1980.

[109] D.A. Dawson. Critical dynamics and fluctuations for a mean-field model of cooperative behavior. *Journal of Statistical Physics*, 31(1):29–85, 1983.

[110] D.A. Dawsont and J. Gartner. Large deviations from the McKean-Vlasov limit for weakly interacting diffusions. *Stochastics: An International Journal of Probability and Stochastic Processes*, 20(4):247–308, 1987.

[111] L. Debnath. *Nonlinear Partial Differential Equations for Scientists and Engineers.* Birkhauser, Boston, MA, 2005.

[112] L. Decreusefond and A.S. Ustunel. Stochastic analysis of the fractional Brownian motion. *Potential Analysis*, 10(2):177–214, 1999.

[113] J.M. Delort, D. Fang, and R. Xue. Global existence of small solutions for quadratic quasilinear Klein–Gordon systems in two space dimensions. *Journal of Functional Analysis*, 211(2):288–323, 2004.

[114] W. Desch, R. Grimmer, and W. Schappacher. Wellposedness and wave propagation for a class of integrodifferential equations in Banach space. *Journal of Differential Equations*, 74(2):391–411, 1988.

[115] Q. Dong, Z. Fan, and G. Li. Existence of solutions to nonlocal neutral functional differential and integrodifferential equations. *International Journal of Nonlinear Science*, 5(2):140–151, 2008.

[116] J.L. Doob. *Stochastic Processes.* John Wiley & Sons, New York, 1990.

[117] J. Duan, P. Holmes, and E.S. Titi. Global existence theory for a generalized Ginzburg-Landau equation. *Nonlinearity (Bristol. Print)*, 5(6):1303–1314, 1992.

[118] N. Dunford and J.T. Schwartz. *Linear Operators, Part I.* John Wiley Interscience, New York, 1958.

[119] J. Dyson, R. Villella-Bressan, and G.F. Webb. Asynchronous exponential growth in an age structured population of proliferating and quiescent cells. *Mathematical Biosciences*, 177:73–83, 2002.

[120] L Edelstein-Keshet. *Mathematical Models in Biology.* Birkhauser, 1988.

[121] R.E. Edwards. *Fourier Series: A Modern Introduction.* Holt, Rinehart and Winston, Austin, TX, 1967.

[122] Y. El Boukfaoui and M. Erraoui. Remarks on the existence and approximation for semilinear stochastic differential equations in Hilbert spaces. *Stochastic Analysis and Applications*, 20(3):495–518, 2002.

[123] K.J. Engel and R. Nagel. *One-Parameter Semigroups for Linear Evolution Equations*. Springer, Berlin, 2000.

[124] M.E. Erdogan and C.E. İmrak. On some unsteady flows of a non-Newtonian fluid. *Applied Mathematical Modelling*, 31(2):170–180, 2007.

[125] S.N. Ethier and T.G. Kurtz. *Markov Processes: Characterization and Convergence*. John Wiley & Sons, New York, 1986.

[126] L.C. Evans. An Introduction to Stochastic Differential Equations Version 1.2. *Lecture Notes, Department of Mathematics, University of California, Berkeley*.

[127] L.C. Evans. *Partial Differential Equations*. Springer, Berlin, 1998.

[128] H. Eyring, S.H. Lin, and SM Lin. *Basic Chemical Kinetics*. John Wiley & Sons, New York, 1980.

[129] K. Ezzinbi and X. Fu. Existence and regularity of solutions for some neutral partial differential equations with nonlocal conditions. *Nonlinear Analysis*, 57(7-8):1029–1041, 2004.

[130] A. Fannjiang and T. Komorowski. Fractional Brownian motions in a limit of turbulent transport. *Annals of Applied Probability*, 10(4):1100–1120, 2000.

[131] H.O. Fattorini. *Second Order Linear Differential Equations in Banach Spaces*. North Holland, Amsterdam, 1985.

[132] H.O. Fattorini. *Infinite Dimensional Optimization and Control Theory*. Cambridge University Press, 1999.

[133] H.O. Fattorini and A. Kerber. *The Cauchy Problem*. Cambridge University Press, Cambridge, UK, 1984.

[134] Z. Feng, W. Huang, and C. Castillo-Chavez. Global behavior of a multi-group SIS epidemic model with age structure. *Journal of Differential Equations*, 218(2):292–324, 2005.

[135] W.E. Fitzgibbon. Global existence and boundedness of solutions to the extensible beam equation. *SIAM Journal on Mathematical Analysis*, 13(5):739–745, 1982.

[136] R. Fitzhugh. Impulses and physiological states in theoretical models of nerve membrane. *Biophysical Journal*, 1(6):445–466, 1961.

[137] W.H. Fleming. Diffusion processes in population biology. *Advances in Applied Probability*, 7:100–105, 1975.

[138] G.B. Folland. *Introduction to Partial Differential Equations*. Princeton University Press, Princeton, NJ, 1995.

[139] A. Friedman. *Stochastic Differential Equations and Applications, Vol. 1 and 2*. Dover Publications, Mineola, NY, 1976.

[140] T. Funaki. A certain class of diffusion processes associated with nonlinear parabolic equations. *Probability Theory and Related Fields*, 67(3):331–348, 1984.

[141] Y.C. Fung. *An Introduction to the Theory of Aeroelasticity*. Dover Publications, Mineola, NY, 2002.

[142] T.C. Gard. *Introduction to Stochastic Differential Equations*. Marcel Dekker, New York, 1988.

[143] C.W. Gardiner. *Handbook of Stochastic Methods for Physics, Chemistry, and the Natural Sciences*. Springer, Berlin, 3rd edition, 1985.

[144] M. Gitterman. *The Noisy Oscillator: The First Hundred Years, From Einstein Until Now*. World Scientific, Singapore, 2005.

[145] A. Glitzky and W. Merz. Single dopant diffusion in semiconductor technology. *Mathematical Methods in the Applied Sciences*, 27(2):133–154, 2004.

[146] M. Goland. The flutter of a uniform cantilever wing. *Journal of Applied Mechanics*, 12(4):197–208, 1945.

[147] J.A. Goldstein. Semigroups and second-order differential equations. *Journal of Functional Analysis*, 4:50–70, 1969.

[148] J.A. Goldstein. On a connection between first and second order differential equations in Banach spaces. *Journal of Mathematical Analysis and Applications*, 30:246–251, 1970.

[149] J.A. Goldstein. *Semigroups of Linear Operators and Applications*. Oxford University Press, UK, 1985.

[150] J.A. Goldstein. *The KdV equation via semigroups*, pages 107–114. Theory and Applications of Nonlinear Operators of Accretive and Monotone Type, Vol. 178. Marcel Dekker, New York, 1996.

[151] A.D. Gottlieb. Markov transitions and the propagation of chaos. *Arxiv preprint math/0001076*, 2000.

[152] T.E. Govindan. Stability of stochastic differential equations in a Banach space. In *Mathematical Theory of Control, Lecture Notes in Pure and Applied Mathematics*, Marcel Dekker, New York, pages 161–181, 1993.

[153] T.E. Govindan. Autonomous semilinear stochastic Volterra integro-differential equations in Hilbert spaces. *Dynamic Systems and Applications*, 3(1):51–74, 1994.

[154] T.E. Govindan. An existence result for the Cauchy problem for stochastic systems with heredity. *Differential and Integral Equations -Athens*, 15(1):103–114, 2002.

[155] T.E. Govindan. Stability of mild solutions of stochastic evolution equations with variable delay. *Stochastic Analysis and Applications*, 21(5):1059–1077, 2003.

[156] T.E. Govindan and M.C. Joshi. Stability and optimal control of stochastic functional-differential equations with memory. *Numerical Functional Analysis and Optimization*, 13(3):249–265, 1992.

[157] C. Graham. McKean-Vlasov Ito-Skorohod equations, and nonlinear diffusions with discrete jump sets. *Stochastic Processes and their Applications*, 40(1):69–82, 1992.

[158] C. Graham. Nonlinear diffusion with jumps. *Annales de l'lHP Probabilities et statistiques*, 28(3):393–402, 1992.

[159] W. Grecksch and C. Tudor. *Stochastic Evolution Equations: A Hilbert Space Approach*. Akademie Verlag, Mineola, NY, 1995.

[160] D.H. Griffel. *Applied Functional Analysis*. Dover Publications, Mineola, NY, 2002.

[161] M. Grigoriu. *Stochastic Calculus: Applications in Science and Engineering*. Birkhauser, 2002.

[162] L.J. Grimm. Existence and uniqueness for nonlinear neutral-differential equations. *American Mathematical Society*, 77(3):374–376, 1971.

[163] G. Grimmett and D. Stirzaker. *Probability and Random Processes*. Oxford University Press, Oxford, UK, 3rd edition, 2001.

[164] G. Gripenberg, S.O. Londen, and O.J. Staffans. *Volterra Integral and Functional Equations*. Cambridge University Press, Cambridge, UK, 1990.

[165] G. Gripenberg and I. Norros. On the prediction of fractional Brownian motion. *Journal of Applied Probability*, 33(2):400–410, 1996.

[166] G. Guatteri and G. Tessitore. On the backward stochastic Riccati equation in infinite dimensions. *SIAM Journal on Control and Optimization*, 44(1):159–194, 2006.

[167] G. Gudehus. On the onset of avalanches in flooded loose sand. *Philosophical Transactions: Mathematical, Physical and Engineering Sciences*, 356(1747):2747–2761, 1998.

[168] R.B. Guenther and J.W. Lee. *Partial Differential Equations of Mathematical Physics and Integral Equations*. Dover Publications, Mineola, NY, 1996.

[169] V.E. Gusev, W. Lauriks, and J. Thoen. Evolution equation for nonlinear Scholte waves. *IEEE Transactions on Ultrasonics, Ferroelectrics and Frequency Control*, 45(1):170–178, 1998.

[170] R. Haberman. *Mathematical Models: Mechanical Vibrations, Population Dynamics, and Traffic Flow: An Introduction to Applied Mathematics*. Society for Industrial and Applied Mathematics, Philadelphia, PA, 1998.

[171] S. Habib, C. Molina-Paris, and T.S. Deisboeck. Complex dynamics of tumors: Modeling an emerging brain tumor system with coupled reaction–diffusion equations. *Physica A: Statistical Mechanics and its Applications*, 327(3-4):501–524, 2003.

[172] T. Hagen. On the semigroup of linearized forced elongation. *Applied Mathematics Letters*, 18(6):667–672, 2005.

[173] T.C. Hagen. *Elongational flows in polymer processing*. PhD thesis, Virginia Polytechnic Institute, Blacksburg, VA, 1998.

[174] J.K. Hale. *Theory of Functional Differential Equations*. Springer, Berlin, 1977.

[175] J.K. Hale and J. Kato. Phase space for retarded equations with infinite delay. *Funkcial. Ekvac*, 21(1):11–41, 1978.

[176] M. Hassani and Y. Ouknine. Infinite dimensional BSDE with jumps. *Stochastic Analysis and Applications*, 20(3):519–565, 2002.

[177] M. He. Global existence and stability of solutions for reaction diffusion functional differential equations. *Journal of Mathematical Analysis and Applications*, 199(3):842–858, 1996.

[178] D. Henderson and P. Plaschko. *Stochastic Differential Equations in Science and Engineering*. World Scientific, Singapore, 2006.

[179] H.R. Henríquez. Regularity of solutions of abstract retarded functional differential equations with unbounded delay. *Nonlinear Analysis*, 28(3):513–531, 1997.

[180] H.R. Henríquez and C.H. Vásquez. Differentiability of solutions of the second order abstract Cauchy problem. In *Semigroup Forum*, volume 64, pages 472–488. Springer, Berlin, 2002.

[181] E. Hernández. Existence results for a class of semi-linear evolution equations. *Electronic Journal of Differential Equations*, 2001(24):1–14, 2001.

[182] E. Hernández. Existence of solutions to a second order partial differential equation with nonlocal conditions. *Electronic Journal of Differential Equations*, 2003(51):1–10, 2003.

[183] E. Hernández, H.R. Henríquez, and M.A. McKibben. Existence of solutions for second order partial neutral functional differential equations. *Integral Equations and Operator Theory*, 62(2):191–217, 2008.

[184] E. Hernández, D.N. Keck, and M.A. McKibben. On a class of measure-dependent stochastic evolution equations driven by fBm. *Journal of Applied Mathematics and Stochastic Analysis*, 2007:Article ID 69747, 26 pages, 2007.

[185] E. Hernández and M.A. McKibben. Some comments on: Existence of solutions of abstract nonlinear second-order neutral functional integrodifferential equations. *Computers and Mathematics with Applications*, 50(5-6):655–669, 2005.

[186] E. Hernández and M.A. McKibben. On state-dependent delay partial neutral functional–differential equations. *Applied Mathematics and Computation*, 186(1):294–301, 2007.

[187] E. Hernández, M.A. McKibben, and H.R. Henríquez. Existence results for partial neutral functional differential equations with state-dependent delay. *Mathematical and Computer Modelling*, 49(5-6):1260–1267, 2009.

[188] E. Hernández and M. Pelicer. Existence results for a second-order abstract Cauchy problem with nonlocal conditions. *Electronic Journal of Differential Equations*, 2005(73):1–17, 2005.

[189] S.C. Hille. Local well-posedness of kinetic chemotaxis models. *Journal of Evolution Equations*, 8(3):423–448, 2008.

[190] T. Hillen and A. Potapov. The one-dimensional chemotaxis model: Global existence and asymptotic profile. *Mathematical Methods in the Applied Sciences*, 27(15):1783–1801, 2004.

[191] E.J. Hinch. The distortion of a flexible inextensible thread in a shearing flow. *Journal of Fluid Mechanics*, 74(Part 2):317–333, 1976.

[192] Y. Hino, S. Murakami, and T. Naito. *Functional Differential Equations with Infinite Delay*. Springer, Berlin, 1991.

[193] M.W. Hirsch and S. Smale. *Differential Equations, Dynamical Systems, and Linear Algebra*. Academic Press, New York, 1974.

[194] A.L. Hodgkin and A.F. Huxley. A quantitative description of membrane current and its application to conduction and excitation in nerve. *Bulletin of Mathematical Biology*, 52(1):25–71, 1990.

[195] P.G. Hoel, S.C. Port, and C.J. Stone. *Introduction to Stochastic Processes*. Houghton Mifflin, 1972.

[196] K. Hoffman. *Analysis in Euclidean Space*. Prentice–Hall, Englewood Cliffs, NJ, 1975.

[197] S.S. Holland. *Applied Analysis by the Hilbert Space Method*. Dover Publications, Mineola, NY, 1990.

[198] S. Hu and N.S. Papageorgiou. *Handbook of Multivalued Analysis*, Volume I: Theory. *Mathematics and its Applications*, Kluwer Academic, 419, 1997.

[199] Y. Hu, J. Ma, and J. Yong. On semi-linear degenerate backward stochastic partial differential equations. *Probability Theory and Related Fields*, 123(3):381–411, 2002.

[200] Y. Hu and S. Peng. Maximum principle for semilinear stochastic evolution control systems. *Stochastics An International Journal of Probability and Stochastic Processes*, 33(3):159–180, 1990.

[201] Y. Hu and S. Peng. Adapted solution of a backward semilinear stochastic evolution equation. *Stochastic Analysis and Applications*, 9(4):445–459, 1991.

[202] R. Huilgol. A second order fluid of the differential type. *International Journal of Nonlinear Mechanics*, 3:471–482, 1968.

[203] A. Ichikawa. Linear stochastic evolution equations in Hilbert space. *Journal of Differential Equations*, 28(2):266–277, 1978.

[204] A. Ichikawa. Stability of semilinear stochastic evolution equations. *Journal of Mathematical Analysis and Applications*, 90(1):12–44, 1982.

[205] A. Ida, S. Oharu, and Y. Oharu. A mathematical approach to HIV infection dynamics. *Journal of Computational and Applied Mathematics*, 204(1):172–186, 2007.

[206] K. Ito and F. Kappel. *Evolution Equations and Approximations*. World Scientific, Singapore, 2002.

[207] Y. Ito. A semi-group of operators as a model of the lateral inhibition process in the sensory system. *Advances in Applied Probability*, 10:104–110, 1978.

[208] A.F. Ivanov, Y.I. Kazmerchuk, and A.V. Swishchuk. Theory, stochastic stability and applications of stochastic delay differential equations: A survey of recent results. *Differential Equations and Dynamical Systems*, 11(1&2):55–115, 2003.

[209] F. Izsak. An existence theorem for Volterra integrodifferential equations with infinite delay. *Electronic Journal of Differential Equations*, 2003(4):1–9, 2003.

[210] D. Jackson. Existence and uniqueness of solutions to semilinear nonlocal parabolic equations. *Journal of Mathematical Analysis and Applications*, 172(1):256–265, 1993.

[211] R. Jahanipur. Stability of stochastic delay evolution equations with monotone nonlinearity. *Stochastic Analysis and Applications*, 21(1):161–181, 2003.

[212] A.H. Jazwinski. *Stochastic Processes and Filtering Theory*. Academic Press, New York, 1970.

[213] K. Jinno and A. Kawamura. Application of Fourier series expansion to two-dimensional stochastic advection-dispersion equation for designing monitoring network and real-time prediction of concentration. *IAHS Publications-Series of Proceedings and Reports-Int. Assoc. Hydrological Sciences*, 220:225–236, 1994.

[214] A. Johnsson and D. Israelsson. Application of a theory for circumnutations to geotropic movements. *Physiologia Plantarum*, 21(2):282–291, 1968.

[215] M.C. Joshi and R.K. Bose. *Some Topics in Nonlinear Functional Analysis.* John Wiley & Sons, New York, 1985.

[216] D. Jou, J. Casas-Vazquez, and G. Lebon. Extended irreversible thermodynamics revisited (1988-98). *Reports on Progress in Physics*, 62(7):1035–1142, 1999.

[217] G. Kallianpur and J. Xiong. Asymptotic behavior of a system of interacting nuclear-space-valued stochastic differential equations driven by Poisson random measures. *Applied Mathematics and Optimization*, 30(2):175–201, 1994.

[218] G. Kallianpur and J. Xiong. Stochastic models of environmental pollution. *Advances in Applied Probability*, 26(2):377–403, 1994.

[219] G. Kallianpur and J. Xiong. *Stochastic Differential Equations in Infinite Dimensional Spaces.* Institute of Mathematical Statistics, Beachwood, OH, 1995.

[220] M. Kanakaraj and K. Balachandran. Existence of solutions of Sobolev-type semilinear mixed integrodifferential inclusions in Banach spaces. *Journal of Applied Mathematics and Stochastic Analysis*, 16(2):163–170, 2003.

[221] D. Kannan. An operator-valued stochastic integral II. *Annales de l'Institut Henri Poincare, Section B*, 8:9–32, 1972.

[222] D. Kannan and A.T. Bharucha-Reid. An operator-valued stochastic integral. *Proceedings of the Japan Academy*, 47(5):472–476, 1971.

[223] D. Kannan and A.T. Bharucha-Reid. On a stochastic integro-differential evolution equation of Volterra type. *Journal of Integral Equations*, 10:1–3, 1985.

[224] G. Karakostas. Effect of seasonal variations to the delay population equation. *Nonlinear Analysis: Theory, Methods and Applications*, 6(11):1143–1154, 1982.

[225] S. Karlin and H. Taylor. *A First Course in Stochastic Processes.* Academic Press, New York, 2nd edition, 1966.

[226] A.G. Kartsatos and K.Y. Shin. Solvability of functional evolutions via compactness methods in general Banach spaces. *Nonlinear Analysis: Theory, Methods and Applications*, 21(7):517–535, 1993.

[227] T. Katō. Quasi-linear equations of evolution, with applications to partial differential equations. *Spectral Theory and Differential Equations*, Springer, Berlin, 448:25–70, 1975.

[228] T. Katō. *Perturbation Theory for Linear Operators.* Springer, Berlin, 1995.

[229] D.N. Keck and M.A. McKibben. Functional integro-differential stochastic evolution equations in Hilbert space. *Journal of Applied Mathematics and Stochastic Analysis*, 16(2):141–161, 2003.

[230] D.N. Keck and M.A. McKibben. On a McKean-Vlasov stochastic integro-differential evolution equation of Sobolev-type. *Stochastic Analysis and Applications*, 21(5):1115–1139, 2003.

[231] D.N. Keck and M.A. McKibben. Abstract stochastic integrodifferential delay equations. *Journal of Applied Mathematics and Stochastic Analysis*, 2005(3):275–305, 2005.

[232] D.N. Keck and M.A. McKibben. Abstract semilinear stochastic Ito-Volterra integrodifferential equations. *Journal of Applied Mathematics and Stochastic Analysis*, 2006(5):Article ID 45253, 22 pages, 2006.

[233] J.H. Kim. On nonlinear stochastic evolution equations. *Stochastic Analysis and Applications*, 14(3):303–311, 1996.

[234] J.R. Kirkwood. *An Introduction to Analysis*. PWS Publishing Company, Boston, MA, 1995.

[235] F.C. Klebaner. *Introduction to Stochastic Calculus with Applications*. Imperial College Press, London, UK, 2nd edition, 2005.

[236] K. Knopp. *Theory and Application of Infinite Series*. Courier Dover Publications, Mineola, NY, 1990.

[237] D. Kolymbas. An outline of hypoplasticity. *Archive of Applied Mechanics (Ingenieur Archiv)*, 61(3):143–151, 1991.

[238] H. Komatsu. Fractional powers of operators. *Pacific Journal of Mathematics*, 19(2):285–346, 1966.

[239] Y. Konishi. On $u_t = u_x x - F(u_x)$ and the differentiability of the nonlinear semi-group associated with it. *Proceedings of the Japan Academy*, 48(5):281–286, 1972.

[240] J. Kopfová. Nonlinear semigroup methods in problems with hysteresis. *Discrete and Continuous Dynamical Systems*, pages 580–589, 2007.

[241] R. Kosloff, M.A. Ratner, and W.B. Davis. Dynamics and relaxation in interacting systems: Semigroup methods. *Journal of Chemical Physics*, 106(17):7036–7043, 1997.

[242] P. Kotelenez. A class of quasilinear stochastic partial differential equations of McKean-Vlasov type with mass conservation. *Probability Theory and Related Fields*, 102(2):159–188, 1995.

[243] E. Kreyszig. *Introductory Functional Analysis with Applications*. John Wiley & Sons, New York, 1978.

[244] V. Krishnan. *Nonlinear Filtering and Smoothing: An Introduction to Martingales, Stochastic Integrals and Estimation.* Dover Publications, Mineola, NY, 2005.

[245] Z.F. Kuang and I. Pázsit. A class of semi-linear evolution equations arising in neutron fluctuations. *Transport Theory and Statistical Physics*, 31:141–151, 2002.

[246] H. Kunita. *Stochastic Flows and Stochastic Differential Equations.* Cambridge University Press, Cambridge, UK, 1997.

[247] T. Kurtz and J. Xiong. Particle representations for a class of nonlinear SPDEs. *Stochastic Processes and their Applications*, 83(1):103–126, 1999.

[248] T.G. Kurtz. Convergence of sequences of semigroups of nonlinear operators with an application to gas kinetics. *Transactions of the American Mathematical Society*, 186:259–272, 1973.

[249] T.G. Kurtz and J. Xiong. A stochastic evolution equation arising from the fluctuations of a class of interacting particle systems. *Communications in Mathematical Sciences*, 2(3):325–358, 2004.

[250] G.E. Ladas and V. Lakshmikantham. *Differential Equations in Abstract Spaces.* Academic Press, New York, 1972.

[251] G.S. Ladde and M. Sambandham. *Stochastic Versus Deterministic Systems of Differential Equations.* Marcel Dekker, New York, 2004.

[252] V. Lakshmikantham and S.G. Deo. *Method of Variation of Parameters for Dynamic Systems.* CRC Press, Boca Raton, FL, 1998.

[253] V. Lakshmikantham and S. Leela. *Nonlinear Differential Equations in Abstract Spaces.* Pergamon, Oxford, UK, 1981.

[254] D. Lauffenburger, R. Aris, and K. Keller. Effects of cell motility and chemotaxis on microbial population growth. *Biophysical Journal*, 40(3):209–219, 1982.

[255] J.R. Ledwell, A.J. Watson, and C.S. Law. Evidence for slow mixing across the pycnocline from an open-ocean tracer-release experiment. *Nature*, 364(6439):701–703, 1993.

[256] J.R. Leigh. *Functional Analysis and Linear Control Theory.* Academic Press, New York, 1980.

[257] D.S. Lemons. *Introduction to Stochastic Processes in Physics.* The Johns Hopkins University Press, Baltimore, MD, 2002.

[258] M.A. Lewis, G. Schmitz, P. Kareiva, and J.T. Trevors. Models to examine containment and spread of genetically engineered microbes. *Molecular Ecology*, 5(2):165–175, 1996.

[259] J. Li. Hopf bifurcation of the sunflower equation. *Nonlinear Analysis: Real World Applications*, 10(4):2574–2580, 2009.

[260] Y. Li and Y. Wu. Stability of travelling waves with noncritical speeds for double degenerate Fisher-type equations. *Discrete and Continuous Dynamic Systems Series B*, 10(1):149–170, 2008.

[261] J. Lightbourne and S.M. Rankin. Partial functional differential equation of Sobolev type. *Journal of Mathematical Analysis and Applications*, 93(2):328–337, 1983.

[262] S.J. Lin. Stochastic analysis of fractional Brownian motions. *Stochastics, An International Journal of Probability and Stochastic Processes*, 55(1):121–140, 1995.

[263] J. Liu, T. Naito, and N. Van Minh. Bounded and periodic solutions of infinite delay evolution equations. *Journal of Mathematical Analysis and Applications*, 286(2):705–712, 2003.

[264] J.H. Liu. Integrodifferential equations with non-autonomous operators. *Dynamic Systems and Applications*, 7:427–440, 1998.

[265] J.H. Liu. *A First Course in the Qualitative Theory of Differential Equations*. Prentice Hall, Englewood Cliffs, NJ, 2003.

[266] J.Y. Liu and W.T. Simpson. Solutions of diffusion equation with constant diffusion and surface emission coefficients. *Drying Technology*, 15(10):2459–2478, 1997.

[267] K. Liu. *Stability of Infinite Dimensional Stochastic Differential Equations with Applications*. CRC Press, Boca Raton, FL, 2006.

[268] K. Liu and J. Zou. Robustness of pathwise stability of semilinear perturbed stochastic evolution equations. *Stochastic Analysis and Applications*, 22(2):251–274, 2005.

[269] M. Lizana. Global analysis of the sunflower equation with small delay. *Nonlinear Analysis*, 36(6):697–706, 1999.

[270] B. Lods. A generation theorem for kinetic equations with non-contractive boundary operators. *Comptes Rendus-Mathématique*, 335(7):655–660, 2002.

[271] P. Lu, H.P. Lee, C. Lu, and P.Q. Zhang. Application of nonlocal beam models for carbon nanotubes. *International Journal of Solids and Structures*, 44(16):5289–5300, 2007.

[272] J. Ma and J. Yong. On linear, degenerate backward stochastic partial differential equations. *Probability Theory and Related Fields*, 113(2):135–170, 1999.

[273] C.R. MacCluer. *Boundary Value Problems and Orthogonal Expansions: Physical Problems from a Sobolev Viewpoint*. Dover Publications, Mineola, NY, 1994.

[274] N.I. Mahmudov. Existence and uniqueness results for neutral SDEs in Hilbert spaces. *Stochastic Analysis and Applications*, 24(1):79–95, 2006.

[275] N.I. Mahmudov and M.A. McKibben. Abstract second-order damped McKean-Vlasov stochastic evolution equations. *Stochastic Analysis and Applications*, 24(2):303–328, 2006.

[276] N.I. Mahmudov and M.A. McKibben. Controllability results for a class of abstract first-order McKean-Vlasov stochastic evolution equations. *Dynamic Systems and Applications*, 15:357–374, 2006.

[277] N.I. Mahmudov and M.A. McKibben. McKean-Vlasov stochastic differential equations in Hilbert spaces under Caratheodory conditions. *Dynamic Systems and Applications*, 15(3/4):357, 2006.

[278] N.I. Mahmudov and M.A. McKibben. On a class of backward McKean-Vlasov stochastic equations in Hilbert space: Existence and convergence properties. *Dynamic Systems and Applications*, 16(4):643, 2007.

[279] N.I. Mahmudov and M.A. McKibben. On backward stochastic evolution equations in Hilbert spaces and optimal control. *Nonlinear Analysis: Theory, Methods & Applications*, 67(4):1260–1274, 2007.

[280] A. Maio and A.M. Monte. Nonautonomous semilinear second order evolution equation in a Hilbert space. *International Journal of Engineering Science*, 23(1):27–38, 1985.

[281] B.B. Mandelbrot and J.W. Van Ness. Fractional Brownian motions, fractional noises and applications. *SIAM review*, 10(4):422–437, 1968.

[282] X. Mao. Approximate solutions for a class of stochastic evolution equations with variable delays. *Numerical Functional Analysis and Optimization*, 12(5-6):525–533, 1991.

[283] X. Mao. Approximate solutions for a class of stochastic evolution equations with variable delays. II. *Numerical Functional Analysis and Optimization*, 15(1-2):65–76, 1994.

[284] X. Mao. Adapted solutions of backward stochastic differential equations with non-Lipschitz coefficients. *Stochastic Processes and Applications*, 58:281–292, 1995.

[285] X. Mao. *Stochastic Differential Equations and Applications*. Horwood Publishing, Chichester, UK, 2nd edition, 2007.

[286] N. Marheineke. *Turbulent Fibers–On the Motion of Long, Flexible Fibers in Turbulent Flows*. PhD thesis, Technische Universitat Kaiserslautern, 2005.

[287] N.G. Markley. *Principles of Differential Equations*. John Wiley & Sons, New York, 2004.

[288] P. Markowich and M. Renardy. The numerical solution of a class of quasilinear parabolic Volterra equations arising in polymer rheology. *SIAM Journal on Numerical Analysis*, 20(5):890–908, 1983.

[289] J.E. Marsden and T.J.R. Hughes. *Mathematical Foundations of Elasticity*. Dover Publications, Mineola, NY, 1994.

[290] R.H. Martin. *Nonlinear Operators and Differential Equations in Banach Spaces*. Krieger Publishing Co., Inc., Malabar, FL, 1986.

[291] M.P. Matos and D.C. Pereira. On a hyperbolic equation with strong damping. *Funkcial. Ekvac*, 34:303–311, 1991.

[292] M.A. Matovich and J.R.A. Pearson. Spinning a molten threadline. Steady-state isothermal viscous flows. *Industrial & Engineering Chemistry Fundamentals*, 8(3):512–520, 1969.

[293] M.A. McKibben. Second-order damped functional stochastic evolution equations in Hilbert space. *Dynamic Systems and Applications*, 12(3/4):467–488, 2003.

[294] M.A. McKibben. Second-order neutral stochastic evolution equations with heredity. *Journal of Applied Mathematics and Stochastic Analysis*, 2004(2):177–192, 2004.

[295] M.A. McKibben. *Discovering Evolution Equations with Applications: Volume 1 - Deterministic Equations*. Chapman & Hall/CRC Press, Boca Raton, FL, 2010.

[296] R.C. McOwen. *Partial Differential Equations*. Prentice Hall, Englewood Cliffs, NJ, 1996.

[297] C.V.M. Mee and P.F. Zweifel. A Fokker-Planck equation for growing cell populations. *Journal of Mathematical Biology*, 25(1):61–72, 1987.

[298] H. Meinhardt and A. Gierer. Applications of a theory of biological pattern formation based on lateral inhibition. *Journal of Cell Science*, 15(2):321–346, 1974.

[299] S. Meleard. *Asymptotic behaviour of some interacting particle systems; McKean-Vlasov and Boltzmann models*, volume 1627 of *Lecture Notes in Mathematics*, chapter Probabilistic models for nonlinear partial differential equations, pages 42–95. Springer, Berlin, 1996.

[300] I.V. Melnikova and A. Filinkov. *Abstract Cauchy Problems: Three Approaches*. CRC Press, Boca Raton, FL, 2001.

[301] M. Miklavcic. *Applied Functional Analysis and Partial Differential Equations*. World Scientific, Singapore, 1998.

[302] T. Mikosch. *Elementary Stochastic Calculus with Finance in View*. World Scientific, Singapore, 1998.

[303] S. Milton and C.P. Tsokos. A stochastic model for chemical kinetics. *Acta Biotheoretica*, 23(1):18–34, 1974.

[304] E. Mitidieri and I.I. Vrabie. Existence for nonlinear functional differential equations. *Hiroshima Mathematics Journal*, 17(3):627–649, 1987.

[305] I. Miyadera. *Nonlinear Semigroups*, volume 109. American Mathematical Society, Providence, RI, 1992.

[306] A. Moameni. On the existence of standing wave solutions to quasilinear Schrodinger equations. *Nonlinearity- London*, 19(4):937, 2006.

[307] M.A. Murad, L.S. Bennethum, and J.H. Cushman. A multi-scale theory of swelling porous media. I. Application to one-dimensional consolidation. *Transport in Porous Media*, 19(2):93–122, 1995.

[308] M.G. Murge and B.G. Pachpatte. Successive approximations for solutions of second order stochastic integrodifferential equations of Itô type. *Indian Journal of Pure and Applied Mathematics*, 21(3):260–274, 1990.

[309] J.D. Murray. *Mathematical Biology*. Springer, Berlin, 2003.

[310] M. Nagasawa and H. Tanaka. Diffusion with interactions and collisions between coloured particles and the propagation of chaos. *Probability Theory and Related Fields*, 74(2):161–198, 1987.

[311] K. Nagel, P. Wagner, and R. Woesler. Still flowing: Approaches to traffic flow and traffic jam modeling. *Operations Research*, 51(5):681–710, 2003.

[312] Y. Naito, T. Suzuki, and K. Yoshida. Self-similar solutions to a parabolic system modeling chemotaxis. *Journal of Differential Equations*, 184(2):386–421, 2002.

[313] S. Nakagiri and J.H. Ha. Coupled sine-Gordon equations as nonlinear second order evolution equations. *Taiwanese Journal of Mathematics*, 5(2):297–315, 2000.

[314] R. Narasimha. Non-linear vibration of an elastic string. *Journal of Sound Vibration*, 8(1):134–146, 1968.

[315] S.K. Ntouyas and P. Ames. Global existence for semilinear evolution equations with nonlocal conditions. *Journal of Mathematical Analysis and Applications*, 210(2):679–687, 1997.

[316] S.K. Ntouyas and P.C. Tsamatos. Global existence for second order functional semilinear integrodifferential equations. *Mathematica Slovaca*, 50(1):95–109, 2000.

[317] D. Nualart and A. Rascanu. Differential equations driven by fractional Brownian motion. *Collectanea Mathematica*, 53(1):55–81, 2002.

[318] B. Oksendal. *Stochastic Differential Equations: An Introduction with Applications*. Springer, Berlin, 5th edition, 1989.

[319] A. Okubo and S.A. Levin. *Diffusion and Ecological Problems: Modern Perspectives*. Springer, Berlin, 2001.

[320] B.G. Pachpatte. *Inequalities for Differential and Integral Equations*. Mathematics in Science and Engineering. Academic Press, New York, 1998.

[321] C.V. Pao. Reaction diffusion equations with nonlocal boundary and nonlocal initial conditions. *Journal of Mathematical Analysis and Applications*, 195(3):702–718, 1995.

[322] E. Pardoux and S. Peng. Adapted solution of a backward stochastic differential equation. *Systems & Control Letters*, 14(1):55–61, 1990.

[323] E. Pardoux and A. Rascanu. Backward stochastic differential equations with subdifferential operator and related variational inequalities. *Stochastic Processes and their Applications*, 76(2):191–215, 1998.

[324] E. Pardoux and A. Rascanu. Backward stochastic variational inequalities. *Stochastics An International Journal of Probability and Stochastic Processes*, 67(3):159–167, 1999.

[325] J.Y. Park and J.J. Bae. On the existence of solutions of strongly damped nonlinear wave equations. *International Journal of Mathematics and Mathematical Sciences*, 23(6):369–382, 2000.

[326] B. Pasik-Duncan, T. Duncan, and B. Maslowski. Linear stochastic equations in a Hilbert space with a fractional Brownian motion. *Stochastic Processes, Optimization, and Control Theory: Applications in Financial Engineering, Queueing Networks, and Manufacturing Systems*, pages 201–221, 2006.

[327] S.K. Patcheu. On a global solution and asymptotic behaviour for the generalized damped extensible beam equation. *Journal of Differential Equations*, 135(2):299–314, 1997.

[328] N.H. Pavel. *Nonlinear Evolution Operators and Semigroups: Applications to Partial Differential Equations*. Springer, Berlin, 1987.

[329] A. Pazy. *Semigroups of Linear Operators and Applications to Partial Differential Equations*. Springer, Berlin, 1983.

[330] J.R.A. Pearson and M.A. Matovich. Spinning a molten threadline. Stability. *Industrial & Engineering Chemistry Fundamentals*, 8(4):605–609, 1969.

[331] R.L. Pego and M.I. Weinstein. Asymptotic stability of solitary waves. *Communications in Mathematical Physics*, 164(2):305–349, 1994.

[332] S. Peng. Backward stochastic differential equations and applications to optimal control. *Applied Mathematics and Optimization*, 27(2):125–144, 1993.

[333] L.C. Piccinini, G. Stampacchia, and G. Vidossich. *Ordinary Differential Equations in Rn: Problems and Methods*. Springer, Berlin, 1984.

[334] D. Pierotti and M. Verri. A nonlinear parabolic problem from combustion theory: Attractors and stability. *Journal of Differential Equations*, 218(1):47–68, 2005.

[335] H.M. Portella, A.S. Gomes, N. Amato, and R.H.C. Maldonado. Semigroup theory and diffusion of hadrons in the atmosphere. *Journal of Physics A: Mathematical and General*, 31(32):6861–6872, 1998.

[336] D.L. Powers. *Boundary Value Problems*. Academic Press, New York, 2nd edition, 1999.

[337] H.K. Preisler, D.R. Brillinger, A.A. Ager, J.G. Kie, and R.P. Akers. Stochastic differential equations: A tool for studying animal movement. In *Proceedings of IUFRO4*, volume 11, pages 25–29. Citeseer, 2001.

[338] C. Prévôt and M. Rockner. *A Concise Course on Stochastic Partial Differential Equations*. Springer, 2007.

[339] J.C. Principe. Artificial neural network. *The Electrical Engineering Handbook*, CRC Press, Boca Raton, FL, 2000.

[340] J. Pruss. Positivity and regularity of hyperbolic Volterra equations in Banach spaces. *Mathematische Annalen*, 279(2):317–344, 1987.

[341] J. Pruss. *Evolutionary Integral Equations and Applications*. Birkhauser, Boston, MA, 1993.

[342] S.M. Rankin III. A remark on cosine families. *Proceedings of the American Mathematical Society*, 79(3):376–378, 1980.

[343] S.M. Rankin III. Semilinear evolution equations in Banach spaces with application to parabolic partial differential equations. *Transactions of the American Mathematical Society*, 336(2):523–535, 1993.

[344] A. Rascanu. Deterministic and stochastic differential equations in Hilbert spaces involving multivalued maximal monotone operators. *Pan-American Mathematical Journal*, 6:83–83, 1996.

[345] T.K. Rawat and H. Parthasarathy. Modeling of an RC circuit using a stochastic differential equation. *Thammasat International Journal of Science and Technology*, 13(2):40–48, 2008.

[346] M. Reed and B. Simon. *Methods of Modern Mathematical Physics*. Academic Press, New York, 1972.

[347] K. Ritz, J.W. McNicol, N. Nunan, S. Grayston, P. Millard, D. Atkinson, A. Gollotte, D. Habeshaw, B. Boag, C.D. Clegg, et al. Spatial structure in soil chemical and microbiological properties in an upland grassland. *FEMS Microbiology Ecology*, 49(2):191–205, 2004.

[348] A.W. Roberts and D.E. Varberg. *Convex Functions*. Academic Press, New York, 1973.

[349] A.E. Rodkina. On existence and uniqueness of solution of stochastic differential equations with heredity. *Stochastics*, 12(3-4):187–200, 1984.

[350] L.C.G. Rogers and D. Williams. *Diffusions, Markov Processes and Martingales, Volume 1: Foundations*. Cambridge University Press, Cambridge, UK, 1994.

[351] J.S. Rosenthal. *A First Look at Rigorous Probability Theory*. World Scientific, New York, 2006.

[352] S.M. Ross. *Introduction to Probability Models*. Academic Press, New York, 7th edition, 2007.

[353] H.L. Royden. *Real Analysis*. Macmillan, New York, 1968.

[354] D. Savić. Model of pattern formation in animal coatings. *Journal of Theoretical Biology*, 172(4):299–303, 1995.

[355] H. Schaefer. Uber die Methode der a priori-Schranken. *Mathematische Annalen*, 129(1):415–416, 1955.

[356] M. Scheutzow. Uniqueness and non-uniqueness of solutions of Vlasov-McKean equations. *Journal of the Australian Mathematical Society*, 43(02):246–256, 2009.

[357] M.J. Schramm. *Introduction to Real Analysis*. Prentice–Hall, Englewood Cliffs, NJ, 1995.

[358] I. Segal. Non-linear Semi-groups. *The Annals of Mathematics*, 78(2):339–364, 1963.

[359] H. Serizawa. A semigroup treatment of a one dimensional nonlinear parabolic equation. *Proceedings of the American Mathematical Society*, 106(1):187–192, 1989.

[360] H. Serizawa and M. Watanabe. Time-dependent perturbation for cosine families in Banach spaces. *Houston Journal of Mathematics*, 12(4):579–586, 1986.

[361] S.E. Serrano, S.R. Workman, K. Srivastava, and B. Miller-Van Cleave. Models of nonlinear stream aquifer transients. *Journal of Hydrology*, 336(1-2):199–205, 2007.

[362] R.E. Showalter. Existence and representation theorems for a semilinear Sobolev equation in Banach space. *SIAM Journal of Mathematical Analysis*, 3(3):527–543, 1972.

[363] R.E. Showalter. A nonlinear parabolic-Sobolev equation. *Journal of Mathematical Analysis and Applications*, 50:183–190, 1975.

[364] R.E. Showalter. Nonlinear degenerate evolution equations and partial differential equations of mixed type. *SIAM Journal of Mathematical Analysis*, 6:25–42, 1975.

[365] R.E. Showalter. *Monotone Operators in Banach Space and Nonlinear Partial Differential Equations*. American Mathematical Society, Providence, RI, 1997.

[366] R.E. Showalter and P. Shi. Plasticity models and nonlinear semigroups. *Journal of Mathematical Analysis and Applications*, 216(1):218–245, 1997.

[367] R.E. Showalter and T.W. Ting. Pseudoparabolic partial differential equations. *SIAM Journal of Mathematical Analysis*, 1(1):1–26, 1970.

[368] S.E. Shreve. *Stochastic Calculus for Finance: The Binomial Asset Pricing Model*. Springer, Berlin, 2004.

[369] R. Situ. On solutions of backward stochastic differential equations with jumps and with non-Lipschitzian coefficients in Hilbert spaces and stochastic control 1. *Statistics & Probability Letters*, 60(3):279–288, 2002.

[370] J.G. Skellam. Random dispersal in theoretical populations. *Bulletin of Mathematical Biology*, 53(1):135–165, 1991.

[371] K. Sobczyk. *Stochastic Differential Equations with Applications to Physics and Engineering*. Kluwer Academic, Dordrecht, 1991.

[372] D. Solow. *How to Read and Do Proofs*. John Wiley & Sons, New York, 1990.

[373] S.K. Srinivasan and R. Vasudevan. *Introduction to Random Differential Equations and Their Applications*. Elsevier, Amsterdam, 1971.

[374] A.S. Sznitman. Nonlinear reflecting diffusion process, and the propagation of chaos and fluctuations associated. *Journal of Functional Analysis*, 56(3):311–336, 1984.

[375] H. Tanabe. *Equations of Evolution*. Pitman, London, 1979.

[376] N. Tanaka. A class of abstract quasi-linear evolution equations of second order. *Journal of the London Mathematical Society*, 62(01):198–212, 2000.

[377] T. Taniguchi. Successive approximations to solutions of stochastic differential equations. *Journal of Differential Equations*, 96(1):152–169, 1992.

[378] D.W. Taylor. *Research on Consolidation of Clays*. Massachusetts Institute of Technology, Cambridge, MA, 1942.

[379] J. Tchuenche. Asymptotic stability of an abstract delay functional-differential equation. *Nonlinear Analysis*, 11(1):79–93, 2006.

[380] J. Tchuenche. Abstract formulation of an age-physiology dependent population dynamics problem. *Matematički Vesnik*, 60(2):79–86, 2008.

[381] R. Temam. *Infinite-Dimensional Dynamical Systems in Mechanics and Physics*. Springer, Berlin, 1997.

[382] G. Tessitore. Existence, uniqueness and space regularity of the adapted solutions of a backward SPDE. *Stochastic Analysis and Applications*, 14(4):461–486, 1996.

[383] J.B. Thomas. *An Introduction to Applied Probability and Random Processes.* John Wiley & Sons, New York, 1971.

[384] L. Tian and X. Li. Well-posedness for a new completely integrable shallow water wave equation. *International Journal of Nonlinear Science*, 4(2):83–91, 2007.

[385] T.W. Ting. Certain non-steady flows of second-order fluids. *Archive for Rational Mechanics and Analysis*, 14(1):1–26, 1963.

[386] K. Tomomi and M. Tsutsumi. Cauchy problem for some degenerate abstract differential equations of Sobolev type. *Funkcialaj Ekvacioj*, 40:215–226, 1997.

[387] C.C. Travis and G.F. Webb. Compactness, regularity, and uniform continuity properties of strongly continuous cosine families. *Houston Journal of Mathematics*, 3(4):555–567, 1977.

[388] C.C. Travis and G.F. Webb. Cosine families and abstract nonlinear second order differential equations. *Acta Mathematica Hungarica*, 32(1):75–96, 1978.

[389] P. Troncoso, O. Fierro, S. Curilef, and A.R. Plastino. A family of evolution equations with nonlinear diffusion, Verhulst growth, and global regulation: Exact time-dependent solutions. *Physica A: Statistical Mechanics and its Applications*, 375(2):457–466, 2007.

[390] C.P. Tsokos and W.J. Padgett. *Random Integral Equations with Applications to Stochastic Systems.* Springer, Berlin, 1971.

[391] C.P. Tsokos and W.J. Padgett. *Random Integral Equations with Applications to Life Sciences and Engineering.* Elsevier Science Ltd., Amsterdam, 1974.

[392] J.A.K.P.P. Van der Smagt and B. Krose. *An Introduction to Neural Networks.* University of Amsterdam, Amsterdam, 1993.

[393] J.L. Vazquez. Perspectives in nonlinear diffusion: Between analysis, physics and geometry. *Proceedings of the International Congress of Mathematicians*, pages 609–634, 2007.

[394] A. Veretennikov. On ergodic measures for McKean-Vlasov stochastic equations. *Monte Carlo and Quasi-Monte Carlo Methods 2004*, 2004:471–486, 2004.

[395] I.I. Vrabie. The nonlinear version of Pazys local existence theorem. *Israel Journal of Mathematics*, 32(2):221–235, 1979.

[396] I.I. Vrabie. *C0-Semigroups and Applications.* North–Holland, Amsterdam, 2003.

[397] I.I. Vrabie. *Differential Equations: An Introduction to Basic Concepts, Results and Applications.* World Scientific, Singapore, 2004.

[398] J.A. Walker. *Dynamical Systems and Evolution Equations: Theory and Applications.* Plenum Publishing, New York, 1980.

[399] P. Waltman. *A Second Course in Elementary Differential Equations.* Dover Publications, Mineola, NY, 2004.

[400] Y. Wanli. A class of the quasilinear parabolic systems arising in population dynamics. *Methods and Applications of Analysis*, 9(2):261–272, 2002.

[401] G.F. Webb. Nonlinear perturbations of linear accretive operators in Banach spaces. *Israel Journal of Mathematics*, 12(3):237–248, 1972.

[402] G.F. Webb. *Theory of Nonlinear Age-Dependent Population Dynamics.* CRC Press, Boca Raton, FL, 1985.

[403] T. Weifner and D. Kolymbas. A hypoplastic model for clay and sand. *Acta Geotechnica*, 2(2):103–112, 2007.

[404] D. Williams. *Probability with Martingales.* Cambridge University Press, Cambridge, UK, 1991.

[405] W. Willinger, M.S. Taqqu, W.E. Leland, and D.V. Wilson. Self-similarity in high-speed packet traffic: Analysis and modeling of Ethernet traffic measurements. *Statistical Science*, 10(1):67–85, 1995.

[406] D. Wrzosek. Global attractor for a chemotaxis model with prevention of overcrowding. *Nonlinear Analysis*, 59(8):1293–1310, 2004.

[407] J. Wu. *Theory and Applications of Partial Functional Differential Equations.* Springer, Berlin, 1996.

[408] J. Wu and H. Xia. Self-sustained oscillations in a ring array of coupled lossless transmission lines. *Journal of Differential Equations*, 124(1):247, 1996.

[409] J. Wu and H. Xia. Rotating waves in neutral partial functional differential equations. *Journal of Dynamics and Differential Equations*, 11(2):209–238, 1999.

[410] J. Wu and X. Zou. Patterns of sustained oscillations in neural networks with delayed interactions. *Applied Mathematics and Computation*, 73(1):55–75, 1995.

[411] B. Xie. Stochastic differential equations with non-Lipschitz coefficients in Hilbert spaces. *Stochastic Analysis and Applications*, 26(2):408–433, 2008.

[412] X. Xu. Application of nonlinear semigroup theory to a system of PDEs governing diffusion processes in a heterogeneous medium. *Nonlinear Analysis: Theory, Methods & Applications*, 18(1):61–77, 1992.

[413] Y. Xu, C.M. Vest, and J.D. Murray. Holographic interferometry used to demonstrate a theory of pattern formation in animal coats. *Applied Optics*, 22(22):3479–3483, 1983.

[414] M. Xuerong. *Exponential Stability of Stochastic Differential Equations*. Marcel Dekker, New York, 1994.

[415] T. Yamaguchi. Nonlocal nonlinear systems of transport equations in weighted Ll spaces: An operator theoretic approach. *Hiroshima Mathematics Journal*, 29:529–577, 1999.

[416] P. Yan and S. Liu. SEIR epidemic model with delay. *Anziam Journal*, 48(1):119–134, 2006.

[417] B.Z. Zangeneh. Semilinear stochastic evolution equations with monotone nonlinearities. *Stochastics, An International Journal of Probability and Stochastic Processes*, 53(1):129–174, 1995.

[418] E. Zeidler. *Nonlinear Functional Analysis: Part I*. Springer, Berlin, 1985.

[419] S. Zheng. *Nonlinear Evolution Equations*. Chapman & Hall/CRC Press, Boca Raton, FL, 2004.

[420] J. Zhu, Z. Li, and Y. Liu. Stable time periodic solutions for damped Sine-Gordon equations. *Electronic Journal of Differential Equations*, 2006(99):1–10, 2006.

Index

A-bounded operator, 364
abstract Cauchy problem, 233
abstract stochastic evolution equation, 219
additive noise, 254
advection equation, 218, 272, 276
advection operator, 337
aeroelasticity
    linear IBVP, 369
almost surely, 74, 81
almost surely exponentially stable, 162
Arzela–Ascoli theorem, 55
asymptotically exponentially pth moment stable, 162

Banach space
    bounded subset of, 55
    closed subspace, 37
    compact subset of, 55
    convex subset of, 55
    definition, 35
    of bounded linear operators
        strongly convergent sequences of, 204
        uniformly convergent sequences of, 204
    precompact subset of, 55
    product space, 38, 206
        graph norm, 206
    sequences
        Cauchy, 35
        convergent, 35
    topology, 37
        closure, 37
        open ball, 37
binomial experiment, 79
Bochner integral, 227
Borel–Cantelli Theorem, 74

boundary conditions (BCs)
    Dirichlet type, 222
    homogenous, 222
    Neumann type, 222
    nonhomogenous, 222
Boussinesq equation, 402
Brouwer Fixed-Point Theorem, 55
Brownian motion, 108
Burger's equation, 376

Cahn–Hillard equation, 378
calculus in abstract spaces, 42
    absolutely continuous (AC), 46
    Banach-space valued integral definition, 228
    continuous functions
        compositions of, 44
        definition, 44
        Intermediate-Value Theorem, 45
        jump discontinuity, 45
        left-sided continuity, 45
        on a product space, 44
        on compact sets, 46
        right-sided continuity, 45
        topological properties, 45
        uniform continuity (UC), 46
    derivative
        definition, 47
        distributional sense, 48
        left-sided derivative, 47
        right-sided derivative, 47
    differentiable function
        properties of, 48
    Frechet derivative, 232
    infinite limits, 43
    one-sided limits, 43
    regularity, 47, 219
        levels of differentiability, 47

Cauchy in mean square, 84
Cauchy problem, 249
    abstract functional-type in Banach
        space
    IVP, 346
    abstract homogenous in Banach
        space
    classical solution, 210
    IVP, 210
    abstract nonhomogenous in Ba-
        nach space
    IVP, 271
    abstract nonlinear in Banach space
    IVP, 404
    abstract quasi-linear functional evo-
        lution equation
    examples, 406
    IVP, 406
    abstract Sobolev-type in Banach
        space
    classical solution, 388
    continuous dependence, 390
    functional right-side, 393
    IVP, 387
    mild solution, 388
    restrictions on operators A and
        B, 387
    abstract time-dependent evolution
        equation
    examples, 404
    abstract time-dependent functional
        evolution equation
    IVP, 405
    variation of parameters formula,
        405
    continuous dependence, 152
    existence and uniqueness, 142
    homogenous in R, 140
    homogenous ODEs in N-space
    classical solution, 181
    strong solution of, 141
Cauchy–Schwarz inequality, 81
Central Limit Theorem, 94
Chebyshev's inequality, 82, 89
chemical kinetics, 171, 285

chemotaxis, 319
classical solution, 181
clay consolidation, 225, 383
Closed Graph Theorem, 207
compact
    set, 11
completeness of the reals
    bounded, 8
    bounded above, 7
    bounded below, 8
    complete, 8
    infimum, 8
    lower bound, 8
    maximum, 7
    minimum, 8
    supremum, 7
    upper bound, 7
conclusion, 1
conditional expectation
    description, 95
    properties, 97
    scenarios, 96
conditional probability, 92
continuity in pth moment, 157
Contraction Mapping Principle, 53
contrapositive, 1
convergence almost surely, 86
convergence in p^{th} moment, 86
convergence in distribution, 86
convergence in mean square, 84
convergence in probability, 86
convergent series in a Banach space,
        205
converse, 1
correlation, 82, 90
correlation coefficient, 82
covariance, 82, 90

differential form, 111
diffusion equation, 221, 315, 336
diffusion-advection equation, 363, 404
distribution function
    definition, 78
    properties, 78
Doob's Martingale Inequality, 105

epidemiological model
  multiple strains, 368
  single strain, 366
equicontinuous family of functions, 55
equivalent norms, 26
event, 70
existential quantifier, 1
expectation
  definition, 79, 89
  properties, 80, 89

F_t-adapted, 102
F_t-measurable, 102
Feller–Miyadera–Phillips Theorem, 213
fiber dynamics, 375
filtration, 102
fixed-point approach, 292
fixed-point theory
  contraction operator, 53
  fixed-point, 53
fluid flow through porous media, 225, 271, 385
flutter, 369
Fourier series, 41
  representation formula, 41
Frechet derivative
  definition, 232
  properties, 232
functional
  examples (see Section 8.2), 340
functions
  composition, 4
  definition, 4
  domain, 4
  image, 4
  independent variables, 4
  mapping, 4
  nondecreasing, 5
  nonincreasing, 5
  one-to-one, 4
  onto, 4
  pre-image, 4
  properties, 5
  range, 4

gas flow in a large container, 276
Gaussian random variable
  definition, 83
  properties, 83
geometric Brownian motion, 152
globally Lipschitz, 288
Gray–Scott model, 317
Gronwall's lemma, 51
groundwater flow, 402
growth condition
  generalized Lipschitz, 290
  global Lipschitz, 288
  global Lipschitz (uniformly in t), 288
  local Lipschitz, 289
  non-Lipschitz conditions, 321
  sublinear growth, 290

Hölder's Inequality, 39, 81, 89, 125
heat conduction, 270
  one-dimensional IBVP, 221
  semi-linear, 282
heat equation
  two-dimensional IBVP, 223
Hilbert space, 39
  basis, 39
    existence of, 39
    representation formula, 39
  complete set, 39
  linearly independent, 39
  orthogonal, 39
  orthonormal set
    definition, 39
    properties, 40
  representation theorem, 40
  span, 39
Hilbert–Schmidt operator, 186
Hille–Yosida Theorem, 212
hypotheses
  H_A, 291
  (H3.1), 140
  (H3.2), 140
  (H3.3), 140
  (H3.4), 159
  (H3.5), 159

(H3.6), 159
(H3.7), 161
(H3.8), 161
(H4.1), 191
(H4.2), 191
(H4.3), 191
(H5.1), 234
(H5.2), 234
(H5.3), 234
(H5.4), 234
(H5.5), 234
(H5.6), 239
(H5.7), 239
(H6.1), 250
(H6.2), 255
(H6.3), 255
(H6.4), 255
(H6.5), 256
(H6.6), 264
(H6.7), 264
(H6.8), 267
(H6.9), 268
(H7.1), 291
(H7.2), 291
(H7.3), 295
(H7.4), 295
(H7.5), 295
(H7.6), 295
(H7.7), 308
(H7.8), 308
(H7.9), 308
(H7.10), 321
(H7.11), 321
(H7.12), 322
(H8.1), 340
(H8.2), 340
(H8.3), 340
(H8.4), 342
(H8.5), 342
(H8.6), 342
(H8.7), 344
(H8.8), 344
(H8.9), 345
(H8.10), 347
(H8.11), 347

(H8.12), 350
(H8.13), 351
(H8.14), 354
(H8.15), 355
(H8.16), 355
(H8.17), 355
(H8.18), 355
(H8.19), 355
(H8.20), 355
(H8.21), 355
(H8.22), 355
(H8.23), 358
(H8.24), 358
(H8.25), 358
(H8.26), 362
(H8.27), 362
(H9.1), 387
(H9.2), 387
(H9.3), 387
(H9.4), 387
(H9.5), 388
(H9.6), 388
(H9.7), 392
(H9.8), 392
hypothesis, 1

implication, 1
independent events, 93
independent increments, 99
independent random variables, 94
independent sigma-algebras, 93
infinite series of real numbers
    absolutely convergent, 23
    Cauchy, 22
    Cauchy product, 24
    comparison test, 22
    convergent , 21
    divergent, 21
    geometric series, 22
    n^{th} term test, 22
    partial sums, 21
    ratio test, 23
initial-value problem (IVP), 49
inner product space
    common examples, 38

inner product, 38
    properties, 38
    relationship to a norm, 38
    separable, 40
Intermediate-Value Theorem, 45
Itó formula
    application of, 151
    in R, 132
    multivariable, 135
Itó integral, 185, 229
    definition, 128
    estimates of, 136
    properties, 128
Itó isometry, 229

Jensen's inequality, 81
joint distribution function, 88

Klein–Gordon equation, 378
Kolmogorov's Criterion, 99

Laplacian operator, 214, 225, 337
Lebesgue Dominated Convergence Theorem (LDC), 126
Lebesgue integral, 183
    definition, 123
    estimates of, 136
    Lebesgue Dominated Convergence, 126
    properties, 125
Lipschitz constant, 288
locally Lipschitz, 289
logistic growth, 317
long-term behavior
    connection to matrix exponential, 182
Lumer–Phillips Theorem, 214
Lyapunov exponent, 162

Markov process, 99
Markov property, 104
martingale, 104
matrix exponential
    contractive, 180
    definition, 179

for a diagonal matrix, 179
    generator, 180
    of a diagonalizable matrix, 179
McKean–Vlasov equation, 407
Mean Value Theorem, 48
measurable set, 73
mild solution, 234
Minkowski's Inequality, 126
Minkowski's inequality, 81, 89
model
    advection equation, 218, 272, 276
    aeroelasticity, 369
    Boussinesq equation, 402
    chemical kinetics, 171
    chemotaxis, 319
    clay consolidation, 225, 383
    diffusion-advection equation, 363
    diffusion equation, 315, 336, 404
        heat conduction, 221
        spatial pattern formation, 284, 316
    effects of random motility on population, 318
    epidemiology, 366
    fiber dynamics, 375
    fluid flow through porous media, 225, 271, 385
    forestation, 406
    gas flow in a large container, 276
    groundwater flow, 402
    heat conduction, 221, 270, 282
    magneto-hydrodynamics, 376, 406
    mathematical physics
        Burger's equation, 376
        Cahn–Hillard equation, 378
        Klein–Gordon equation, 378
        Schrodinger equations, 376
        semiconductors, 377
        Sine–Gordon equation, 377
    neural networks, 283, 316, 404
    non-Newtonian fluids, 385
    nonlinear diffusion, 402
    nonlinear waves, 403
    Ornstein–Uhlenbeck Process, 165

pharmacokinetics, 174, 245, 281, 311, 335
phase transitions, 378
pollution, 218, 337, 365, 406
soil mechanics, 225, 383
spring mass system, 175
spring-mass system, 246, 312, 364
thermodynamics, 225
traffic flow, 218, 376
transverse vibrations in extensible beams, 371, 403
wave equation, 272, 282, 314, 361, 406
moment, 82
multiplicative noise, 259

N-Dimensional Gaussian random variable, 90
N-space
    algebraic operations, 25
    complete, 29
    equivalent norms, 26
    Euclidean norm, 25
    geometric structure, 25
    inner product
        Cauchy-Schwarz inequality, 27
        definition, 26
        properties, 27
    norm properties, 26
    open N-ball, 28
    orthonormal basis, 27
    sequences
        Cauchy, 28
        Cauchy Criterion, 29
        convergent, 28
        definition, 28
    topology, 28
    zero element, 25
natural filtration, 103
Navier-Stokes equations, 406
negation, 1
neural networks, 283, 316, 404
non-Newtonian fluids, 385
nonlinear diffusion, 402
norms

graph norm for a product space, 206
L^2-norm, 34
sup norm, 34

operators
    A-bounded, 364
    accretive, 213
    bounded operator, 201
    closed operator, 207
    composition of, 204
    densely-defined operator, 208
    dissipative, 213
    equality, 203
    graph of an operator, 207
    Laplacian, 225
    linear operator
        definition, 201
        invertible, 205
        one-to-one, 205
        onto, 205
        resolvent operator, 212
        resolvent set, 211
        Yosida approximation of, 212
    m-dissipative, 213
    monotone, 213
    nonexpansive, 213
    operator norm, 201
        estimate for, 202
    restriction to subset, 203
    space of bounded linear operators, 201
    unbounded operator, 202
ordinary differential equations (ODEs)
    first-order linear, 49
    general solution, 48
    higher-order linear, 50
    initial-value problem (IVP), 49
    separable, 48
    systems of, 171
    variation of parameters formula, 50
    variation of parameters method, 49
Ornstein–Uhlenbeck Process, 165
outward unit normal vector, 366, 406

P-null set, 74
partial differential equations (PDEs)
    complex-valued, 376
        separation of variables method, 222, 372
pharmacokinetics, 174, 245, 281, 335
pollution, 218, 337, 365, 406
precompact set
    definition, 55
Principle of Uniform Boundedness, 204
probability density function, 79
probability measure
    definition of, 73
    properties, 73
probability measure induced by a random variable, 304
probability space
    complete, 75
    definition of, 73
product space, 38
progressively measurable, 103
Prokorov's Theorem, 304

quadratic variation, 110

random characteristic function, 77
random operator-valued step function, 231
random simple function, 77
random variable
    binomial, 79
    Cauchy in mean square, 84
    characteristic function, 77
    continuous, 76
    convergence in mean square, 84
    correlation, 82
    correlation coefficient, 82
    covariance, 82
    discrete, 76
    distribution function, 78
    expectation, 79
    Gaussian, 83
    Hilbert-space valued, 216
    joint distribution function, 88
    limit theorems, 84

moment, 82
    probability density, 79
    properties, 76
    R_n-valued, 88
    real-valued, 76
    simple function, 77
    space of, 83
    uniform, 82
    variance, 81
reaction-convection-diffusion equation
    epidemiological models, 339
    genetically engineered microbes model, 339
    pollution model, 337
    semiconductor, 339
reaction-diffusion equations, 284
real linear space
    definition, 33
    linear subspace of, 33
    measuring distance, 34
    norm, 33
real number system
    absolute value, 6
        properties, 7
    Cauchy–Schwarz inequality, 7
    Minkowski inequality, 7
    order features, 5
    topology, 9
        Bolzano–Weierstrass Theorem, 11
        boundary, 9
        boundary point, 9
        closed, 9
        closure, 9
        compact, 11
        dense, 10
        derived set, 9
        Heine–Borel Theorem, 11
        interior, 9
        interior point, 9
        limit point, 9
        open, 9
resolvent operator
    definition, 212
    properties, 212

robustness, 163

sample paths, 100
sample space, 70
Schrodinger equations, 376
semigroup of bounded linear operators
    C0-semigroup
        properties, 210
    contractive
        approximation of, 215
        definition, 209
    example
        Sobolev IBVP, 226
    generator
        bounded perturbation of, 215
        definition of, 209
        properties, 210
    semigroup property, 209
sequences
    real-valued sequence
        below, 12
        bounded, 12
        bounded above, 12
        Cauchy Criterion, 20
        Cauchy sequence, 19
        convergent, 13
        decreasing, 12
        divergent, 13
        increasing, 12
        limit, 12
        limit inferior, 18
        limit superior, 18
        monotone, 12
        nondecreasing, 12
        nonincreasing, 12
        squeeze theorem, 14
        subsequence, 12
sets
    bounded domain in N-space
        smooth boundary, 48
    complement, 2
    concave, 321
    convex, 321
    countable, 40
    elements, 2

family of sets, 3
    intersection, 3
    subset, 2
    union, 3
sigma-algebra
    Borel class, 72
    definition of, 72
    generated by a set, 72
Sine-Gordon equation, 377
Sobolev spaces
    continuous functions, 34
        pointwise convergence, 35
        Taylor series, 37
        uniform boundedness, 35
        uniform convergence, 35
        uniformly convergent series, 35
        Weierstrass M-Test, 36
    locally p-integrable functions, 34
    n-times continuously differentiable
        functions, 34
    p-integrable functions, 34
Sobolev-type evolution equation, 387
soil mechanics, 225, 383
Space U, 122
spatial pattern formation, 284, 316
spring mass system
    coupled springs, 177, 365
    harmonic oscillator, 176
square matrices
    algebraic operations, 30
    eigenvalue, 31
        multiplicity, 31
        of an inverse matrix, 31
    invertible, 31
    sequences
        convergent, Cauchy, 32
    space of, 30
        norms in, 31
    terminology
        diagonal, 30
        identity matrix, 30
        symmetric, 30
        trace, 30
        transpose, 30
        zero matrix, 30

stability, 162
Standing Assumptions
  (**S.A. 1**), 112
  (**S.A. 2**), 112
  (**S.A. 3**), 112
stationary increments, 99
stochastic differential equation (SDE), 120
stochastic differential equations (SDE)
  blow-up of solutions, 287
  nonuniqueness of solutions, 286
  systems of
    matrix form, 172
stochastic process
  continuity, 101
  covariance, 100
  definition, 97
  equivalence of, 99
  finite-dimensional joint distribution, 99
  Gaussian, 99
  independent increments, 99
  Markov, 99
  martingale, 104
  mean, 100
  mean square limit, 100
  notation, 98
  stationary, 99
  stationary increments, 99
  variance, 100
  version, 99
strong solution, 141, 192
sublinear growth, 290

Taylor series, 37
thermodynamics, 225
Thomas model, 317
tight family, 304
time-dependent evolution equation, 404
traffic flow, 218
transverse vibrations in extensible beams, 371
Trotter–Kato Approximation Theorem, 215

uniform random variable, 82
universal quantifier, 2

variance, 81, 89
variation of parameters formula, 234

wave equation, 361
  homogenous
    classical solution of, 273
    equivalent system formulation, 273
    IBVP, 272
    with viscous damping, 274
  quasi-linear, 406
  semi-linear forcing term, 282
  Sobolev-Type, 386
  system of coupled semi-linear, 314
  weakly-coupled and damped with nonlinear dispersion, 282
  with nonlinear dissipation, 403
weak convergence of probability measures, 304
Weierstrass M-Test, 36
Wiener process
  definition, 108
  M-dimensional, 112
  modeling with, 110
  properties, 109

Yosida approximations, 237

zeroth-order approximation, 239

Printed and bound by CPI Group (UK) Ltd, Croydon, CR0 4YY

26/10/2024

01779752-0001